Gordon MacPherson and
Jon Austyn

Exploring Immunology

Related Titles

Chapel, H., Haeny, M., Misbah, S., Snowden, N.

Essentials of Clinical Immunology

ISBN: 978-1-4051-2761-5

Coico, R., Sunshine, G.

Immunology
A Short Course

ISBN: 978-0-470-08158-7

Elgert, K. D.

Immunology
Understanding The Immune System

ISBN: 978-0-470-08157-0

Williams, A. E.

Immunology
Mucosal and Body Surface Defences

ISBN: 978-0-470-09004-6

Strober, W., Gottesman, S. R.

Immunology
Clinical Case Studies and Disease Pathophysiology

ISBN: 978-0-471-32659-5

Gordon MacPherson and Jon Austyn

Exploring Immunology

Concepts and Evidence

Illustrations by Ruth Hammelehle

WILEY-BLACKWELL

The Authors

Dr. G. Gordon MacPherson
Sir William Dunn School of
Pathology, Univ. of Oxford
South Parks Road
Oxford OX1 3RE
and
Oriel College
Oxford OX1 4EW
United Kingdom

Prof. Jonathan M. Austyn
Nuffield Department of
Surgical Sciences, University
of Oxford, John Radcliffe Hospital
Headington, Oxford OX3 9DU
and
Wolfson College
Oxford OX2 6UD
United Kingdom

Illustrations by
epline
Ruth Hammelehle
Marktplatz 5
73230 Kirchheim/Teck
Germany

Library of Congress Card No.: applied for

British Library Cataloguing-in-Publication Data
A catalogue record for this book is available from the British Library.

Bibliographic information published by the Deutsche Nationalbibliothek
The Deutsche Nationalbibliothek lists this publication in the Deutsche Nationalbibliografie; detailed bibliographic data are available on the Internet at http://dnb.d-nb.de.

Wiley-Blackwell is an imprint of John Wiley & Sons, formed by the merger of Wiley's global Scientific, Technical, and Medical business with Blackwell Publishing.

Typesetting Thomson Digital, Noida, India
Printing and Binding Himmer AG, Augsburg, Germany
Cover Design Adam Design, Weinheim, Germany

Printed in the Federal Republic of Germany
Printed on acid-free paper

Print ISBN Hardcover: 978-3-527-32430-9
Print ISBN Softcover: 978-3-527-32412-5

To the memory of Ralph Steinman, 1943–2011
A wonderful early mentor to one of us;
an outstanding friend and colleague to both

Contents

Preface

We hope this book will be of value to readers at different levels of professional development. First and foremost, it is intended for undergraduates starting their studies of immunology, such as biomedical, medical and veterinary students (and perhaps those from other disciplines with an interest in the subject). Second, it should be of value to graduate students entering research where there is an immunological content. Last, but not least, we also hope it will be useful to more senior scientists and clinicians wanting to learn a bit more about immunology. In writing this book we have tried to help all of these groups to understand how the immune system works, not just by understanding its basic scientific and clinical concepts, but also by appreciating some of the evidence that has increased our knowledge about how and why immune responses have evolved. We also hope to stimulate readers to appreciate immunology as a science that underpins much of current clinical practice and biomedical research.

This book shows how immune defence has evolved in the face of the continual selective pressure exerted by viruses, bacteria and other pathogens, and *vice versa*. It explains how defence against different types of infectious agents requires different immune mechanisms and shows how pathogens have also evolved to evade these mechanisms. It describes many of the different anatomical, cellular and molecular components involved in immune responses. The book also shows how many immunological mechanisms that have evolved to protect against infection can also cause disease, and how, increasingly, we can use our knowledge of immunology to prevent or treat disease.

We aim to show that the continued development of our understanding of immunology is based crucially on a combination of experimental analysis and clinical observation. Readers will be introduced to some of the techniques currently used in immunological research, as well as a few of historical interest. They will also be introduced to selected clinical case studies to aid in understanding how and why different types of infectious and immunologically-mediated diseases are caused, and we hope that these may be of particular value to undergraduate medical students. By taking this approach, we hope that readers will gain an integrated understanding of both the basic and clinical aspects of key areas of immunology, all too often treated separately in other texts. We expect that readers will end up understanding not only the role of immune responses in infection and disease, but also how our current understanding of immunity is still deficient in so many areas. We are strongly of the opinion that the only way to increase our understanding, with the goal of more effective immune intervention, is through a balanced combination of clinical and experimental approaches.

The first chapter provides a straightforward overview of immunity and immune responses, both beneficial and harmful. The second chapter covers in more depth the nature of pathogenic organisms and their interactions with the immune system, including mechanisms that pathogens have evolved to defend themselves against immune responses. The next four chapters collectively cover in more detail the induction and regulation of immune responses at the whole organism, tissue, cellular and molecular levels. The final chapter deals with the less welcome aspects of immunity, including allergies and autoimmune diseases, and also introduces some general principles of transplant reactions and tumour immunity.

Acknowledgements

Many colleagues have assisted us in the development of this book at informal and formal levels. Those who have contributed by giving us suggestions and reviewing draft chapters include Helen Chapel, Paul Crocker, Tim Elliott, Simon Hunt, Sarah Marshall, Alan Mowat, Caetano Reis e Sousa, Adrian Smith and Tony Williams. We are grateful to all of these colleagues for their detailed inputs. We are also grateful for the advice and assistance given to us by the team at Wiley, Gregor Cicchetti, Andreas Sendtko and Anne Chassin du Guerny in particular, for the sterling work of our Graphic Designer Ruth Hammelehle who managed to convert our roughly penciled sketches into meaningful figures, and to Nitin Vashisht at Thomson Digital. Finally, we would not have been able to write this book without the patience, encouragement and support of our wives, Shelley and Karen.

A Note to the Reader

Chapters

We endeavoured to write this book so that it can either be read in its entirety from beginning to end or, alternatively, the chapters can be read separately in any order. Those with little background in immunology can begin by reading the Introduction and Chapter 1 which provides an outline of 'The Immune System'. Readers who would like an accessible overview of the main areas of immunology might read Chapter 2 on 'Infection and Immunity', and/or Chapter 7 on 'Immunity, Disease and Therapy'. These readers, and those who already have specialist knowledge, as well as all those interested in further detail can then turn to Chapter 3 on 'Functional Anatomy of the Immune System', Chapter 4 on 'Innate Immunity', Chapter 5 on 'T Cell-Mediated Immunity' and/or Chapter 6 on 'Antibody-Mediated Immunity'.

Boxes

The information provided in boxes provides examples of some key techniques used in immunology, explains some aspects described in the text in a little more detail, and encourages the reader to consider broader related areas such as evolutionary aspects.

Case Studies

We have included some selected case studies in Chapters 2 and 7 to highlight the typical clinical presentation of different types of infectious and immunologically-mediated diseases, and to relate these to their underlying pathogenesis or treatment. A few others are also included in Chapters 3–6.

Questions in the Text

Scattered through the book are questions related to the topics under consideration. These are designed to make readers think critically and in more depth about important problems, and are designed to be of use for self-learning and in tutorials and seminars. As far as we know, many of these questions do not have definitive answers. (If they do, we will not apologise for our own lack of understanding since we believe that it is only through testing ones knowledge that one really begins to learn, and we are certainly not ashamed to reveal our own incomplete grasp of the subject!). Some of the questions may prove particularly challenging for the novice reader and it may be best for such readers to read a chapter in its entirety before attempting to answer many of the questions.

Learning Outcomes

At the end of each chapter we have highlighted some key questions which will enable readers to review their understanding of each topic. We have also included a few questions ('General' and 'Integrative') which will, we hope, encourage the reader to think about the broader areas covered in that chapter, and to considers these areas in relation to others covered elsewhere in the book.

Further Study Questions

At the end of Chapters 3–6 respectively are a few 'Further Study Questions' designed to stimulate further investigation by the reader. They could for example form the basis of extended essays, perhaps to be set by tutors. Because many of these questions are deliberately open-ended, we also have also provided a few hints as to how they might be approached.

Website

The accompanying website www.wiley-vch.de/home/immunology is based on the figures and legends from the book. These have been arranged and modified so that each set tells a coherent story that will be particularly useful for revision purposes.

Further Reading

We have deliberately avoided giving detailed bibliographical references. It is our experience that these are rarely used by students, and the speed of change in immunology means that many will be out-dated by the time the book is read. As a starting point, in terms of publications, the interested reader may wish to consult specialist articles in the *Annual Reviews* and *Advances* series, as well as *Immunological Reviews*, for comprehensive information on many topics. Journals in the *Trends* (e.g. *Immunology Today)* and *Nature Reviews* series provide particularly timely updates on key areas, while the *Current Opinion* series also provides some helpful pointers for readers who are ready to engage with the current primary literature. Some of the top primary research journals such as *Nature Immunology, Immunity* and the *Journal of Experimental Medicine* also contain valuable review articles. You will find many others. We consider that becoming proficient in navigating the immunological literature is an essential part of the learning process, and we hope that after reading this book you will continue to enjoy exploring immunology!

Introduction

1
Introduction

Why explore immunology? Because it is clinically important, it is academically challenging and exciting, and it holds out the promise of important advances in understanding and treating disease. In this book we aim to give the reader an understanding of how the immune system works, of how our understanding of its working has developed from both clinical and experimental evidence, and where some major gaps and problems in our understanding lie. In this short introduction we will give you some idea of the different areas of immunology that will be discussed in more detail in the following chapters. Throughout most of the book we will use real clinical and experimental examples to illustrate the importance of immunology to all those with an interest in the biomedical and biomedical sciences.

2
Immune Responses

Recovery from influenza and resistance to re-infection, vaccines, hay fever, asthma, treatment of rheumatoid arthritis, rejection of transplanted kidneys, the diagnosis of leukaemia and a potential cure for cancer – all of these involve immunology. Immunology is the study of the immune system and of its multiple, complex responses. Most immunologists consider that the immune system has evolved to defend the host against infectious agents, some of which have the potential to cause real harm. These agents, ranging in type from viruses, bacteria and fungi to worms, are called pathogens. However the immune system can respond to almost any foreign molecule including proteins, carbohydrates, nucleic acids – even molecules such as dinitrophenol that have never existed in nature. For our purposes, at this stage, these types of molecule can be loosely and collectively termed antigens.

The immune system constantly patrols our entire body for any signs of infection. If it occurs, rapid changes are induced at the local site of infection, typically leading to inflammation which we all know so well – think of the redness, swelling , sense of heat and pain (and pus) if you have a boil on your skin or a stye in your eye. At the same time, other changes are more slowly induced away from the site of infection in specialised organs of immunity such as the lymph nodes – these include the painful swellings in your neck you will sometimes feel f you have a really bad sore throat. All these changes are caused by specialised cells of immunity, some of which are selectively involved in different types of infection and at different sites of the body. These cells use a vast array of specialised molecules to talk to neighbouring and more distant cells, and some of these can also cause changes in organs very distant from the site of infection – if you have a fever it is because some of them act on part of your brain. All these different organs, cells and molecules work together to bring about a highly coordinated and tightly regulated series of events, an immune response, which endeavours to bring about the elimination of the infectious agent.

The immune responses is a reactive, homeostatic response to changes in the host's internal and external environment. Importantly, in vertebrates such as ourselves, some types of immune response ensure that when an infectious agent has been eliminated the individual is protected against subsequent encounters with the same agent – if you have had measles as a child you will (almost certainly) never get it again. This phenomenon, termed immunological memory, thus ensures that a state of immunity is generated that can last a lifetime.

3
Infection and Immunity

3.1
Life in a Micro-Organism-Rich World

The human body contains some 10^{13} cells. The human large intestine contains about 10^{14} bacteria. We are constantly interacting with viruses, bacteria, fungi and smaller parasites. In the developing world, interaction with larger parasites such as worms is a continuous, unremitting part of life. Most of these interactions are harmless, indeed some are beneficial – the commensal bacteria that fill our large intestines and coat many other sites exposed to the external world serve to protect us from infection by harmful microbes.

All multicellular organisms provide a potential niche for micro-organisms to colonize and even bacteria can be infected

Exploring Immunology: Concepts and Evidence, First Edition. Gordon MacPherson and Jon Austyn.

by viral pathogens, the bacteriophages. Our bodies are potentially a rich source of food for microbes and parasites, and without an immune system we would be eaten to death. While many microbes can co-exist peacefully with their hosts, some have the potential to cause damage or even death. It follows that evolutionary pressures will select for mutations in hosts that increase their ability to resist potentially damaging infections and to tolerate colonization with harmless bacteria that may prevent attack by pathogens. In parallel, micro-organisms capable of surviving within a multicellular host will mutate to be able to increase their chances of survival and thus of spreading their gene pools to other hosts. However, micro-organisms evolve much faster than mammals, for example, and thus another challenge for any host immune system is to try to anticipate in advance what possible mutations might arise in micro-organisms in the future, as the host itself will not be able to evolve nearly so quickly. Remarkably, in many cases the immune system does this very successfully.

3.2 Infectious Disease

Fortunately, almost all of us have some capacity to defend ourselves against infection by potential pathogens. Most of us are able to recover from repeated bouts of influenza, year after year. And yet over 14 million people world-wide are infected with tuberculosis of whom around 1.6 million will die of this disease every year. More than a million people die of malaria each year about 1 child every 30 seconds in Africa alone. Acquired immunodeficiency syndrome (AIDS) cases continue to increase dramatically in the developing world. Closely related to the above is the problem of drug resistance, which is on the increase in all forms of infectious disease. There are some strains of Mycobacterium tuberculosis (the causative bacterium ofTB) that are resistant to all current antibiotics. Some strains of the malarial parasite are becoming resistant to almost all drugs. Infection with drug (methicillin)-resistant Staphylococcus aureus (MRSA) is a major problem in hospitals, and the human immunodeficiency virus (HIV) is becoming resistant to many of the drugs in clinical use. We are running out of antibiotics – no new class of antibiotic has come into clinical use in the last 20 years (there are some hopeful signs). Microbes will always evolve new ways of defending themselves against antibiotics, so we have an urgent need to understand how to manipulate the immune system to generate even better protective responses against continually-evolving pathogens. This is one of the main focuses of immunological research today.

4 Immunopathology and Immunotherapy

4.1 Immune-Mediated Disease

Pathogens, by definition, can cause disease. Disease can also be caused by the inability to make an effective immune response to a pathogen; this is termed an immunodeficiency disease. Happily such diseases are rare, but we have all seen images of children having to be enclosed in sterile plastic bubbles because of their defective immunity. Disease can further be caused by the immune system making an inappropriate response to a normal component of the host. Most of us know people who are diabetic or who have the crippling joint destruction seen in some forms of arthritis – these are different forms of autoimmune diseases. We also all know people who suffer from allergies, caused by apparently harmless antigens such as pollen in hay fever. Or those who have reactions (sensitivities) to metals such as nickel in jewelry sensitivities to metals such as nickel; we will term these conditions immune-related sensitivities (they are sometimes called allergies and hypersensitivity diseases). Many of us will have heard that these diseases are on the increase in developed countries, asthma being a prime example. Fewer will be aware that this increase is not in general seen in the developing world. We are still puzzling to understand why these changes occur and how we might prevent or explain them.

Thus, it is essential that we deal effectively with pathogens, but it is inconvenient, even life-threatening, when the immune system – often for completely unknown reasons – makes a powerful response against something that is inherently harmless, such as pollen, or against part of the body itself. It is another major goal of immunological research to discover how to switch off these unwanted responses selectively, rather than having to use non-specific drugs, many of which such as steroids have serious side-effects and often render the patient at high risk of infection.

4.2 The Challenge of Transplants, the Problem of Cancer

A different setting where scientists and clinicians really want to know how to manipulate immune responses is after organ transplantation. Many of us will know or know of people who have had kidney transplants because their own organs have failed. Most of us will know that often the only way that transplants can be accepted is by using powerful chemical immunosuppressive agents such as cyclosporin or tacrolimus. Usually transplants come from other people, generally people who are more or less genetically different from the patient. Not surprisingly the immune system of the patient recognizes that these foreign organs are not normal parts of the body and tries to eliminate them. Indeed, the dramatic power of the immune system is manifested by the subsequent rejection episodes that may follow. If we try to turn these off, using available chemical agents, the patient becomes increasingly susceptible to infections, and also to certain cancers some of which may actually be triggered by viral infections.

What about malignant tumours (cancer)? A disconcerting fact is that potentially malignant cells are probably appearing within our bodies almost continuously – and yet two in every three people will never develop a malignant tumour. Back in the 1950s, it was first suggested that the immune system is continuously surveying the body and trying to eradicate malignant cells as they arise, a phenomenon that was thus called immune surveillance. Quite a challenge, given that these cells derive from cells that naturally belong to the body! Nevertheless, there is some evidence that this idea is correct, and yet malignant tumours do develop in many people particularly as they get older and past reproductive age. We are still puzzling to understand why this happens in some and not others; while some parts of the puzzle seem to have been solved, such as the links between tobacco smoking and lung cancer, others have certainly not.

Another remarkable idea to emerge in quite recent years is that it may eventually become possible to vaccinate people against cancers. One recent dramatic advance has come with the development of a vaccine against the strain of human papilloma virus (HPV) that is associated with cervical carcinoma; by preventing infection by HPV this vaccine protects women against the development of the cancer. This form of vaccination is called prophylactic because it is given before infection actually occurs. An even bigger challenge will be to discover if and how we can vaccinate people who have actually developed a cancer, and also people who already have an infectious disease (e.g. HIV infection). Since this form of vaccination is designed as a therapy to treat an already existing condition it is termed therapeutic vaccination.

4.3 Immunological Interventions in Disease

Vaccines are, of course, the best-known and most effective examples of immune intervention. Smallpox has been eradicated globally, poliomyelitis exists in only a few areas of a few countries and tetanus is 100% preventable by vaccination. We also have highly successful vaccines against some other infections, such as diphtheria, measles, mumps and rubella. Why do we not have effective vaccines against HIV, tuberculosis, malaria and many other major infectious diseases? It is not for want of trying. All over the world vast sums of money are put into vaccine development, yet researchers still say that successful vaccines against these infections are 5, 10 or even more years away, which in reality means that we have no idea if or when they will actually be available.

Increasing numbers of immune-based treatments are being developed. Apart from the development of some really successful vaccines against infection, the production of therapeutic antibodies is probably the most dramatic success story of immunology to date. For example, some forms of rheumatoid arthritis that are totally resistant to all standard treatments can be halted in their tracks by a genetically-engineered antibody generated against a protein involved in stimulating inflammation. Modern molecular approaches are starting to make real differences and this last example demonstrates how an understanding of the basic mechanisms of immunology can be used to design new therapies using tools derived from the immune system itself.

4.4 Using Immunological Tools for Diagnosis

Leukaemia is a malignant tumour of blood cells. In leukaemias that arise from the lymphocytes – normally helping to defend us against infection – the leukaemic cells can originate from either T or B lymphocytes. The success of treatment depends on giving the best drugs, and these differ for T- and B-derived leukaemias. T and B cells differ in the proteins they carry on their surfaces and highly specific, artificially generated monoclonal antibodies have been developed which bind to molecules expressed by only one or the other. This makes it straightforward to distinguish between the leukaemias and to give the optimal treatment.

Another example comes from breast cancer. In assessing the likely outcome for a patient diagnosed with such a tumour it is crucial to know if the tumour has spread to the glands (actually, the lymph nodes) in the armpit. Using basic histological techniques it can be very difficult to identify small numbers of tumour cells in the node. If however a section of the lymph node is labelled with an antibody to a molecule (cytokeratin) found only in the epithelial cells from which the tumour originates and is stained using special techniques, tumour cells can be made to stand out as bright red on a blue background under the microscope. This enables the clinician to determine if the tumour has spread to the node and to adjust therapy accordingly.

5 Exploring Immunology

Is immunology difficult? Well, immunology is a complex and incompletely understood science that many find difficult to understand. We are tempted to say that if you find it easy you are not doing it properly – there are many areas that still puzzle all immunologists. We hope that we have already shown that immunology is a field of study that impinges importantly on all areas of medicine and biomedical science.

Another difficulty for anyone trying to explain immunology is that it is not possible, at least to our minds, to teach or learn immunology in a linear, incremental way. In comprehending immune responses, understanding one part depends on understanding others, and the understanding of these others depends on understanding the first part. It all seems confusing at first and it is only after covering much of immunology that it starts to form a coherent picture. If you find it confusing at first, in Chapter 1 we give you a brief, straightforward overview of the immune system and immune responses, before going into more detail in the rest of the book. Finally, when you have finished reading this book , we hope that you will find yourself fully equipped to continue and develop your interest in this clinically and scientifically important, sometimes frustrating, but always intriguing field by further exploring immunology.

1
The Immune System

1.1
Introduction

All living things – animals, plants and even bacteria – can act as hosts for infectious organisms and thus have evolved mechanisms to defend themselves against infection. Infection can be by other living things, non-living things (viruses) and possibly even molecules (prions). Since it is so crucial to our own survival, much of our understanding of immunity has come from studies in humans – particularly in relation to the causes and prevention of disease – but deep insights have also come from experimental studies in animals such as mice. For these reasons, in this book we concentrate on the immune systems of humans and mice. These, along with other more recently evolved organisms (e.g. birds and amphibians), have the most complex and sophisticated immune systems, but the origins of these can in many instances be traced back to the most distant and ancient species in evolutionary history.

In this chapter we provide an overview of immunology in which we introduce the key players in immunity, largely focussing on the immune systems of humans and mice. We start by briefly considering how infection can be sensed by the host organism and how it is possible for a host to recognize many very different infectious agents (Section 1.2). We then introduce the tissues and specialized organs where immune responses occur (Section 1.3). To eliminate different infections effectively, immune responses need to be tailored to particular types of infection. This requires a variety of cells and molecules that can interact coherently to generate the mechanisms that are needed to eliminate each type of infection. As these mechanisms help to bring about or "effect" the elimination of infectious agents they are termed effector mechanisms.

Defence against infection is divided into two main forms termed innate immunity and adaptive immunity. Innate defence mechanisms are present in different forms in all multi-cellular organisms, including plants. Adaptive defence mechanisms have evolved more recently in vertebrates. In vertebrates, the interaction of innate and adaptive immune mechanisms is essential for the generation of effective immunity to infection.

To introduce the mechanisms of immunity we start by describing the different types of immune cells and their function in innate and adaptive immunity to infection (Section 1.4). We then introduce the major classes of molecule involved in functions such as the detection of infection, the recruitment of cells to infected sites, communication between cells and tissues, signalling within cells, and, usually, elimination of the infectious agent (Section 1.5).

The immune system that humans have evolved is, however, not perfect and we discuss some of these imperfections at the end of this chapter (Section 1.6). The immune system is a very effective killing machine, and if it goes wrong it can cause severe disease and even death of its host. To cover these latter areas we first consider how the immune system is able to discriminate between what needs to be eliminated and what does not – particularly in the case of adaptive immunity, which has evolved to recognize molecular structures largely at random. We then introduce the different ways in which the immune system can cause damage if it becomes directed not to infectious agents, but to otherwise harmless targets, including many inert substances around us and within the tissues of the host itself. We discuss the problems of transplants (some of which can even attack their hosts) and why the immune system fails to reject malignant tumours (cancer). Finally, we turn from problems to solutions and introduce two areas in which either the intact immune system and components of immunity can be harnessed for our own benefit and from which tools can be derived to treat disease.

By the end of this chapter you should have insight into of the basic properties and functions of the immune system, and will understand the principles of its roles in defence against infectious disease. You will start to have an appreciation of why it is pivotal to life, disease and death, and how it is important not only in prevention of disease but in its causation. This chapter will lead you on to the following chapters where different areas of immunity are discussed in greater depth.

1.2
Host Defence Against Infection

We need immune responses to defend ourselves against infection. Many different kinds of organism have the potential to infect us and, if they do so, can cause us harm in many

Exploring Immunology: Concepts and Evidence, First Edition. Gordon MacPherson and Jon Austyn.

different ways. To deal with all these potential threats we, as hosts for infectious agents, need a variety of different kinds of host defence mechanisms. Indeed this applies for any living organism.

1.2.1 Infectious Agents

To understand how the immune system works in infection we need to know who the aggressors are. Potentially infectious agents include the following:

- Viruses, which are non-living entities. Common examples are influenza virus, human immunodeficiency virus (HIV) and herpes simplex virus (HSV, which can cause cold sores or genital ulcers).
- Bacteria, are single-celled prokaryotic organisms. Examples include *Staphylococcus* and *Streptococcus* that cause acute infections such as abscesses and sore throats, and *Mycobacteria* that cause chronic infections such as tuberculosis and leprosy.
- Fungi, which are unicellular, such as Candida that causes thrush, or multicellular.
- Parasites, which are eukaryotic organisms. Some are single-celled protozoa that cause diseases such as malaria, others are large, multicellular organisms (metazoa) such as tapeworms.

In this book, for convenience, we will sometimes refer to smaller infectious agents, including viruses, as microbes because they are microscopic in size. However, many parasites, the metazoa, are often far from microscopic in size.

1.2.2 Host Defence

All organisms possess mechanisms to defend themselves against infection, and immunity is a specialized form of host defence. In mammals, defence mechanisms can be passive or active. Passive defence comes in the form of natural barriers that hinder infection. Examples are skin, which prevents access of microbes to the underlying tissue, and gastric acid in the stomach which, not surprisingly, can kill many microbes that might be ingested with food. Their existence is quite independent of the presence of infection. Active defence is brought about by immune responses that involve a diversity of different effector mechanisms that are induced by the presence of infection and which may eliminate the microbe. Thus, all forms of active immunity depend on specific recognition of molecules present in the infecting agent. This is turn leads to a response, involving the interaction of cells and molecules to produce different effector mechanisms that can often eliminate the infection.

Immunity is itself divided into two different forms – innate and adaptive. Innate responses occur rapidly and can generate effector mechanisms that are effective within minutes or hours of infection. In contrast, adaptive immunity takes much longer to become effective, usually over a few days. In immunity to most forms of infection, however, both innate and adaptive immunity are essential. A major advantage of adaptive immune responses, not seen with innate immunity, is that they generate memory – a second infection with the same microbe elicits a stronger, faster and usually more effective response. See Figure 1.1.

1.2.3 Immune Recognition

Different types of cells and molecules are involved in the initiation of innate and adaptive immune responses although, as mentioned above, their interaction is essential in defence against most infectious agents. So what do the innate and adaptive arms of immunity do in general terms? Broadly speaking we can view some components of the innate immune system as being involved in the detection of

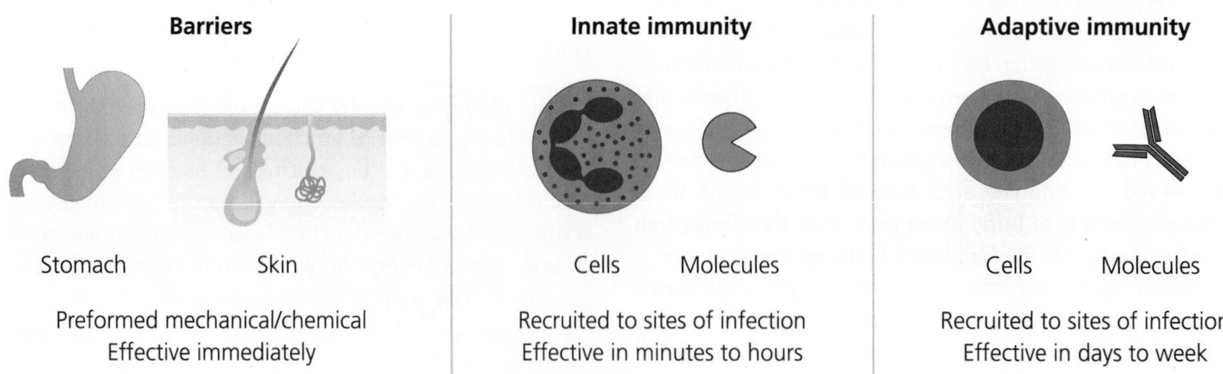

Fig. 1.1 Mechanisms of defence against infection. Natural barriers. These stop infectious agents entering the host or provide a hostile environment. Physical barriers to infection include the epithelia of the skin, lung and airways, and the gastro-intestinal and urogenital tracts. Cells in these barriers may also secrete agents that kill infectious agents. **Innate immunity.** This is the first form of immunity induced by infectious agents. Cells and molecules such as phagocytes and complement can make rapid responses that may eradicate the infection. **Adaptive immunity.** Later adaptive responses may be generated if the infectious agent is not killed by innate immunity. Cells and molecules such as lymphocytes and antibodies take longer to become effective, but adaptive immunity can also lead to a state of long-lasting resistance to re-infection termed **immunological memory** (not shown).

"harmful" things that represent "danger" to the organism, such as general classes of microbes that may have infected the host. Other components then endeavour to eliminate the microbe. In contrast, the adaptive immune system can discriminate very precisely between individual microbes, even of the same type, but can generally only make a response if it has been informed by the innate system that what is being recognized is "dangerous". If so, adaptive responses may then help to eliminate the microbe, if it has not already been eradicated during the earlier innate response. Recognition of infectious agents is essential for any form of immunity and thus for host defence against them. Generally speaking, the types of receptors used for recognition differ in innate and adaptive responses. See Figure 1.2.

1.2.3.1 Recognition in Innate Immunity: Pattern Recognition Receptors

The key components of the innate immune system include cells such as phagocytes and soluble molecules such as complement. These work together to sense the presence of infection. The recognition of potentially dangerous microbes usually leads to the generation of inflammation, familiar to us all. One way of viewing this is that the innate immune systems of multi-cellular organisms can generate "alarm" signals in response to danger, and that some of these signals cause inflammation. Alarm is not a conventionally used term, but is one that we find helpful and therefore will use it from time to time in this book. Inflammation enables effector cells and molecules to be targeted to the site of infection. As noted

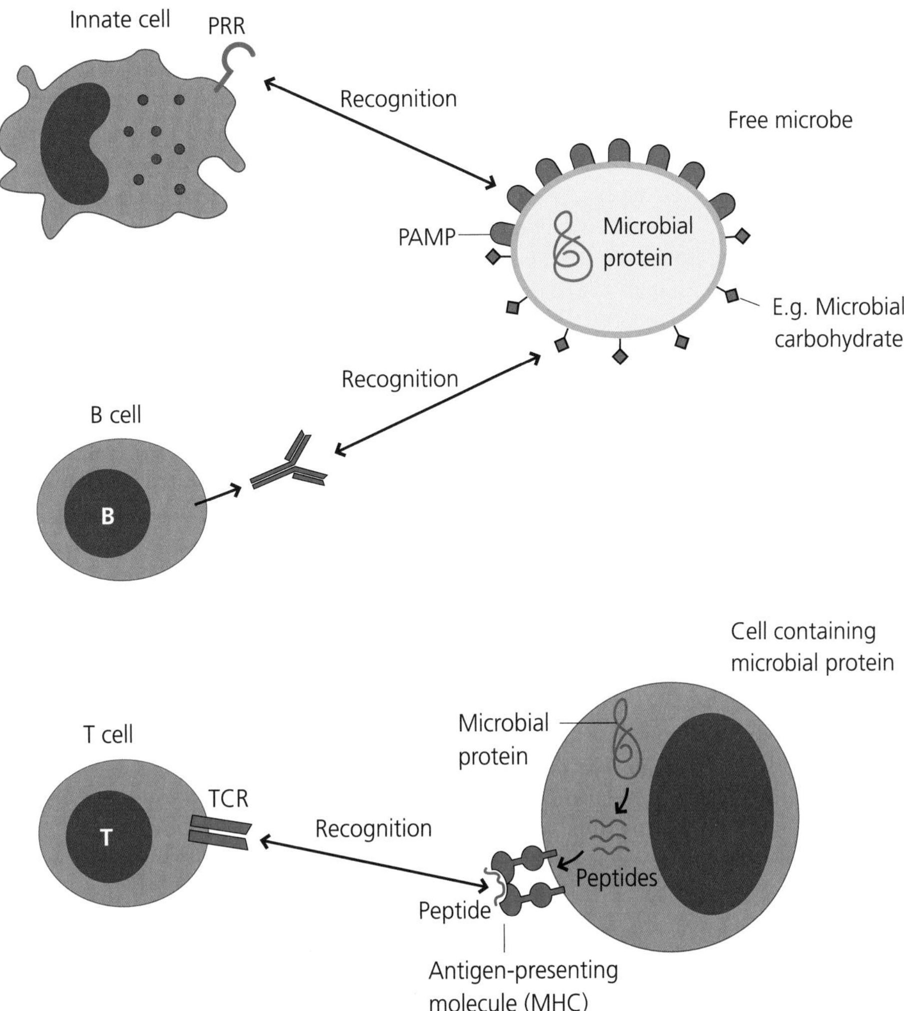

Fig. 1.2 Immune recognition. Innate immunity. PRRs directly or indirectly recognize conserved features of infectious agents called PAMPs. PRRs are widely expressed throughout the innate immune system. **Adaptive immunity.** The two main types of lymphocytes, B cells and T cells, have highly discriminatory receptors for microbial components or antigens, BCRs and TCRs respectively. These recognize antigens in totally different ways. BCRs can be secreted as soluble antibodies and bind to different types of antigen, such as carbohydrates on glycoproteins in their unfolded, native form. In contrast, TCRs generally recognize small peptides, generated by degradation of microbial proteins, in association with specialized presenting molecules (MHC molecules) on the surface of other cells (i.e. as peptide–MHC complexes).

above, other signals generated during innate responses can also determine whether, and in what way, the lymphocytes of adaptive immunity will respond.

The recognition of infectious agents in innate immunity is mediated by germline-encoded receptors called pattern recognition receptors (PRRs). These receptors generally recognize conserved features of infectious agents that are often shared by different classes of microbes, these microbial features are called pathogen-associated molecular patterns (PAMPs). PAMPs directly or indirectly stimulate innate immune responses by acting as agonists for PRRs. An agonist is anything that stimulates a response through a receptor, as opposed to an antagonist that inhibits it. PAMPs may bind directly the PRRs, therefore acting directly as ligands for these receptors, but some PAMPs can trigger responses by binding to a different molecule that then associates with a PRR, so it is useful to use the general term agonist. This also allows us to discriminate clearly between components of microbes that trigger innate responses, and molecular structures which are recognized in adaptive immunity that are termed antigens (below).

The cells responsible for initiating activation of the innate immune system are widely distributed in tissues and organs, and they possess many copies of different types of PRR that trigger rapid responses. This allows very rapid activation and deployment of the effector mechanisms of innate immunity. In many cases the innate system can eliminate the infectious microbe, often without any symptoms occurring (i.e. subclinically), and if there has been damage to the tissues at the site of infection the innate system will initiate repair and healing. Importantly, activation of the innate system is also essential for the triggering of adaptive immune responses

1.2.3.2 Recognition in Adaptive Immunity: Antigen Receptors

The key components of the adaptive immune system are the lymphocytes. It is convenient at this stage to divide these into two main groups (other types do exist). One group is the T lymphocytes (T cells) which have evolved to interact with other cells. The other is the B lymphocytes (B cells) which are the precursors of cells that can make soluble antibodies. The recognition of molecules from infectious agents by lymphocytes is mediated by their specialized antigen receptors, which are not present on cells of innate immunity. An antigen can be defined as a molecular structure against which a specific adaptive immune response can be made. In contrast to PRR agonists in innate immunity (above), the antigens which stimulate lymphocyte responses are generally unique to particular infectious agents no matter how closely they are related.

Collectively, lymphocytes express a vast range or repertoire of antigen receptors of different specificities, but each lymphocyte expresses multiple copies of a receptor of only a single given specificity. The term specificity relates to the particular antigen(s) that each lymphocyte is able to recognize. These receptors are generated by rearrangement of germline DNA, a process that is not known to occur for any other type of molecule. Their specificity is generated largely at random and in advance of any infection. Lymphocyte recognition of antigen is thus anticipatory. Lymphocyte antigen receptors are highly discriminatory and distinguish between even very small differences in antigens, such as an amino acid substitution in a peptide or a specific side chain in an organic molecule.

The vast repertoire of T and B cell antigen receptors means that a lymphocyte expressing a particular receptor is exceedingly rare. In addition, lymphocytes are normally small, relatively inactive cells. Thus, adaptive immune responses require the activation and proliferation of specific clones of B cells or T cells in order to reach a critical mass that can deal with the infectious agents and this takes time. Adaptive responses also need to generate the particular effector mechanisms that are most suited to eliminate the infection and this also takes time. Hence, adaptive immune responses are generally slower to be triggered than innate responses. However, following infection, increased numbers of antigen-specific lymphocytes remain in the body and these provide stronger, more rapid responses should re-infection occur.

1.2.3.3 Types of Recognition in Innate and Adaptive Immunity

Earlier, we suggested that some components of innate immunity can be viewed as recognizing danger, and that in general they are able to discriminate between harmful and harmless stimuli. Danger can be represented by the presence of an infectious agent or by signs of cell damage or stress that may or may not be associated with infection. In contrast, the antigen receptors of lymphocytes enable them to discriminate very precisely between what is "self" (any normal component of the host) and "non-self" (such as a component of a microbe), but not between harmless and harmful. Therefore, it is primarily the signals generated by the innate system that inform lymphocytes as to whether the antigens that they are recognizing originate from harmful or harmless agents, and hence determine whether or not these lymphocytes become activated and also precisely how they need to be activated in the context of the specific danger that is posed. If they are instructed that there is no danger (or not instructed that there is danger), then lymphocytes become unresponsive, or "tolerant", to what they are recognizing (Section 1.6.1).

1.2.4 Stages of Immunity

Immune responses to infectious agents involve a sequence of events that usually occur in distinct anatomical compartments. Immunologists divide tissues into lymphoid tissues, on one hand, and non-lymphoid or peripheral tissues, on the other. As we shall see (below), lymphocytes are produced in primary lymphoid tissues, but tend to localize in specialized sites, the secondary lymphoid tissues. In contrast, the cells of innate immunity are distributed throughout all tissues and organs. Once the natural barriers are breached by an infectious agent, innate agonists such as PAMPs, derived from the infecting agent, trigger the activation of innate immunity. Infections most commonly start in peripheral, non-lymphoid

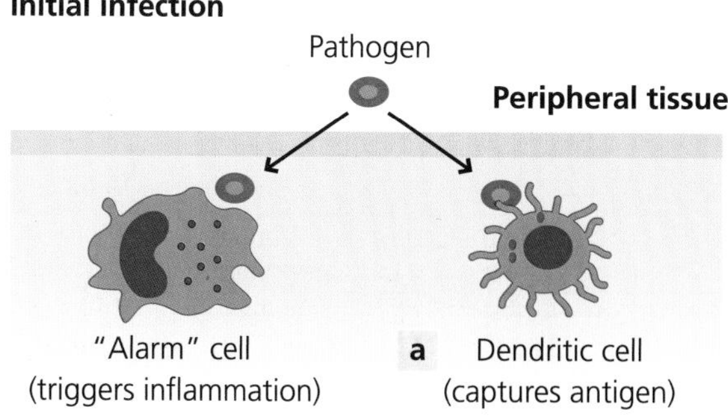

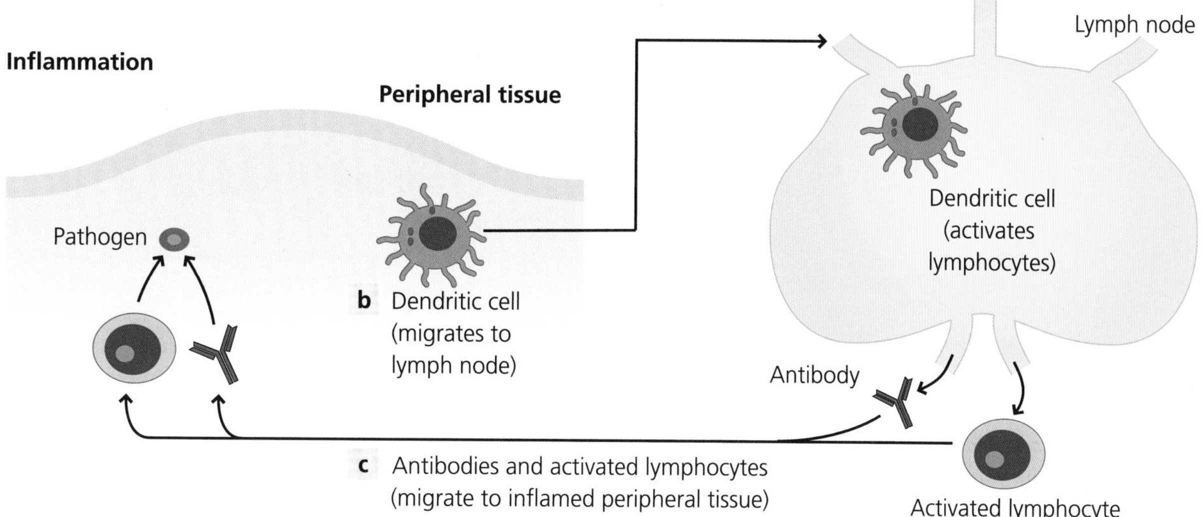

Fig. 1.3 Stages of immunity. Innate and adaptive immunity are closely interlinked. Specialized local ("alarm") cells of innate immunity can sense the presence of infectious agents. Consequent inflammation enables blood-borne innate effector cells and molecules to enter the tissue. (a) Dendritic cells (DC) at the site of infection sense the presence of an infectious agent and capture molecules (antigens) from it. (b) They migrate into secondary lymphoid tissues and activate lymphocytes that are specific for the infectious agent. (c) Some lymphocytes then make **antibodies** that circulate in blood to the site of infection and attack the infectious agent; other lymphocytes enter sites of infection and help or recruit other cells to kill the infectious agent, or directly kill infected cells.

tissues, and it is the resident innate cells such as macrophages and mast cells in these tissues that first recognize innate agonists. The subsequent inflammatory response is then crucial for the rapid recruitment of other innate effector cells and molecules, such as specialized phagocytes and complement, into the site of infection to help eliminate the microbes that are present there.

Adaptive immunity is also initiated by the infection and becomes evident a few days to a week later. For this to happen, antigen needs to be transported to secondary lymphoid tissues (e.g. to lymph nodes or spleen; below) during the innate phase so that it can be recognized by lymphocytes that tend to congregate in these organs. The molecules responsible for antigen recognition are the respective antigen receptors of the two main groups of lymphocytes: T cell receptors (TCRs) and B cell receptors (BCRs, that can later be secreted as antibodies). The activated lymphocytes then either become effector cells themselves (e.g. some T cells develop into cytotoxic cells that can kill virally infected cells, or B cells develop into plasma cells that secrete antibodies) or they can recruit and activate other effector cells at the site of infection, often including cells of innate immunity. Together, these different effector mechanisms, which are recruited into sites of infection because of inflammation, in most cases control and ultimately clear the infection. Finally, as noted above, in most forms of adaptive immunity, memory lymphocytes are generated in lymphoid tissues. Once induced, memory T and B cells localise to non-inflamed peripheral tissues or secondary lymphoid organs and, in the case of re-infection with the same organism, generate rapid, stronger adaptive responses. See Figure 1.3.

1.3 Anatomical Basis of Immunity

Immune responses occur in complex living organisms. Infection has driven the evolution of tissues and organs in which these responses take place. To understand immune responses requires an understanding of the structure of these tissues and organs in relation to their immune functions. For the

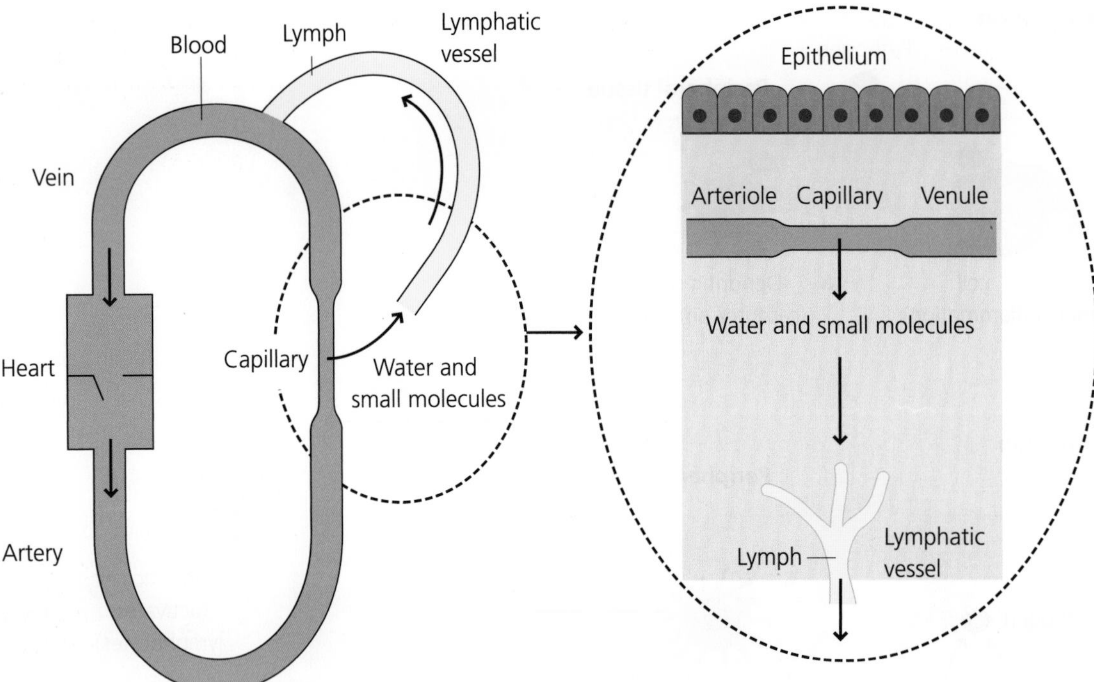

Fig. 1.4 Blood and lymph. Blood transports water, oxygen and small molecules that diffuse into extravascular tissues across endothelial cells of small blood vessels. Blood also carries leukocytes (white blood cells), but these cells, and larger molecules such as antibodies, can only enter tissues in large amounts at inflammatory sites (Figure 1.3). Extravascular fluid collects into lymphatic vessels and ultimately re-enters blood. Lymph also transports cells from sites of infection into lymph nodes, enabling adaptive immunity to be triggered (Figure 1.3).

benefit of readers who have little or no background in anatomy and histology we will first give a very general introduction to this area.

1.3.1 Peripheral Tissues

For immunologists, peripheral tissues refer to most tissues and organs of the body. They exclude primary and secondary lymphoid organs (below). A generalized tissue or organ is covered by an epithelium, a continual layer of cells that is often in contact with the external environment. Examples are the outer layers of the skin, and the lining of the gastro-intestinal, respiratory and urogenital tracts. Underlying the epithelium there is usually a basement membrane and under this lies loose connective tissue containing cells such as macrophages, mast cells and fibroblasts, and structural components such as collagen fibres (e.g. Figure 1.10).

All tissues and organs are vascularized they have a blood supply. It is crucial for water and solutes, such as nutrients and dissolved oxygen, to be able pass through the linings of blood vessels into the surrounding tissues, so that the cells in these extravascular sites can be nurtured and maintained. This extravascular fluid with its solutes then needs to be collected, as lymph, and returned back into the blood. Therefore, all tissues and organs contain vessels of two types: blood vessels and lymphatics (see Figure 1.4).

Blood leaving the heart is brought to tissues by arteries, which subdivide to form arterioles and ultimately capillaries. The latter are the sites where molecules normally diffuse into and from extravascular tissues. Capillaries join to form venules, these are the vessels where cells migrate into tissues under both normal and inflamed conditions. The capillarity then join to form venules that join to form veins, which ultimately return the blood to the heart. Lymphatics originate from blind endings in tissues. They collect molecules as well as motile cells from the extravascular spaces in tissues and transport them to lymph nodes in afferent lymph. Lymph nodes serve as filters and are sites where adaptive immunity can be initiated. Efferent lymph, leaving the lymph nodes, then carries cells and molecules into larger vessels. The main one is the thoracic duct, which then drains back into the blood.

During immune responses the blood acts as a delivery system. Blood vessels and lymphatics are lined by flattened endothelial cells. The blood transports leukocytes, which are the white blood cells as well as erythrocytes, which are the red blood cells (RBCs). If molecules on the plasma membranes of leukocytes recognize complementary molecules on the endothelium, they can adhere to it and may subsequently migrate into the extravascular spaces of the local tissue. As noted above, some small molecules can normally diffuse freely out of the blood in capillaries and venules, but most macromolecules are retained in the blood. Such macromolecules can, however, be delivered selectively to sites of inflammation because gaps open up between venule endothelial cells at these sites (e.g. Figure 1.6).

Some tissues are called "mucosal". In general, but not always, this is because their epithelia contain goblet cells

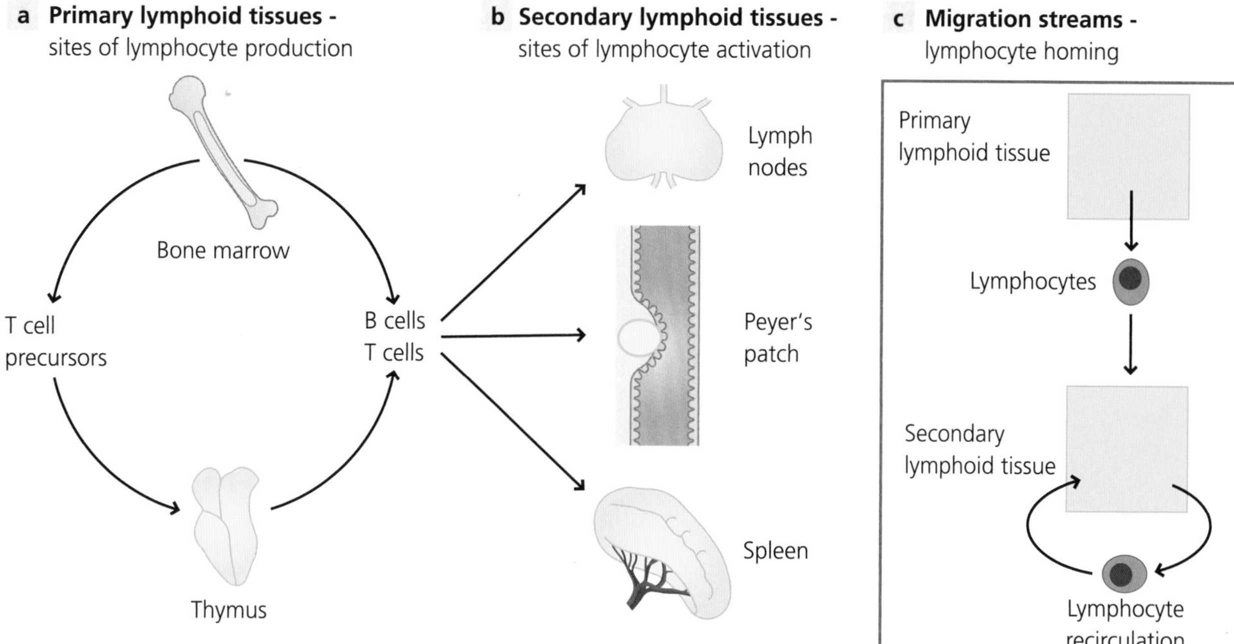

Fig. 1.5 Primary and secondary lymphoid tissues. (a) Primary lymphoid tissues are sites where lymphocytes are produced. In adult humans and mice the main sites are the *bone marrow* and the *thymus*, where B cells and T cells, respectively, undergo most or all of their development. (b) **Secondary lymphoid tissues** or organs are sites where adaptive responses are induced and regulated. The main tissues are lymph nodes, the spleen and specialized MALTs such as the Peyer's patches in the small intestine. All have highly organized areas containing T cells and B cells. Lymphocytes recirculate between blood and lymph and can monitor different anatomical compartments for infection and mount appropriate responses.

which secrete mucus onto the epithelial surface; mucus acts as a lubricant and as a barrier to infection. Mucosal tissues include the gastro-intestinal tract, the respiratory and urogenital tracts, the eye, and the lactating mammary gland. From an immunological point of view mucosal tissues are important because they are associated with immune responses that differ in important respects from those associated with other peripheral tissues.

1.3.2 Lymphoid Tissues

Lymphoid tissues are those whose major functions are associated with adaptive immune responses. Primary lymphoid tissues are sites where lymphocytes are produced. In adult humans they include the bone marrow and a specialized organ called the thymus, which is the organ in which T cells develop. The bone marrow is also the site where other leukocytes are generated, such as monocytes (which can develop into macrophages) and granulocytes (which are themselves of different types). The process of generating blood cells is called haematopoiesis (below), so the bone marrow can also be termed a haematopoietic organ. In contrast, the secondary lymphoid tissues are the sites where adaptive immune responses are initiated and regulated. These tissues include lymph nodes, which are distributed throughout the body, and the spleen. There are also specialized secondary lymphoid tissues associated with mucosaecalled mucosal-associated lymphoid tissues (MALTs), such as Peyer's patches in the small intestine and the tonsils and adenoids in the throat and nose. See Figure 1.5.

1.3.3 Inflammatory Sites

Usually, recognition of infection in a peripheral site by the innate immune system leads to local inflammation. This is clinically evident as reddening, heat, swelling and pain at the site of infection. One of the main functions of inflammation is to ensure that the cells and molecules needed to deal with the infecting agent are delivered to the right place at the right time. These include, for example, specialized phagocytes and complement components and, often later, lymphocytes and antibodies.

If the infectious agent can be eliminated rapidly (within days or a week or so) the response is known as acute inflammation, and repair and healing of the tissue follows. Typical examples known to almost everyone are abscesses (boils) and sore throats. If, however, the infectious agent persists, continuing inflammation, termed chronic inflammation, leads to continuing tissue damage. This is typically seen with infections such as tuberculosis which may persist for months or years (See Figure 1.6).

More distant sites can also be affected by inflammation at sites of infection, these changes are called the systemic effects of inflammation. Thus, the bone marrow can be stimulated to produce more leukocytes, the liver to produce larger amounts of soluble effector molecules that can

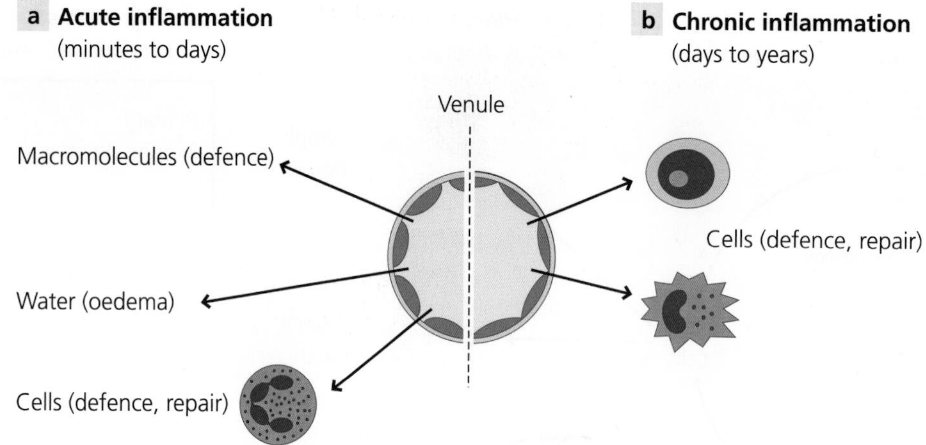

Fig. 1.6 Acute and chronic inflammation. (a) Acute inflammation. Infectious agents such as extracellular bacteria trigger inflammation that starts quickly and lasts a relatively short time. The main features are recruitment of cells and molecules such as neutrophils and complement, and the accumulation of extravascular fluid (oedema). (b) **Chronic inflammation.** Infectious agents that are not quickly eliminated and which trigger adaptive immune responses can lead to chronic inflammation that can last much longer. The main features are recruitment of blood monocytes, which become macrophages, and activated lymphocytes, especially T cells, and sometimes the development of long-lasting granulomas.

contribute to eliminating the infectious agent and the hypothalamus in the brain to induce fever.

1.4 Cellular Basis of Immunity

The primary function of the immune system is to eliminate infectious agents that have breached the natural barriers. This is brought about by the integrated actions of different cells and molecules that, directly or indirectly, lead to recovery from infection. It is the effector cells and molecules that actually bring about the elimination of infectious agents. To simplify understanding; we focus first on the cellular basis of immunity, before turning to molecules involved in immunity in Section 1.5. This is an artificial distinction, since cells and molecules are usually both involved in immune responses, as will be seen in the rest of this book.

1.4.1 Origin of Immune Cells

1.4.1.1 Haematopoiesis

Haematopoiesis is the name given to blood cell development. In the adult mammal this takes place primarily in the bone marrow (also a primary lymphoid tissue), but in the foetus this happens in the liver. The earliest cells in haematopoiesis are the multi-potential stem cells, which can give rise to any and all blood cells. When cells divide each of the progeny cells is usually of the same type. Stem cells are unusual in that the one of the progeny cells is another stem cell, thus maintaining this population, while the other can start to differentiate into a different cell type, ultimately producing all the cells of different lineages.

Initially stem cells divide to form cells that can either give rise to different types of "myeloid" cells or "lymphoid" cells – these are the common myeloid progenitor(CMPs) and common lymphoid progenitors (CLPs). In turn, the former precursors can ultimately produce monocytes and macrophages, different types of granulocytes, and other cells including erythrocytes (RBCs), megakaryocytes (the precursor of blood platelets) and some DCs. In contrast, the latter give rise to the lymphocytes (T cells and B cells) as well as natural killer (NK) cells and other DCs. The development of cells in each pathway is regulated by growth factors that are more or less restricted in their activities. Thus, interleukin (IL)-3 and stem cell factor (SCF) stimulate stem cells which can develop into many different lineages, granulocyte macrophage colony-stimulating factor (GM-CSF) preferentially stimulates development of monocytes, granulocytes and some DCs, and erythropoietin stimulates red cell development. See Figure 1.8.

1.4.1.2 Lymphopoiesis

The term lymphopoiesis refers to the generation of lymphocytes, in contrast to myelopoiesis which refers to generation of myeloid cells (above), and thus can be viewed as a subset of haematopoiesis. In the adult, common lymphoid precursors are present in the bone marrow. B cells start to differentiate in the bone marrow and are released as partially mature (transitional) B cells into the blood. T cell precursors, however, migrate directly to the thymus and T cells complete their development in this organ (a primary lymphoid tissue). It is important to realize that lymphocytes can also undergo further rounds of division and development outside the primary lymphoid organs. This normally happens after they have recognized the antigens for which they are specific, such as molecular components of infectious agents and occurs in organs such as lymph nodes, spleen and Peyer's patches (secondary lymphoid organs). The phenomenon of clonal expansion is an essential feature of adaptive immune responses for it enables clones of specific lymphocytes to

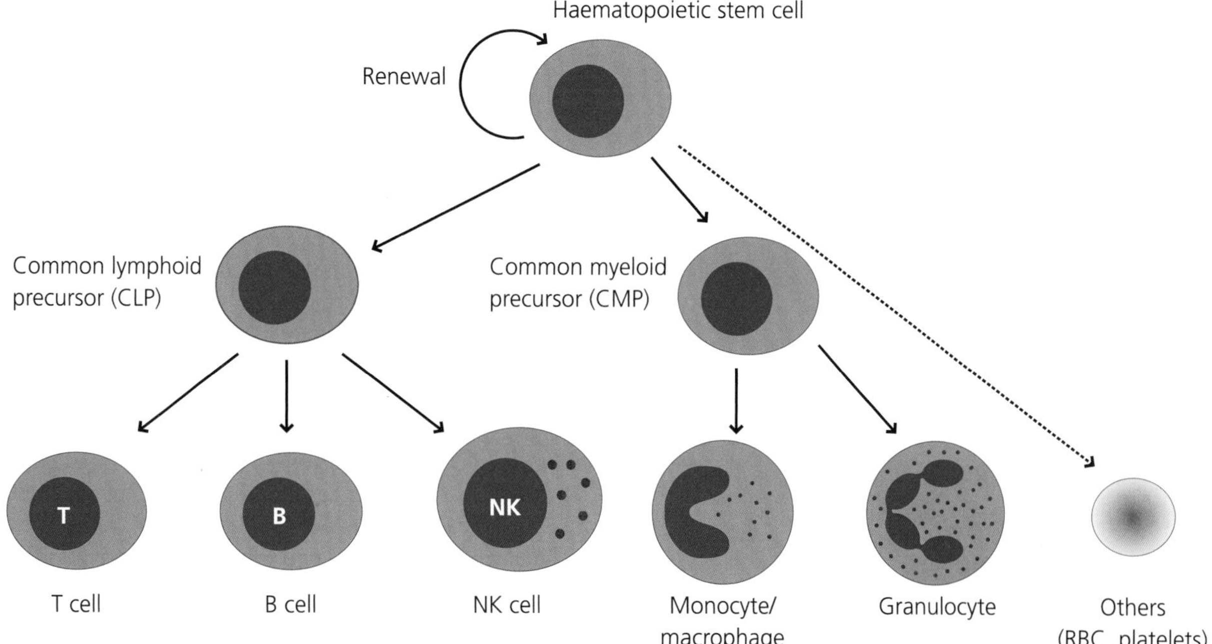

Fig. 1.7 Haematopoiesis. All blood cells are formed from haematopoietic stem cells. On division they form another stem cell (self renewal) and a more committed precursor for a blood cell. In turn, the CLP gives rise to different types of lymphocyte as well as natural killer (NK) cells, while the CMP can generate monocytes and macrophages, granulocytes, and other cell types. The production of different blood cells is regulated by different growth factors and is largely under feedback control.

expand and reach a critical mass that can now help to deal effectively with eliminating the infection.

1.4.1.3 Cell-Mediated Immunity

The principal mechanisms by which cells can help to defend against infection are the following:

i) Killing the microbe directly (e.g. bacteria are often taken up and killed intracellularly by phagocytes).
ii) Killing the cells that harbour microbes (e.g. killing of virally infected cells to prevent the release of new viruses).
iii) Preventing access to, or expelling the microbe from, the body and providing defence against larger parasites such as worms in the gut.

These cellular responses, which contribute to cell-mediated immunity as a whole, need to be tightly regulated and coordinated and, quite often, different cells need to collaborate to help to bring about an overall response that lead to the elimination of an infectious agent. See Figure 1.8.

1.4.1.4 Introduction to Cells of the Immune System

The different cells of the immune system can be classified according to overlapping principles. The most important classification is functional – what do they do? This relates to other aspects – their developmental origins and relationships, and their anatomical distribution in the body. Cells are also classified by their morphology, but this can be misleading – naïve B and T cells look more or less the same under the light microscope – and it is more accurate to combine morphology with other approaches such as identification of the surface molecules that the cells express. Based on the these criteria we will now briefly introduce the major groups of cells involved in immune responses before discussing each in more detail. We will divide the cells into those primarily thought of as belonging to innate immunity and those of adaptive immunity, while emphasizing that in reality there is much overlap between the two groups.

1.4.1.5 Cells of Innate Immunity

Macrophages These are divided into two main types:

- Resident macrophages are present in steady-state tissues (i.e. before infection occurs) and can detect the presence of microbes. In turn they can help to trigger inflammation.
- Recruited (or elicited) macrophages are not tissue-resident cells, but they develop from circulating precursors called monocytes that can be recruited into sites of infection. After development into macrophages they can act as effector cells to help eliminate the infection.

Macrophages are one of the two main types of specialized phagocyte that can engulf and internalize (phagocytose), and subsequently kill microbes such as bacteria.

Mast Cells These cells also reside in steady-state tissues and can detect the presence of microbes. Mast cells contain granules that are discharged when they are stimulated, and the granule contents can contribute to triggering of local inflammation.

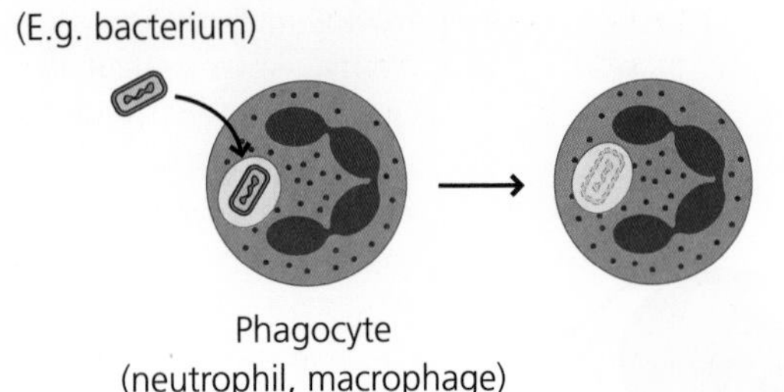

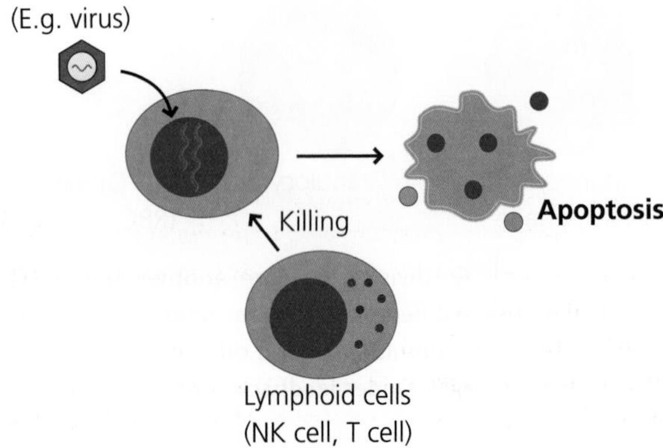

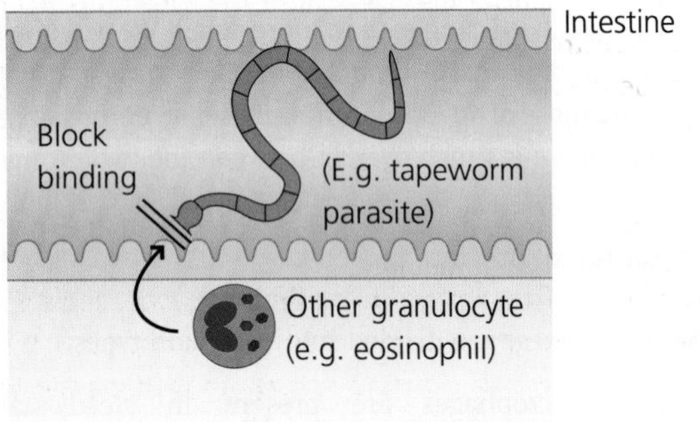

Fig. 1.8 Mechanisms of cell-mediated immunity. There are three main ways in which cells can eliminate infectious agents or protect against infection. (a) **Direct killing**. Specialized phagocytes, mainly macrophages and neutrophils, can internalize microbes such as bacteria and destroy them intracellularly. (b) **Killing of infected cells**. NK cells and cytotoxic T cells can kill infected cells, preventing the replication and release of microbes such as viruses. (c) **Maintaining natural barriers**. Larger parasites, such as intestinal worms, cause damage to the epithelium, risking entry to the body. Cells such as eosinophils can help to repair the damage and, in some cases, may help to expel the organisms or kill it by secreting toxic substances.

Granulocytes These are the major populations of blood leukocytes, the circulating white blood cells. They are called granulocytes because they contain cytoplasmic granules that are visible under the light microscope. They have nuclei that possess two or more lobes, and are thus also called polymorphonuclear leukocytes to distinguish them from monocytes and lymphocytes, the mononuclear leukocytes, which do not have lobulated nuclei. Granulocytes are divided into three groups derived from a common precursor: neutrophils, which are abundant in blood and are a very important type of phagocyte; the rarer eosinophils, and the basophils that are somewhat related to mast cells in function. A crucial feature of granulocytes is that they normally circulate in the blood but can be recruited selectively into inflammatory sites in response to different types of infection. (Note that although mast cells also contain granules, they do *not* circulate in the blood in a mature form and are therefore by definition *not* included in the term granulocyte; they may also originate from a distinct progenitor.)

NK Cells NK cells are developmentally related to lymphocytes, but differ in many aspects from them. NK cells are present in tissues as resident cells and can also be recruited to sites of inflammation. They can kill other cells, such as virally infected cells (i.e. they have cell-killing (cytotoxic) activity) and they also regulate immune responses.

1.4.1.6 Cells of Adaptive Immunity

Lymphocytes These are the primary cells involved in adaptive immune responses. As noted earlier, they have highly discriminatory antigen receptors and are divided into two main groups: B cells and T cells. B cells are the precursors of plasma cells that secrete antibodies. T cells are themselves divided further into two main groups: CD4 T cells that function principally as regulator and coordinator cells in adaptive immune responses, and CD8 T cells that can develop into cytotoxic cells with the capacity to kill cells infected with viruses or other microbes. Cytotoxic T lymphocytes (CTLs) and NK cells, noted above, are the two main types of cytotoxic cells in the immune system.

1.4.1.7 Cells that Directly Link Innate and Adaptive Immune Responses

DCs are another cell type, which actually exist as several subgroups. Most of these are involved in the initiation and regulation of adaptive immune responses. They can be viewed as linking innate and adaptive immunity because they respond quite rapidly to the presence of infection and subsequently change their properties so they can activate lymphocytes, particularly the CD4 T cells.

1.4.2 Phagocytes

All cells in the body are capable of sampling their extracellular milieu in a general process called endocytosis. They may do so

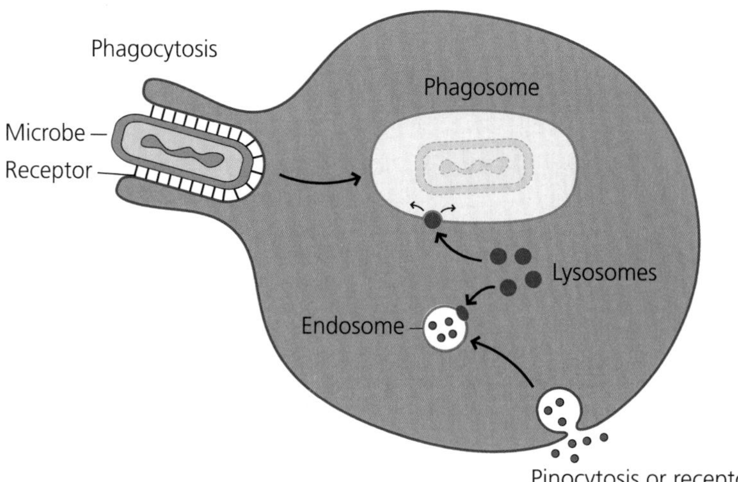

Fig. 1.9 Mechanisms of endocytosis. **Phagocytosis**. Phagocytes can internalize particles such as bacteria into vesicles called phagosomes. **Pinocytosis**. All cells engulf small quantities of fluids and their dissolved solutes. These are taken up in membrane-bound vesicles called (primary) endosomes. In all cases, the primary endosomes or phagosomes then fuse with other intracellular vesicles called lysosomes to form secondary endosomes or phagolysosomes in which the internalized contents can be degraded. This can lead to killing of phagocytosed microbes.

by internalizing small samples of fluid and solutes into membrane-bound vesicles called endosomes. This is the process of fluid-phase endocytosis or pinocytosis. It is used by cells, for example to obtain nutrients, and in some cases cell surface receptors extract the required nutrient molecules in a related process called receptor-mediated endocytosis. Some specialized cells are also capable of internalizing particles through the process of phagocytosis; the particles are contained in membrane-bound vesicles called phagosomes. In all forms of endocytosis, including phagocytosis, membrane-bound vesicles called lysosomes, containing catabolic enzymes, can fuse with the endocytic vesicle or the phagosome and release the enzymes into the vesicle. Generally speaking this usually leads to the degradation of the vesicular contents. Note that the term endocytosis can be used in a general sense to include phagocytosis (as we do), but is sometimes used for all processes other than phagocytosis. See Figure 1.9.

Phagocytes are specialized cells that are able to internalize particles such as bacteria and small protozoa, and are able to kill microbes intracellularly. The two main classes of phagocytes are macrophages and neutrophils. Other cells such as eosinophils are also weakly phagocytic, but this is not known to be one of their main purposes.

1.4.2.1 Macrophages

Macrophages reside in almost every tissue of the body and are important components of *both* innate and adaptive immune responses. Circulating precursors of macrophages, which are produced in the bone marrow, are called monocytes. When a monocyte enters a tissue it may develop into a form called mature macrophage.

Macrophages in tissues are generally long-lived cells (tattoos, which last for a lifetime, represent ink particles inside macrophages). Resident macrophages in connective tissues and solid organs play crucial roles in the remodelling and growth of organs during development and in regulation of normal tissue functions (homeostasis). Tissue-resident macrophages possess specialized receptors that recognize cells that have been programmed to die through the process of apoptosis and assist in tissue remodelling without scarring (e.g. during embryogenesis).

Infection most often begins in peripheral tissues and resident macrophages play very important roles in the initiation of innate immune response. They possess receptors (PRRs) that can recognize different types of infectious agent, enabling them to distinguish between broad classes of viruses, bacteria and fungi for example. Some of these receptors enable the macrophage to phagocytose the infectious agent so it can be destroyed intracellularly. Importantly, other PRRs signal to the nucleus to change gene expression, leading, for example, to the secretion of small, hormone-like proteins (cytokines; below) which act on other cell types. Crucially, some of the cytokines that resident macrophages secrete early in infection (as alarm signals; Section 1.2.3.1) then help to stimulate local inflammation by modifying the structure and function of the endothelial cells of blood vessels. Later they may also stimulate more distant tissues, inducing systemic inflammatory responses. See Figure 1.10.

Once the inflammatory process has been initiated, large numbers of monocytes can be recruited from the blood to the site of infection. These cells develop into inflammatory macrophages with a much larger repertoire of functions than the resident cells. (Depending on how they are being studied these cells are sometimes also called elicited macrophages.) For example, they are much more phagocytic and secrete an enormous diversity of molecules that help to amplify the inflammatory response, bring in other types of effector cells to the area and eliminate the infectious agent. These and other macrophages may also contribute to the repair and healing of the tissue.

In some circumstances macrophages can acquire new and highly potent antimicrobial killing mechanisms. This is particularly the case once the later adaptive immune responses have been initiated, although it can happen earlier. A key macrophage-activating agent, produced in such responses is

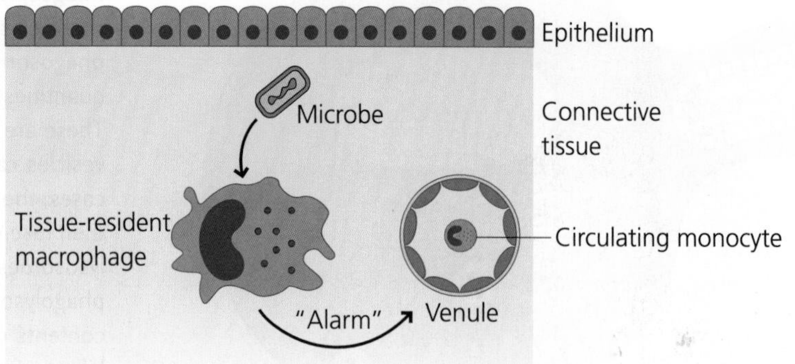

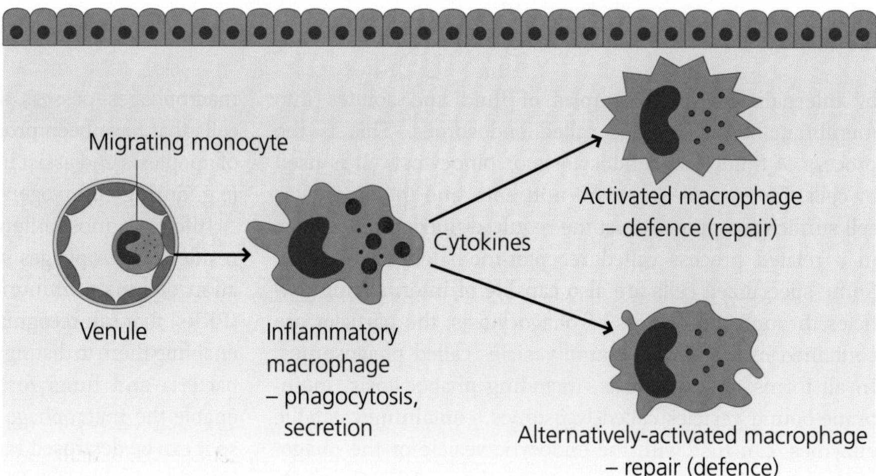

Fig. 1.10 Types of macrophages. Macrophages are plastic cells that can develop different specialized functions. **Tissue-resident macrophages** are long-lived cells present in connective tissues. In the steady state they are involved in homeostasis, helping to maintain the architecture of the tissue, and can also sense infectious agents and help to trigger inflammation (e.g. Figure 1.3). **Inflammatory macrophages**. Local inflammation recruits blood monocytes into tissues. In response to infectious agents these can become highly secretory cells that produce many innate effectors such as complement components. **Activated macrophages**. Innate and adaptive responses can help turn macrophages into potent anti-microbial cells. **Alternatively activated macrophages**. Macrophages can also develop functions involved in wound repair and healing, and may help to dampen down the inflammation.

the cytokine interferon (IFN)-γ. Activated macrophages do not have the ability to produce the wide variety of different mediators typical of elicited macrophages, but they concentrate on killing the microbes they have phagocytosed. For this they produce, for example, nutrient-depleting molecules that prevent growth and replication of the microbes, highly toxic agents that poison them, including reactive oxygen intermediates/species (ROIs/ROS), and a diversity of enzymes that digest them. Activated macrophages are very important in defence against certain bacteria, such as the mycobacteria, some of which cause tuberculosis and leprosy. In other circumstances different types of macrophage activation (e.g. alternative activation) or deactivation can be triggered, perhaps as sequential stages, depending on the local environmental conditions (e.g. which cytokines are present). You can see that the macrophage is therefore a highly adaptable cell type.

1.4.2.2 Neutrophils

Neutrophils are the most abundant leukocytes in human blood, comprising approximately 70% of total white blood cells. Neutrophils are crucial in defence against pyogenic (pus-forming) bacteria such as *Staphylococcus* and *Streptococcus* which typically cause abscesses and similar infections involving acute inflammation. Neutrophils are released from the bone marrow as mature cells and have very short lifespans (1–2 days). They are not present in normal tissues, but are rapidly recruited to sites of acute inflammation.

Neutrophils are highly phagocytic. They constitutively possess a wide variety of antimicrobial killing mechanisms (in contrast to macrophages, in which these are usually induced). Many of the molecules needed to generate killing mechanisms are preformed and stored in different types of cytoplasmic granules. When a microbe has been phagocytosed, some of these granules fuse with the phagosome, delivering their toxic contents – some of which are similar to those of activated macrophages (e.g. ROIs) – onto the microbe. Neutrophils may also have the capacity to kill microbes extracellularly. In addition to their anti-microbial agents the granules also contain proteolytic enzymes that breakdown (liquefy) the surrounding tissue. When neutrophils encounter bacteria that are able to resist killing, they die and in some infections, typically with *Staphylococcus aureus*, the liquefied tissues form the pus that discharges from abscesses. See Figure 1.11.

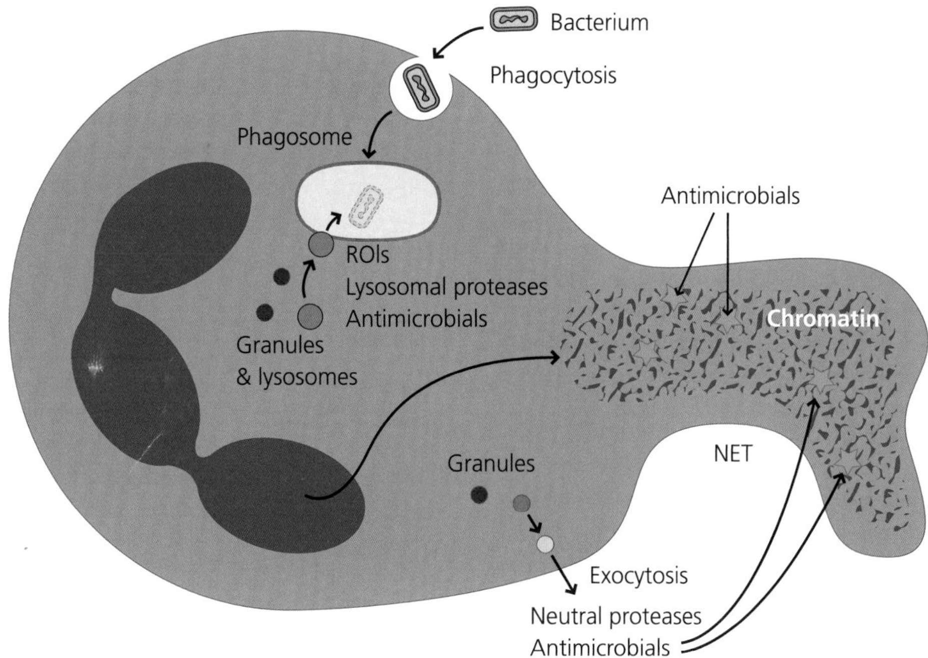

Fig. 1.11 Killing by neutrophils. Neutrophils can kill microbes intracellularly and perhaps extracellularly. **Intracellular killing**. After phagocytosis of microbes the phagosomes fuse with cytoplasmic granules that contain or can produce anti-microbial agents. The latter include reactive oxygen intermediates and other compounds that may be toxic to the microbes and proteases which may degrade them. **Extracellular killing**. These cells can also extrude **neutrophil extracellular traps (NETs)** composed of chromatin that are thought to trap extracellular microbes. The contents of other granules may subsequently discharge to release anti-microbial agents and proteases to help kill them, although this is not yet proven.

1.4.3 Mast Cells, Eosinophils and Basophils

As for the different phagocytes (Section 1.4.2), the functions of mast cells, eosinophils and basophils significantly overlap, although each, too, has its own specialized properties. Mast cells can be viewed as being resident cells in tissues, whereas the basophils and eosinophils are typically recruited to inflammatory sites. They are commonly found in the anatomical barriers of the body e.g. immediately under mucosal epithelia or in the skin.

1.4.3.1 Mast Cells

Mast cells are resident cells in mucosal tissues and loose connective tissues. They function, at least in part, by sensing infection or tissue damage and then triggering inflammation. Mast cells possess cytoplasmic granules that contain histamine and cytokines. Mast cells may be activated through their PRRs and by small molecules such as complement fragments. When activated these cells very rapidly release their stored mediators, but also synthesize cytokines and lipid mediators. These secreted molecules are generally involved in the induction of acute inflammation. The lipid mediators also induce contraction of smooth muscle, such as in the lung and airways and in the gut. In the latter case, muscle contraction may help expel parasites. Mast cells are, however, better known for the roles they play in pathological, allergic responses. See Figure 1.12.

1.4.3.2 Eosinophils

These granulocytes, normally representing 1–3% of all blood leukocytes, usually possess a bi-lobed nucleus and large cytoplasmic granules of different types that stain orange with the

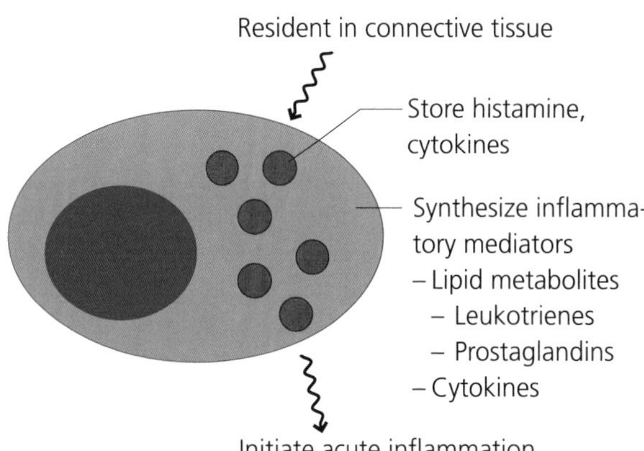

Fig. 1.12 Mediators produced by mast cells. Mast cells are present in all connective tissues. Mast cells have large cytoplasmic granules that store pre-formed inflammatory mediators such as histamine and cytokines, and which are discharged when the mast cell is activated (degranulation). Activated mast cells also synthesize inflammatory mediators including lipid metabolites and other cytokines when activated. Mast cells can be activated by pathogen molecules or by molecules produced during acute inflammation. They can also be activated during allergic reactions such as hay fever.

acidic dye, eosin (hence their name). They are recruited to certain types of inflammatory sites, particularly those associated with parasitic infections and allergic responses. As for mast cells, the contents of their granules are released when eosinophils are activated by cytokines. Other mediators can be synthesized later. Several functions of eosinophils overlap with those of mast cells (above). These include inducing the increased production of mucus, which may help prevent attachment of infectious agents to the underlying epithelial cells. Eosinophils may have a role in healing certain types of tissue damage, such as that caused by intestinal infestation of worms. Thus, eosinophils may be important in defence against parasites such as intestinal worms. See Figure 1.13.

1.4.3.3 Basophils

Basophils possess bi-lobed nuclei and their large cytoplasmic granules stain dark blue with basic dyes (hence the name). Basophils circulate in the blood in very small numbers (normally less than 1% of leukocytes) and are normally absent from tissues. However, they can be recruited into inflammatory sites, typically together with eosinophils (above). They are functionally similar to mast cells, although there are some differences, and basophils are also potent stimulators of inflammation.

Neutrophil

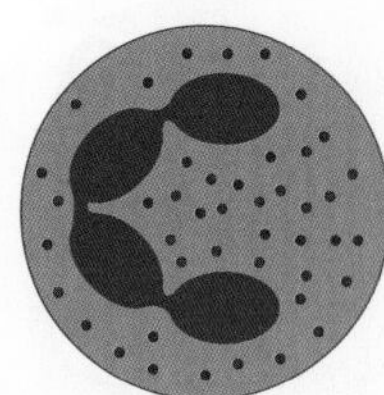

- Most abundant leukocyte
- Defence against extracellular pyogenic bacteria and fungi
- Numbers raised in pyogenic infection
- Granules of two main types

Eosinophil

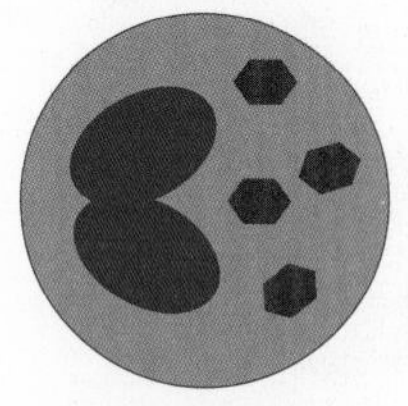

- Scarce leukocyte
- Defence against metazoan parasites
- Numbers raised in parasite infection (metazoan worms)
- Numbers raised in allergies
- Crystalloid cytoplasmic granules

Basophil

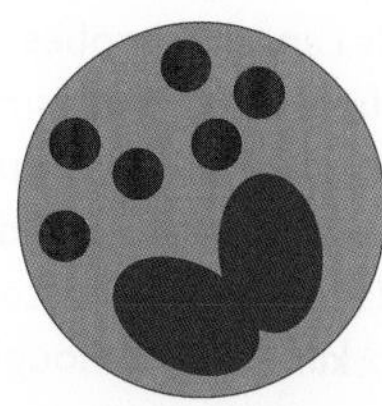

- Rare leukocyte
- Large dense granules
- Similarities to mast cells
- Uncertain role in defence

Fig. 1.13 Types of granulocytes. Granulocytes are leukocytes that contain different cytoplasmic granules with characteristic staining properties. All can be recruited to different sites of inflammation. **Neutrophils** are the most abundant leukocytes in blood. Large numbers can be recruited rapidly to sites of acute inflammation (Figure 1.6) and their production from the bone marrow can be greatly increased during some bacterial infections. They are highly active phagocytes (Figure 1.11). **Eosinophils** are normally present in low numbers, but can be recruited particularly into mucosal tissues. Larger numbers can be produced from the bone marrow during other types of infection, especially with large parasites against which they may provide resistance or defence, and in allergic responses such as asthma. **Basophils** are normally very rare in blood. They, too, can be recruited to inflammatory sites where they typically accompany eosinophils. They have very similar properties to mast cells but their roles in host defence are poorly understood.

1.4.4 Natural Killer Cells

All viruses need to infect cells in order to replicate. To limit viral replication and the release of new virions, immune cells have evolved that can kill virally infected cells and perhaps some tumour cells. The ability of an immune cell to kill another cell in this way is termed cellular cytotoxicity and the cell that does the killing is generally referred to as a cytotoxic cell. Such cells kill by triggering a process called apoptosis in the cells they recognize. All cells have the potential to undergo apoptosis (e.g. during tissue remodelling) and some specialized cells of the immune system have evolved mechanisms to latch on to this process; they therefore kill by enforced suicide rather than murder of their targets. The advantage of triggering this form of cell death is that the contents of the dead and dying cell can be rapidly eliminated, usually without triggering inflammation. One type of cell that can do this, although it may not be its principal function in host defence, is the NK cell. natural killer (NK) cells were first recognized by their ability to kill tumour cells in culture as soon as they were isolated from blood. They are present in normal blood and organs such as spleen and liver, but other NK cells can be recruited to inflammatory sites. Owing to their spontaneous cytotoxic capacity, they were termed natural killer cells – as opposed to cytotoxic T cells (CTLs) that need to be activated before they can kill (Section 1.4.5.2).

NK cells resemble lymphocytes, but in their resting state they contain large cytoplasmic granules. These granules contain specialized molecules, including one called perforin that can insert into membranes of the target cell and polymerize to form pores that allow entry of other molecules (particularly specialized enzymes called granzymes) that can trigger death of the target cell. This may be important in defence against some viral infections (e.g. herpes viruses). As the molecules contained in granules are pre-formed, NK cells can kill very rapidly. NK cells additionally possess membrane molecules such as Fas ligand which binds to death-inducing Fas on target cells and can also trigger apoptosis. (In contrast, these components normally have to be induced over time in CTLs.) Crucially, NK cells additionally secrete a variety of cytokines that induce or regulate the activity of other immune cells. See Figure 1.14.

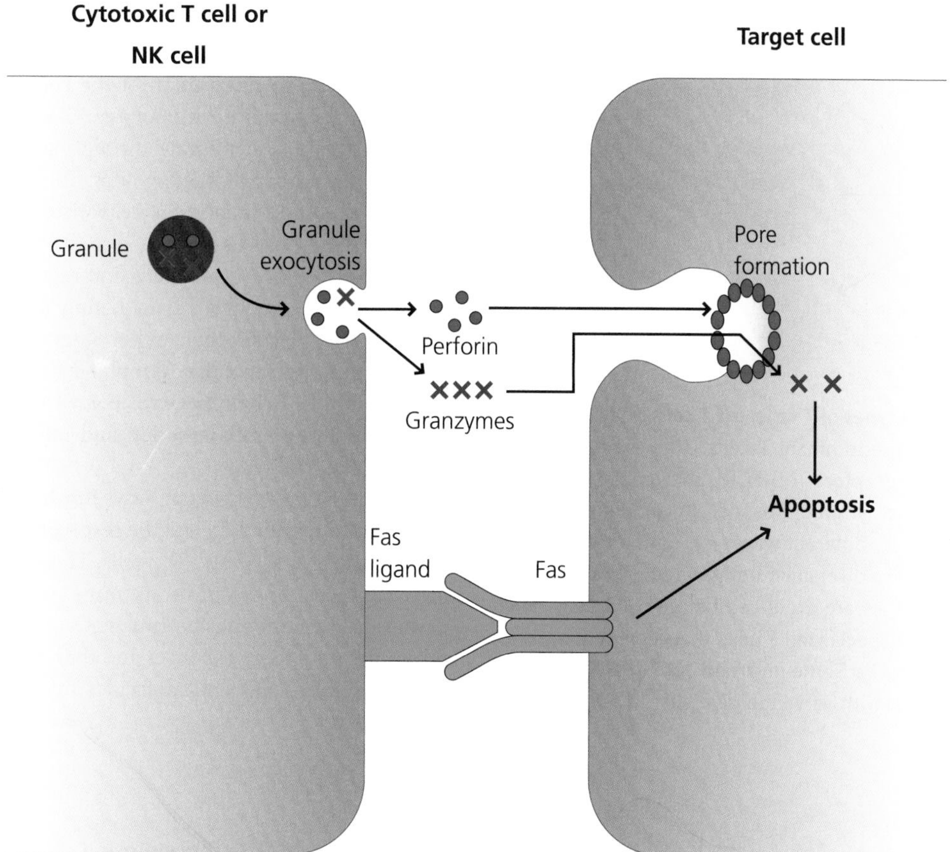

Fig. 1.14 Mechanisms of cellular cytotoxicity. NK cells and cytotoxic T cells can kill infected cells by inducing apoptosis in a process called cellular cytotoxicity. **Granule-dependent mechanisms**. Pre-existing (NK cells) or newly generated (CTLs) granules store molecules which can trigger apoptosis. These include perforin that can polymerize to form pores in target cell membranes and enables molecules such as granzymes to enter the cytosol where they trigger apoptosis. **Granule-independent mechanisms**. Cytotoxic T cells and NK cells can also express cell surface molecules such as Fas ligand which, when it bind to Fas on the target cell, initiates apoptosis in the target cell.

So how do NK cells recognize and kill their target cells? As we shall see later (Chapter 4) it turns out that NK cells possess a large set of specialized receptors that belong to different structural families and which are entirely different from PRRs or lymphocyte antigen receptors. Essentially, NK cells monitor the levels of specialized molecules that are present on other cells (particularly major histocompatibility complex (MHC) molecules; below). If these levels are normal, killing by the NK cells is inhibited by inhibitory receptors. However, if they are significantly reduced, typically because a virus has infected a cell, activating receptors can become engaged that overcome the repressed state and lead to killing by the NK cells. The ability of NK cells to kill other cells that lack (or have very low levels) of normal components was originally termed the missing self hypothesis.

1.4.5 Lymphocytes

Lymphocytes are the cells that mediate adaptive immunity. In the resting state they are inactive "naïve" precursor cells that need to be stimulated by antigen, and usually by other signals before they become fully activated. The mechanisms of activation are complex, but the dangers of inappropriate activation (e.g. leading to autoimmune diseases; Chapter 7) are so great that there need to many points at which activation can be regulated. Once it has been activated a lymphocyte can become an effector cell, which is usually short-lived, or a long-lived memory cell capable of being reactivated should an infection be reoccur.

In this section we will introduce two major classes of lymphocytes, and briefly describe their properties and functions. As noted earlier, these comprise the T cells – which are in fact of two main types, CD4 T cells and CD8 T cells – and the B cells. These are the "conventional" types of lymphocyte that play central roles in adaptive immunity. Conventional T cells are also called αβ T cells. There are in addition other types of less conventional (or "unconventional") lymphocytes, including γδ T cells and NKT cells (not to be confused with NK cells!) which are discussed in Chapters 5 and 6. See Figure 1.15.

1.4.5.1 CD4 Helper and Regulatory T Cells

CD4 T cells, which when activated become conventional helper T (T_h) cells or regulatory T (T_{reg}) cells, are one of the

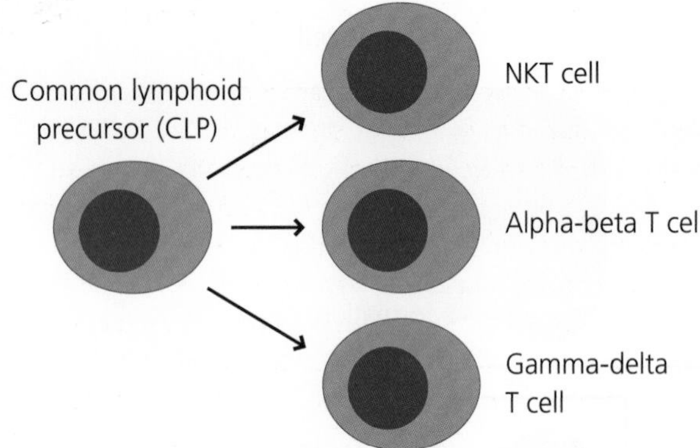

Fig. 1.15 Different types of T cells. αβ T cells. In humans and mice these are the predominant T cells. They express highly diversified antigen receptors (αβ TCR); they pass repeatedly through all secondary lymphoid tissues (recirculation). **γδ T cells.** In foetal mice, different waves of non-conventional T cells with much less diversified antigen receptors (γδ TCR) populate epithelia; they are produced before αβ T cells. **NKT cells.** These are very specialized T cells that represent another developmental pathway. Some **invariant NKT (iNKT)** cells express αβ TCR, but with very little diversity.

two main types of T cells. They are so-called because they possess a molecule called CD4 (in multiple copies, of course) that is involved in recognition of the cells with which they interact. As for all naïve lymphocytes, CD4 T cells circulate in the bloodstream and migrate through secondary lymphoid tissues. They are small, resting cells. To function, they need to be activated and in primary responses this occurs in the secondary lymphoid tissues. In secondary lymphoid tissues DCs (below) activate CD4 T cells and regulate their differentiation into cells capable of mediating different functions. These effector T cells then either interact with other cells locally within the secondary lymphoid tissues (e.g. they can help to activate B lymphocytes; below) or they migrate to peripheral sites of inflammation and infection and interact with different cell types.

There are at least four main ways in which CD4 T cells can be instructed to function, and the respective T cell subsets are termed T_h1, T_h2, T_h17 and T_{reg} cells. These subsets stimulate different types of adaptive immune response, either by recruiting or modifying the functions of the cells of innate immunity, initiating responses in other adaptive cells or by bringing in new cells and molecules into the response. See Figure 1.16.

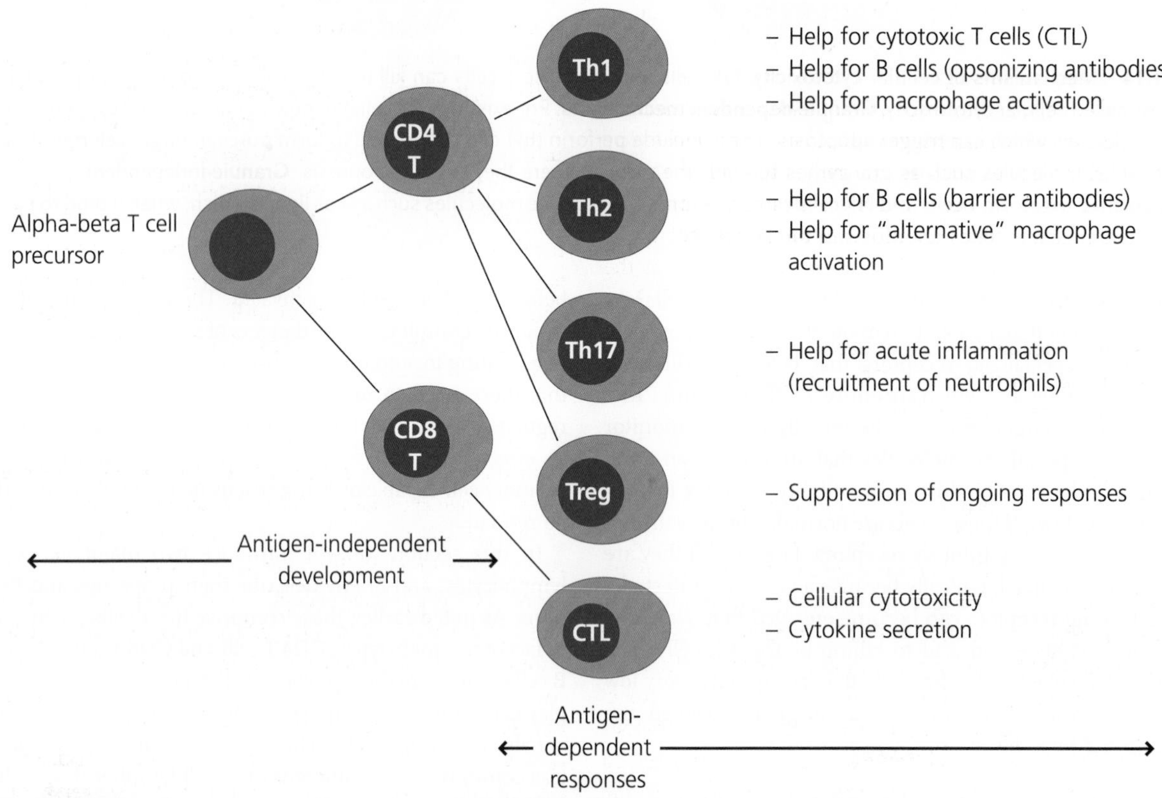

Fig. 1.16 Subsets of conventional αβ T cells. Two main αβ T cell subsets are produced in the thymus: CD4 and CD8 T cells. Each can acquire specialized (polarized) functions during immune responses. **CD4 T cells.** These may develop into T_h1, T_h2 or T_h17 cells depending on the type of infection. They regulate the functions of other cell types or help to recruit and orchestrate different effector mechanisms, some of which are indicated. CD4 T cells can also develop into T_{reg} cells that can suppress immune responses. **CD8 T cells.** These cells can develop into cytotoxic cells (CTLs) and may also adopt polarized functions somewhat resembling those of CD4 T cells although these are less well understood.

In general terms the functions of these activated CD4 T cells are as follows:

- T_h1 cells stimulate anti-microbicidal and cytotoxic effector functions of immunity (e.g. they activate macrophages and recruit and activate cytotoxic cells, both NK cells and CD8 T lymphocytes). They also instruct B cells to develop into plasma cells and secrete certain types of antibodies (which are of different classes, Section 1.5.5.3) that can interact with some of these cells. T_h1 responses may be particularly important for later defence against some types of bacterial and viral infections (e.g. tuberculosis, influenza).
- T_h2 cells stimulate the barrier functions of immunity (e.g. they recruit and maintain eosinophils, cause macrophages to become alternatively activated, and induce the production by plasma cells of other types of antibodies that can interact with these cells; Section 1.5.5.3). This type of response may be particularly important for host defence or resistance against parasitic infections such as worm infestations
- T_h17 cells are particularly efficient at recruiting neutrophils to the site of infection; hence, neutrophils (like other innate cells) can be recruited in both the innate and adaptive responses of immunity. This type of response may be particularly important for defence against other types of bacteria, particularly bacteria that cause acute inflammation such as *Staphylococcus* and *Streptococcus*, and perhaps some fungi.
- T_{reg} cells suppress the responses of other cells, including DCs and/or other lymphocytes. This type of response may be particularly important in switching off immune responses when an infectious agent has been eliminated and in ensuring that harmful responses are not made against innocuous agents (including components of the body itself).

These are probably the best understood subsets, although there are in fact other types of CD4 T cells with specialized functions that are noted in Chapter 5.

1.4.5.2 CD8 T cells

CD8 T cells represent the other main type of conventional T cell (see Figure 1.16). These T cells possess multiple copies of a molecule called CD8 that is involved in recognition of the cells with which they interact. Following activation, these T cells can become cytotoxic T cells capable of inducing apoptosis in cells they recognize (note, in contrast, that NK cells do not need to be activated before they can be cytotoxic; above). CD8 T cells, like CD4 T cells, circulate in the bloodstream and migrate through secondary lymphoid tissues. To function they also need to be activated and this occurs in the secondary lymphoid tissues. When CD8 T cells are fully activated, they become cells with potent cytotoxic activity. These cells leave the secondary lymphoid tissues, and enter peripheral sites of inflammation and infection. Here they can kill virally infected cells. They do so in two main ways – through perforin and granzymes, and through Fas ligand–Fas interactions (which are very similar to those used by NK cells).

As cytotoxic cells, CD8 T cells are a central component of adaptive immunity to viruses such as the influenza virus. However, they can also produce cytokines that (i) are directly toxic, such as tumour necrosis factor (TNF)-α, which induces apoptosis by binding to death-inducing receptors on other cells, or (ii) can modulate or enhance the functions of innate cells (e.g. IFN-γ, the major macrophage-activating cytokine), thus providing additional mechanisms by which they may help to eliminate infectious agents. Note that polarized subsets of cytokine-secreting CD8 T cells have also been postulated and named in accordance with CD4 T cell subsets (T_c1, T_c2, etc.).

1.4.5.3 B Lymphocytes

B lymphocytes are also small, resting cells, morphologically indistinguishable from T cells. When they are appropriately activated their primary function is to develop into plasma cells, which can be regarded as antibody-synthesizing factories, or to become memory cells. In many cases B cells need help from T cells to become activated and to develop into plasma cells, and the T cells also control the type of antibody that they make (see Figure 1.16). This type of response, which is typically made in response to protein antigens by B cells in specialized sites of secondary lymphoid tissues (termed follicles), is therefore termed a T-dependent (TD) response. However, there are different types of B cells. Another subset of B cells, located in a specialized site of the spleen (called the marginal zone) can produce antibodies to other types of antigen, such as polymeric carbohydrates, without needing any help from T cells and this is an example of a T-independent (TI) response. Finally, a different type of non-conventional B cell (called B-1 cells in the mouse) can produce so-called natural antibodies in the apparent absence of any antigenic stimulation. The functions of different types of antibody are described in detail in Chapter 6. See Figure 1.17.

1.4.5.4 Memory Lymphocytes

A central feature of adaptive immunity is the phenomenon of immunological memory. This is a crucial property of lymphocytes and a function of adaptive immune responses in general. As we mentioned earlier, once an initial infection has been overcome by a primary adaptive immune response, a long-lasting state of resistance to re-infection by the same organism ("immunity") can often be achieved. Good examples are immunity to childhood infections such as measles, and the long-lasting immunity that can be induced by vaccination against smallpox and tetanus. Immunity in these cases may last for many years and is due to the development, in the course of the adaptive response, of populations of memory lymphocytes, both T and B cells.

Once an infectious agent has been cleared, most of the expanded populations of effector T cells, or plasma cells, die. However, populations of antigen-specific memory T cells, both CD4 and/or CD8 T cells, and memory B cells persist.

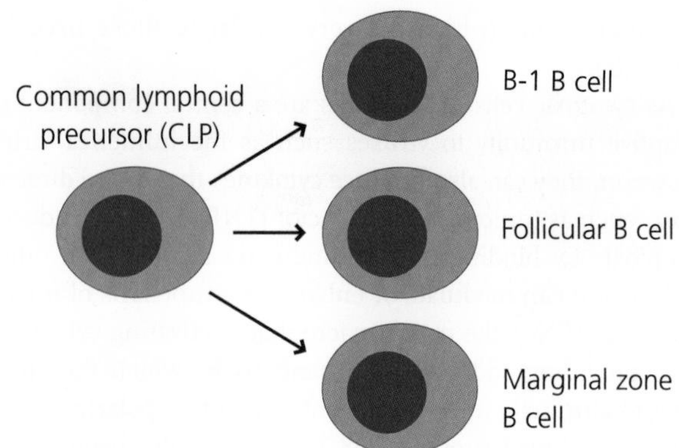

Fig. 1.17 Different types of B cells. Different types of B cell develop during foetal and neonatal life (*cf.* Figure 1.15 for T cells). All can develop into plasma cells and make antibodies. They may respond to different types of antigen. **Follicular B cells**. The predominant, conventional, B cells in humans and mice have highly diversified antigen receptors (BCRs) and pass repeatedly through all secondary lymphoid tissues (recirculation). **B-1 cells**. In mice, non-conventional B cells with much less diversified BCRs populate the peritoneal and pleural cavities; they are first produced before conventional B cells. **Marginal zone B cells**. These are sessile cells in splenic marginal zone that represent an alternative pathway of development to follicular B cells.

The numbers of these antigen-specific clones of lymphocytes are now higher than before the infection occurred and memory cells can be more rapidly reactivated if the infectious agent is encountered again. Hence, secondary responses are much larger and more rapid, and usually lead to highly efficient clearance of the infectious agent. See Figure 1.18.

1.4.6 Dendritic Cells (DCs)

There are different types of DCs and their nomenclature can be confusing. We will focus primarily on "classical" DCs whose primary function is to trigger and regulate most types of adaptive immune responses, particularly the responses of CD4 T cells. Such DCs, which are transiently resident in peripheral tissues sense different types of infectious agents using the PRRs that are also expressed by innate immune cells. They can also internalize infectious agents (viruses, bacteria, etc.) or smaller antigens derived from them (e.g. bacterial toxins, components of parasites). See Figure 1.19.

Depending on the particular type of infection that has occurred, and in response to other stimuli such as the particular cytokines that are produced in inflammatory sites, these DCs then change their properties. They migrate in increased numbers from peripheral tissues into secondary lymphoid tissues (e.g. from the skin into the lymph nodes that drain that site), transporting the antigens they acquired in peripheral sites of infection in forms that can be recognized by T cells, i.e. peptide-MHC complexes. In these lymphoid organs the DCs can interact with antigen-specific T lymphocytes, particularly CD4 T cells, and, because they now express specialized molecules that are needed for full activation of naïve T cells (a process called costimulation), they can initiate many types of adaptive immune responses. Depending on the type of infection, DCs also cause activated CD4 T cells to adopt specific functions appropriate to the type of microbe that needs to be eliminated. Having done their job, DCs die in secondary lymphoid tissues.

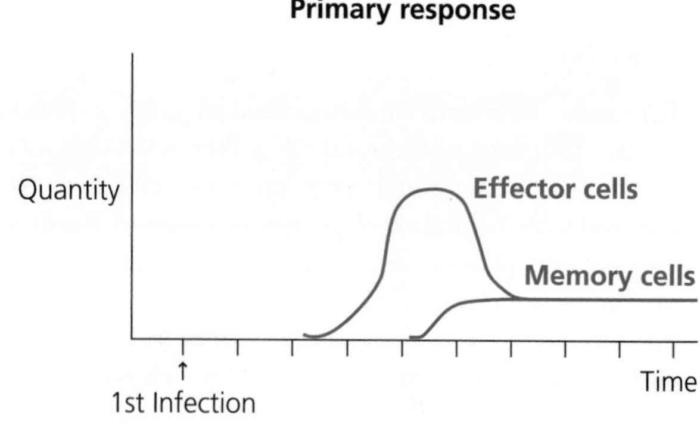

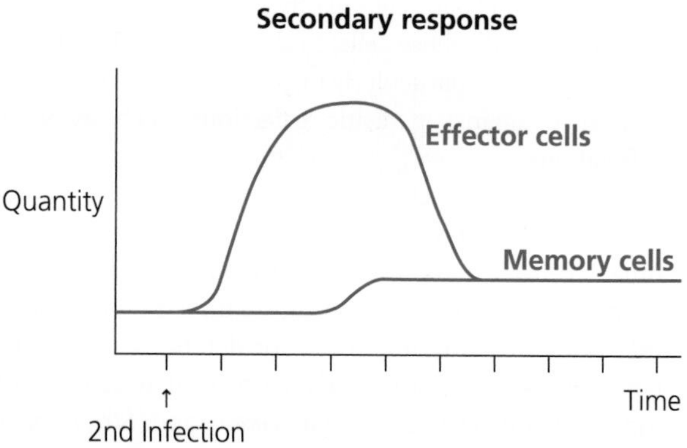

Fig. 1.18 Effector and memory lymphocyte responses. Primary responses. In response to a new infection, effector lymphocytes are produced after a lag (when the cells are being activated and expanded) and their numbers slowly decrease as the microbe is cleared. A population of memory lymphocytes then appears and persists, often for a lifetime. **Secondary responses.** If the same infection occurs again, memory lymphocytes can rapidly develop into effector cells and help to clear the infection more efficiently. They then revert back to memory cells (or are replaced by new ones) and can respond again similarly if the infection happens at any time in the future. These principles apply for both T cell and B cell responses. The efficiency of secondary responses is seen even more clearly in antibody responses, since more appropriate types of antibodies of higher affinity can be produced, compared to those secreted in a primary response (not shown; see Chapter 6.)

Plasmacytoid DCs (pDCs) are a different cell type. They are called plasmacytoid because they possess much rough endoplasmic reticulum and, in this respect, they resemble plasma

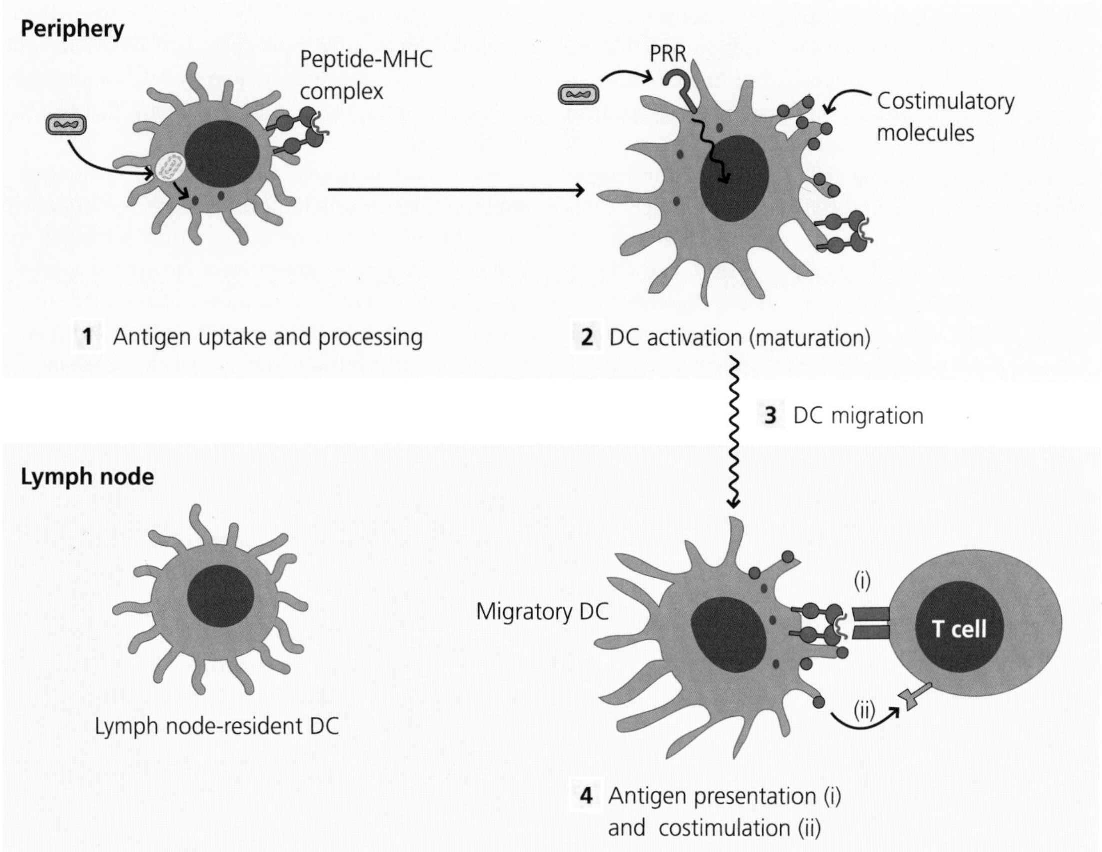

Fig. 1.19 Functions of "classical" dendritic cells. Classical DCs are found in nearly all peripheral tissues; in skin epidermis they are called **Langerhans cells.** (1) When infection occurs, these cells internalize and degrade microbial antigens to forms that can be recognized by T cells (peptide–MHC complexes). (2) Sensing of the pathogen through PRRs also causes them to increase expression of specialized costimulatory molecules that are needed to activate T cells. (3) They then migrate into secondary lymphoid tissues (e.g. from skin to draining lymph nodes via afferent lymph). (4) These migratory DCs can now activate antigen-specific T cells, triggering adaptive immunity. Lymph nodes additionally contain a distinct population of resident DCs that play less understood roles in T cell activation or regulation than the migratory DCs.

cells. They are very potent secretors of cytokines collectively called type I IFN and may play a role in anti-viral defence. When stimulated via PRRs they rapidly secrete a variety of pro-inflammatory cytokines which probably also play a role in the regulation of T cell activation, but their real functions in immune responses are still unclear.

Finally, there is another cell type called the follicular DC (FDC) that we discuss in other chapters. Despite their name, these are a totally different type of cell that must not be confused with either classical DCs or pDCs. FDCs are localized in specialized sites of secondary lymphoid organs (the follicles) where they play central roles in B cell responses, as opposed to the several types of classical DCs that are involved in T cell responses (above).

1.4.7
Coordination of Immune Responses

To conclude this section on the cellular basis of immunity, we will summarize a few key principles.

As a first general rule, the type of innate or adaptive response that occurs after infection is tailored to the type of infectious agent that initiated it. Thus, the immune system generates an immune response that is most appropriate for eliminating the infectious agent. For example, we have seen above that phagocytes are involved in direct killing of bacteria, cytotoxic cells can kill cells that have been infected by viruses, and eosinophils and basophils may help in resistance to worm infestations. Each of these cell types uses very different mechanisms to perform its functions.

A second general rule is that cells that are involved in any immune response do not act independently, they are regulated by, and can regulate the functions of, other cells. The production of innate cells from the bone marrow may be increased or decreased during both innate and adaptive responses, their accumulation at sites of infection is regulated by inflammation and, importantly, their functions can be modulated. In this way each cell can help to fine tune the type of immune response that needs to be induced at different times after infection.

It is a third general rule that all immune responses need to be initiated, particularly in response to infection, and that the overall response needs to be properly coordinated and regulated. Just as very different types of response can occur in innate immunity, the same is true for adaptive responses. In adaptive immunity, specialized cells are required to trigger lymphocyte responses and a specialized set of lymphocytes then controls the overall type of response that follows. The most important controlling cells are DCs, and CD4 T cells that have helper or regulatory functions. Together these cells are responsible, with assistance from other cell types, for co-ordinating the adaptive immune response. (To use a business analogy, DCs might be viewed as the executives, CD4 T cells as middle management and all the other cell types we have mentioned as the workers. To a significant extent DCs assess the risk, take strategic decisions and tell the CD4 T cells what to do; in turn, CD4 T cells instruct the different types of worker to help achieve the task.)

Finally, of course, nothing could happen without molecules. Some key molecules of immunity are considered in the next section.

1.5 Molecular Basis of Immunity

The primary function of the immune system is to eliminate infectious agents that have breached the natural defences. This is brought about by the integrated actions of different cells and molecules that, directly or indirectly, lead to their elimination. Having outlined the cells of immunity, we now examine the molecular basis of immunity. We describe in a little more detail some of the molecules mentioned above, and introduce others that also play key roles in immunity.

1.5.1 Cell-Associated and Soluble Molecules of Immunity

Some of the most important classes of molecules expressed by immune cells are those with the following functions:

i) Controlling the positioning of immune cells within the body; for example, they enable these cells to localize within normal tissues or to reach the sites of infection.
ii) Enabling the recognition of infectious agents and other signals, so that the cell can make an appropriate response.
iii) Communicating with other cells in the vicinity or in more distant tissues, to help bring about a coordinated response.
iv) Directly or indirectly acting as effector molecules; for example, by helping to kill the infectious agent (inside or outside of cells) or to kill the cells that may harbour infectious agents (cellular cytotoxicity).

In addition, many cell-associated molecules, which include receptors for soluble molecules, are linked to intracellular signalling cascades that frequently lead to changes in gene expression or other cellular functions that are needed in specific responses.

Many molecules involved in immunity belong to different superfamilies in which there is a common underlying structure, although functions of the family members may differ widely. For example, molecules of the immunoglobulin superfamily contain different numbers of so-called immunoglobulin domains possessing a characteristic structure. For the purposes of this introduction we will, however, use *function* to group the molecules of immunity into six main types (Figure 1.20):

- **Adhesion molecules** which, for example, enable leukocytes to interact with endothelial cells during migration out of the blood into specific sites, components of connective tissues such as the extracellular matrix, and other cells.
- **Innate agonist receptors**, particularly PRRs that activate the innate immune system.
- **Antigen receptors** which enable lymphocytes to recognize infectious agents or to detect the presence of infectious agents inside other cells.
- **Cytokines** which enable cells to signal to other immune and tissue cells in the same vicinity or at a distance; a subset of these are called chemokines that are involved in directing the migration of cells (e.g. to sites of infection).
- **Effector molecules** that directly kill the infectious agent or enable it to be targeted to specific cells for elimination.
- **Intracellular signalling molecules** without which immune cells (or, indeed, any other cell type) would be unable to function.

1.5.2 Adhesion Molecules

In all cases, cells that enter tissues from the blood must cross the barrier of endothelial cells that line blood vessels and enter extravascular sites. This process is called extravasation or transendothelial migration. Clearly, any cell needs to discriminate between one site and another (e.g. peripheral versus secondary lymphoid tissues, or normal versus inflamed tissues). Cells also need to able to migrate within tissues and to retain their positions in these tissues. These functions are dependent on adhesion molecules.

The selective homing of different cells to different sites at different times is controlled by a remarkable recognition system. In part, this involves adhesion molecules of leukocytes, which are of two main classes: selectins and integrins. These bind to counter-ligands on endothelial cells that are collectively termed vascular addressins because, together, they provide "addresses" to help leukocytes find their way through and into different types of tissue. A third set of molecules involved in cell homing are not adhesion molecules, but are receptors for a specialized set of cytokines called chemokines (below).

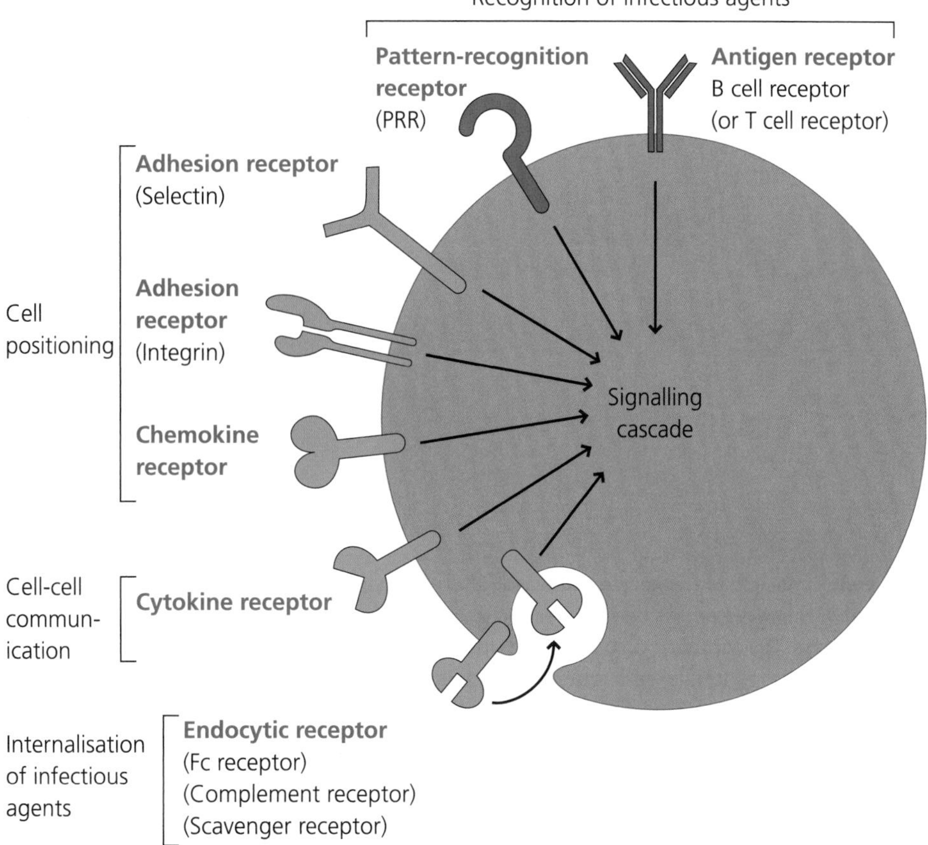

Fig. 1.20 Examples of molecules involved in immune responses. Recognition of infectious agents. PRRs are expressed by many different types of cell, particularly innate cells, and enable them to recognize different classes of infectious agent. Antigen receptors (TCRs and BCRs) are expressed only by lymphocytes and enable them to recognize specific components of infectious agents. **Endocytic receptors.** These facilitate uptake of molecules and particles which are sometimes coated with other immune components (e.g. antibodies or complement). **Communication molecules.** Cytokine receptors bind cytokines that are produced by the same or other cells and change cellular functions. Some cytokines, called chemokines, generally alter cell migration after binding to chemokine receptors. **Cell-positioning molecules.** These are needed for cells to migrate to and from normal or inflamed tissues, or to adhere to and communicate with other cells. **Signalling molecules.** Binding of ligands to receptors typically triggers biochemical cascades within the cell involving intracellular signalling molecules. These form pathways that ultimately change the behaviour of the cell (e.g. movement) or change gene expression so the cell can acquire new roles in immunity.

Collectively, the permutations and combinations of different selectins, integrins. chemokines and chemokine receptors provides a highly discriminatory system to ensure that the right cells go to the right places at the right time – a system that one might term the "post code" (or "ZIP code") principle. See Figure 1.21.

1.5.2.1 Selectins

Leukocytes in the blood can transiently attach to and roll along the endothelium of venules. Rolling is usually mediated by selectins (lectins are proteins that bind to carbohydrate ligands). Rolling gives the cell time to interact with molecules expressed on the endothelial surface that define the type of endothelium, and hence to identify the tissue in which it finds itself (e.g. skin or gut) and the state of that tissue (e.g. normal or inflamed). These complementary molecules, which are constitutively expressed or inducible at different sites, contain carbohydrate residues that can act as ligands for the selectins.

1.5.2.2 Chemokines and Chemokine Receptors

Chemokines are a specialized class of cytokines (Section 1.5.4) that can stimulate cells to migrate directionally along a concentration gradient of the chemokine. Directed cellular migration is called chemotaxis. Chemokines can be produced in tissues by many different types of cell (e.g. by macrophages and mast cells in response to infection). As well as promoting directional cell migration within the tissues where they are produced, chemokines may be transferred across the endothelium to the luminal (blood) side and become attached to the endothelial cells. Different types of leukocytes such as

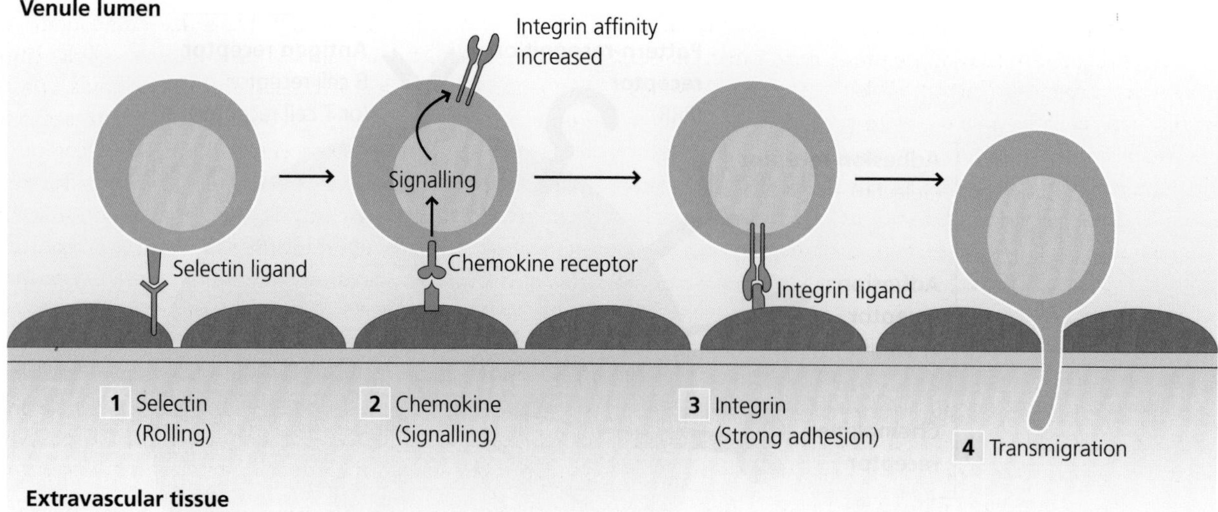

Fig. 1.21 The "post code" principle for leukocyte extravasation. Leukocytes from blood must enter specific tissues and organs (extravasation) to mount immune responses during infection (peripheral tissues in innate responses and secondary lymphoid tissues in adaptive responses). The combination of different types of molecule expressed by vascular endothelial cells resembles a post code (ZIP code) that can be read by the leukocyte. **(1) Selectins.** Selectins, through weak binding to their carbohydrate ligands, mediate transient attachment and rolling of leukocytes along the endothelium. **(2) Chemokine receptors.** Chemokines produced at sites of inflammation are transported across the endothelium to the luminal side where they are displayed for recognition by leukocytes with corresponding receptors. **(3) Integrins.** Intracellular signalling from chemokine receptors leads to activation of integrins, which can now bind strongly to counter-receptors on the endothelial cells, leading to firm attachment. **(4) Molecules involved in transmigration.** Other types of molecule (not shown) enable the leukocyte to cross between or directly through the endothelial cells to enter extravascular tissues.

neutrophils, or later in the response activated T cells, possess different types of chemokine receptors. The corresponding chemokines, which are attached to the endothelial cells, bind to chemokine receptors and, in turn, activate some of the integrins on the leukocyte. Once activated, integrins bind very strongly to their ligands (below).

1.5.2.3 Integrins

When leukocytes receive appropriate signals via chemokines they stop rolling and attach firmly to the endothelial cells. The rolling of leukocytes can be arrested because of the strong binding that occurs between the integrins and their counter-ligands on the endothelium. Integrins exist in two states either high or low affinity. The integrins of circulating leukocytes are in the low-affinity state and incapable of strong binding. The switch to the high-affinity state occurs in response to the leukocyte receiving signals through its chemokine receptors.

Once within the tissue, different integrins can also enable the cell to attach to components of the extracellular matrix and, later, to the cells with which they may need to physically interact during the immune response. Within the tissue, a gradient of chemotactic molecules – including chemokines produced at the site of infection – is established. These permit the directional migration of the leukocyte towards the site of infection itself. Other types of integrin are not involved in migration but have different functions.

1.5.3 PRRs and Antigen Receptors

All immune responses, both innate and adaptive, are initiated by recognition of molecules, and particularly components of infectious agents. Some components may act as agonists in innate immunity, while others are recognized as antigens by the lymphocytes of adaptive immunity. We will highlight the functions of four different types of cell-associated recognition systems in the immune system.

- PRRs play central roles in the initiation and regulation of innate immunity, and in part regulate adaptive responses.
- BCRs on the surface of B cells can recognize antigens directly; after a B cell develops into a plasma cell, BCRs can be secreted as antibodies. Other molecules are involved in signalling from BCRs to the B cell when antigen recognition occurs, while others regulate B cell activation and responses.
- TCRs of T cells remain cell-associated and are not secreted. TCRs *cannot* recognize antigen directly. Instead the antigen (usually a peptide) must be bound to a MHC molecule. As for B cells, other molecules are additionally involved in T cell activation and responses.
- MHC molecules bind peptides and enable conventional T cells to recognize antigens.

1.5.3.1 Pattern Recognition Receptors (PRRs)

PRRs are widely distributed on immune and other cells. Other types exist in soluble forms and are sometimes termed pattern recognition molecules (PRMs). A key feature of PRRs is that they are generally involved in responses to components of infectious agents that are not synthesized by their host (e.g. humans). Some PRRs may also be involved in responses to components of the host itself, but which are usually only produced in times of damage or stress as during an infection. PRRs are phylogenetically ancient and are also present in invertebrates. One of the best understood types of PRRs, the Toll-like receptors (TLRs) were first discovered in insects.

Cell-associated PRRs are present in different cellular compartments, on the cell surface, in endosomes or in the cytosol. As such they have the potential to recognize (directly or indirectly via an associated molecule) infectious agents that are present outside the cell, have been internalized (e.g. by phagocytosis in macrophages) or have infected cells. The outcome of recognition by PRRs differs depending on the cell type that is involved. The best understood PRR functions are those involved in innate immunity, the induction of inflammation and the initiation of adaptive immunity. Thus, PRRs expressed by phagocytes can promote phagocytosis or stimulate the production of cytokines that induce inflammation (below) whereas. PRRs expressed by mast cells may lead to the release of their granule contents. Whereas PRRs expressed by DCs stimulate cellular responses, including the secretion of cytokines, which help to regulate T cell responses. See Figure 1.22.

PRRs are also expressed on epithelial cells, leading to secretion of anti-microbial agents; endothelial cells, helping to regulate the recruitment of different types of leukocyte in inflammatory responses; and even on lymphocytes, where their functions are not fully understood.

1.5.3.2 B Cell Receptors (BCRs)

B lymphocytes have evolved to recognize components of infectious agents that are derived from the extracellular compartments of the body, including the tissue fluids. These infectious agents include, for example, free viruses and bacteria in the blood or extracellular fluid. This recognition is initially mediated by membrane-bound BCRs. BCRs have evolved to recognize complementary parts of the three-dimensional structure of foreign antigens, such as part of a molecule on the surface of a virus or bacterium – these are termed conformational epitopes. The same is true when BCRs are secreted as antibodies (below). The only difference between a BCR and its corresponding antibody is whether or not it contains a small molecular portion that attaches it to the cell membrane, and sometimes an additional component that can help the soluble molecule polymerize into dimers or pentamers.

Generation of Diversity of Lymphocyte Antigen Receptors A vast number of antigen receptors of different structures – and hence of different antigen specificities – can be assembled during early development of B cells in the bone marrow, and of T cells in the thymus (below). In general, each B cell possesses multiple copies of BCRs of only one specificity; the same applies for T cells. The process of generating the repertoires of lymphocyte antigen receptors occurs through a remarkable process of diversification that is not known to occur in any other cell type. Unlike all other genes in the genome, those that encode lymphocyte antigen receptors exist in germline DNA not as functional genes, but as gene segments. To create a functional antigen receptor gene, these segments are rearranged at the DNA level in lymphocytes. In humans and mice this process occurs by a genetic mechanism termed somatic recombination.

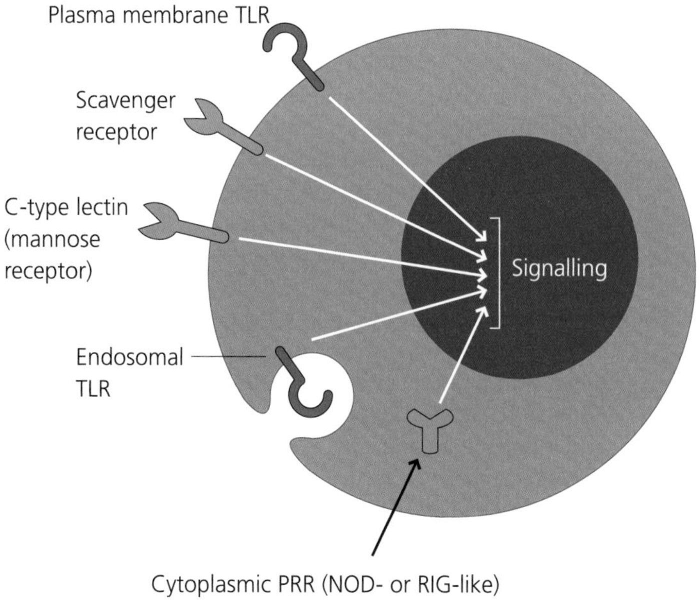

Fig. 1.22 Types and locations of pattern recognition receptors. PRRs are localized in different cellular compartments. Widely expressed components of different classes of infectious agent act as agonists for different PRRs and stimulate different cellular responses (e.g. phagocytosis, degranulation, cytokine secretion). **Plasma membrane PRRs.** These PRRs include some TLRs, C-type lectin receptors and scavenger receptors. Typically they are involved responses to bacteria, fungi or protozoa. **Endosomal PRRs.** Some TLRs are expressed on endosomal membranes. Typically they respond to nucleic acids (RNA, DNA) from viruses and bacteria. **Cytoplasmic PRRs.** These include RNA helicases, which for example recognize nucleic acids of viruses, and NOD-like receptors which recognize components of bacteria that may have escaped into the cytoplasm. Many PRRs signal to the nucleus to change gene expression, but others modulate other cell functions by acting on cytoplasmic components (such as actin, leading to changes in cell shape (not shown).

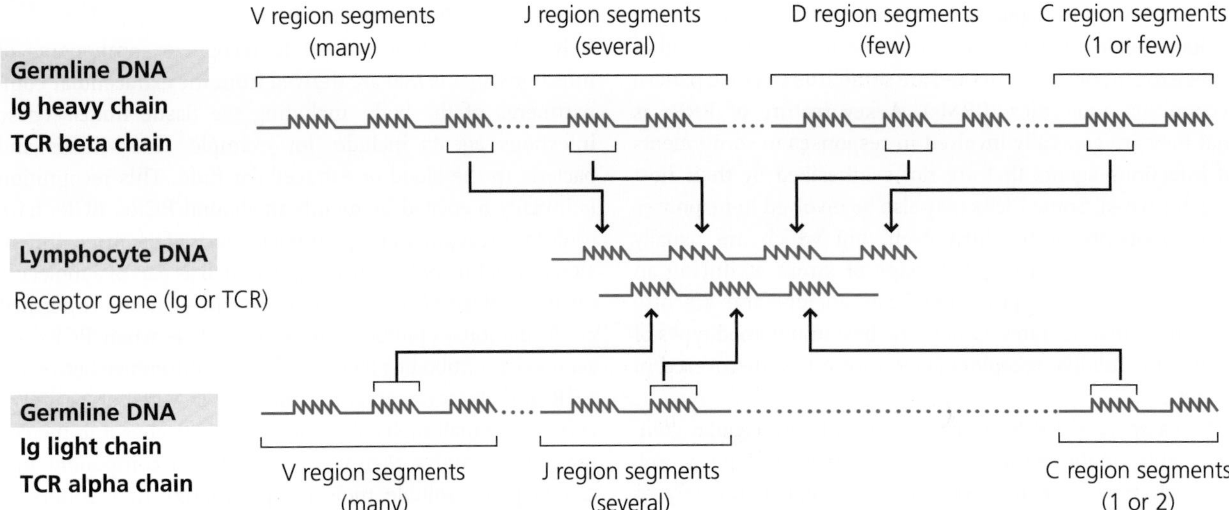

Fig. 1.23 Rearrangement of antigen receptor genes. BCRs and TCRs are comprised of two different molecules – immunoglobulin (Ig) heavy (H) and light (L) chains, and TCR α and β chains respectively. Each chain consists of variable and constant regions. Each chain is encoded in a particular region (locus) of a different chromosome but, in germline DNA, the receptor genes are not present as functional genes. In lymphocytes, functional antigen receptor genes are assembled by rearrangement of gene segments. **IgH and TCR β genes.** There are three groups of gene *segments*: V, D and J in the variable region locus. Largely at random, one D segment joins to one J segment and the DJ segment then joins to a V segment. This encodes the variable (V) *region* of the IgH chain or the TCR β chain respectively. **IgL and TCR α genes.** A similar principle applies but D segments are absent. One V segment joins to one J segment, and the assembled VJ segment encodes the V region of the IgL chain or the TCR a chain. **C (constant) region segments.** Downstream of the V, D and/or J segments, there are one or more C region segments. The newly assembled VDJ or VJ segments become juxtaposed to the closest C region segment and a functional gene is created. In the case of the IgH locus there are several different C regions enabling different types of BCR to be assembled with the same V region; these can be secreted as antibodies with different corresponding functions but with the same antigen specificity.

In their membrane-bound form, attached to B cells, BCRs are composed of two pairs of molecules, termed heavy (H) chains and light (L) chains. These associate to form a "Y"-shaped molecule (e.g. Figure 1.24). The top two regions of the "Y" – called the variable regions – are responsible for antigen recognition. Each of these is formed from a variable region of a H chain combined with the variable region of a L chain and, together, these determine the antigen specificity of the molecule. The stem of the Y – called the constant region – is attached to the cell in a BCR. It also controls all aspects of the biological and immunological functions of the corresponding antibody when it is secreted. These molecules are encoded by the immunoglobulin genes that are formed after DNA rearrangement of the corresponding segments.

To make a variable domain of a BCR, different DNA segments are recombined. These are called V, D and J segments for H chains, and V and J segments for L chains. There are multiple V, D and J segments, and joining happens largely at random, but selecting only one of each. Where the joins are made, specialized mechanisms can introduce further variations in the DNA, thus changing the structure of the V region gene that is produced and hence the antigenic specificity of the molecule it encodes. Very similar process are involved in generation of TCRs (below). For BCRs, the assembled variable region genes are then attached to one of a few constant region genes that encode part of the constant region of the antibody. Depending on which C region gene is used, a different type or class of antibody is produced. Assembly of a functional BCR is crucial for B cell development – if a functional BCR is not produced, the developing cell dies; the same is true for T cells (below). See Figure 1.23.

Functions of BCRs When they are first produced, B cells are small, resting cells. These mature, but naïve (antigen-inexperienced), B cells are not able to do much by themselves. To develop into plasma cells that secrete antibodies, or to develop into memory B cells, they first need to be activated. Recognition of antigen by the BCR is important for B cell activation. BCRs on the surface of B cells are associated with a molecular complex, CD79, that delivers signals to the B cell when antigen recognition occurs and helps to activate the B cells. Another important function of BCRs is to trigger internalization and degradation of the antigens they recognize. These antigens are subsequently presented at the cell surface (as peptides bound to MHC molecules; below) to enable recognition by specific T cells that can then instruct the B cells what to do next, particularly in the case of antibody responses against protein antigens. See Figure 1.24.

However, regulation of lymphocyte responses is, in general, a tightly controlled process that depends in part on what has happened (or still is happening) in the innate response. If, for example, complement has been activated, one particular complement component (bound to an antigen) can bind to a receptor

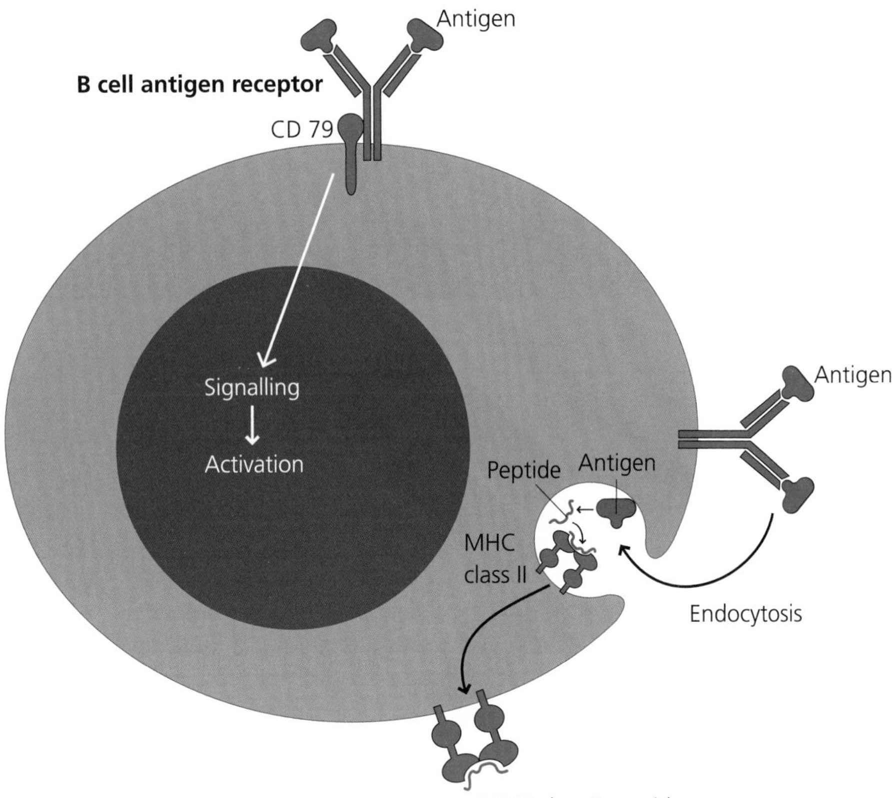

Fig. 1.24 Functions of B cell antigen receptors. All B cells express membrane-bound BCRs. **Antigen recognition**. BCRs recognize structural features on antigens known as conformational determinants or B cell epitopes. These are three-dimensional surfaces of folded microbial proteins, carbohydrates or glycolipids, for example. **Intracellular signalling**. BCRs are associated with a molecular complex, CD79, which triggers intracellular signalling following antigen recognition. These signals can help to activate a naïve B cell, after which it may develop into a plasma cell that secretes its BCRs as antibodies. **Antigen internalization**. Antigens bound to BCRs can be internalized and degraded. Peptides from protein antigens can be expressed as peptide–MHC complexes at the cell surface, enabling T cells to recognize the B cells and change their functions (e.g. to instruct them to make different types of antibodies).

complex (CD19/CD21/CD81) on the surface of B cells and deliver powerful signals to increase B cell activation. Other B cell molecules, such as CD40 and CD40 ligand, are involved during communication with activated CD4 T cells, which in many adaptive responses help to instruct the B cells as to what type of antibody they should secrete. See Figure 1.26.

During antigen-specific responses of B cells, further changes can be introduced into their immunoglobulin genes. First, B cells can change the constant domains of their BCRs and antibodies. A new constant region gene (attached to the pre-assembled variable region genes) may be selected to produce a different type (or class) of antibody with a different function (Section 1.5.5.3). This is called class switching. Second, B cells can also change the variable domain of their BCRs and antibodies. This is because mutations can be introduced into the variable region genes in a process called somatic hypermutation. Potentially this can increase the strength of recognition (affinity) of the BCR and of the respective antibodies that can be produced when the B cell develops into a plasma cell. The general increase in average affinity of antibodies that is seen after repeated antigenic stimulation is called affinity maturation and is a direct consequence of somatic hypermutation. It is important to stress that class switching and somatic hypermutation do *not* occur in TCRs.

1.5.3.3 T Cell Receptors (TCRs)

In contrast to B cells, T cells have evolved to recognize other cells that may *contain* foreign antigens. These include, for example, specialized cells such as phagocytes that have internalized bacteria or cells that have been infected by viruses. This recognition is mediated by membrane-bound TCRs. There are two different types of TCRs and the two main populations of T cells that express them generally have very different functions. The TCRs of conventional T cells in humans and mice are composed of a pair of molecules called the **α chains** and **β chains**, which combine to form an αβ TCR; these T cells are therefore also called αβ T cells. (Remember there are other types of T cells, but these are discussed in Chapter 5.) Conventional αβ TCRs have evolved to recognize peptide–MHC complexes on the surface of other cells. See Figure 1.25.

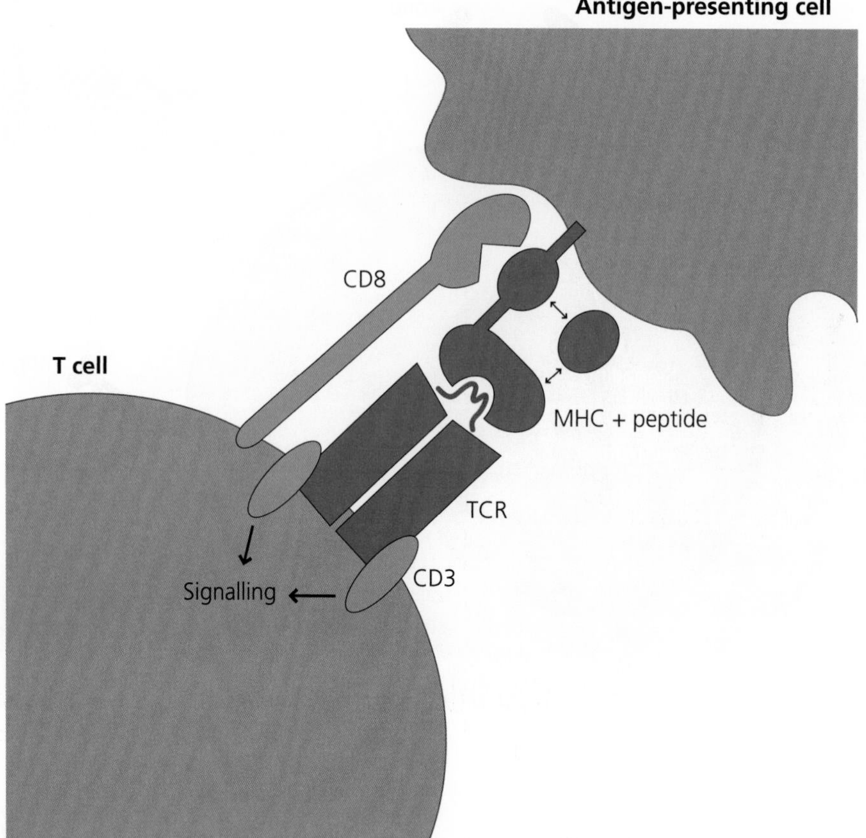

Fig. 1.25 Structure and function of T cell receptors. Conventional T cells express plasma membrane αβ TCRs; these are structurally related to immunoglobulin molecules, but functionally very different. **Antigen recognition.** αβ TCRs do not recognize whole proteins. Small peptides are produced by degradation of microbial proteins present in different cellular components. These peptides then bind to MHC molecules that transport them to the cell surface. Conventional αβ TCRs can then recognize these peptide–MHC complexes (but not either alone). The peptide represents a sequential determinant or (in association with an MHC molecule) a T cell epitope. **Intracellular signalling.** TCRs are associated with a molecular complex, CD3, which triggers intracellular signalling when TCR recognition occurs. These signals, to which CD8 molecules (or CD4 molecules; not shown) also contribute, represent signal 1 that is necessary for T cell responses to occur, but which is usually not alone sufficient for naïve T cells to become activated.

A vast number of αβ TCRs of different structures, and hence antigenic specificities, can potentially be assembled during development of T cells in the thymus. However, as for BCRs, these differ only in the variable antigen recognition domain at the top of the molecule, while the stem remains essentially constant. In general, every T cell can be thought of as having multiple copies of TCRs of only one specificity. TCR genes are generated through DNA rearrangement in a manner highly analogous to that described for BCR genes (above) – involving V, D and J segments for the β chain, and V and J segments for the α chain. However, the rearranged TCR variable region effectively associates with only one type of C region gene that cannot produce a secretory protein. Hence, the TCR of T cells is always membrane-bound (it cannot be secreted like antibodies) and T cells cannot undergo class switching (they keep the same TCR for life). In addition, mutations cannot be introduced in TCR variable regions, so T cells cannot undergo somatic hypermutation, and affinity maturation during T cell responses does not occur.

Functions of Co-Receptors on T Cells As we have seen there are two main subsets of conventional αβ T cells; CD4 and CD8 T cells. It is impossible to discriminate between the two subsets of T cells on the basis of the TCRs that they express. However, these subsets have very different functions. As already described in Section 1.4.5, CD8 T cells can develop into cytotoxic cells that can kill cells containing foreign antigens (e.g. virally infected cells). In contrast, CD4 T cells secrete a variety of cytokines and act as helper or regulatory cells by interacting with other cell types. For example, as noted above, CD4 T cells can provide help to B cells, particularly during responses to protein antigens, and can also regulate the functions of other cells such as macrophages. It is the expression of CD4 or CD8 that differentiates the different subsets of T cells. These molecules help the T cells to distinguish between cells containing different types and origins of antigen (see MHC; below) and also deliver signals to the T cells which help to trigger T cell activation.

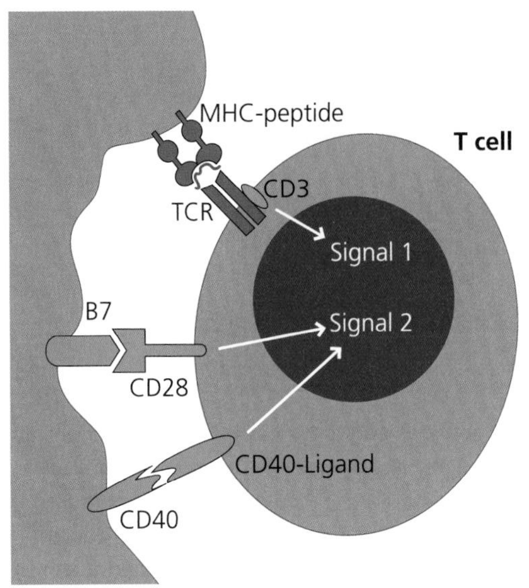

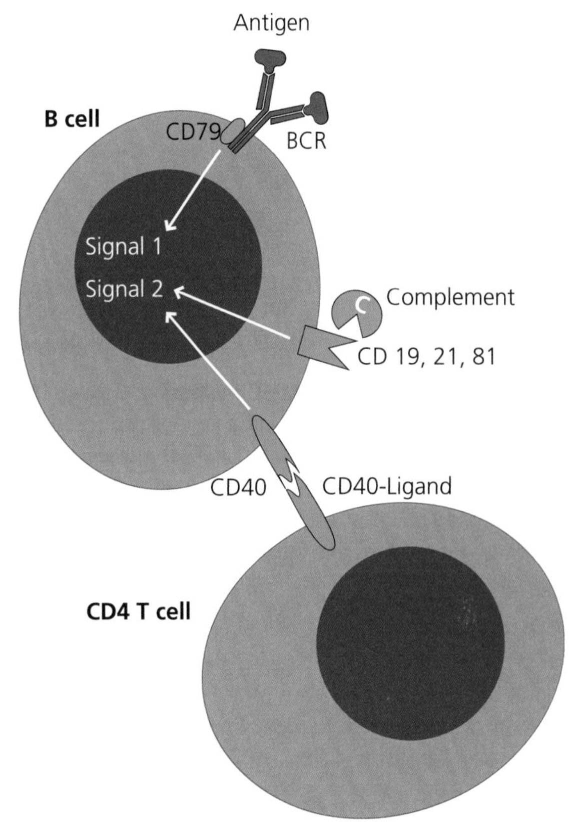

Fig. 1.26 Costimulation of lymphocytes. For naïve T and B cells, antigen recognition through TCRs and BCRs respectively (Signal 1), is not normally sufficient to induce full activation. Instead, they need additional signals (Signal 2) to provide costimulation. For **T cells** the most important are provided when B7 molecules (CD80, 86), typically expressed by activated DCs, bind to CD28 on T cells; additional signals can be provided by CD40 and CD40 ligand interactions. For **B cells**, binding of CD40 ligand on activated CD4 T cells to CD40 on the B cell is one of the most important for production of antibodies against protein antigens; other signals that are delivered when an activated complement component binds to a complex containing CD19 greatly enhance B cell responses. Cells that do not receive sufficient costimulation do not undergo full activation and may become unresponsive (anergy) or die. This can result in antigen-specific unresponsiveness and tolerance.

Functions of TCRs and Costimulatory Molecules To develop into the different types of T_h or T_{reg} cells, or into cytotoxic cells, naïve T cells first need to be activated. Recognition of antigen by the TCR is essential for T cell activation. TCRs on the surface of T cells are associated with another molecular complex called CD3 (compare CD79 on B cells; above) that delivers signals to the T cell when recognition occurs and helps to activate the T cells. This alone, however, is insufficient to fully activate a naïve T cell. In addition to the TCR–CD3 complex, T cells also possess molecules that deliver additional, crucial, activating signals to the cell when recognition occurs. When these interact with complementary molecules on other cells, particularly DCs that have been stimulated by infectious agents, T cell activation can occur. These molecules are termed costimulatory molecules. Examples of these include some members of the B7 family of molecules, which interacts with CD28 on T cells, and CD40 and CD40 ligand. Both antigen recognition *and* costimulation are needed for initial T cell activation; this principle generally also applies for B cell activation. A most important point to remember is that once a T cell has been activated, it no longer needs costimulation and can make a response when it recognizes any other cell containing antigen. The same is true for some memory T cells. See Figure 1.26.

1.5.3.4 Major Histocompatibility Complex (MHC) Molecules

Antigen recognition by B cells can occur directly, because their BCRs have evolved to recognize antigens *outside* cells (e.g. components derived from viruses or bacteria in the tissue fluids). However, antigen recognition by T cells *cannot* occur directly, because the antigens for which TCR are specific are *inside* other cells (e.g. a virus in the cytosol of an infected cell or a bacterium in the phagosome of a macrophage).

If recognition of antigen by T cells cannot occur directly, it is necessary for any cell containing antigen to "show" it, or "present" it, to the T cells instead. This is the function of the classical MHC molecules which are expressed by almost every cell of the body. The function of these MHC molecules is to

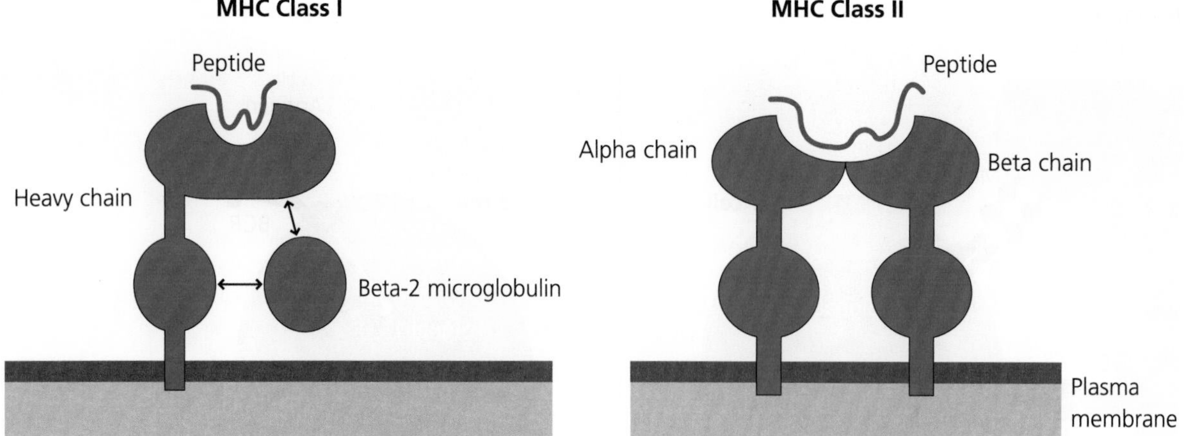

Fig. 1.27 Types of classical MHC molecules. Classical MHC molecules bind peptides and these complexes are recognized by "conventional" $\alpha\beta$ T cells. **MHC class I molecules** contain a heavy chain, non-covalently associated with an invariant molecule, β_2-microglobulin. In contrast, **MHC class II molecules** are heterodimers composed of α and β chains (not to be confused with the molecules that comprise the TCRs). Both types of classical MHC molecules contain an antigen-binding groove that can potentially bind many different peptides (one at a time). They are also highly polymorphic, differing from each other mostly in the amino acid residues that form the antigen-binding groove which thus determine the precise nature of the peptides that can be bound.

bind representative samples (usually peptides) of molecules (usually proteins) that are synthesized within a cell, or which have been internalized by the cell, and to transport them to the cell surface. (So called non-classical MHC molecules have different functions as discussed in Chapter 5.) One molecule of peptide can bind to one classical MHC molecule. Every cell is able to degrade proteins that it synthesizes within the cytoplasm, while a more limited number of cells (e.g. macrophages) can degrade proteins that have been internalized in phagosomes. If these samples (generally, small peptides) happen to be derived from "foreign" infectious agents, complexes of foreign peptides and MHC molecules are displayed at the cell surface. In turn, $\alpha\beta$ TCRs can bind to these foreign peptide–MHC complexes, thus permitting the T cells to recognize antigens that are present inside cells.

There are two different types of classical MHC molecules; MHC class I and MHC class II. During their biosynthesis MHC class I and II molecules are routed through different cellular compartments. Hence, they can bind peptides derived from different sites. As a general rule MHC class I molecules, which are widely expressed on different cell types, bind peptides that have been produced within the cytoplasm (e.g. where viruses often infect and replicate). In contrast MHC class II molecules, which are expressed on a more limited number of cell types, bind peptides that have been produced within endosomes or phagosomes (e.g. where pathogens can be internalized). See Figure 1.27.

MHC Molecules and T Cell Recognition How do the different subsets of CD4 and CD8 T cells discriminate between the different MHC molecules and bound peptides? The answer is quite simple and very elegant. CD8 T cells use the CD8 molecule (in conjunction with their TCR) to recognize peptide–MHC class I complexes. Any cell containing foreign antigen in the cytosol can therefore potentially be killed by CD8 T cells after they develop into CTLs. In contrast CD4 T cells, which can develop into T_h1 or T_h2 cells for example, use the CD4 molecule (in conjunction with their TCRs) to recognize peptide–MHC class II complexes. Any cell expressing MHC class II that contains foreign antigen in endosomes or phagosomes can therefore be recognized and its functions can be modulated by the CD4 T cells. See Figure 1.28.

Function of MHC Molecules in Presenting Antigens T cell-mediated immunity is crucial in defence against infection, and the ability of MHC molecules to bind and present peptides from infectious agents to T cells is essential for T cell activation and thus for host defence in general. Infectious agents are usually multiplying very rapidly, generating mutants, and any mutation that modifies a peptide so that it cannot bind to a MHC molecule will give the agent a selective advantage by facilitating evasion of the immune response. Since we cannot mutate our MHC molecules fast enough to keep up with the high rate of mutations in many microbes, we need other ways to maintain the efficacy of our MHC molecules in defence against infection.

How is this efficacy ensured? First, MHC molecules are highly promiscuous peptide receptors: one MHC molecule can potentially bind thousands of possible peptides (although one at a time). Thus, a few MHC molecules can cover a very large number of pathogen-derived peptides, increasing the chances that a given pathogen might be recognized by the respective antigen-specific T cells. However, the possibilities are not unlimited, so any given MHC molecule may not be able to bind a certain peptide. Second however, any individual expresses multiple MHC molecules and these are co-dominantly expressed. This means that the cells of any individual express a set of MHC molecules from both maternal and paternal chromosomes, thus increasing the number of

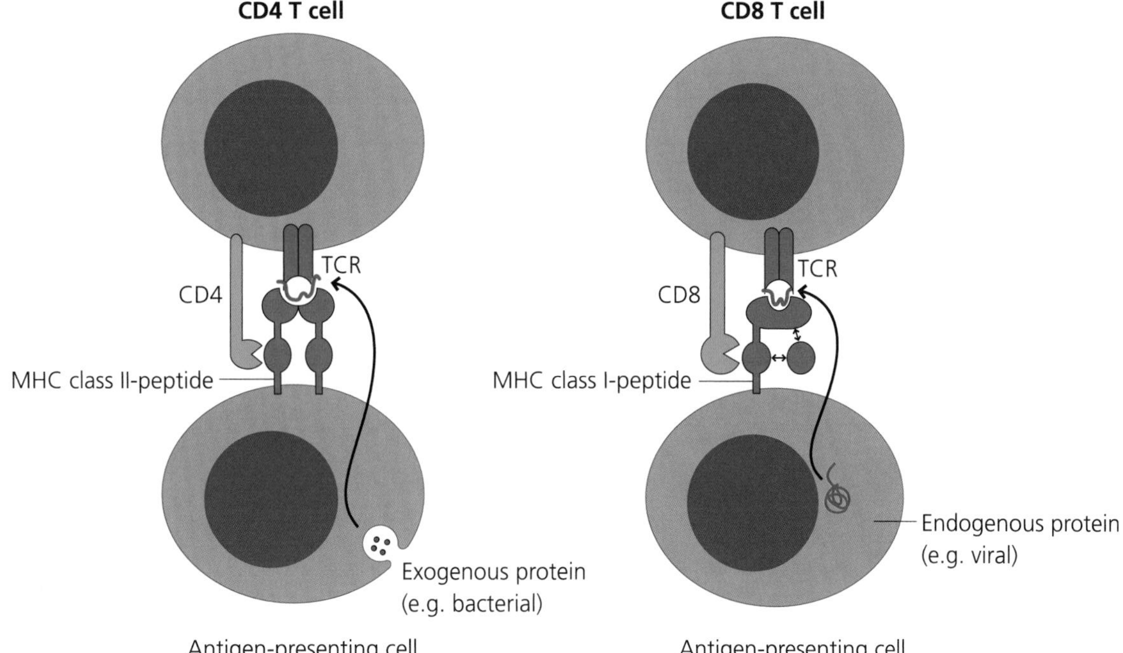

Fig. 1.28 Antigen recognition by CD4 and CD8 T cells. Conventional αβ T cells express either CD4 or CD8. **CD4 T cells** recognize peptide–MHC class II complexes. Their CD4 molecules bind to conserved regions of MHC class II molecules, helping their TCRs to recognize peptide–MHC class II complexes. The peptides bound to MHC class II molecules are derived from "exogenous" antigens from the extracellular milieu (e.g. microbial components that have been endocytosed). Only a few specialized cell types express MHC class II molecules constitutively. Activated CD4 T cells are often known as helper T cells. **CD8 T cells** recognize peptide–MHC class I molecules. The CD8 molecules bind to conserved regions of MHC class I molecules and perform an analogous function to CD4 molecules (Section 1.5.3.4) in permitting T cell recognition. However peptides bound to MHC class I molecules are typically derived from "endogenous" antigens from the intracellular milieu (e.g. components of infecting viruses in the cytoplasm). Almost all cell types express MHC class I molecules. Fully activated CD8 T cells are generally known as cytotoxic T lymphocytes (CTLs).

MHC molecules that are present in any individual. Third, MHC genes are very polymorphic, meaning that sometimes 100 or more alleles of a given MHC molecule exist within the population (e.g. humans). This means that the chances of any given individual sharing exactly the same set of MHC molecules as another is extremely small, except in the case of identical twins. Inevitably, because of the particular set of MHC molecules they possess, some individuals may be able to respond better to some infectious agents than others. Hence, the extensive diversity of MHC molecules within a population is extremely important to ensure survival of the species as a whole – even if some individuals are killed by any given infectious agents, it is likely that others can survive and reproduce.

1.5.4
Cytokines and Cytokine Receptors

Cytokines are small, hormone-like proteins produced by many different cell types, including the cells of immunity. They work by binding to cytokine receptors which are expressed by many different cell types, again including the cells of immunity. In the immune system, cytokines help to bring about an integrated and coordinated response appropriate to the type of infection that has occurred. Cytokines can act locally on the same or different cells, in an autocrine or paracrine manner; some also can act on more distant cells or tissues in an endocrine manner. Cytokine receptors, as well as cytokines themselves, can be grouped into different structural families; for example, some cytokine receptors comprise multiple subunits, one of which can be shared between all in the family. Some pairs or groups of cytokines tend to be involved in "cross-talk" between particular cell types of immunity, such that a cytokine produced by one cell may trigger a second cytokine to be produced by another cell that then feeds-back to enhance or inhibit the functions of the first.

Any given cytokine may have multiple functions, depending on which cell types express the corresponding cytokine receptor. The overall effects of cytokines are mediated by the different responses that the respective cell types make. Cytokines are involved in both innate and adaptive immunity, often both. Below, to provide some examples we will briefly consider these responses separately, but stress that this is a highly artificial and over-simplified distinction. We should also stress that some cytokines which play crucial roles in immunity are also produced by "non-immune" cells. See Figure 1.29.

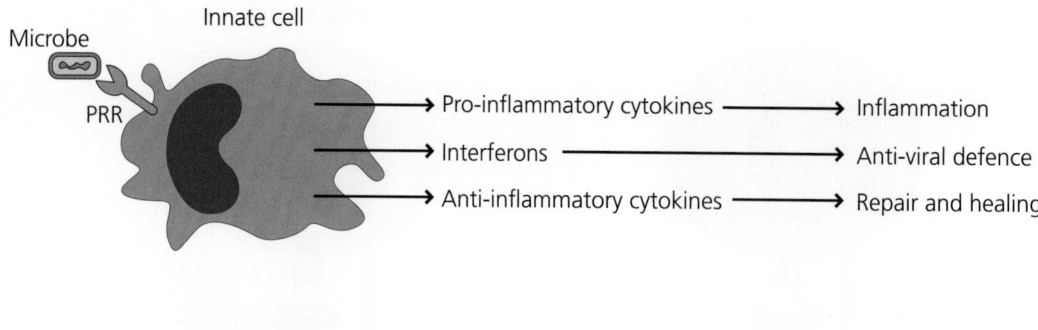

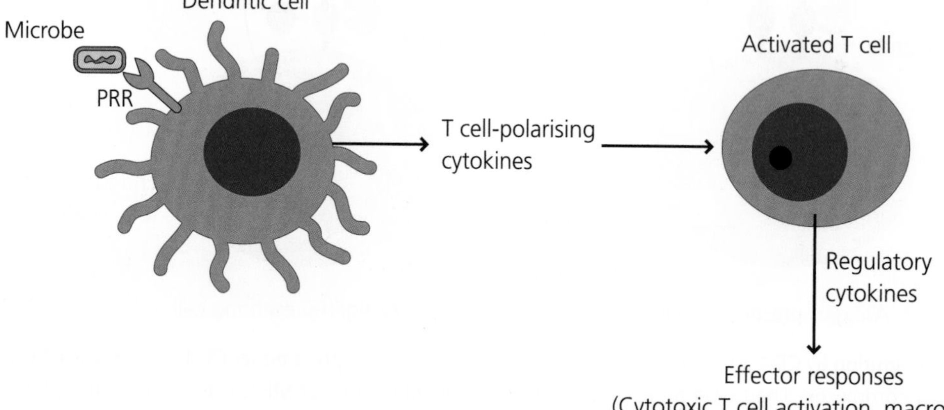

Fig. 1.29 Examples of cytokine functions. All immune (and many other) cell types can produce cytokines, which represent the main method of communication between cells that are not in direct contact with each other. Cytokines are proteins that act on other cells to alter their properties and functions. They can act on the cell that produces them (**autocrine**), on local cells (**paracrine**) or distantly (systemically – **endocrine**). They can be involved in activation or inhibition of cell functions. Some general functions of cytokines that play particularly important roles in innate and adaptive immunity or which help link these two arms are indicated. These are, however, gross over-simplifications. Not shown is another class of specialized cytokines, chemokines, that play important roles in regulating cell migration and localization, and which can be produced by all cell types indicated.

1.5.4.1 Cytokines in Innate Immunity

Cytokines are produced by all innate cells including macrophages, granulocytes, mast cells and NK cells. Some cytokines are particularly efficient at stimulating inflammatory responses and are therefore termed pro-inflammatory cytokines. For example, some cytokines induce local endothelial cell responses, leading to increased permeability and leukocyte recruitment. Some also act on distant tissues, including bone marrow, liver and hypothalamus, and contribute to systemic inflammatory responses. Other cytokines are particularly efficient at inhibiting inflammatory responses; these are termed anti-inflammatory cytokines. Some may also help to stimulate wound repair, healing and tissue remodelling, including the stimulation (or inhibition) of growth of new blood vessels (angiogenic effects). Some of these cytokines, and others produced by DCs, also play very important roles by acting on the cells of adaptive immunity and polarizing their responses (e.g. to induce different types of CD4 T cell responses).

A specialized group of cytokines is the interferons (IFNs), all of which have anti-viral activity. Type I IFNs are very powerful anti-viral agents; these include the α- and β-IFNs. They are secreted by virally-infected cells and some activated leukocytes, such as macrophages and pDCs. They act on other cells to induce an anti-viral state by inducing molecules involved in inhibition of viral protein synthesis. Type I IFNs also regulate other immune cells. Type II (γ) IFN is a relatively weak anti-viral agent; however, more importantly, it is a highly potent macrophage activator that induces strong microbial killing activity in the cells. It is secreted by NK cells and activated T cells.

1.5.4.2 Cytokines in Adaptive Immunity

Many cytokines are produced by activated lymphocytes. Some of the best understood, functionally, are probably those that are produced by CD4 T cell subsets (though CD8 T cells and even B cells also secrete cytokines). Different sets of cytokines are produced by different polarized subsets of CD4 T cells. For instance, a signature cytokine secreted by T_h1 T cells is IFN-γ and, by T_h2 cells, IL-4. Together, these sets of cytokines help to orchestrate different types of adaptive immune responses. For example, different sets of cytokines are involved in stimulating B cells (once activated) to secrete different types of antibodies that can interact selectively with phagocytes and NK cells or with mast cells and eosinophils.

Generally speaking, cytokines can be thought of as soluble molecules that bind to their respective cellular receptors. However, certain cell surface molecules that are involved in communication between cells may also be structurally related to certain cytokines and their receptors, and are thus said to belong to a cytokine–cytokine receptor family. A case in point is the TNF–TNF receptor family. Many molecules in this family play crucial roles in life or death decisions of cells. For example, the prototypic member of this family, TNF-α, can function as a pro-inflammatory cytokine in some cases, but trigger apoptotic cell death in others. Fas and Fas ligand, introduced earlier as molecules that can be used by cytotoxic cells to kill other cells, are also, structurally, members of this family.

1.5.4.3 Chemokines and Chemokine Receptors

As discussed earlier, a specialized subset of cytokines is involved in the localization and directional migration (chemotaxis) of immune cells into and within different anatomical compartments (e.g. at sites of inflammation or within secondary lymphoid tissue). These chemoattractant cytokines are hence termed chemokines. Some chemokines also have additional functions (e.g. they may have direct anti-microbial activities). Chemokines bind to chemokine receptors that are widely expressed on different types of cell. Often, several different chemokines can bind any given chemokine receptor and any given chemokine receptor can bind different sets of chemokines – the number of possible permutations and combinations is immense.

As for cytokines in general, different chemokines are produced by the different cells of innate and adaptive immunity, and help to orchestrate immune responses. For example, chemokines produced by polarized subsets of T cells can recruit more of the same subset of T cells into sites of infection, as well as other cells with which they need to interact (e.g. monocytes or granulocytes).

1.5.5 Effector Molecules of Immunity

In previous sections we have already mentioned several types of effector molecules that help to bring about the elimination of infectious agents. For example, these include perforin and granzymes that are used by cytotoxic cells to kill infected cells (Section 1.4.4). Other molecules such as defensins have direct anti-microbial functions in that they are directly toxic to microbes (Chapter 4); arguably these might also includes some of the ROIs and other toxic molecules that can be produced by phagocytes. In this section however, we will focus on two other sets of very important effector molecules in immunity: complement and antibodies. Complement is a multi-molecular system of proteins in blood and tissue fluids; it is primarily a component of innate immunity. Antibodies, secreted by B cells when they develop into plasma cells, are primarily components of adaptive immunity. However, both complement and antibodies are involved in each type of immunity (below).

1.5.5.1 Shared Functions of Complement and Antibodies

Complement and antibodies each have specialized features and functions, but some are shared between them. These include:

- **Recruitment to sites of infection.** Both are types of soluble molecules that circulate in the blood, but which can be recruited to sites of infection because local inflammatory responses increase the permeability of the endothelium, allowing them to cross into extravascular tissue spaces.
- **Formation of immune complexes.** Both can bind to soluble antigens to form complexes that may induce inflammation.
- **Opsonization.** This is the process of coating microbes, to enhance their uptake and subsequent elimination by phagocytes. Microbes coated with complement or antibodies are directed to specialized receptors, called opsonic receptors. These are respectively the complement receptors and the Fc receptors (FcRs); there are several types of each.
- **Enhancing adaptive immunity.** Components of complement and antibodies, when bound to antigens, can enhance different types of adaptive immune responses.

1.5.5.2 Complement and Complement Receptors

The complement system is a cascade in which activation of one component can lead to activation of the next – and because one activated molecule can activate several or many others, there is built-in amplification in the system. The complement system can be activated by three different pathways. One, the mannose-binding lectin (MBL) pathway, is initiated by a soluble PRR, MBL, binding to the surface of microbes. The second, the alternative pathway, is triggered spontaneously and serves to amplify the other two pathways. The third is the classical pathway, activated by complement binding to antibodies on the surface of microbes. In addition to their roles in opsonization of microbes and enhancing B cell responses (above), binding of complement to some microbes can lead to the assembly of other components that form a pore in the membrane, leading to direct lysis of the microbe. Other complement components diffuse away from the site of activation and are involved in stimulation of inflammation (e.g. by binding to different types of complement receptors on endothelial cells or mast cells). Complement also plays an important role in helping to solubilize and clear immune complexes from the blood. As with all cascade systems, the complement pathway is tightly regulated to help limit damage to host cells; in many cases this regulation is mediated by enzymes that cleave and inactivate active components. See Figure 1.30.

1.5.5.3 Antibodies and Fc Receptors

Different types or classes of antibodies are classified according to the type of constant region that they contain. In humans and mice these are called IgM, IgG, IgA, IgE and IgD (there are other types in species such as fish). All can be secreted as soluble antibodies although IgD exists mainly as a cell-bound form. In addition to these main classes of

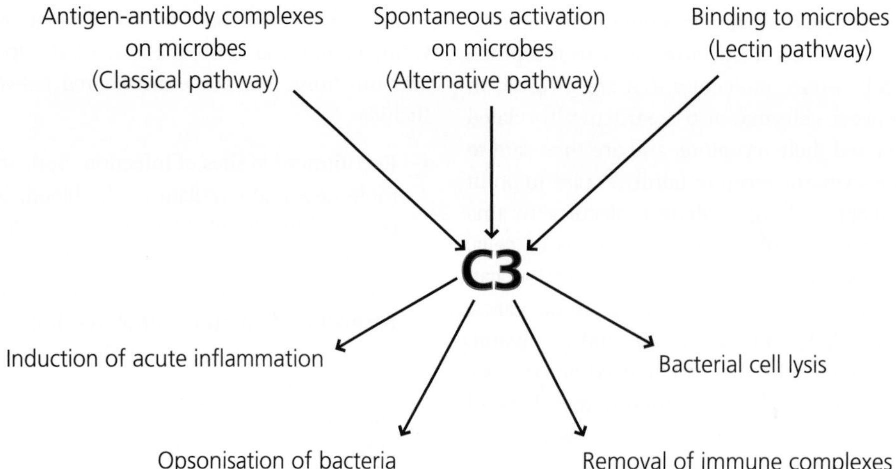

Fig. 1.30 Activation and functions of complement. The complement system comprises many different proteins, some of which form a proteolytic cascade that becomes activated during innate and adaptive responses. Complement can be activated by three different routes. Typically the lectin pathway is triggered directly by some microbial structures, the classical pathway is triggered by some types of antibodies, and the alternative pathway acts as an amplification loop for the former two. **Convergence at C3.** Activation of complement by any of these pathways leads to activation of the central C3 component, after which the pathways are identical. **Functions of complement.** Complement activation has four main outcomes. (i) Binding of complement components to the surface of microbes can opsonize them, promoting phagocytosis via complement receptors. (ii) Small complement fragments help to trigger local inflammatory responses. (iii). Activation can help to solubilize or eliminate large antibody–antigen immune complexes. (iv). The late complement components can assemble into pores in microbial membranes and (in some cases) kill them.

antibody, there are also subclasses of some of them, such as the different forms of IgG that have different functions. The different functions of different types of antibody are determined exclusively by their constant regions.

Different classes of antibodies are generally found in different compartments of the body. For example, IgM and IgG are commonly found in blood and at sites of inflammation, and IgG can also cross the placenta to the foetus. In contrast, IgA is typically secreted into mucosal sites (e.g. the gut) and into maternal milk of mammals from where it can be delivered to neonates. IgE is present at low levels in blood, but binds with high affinity to mast cells in tissues.

Different classes of antibody also have very different functions. As noted (above) some antibodies can help to opsonize microbes, directly or indirectly by activating complement. IgM, which generally has a pentameric structure (five "Y"-shaped monomers linked together) cannot act as an opsonin on its own, but is particularly efficient at activating complement. Some types of IgG are very good opsonins and in some cases they also enhance the killing of the microbes by phagocytes. IgA, which in mucosal tissues is largely dimeric, and some types of IgG are very efficient at "neutralization" – the capacity to bind to a virus, bacterium or bacterial toxin and prevent it from infecting or acting on cells. IgE is a very good "sensitizer" of mast cells – when antigen binds to IgE that is attached to mast cells or eosinophils, it stimulates degranulation and the mediators that are released help to stimulate inflammation and may be toxic to microbes or larger parasites. See Figure 1.31.

1.5.6
Cell Signalling Components

In general, receptor molecules associated with cells are linked to intracellular components that can transduce signals. Signal transduction is initiated when a receptor recognizes its ligand. This stimulates different intracellular biochemical cascades (signalling pathways). Some of these signals ultimately act on the cytoskeleton so the cell can change shape, internalize molecules or particles, or move. Others increase or decrease the susceptibility of a cell to apoptosis, thus regulating cell survival. Still others can signal to the nucleus, activate transcription factors and alter gene expression, and hence the proteins that are synthesized within that cell.

Any given receptor may be expressed on more than one cell type and similar receptors may often deliver signals in similar ways. However, the outcome can be very different depending on the type of cell that is involved. Part of the difficulty in understanding signal transduction is that the names of components are very complex. However, some of the principles are fairly straightforward. Many signal transduction pathways involve enzymes such as tyrosine kinases which introduce phosphate groups onto intracellular tyrosines in other proteins, and tyrosine phosphatases which counter the effect of the kinases by removing phosphate groups. Phosphorylated tyrosine residues can be recognized by the so-called SH2 domains of other signalling components. An example is the JAK–STAT pathway that is used by many cytokine receptors (above) JAKs (Janus kinases) are tyrosine kinases and STATs

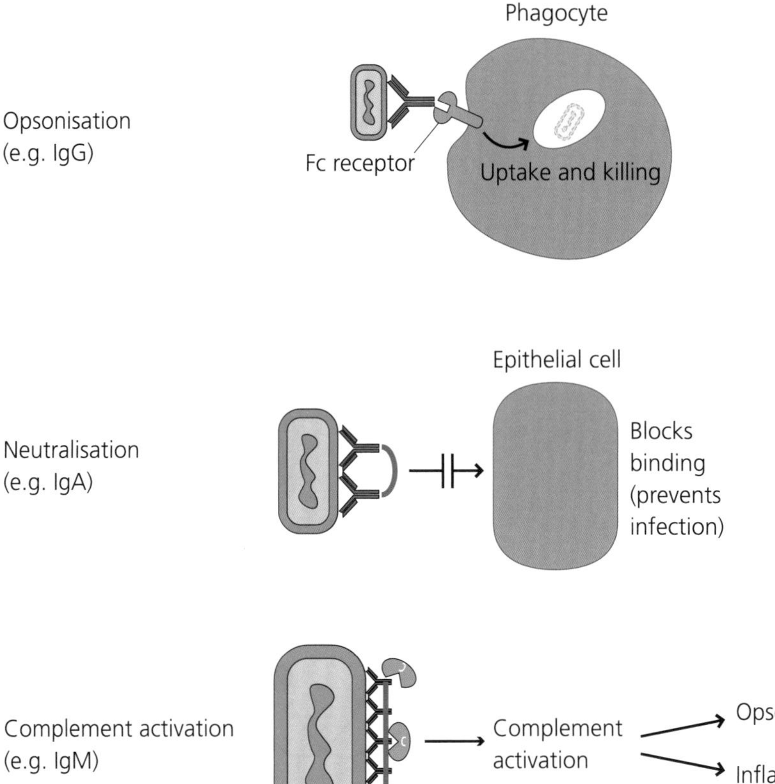

Fig. 1.31 Some functions of antibodies. Antibodies are of different classes: IgD, IgM, IgG, IgA and IgE. These all comprise the typical immunoglobulin monomer (a "Y"-shaped molecule), but in some cases they can form multimers, particularly IgA and IgM. Some classes, such as some types of IgG, are important opsonins, binding to microbes and targetting them to phagocytes through specific FcRs. Several classes, such as IgA in the gut, can bind to microbes and inhibit their attachment to host cells; by thus preventing infection they neutralize the microbe. Particular classes, including IgM, are very efficient at activating complement, thus indirectly leading to opsonization or inflammatory responses (mediated by activated complement components).

(signal transducers and activators of transcription) contain a SH2 domain that allows them to bind to the JAKs. In turn, the STATs are phosphorylated, dimerize and translocate to the nucleus where they act as transcription factors to regulate gene expression. Some signal transduction modules also use G-proteins that act as molecular switches. Examples of G-protein-coupled receptors (GPCRs) are the chemokine receptors (above). Signalling pathways may also contain adaptors that help one component bind to another and scaffolds on which several components can be attached to facilitate their coordinated functioning, particularly in specific intracellular locations. See Figure 1.32.

1.6 Immune Responses and Disease

As will become evident in other chapters, the absolute necessity for an immune system is shown by the life-threatening infections that can occur when it is defective. In addition, the immune system has enormous power to attack the body itself or to make damaging, sometimes fatal, responses to otherwise harmless agents. The immune system is also very complex. This is evident from the problems we have in understanding and preventing the rejection of transplants or the development of cancer. The immune system has, however, huge potential, only partially realized as yet, in the development of strategies for the treatment of disease.

In this last section we first outline some problems of immunity in terms of one particular problem the immune system has faced during its evolution and the different problems that result from defective, aberrant or unwanted immune responses. Finally, we introduce two therapeutic approaches designed to manipulate immunity for our own benefit.

1.6.1 The Adaptive Immune System Needs to be Educated

1.6.1.1 The Problem of Self–Non-Self Discrimination

The immune system has evolved to have a defensive role against any infectious microbe, but should not cause harm to the body. In order to do this, the immune system must be able to discriminate between harmless components of the body (self) and components such as food and commensal bacteria, which should not be attacked, and those foreign agents such as potential pathogens, which do need to be attacked. Although this is often called self–non-self discrimination, more accurately it could be called harmful–harmless discrimination. The normal non-reactivity of the immune system towards harmless antigens is called tolerance.

1.6.1.2 Immunological Tolerance

How does the immune system discriminate between harmful and harmless? We have briefly touched on this earlier (Section 1.2.3.3). The evolution of innate immunity has led to the development of PRRs, some of which recognize components of infectious agents that are not synthesized by the host. In this sense, PRRs, by their very nature, may provide perfect discrimination between harmless self and harmful infectious non-self. Thus, non-reactivity in innate immunity is essentially passive. However, there is a big problem in terms of adaptive immunity. The specificity of all lymphocyte antigen receptors, the TCRs and BCRs of lymphocytes, is largely generated at random (Section 1.5.3.2). This means that many of these receptors will potentially recognize self and there is very good evidence that this does occur, as we shall see later (Section 1.6.4). Such receptors are termed autoreactive receptors. They include autoantibodies that can cause much damage and lead to autoimmune diseases. Hence, there have to be active mechanisms of tolerance to prevent lymphocytes bearing these receptors from being produced in the first place and/or to inhibit the responses of any lymphocytes that might develop with such receptors. Both mechanisms occur, but at different times and in different places.

The first mechanism for inducing tolerance occurs during the early development of B cells in the bone marrow and of T cells in the thymus. If a developing B cell or T cell expresses an antigen receptor that can bind to any self component with sufficient strength (affinity), that cell can be either killed or inactivated; these two outcomes are termed clonal deletion and anergy, respectively. In the case of B cells, unlike T cells, they have the further option of changing their antigen receptors if they recognize self antigens; this is termed receptor editing. Since these processes occur in central or primary lymphoid tissues (as opposed to peripheral sites or secondary lymphoid tissues) these forms of tolerance are known as central tolerance. This process is, however, imperfect. For example, it is impossible to ensure that every single one of the self components that is produced throughout life is represented in the tissues where lymphocytes develop; think, for example, of the new hormones and proteins that are produced during puberty or during pregnancy and lactation. Hence, despite the efficiency of central tolerance, some mature autoreactive T cells and B cells are inevitably released from bone marrow and thymus. See Figure 1.33.

The second mechanism of tolerance induction occurs in the periphery (peripheral tolerance). It differs somewhat for T cells and B cells. In the case of naïve T cells, normal activation

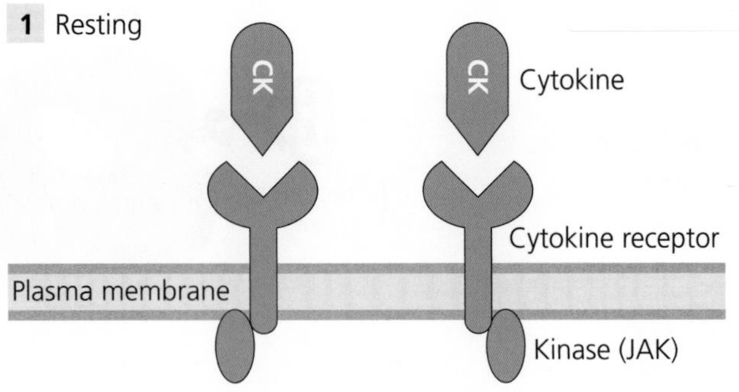

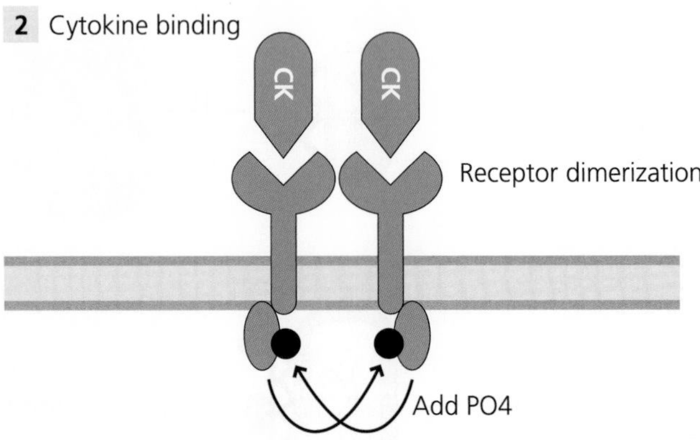

3 Recruitment of transcription factor

CK

CK

Transcription factor (STAT)

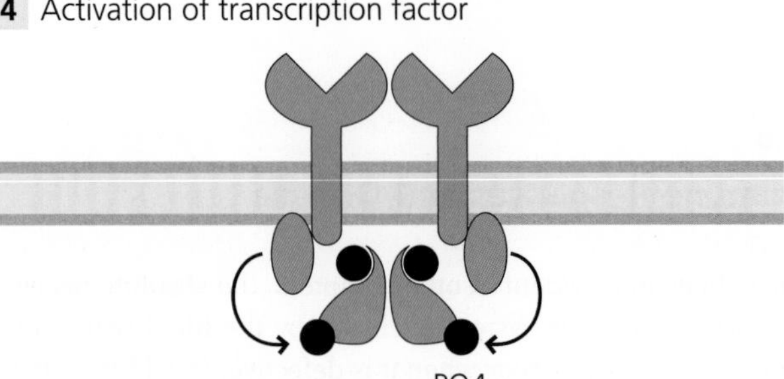

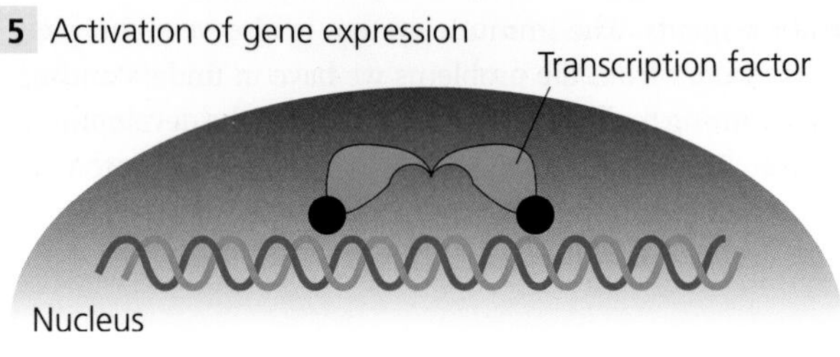

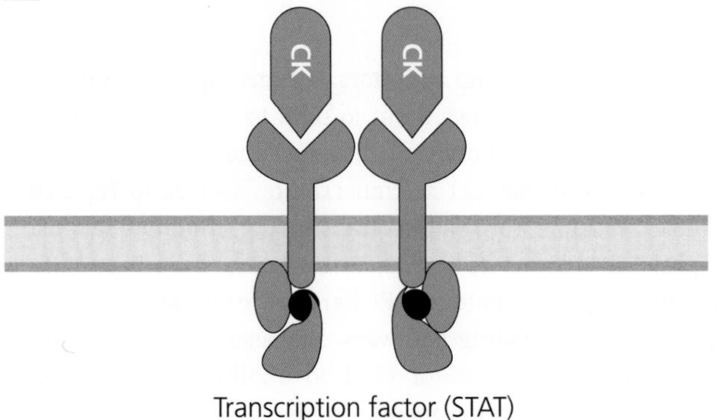

Fig. 1.32 An example of an intracellular signalling pathway. A number of cytokine receptors signal through the JAK–STAT pathway. These receptors are associated with tyrosine kinases called JAKs. Binding of a cytokine leads to dimerization of the receptors and juxtaposition of the JAKs which phosphorylate and activate each other. This enables STATs to bind (they have SH2 domains which recognize phosphorylated tyrosines). The JAKs phosphorylate the STATs, which then translocate as a dimer to the nucleus, bind DNA and act as transcription factors to change gene expression. This is a very simple signalling pathway that does not involve other molecules as adaptors or scaffolds to help link or localize different components of the pathway.

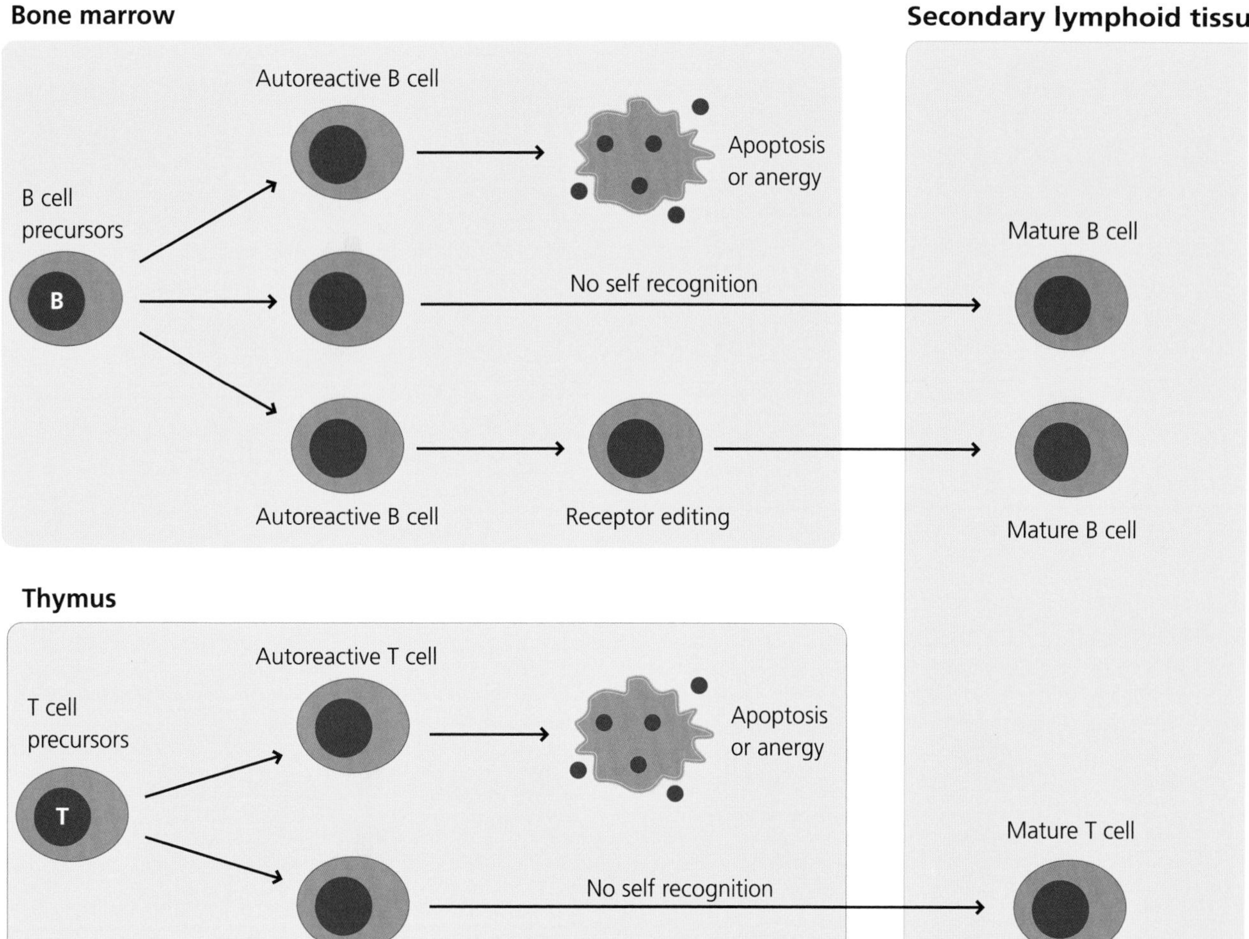

Fig. 1.33 Some mechanisms of tolerance induction. As B cells and T cells generate their respective antigen receptors largely at random (Figure 1.23) many have a high risk of recognizing components of the host itself (e.g. structural components of tissues or other cells of the body) – such receptors are termed autoreactive. Generally speaking, during lymphocyte development, cells with autoreactive receptors are killed through the induction of apoptosis (clonal deletion) or rendered unresponsive (anergy). Both apply to B cells developing in the bone marrow; the former particularly applies to T cells developing in the thymus. In addition, B cells can try to make another receptor that is not autoreactive (receptor editing). The remaining cells mature, expressing receptors are generally not able to recognize self components – hence they are tolerant – but they can potentially recognize foreign antigens should they be encountered in the future.

needs two sets of signals, including specialized costimulatory molecules that are generally expressed by DCs but not by most other cell types. Hence, if a T cell recognizes a self antigen on most other cell types that do not express costimulatory molecules, activation cannot occur; instead the cell becomes unresponsive or dies. In addition, regulatory T cells, of different types, can develop to suppress immune responses. In the case of B cells, many responses need T cell help (e.g. in the production of antibodies against protein antigens). In the absence of such help, a B cell that has recognized antigen does not become activated or in other cases is anergized and subsequently dies. This implies that T cell tolerance needs to be more complete than B cell tolerance and this is what is actually observed.

1.6.2 Immune Responses Against Infection Can Cause Damage

1.6.2.1 Immunity is Generally Beneficial

Having an immune system is of course essential for defence against infection. The rapid activation of innate and adaptive immunity often means that even if we become infected the infectious agent can be cleared without us even being aware of this happening; this is termed a subclinical infection.

1.6.2.2 Normal Immunity Can Cause Problems

Sometimes, however, immune responses can lead to unpleasant symptoms (think of how you feel if you are fighting a cold or flu) and may even cause long-lasting tissue damage, as in

1 Normal individual, non-pathogenic microbe

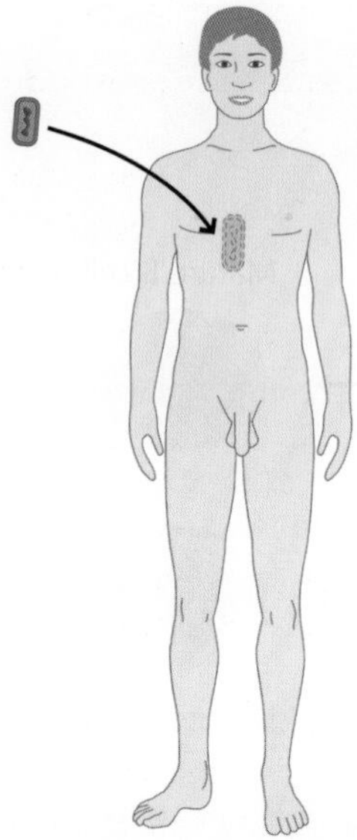

2 Immunodeficient individual, non-pathogenic microbe

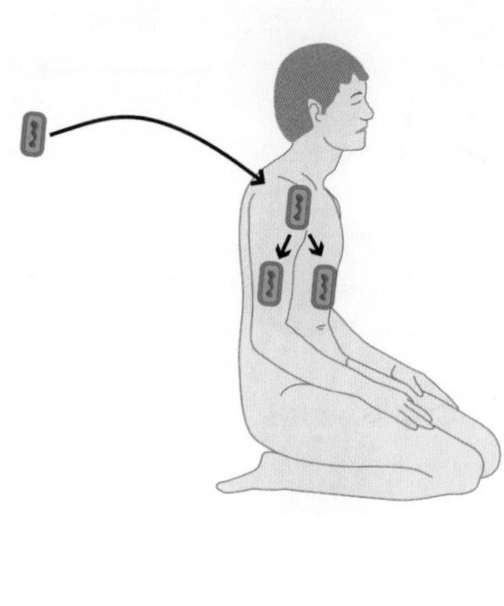

3 Normal individual, pathogenic microbe

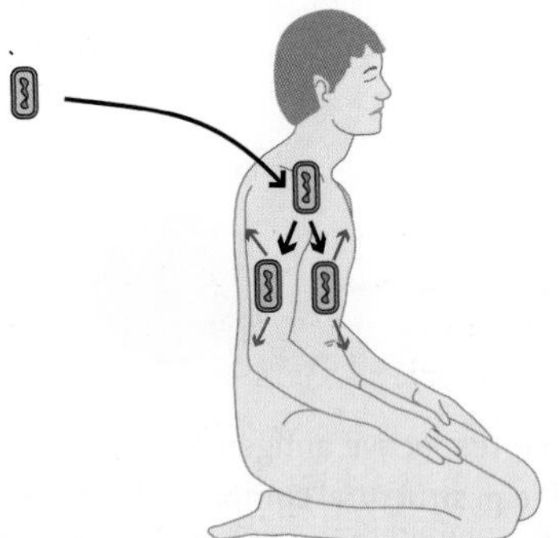

4 Normal individual, pathogenic microbe

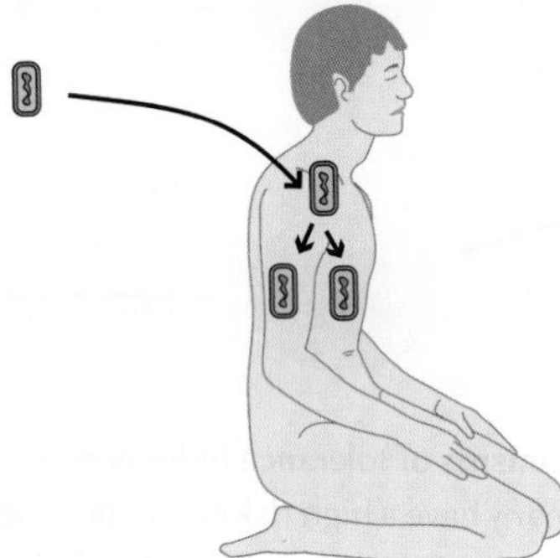

Fig. 1.34 Infection, disease and tissue damage. (1) In healthy (normal) individuals with functional immunity, many non-pathogenic microbes are eliminated without causing disease, subclinically. (2) In immunodeficient individuals these same microbes may cause disease (Section 1.6.3). (3) In normal individuals infected with some pathogenic microbes, secretion of toxic molecules or other mechanisms can cause disease. (4) In some normal individuals, the immune response to the microbe is the actual cause of clinical disease – collateral damage.

the lung of someone with tuberculosis. This is because the highly potent mechanisms that are involved in fighting infections can cause side-effects and sometimes significant collateral damage ("friendly fire"). These are examples of how normal immune responses to "infectious non-self" antigens can cause problems. See Figure 1.34.

A further problem arises because viruses and microbes can evolve so much faster than the host's immune system. The efficiency of the many components of immunity in dealing with infections means that the immune system itself helps to drive the evolution of infectious agents. In other words, it exerts considerable pressure to select infectious agents with mutations that allow them to evade immune responses. In this way an infectious agent, that would otherwise be cleared, may actually become a pathogen. Pathogens are successful because they evade, or subvert, immune responses and the diseases that they cause can sometimes be due to collateral damage caused by host immunity.

We discuss the benefits of immunity in relation to these problems and provide further examples of them, particularly in Chapter 2, and we will also meet them again from time to time in Chapters 3–6. For now we will introduce the ways in which defects in immunity lead to disease and how misdirected or unwanted immune responses can cause harm; these are discussed in more detail in Chapter 7.

1.6.3
Defects in Immunity Can Cause Serious Infections

The importance of an effective immune system is shown most clearly by the increased frequencies of infectious diseases seen in patients with defective immunity.

1.6.3.1 Primary Immunodeficiency Diseases

Any genetic defect that alters the function of a component that is normally involved in elimination of an infectious agent may lead to serious infections caused by that agent. These are called primary immunodeficiency diseases (Figure 1.34). Commonly, primary immunodeficiencies are associated with mutations in the genes that encode soluble or cell-associated components of immunity, although in other cases they may, for example, result from mutations in metabolic pathways. At least 180 different mutations that lead to primary immunodeficiencies have been identified to date; there are likely to be many more. Primary immunodeficiencies may result from defects in innate or adaptive components, or both, including the following general types.

i) Defects in phagocyte functions.
ii) Defects in complement activation, function or regulation.
iii) Defects in the generation of lymphocyte antigen receptors, leading to failure of development of T cells and/or B cells.
iv) Defects in the ability of cells to cooperate (e.g. for T cells to help B cells make antibodies).

Not surprisingly, primary immunodeficiencies can therefore lead to severe, persistent, unusual or recurrent (SPUR) infections, and the type of infection that occurs reflects the normal functioning of the component or cell that is involved. We use this as evidence for the involvement of different immune components in defence against certain types of infectious agents in Chapters 2–6.

1.6.3.2 Secondary (Acquired) Immunodeficiency Diseases

Immunodeficiencies that are not caused by a genetic defect are termed secondary (or acquired). In general these immunodeficiencies do not appear in early childhood. The individual is born with a normal immune system, but a later event causes damage leading to defective immunity. In some cases the cause(s) is unknown, but in other cases this may, for example, result from aggressive treatment for cancer or as a complication of the advanced disease itself. Secondary immunodeficiencies can also be caused by infection with pathogens. A well-known example is HIV, which, because it leads to destruction of key immune components (CD4 T cells) renders the infected individual susceptible to other types of infection that can be life-threatening as the individual progresses to acquired immunodeficiency syndrome (AIDS).

1.6.4 Immune Responses Can Sometimes be Made Against the Wrong Antigens

There are circumstances where immune responses attack the body itself, or are made against otherwise harmless agents, causing disease and sometimes leading to life-threatening conditions. These conditions are of two main types.

- Aberrant responses against self antigens, which cause autoimmune diseases.
- Aberrant responses against apparently innocuous non-self antigens that result in allergies and other immune-related sensitivities (sometimes termed hypersensitivity diseases).

All of these conditions are primarily caused by aberrant adaptive immune responses, which may, however, reflect inappropriate activation of the innate immune system. With one exception (i.e. allergies), the different types of immunological mechanism involved in autoimmune diseases and other immune-related sensitivities overlap. One convenient way of distinguishing between these two types of aberrant responses is to consider whether the antigen that is ultimately recognized is "intrinsic" (i.e. a self antigen of the host itself) or "extrinsic" (i.e. a non-infectious, non-self antigen from the environment of the host, such as pollen in those who suffer from hay fever). In most cases we do not fully understand why such responses are triggered, although often we can appreciate the underlying mechanisms that lead to disease. These conditions are considered more fully in Chapter 7. See Figure 1.35.

1.6.4.1 Autoimmune Diseases

Despite the varied mechanisms of immunological tolerance (above), immune responses are sometimes triggered against normal self components, leading to autoimmune disease. Perhaps one of the simplest mechanisms to understand is when normal mechanisms of tolerance induction are defective. This could be due to a defect in central and/or peripheral tolerance of T cells. An example of the former is if there is a failure to express self antigens, to which T cells are normally tolerized, in the thymus. An example of the latter is if regulatory T cells are not induced. Both types of defect lead to very serious autoimmune diseases and widespread tissue damage and are further discussed in Chapter 5.

Other types of autoimmune disease occur through different mechanisms, considered in Chapter 7. For most of these we understand the mechanisms causing the disease, but we do not understand why it occurs. They can be classified as follows, although more than one mechanism may be involved in any given disease:

- The activation of autoreactive B cells and the secretion of antibodies against normal self components, termed autoantibodies. These can lead to cytotoxic effects and tissue damage that is mediated by the cells with which they interact. Alternatively, they can directly inhibit the normal functioning of other cells of the body (e.g. because they block the binding of molecules to their receptors). An example of the latter is myasthenia gravis where an

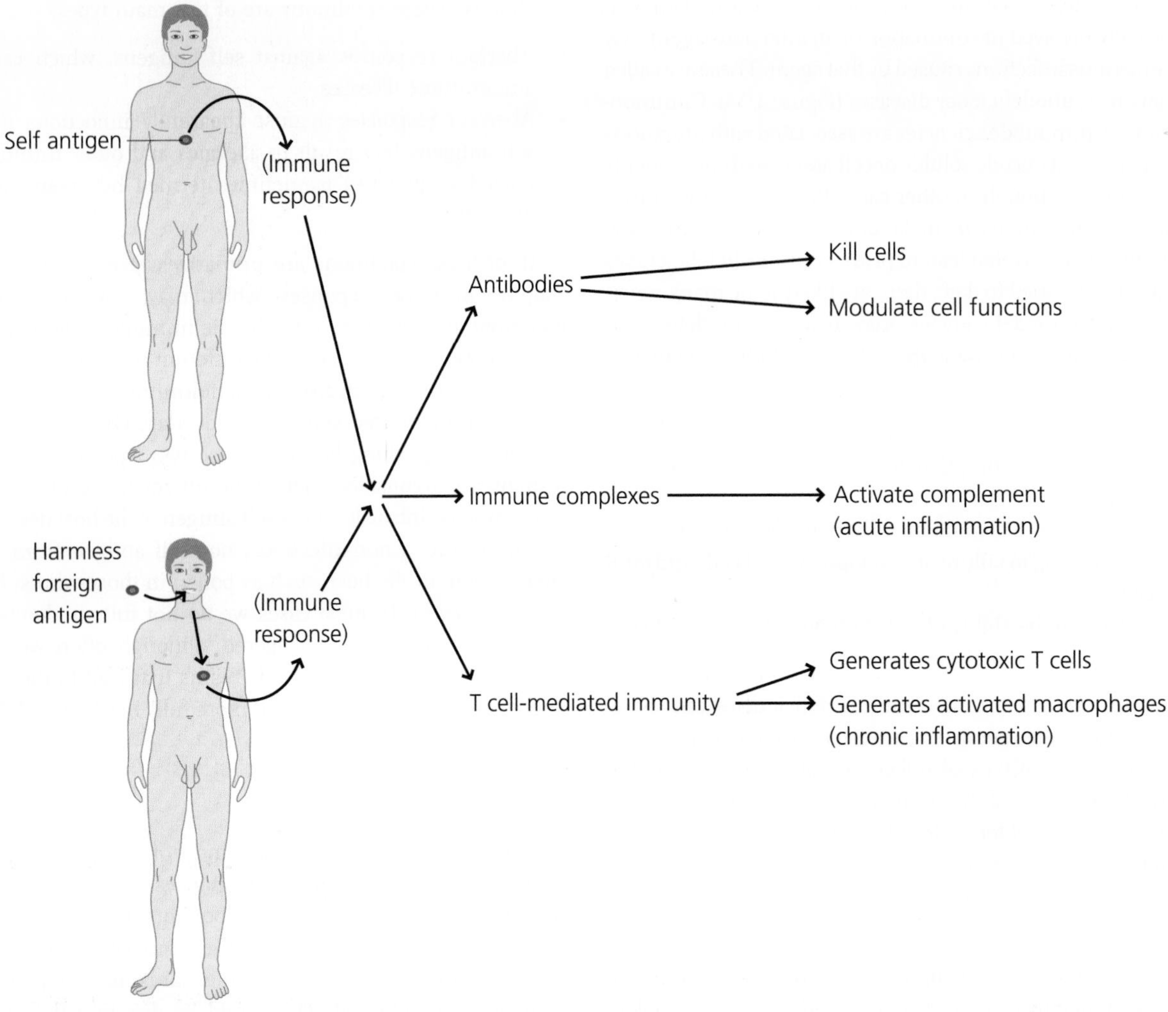

Fig. 1.35 Autoimmune diseases and immune-related sensitivities. Disease can result from activation of adaptive immune responses against otherwise harmless ("intrinsic") components of the body itself or against non-infectious ("extrinsic") foreign agents from the environment. The effector mechanisms causing the tissue damage may be mediated by antibodies of different types, large immune complexes, T cells that aberrantly activate macrophages or cytotoxic T cells, or a combination of these. Symptoms can be mild (e.g. skin rashes) to life-threatening (e.g. anaphylaxis following a wasp sting in a sensitized individual). The clinical pattern of disease usually reflects the distribution of the antigen being recognized and may be localized to particular organs or become widespread throughout the body (**systemic**).

antibody blocks the transmission of nerve impulses to muscles, leading to progressive paralysis.

- The production of immune complexes composed of antigens bound to antibodies and/or complement either in specific tissues or in the circulation. If circulating immune complexes are not effectively removed they can induce inflammation that may prevent normal functioning of certain organs. An example of the latter is the kidney damage seen in serious cases of systemic lupus erythematosus.
- The induction of T cell-mediated responses against self antigens. This is responsible for autoimmune diseases such as insulin-dependent (Type I) diabetes. Here the T cell responses involve the activation of cytotoxic T cells or activated macrophages that are, in turn, the main mediators of damage.

1.6.4.2 Immune-Related Sensitivities (Hypersensitivity Diseases)

In some cases, for reasons that are often not understood, an apparently innocuous antigen, which is not related to an infectious agent, can trigger aberrant responses that cause tissue damage. Perhaps the best well-known example of this is allergy (for which there is no known equivalent mechanism of autoimmunity). Examples are hay fever and food sensitivity. Allergy is caused when a specific type of antibody, IgE, is made against an extrinsic antigen, such as a component of pollen or peanuts. IgE then binds to mast cells with no deleterious effect. However, if the antigen is encountered again, it binds to the IgE that is coating the mast cells and triggers explosive degranulation of the cells. The mediators produced then cause the symptoms and side-effects of allergy that can, in severe cases, include obstruction of the airways, cardiovascular shock and death.

The mechanisms underlying other types of immune-related sensitivities are shared with those of autoimmune diseases. As such they include diseases caused by antibodies, immune complexes and T cells. Examples of antibody-mediated disease include drug sensitivities, such as where IgG antibodies are made against penicillin. Immune complex-mediated diseases include a wide variety of "occupational" sensitivities such as Farmer's lung, in which inhaled fungal spores in mouldy hay cause immune complexes to form and trigger inflammation in the lungs. T cell-mediated diseases include contact sensitivities, in which T cells are involved in directing damaging inflammatory responses against small molecules such as metals in jewellery. The presence of pre-formed antibodies against extrinsic antigens generally leads to rapid responses called immediate-type hypersensitivities. In contrast, T cell-mediated responses against extrinsic antigens (as well as certain pathogens such as tuberculosis) are often called delayed-type hypersensitivity(DTH) responses as it takes a day or so after antigen administration for the response to become apparent.

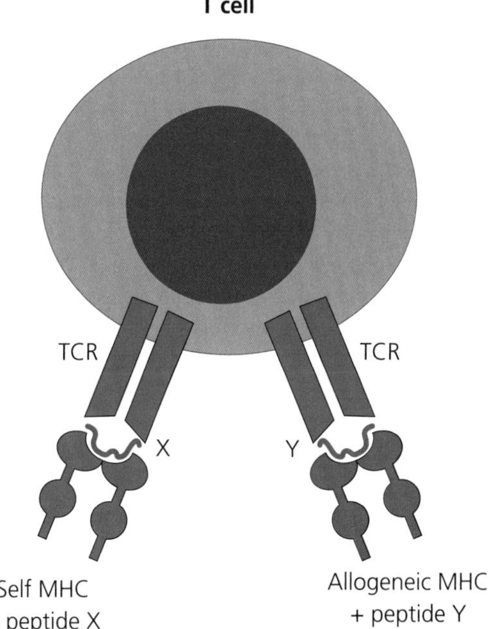

Fig. 1.36 Alloreactivity. Transplants between non-identical members of the same species (allografts) are rejected with surprising vigour because a very high frequency of T cells from any individual can recognize MHC molecules of a genetically different (allogeneic) individual. This very high frequency represents cross-reactive recognition; a T cell that can potentially recognize a foreign (e.g. microbial) peptide bound to a self peptide–MHC may also recognize an allogeneic MHC molecule(s) binding a different peptide(s).

1.6.5 Transplantation Reactions

The immune responses that occur when a foreign organ is transplanted and which often lead to rejection of the transplant are quite normal immune responses. However, in this setting they are unwanted. There are different types of rejection that occur for different immunological (and non-immunological) reasons.

1.6.5.1 Acute Rejection

Acute rejection of a transplant typically occurs within days or weeks of transplantation of a graft from a genetically different (allogeneic) individual of the same species. It is initiated by the activation of the recipient's T cells against foreign molecules present in the transplant that are termed transplantation antigens. The most important transplantation antigens are the foreign MHC molecules of the grafted tissue. T cells which would otherwise recognize peptide–MHC complexes during responses against infectious agents mount a response against the transplant because their antigen receptors (TCRs) can cross-react with foreign MHC molecules (and their bound peptides). This form of recognition is termed alloreactivity. It turns out that there is a relatively large number of alloreactive T cells in any individual, so a very substantial T cell response can be made against the foreign organ. As CD4 T cells control the responses of other cell types, allograft rejection usually involves the activation of different components of both innate and adaptive immunity. See Figure 1.36.

1.6.5.2 Other Types of Rejection

Other types of rejection are the much faster phenomenon of hyperacute rejection, when a transplant is rejected within a few minutes or hours, and the much slower phenomenon of chronic rejection that can occur even years after transplantation. The bases of these types of rejection are covered in Chapter 7. However, we will note for now that hyperacute rejection is triggered by the presence of pre-formed antibodies against the graft (e.g. because of an immune response to a previous transplant). A similar mechanism can often lead to rapid rejection of transplants between species (xenografts), such as if a pig kidney is transplanted to a human. We still do not know exactly why chronic rejection occurs.

1.6.5.3 Graft Versus Host Disease

The ability to replace defective or damaged cells or tissues by transplanting stem cells from normal individuals holds huge therapeutic potential. Bone marrow transplantation is by far the most successful example of this approach in current use. As stem cells in bone marrow can replace any and all haematopoietic cell types, this procedure is widely used for the treatment of many primary immunodeficiencies in which a defective component needs to be replaced (Section 1.6.3.1). It is also commonly used after treatment for leukaemia, because the drugs that are used to kill the leukaemic cells also kill the normal stem cells, so these can be replaced from a bone marrow transplant. One problem with bone marrow transplantation, however, is graft versus host disease (GVHD). Recipients who receive a bone marrow transplant are either incapable of rejecting the transplanted cells because of their underlying immunodeficiency or they have been immunosuppressed to prevent rejection of the transplant. The bone marrow, however, contains mature T cells from the donor. These cannot be rejected but can recognize the recipient's MHC antigens because of alloreactivity (Section 1.6.5.1). Hence, these donor T cells can become activated and generate an immune response against host tissues, often including the skin and gut. This is GVHD – the opposite of the host versus graft responses that occur during different types of graft rejection (above).

1.6.6 Tumours Can Evade the Immune System

Tumours are caused by the uncontrolled growth of cells. Benign tumours remain localized, but malignant tumours (cancers) can spread to other parts of the body (metastasis) – this is why they can be lethal. Tumours of course originate from self cells. This is seriously problematic because of the powerful mechanisms of tolerance which generally ensure that immune responses are not made against self. In many cases, however, tumours can express antigens that can potentially be recognized by the immune system. These include viral antigens if an oncogenic virus has provoked the abnormal growth of cells, as well as components of the host that have become mutated, over-expressed or abnormally expressed (e.g. those that are normally expressed only in the foetus). Hence, some tumours can potentially be recognized by the immune system. The antigens of tumours that can be recognized by T cells and antibodies are called tumour antigens.

A further problem arises because tumours are very good at evading immunity. Thus tumours can down-regulate molecules that are needed for immune recognition, such as the MHC molecules that are normally needed for T cell recognition. In addition, tumours can secrete molecules that have powerful inhibitory effects on cells of the immune system such as DCs or they can trigger the production of cells that suppress immune responses such as regulatory T cells. We have mentioned previously that the immune system itself can select variants of microbes that are resistant to defence mechanisms and exactly the same can happen with tumours. Malignant tumours have a very high mutation rate and any tumour cell that mutates in a way that enables it to resist defence mechanisms will have a selective advantage over the rest of the tumour. This is an example of Darwinian selection occurring within a single organism. See Figure 1.37.

1.6.7 Immune-Based Therapies

We conclude this section by briefly introducing two therapeutic approaches that are allowing us to modulate immunity for our own benefit. These are vaccines, which are hugely successful in preventing some disease, but which are ineffective or do not exist for others, and therapeutic antibodies which are showing real benefit in the treatment of disease. The development of vaccines originated from empirical observations made in the days before anything was known about the immune system itself, while the development of therapeutic antibodies has only been made possible from our increasingly sophisticated understanding of immunity.

1.6.7.1 Vaccines

Protection against infection can be either active, involving direct activation of the host immune system, or passive, where antibodies or immune cells are transferred to the host. One of the most effective ways of generating active resistance to infection is to have recovered from an actual infection. To infect an uninfected person deliberately in an attempt to trigger immunity is, however, risky but has been done: infection of children with virulent smallpox (variolation) was widely practiced until Jenner developed vaccination with what we now know to be the closely-related cowpox virus. Alternatively, resistance can be acquired passively. Thus, the human foetus acquires maternal IgG by transfer across the placenta and the neonate obtains maternal IgA in the early milk (colostrum). Antibodies (e.g. generated in horses) can also be given by injection to protect against potential infection, as in wounded individuals who have not been vaccinated against tetanus. See Figure 1.38.

Vaccination is an active process that is designed to mimic, as closely as possible, the infectious agent against that one wishes to induce protection and to elicit the most effective form of immunity against that agent. The enormous potential of manipulating the immune system for the treatment of disease in this way is evident from the remarkable success that vaccination has had, for example, in the global eradication of smallpox. To date, most vaccines have been designed to trigger immunity to infectious diseases in individuals before they have been infected (prophylactic vaccination). However more recently there have been increasing attempts to design vaccines for treatment of individuals with disease (therapeutic vaccination). It is crucial to understand that an effective vaccine needs two things:

- An antigen(s), against which a response is to be generated.
- An adjuvant that triggers an effective immune response.

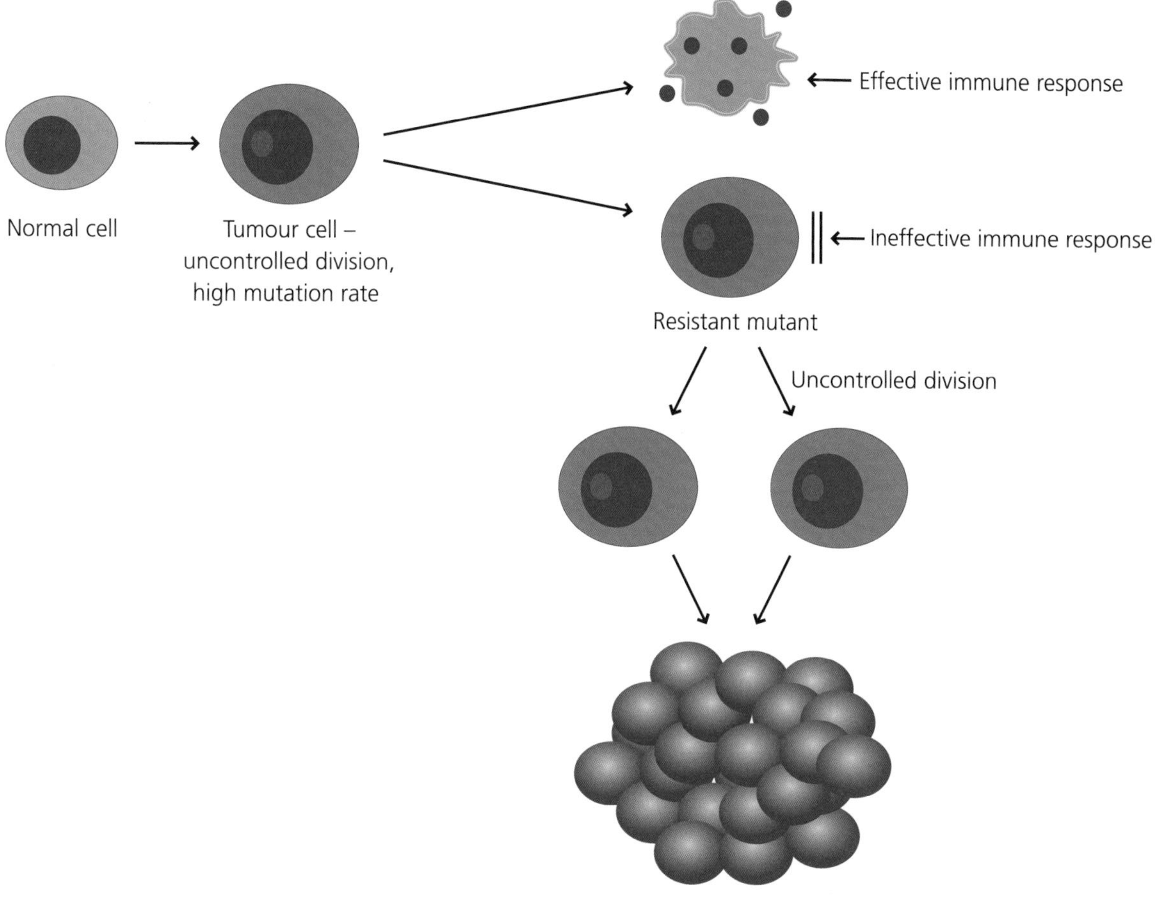

Fig. 1.37 Immune evasion by tumours. Malignant tumours (cancers) are clones of cells that have very high mutation rates. This means that many of them will express mutant proteins that can give rise to peptides not present in normal cells, which may be potentially antigenic. These peptides may induce an adaptive response to the tumour. The high mutation rate also means that new tumour variants are continually forming and, inevitably, some of these will be able to evade or avoid the immune response. (They may, for example, lose a tumour antigen, decrease MHC expression, secrete anti-inflammatory cytokines or induce regulatory instead of effector T cells). These mutant subclones will have a selective advantage and will outgrow the parental clone. Thus, over time the tumour will develop multiple means of avoiding the immune response and a cancer may develop. This is a good example of Darwinian selection in action within an individual organism.

In most cases of prophylactic vaccination against infectious diseases, protection is achieved by giving vaccines that generate active immunity. These can be live, attenuated organisms, dead organisms or parts (subunits) of organisms. These vaccines, which contain the requisite antigens and which often have intrinsic adjuvant activity, are discussed further in Chapter 2. Some vaccines are also being developed to protect against tumours. Thus, girls are now being vaccinated against the human papilloma virus (HPV) that causes carcinoma of the uterine cervix and, by preventing viral infection, initiation of the tumour is prevented.

What of therapeutic vaccines? There are situations (e.g. malignant tumours and chronic infections such as HIV) where in theory vaccines could be given to patients suffering from the disease with a view to eradicating the infection or tumour. In general, it seems much more difficult to use immune responses to treat ongoing disease than to prevent it and although many trials of therapeutic vaccination are underway, so far there are relatively few signs of real benefit. Nevertheless, progress is being made as we describe in Chapter 5 (adoptive cell therapy).

1.6.7.2 Monoclonal and Therapeutic Antibodies

Antibodies, because of their specificity, have the potential to be potent, highly selective drugs able to modulate cells and molecules *in vivo*. Until fairly recently this potential could not be realized except in a very few cases because antibodies could not be made reproducibly or in sufficient quantities. This form of treatment was revolutionized by the development of monoclonal antibodies. To make monoclonal antibodies, usually, antibody-secreting (B) cells from an

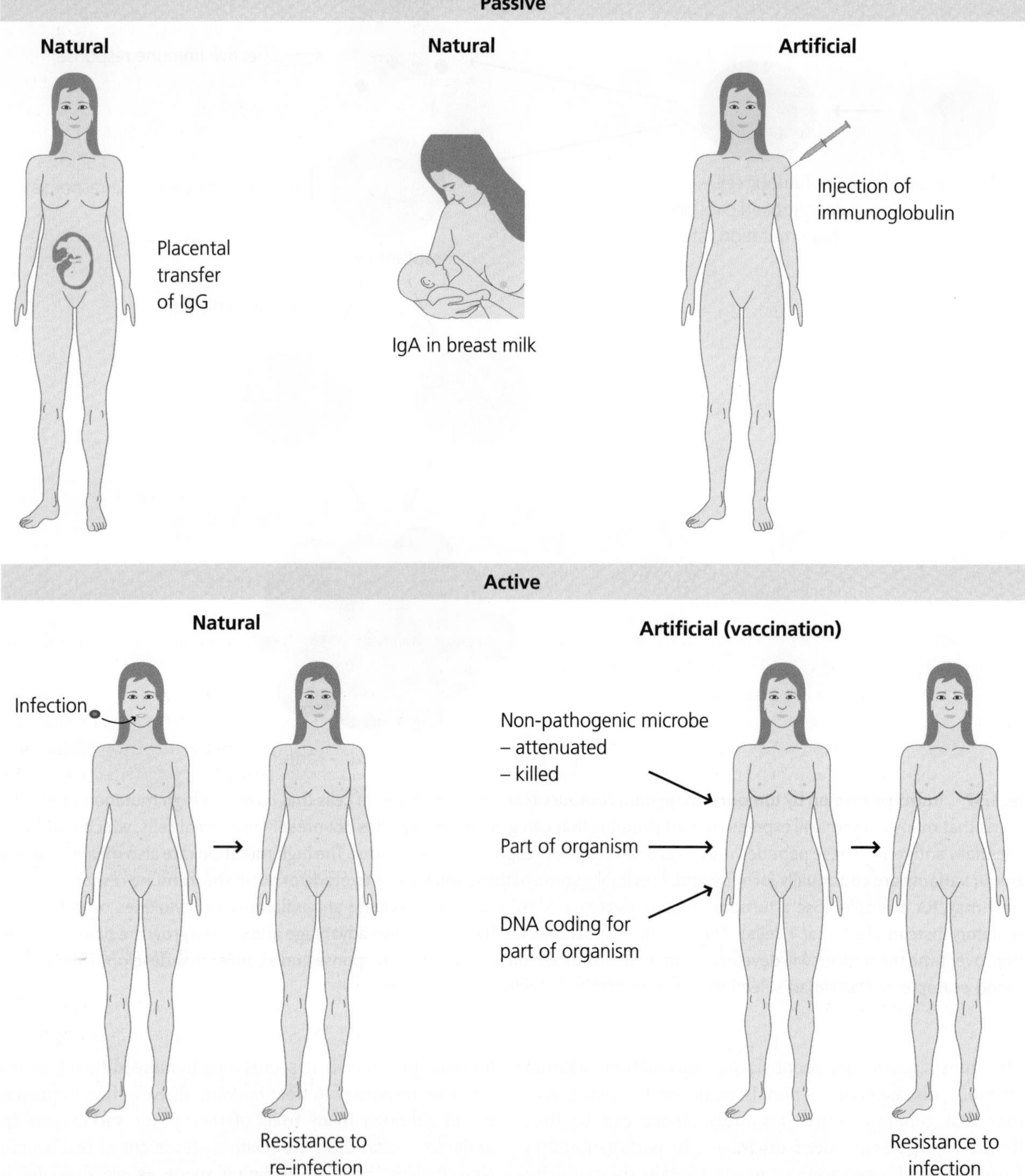

Fig. 1.38 Resistance to infection. Resistance to infection can be acquired passively or actively, through natural or artificial means. For example, passive immunity may be acquired naturally by a foetus or neonate through transfer of maternal antibodies across the placenta or in milk. It may also be delivered artificially, such as by giving pooled human immunoglobulin to antibody-deficient patients. Active immunity generally follows naturally after recovery from an infection. It may also be stimulated artificially during vaccination with a vaccine designed to induce a protective response.

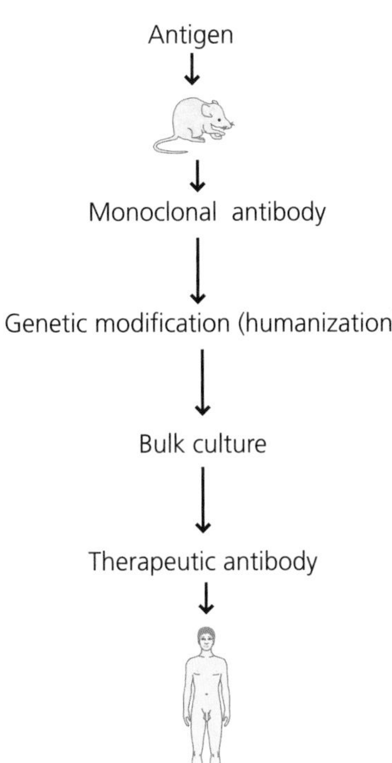

Fig. 1.39 Therapeutic antibodies. The development of monoclonal antibody technology has enabled the production of large quantities of homogenous antibodies with a defined antigenic specificity that can be used therapeutically to treat human disease. A major problem with such antibodies is that they are foreign proteins (they are typically produced in mice). The immune responses induced against them (e.g. the production of anti-antibodies) can lead to their very rapid destruction. To avoid this, genetic engineering has been used to create antibodies in which the only non-human parts are the hypervariable regions that form the antigen-binding site itself and hence are much less immunogenic. These can be used, for example, to block the activity of pro-inflammatory cytokines such as TNF-α in some autoimmune diseases, or to target molecules expressed selectively by tumour cells in certain cancers.

immunized mouse are fused with tumour cells of B lymphocytes called myeloma cells. The resulting fused cells are termed hybridomas. These are monospecific and immortal, and can be grown as clones on industrial scales. Thus, large amounts of a monoclonal antibody can be produced. Since some antibodies can inhibit or stimulate cellular responses they have huge potential as therapeutic agents. The use of monoclonal antibodies has revolutionized immunotherapy and there are now over 20 such antibodies in clinical use and many more under development. These therapeutic antibodies are being used increasingly to treat certain autoimmune diseases, to help prevent transplant rejection and to eliminate tumour cells in the treatment of some cancers. See Figure 1.39.

Learning Outcomes

By the end of this chapter you should be able to understand, briefly explain and discuss some aspects of the following topics – the relevant sections of the chapter are indicated.

- Host defence against infection (Section 1.2)
  - What are the major differences between innate and adaptive responses?
- The anatomical basis of immunity (Section 1.3)
  - What are the major differences between primary and secondary lymphoid tissues?
  - Why is inflammation important in host defence against infection?
- The cellular basis of immunity (Section 1.4)
  - What are leukocytes?
  - What are phagocytes and why are they important?
  - What are granulocytes?
  - What are lymphocytes and how many types of conventional lymphocyte are there?
  - What is immunological memory?
- The molecular basis of immunity (Section 1.5)
  - What types of molecules are involved in the migration of leukocytes from the blood into sites of infection?
  - How does recognition of infectious agents differ in innate and adaptive responses?
  - What mechanisms are involved in the generation of receptors used for antigen recognition in adaptive immunity?
  - What are cytokines and what do they do?
  - What is complement?
  - What do antibodies do?
- Immune responses and disease (Section 1.6)
  - How does the immune system normally avoid making harmful responses to self or harmless foreign molecules?
  - What is an immunodeficiency disease?
  - What different types of antigen are involved in autoimmune diseases and allergies?
  - How and why might transplants be rejected?
  - How might the immune system be able to recognize tumours?
  - How might immunity be manipulated for therapeutic purposes?

2 Infection and Immunity

2.1 Introduction

Our planet Earth is estimated to be around 4.5 billion years old. Life on this planet is thought to have emerged at least 3.8 billion years ago. The emergence of entirely new species, and the extinction of many others, has been driven by natural selection. So, we believe, has the immune system.

Each new species that evolves represents a new ecological niche that can be used by other species (e.g. as food or shelter, or as a vehicle for transmission of their genes) and which can be considered as parasites in the most general sense. The emergence of these species, both the hosts and the parasites, has been driven by natural selection. Parasites and their hosts are involved in an evolutionary arms race. In most cases it is not to the parasite's advantage to kill its host because this would prevent spreading of parasite genes. If, however, the host evolves new defence mechanisms to prevent or limit infection, any parasite that evolves ways of overcoming these mechanisms will have a selective advantage and will reproduce more successfully. Reciprocally, but generally much more slowly, a host that mutates to resist parasites more effectively will also have a selective advantage. Thus, it is host–parasite interactions that drive the evolution of immune defence mechanisms in the host.

Viruses and microscopic organisms such as bacteria and small protozoa, that may collectively be termed microbes, evolve much faster than their mammalian hosts. If a microbe mutates so that it is "invisible" to the host's immune system it would take many generations before the host could evolve effective defence. If, however, the host had evolved anticipatory mechanisms that could recognize any variants that might occur in a microbe, it would be well placed to cope with many mutants. This is precisely what the adaptive immune system in mammals has achieved by randomly generating lymphocytes that express a vast diversity of antigen receptors.

In this chapter we examine immunity to infection in mammals. We start by introducing the different types of organism that can infect us and explain that, while the vast majority of these organisms do not cause disease, some of them, the pathogens, can cause serious disease and death in normal individuals (Section 2.2). We note that we have mutually beneficial relationships with the commensal bacteria that continuously inhabit our bodies. We introduce the concept that defects in immunity, the immunodeficiencies, can result in opportunistic infections by microbes that are otherwise rapidly eliminated by those with functioning immune systems. We emphasize that the types of infection that occur in an immunodeficient host provide evidence for the normal functioning of that component in immunity.

We then explain the function of each major component of immunity in defence against infection, using studies of natural human defects and the responses of genetically modified mice as evidence for these functions (Section 2.3). We next outline how some of the main classes of infectious agent can cause disease and explain how the different components of immunity, considered in isolation earlier, work together to try to eliminate them (Section 2.4). To illustrate the different interactions between hosts and different classes of pathogens we use selected case studies to describe the type of disease that each can cause. We discuss why disease happens, how different mechanisms of immunity interact during different types of response, and how these diseases can be prevented or treated by strategies based on our knowledge of immunity.

Finally, we discuss briefly the vaccines that have successfully eradicated some of infectious diseases, but which are not available for so many others, and outline some new approaches for vaccine design that are being developed for the future (Section 2.5).

By the end of this chapter you will have an appreciation of the diversity of potentially infectious agents that exist, why some of them cause disease, and how normal mechanisms of host defence protect against so many (and vaccines against a few).

2.2 Pathogens and Infectious Disease

2.2.1 What is a Pathogen?

Any given species has very few pathogens. Probably more than 99.9% of all potentially infectious agents are in fact non-

Exploring Immunology: Concepts and Evidence, First Edition. Gordon MacPherson and Jon Austyn.

pathogenic in normal individuals, and only a few represent opportunistic infectious agents that can cause disease in immunocompromised individuals (Section 2.2.4).

A pathogen is any infectious agent that causes disease. This is, however, too simple a definition. Some pathogens such as rabies virus will virtually always cause disease if they get into the tissues of a mammalian host (e.g. through a bite), but it is highly unlikely that swallowing rabies virus would cause any harm at all. Others, which we often think of as highly pathogenic, such as *Mycobacterium tuberculosis* (the bacterium that causes tuberculosis), will cause clinical disease only in a minority of healthy people who inhale them. Some, that we think of as normally harmless can cause disease if they are in the wrong place; in other words, they become pathogenic. For example, *Escherichia coli*, a normal bacterial inhabitant of the large bowel, can cause acute inflammation of the bladder (cystitis) if it gets into the urinary tract.

Other micro-organisms are relatively or completely harmless to most of us, but can cause disease in people with genetic defects of immunity; in other words, primary immunodeficiencies. *Pneumocystis jirovecii* is a fungal organism very widespread in the environment. We must all be inhaling it at frequent intervals with no harmful outcomes. It is, however, a major cause of illness in patients infected with human immunodeficiency virus (HIV) who progress to a secondary immunodeficiency – acquired immunodeficiency syndrome (AIDS) – as their CD4 T cells are eliminated. The kind of infection that *Pneumocystis* causes in such cases is called opportunistic. There is an implication of this observation that does not seem to be widely appreciated. Patients with immunodeficiency may become clinically infected by organisms that are harmless to others. So this can only mean that in immunocompetent individuals the immune system is working continually, subclinically, without us being aware at all that defence mechanisms are or have been in action. This suggests that we should radically change our view of the immune system. It is not a system that comes into operation only when we encounter a pathogen; it is a physiological system that is as much a part of normal homeostasis as temperature regulation or the regulation of food intake.

Q2.1. How far is it valid or useful to think of the immune system as a sixth sense?

To be a successful pathogen, a microbe must have evolved mechanisms that enable it to evade or avoid host defence (here we use microbe to cover all infectious agents, whatever their size). These mechanisms allow it to infect the host and cause disease (Section 1.2), whereas similar but non-pathogenic agents lacking such mechanisms cannot infect and cause disease. The molecules expressed by pathogens that allow them to infect and survive in their hosts, and which enable them to cause disease, are known as virulence factors. Pathogens use many different mechanisms to subvert immunity and by examining these mechanisms we can learn much about immunity to infection. One authority on vaccinia virus has suggested that we will learn more about immunity to vaccinia from deciphering the vaccinia genome, and hence identifying components the virus might use for subversion or evasion, than by studying immune responses to the virus itself. So what are the different types of pathogen to which humans with otherwise normal immunity are susceptible, and how do they establish infections?

2.2.2 What Types of Pathogen Can Cause Disease?

A remarkable diversity of organisms can be disease causing, or pathogenic, although only a relatively small number of each is pathogenic in any given species such as humans. Perhaps the smallest – although these are not true organisms – are viruses, which include many different types, many probably yet to be discovered. Then there are bacteria, a vast number of different types of single-celled prokaryotic organisms that can live either outside or within the host's cells (prokaryotes lack a membrane-bound nucleus). Increasing in complexity there are also single-celled and multi cellular eukaryotic yeasts and fungi (eukaryotes possess a membrane-bound nucleus). In addition there is a huge group of organisms classed as true parasites which range in size from microscopic (e.g. protozoa, the malaria parasite) to truly enormous (metazoa, such as helminths such as tapeworms in the gut). We may even have to consider molecules as pathogens if it is the case that diseases such as bovine spongiform encephalopathy (BSE) in humans or scrapie in sheep, are caused by misfolded protein molecules called prions. It is astonishing, but essential, that the immune system has evolved a variety of mechanisms that can potentially attack, and often successfully eradicate, so many different types of infectious agent, differing dramatically in sizes (nanometres to metres) and locations after infection (outside and inside cells, and infecting a variety of different tissues). See Figure 2.1.

Yet, even if many of these infectious organisms are successfully eradicated in due course, many can still cause disease before this happens; the symptoms we suffer when we have a cold or flu are examples of this type of disease. We will now examine the main classes of infectious agents, some of which are pathogenic, in a little more detail.

2.2.2.1 Viruses

Viruses are subcellular particles that contain nucleic acid (DNA or RNA) in a protein coat. They may or may not have an external lipid envelope derived from host cells. They rely entirely on host cells for their replication. As for other pathogens there are many ways in which they can enter the body; from the air, food or water (e.g. the oro-faecal route for polio), by sexual transmission, or through living vectors (e.g. from an insect bite for yellow fever, or a mammalian bite for rabies). Most or all infectious agents including viruses survive by following a cycle from their reservoirs (natural sources of the agent), through their vectors (which enable them to be transmitted to the hosts through different routes) to their hosts (you or us) and back again. See Figure 2.2.

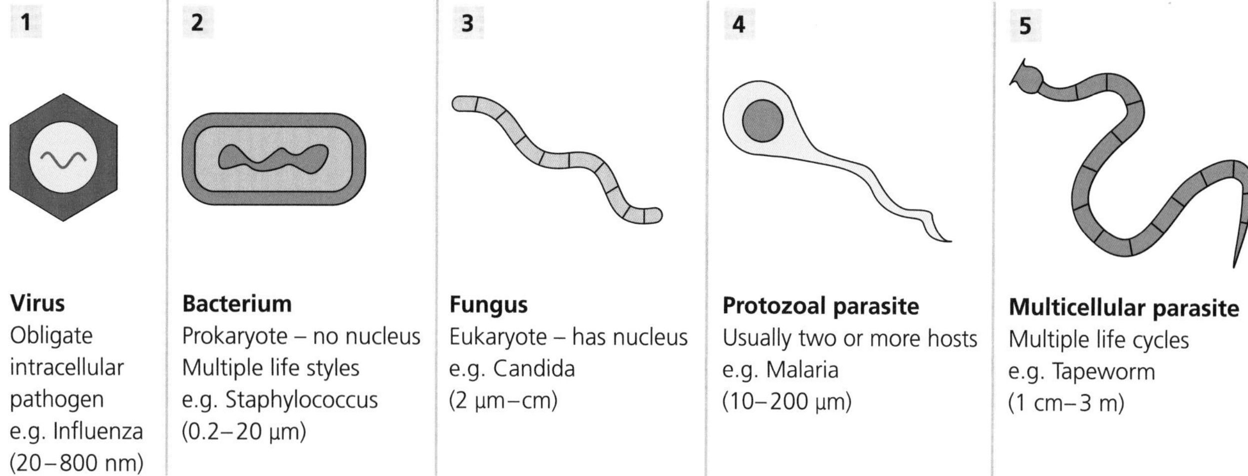

Fig. 2.1 Major classes of pathogen. Very few potentially infectious agents are pathogenic. A pathogen invades a host to gain shelter, feed or reproduce , and so that it can survive to infect other hosts. (1) The smallest pathogens are viruses, which use the machinery of the host's cells to reproduce. (2–4) Next in size are bacteria, fungi and single-celled protozoa; some live outside cells, others preferentially live inside cells. (5) The largest parasites are multi-cellular metazoans which are too big to invade cells, but which can live in body cavities, such as intestinal worms. All these can trigger immune responses. Not surprisingly different types of immunity are needed to deal with these different infectious agents.

Important barriers to entry of viruses are of course the epithelia – the sheets of cells that cover body surfaces, such as the skin, and the linings of the respiratory and intestinal tracts, and including structures such as secretory glands in the lactating breast and absorptive tubules in the kidney. Some viruses, such as rabies, can bypass the outer epithelial layers of the body through a bite. (In this case the virus even changes the behaviour of its host to increase the chances of

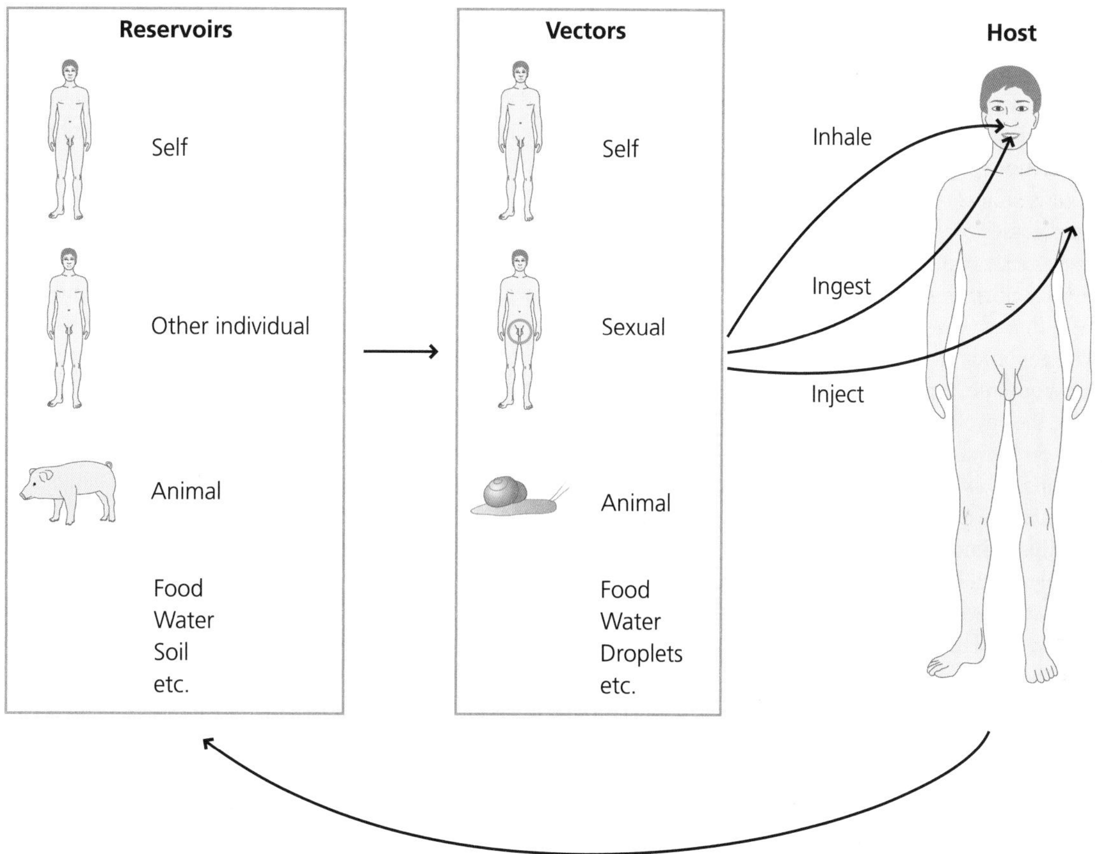

Fig. 2.2 Transmission of infectious diseases. Infectious agents are transmitted to their host from reservoirs, which may include the host itself. The vectors of transmission can be physical (e.g. droplets) or other biological species. Infectious agents can enter the host by different routes, usually by ingestion, inhalation or injection (sexual transmission is also frequent).

transmission to others; it makes them aggressive and more likely to bite.) Most viruses, however, need to infect the first epithelial layer they come to (e.g. in the respiratory tract or intestine). Many viruses, such as influenza, only infect this first epithelial layer. The damage viruses cause to epithelia may then facilitate secondary infections by bacteria; this often happens in the common cold.

Q2.2. If influenza only infects the respiratory epithelium, why do we feel so generally unwell when we have flu?

Other viruses have a life history in which they sequentially infect different cell types. For example, the polio virus first infects the intestinal epithelium and then may infect cells in lymph nodes draining the intestine to which it travels in the lymph. In a small minority of cases, the virus also infects motor neurons in the spinal cord leading to muscular paralysis and thus preventing breathing; in the 1950s, this necessitated patients spending the rest of their lives in iron lungs to enable artificial respiration. Some, such as the influenza virus, are "one-hit" viruses – they cause an infection, but are rapidly cleared by the immune system. Others, once they have infected, may stay with us for long periods or a lifetime. Herpes simplex virus (HSV) infects epithelial cells in the oropharynx or genital tract, but then infects sensory neurons and travels to dorsal root ganglia in the spinal cord. Although you are unaware of it, once you have been infected, the virus will stay there for the rest of your life, often hidden from the immune system; this is called latency. If you have ever had a cold sore you will always have the virus. From time to time the virus becomes re-activated and travels back down the nerves to the skin, causing the typical lesions of cold sores. The same is true of chickenpox – once infected you always have the virus, but in this case if it is reactivated it causes the painful disease known as shingles.

To infect a cell, a virus has first to attach to the cell. It usually does this by binding to specific molecules on the cell surface. Thus, for influenza, haemagglutinin (HA; a molecule in the viral envelope) binds to sialic acid (a carbohydrate that is part of many glycoprotein molecules on the surface of cells). In the case of HIV, the envelope protein gp120 can bind to a molecule called CD4 on human T cells, as well as dendritic cells (DCs) and macrophages (which, in human, also express CD4). This binding specificity determines which cells a particular virus can infect; this is known as viral tropism. For example, polio virus can only infect human and some primate cells. However if polio virus RNA is injected into chicken cells, fully infectious virus can be formed because, once inside the cell, it can use the cellular machinery to replicate. The resultant viruses are able to infect human cells, but still cannot infect other chicken cells because they do not have the appropriate receptor. This shows that the tropism (host range) of this virus is determined by its ability to bind to molecules expressed by the host cell. Similarly, mice are normally totally resistant to polio. If, however, transgenic mice are constructed that express the human polio virus receptor, the mice can be infected and their cells release fully infectious virus. See Figure 2.3.

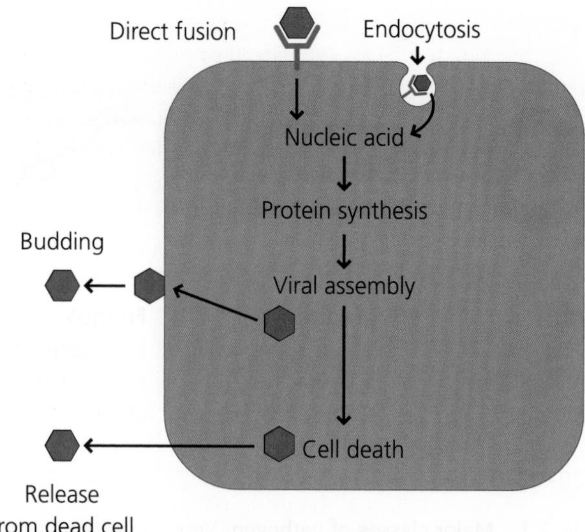

Fig. 2.3 Life cycles of pathogenic viruses. Following transmission, viruses bind to specific molecules expressed by the host cell (this determines the cells that can be infected i.e. the tropism of the virus) and are then internalized by endocytosis or by direct fusion with a cell membrane. The viral nucleic acid is released and viral proteins are synthesized. Some proteins are involved in replication of the viral nucleic acid and others are used for assembly of new viral particles. Viruses can then be released from the cell either by budding from the membrane or by killing the cell.

The attached virus may bind to the plasma membrane, after which its envelope fuses directly with the membrane, or may enter the cell by endocytosis. In endocytosis the cell membrane associated with the bound virus invaginates, and closes to form an endocytic vesicle containing the virus and some extracellular fluid. (In contrast, the uptake of a small particle such as a bacterium is called phagocytosis, and the vesicles formed are termed phagosomes.) Proteins in the virus enable the virus to fuse with the endosomal membrane and the viral nucleic acid can now enter the cytoplasm. For example, in influenza, the low pH in the endosome induces a conformational change in the viral HA, which permits it to fuse with the endosomal membrane. The viral nucleic acid hijacks the cell's metabolic machinery and instructs the synthesis of viral proteins. In many cases, so-called early proteins are involved in the regulation of metabolism and the replication of the viral genome, followed by the synthesis of late proteins which are assembled to form the viral particles. The virus is released from the cells either by budding, in which case part of the infected cell membrane forms the viral envelope, or by causing the death of the cell, and allowing release of new viruses into the extracellular spaces from where they can infect new cells or new hosts.

2.2.2.2 Bacteria

Bacteria are small, self-contained organisms, generally capable of reproducing independently of host cells. They are present everywhere throughout our environment but almost all of them are harmless, and indeed some are very beneficial to their hosts. Bacteria are classified by their morphology and biochemical characteristics (e.g. cocci are round and bacilli are elongated). You may also see bacteria described as Gram-positive or Gram-negative. This refers to their staining properties with particular dyes and reflects the biochemistry of their cell walls. However, from the point of view of pathogenesis and immunity, it is most useful to classify bacteria in terms of the mechanisms by which they cause disease. This approach is used in the following sections. Unlike viruses, there is no generally applicable life history because bacteria have very diverse patterns of behaviour. Acquisition of bacterial infection is however by much the same mechanisms as for viruses, as shown by the following examples. They may be acquired from droplets in the air, as in bacteria that cause sore throats (*Streptococcus pyogenes*) or tuberculosis (*Mycobacterium tuberculosis*); from contaminated food and drinking water, as in bacteria that cause food poisoning (*Salmonella*) and diarrhoeal diseases (*Vibrio cholerae*; cholera); through sexual contact (*Treponema pallidum*; syphilis); or through cuts and abrasions (*Clostridium tetani*; tetanus) or insect bites (*Yersinia pestis*; plague). Often indeed they are derived from the hosts themselves: urinary tract infection originating from normal bacteria of the large intestine (*Escherichia coli*; cystitis) is an example we noted earlier.

Some bacteria do not need to infect or cross an epithelial surface to cause infection. Following inhalation, *Streptococcus pneumoniae*, the cause of lobar pneumonia, inhabits the smallest air spaces (alveoli) in the lung. After ingestion, *Vibrio cholerae* attaches to intestinal epithelium but does not invade. Some bacteria, however, do invade epithelia. Following ingestion, *Shigella disenteriae* (the cause of dysentery) invades intestinal epithelium but does not go further, while other organisms such as *Salmonella typhi*, the cause of typhoid fever, go even further. The latter first invades intestinal epithelium, but it then crosses the endothelium (the layer of cells that lines blood vessels, as well as lymphatic vessels) to enter the bloodstream and infect other sites such as the liver. Many bacteria that have crossed an epithelial surface nevertheless remain largely extracellular and for this reason they are termed extracellular bacteria. Others, however, particularly those mycobacteria that cause tuberculosis and leprosy, live preferentially inside host cells, often in fact primarily within macrophages, cells that normally play important roles in defence against this type of organism, The bacteria can therefore be termed intracellular bacteria. They, and some protozoal parasites such as *Leishmania*, can also be referred to as facultative intracellular parasites – the term parasite being used here in a generic rather than specific sense. See Figure 2.4.

2.2.2.3 Fungi

Fungal infections are mainly acquired from the local environment. Their spores are everywhere and mostly they cause superficial infections: athlete's foot, Dhobie itch (groin) and thrush (vulva and mouth) are common examples. Occasionally fungal infections may be widespread throughout the body (systemic), but this is rare in normal individuals although it can be a major problem in patients with immunodeficiencies such as AIDS.

2.2.2.4 Parasites

As noted above, a wide variety of different organisms are grouped together and classified as parasites, although the term is sometimes used more loosely to include certain bacteria as well (above). In fact all "true" parasites are, by definition, eukaryotes. The diversity of such parasites ranges from single-celled eukaryotic organisms such as protozoa, to multi-cellular metazoa such as worms. Parasitism refers to a symbiotic relationship in which the infectious organism lives on or in its host and obtains nourishment from it, often to the detriment of the host. Another form of symbiosis is commensalism, in which one partner benefits without apparently affecting the other. For example commensal bacteria may benefit the host by competing with other, potentially pathogenic bacteria. Nevertheless, it is clear that the commensal bacteria also benefit from this relationship (e.g. the host provides shelter and a source of nutrients) and as such this relationship actually reflects another form of symbiosis called mutualism in which both partners benefit. It is however difficult to see how true parasites could benefit their hosts in any way.

Parasites have hugely diverse life histories – sometimes undergoing remarkably different morphological (and sexual) changes at different stages – and it would be unwise to try to generalize as to how they are all acquired or how they all cause disease. Nevertheless some protozoal infections are transmitted to humans through insect bites. For example, the protozoa that cause malaria (*Plasmodium*) are transmitted by mosquitoes, those that cause African sleeping sickness and Chagas' disease (*Trypanosoma*) by tsetse flies and assassin bugs, respectively, and *Leishmania* by sand flies. Others are acquired orally, such as the protozoa that cause amoebic dysentery (*Entamoeba hystolitica*) and toxoplasmosis (*Toxoplasma gondii*). After infection of their hosts, some protozoa live inside particular cells, such as *Leishmania*, which preferentially infects macrophages as noted earlier. Others can infect a variety of different cell types even at different stages of their life history. As we shall see (e.g. Section 2.4.5), the malarial parasite (*Plasmodium falciparum*) after infecting hepatocytes in the liver undergoes repeated cycles of infection of red blood cells. See Figure 2.5.

In terms of routes of transmission, similar principles to the above often also apply to worm infections. For example, the worms that cause filariasis (*Onchocercidae*) are transmitted by biting insects, whereas those that cause trichinosis (*Trichinella spiralis*) are acquired orally from food. Some have even evolved mechanisms to infect directly through the skin: after release by snails into the water, the parasites that cause schistosomiasis (*Schistosome cercariae*) can directly penetrate the exposed skin of humans that come into contact with it. In general, once they have infected, some parasites at different stages of their life cycle can invade a variety of different tissues, such as

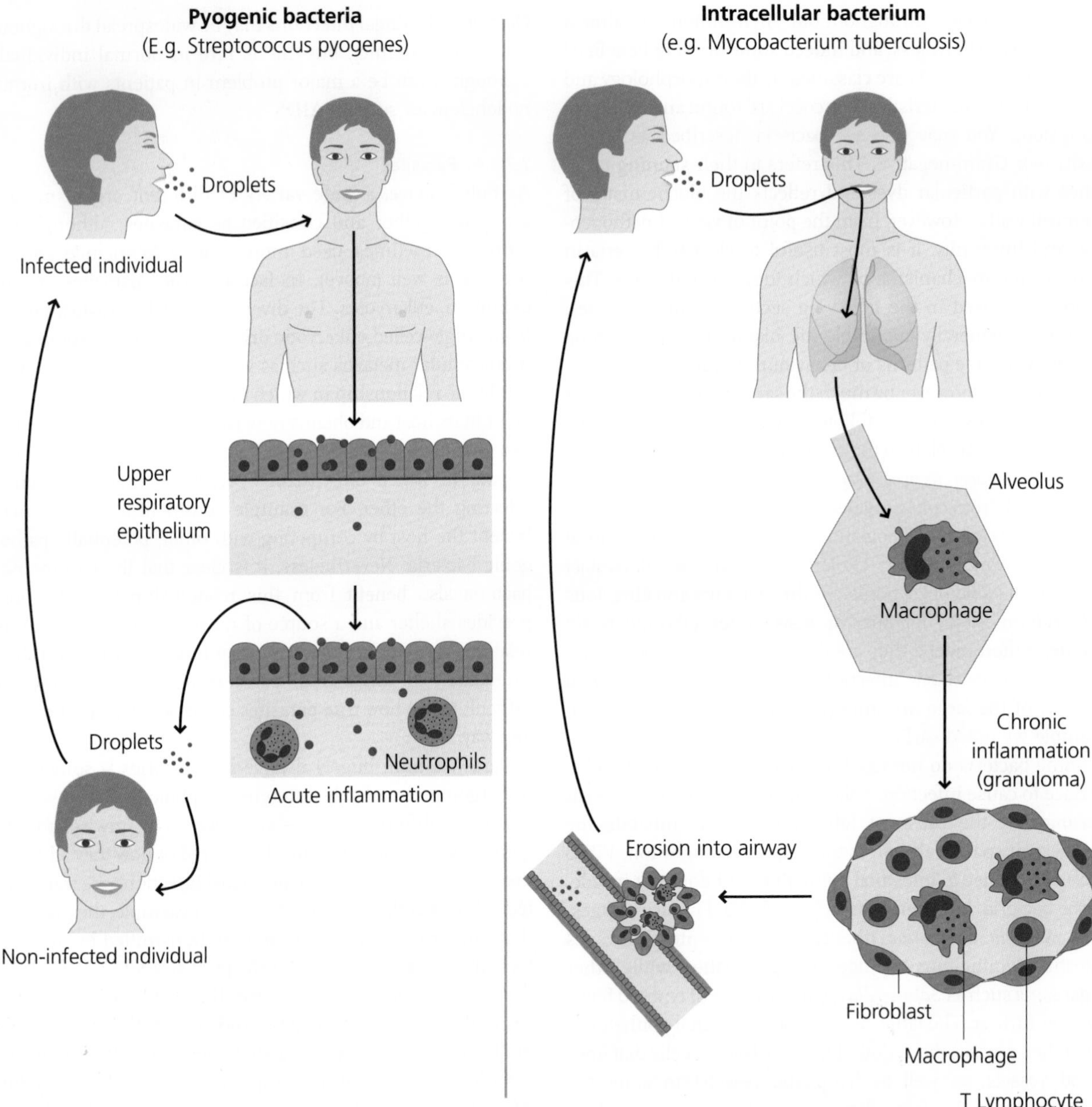

Fig. 2.4 Life cycles of pathogenic bacteria. Different types of bacteria cause very different types of diseases. Some pyogenic (pus-forming) bacteria such as *Streptococcus* can live and replicate extracellularly and may spread destructive infections to a variety of tissues. If not treated they can cause acute inflammation that resolves within a few days or can cause death, again in a few days. Other bacteria such as *Mycobacterium tuberculosis* which causes tuberculosis can only survive and replicate intracellularly. The immune response against the infected cells often causes chronic inflammation that typically lasts for months or years if not treated successfully.

T. spiralis which infects the mucosae of the intestine, liver and skeletal muscles. Others, such as tapeworms, being so much larger, remain in the lumen of the intestine.

2.2.3 Infection and Disease

It is important to make a distinction between infection and disease. If a room full of immunocompetent individuals were subjected to an aerosol containing a pathogenic bacterium (e.g. *Mycobacterium tuberculosis*), everyone would inhale the bacterium, but only a small proportion, possibly around 10%, would develop clinical disease. The reminder would have been actively infected, but would have controlled the bacteria without any clinical sign of disease. In this latter case people were actually *infected* with the organism, in the sense that it gained access to their bodies, but with fully functioning immune systems they were able to recover without clinical signs of *disease*. Therefore, infection does not necessarily lead to disease. See Figure 2.6.

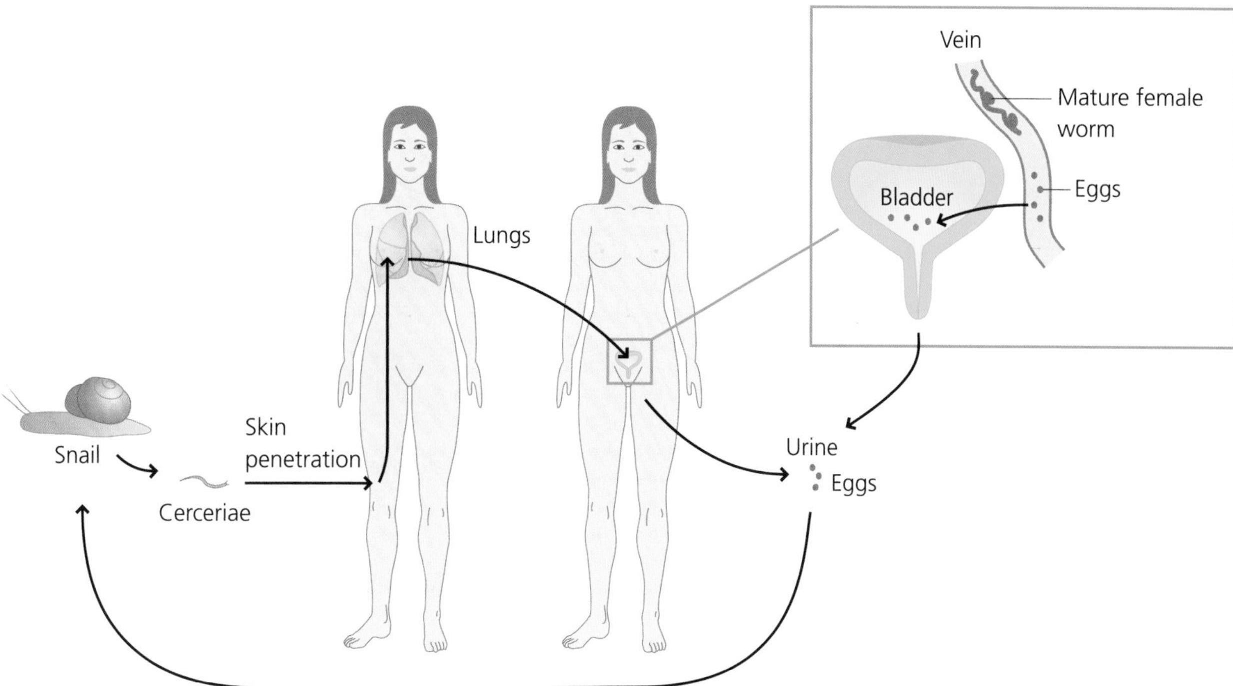

Fig. 2.5 Generalized life cycles of parasites. Parasites display a huge variety of life cycles. In many cases there is an animal reservoir and a different animal vector. Often, as in schistosomiasis, the parasite exists in different forms in the reservoir, the vector and the human host. Schistosomes are metazoan parasites that live in water snails and are transmitted by their cerceriae forms that penetrate the skin of humans in the water. They migrate to the lungs, mature and adult sexual forms live in blood vessels in different sites. Some species live in blood vessels around the bladder and the eggs may then enter the bladder, and re-enter water via urine, to infect more snails.

What properties does a microbe need to possess if it has the potential to cause disease in normal individuals?

i) It needs to be able to access the host and survive.
ii) It must be able to evade or subvert the defence mechanisms that the normal individual possesses.
iii) It needs to be able to cause, directly or indirectly, the tissue damage associated with disease. Direct causation may be, for example, by toxin secretion. Indirect causation often represents collateral damage, since the immune response against the microbe is the actual cause of tissue damage.

It is not uncommon for an individual who has recovered clinically from an infection and is hence asymptomatic, or who has never had clinical symptoms, to be infectious to others. These individuals are known as carriers. The most famous example is Typhoid Mary – a cook in New York at the start of the twentieth century – who, after recovering from typhoid, retained live bacteria in her gall bladder from which they were excreted into her faeces. It is said that she managed to infect over 40 people, several of whom died, presumably because of inadequate personal hygiene. Many other examples of carrier states exist: hepatitis B, hepatitis C, HIV, methicillin-resistant *Staphylococcus aureus* (MRSA), syphilis, meningococcus (*Neisseria meningitidis*) and gonorrhoea (*Neisseria gonorrhoeae*) are some of the better known.

2.2.4 Immunodeficiency Diseases

Some individuals suffer from an unusual frequency or pattern of infectious disease, often becoming apparent in very early life. These conditions may represent an inability to mount an effective immune response against an infection and may be inherited or acquired. These are the immunodeficiency diseases. If a child is born with a mutation in a gene encoding a molecule important in immunity to infection, the defect may result in the child becoming infected with common pathogens more frequently than normal individuals. They may also become infected with microbes that do not give rise to clinical infections in individuals with normal immunity. Often these are evident as severe, persistent, unusual or recurrent (SPUR) infections. These genetic defects, which can be inherited in an autosomal or X-linked manner, are known as primary immunodeficiencies. They differ from infections that are secondary to another cause, such as in individuals with advanced cancer or who have been aggressively treated for cancer, or, as we have noted earlier individuals who have been infected with HIV and have progressed to AIDS. These latter conditions are called acquired or secondary immunodeficiencies.

Primary immunodeficiencies can result from defects in genes involved in innate immunity or, more frequently, in adaptive immunity. Defects in innate immunity include mutations in genes that encode proteins involved in the recruitment

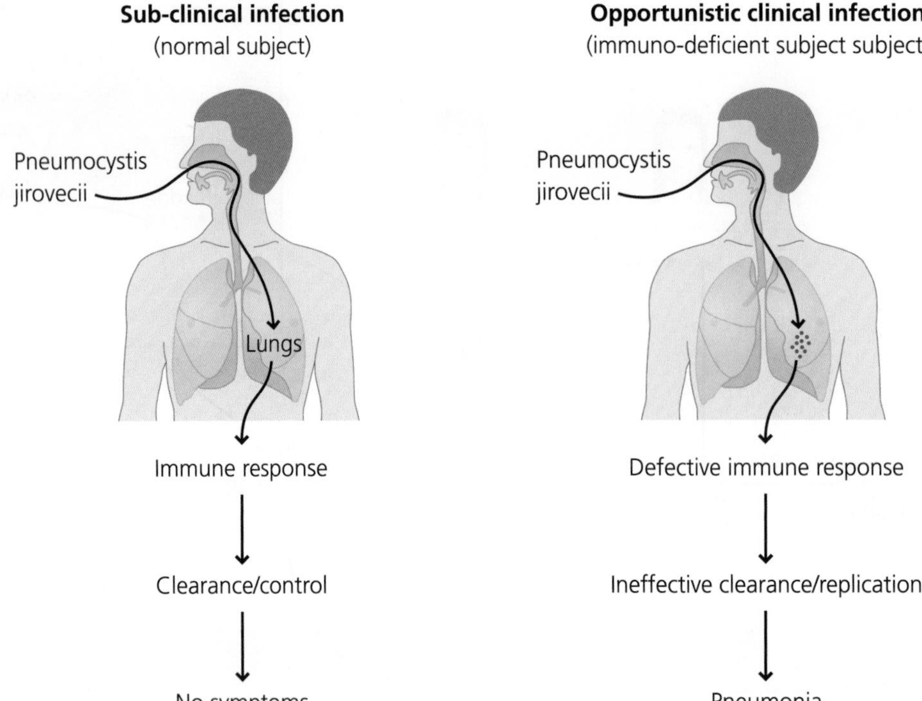

Fig. 2.6 Clinical and subclinical infection. Most microbes that enter the body do not cause any symptoms: they are dealt with by the immune system silently so the infection is subclinical. Other microbes when they infect will almost always cause symptoms: the infection is clinically evident. Some organisms such as the fungus *Pneumocystis jirovecii* do not cause clinical infection in normal humans, but can do so if the immune system is defective, as in AIDS; these infections are opportunistic. Opportunistic infections tell us that in normal individuals the immune system is working continually to eliminate or control many microbes.

of phagocytes to inflammatory sites or in the ability of phagocytes to kill bacteria, e.g. leukocyte adhesion deficiency (LAD) and chronic granulomatous disease (CGD), respectively, as well as in structural or regulatory components of the complement system (complement deficiencies). Defects in adaptive immunity include mutations in genes that encode proteins which are essential for normal lymphocyte development, so that T cells and/or B cells are absent or defective, or which are needed for T cells to help B cells make different types of antibodies, e.g. severe combined immunodeficiency (SCID) and hyper-IgM (HIGM) syndrome, respectively.

Around 200 different primary immunodeficiencies are now recognized and the genetic bases of more than half of them have been identified to date. These immunodeficiencies are very rare, but are highly informative because they provide evidence for the role of those genes – and of the mechanisms they regulate – in normal immunity. We shall meet again some of those mentioned above, and others, later in this chapter, and we discuss some of them in more detail in Chapters 3–6, sometimes illustrating them with case studies.

Q2.3. If immunodeficiencies lead to so much death and disease, why has evolution not selected for their elimination?

Secondary (acquired) immunodeficiencies are similarly varied in their nature. Some, such as AIDS, are well understood, even if we can do little about them. Others, such as those accompanying infections such as measles, are much less well understood.

2.2.5
Exploring Immunity to Infection

Before we start discussing the ways in which we combat real infections, it is important to think about evidence: how is it that we can know which molecules, cells or tissues are important in defence against any particular pathogen? This type of evidence can be obtained from different settings.

First, as we outlined above (Section 2.2.4), we have studies of human disease. The roles of different defence mechanisms in combating infection are often most clearly demonstrated by individuals who for one reason or another cannot generate a particular mechanism. We can then ask, what kind of infections do they acquire? This in turn provides evidence for the normal role of that mechanism in host defence against the particular type of infectious agent. Thus, the value of studies of these patients in understanding immunity to infection cannot be over-emphasized. In some cases, however, the deficiency may have been identified in only one or two extremely rare individuals, so that care is still needed in extrapolating some of these observations to the whole population.

Second, we have studies in experimental animals. Animal experiments are crucial in that they allow the dissection and analysis of immune responses in ways that can never be

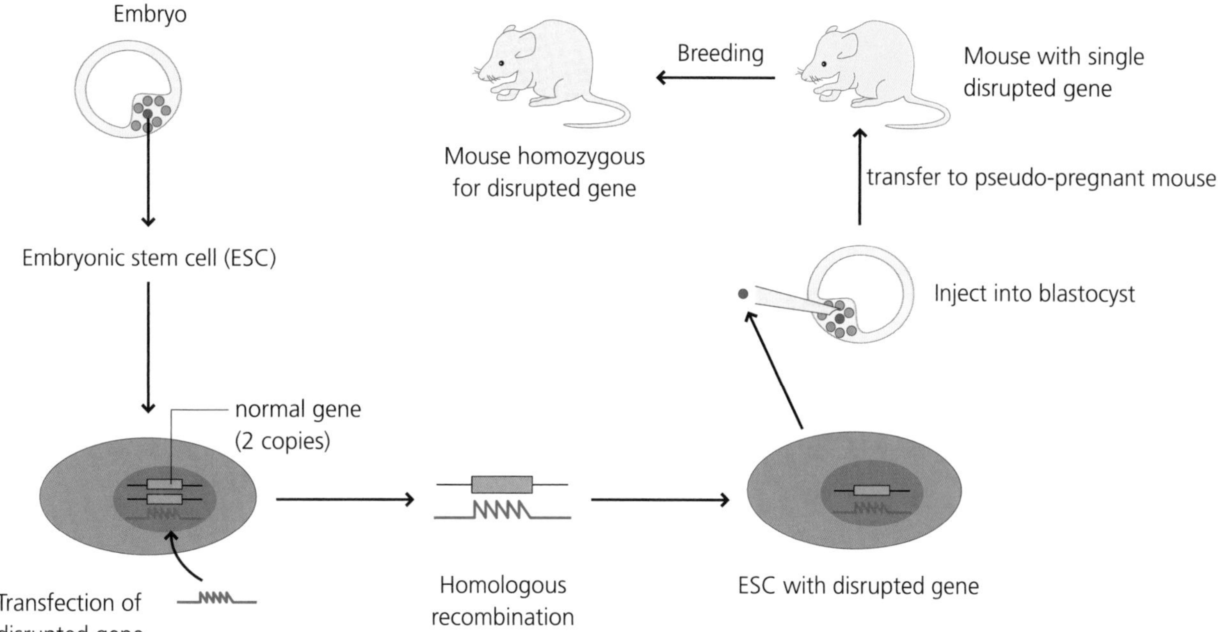

Fig. 2.7 Gene-targeted, knock-out mice. Embryonic stem (ES) cells from the blastocyst (an early stage of embryogenesis) have the potential to develop into any cell or tissue of the body. Techniques have been developed to maintain these cells indefinitely in culture. DNA containing a gene whose sequence has been disrupted can then be inserted (transfected) into ES cells. In tissue culture the stem cells divide and, on rare occasions, the inserted gene will replace the normal gene through the process of homologous recombination. The altered ES cell can then be injected into an early mouse embryo, inserted into a "pseudo-pregnant" female, and may enter the germline. One copy of the defective gene may then be expressed in the progeny of the resulting mouse. By interbreeding mice expressing one copy of the defective gene, mice homozygous for the defective gene may be selected. These mice will not express the protein encoded by the normal gene because that gene has now been "knocked out".

achieved in humans. A good example of this comes from the use of genetically modified mice (e.g. mice in which a gene has been "knocked out" so that its product can no longer be expressed). Thus if a patient with increased susceptibility to a particular kind of infection is found to have a defect in the production of a protein, say a receptor for a cytokine such as the interferon (IFN)-γ receptor, this defect is a good candidate for the molecular basis of the increased susceptibility, but of course there may be other defects that have not been identified. If, however, the gene coding for that protein is inactivated in a mouse (and only that gene is inactivated) and the mouse shows increased susceptibility to a similar infection, this is much stronger evidence that the candidate gene is in fact crucial in the development of immunity to that kind of infection. A word of caution, however: mechanisms of immunity that have evolved in humans are sometimes different from those that are found in mice, so sometimes gene knock-outs in mice do not fully recapitulate the human diseases. See Figure 2.7.

Another, third, approach is more active. Experimentally, resistance to infection can often be transferred to a non-immune animal by transfer of the appropriate immune effector mechanism. This is known as adoptive transfer – a most important approach in immunological research. Thus, cells such lymphocytes or lymphocyte subpopulations, or serum containing antibodies, can be transferred from an immune animal to a normal animal and the recipient tested for immunity. Although this approach is most often used experimentally, in some instances adoptive transfer is used to treat or prevent human disease; for example, antibodies raised in horses or humans have been used to treat tetanus, diphtheria and hepatitis. Hence, the effectiveness of these latter treatments is very strong evidence for the potential of antibodies to mediate defence against these the causative microbes. See Figure 2.8.

2.3 Host Defence Against Infection

In this section we will give an overview of the different mechanisms that are used to defend against infectious disease. To illustrate how each component of the immune system can contribute to host defence, we highlight specific types of infectious agent that can be eliminated by each. This approach is of course an over-simplification, as multiple mechanisms are usually involved in immune responses against any given potential pathogen, but it enables us to establish general principles, after which the exceptions can be better

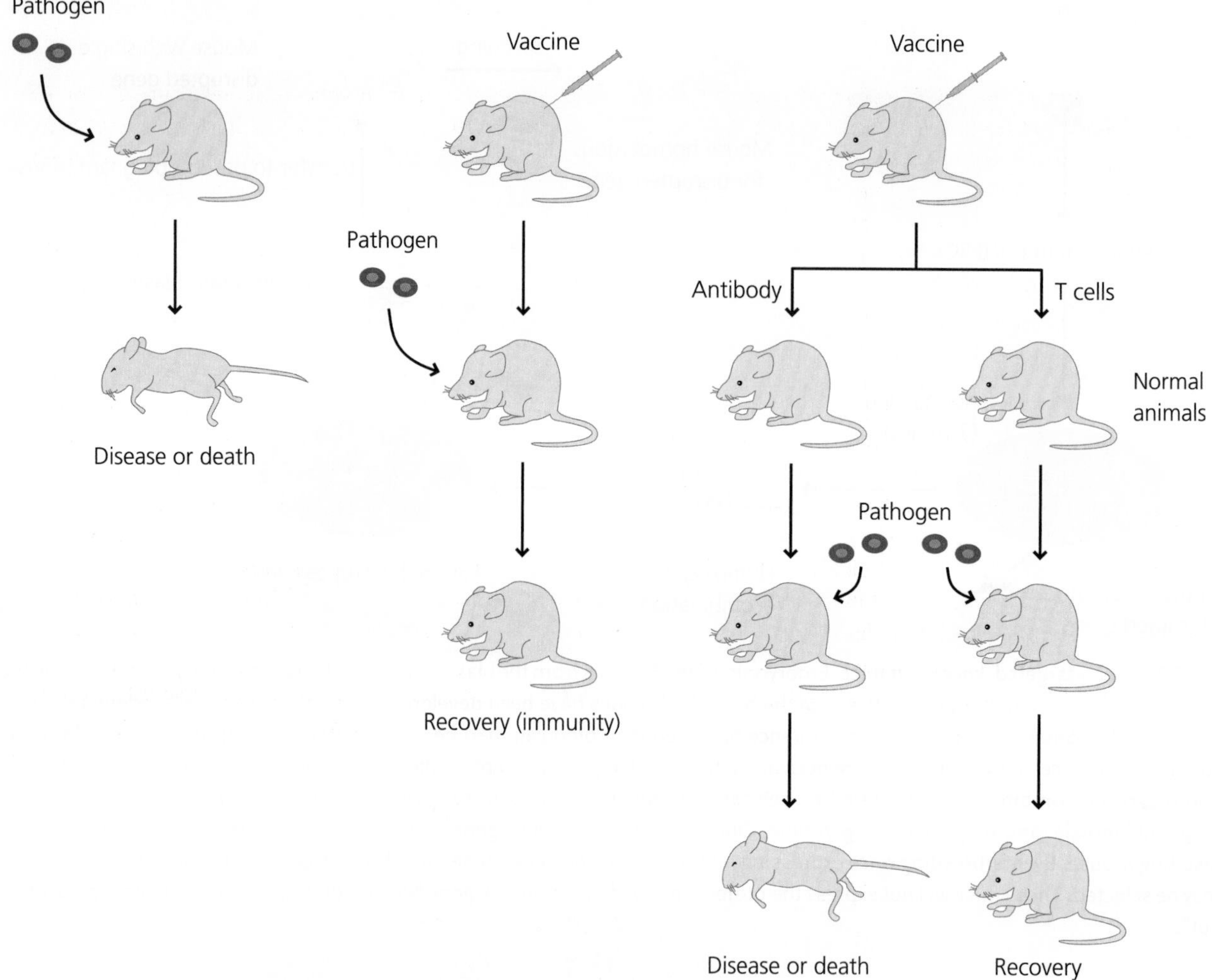

Fig. 2.8 The principle of adoptive transfer of immunity. If an animal is vaccinated against a pathogenic microbe (or has recovered from an infection) it is usually resistant to re-infection with the same microbe. To find out which part of the immune system is mediating this resistance, different immune components such as antibodies in serum or T cells isolated from lymphoid tissues can be transferred to a normal animal; this procedure is known as adoptive transfer. The recipient can then tested for resistance to the microbe. In the case shown, T cells, but not antibodies, are needed for defence against this particular pathogen.

understood. Then, in Section 2.4, we bring these mechanisms together, and emphasize the ways in which innate and adaptive mechanisms can be integrated to provide effective protection. One important principle that should be stressed is that because the different types of pathogen vary so much in size and life history, very different mechanisms are needed to deal with them. Another is that because hosts and pathogens have co-evolved they have generally reached a "balanced" relationship despite disease. See Box 2.1.

2.3.1 Mechanisms of Innate Immunity

2.3.1.1 Phagocytes and Phagocytosis

Phagocytosis is the internalization of particles by cells. Macrophages, and neutrophils, polymorphonuclear neutrophils (PMNs), are specialized phagocytes and are the most important phagocytes in defence against infection. They are able to phagocytose small micro-organisms such as bacteria, small parasites (protozoa) and some fungi, and they also internalize viruses. This is an important mechanism leading to the intracellular elimination of some types of infectious agent. Its importance is shown by patients in whom phagocytosis is defective or who cannot recruit phagocytes to inflammatory sites and who consequently suffer from an increased incidence of infection; LAD (see Chapter 4) is an example of the latter type of defect. See Figure 2.9.

In phagocytosis, the micro-organism is recognized by cell surface receptors, which often include certain pattern recognition receptors (PRRs) (Section 1.2.3.1). Subsequently it is internalized into a membrane-bound vacuole called a phagosome. The phagosome then fuses with cytoplasmic organelles, called lysosomes, which contain pre-formed anti-microbial toxins and digestive enzymes. Often the micro-organism can be killed and degraded in this way. Neutrophils and, to a lesser but no less important extent,

Box 2.1: Balance in Host–Pathogen Relationships

Complex hosts, and the microbes that infect them, have co-evolved over very long periods of time. It is to the microbe's advantage that it maintains a balanced relationship with its host. If host defence predominates, the microbe will disappear (as has happened with smallpox following vaccination). On the other hand, if the microbe predominates it will reduce the supply of hosts to a point where it cannot transmit its genes to other individuals (whether this has contributed to the disappearance of species is not known).

Perhaps the most successful microbes are those which we do not know inhabit our bodies. These include many silent viruses and the commensal bacteria that we possess in such large numbers in our large intestines. These are usually transmitted directly to foetal or neonatal progeny, and under normal circumstances never cause disease.

Even within a species, co-evolution can mould the relationships between microbes and their hosts. Human disease provides some good examples. Where a pathogen is introduced *de novo* into a population, the severity of the disease is often much greater than in the population that provides its normal hosts. Smallpox, introduced into Mexico by the Spanish conquistadores, caused a much more severe disease in the natives than in the Spaniards, and contributed to the conquest of Mexico by a relatively small force of invaders. Yellow fever caused an epidemic in Memphis in 1878 and the mortality amongst African Americans was around 10%; in Caucasians it was 70%, reflecting the co-evolution of the virus and its hosts in Africa rather than Europe.

Zoonoses (infections in humans caused by microbes whose natural host is another species) can also cause very severe disease. The viruses of HIV, Ebola fever, green monkey disease and Lassa fever cause much more severe disease in humans than in their natural primate hosts. The new host, not having had time to co-evolve with the pathogen, cannot maintain a balanced relationship.

These observations tell us that microbes and their hosts have co-evolved in ways that maximize the probability of the pathogen being able to infect other hosts and that, for the most successful infectious agents, usually this is brought about by the infection not causing death or severe disease in their natural hosts.

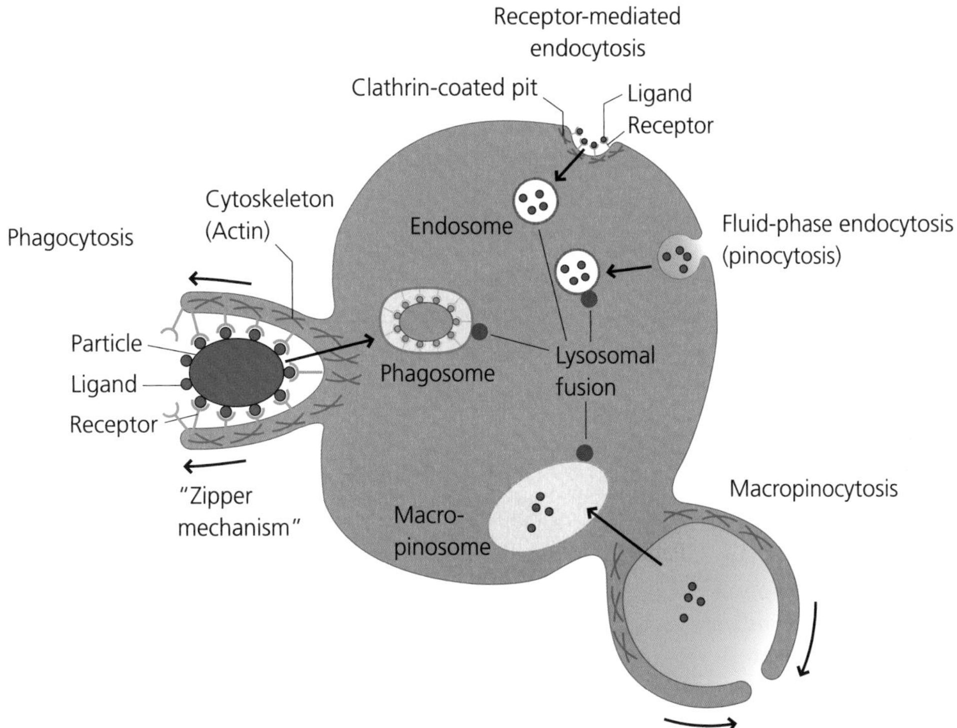

Fig. 2.9 Endocytosis and phagocytosis. All cells can sample their extracellular milieu through the process of endocytosis. All can take up molecules in the fluid-phase by pinocytosis or through receptor-mediated endocytosis, into endosomes. Specialized cells such as neutrophils and macrophages can also internalize particles by phagocytosis. During phagocytosis, sequential interactions of cell surface receptors and ligands on the particle may result in a zippering process involving the actin cytoskeleton that leads to the particle being enclosed within the cell in a phagosome. Some cells (especially some DCs) can extend large sheets of cytoplasm that fuse to enclose large volumes of fluid; this is macropinocytosis. The internalized vesicles may then fuse with lysosomes that contain degradative enzymes. As the endolysosomes mature they become increasingly acidified, resulting in activation of the degradative enzymes (e.g. acid proteases) that degrade their contents. Sometimes internalized receptors are recycled back to the surface to be re-utilized (not shown).

macrophages use phagocytosis as a crucial mechanism of host defence. This ensures that many infectious agents are not pathogenic to normal individuals because they can be so rapidly phagocytosed and eliminated early after infection. It should be emphasized that phagocytosis by macrophages is also an essential part of normal homeostasis, such as in the removal of aged erythrocytes and the elimination of dying (apoptotic) cells during tissue remodelling.

2.3.1.2 Opsonization, Complement and Natural Antibodies

Phagocytosis is an essential first stage in the killing of many viruses and other microbes. Because they are so small, many viruses are actually internalized through receptor-mediated endocytosis. Not surprisingly, many pathogens have evolved means of avoiding phagocytosis (e.g. by changing the structures of their external surfaces). In turn, vertebrate hosts have evolved additional mechanisms that are nevertheless still able to direct such pathogens into the phagocytic pathway. This is the process of opsonization. For example, microbes can be coated by specialized host molecules such as complement components or, usually later after infection, with antibodies. These can subsequently be recognized by cell surface receptors on phagocytes to stimulate phagocytosis. See Figure 2.10.

The complement system consists of a large number of soluble molecules that circulate in the blood. Together they constitute a proteolytic cascade in which initial activation, by three different routes, results in amplification of the numbers of activated molecules generated at each stage. The deposition of certain activated complement components on a microbe enables them to act as opsonins: their subsequent recognition by complement receptors enables phagocytosis of the opsonized microbe. Other complement components are involved in the induction of acute inflammation. The components activated later in the complement pathway may also assemble into pores (the membrane attack complex) on bacterial surfaces leading to lysis of some types of bacteria. The complement system appears to be particularly important in defence against extracellular, pyogenic bacteria because individuals with genetic defects of the so-called central and late components which are common to all three pathways show an increased incidence of such infections. Pyogenic means pus-forming, and the pyogenic bacteria are typically those such as *Staphylococcus aureus* and *Streptococcus pyogenes* that cause acute inflammation and the formation of pus at superficial sites. See Figure 2.11.

Q2.4. Another important function of complement is to clear immune complexes from the bloodstream. Individuals lacking some particular complement components often present with skin rashes and, in some cases, severe kidney damage. Why might this be?

Antibodies can also act as opsonins on their own, or after activating complement and subsequently leading to uptake via complement receptors (as above). Even before adaptive responses are mounted, so-called natural antibodies are produced that have weak affinity for a variety of infectious agents and which can generally activate complement. The deposition of other types of antibodies on a micro-organism can enable its subsequent recognition by another set of phagocyte receptors, termed Fc receptors (FcRs), some of which induce phagocytosis.

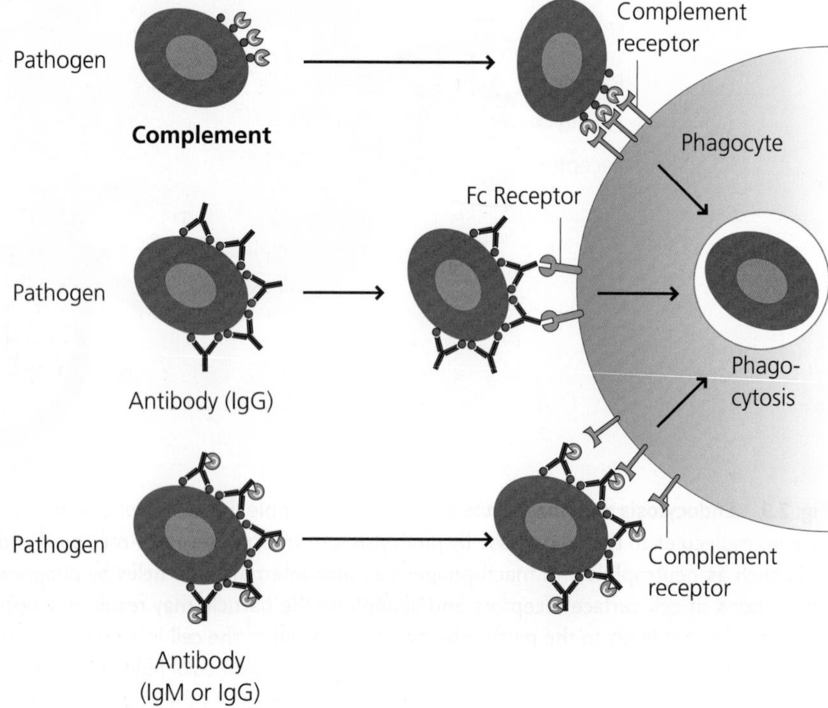

Fig. 2.10 Opsonization. Opsonization prepares a particle for phagocytosis by coating it with molecules for which phagocytes have receptors that can mediate internalization. For example, some bacteria can be phagocytosed directly via PRRs (not shown) and are often non-pathogenic. Pathogenic bacteria may have developed capsules to protect themselves from uptake in this way, but can be taken up after they have been opsonized. The main opsonins are antibodies and certain complement components that are bound directly to the surface of microbes, or to antibodies that are attached to them (mainly IgM and some classes of IgG). A variety of different Fc and complement receptors can mediate internalization of antibody- and complement-opsonized particles, respectively.

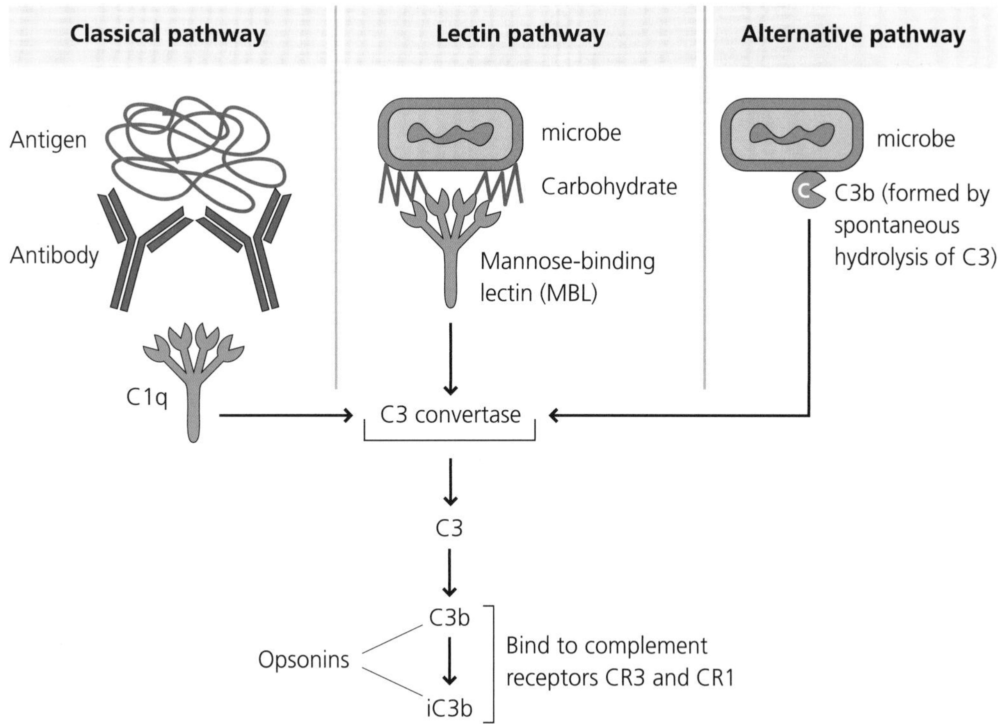

Fig. 2.11 Complement activation and production of opsonins. Complement comprises a large group of proteins, some of which can be activated sequentially in a cascade. Three main pathways are involved in activation. The classical pathway typically starts with C1q binding to antibodies on the surface of a particle. The lectin pathway often involves mannose-binding lectin (MBL) binding to carbohydrates on a particle. The alternative pathway utilizes the continual activation of C3 on particle surfaces and serves to amplify the other pathways. These pathways all come together with the activation of component C3. Its products C3b, and particularly iC3b, act as potent opsonins through binding to complement receptors such as CR3 and CR1.

2.3.1.3 Neutrophils, Extracellular Bacteria and Fungi

Neutrophils are the most common leukocytes in blood, and hence they can be rapidly recruited to sites of infection and inflammation. Substantially more neutrophils can be mobilized from the bone marrow as the infection progresses. Extracellular pyogenic bacteria, which can multiply rapidly outside host cells, require a fast host response. Owing to their large numbers and rapid mobilization, neutrophils play an essential role in the elimination of such bacteria as well as some types of fungi. Neutrophils have several mechanisms for killing bacteria. Particularly important is the intracellular production of reactive oxygen intermediates (ROIs), including hydrogen peroxide (good if you want to bleach your hair) and hypochlorous acid (the main component of domestic bleach). These ROIs are secreted into the phagocytic vacuole and can be highly toxic to many bacteria. Lysosomal proteases, which are secreted into the vacuole, are important in digesting bacteria but may also have a direct role in killing. In neutrophils, ROIs may facilitate this killing by actually helping to increase the acidity of the vacuole. In addition, neutrophils extrude chromatin and certain enzymes that may trap and digest bacteria extracellularly in more-recently identified structures termed neutrophil extracellular traps (NETs). Having done their job, the neutrophils die. Other mediators they produce help to digest and liquefy the local tissue (in preparation for repair and reconstruction) and this liquid, together with dead and dying neutrophils, forms pus. See Figure 2.12.

We know neutrophils are important in particular types of infection because patients with reduced numbers of neutrophils or genetic defects, such as the inability to generate ROIs, become highly susceptible to infection with some extracellular pyogenic bacteria and fungi; CGD (see Chapter 4) is an example of this type of defect. Some of the infectious agents that more readily cause disease in these patients are not pathogenic in immunocompetent individuals. (That they do infect people with defective neutrophils tells us that, in immunocompetent people, neutrophils are working away all the time without us being aware of it.) In particular, these patients may have an increased incidence of infection with pyogenic bacteria such as *Staphylococcus aureus*. It is important to realize that patients with neutrophil defects do not show an increased susceptibility to other infectious agents such as intracellular bacteria (e.g. mycobacteria) and viruses. This is good evidence that neutrophils do not normally play major roles in defence against these types of organism.

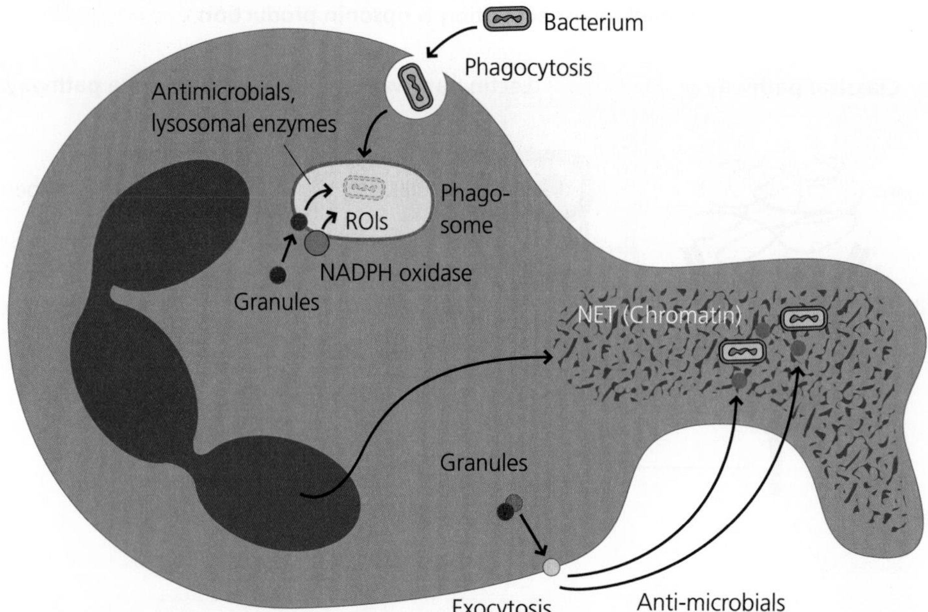

Fig. 2.12 Neutrophil microbicidal mechanisms. Phagocytosed bacteria are subjected to a variety of potential killing mechanisms within the phagosome. These include reactive oxygen intermediates, anti-microbial proteins and peptides, and lysosomal enzymes. Neutrophils may also be able to secrete or release anti-microbial mechanisms that act extracellularly. These include anti-microbial proteins and peptides such as defensins, and NETS consisting of extruded sheets of nuclear material that may serve to trap bacteria, facilitating their killing.

Q2.5. Why might neutrophils play a lesser role in host defence against *intracellular* than *extracellular* bacteria?

2.3.1.4 Macrophages, Intracellular Bacteria and Viruses

In contrast to neutrophils, macrophages are normally resident within almost every tissue of the host. Microbial recognition by PRRs on tissue-resident macrophage is an important stimulus in the initiation of inflammation because they stimulate secretion of cytokines and chemokines. When infection causes inflammation, monocytes are recruited to the site of infection and develop into macrophages or, under influences that are still unclear, into dendritic cells (DCs).

Unlike neutrophils, normal tissue-resident macrophages have limited anti-microbial defences, probably because they are usually pre-occupied with other homeostatic duties (tissue remodelling and such-like). Potent anti-microbial activities, including the production of ROIs, can, however, be induced in macrophages (perhaps mainly in the newly recruited macrophages) by cytokines produced by other cells, particularly IFN-γ. This cytokine may be produced, at early stages of infection, by natural killer (NK) cells and, later, by a subset of activated (T_h1) T helper cells. The importance of such signals is highlighted by individuals who, for example, have genetic defects in the IFN-γ receptor and who become particularly susceptible to infection by intracellular bacteria. Many of these bacteria are considered to be commensals in normal individuals, again showing the importance of subclinical activity of the immune system.

Phagocytosis by macrophages can, however, be a double-edged sword. If a pathogen can survive inside a macrophage, the longevity of these cells can provide a safe haven. Such pathogens have evolved a variety of mechanisms to permit their survival inside macrophages. For example, *Listeria* can escape from the phagosome into the cytosol, avoiding the killing mechanisms which are directed into the phagolysosome, and pathogenic *Mycobacterium* prevent the fusion of phagosomes with lysosomes. In addition to this, the activated macrophage is a rather nasty customer – it is actively secretory, releasing molecules such as hydrogen peroxide which are toxic to many neighbouring cells, and proteolytic enzymes such as collagenase and elastase that break down connective tissue. Thus activated macrophages, as well as being the only cells that can kill pathogens such as *Mycobacterium tuberculosis*, are also the cells responsible for the tissue destruction that is so typical of chronic inflammation, as we shall see later (Section 2.4.2.3). See Figure 2.13.

Q2.6. Why might macrophages play a lesser role in defence against *extracellular* bacteria?

2.3.1.5 Natural Killer (NK) Cells and Viruses

Another cell involved in innate immunity is the natural killer (NK) cell. Some NK cells are present in the blood and blood-filtering organs such as the spleen, liver, lungs and bone marrow, and others can also be recruited to inflammatory sites. They can contribute to early responses against some bacteria because they can be triggered to secrete cytokines. For example, the cytokine IL-12 produced by other cells, such as macrophages or DCs, can trigger these cells to secrete IFN-γ. Another important feature of NK cells is that they can

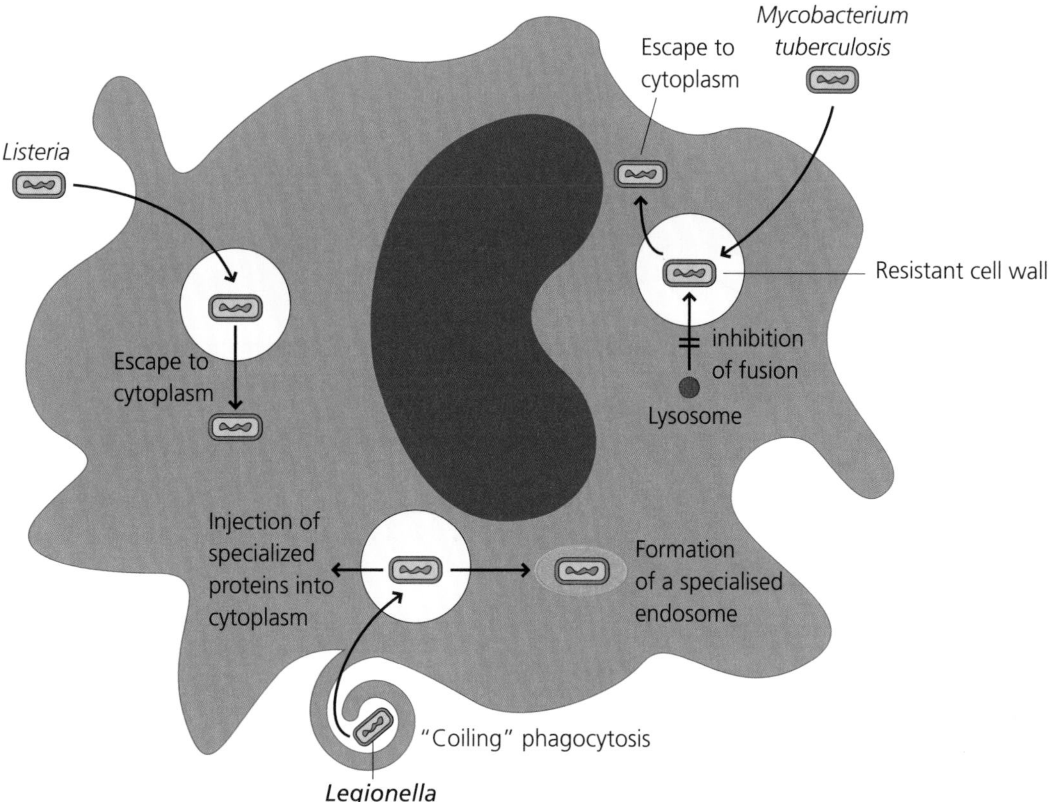

Fig. 2.13 Microbial macrophage evasion mechanisms. Many pathogens have evolved mechanisms to help them avoid being killed by macrophages. Some general mechanisms used by three different types of bacteria are shown for *Listeria*, *Mycobacterium tuberculosis* and *Legionella* (the causative agents of diseases such as listeriosis, tuberculosis and Legionnaire's disease respectively); the relevance of the coiling mechanism of phagocytosis that is induced by the latter is not known. Many other evasion mechanisms have been discovered; for example, some bacteria may interfere with signalling pathways such as that stimulated by IFN-γ, thus inhibiting macrophage activation (not shown). Many others are probably yet to be discovered.

directly kill other infected host cells; unlike cytotoxic T cells, which need to be activated before they become cytotoxic (Section 2.3.2.2), NK cells show spontaneous cytotoxicity.

Viruses, once intracellular, can evade many host defence mechanisms including killing of infected cells by cytotoxic T cells. In some cases, however NK can recognize these virally infected cells and kill them. If the NK cell kills an infected cell before new virus has been released, it can limit viral replication and spread. Killing of virally infected cells by NK cells involves both the delivery of pre-formed granule contents or interaction with death-inducing receptors, leading to apoptosis of the target cell. Clinical and experimental evidence suggests that NK cells play particularly important roles in host defence against some viruses. This is demonstrated by very rare individuals with reduced numbers of NK cells or whose NK cells are defective at killing. Such patients suffer very severe, potentially fatal, infections with herpes viruses, but, if they can survive until the adaptive immune response kicks in, they may survive as well as normal individuals. See Figure 2.14.

Q2.7. Why might NK cells play lesser roles in host defence against viruses *other than* the herpes viruses?

Q2.8. Might you expect NK cells to play a role in host defence against *tumour cells* as well as viruses?

2.3.1.6 Mast Cells and Basophils, Extracellular Bacteria and Parasites

Mast cells are present in all loose connective tissues. They express PRRs and can rapidly produce pro-inflammatory cytokines that trigger inflammatory responses when they sense infection (i.e. these cytokines can be viewed as being alarm signals that are produced in response to the danger posed by infectious agents). Mice that lack mast cells may be particularly susceptible to infection by extracellular bacteria because mast cells are needed to recruit neutrophils (Section 2.3.1.3) which probably deal best with this type of pathogen. Mast cells also store pre-formed mediators of inflammation such as histamine, and can later synthesize lipid metabolites such as prostaglandins and leukotrienes which are important in mediating the vascular changes in inflammation (venular vasodilatation and increased permeability). See Figure 2.15.

Mast cells are also present in mucosal tissues. Here they may play an important part in immunity to parasites,

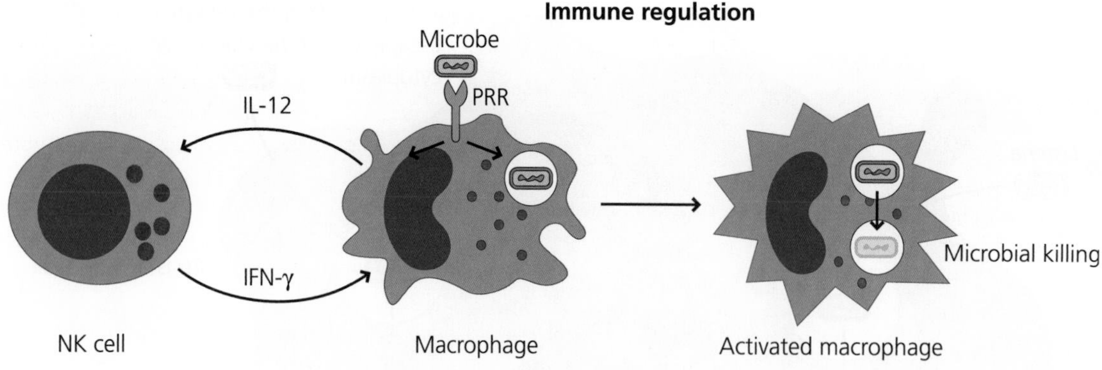

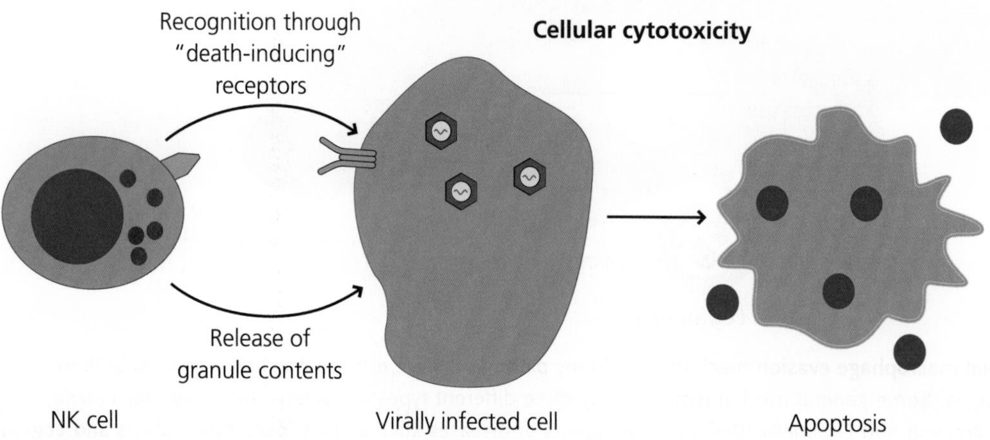

Fig. 2.14 Some functions of natural killer cells. NK cells are developmentally related to lymphocytes but are considered as belonging to the innate arm of immunity. Two main types of function are highlighted. **Immune regulation.** NK cells can regulate the functions of other cell types. For example, a feedback loop triggered by IL-12 secretion from macrophages can stimulate IFN-γ production by NK cells which in turns helps macrophage activation. **Cellular cytotoxicity.** NK cells can kill other cells. Different recognition systems enable NK cells to deliver pre-formed granule contents or to ligate "death-inducing" receptors and induce apoptosis (e.g. in cells infected with certain types of virus).

especially worms. Some of the lipid-derived mediators they produce increase the production of mucus, thus potentially adding a more effective barrier to access of worms to the gut wall. They may also cause smooth muscle contraction which might help to expel the worms in the faeces. Certainly mice lacking mast cells cannot clear some intestinal worms as effectively as normal mice under experimental conditions. However, given that we generally do not manage to get rid of most worms, their efficacy may be called into doubt.

A rare leukocyte, the basophil, somewhat similar to the mast cell, can be recruited from the blood to sites of some types of inflammation and infection. We still know relatively little about what these cells do. However, because basophils share many features with mast cells, they may be recruited to help to amplify the local responses that are typically stimulated by mast cells. Some more recent research also suggests they may play a role in polarizing CD4 T cell responses towards the T_h2; type that typically occurs during parasitic infections (Section 2.3.2.1).

2.3.1.7 Eosinophils, Parasites and Tissue Repair

Eosinophils are normally rare in most connective tissues but are commonly found in mucosal tissues. Details of their roles in immunity to infection remain unclear. They have cytoplasmic granules that contain mediators that are known to be toxic to larger parasites, at least *in vitro*. Some of the functions of eosinophils overlap with those of mast cells (above). Both cell types, and additionally basophils, appear to respond to blood sucking ticks by degranulation at skin sites where ticks are feeding, although whether or how they might play a defensive role is not entirely clear.

Eosinophils (and indeed mast cells) additionally store or synthesize molecules that are important for tissue remodelling (e.g. growth factors for fibroblasts) and are present in tissues undergoing repair. Indeed, the T_h2 types of adaptive immune response in which eosinophils (and basophils) are recruited may assist in the resolution of inflammation (production of anti- rather than pro-inflammatory cytokines), tissue repair and wound healing. In some cases, however,

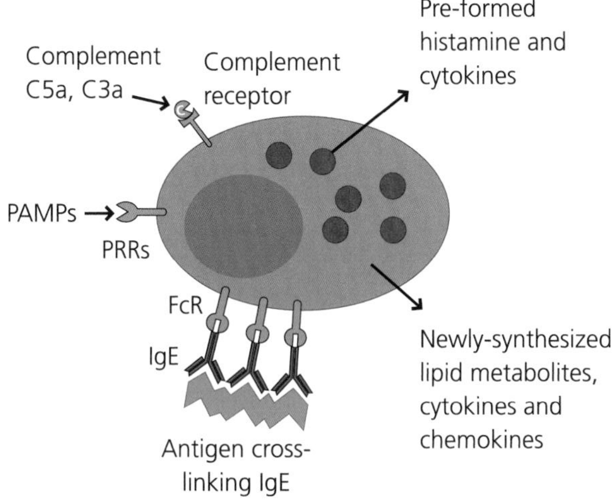

Fig. 2.15 Mast cell activation, degranulation and secretion. Mast cells reside in connective tissues and can be activated in several ways. They express receptors for complement components C3a and C5a, and a selection of PRRs such as Toll-like receptors (TLRs) that recognize microbial pathogen-associated molecular patterns (PAMPs). They can also be activated by cross-linking pre-formed IgE bound to FcRs on their surface. Activation results in very rapid degranulation with the release of pre-formed granule components such as histamine and cytokines. Mast cells can then synthesize and secrete other pro-inflammatory molecules such as lipid mediators (leukotrienes, prostaglandins) and more cytokines. Mechanical damage to mast cells can also lead to the release of histamine.

the excessive laying down of collagen can lead to loss of function in tissues, as can happen in chronic asthma. See Figure 2.16.

2.3.2 Mechanisms of Adaptive Immunity

Lymphocytes are crucial for adaptive immune responses, and specifically the conventional T cells and B cells that are introduced in Chapter 1. Here, we focus on some key functions of these CD4 and CD8 T cell populations, and of the different classes of antibodies that are produced when B cells develop into plasma cells.

2.3.2.1 CD4 T Cells

The need for CD4 T cells is well-demonstrated in AIDS patients, whose CD4 T cells are eventually killed by the HIV virus. Not only do these patients suffer from an increased incidence of infections such as tuberculosis, they develop potentially fatal opportunistic infections with microbes that are harmless to normal individuals. Normally, however, CD4 T cells are highly adaptable and they can take on specific sets of functions that help best to deal with particular types of infectious agents. In other words, CD4 T cell responses can become polarized. In extreme cases, the polarized cells can be considered as representing discrete subsets of which the best known are T_h1, T_h2, T_h17 and regulatory (T_{reg}) cells. These subsets are most clearly evident in experimental studies of T cells in culture or immunized laboratory mice, from where most evidence for their existence has been obtained. Clinically, however in most human infectious diseases, a more mixed pattern of responses is typically seen.

Q2.9. Why might it be advantageous to the host not to polarize fully all CD4 T cell responses?

T_h1 Cells: Intracellular Bacteria and Viruses T_h1 cells are typically induced following infections by intracellular bacteria and certain viruses. Some of the cytokines they secrete, such as IFN-γ, can activate and induce potent anti-microbial mechanisms in macrophages (above). Furthermore T_h1 cells are often needed to help trigger cytotoxic T cell responses (below). In addition, T_h1 cells can activate B cells and regulate the production of specific types of antibodies. The latter particularly include IgG subclasses that are the most effective opsonizing antibodies, and which target infectious agents to the anti-microbial functions of neutrophils and macrophages. The importance of T_h1 cytokines is highlighted by patients with genetic defects (e.g. of IFN-γ or its receptor; above) and from studies in mice in which the cytokine genes (or their

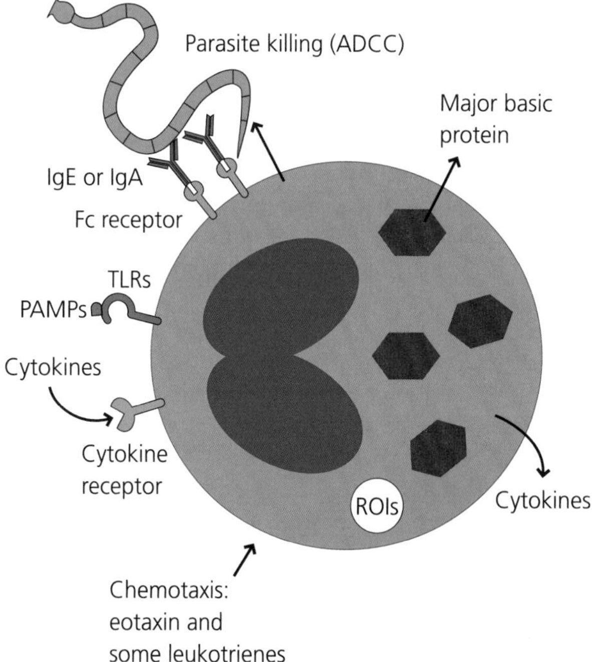

Fig. 2.16 Eosinophil properties and functions. The functions of eosinophils are not fully understood. They are recruited into inflammatory sites such as those associated with parasitic infections and asthma. They express TLRs, and respond to cytokines such as IL-4, the chemokine eotaxin and some leukotrienes, for example. When activated they can secrete several cytokines including IL-4 and IL-13, and major basic protein which may be involved in defence against parasites. They can also mediate antibody-dependent cell-mediated cytotoxicity (ADCC) against some IgE- or IgA-coated parasites.

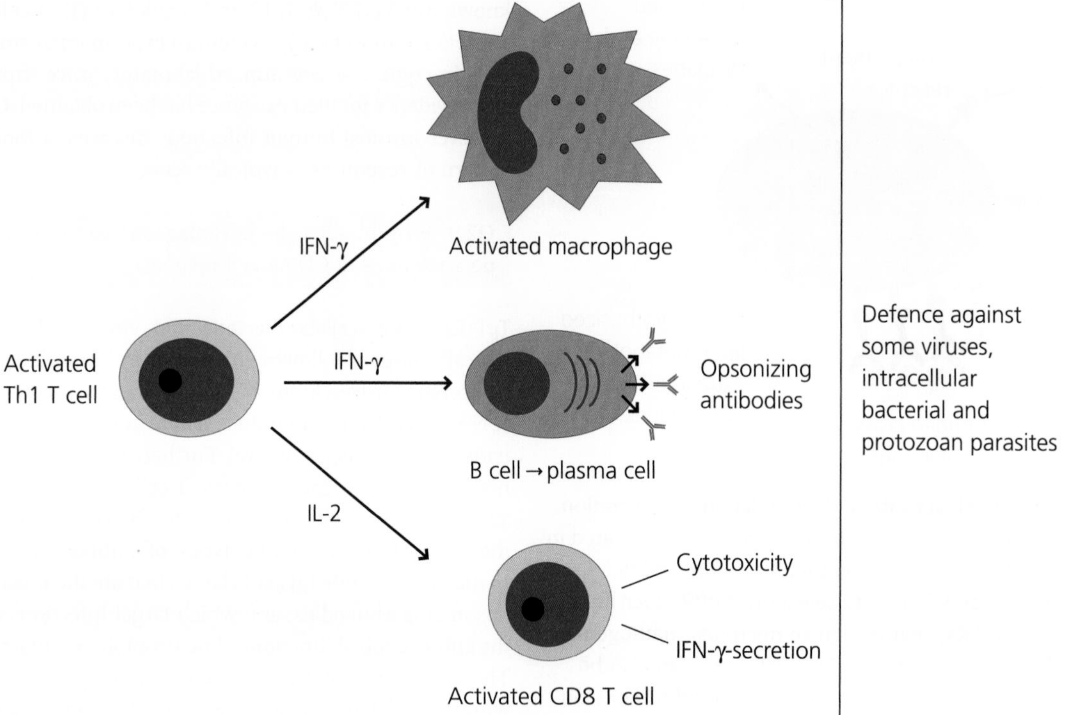

Fig. 2.17 Functions of T_h1 CD4 T cells. T_h1 cells secrete IFN-γ which activates macrophages and is involved in B cell differentiation into plasma cells that secrete opsonizing antibodies (e.g. those that can bind to FcRs of activated macrophages). They also secrete IL-2 which may be important in the activation and clonal expansion of CD8 cytotoxic T cells at sites of infection. T_h1 cells express the T-bet transcription factor (not shown).

receptors) have been deleted, rendering them susceptible to infection with these micro-organisms. Tellingly, vaccinia virus, for example, encodes a protein that binds to and inhibits the activity of IFN-γ; the virus would not continue to carry this genetic burden if it did not confer a selective advantage. See Figure 2.17.

T_h2 Cells: Larger Protozoa and Worms T_h2 cells are more often induced during certain stages of fungal and parasitic worm infections, and for some of us, in allergies (e.g. hay fever). T_h2 cells are largely involved in high-level antibody production, particularly IgE, in mice. Their cytokines also stimulate eosinophil and mast cell production and/or survival and activity. These latter cells express FcRs for IgE and are thought to have important roles in defence against larger parasites such as helminth worms. Binding of worm antigens to IgE-coated mast cells stimulates degranulation, leading to smooth muscle contraction that may aid worm expulsion from the gut. In addition, eosinophils may be targeted to larger parasites coated with IgE and deliver highly toxic granule contents (including a neurotoxin) onto them. In patients that have been treated by drugs to clear parasitic infections, resistance to re-infection correlates with levels of anti-parasite IgE. See Figure 2.18.

T_h17 Cells: Extracellular Bacteria and Fungi The functions of T_h17, cells so named because they secrete IL-17, are less completely understood than those of T_h1 and T_h2 cells. They were originally identified as mediators of autoimmunity, but clearly this is not a role that would be selected for in evolution. Studies in IL-17-deficient mice suggest that they stimulate recruitment and activation of neutrophils, and hence these cells may indirectly play an important role in defence against extracellular bacteria and some fungi. See Figure 2.19.

Regulatory (T_{reg}) Cells: Turning Off Responses What stops immune responses when infectious agents have been eliminated? Lack of antigen is obiviously a major factor, but it is clear that some CD4 T cells have a crucial role in regulating the activity of other T cells or of DCs. These cells are known as regulatory T cells (T_{reg}). Some are formed in the thymus (natural T_{reg}), while others are induced during adaptive immune responses. The importance of these cells is shown in many mouse models, and crucially in a very rare human disorder, IPEX (Immune dysregulation, Polyendocrinopathy, Enteropathy X-linked) in which patients have a mutation in both copies of the FoxP3 gene – essential for the development of T_{reg}. These children suffer from multiple autoimmune disorders and usually die in their first year, suggesting that, T_{reg} normally play a crucial role in preventing T cells from attacking components of the body itself (a form of peripheral tolerance; Chapter 5). Given the ability of T_{reg} to suppress immune responses, it is not surprising that many chronic infections are associated with increased T_{reg} activity, the hypothesis being that the microbes are inducing T_{reg} to enhance their own survival. A similar phenomenon is seen

IgE
B cell → plasma cell
IL-4
IL-4,
IL-13
Alternatively-activated
macrophage
Activated Th2 T cell
IL-4,
IL-5
Mast cell or basophil
IL-4,
IL-5
Defence against
some extracellular
bacteria and some
metazoan parasites
Eosinophil

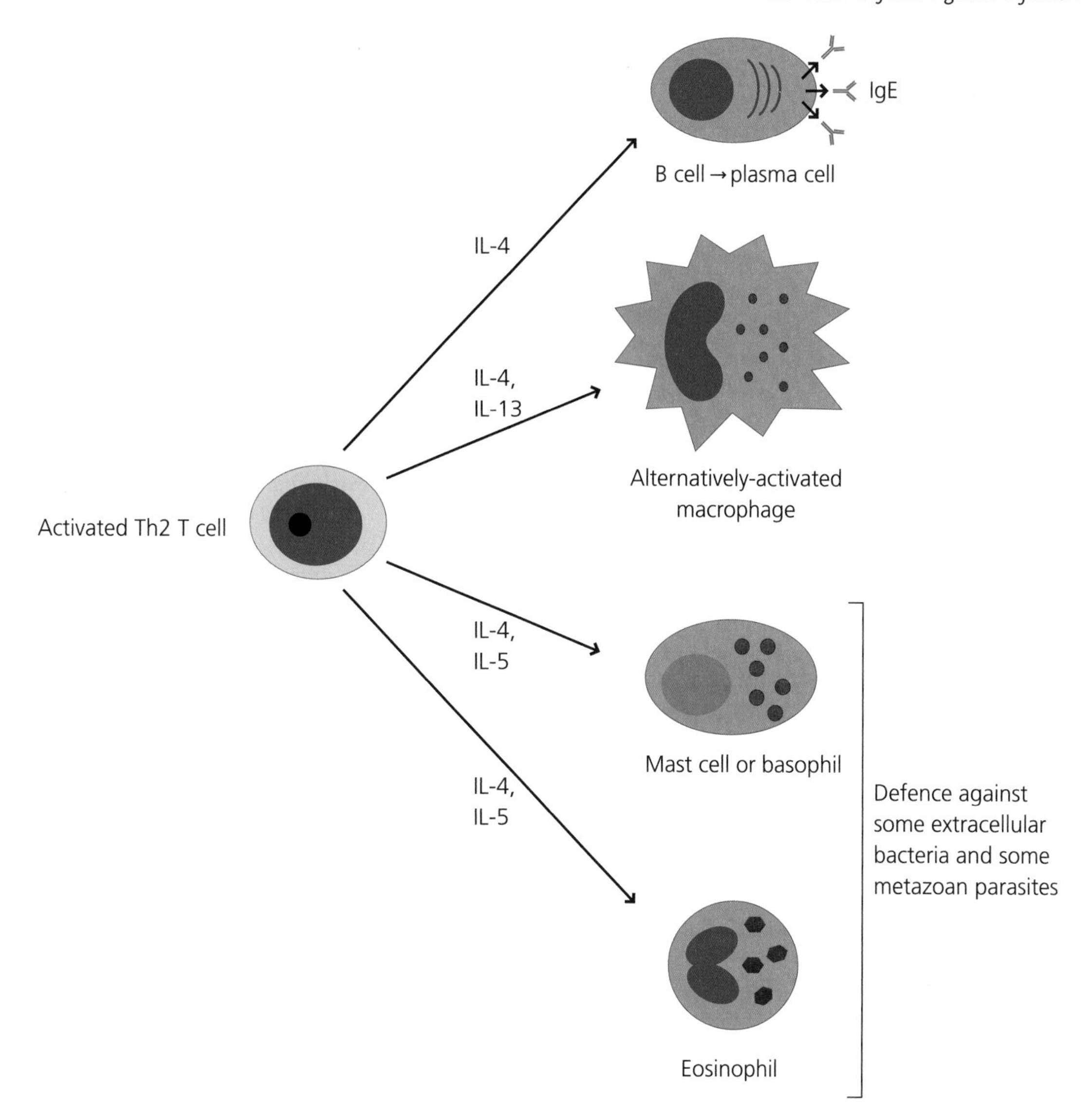

Fig. 2.18 Functions of T_h2 CD4 T cells. T_h2 cells secrete IL-4, IL-5 and IL-13 which are involved in helping B cells to develop into plasma cells that secrete IgE (and some subclasses of IgG). They are also involved in mast cell activation, and eosinophil and basophil production or recruitment (all of which have FcRs for IgE). These cytokines are also involved in alternative activation of macrophages that may be important in helping tissue repair. T_h2 cells express the transcription factor GATA-3.

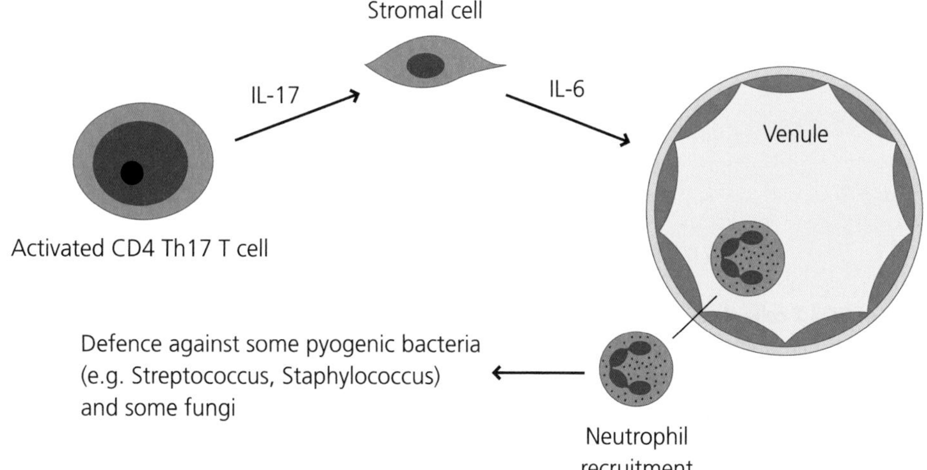

Fig. 2.19 Functions of T_h17 CD4 T cells. T_h2 cells secrete IL-17 that acts on stromal cells to stimulate secretion of IL-6, which is involved in neutrophil recruitment to sites of acute inflammation, particularly in pyogenic bacterial and some fungal infections. T_h17 cells express the transcription factor RORγt.

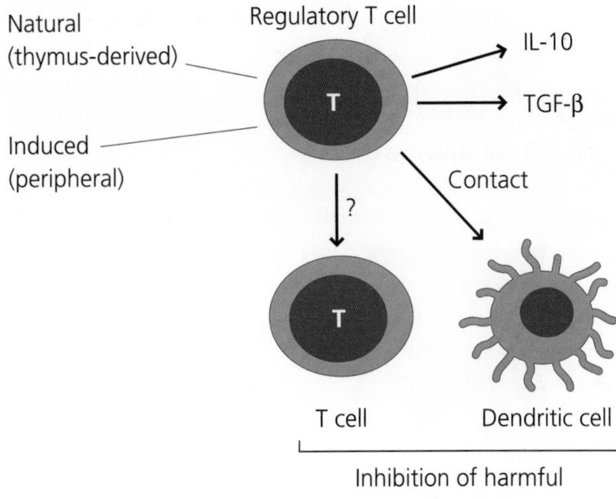

Fig. 2.20 Functions of regulatory CD4 T cells. Some "natural" CD4 regulatory T cells (T_{reg}) are formed in the thymus. They generally express CD25 on one of the high-affinity IL-2 receptor chains. Naïve CD4 T cells activated with insufficient co-stimulation, and perhaps also in the presence of transforming growth factor (TGF)-β, may become T_{reg} in peripheral tissues. Other populations of T_{reg} also exist, but generally all seem to regulate immune responses by acting on other T cells or on different cells such as DCs. Natural T_{reg} express the FoxP3 transcription factor (not shown).

in some patients with malignant tumours (cancer); any tumour mutant that could induce T_{reg} would have a survival advantage. See Figure 2.20.

2.3.2.2 CD8 T Cells

CD8 T cells play very important roles in defence against viruses. When activated they can become cytotoxic and can kill other cells that they recognize and with which they are in contact. If a CD8 T cell can kill a virally-infected cell before new, infectious virus has been assembled, the spread of infection can be stopped. They are also potent secretors of some cytokines. Secretion of IFN-γ, itself an anti-viral molecule, can of course activate macrophages and also make them more resistant to viral replication. CD8 T cells may also play important roles in defence against intracellular bacteria and protozoa. Killing an infected cell may release the microbe and permit phagocytosis by macrophages which, when activated, may kill the organism. See Figure 2.21.

There is overwhelming evidence that CD8 T cells are important in defence against viral infection. Perhaps the most compelling, although indirect, evidence lies in the number of ways in which different viruses have evolved means of interfering with the major histocompatibility complex (MHC) class I processing and presentation pathways required for CD8 T cell function. The viruses would not carry this extra genetic load if it did not give them a selective advantage (Section 5.2.7). Additionally, the numbers of CD8 T cells increase massively during many viral infections. In fact, antigen-specific cytotoxic T lymphocytes (CTLs) may represent almost as much as 50% of the entire circulating T cell pool at the height of certain infections. That CD8 T cells are important in other infections is suggested from studies of mice deficient in perforin, a granule component essential for the killing of cells. These mice, if infected with the bacterium

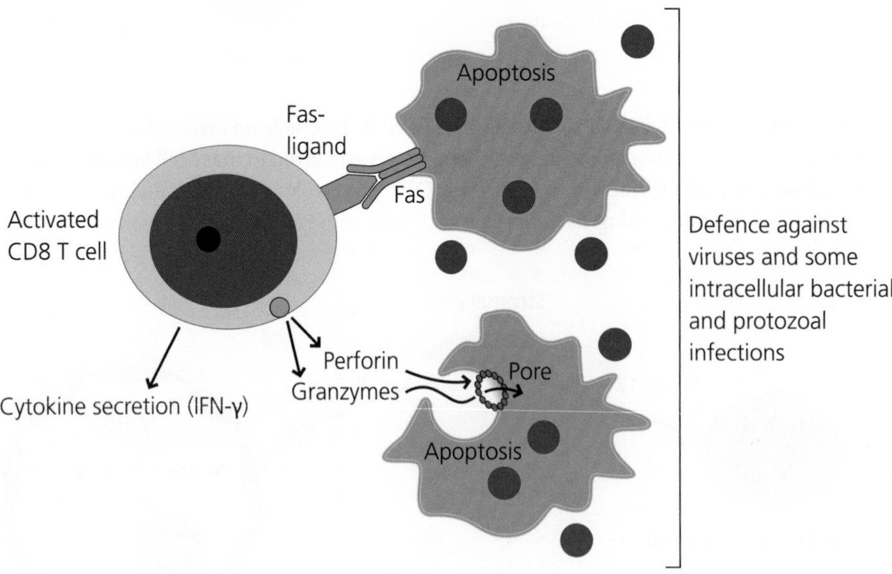

Fig. 2.21 Functions of CD8 T cells. Activated CD8 T cells can express Fas ligand on their surface; Fas ligand can bind to Fas on another cell and this may initiate apoptosis. Activated CD8 T cells are also potent secretors of IFN-γ. It is possible for naïve CD8 T cells to be activated in a manner analogous to T_h2 differentiation in CD4 T cells, but the functions of these cells are unclear. Activated CD8T cells can also acquire conspicuous cytoplasmic granules. These contain molecules involved in cell killing (cellular cytotoxicity) such as perforin and the granzymes. When such a CTL binds to its target, the granule contents are secreted into the immunological synapse between the two cells. Perforin monomers polymerize in membranes of the other cell forming a pore though which granzymes enter the cytoplasm and also initiate the apoptotic sequence.

Listeria, which is able to enter the cytoplasm of the cells it infects, cannot recover from the infection.

2.3.2.3 Antibodies

B lymphocytes are crucially important in normal immunity because they can develop into plasma cells which make different types of antibodies, depending on the type of infection that has occurred. The different types of B cells, and the different types of antibodies that they can make as plasma cells, are introduced in Chapter 1. While they are similar in overall shape, the structure of each class (and subclass) of antibody is unique. It is these structures that determine their distinct properties, such as in which compartments they are distributed throughout the body. Here, we will briefly review some of the principal effector functions of the main classes of antibodies; these are described in more depth in Chapter 6.

IgM Membrane-bound IgM is a typical "Y"-shaped monomer of two heavy (H) plus two light (L) chains anchored to the membrane. However, secreted IgM is composed of five immunoglobulin units which form a pentamer. This greatly increases the binding strength of the whole molecule (avidity). Is IgM important in defence? It is the first antibody to be made following pyogenic infections and thus can act before IgG has been produced. Since specific FcRs for IgM have not been identified, it may help defence against pyogenic bacteria primarily by activating complement, resulting in opsonization for example. Additionally many bacteria are coated with polysaccharide capsules (e.g. *Streptococcus pneumoniae*) and recovery from these infections is dependent on opsonization. Specialized marginal zone B cells in the spleen make IgM to these polysaccharides and if an individual has their spleen removed they become susceptible to infection with this type of bacteria. If, however, an individual has normal or raised levels of IgM, but is unable to make IgG, they are highly susceptible to infection with some pyogenic bacteria; HIGM syndromes (See Chapter 6) are one example of this type of defect. This shows that IgM on its own is not sufficient for the elimination of many extracellular bacteria, even though it is a very effective opsonin because it can activate complement. See Figure 2.22.

IgG IgG is the most abundant class of antibody and is secreted as a monomer. IgG itself comprises several different subclasses (four in humans and mice) with different Fc regions. These have some significant differences in function, such as the relative abilities to activate complement and trigger inflammation or to directly opsonize microbes (via FcRs). Is IgG important in defence? IgG deficiency can exist as a single deficiency or as part of a more general immunoglobulin deficiency (see Chapter 6). The commonest presenting features are bacterial respiratory tract infections. That treatment of these deficiencies by passive transfer of IgG from normal individuals is highly effective gives further strong evidence of its importance in certain types of bacterial infection. See Figure 2.23.

IgA IgA is "the" mucosal antibody. It is secreted by plasma cells in mucosal tissues and is transported across the mucosal epithelium to the luminal surface as a dimer, where it acts by blocking (neutralizing) binding of pathogens and toxins to epithelial cells. Is IgA important? Children in developing countries who are not fed with breast milk are much more susceptible to infectious diarrhoeal disease. This is because IgA in milk confers protection against these infections. People with an IgA deficiency (this is one of the commonest primary immunodeficiencies, around 1:300–400 in developed countries) may show increased susceptibility to respiratory tract and intestinal infections. Vaccinologists are striving to develop vaccines that can be given orally to stimulate IgA against mucosal infections such as cholera and typhoid fever. See Figure 2.24.

IgE IgE is secreted in T_h2-biased responses. Levels of serum IgE are raised in parasitic infections, particularly intestinal worms. Although IgE has been shown to play a role in immunity to intestinal parasites in mice, selective IgE deficiencies have not been identified in humans and the role of IgE in human parasitic infections is uncertain. Much more important in human disease, at least in developed countries, is the key role of IgE in allergic disorders. See Figure 2.25.

IgD IgD is still somewhat a mystery molecule. It differs from all other immunoglobulins in that it is almost always expressed on the cell surface of naïve B cells, but is not normally secreted in significant amounts and is found only in very small amounts in the plasma. For decades it has been thought to play primarily a signalling function in B cell responses. The first evidence is now beginning to emerge that IgD might also play a direct role in defence against infection, as is further discussed in Chapter 6.

2.4 Infection and Immunity in Action

It should now be clear that pathogens cause disease, but that the vast majority of potential infectious agents are not pathogenic in any given species either because they fail to infect or, if they do, they are rapidly eliminated by the mechanisms of immunity. This section provides an overview of the different ways in which pathogens (introduced in Section 2.2) can cause disease, and of the ways in which the individual mechanisms of innate and adaptive immunity (introduced in Section 2.3) work together and collectively attempt to combat different classes of infectious agent.

2.4.1 How Do Pathogens Cause Disease?

Immune defence mechanisms have been selected during evolution to defend against different kinds of infectious agents. Although a relatively small but significant number of pathogenic micro-organisms can cause disease, the

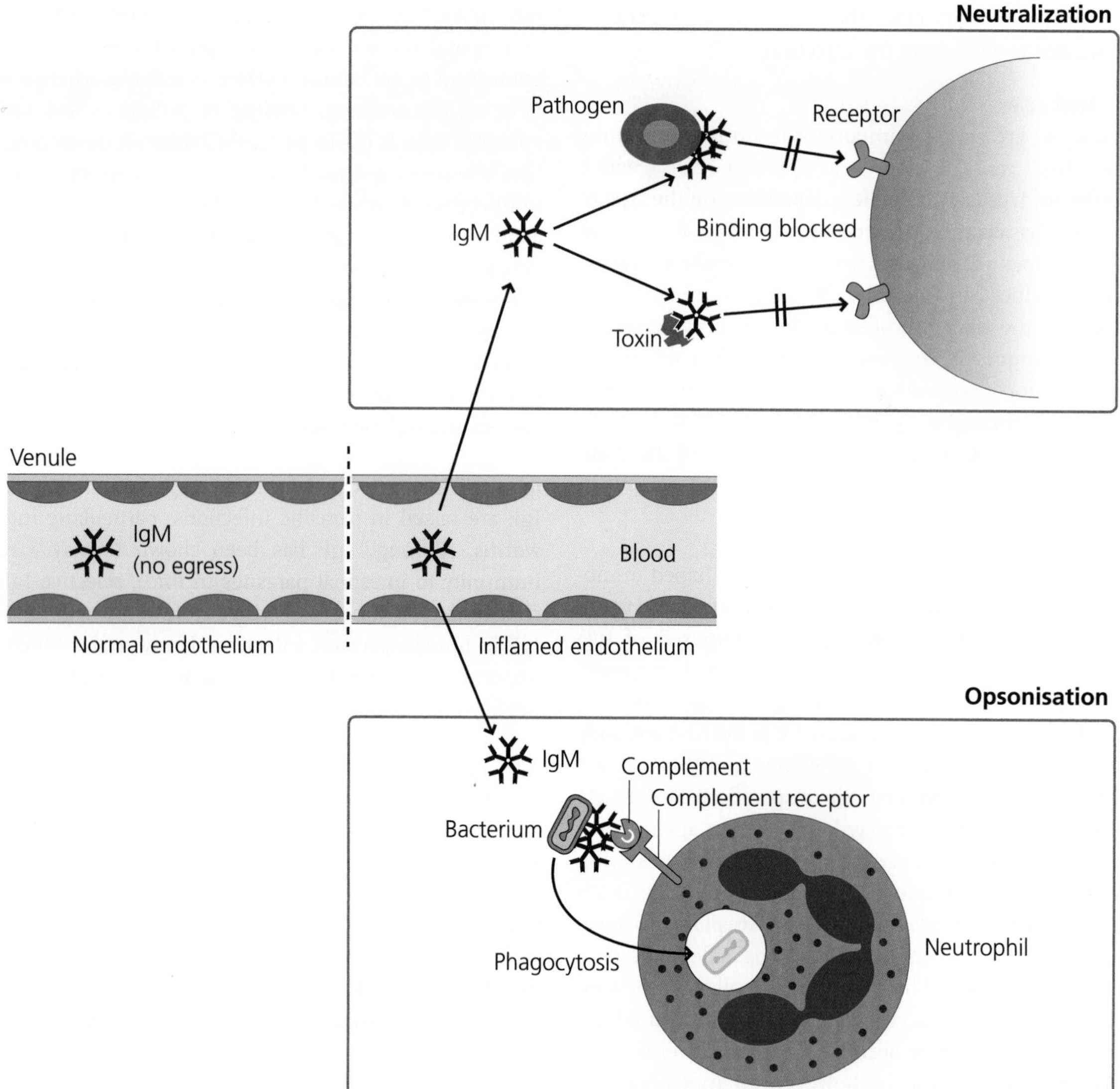

Fig. 2.22 Functions of IgM. Pentameric IgM is recruited to sites of inflammation. IgM is usually of low affinity but its multiple antigen-binding sites give it high avidity for microbes expressing multiple copies of an epitope. Hence, it is a potent neutralizing antibody for bacteria and viruses. It cannot opsonize microbes directly, but is a potent complement activator, and this permits it to act indirectly as an opsonin and to trigger acute inflammation.

number of ways in which they actually do this is rather limited. It is therefore possible to define a relatively small number of different mechanisms of pathogenicity. Not surprisingly, the defence mechanisms acting in response to infection by these different groups of pathogens tend to be rather similar and can themselves be grouped, making understanding of the principles involved relatively straightforward. Of course within each group of defence mechanisms there is much variety, but this should not distract from appreciating the underlying principles that apply to each.

2.4.1.1 Mechanisms of Pathogenicity

Initially it is useful to consider two general ways in which pathogens may cause disease: (i) they can kill or interfere directly with cell function (e.g. viral infections of host cells or production of bacterial toxins) and/or (ii) the host's innate or adaptive immune response against the pathogen can damage its own cells, tissues or organs (in other words, collateral damage or friendly fire). In all cases it is important to remember that pathogens are pathogens because they have evolved mechanisms to evade or redirect host immune responses to their own benefit.

2.4.1.2 How Do We Know a Particular Pathogen Causes a Particular Disease?

An association between two events does not identify a causal relationship between them. That a particular micro-organism is associated with a particular disease may be coincidental and there are several examples where identification of an

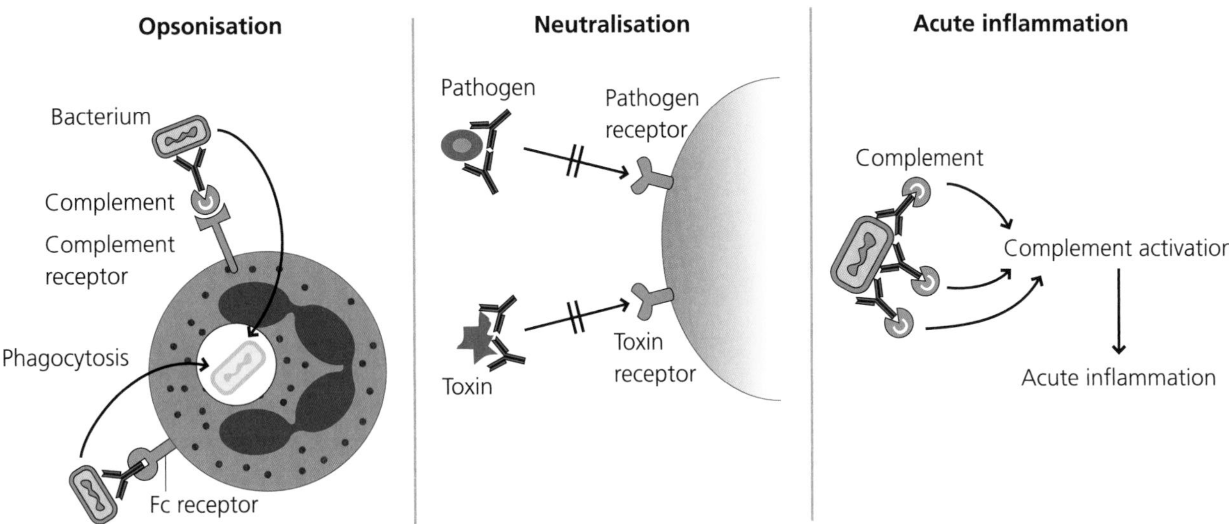

Fig. 2.23 Functions of IgG. IgG is made later than IgM and in small quantities in primary immune responses, but is made rapidly in large quantities in secondary responses. IgG can enter non-inflamed extravascular tissues. It can also cross the human placenta and confer resistance to infection on the foetus and early neonate. IgG can neutralize pathogens and opsonize them directly or indirectly after activating complement; by activating complement it can also initiate acute inflammation. IgG can also mediate antibody-dependent, cell-mediated cytotoxicity (ADCC e.g. by monocytes; not shown).

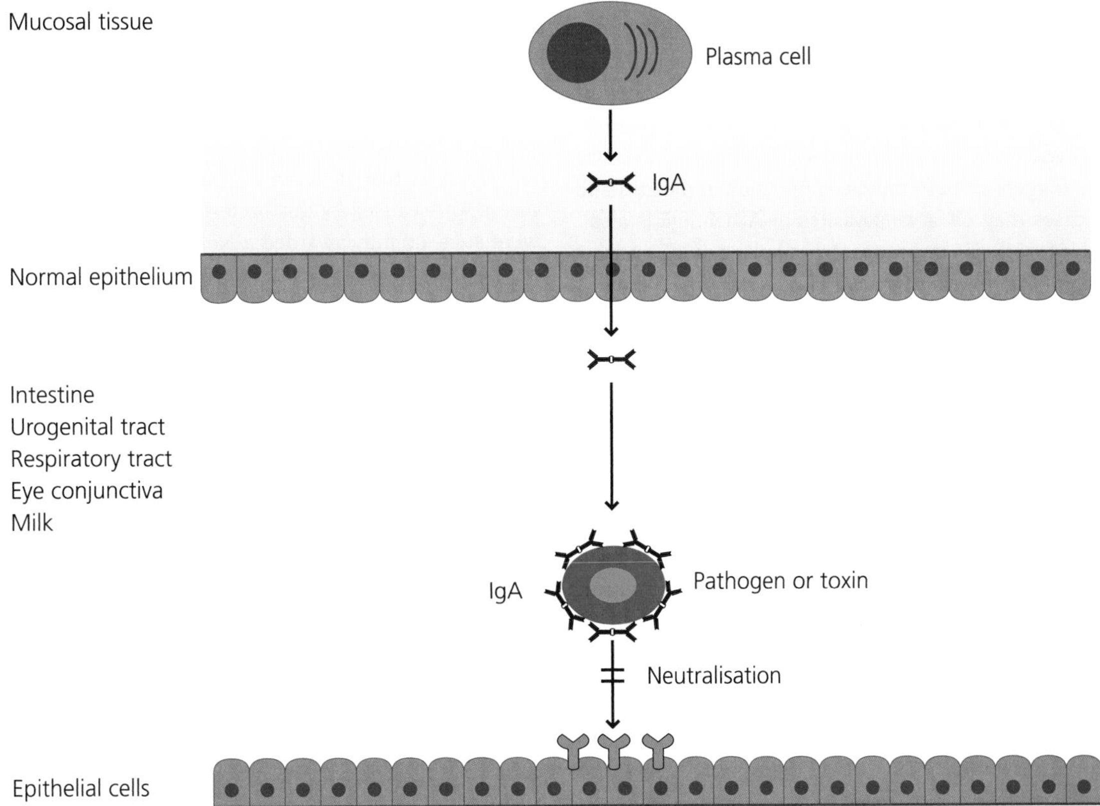

Fig. 2.24 Functions of IgA. Plasma cells in mucosal tissues typically secrete dimeric IgA. This binds to a specific receptor that transports it across mucosal epithelial cells. On the luminal surfaces of mucosal tissues it acts as a neutralizing antibody, blocking attachment of pathogens and toxins to the epithelial cells.

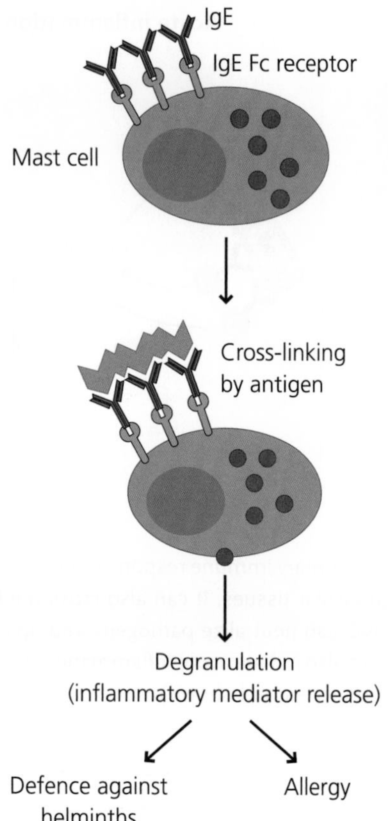

Fig. 2.25 Functions of IgE. IgE, secreted by plasma cells in connective tissues, binds to high-affinity FcRs on mast cells in the absence of antigen. It is present at very low levels in the blood of normal individuals, but its levels are raised in parasitic infections, particularly with helminths. IgE may play a role in defence against these parasitic infection by helping to maintain epithelial barriers, assisting expulsion of worms (e.g. because histamine triggers smooth muscle contraction in the gut) and in some cases may kill larval parasites by ADCC. IgE is also responsible for allergic responses: cross-linking of IgE on mast cells leads to the release of many inflammatory mediators.

association has been wrongly interpreted as showing the organism caused the disease. In 1918, an epidemic of an infectious disease, influenza, spread throughout most of the world. The causal organism was unknown but an eminent American microbiologist became convinced that the infection was due to a bacterium, now known as *Haemophilus influenzae*, because he could isolate the bacterium from the sputum of many cases of influenza. In fact, the bacterium was probably there as a result of secondary infection (a result of the damage caused to the airway epithelium by the primary agent of disease). This theory may have hindered significantly the identification of the real causative organism, the influenza virus.

The great German bacteriologist, Robert Koch, in 1878, set down conditions that had to be satisfied if a bacterium was to be identified as the cause of a disease. The conditions, known as Koch's postulates (Figure 2.26), are still valid today and apply not just to bacterial infections, but also infections with viruses and other type of pathogen.

- The organism must be able to be isolated from all cases of the disease.
- The organism must be able to be grown *in vitro* in pure culture.
- Infection of a suitable host with the organism must give rise to a disease with the features of the original disease.

There are of course some difficulties with applying these postulates to all infectious diseases. Thus, the leprosy bacterium, *Mycobacterium leprae*, is present in all cases of leprosy, but has never been cultured successfully *in vitro* (there is large prize awaiting the first person to culture the organism) and the only other animal known that can be infected by it is the Central American Armadillo. Similarly, the spirochaete, *Treponema pallidum*, has never been cultured *in vitro* and is not known to infect any animals except humans, but is universally accepted as the cause of the venereal disease, syphilis. Even for AIDS, although there is good evidence that non-human primates can be infected with the relative of HIV, simian immunodeficiency virus (SIV), leading to an AIDS-like illness, it is clearly not ethical to deliberately infect any human with a pure culture of HIV. Indeed, in some parts of the world, until recently, certain governments maintained that HIV was not the causative agents of AIDS and the experiment could not be done to prove definitively otherwise. Needle-stick accidents, in which health workers with no risk factors for HIV infection developed AIDS after injuring themselves with a needle contaminated with material from an AIDS patient, are as close as we have come to the definitive experiment. On the other hand, it was only when Barry Marshall ingested a pure culture of *Helicobacter pylori* and developed inflammation of the stomach that it became generally accepted that peptic ulcers were caused by this bacterium. He and Robin Warren were subsequently awarded a Nobel Prize for their work on peptic ulcer pathogenesis, but this approach is not to be generally recommended. See Figure 2.27.

In the remainder of this section we will first focus on the respective mechanisms of pathogenicity of the major groups of bacteria and viruses, and host defence against them. We then briefly discuss fungi before turning to one particular example of a protozoan parasite. Here, we will not consider larger, metazoan parasites such as worms, many of which do not cause overt disease in their hosts because they are effectively contained in topologically external compartments such as the lumen of the gut (although they may compete with their hosts for nutrients).

2.4.2 Pathogenesis of Bacterial Infections

2.4.2.1 Bacteria that Secrete Toxins

In these infections, the disease is caused solely, or very largely, by protein toxins secreted by the infecting bacteria. These are known as exotoxins, in contrast to endotoxins which are part of the bacterial cell wall and which are not secreted

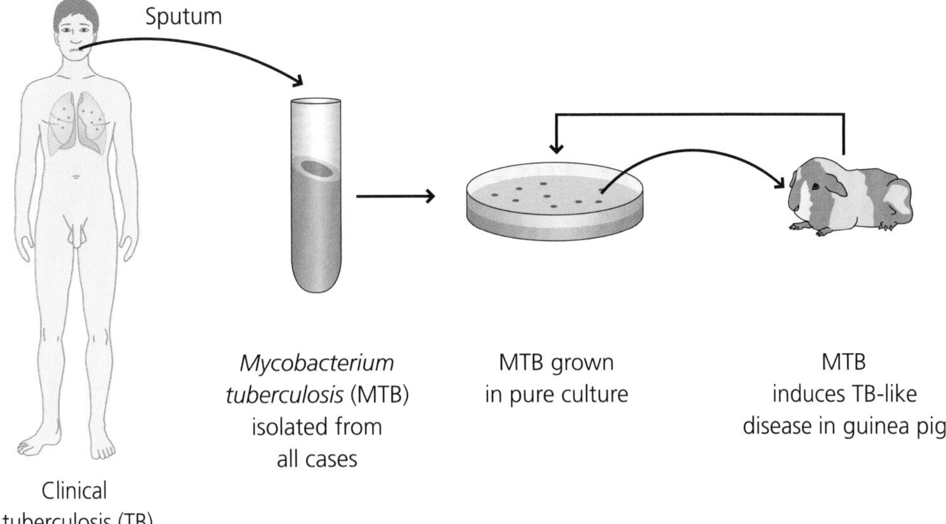

Fig. 2.26 Koch's postulates. Koch defined conditions that needed to be fulfilled to show that a particular microbe is the cause of a particular disease. The microbe must be able to be isolated from all cases of the disease and grown as a pure culture *in vitro*. The isolated microbe must cause a similar disease in a suitable experimental model. It must also be able to be re-isolated and cultured from the infected animal. Tuberculosis is shown as one example.

(Section 2.4.2.5). This kind of infection is a major cause of illness and death in the developing world, in the form of diarrhoeal diseases such as cholera (caused by *Vibrio cholerae*). Tetanus and botulism have a similar pathogenesis, and the symptoms of whooping cough (pertussis infection) and diphtheria are also due mainly to toxins. These toxins work by binding to receptor molecules on host cells (e.g. cholera toxin binds to intestinal epithelial cells) and killing the cells or perturbing their metabolism. In these infections, antibodies are the main if not the only defence mechanism. Antibodies, by binding to the toxin, can prevent it attaching to the host cell, thus neutralizing it and completely preventing disease. See Figure 2.28.

Q2.10. What might be the selective advantage to Salmonella of inducing intestinal inflammation?

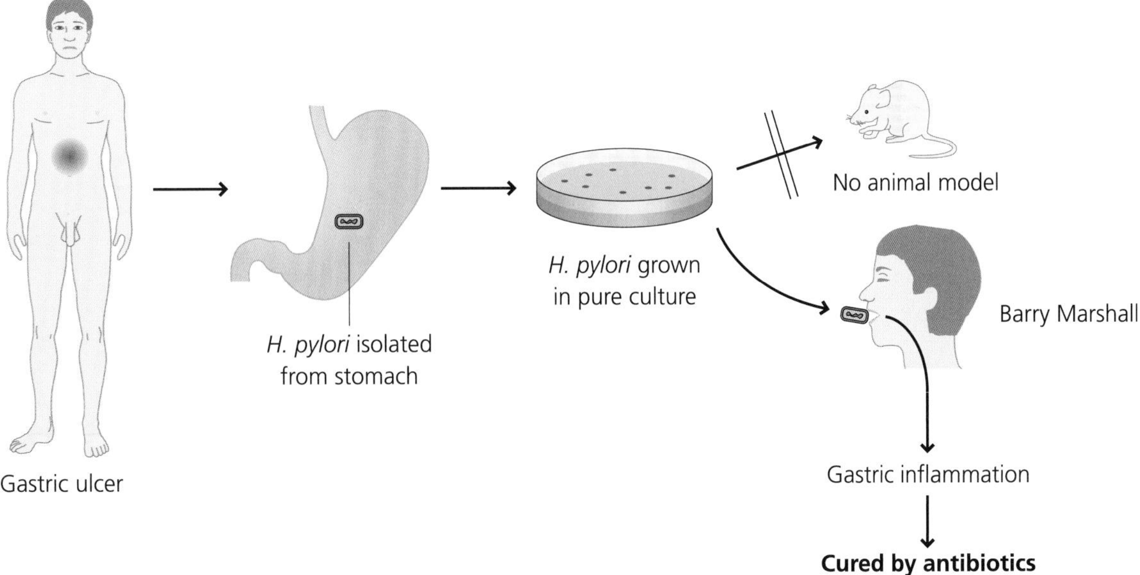

Fig. 2.27 *Helicobacter pylori* and peptic ulcers. Stomach (peptic) ulcers were not thought to have an infectious origin. Australian workers, however, isolated a novel bacterium, *Helicobacter pylori*, from the stomachs of ulcer patients. They could grow it in pure culture but they did not have an animal in which it would grow and cause disease. To convince the sceptics, one of the workers swallowed a pure culture of the bacterium and developed stomach inflammation, gastritis. He, and subsequently other patients with peptic ulcers, were cured by antibiotic treatment.

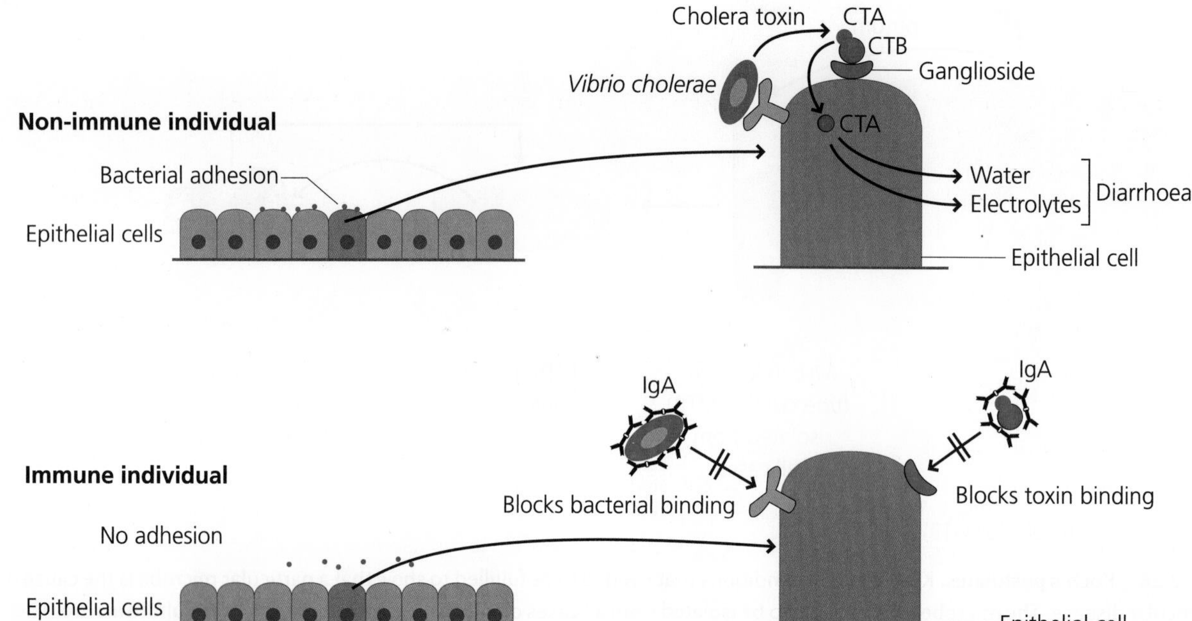

Fig. 2.28 Pathogenesis of cholera. If water containing *Vibrio cholerae*, the cause of cholera, is drunk, it is likely the bacteria will adhere to intestinal epithelial cells. The bacteria secrete toxin composed of two parts. CTB binds to a ganglioside on the cell surface, facilitating the internalization of CTA. The latter induces the intracellular synthesis of large amounts of cyclic AMP, resulting in massive chloride secretion into the intestinal lumen. Water and other electrolytes follow the chloride ions, causing profuse diarrhoea. In immune individuals, IgA antibodies block the binding of both the toxin and of the bacteria to the intestinal cells.

Immunization against tetanus consists of giving a modified form of the toxin (toxoid) which is non-toxic, but retains its antigenicity. This induces an IgG anti-tetanus toxin antibody response. The antibodies bind to the toxin and block its binding site for the receptors on nerves. Immunity lasts for at least 10 years, but depends on having pre-formed antibody; after re-infection there is not enough time for an effective secondary antibody response to occur between the toxin being released and it reaching a nerve. See Figure 2.29 and Case Study 2.1.

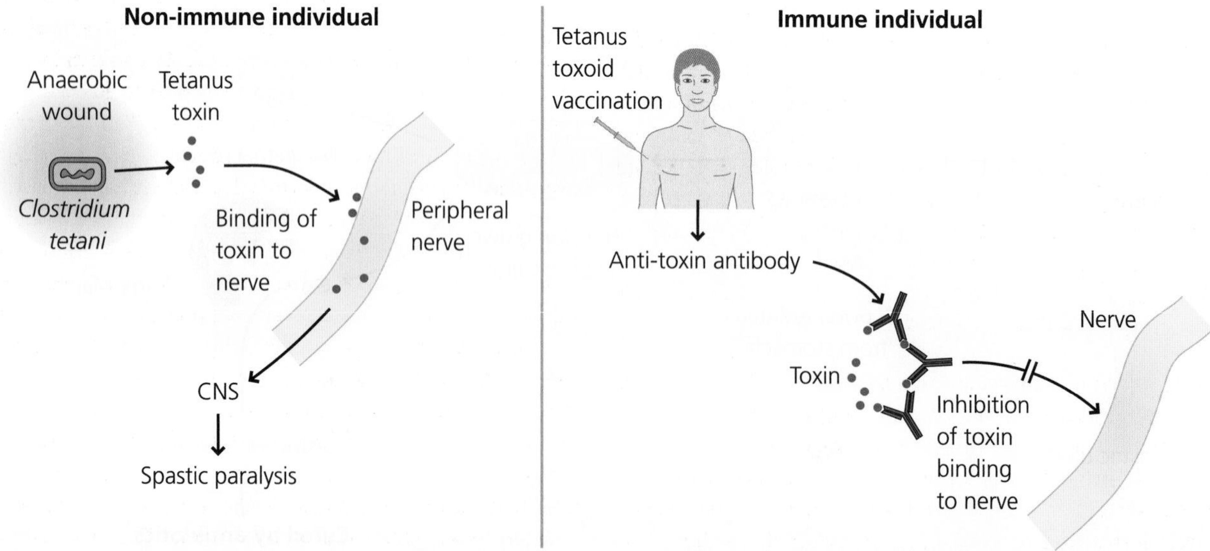

Fig. 2.29 Pathogenesis and immunity in tetanus. Tetanus spores germinate in the anaerobic environment of dead tissue. The bacteria secrete a toxin which, in non-immune individuals, diffuses to a nearby peripheral nerve, binds to receptors and is internalized. It travels by retrograde axonal transport to the cell body of the nerve in the CNS, where it blocks secretion of inhibitory neurotransmitters, leading to uncontrolled activation and spastic paralysis. Individuals who been vaccinated against tetanus have pre-formed antibodies that can bind the toxin and prevent its binding to the nerve.

Case Study 2.1: Tetanus

Clinical: A 57-year-old English woman was digging her well-manured rose bed. She stabbed her boot with the gardening fork, mildly injuring her foot. She washed and dressed the wound, and thought nothing more of it. A few days later she developed marked rigidity of her limbs and was admitted to hospital. When questioned by the physician, she could not remember when she had last been immunized against tetanus. She was curarized to induce complete muscle paralysis, put on a respirator and treated with antibiotics. After 14 days she made a full recovery and was allowed home.
Explanation: Horse manure contains many Clostridium tetani spores. These spores are highly resistant to inactivation and may remain in the manure for long periods. The woman's wound contained dead tissue, which is anaerobic. Some spores had entered this dead tissue and germinated in the anaerobic conditions. These bacteria really are saprophytes – using the dead tissues as food. The bacteria began to secrete tetanus toxin that diffused to nearby nerves, bound to receptors on these nerves and was internalized. The toxin was transported by retrograde axonal transport to the spinal cord, where it blocked the release of inhibitory neurotransmitters at motor neuron synapses. This led to hyperactivity of motor neurons and the resultant spastic paralysis. By paralysing the woman and giving her artificial respiration she could be kept alive until the toxin was metabolized and inhibition restored.
Footnote: This is not such a rare event even in the UK. This actually happened to a woman known to one of the authors.

Q2.11. In contrast to most infections, patients who recover from tetanus are usually not immune to re-infection and need to be actively immunized. Why might this be?

2.4.2.2 Bacteria that Cause Acute Inflammation

Acute inflammation is a condition of rapid onset, characterized by local pain, increased dilatation of the blood vessels (vasodilatation) and leakiness leading to tissue swelling (oedema). It is also often accompanied by systemic symptoms such as fever and loss of appetite, which normally last for a few hours or days. An abscess (boil, stye), an acute middle ear infection, meningitis and pneumonia are all examples of acute inflammation. The bacteria that cause acute inflammation are typified by *Staphylococcus aureus*, which causes abscesses or boils including styes on the eyelid; *Streptococcus pyogenes*, which cause some sore throats and is one of the "flesh-eating" bacteria; and *Neisseria meningitidis*, one of the causes of meningitis. These bacteria can exist and multiply in extracellular tissues and are logically called extracellular bacteria. This type of infection is often typified by the presence of pus at the site of infection. As mentioned earlier, this type of infection is termed pyrogenic, not to be confused with the term pyrogenic which means "fever-inducing". See Case Study 2.2

Pathogenesis and Recovery In Case Study 2.2, an innate immune response occurred in the lungs that led to the recruitment of neutrophils. In the case of a related, but non-pathogenic, bacterium the neutrophils might have been able to clear the infection. In this case however, the bacteria avoid destruction by evading the neutrophils. At the same time, however, an adaptive immune response was starting. Capsular polysaccharide was carried to secondary lymphoid organs. After 6–7 days the antigen-specific B cells had started to secrete anti-capsular antibodies. At first this is in the form of IgM since it is a primary antibody response; IgG is made in significant quantities only in a T-dependent (TD) secondary response. See Figure 2.30.

Q2.12. Normal individuals possess IgM and IgG antibodies against capsular polysaccharides of *Streptococcus pneumoniae*. Given that such polysaccharides are T-independent (TI) antigens (Section 1.4.5.3), how might IgG be synthesized?

The IgM was secreted into the blood and, in the lungs, the increased venular permeability permitted the antibodies to enter the alveoli and to bind to the bacterial capsules. Neutrophils do not express FcRs for IgM, but IgM is a very efficient activator of complement. Hence, complement components C3b and its inactivated form iC3b were deposited on the bacterial cell capsules, close to where IgM was bound. Recognition by complement receptors on the neutrophils now enabled the cells to phagocytose and subsequently kill the bacteria. This is one of the ways in which the immune system has evolved in attempting to overcome the evasion mechanisms that bacteria have evolved.

The dead and dying neutrophils, and damaged and liquefied tissues, formed much exudate in the lungs. Monocytes were also entering the alveoli in large numbers and they differentiated into inflammatory macrophages which phagocytosed the debris. The lungs are unique in that all this debris can be disposed of by coughing; this explains the above patient's profuse sputum, and the red-brown colour represents breakdown products of haemoglobin. This very rapid and efficient garbage disposal system means that the lung structure and function can be fully restored to normal with little or no collagen deposition, and thus no scarring. This is known as resolution, which is unusual in acute inflammation since in most sites, acute inflammation leads to significant scarring.

Vaccination against *Streptococcus pneumoniae* has proved difficult because there are many different forms of capsular polysaccharide (at least 23). A vaccine containing as many as 23

Case Study 2.2: Lobar Pneumonia

- Clinical: A 23-year-old man is on an expedition in the Himalayas. One night he develops a dry cough and complains that he is feeling cold. He is shivering violently. Next morning he feels very unwell, has lost his appetite and his temperature has increased to 39.5 °C. His pulse rate is 95 beats/min. He produces blood-stained sputum. He continues to be unwell, and the next day when he coughs he has a sharp, catching pain in the right side of his chest. A medical student in the expedition percusses his chest (lays one finger on the chest and taps it hard with another finger) and finds that the sound over the lower right side is very dull in comparison with the rest of his chest. The sledge carrying the antibiotic supply has been lost down a crevasse. Over the next 3 days the patient becomes drowsy, he remains feverish, his sputum continues to be blood-stained and the expedition is worried that he is going to die. On the seventh morning after starting to feel unwell he wakes up, says he feels much better, starts to sweat profusely and asks for breakfast. He starts to cough up much sputum that is red-brown in colour. He makes a rapid recovery and completes the rest of the expedition in good health.
- Explanation: The patient had inhaled *Streptococcus pneumoniae*, possibly transmitted from another member of the expedition who was a symptomless carrier, and developed lobar pneumonia due to the properties of the bacteria. The inhaled bacteria passed to the lung alveolae and started to multiply extracellularly. Toxins released by the bacteria initiated an acute inflammatory response. Increased venular permeability allowed the passage of blood plasma and erythrocytes into his alveoli, accounting for the blood-stained sputum. Cytokines released from resident and recruited cells (mainly macrophages) acted on his central nervous system (CNS) to cause his fever and other symptoms. Neutrophils were recruited in large numbers into the alveoli. *Streptococcus pneumoniae* however, has, a thick polysaccharide capsule – this is its main virulence determinant – and the neutrophils have no receptors for the polysaccharide so they can multiply unchecked. Their rate of multiplication and the inflammatory exudate enables the bacteria to colonize a whole lobe of the lung, this is why it is called "lobar". The affected lobe is now filled with fluid, this is why it is dull to percussion. After several days the bacteria induced an effective antibody response that was responsible for his recovery.

variants is partially protective but, as noted above, these trigger TI responses. Recently attempts have been made to make conjugate vaccines containing proteins such as tetanus toxoid as carriers to stimulate T cell help (Section 2.5.2.2).

2.4.2.3 Bacteria that Cause Chronic Inflammation

Chronic inflammation is a much longer lasting process than acute inflammation. It is typically characterized by a mononuclear infiltrate of immune cells, particularly macrophages and activated lymphocytes; these cells are termed mononuclear because they have a round or oval, or kidney-shaped, unlobulated nucleus, in contrast to polymorphonuclear leukocytes such as neutrophils. Bacteria that stimulate chronic inflammation such as *Mycobacterium tuberculosis* (the cause of tuberculosis), and protozoa such as *Leishmania*, live inside host cells, primarily macrophages (Section 2.3.1.4). They can replicate in these cells and kill them. Molecules released by the dying macrophages induce local inflammation, but this is not the main cause of clinical disease in tuberculosis, it is a bit more complicated. it is a bit more complicated. The disease, in this case tissue damage, is primarily the result of the host's own immune response against the bacteria. (See Case Study 2.3).

Pathogenesis and Recovery Initial infection. At some time before she developed symptoms this girl (Case Study 2.3) had inhaled some tuberculosis bacteria (*Mycobacterium tuberculosis*). These were phagocytosed by macrophages in her lung alveoli and connective tissues. Although the differences between pathogenic and non-pathogenic mycobacteria are not fully understood, it is clear that the pathogenic bacteria are able

Case Study 2.3: Tuberculosis

Clinical: A 15-year-old school girl develops a cough and a fever, which do not subside after 2 weeks. She is given a chest X-ray that shows areas of consolidation (filling-in of the air spaces). A Mantoux test is positive (this is a test for a T cell response against antigens of the mycobacteria; below). Examination of her sputum shows the presence of bacteria typical of tuberculosis. She is treated with antibiotics and makes a complete recovery.

Explanation: The girl had inhaled *Mycobacterium tuberculosis* from an infected individual. The infection stimulated the development of chronic inflammation in her lungs as a side-effect (collateral damage) of a T cell-mediated immune response against to the infection. This chronic inflammation caused tissue damage that appeared as areas of consolidation in the X-ray. Antibiotics were used to kill the bacteria.

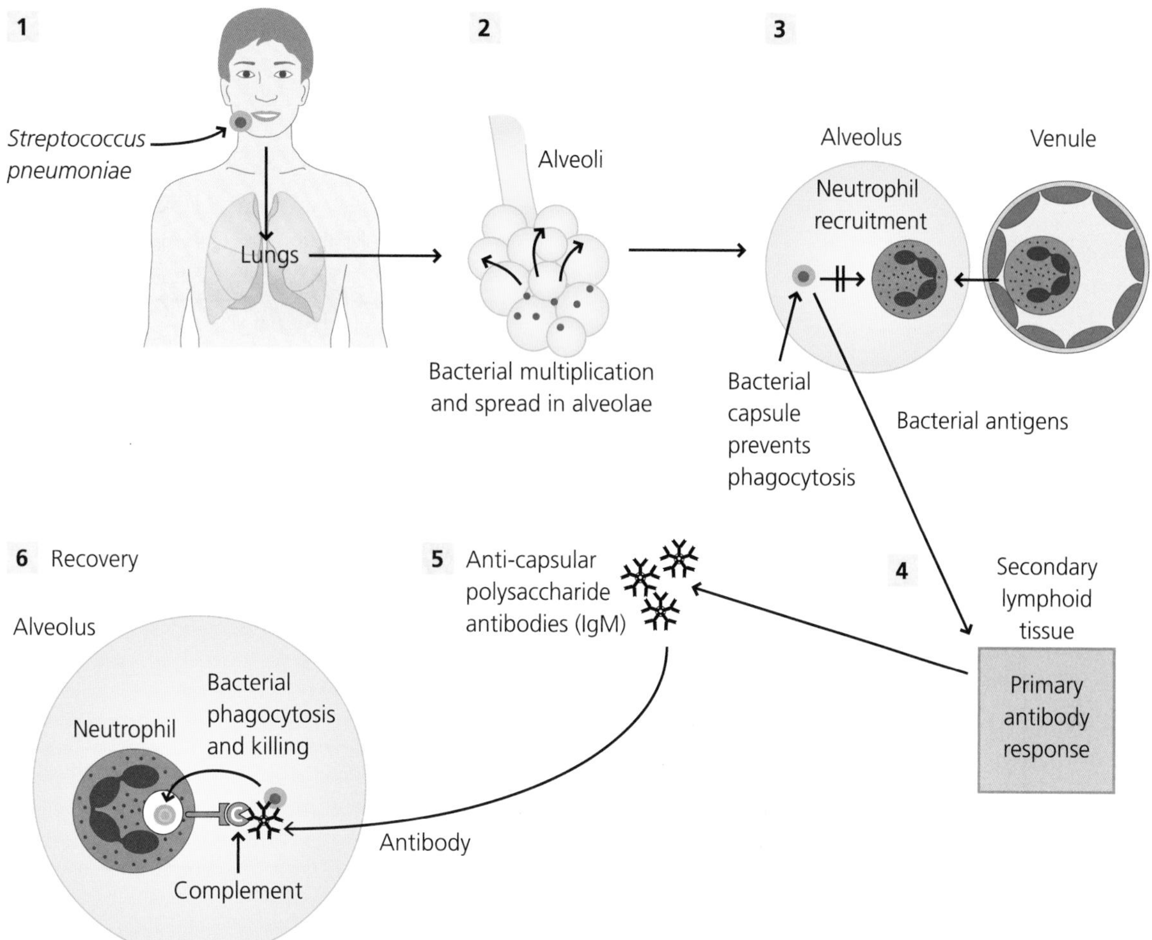

Fig. 2.30 Lobar pneumonia. (1) The bacterium *Streptococcus pneumoniae* is inhaled and travels to the lung alveoli. (2) It multiplies in the alveolar air spaces and release toxins that cause acute inflammation. (3) The inflammation recruits neutrophils to the alveolar spaces, but they cannot phagocytose the bacteria because it has a polysaccharide capsule. (4) Bacterial antigens are carried to secondary lymphoid organs where they initiate an adaptive immune response to the capsular polysaccharide. (5) IgM is synthesized and travels in the blood to the inflamed area. Here it can enter the alveoli because of the increased venule permeability consequent to the inflammation. (6) IgM binds to the bacteria, but cannot act directly as an opsonin since neutrophils do not express FcRs for IgM. However, complement also enters the alveoli and is activated and binds to the IgM on the bacteria. Neutrophils have receptors for complement components C3b and iC3b, permitting phagocytosis and killing of the bacteria. The inflammatory exudate containing bacteria, neutrophils and other molecules is coughed up, often leading to complete recovery of lung function.

to interfere with the killing mechanisms in macrophages. *Mycobacterium tuberculosis* evades the immune response in several ways – these include prevention of phagosome maturation and subsequent lysosomal fusion, and interference with the synthesis of reactive nitrogen intermediates. Additionally, *Mycobacterium tuberculosis*, but not non-pathogenic mycobacteria such as Bacillus Calmette-Guérin(BCG) (a vaccine against tuberculosis prepared from a strain of the attenuated live bovine tuberculosis bacillus *Mycobacterium bovis*) can escape from the phagosome into the cytoplasm thus evading the normal route of MHC class II presentation to CD4 T cells (introduced in Chapter 1 and discussed in Chapter 5). See Figure 2.31.

In Case Study 2.3, the bacteria multiply in the macrophages and kill them. Molecules released by the dying macrophages (e.g. proteases) cause local tissue damage (inflammation). Other molecules cause changes in the expression of adhesion molecules on the luminal surface of local venular endothelium that permit blood monocytes to adhere and cross into the tissues, and develop into inflammatory macrophages. These again phagocytose the bacteria but cannot kill them and are themselves killed, increasing the degree of inflammation. As this process occurs over days, weeks or months, it is known as chronic inflammation.

At the same time that macrophages are failing to kill the bacteria, bacteria and bacterial antigens are being transported via the lymph to lymph nodes draining lung. These bacteria could be transported free within the lymph or within migrating cells (DCs and perhaps some monocytes or macrophages). In the node, two processes are occurring. The bacteria continue to multiply in macrophages setting up chronic

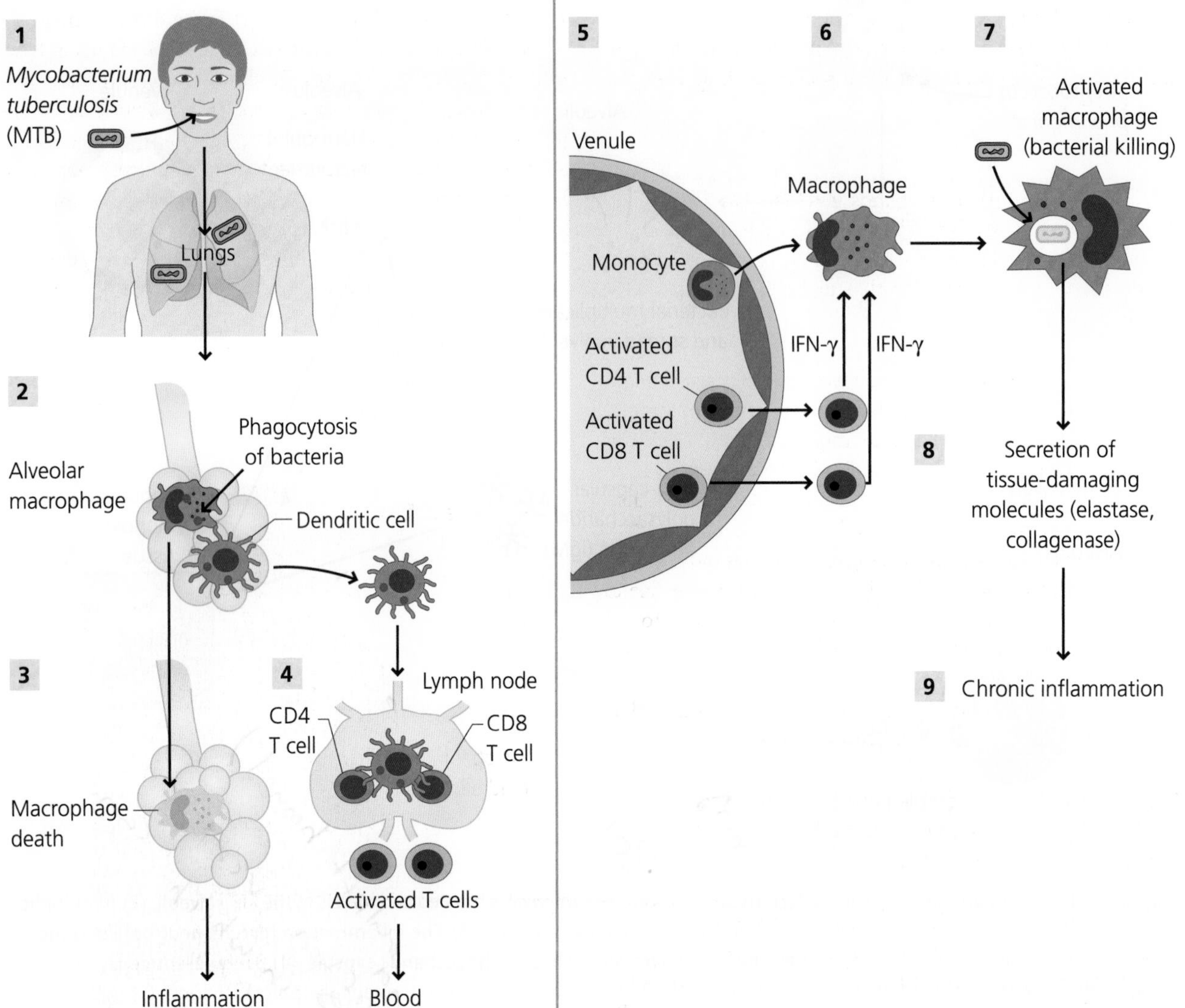

Fig. 2.31 Pathogenesis of tuberculosis. (1) *Mycobacterium tuberculosis* enters the body by inhalation. (2) In the lung alveoli it infects and kills alveolar macrophages, leading to local inflammation (3). Mycobacterial antigens are transported to the draining lymph nodes, probably by DCs which have been activated in the lung via PRRs. (4) The DCs activate CD4 and CD8 T cells, which migrate to the inflamed lung tissues (5). (6) The local inflammation also recruits blood monocytes, and IFN-γ activates these macrophages, increasing their ability to kill the bacteria, most probably by hydrogen peroxide and reactive nitrogen intermediates. (7) The bacteria, however, possess many mechanisms that help them to avoid being killed, and the infection becomes chronic. (8) The macrophages are also secreting tissue-damaging enzymes that result in the breakdown of connective tissues, leading to chronic inflammation (collateral damage). (9) When the infection is controlled, the resulting repair is likely to cause much fibrosis and scarring. Some bacteria are, however, likely to remain in the tissues in a dormant state.

inflammation in the node. Additionally, however, an adaptive immune response to the bacteria starts. DCs degrade proteins expressed by the bacteria that they phagocytosed in the periphery, and small peptides from these proteins, bound to MHC molecules, are recognized by T cell receptors (TCRs) on CD4 and CD8 T lymphocytes present in the node. These T cells become activated and start to divide, forming clones of T cells expressing the same TCR specificity (clonal expansion). The CD4 T cells also start to polarize their responses and develop into T_h1 cells as a consequence of the signals they receive, including those produced by the DCs in response to recognition of bacterial antigens via their PRRs.

The activated T cells are then released from the node into the blood and migrate back to the site of infection in the lung. They can do this because they now express adhesion molecules that bind to complementary molecules on endothelial cells in the inflamed area. The activated T cells secrete cytokines that have multiple effects. Crucial, however, is the IFN-γ that binds to receptors on the macrophages, inducing activation and the expression of a variety of novel genes including those involved in production of ROIs. These activated macrophages can now kill the bacteria they have phagocytosed, and the infection is controlled. Surprisingly, however, not all bacteria are killed, some remain dormant in the tissues for many years and do not cause clinical disease unless the individual becomes immunosuppressed (e.g. after HIV infection and AIDS or anti-inflammatory treatment for diseases such as rheumatoid arthritis).

Q2.13. If for some reason, the antigen that stimulates such an immune response is not derived from a pathogen, but is a self antigen (e.g. a molecule in β cells of the Islets of Langerhans in the pancreas that produce insulin), what might ensue?

Activated macrophages are not, however, totally a good thing. As well as being able to kill bacteria, they become active secretory cells, producing enzymes and other molecules that kill cells and break down connective tissue, so that in the area of infection there is a lot of tissue damage. In contrast to acute inflammation, however, pus never forms. Instead, because the inflammatory process is relatively slow, the healing response has more time to develop; much collagen is laid down (fibrosis) giving rise to structures termed granulomas. In tuberculosis and other forms of immune-mediated chronic inflammation the granulomas contain many T cells; these are the cells that are driving the inflammation. The solidification of areas of the lung caused by the granulomas gives rise to the opacities seen on X-ray.

Q2.14. Can you suggest a hypothesis to explain how, following a primary infection, the tubercle bacteria are maintained in a dormant state?

The Mantoux test consists of injecting some *Mycobacterium tuberculosis* proteins into the skin. In a positive test, an area of inflammation appears at the site of injection within 24–48 h. Why does this occur? During the adaptive response greatly increased numbers of activated and memory T cells specific for the bacterial proteins are generated by clonal expansion. Large numbers of these T cells migrate to the site of injection, recruit and activate more macrophages, and stimulate chronic inflammation. This kind of response is known as a Type IV delayed-type hypersensitivity (DTH) reaction (Chapter 7). If a person has not previously had, or been immunized against, tuberculosis there are not enough antigen-specific T cells present to mount a DTH response so they will not respond to a Mantoux test. Thus, in tuberculosis, the inflammation is not due to direct effects of the bacteria. It is due to mediators released by host cells, primarily macrophages, which have been activated by T cells.

Q2.15. Robert Koch – he of the postulates fame (Section 2.4.1.2) – recognized the importance of DTH responses in tuberculosis and attempted to develop a cure for tuberculosis by inoculating diseased patients with killed mycobacteria. Unfortunately a significant number rapidly became seriously ill and died. Why might this have happened?

2.4.2.4 Bacteria that Cause Intestinal Inflammation

Bacteria such as *Salmonella* mostly cause direct damage to intestinal epithelial cells, which in some cases results in local inflammation. In most cases the clinical effects are diarrhoea and vomiting. Many of these bacteria are able to alter the properties of host epithelial cells to their own advantage. Thus, a *Salmonella* bacterium, having bound to the cells, inserts a hollow needle-like tube (called a Type 3 secretory system) into the cytoplasm of the cell. Bacterial proteins are then injected into the epithelial cell and these can radically change the properties of the cells, e.g. by making them phagocytic, inducing apoptosis and interfering with intracellular signalling. See Figure 2.32.

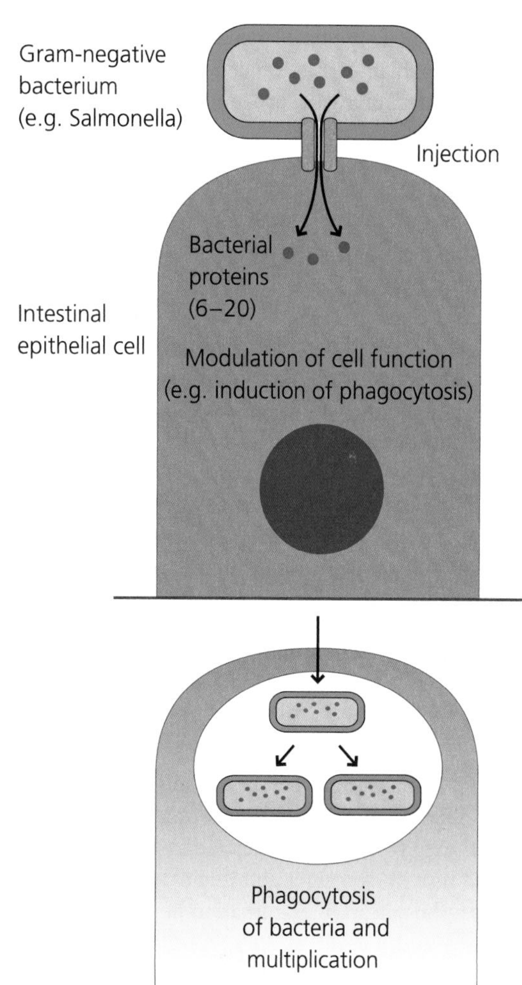

Fig. 2.32 Bacterial Type III secretory systems. After binding to an intestinal epithelial cell, Gram-negative bacteria such as *Salmonella* construct hollow needles which insert into the cytoplasm of the host cell. Bacterial proteins are injected into the host cell, inducing changes that help the bacteria to infect and maintain itself in the host. One example is that the intestinal cell can become phagocytic, now enabling the bacteria to enter the cell where they are able to survive by interfering with phagolysosomal function.

Q2.16. What might be the selective advantage to Salmonella of inducing intestinal inflammation?

Normally the gut is protected from infection by many pathogens by antibodies (particularly IgA) that are secreted directly into the large intestine and which act to prevent attachment of invading bacteria to the gut wall. In other words, IgA neutralizes the bacterial infection. Some bacteria, however, such as those that cause typhoid fever (*Salmonella typhi*) invade through the intestinal wall and can spread to other organs such as the liver, and cause damage in these sites. The mechanisms involved in initial recovery from these infections probably include neutrophils and activated macrophages, and, later, antibodies. However, resistance to infection with bacteria such as those that commonly cause bacterial food poisoning (*Salmonella typhimurium*), and resistance to re-infection, is primarily antibody-mediated (usually IgA). Neutrophils play an important part in early defence against *Salmonella*; however, unlike infections with pyogenic bacteria, they are not sufficient for defence.

Q2.17. Although diarrhoea and vomiting can lead to serious dehydration these responses are often considered also to be beneficial to the host. Why?

2.4.2.5 Bacteria that Cause Septic Shock

Both Gram-negative and Gram-positive bacteria contain molecules in their cell wall that are able to induce strong activation of the innate immune system by acting as agonists for Toll-like receptors (TLRs, a subset of PRRs). In the case of Gram-negative bacteria one such molecule is lipopolysaccharide (LPS) or endotoxin, for Gram-positive organisms the corresponding molecule is lipoteichoic acid. These molecules are structural components of the bacterial cell wall and are not secreted or released from living bacteria. They are, however, released from dead or dying organisms, and if they get into the bloodstream, they can induce potentially fatal cardiovascular shock – commonly known as septic or endotoxic shock. See Case Study 2.4.

Septic shock can have other consequences for example decreased organ perfusion can lead to organ failure (e.g. kidney failure). The major problem is that the low blood pressure (hypotension) is refractory to treatment and that all the usual drugs used to raise blood pressure are ineffective. In animal models, if an antagonistic monoclonal antibody to TNF-α is given early (before hypotension has set in), the hypotension can be prevented. This treatment is, however, ineffective in human septic shock, probably because by the time it is given, the TNF-α has done its work. Most septic shock is caused by LPS from Gram-negative bacteria but can however also be caused by lipoteichoic acid from Gram-positive bacteria.

2.4.2.6 Bacteria that May Cause Malignant Tumours

Unlike viruses, there is relatively little direct evidence that links bacterial infection with an increased risk of developing malignant tumours. One such linkage, however, is that of *Helicobacter pylori* with carcinoma of the stomach. Stomach (gastric) ulcers are a common disease in the developed world. It is only recently, however, that it has been shown that these ulcers are caused by a bacterium, *Helicobacter pylori* Fig 2.27. Many gastric carcinomas are associated with gastric ulcers and there is a strong correlation between the presence of *Helicobacter pylori* in the stomach and the development of gastric carcinoma. Bacterial infection is, however, a major complication of many malignant tumours and is a frequent cause of death in patients with widespread metastases.

Case Study 2.4: Septic Shock

Clinical: A 15-year-old boy on a camping trip develops severe abdominal pain, starting centrally but later localizing to the right lower abdomen. He has a mild fever. It has snowed heavily overnight and it is many hours before he arrives at hospital. On examination he has a rigid, tender abdomen and a fever of 39.5 °C. Blood tests show an elevated neutrophil count. His blood pressure is slightly elevated. He is diagnosed as having acute appendicitis. While being prepared for surgery he becomes pale and lethargic. His blood pressure has fallen dramatically and he has difficulty with breathing. Despite all attempts at resuscitation he dies. At post-mortem his peritoneal cavity is found contain cloudy fluid, and his appendix is swollen, reddened and shows a rupture. His blood was cultured for bacteria and showed the presence of Gram-negative rod-like bacteria.

Explanation: The boy had developed acute appendicitis. The acute inflammation had weakened the wall of the appendix and it had ruptured, allowing the inflammatory exudate to enter the peritoneal cavity, causing acute peritonitis (this accounts for his rigid abdomen). Some bacteria had entered the bloodstream (septicaemia). LPS was released from dead or dying bacteria. This was recognized by TLR4 (Section 4.2.2.2) on circulating monocytes. TLR4 signalling resulted in the activation of cytokine secretion, including tumour necrosis factor (TNF)-α. TNF-α acted on veins to induce the synthesis of nitric oxide by endothelial cells. The nitric oxide acted on the veins to cause smooth muscle relaxation, leading to vasodilatation and pooling of blood in the venous circulation. The fall in input of blood to the heart led to decreased cardiac output and death.

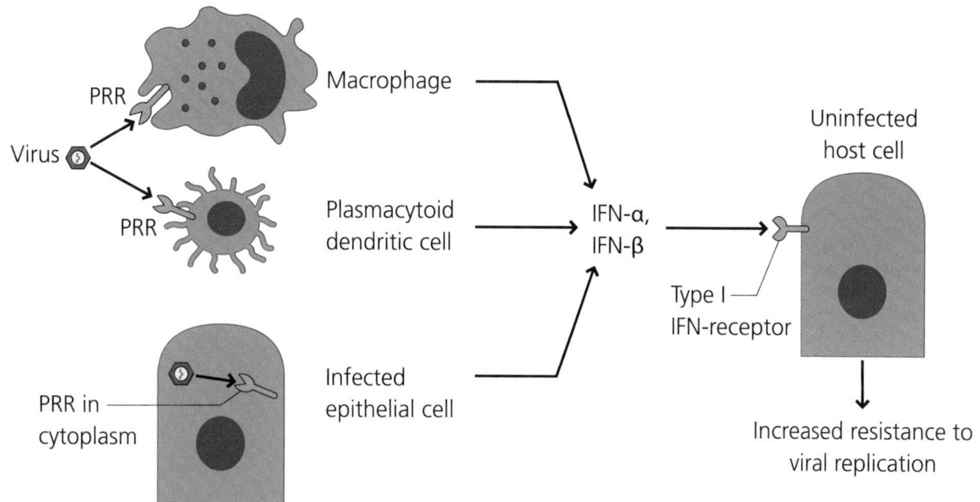

Fig. 2.33 Initial resistance to viral infection. Viruses need to evade or subvert a series of defence mechanisms if they are to set up an infection. Natural barriers such as stomach acid can inactivate many viruses. If they pass these barriers, the viruses activate the innate immune system via PRRs expressed by alarm cells such as macrophages, pDCs and other cell types. The most important effect of this activation is the synthesis of Type I IFNs e.g. IFN-α and IFN-β, which act on other cells to make them resistant to viral replication. The infected cells may also produce Type I IFNs to help protect other cells in the vicinity.

Q2.18. How might you try to show that *Helicobacter pylori* is the cause of gastric cancer, rather than just being associated with it?

2.4.3 Pathogenesis of Viral Infections

Viruses are obligate intracellular pathogens – they can only replicate inside cells. The mechanisms by which they cause disease overlap in many ways with bacteria. Viruses can be roughly divided into two major groups according their overall life history. Some viruses only infect locally, at the sites they first enter the body. Examples include the common cold and influenza viruses in the respiratory tract, and norovirus in the intestine. Others spread from the initial site of infection into different regions of the body and may cause disease in distant organs (systemic infection). Examples of these include the viruses that cause hepatitis, measles and smallpox.

In contrast to the defence mechanisms that operate during bacterial infections, those in viral infections are more homogenous. Initial resistance in a non-immune individual is largely via innate mechanisms such as IFNs that induce resistance in neighbouring cells (Chapter 4). Type I IFNs, as opposed to IFN-γ are cytokines that are secreted in response to viral infection by infected cells, as well as macrophages and NK cells, and at especially high levels by a specialized subset of plasmacytoid DCs (pDCs). See Figure 2.33.

However, recovery from most viral infections depends on the activity of CD8 T cells, which often need help from CD4 T cells to be activated. Activated cytotoxic CD8 T cells (CTLs) can kill virally infected cells. If this happens before infectious virus has been assembled in the cell, the spread of infection can be stopped. Antibodies probably play little part in recovery from infection. However, in individuals who are immune to a virus as a result of vaccination or recovery from a previous infection, it is almost always antibodies that mediate resistance, usually by preventing binding to host cells (neutralization) or coating them to enable their phagocytosis (opsonization). See Figure 2.34.

2.4.3.1 Viruses that Cause Direct Cell Death

In some diseases, the clinical effects are due solely to cell death caused by the virus. For example, in polio the virus kills the motor neurons in the spinal cord that initiate muscle contraction. In other infections such as HIV, the loss of a particular subset of T lymphocytes (CD4 T cells) may be caused by the virus directly killing the T cells (but could also reflect killing of infected CD4 T cells by other cell types). See Case Study 2.5.

Pathogenesis The source of the virus in Case Study 2.5 could have been a child who had recently been immunized with the live but attenuated (weakened) polio vaccine, called the Sabin vaccine. Children immunized with the Sabin virus excrete live virus in their faeces and, very rarely, some of the viruses can revert to being virulent. The patient above may have swallowed some virulent virus that passed through his stomach and bound to the human polio virus receptor (PVR) on his intestinal epithelial cells. Peyer's patches may be important sites of initial infection. Polio virus is non-enveloped and either gains direct entry into the epithelial cell or is endocytosed by it (which of these actually happens is not certain). It replicates in these cells and following viral assembly the cell lyses, permitting release of infectious virus. The virus then probably enters lymph and is carried to the mesenteric nodes, where it may undergo another round of replication, before being released into the bloodstream. It eventually reaches the CNS and infects anterior horn motor neurons, killing them

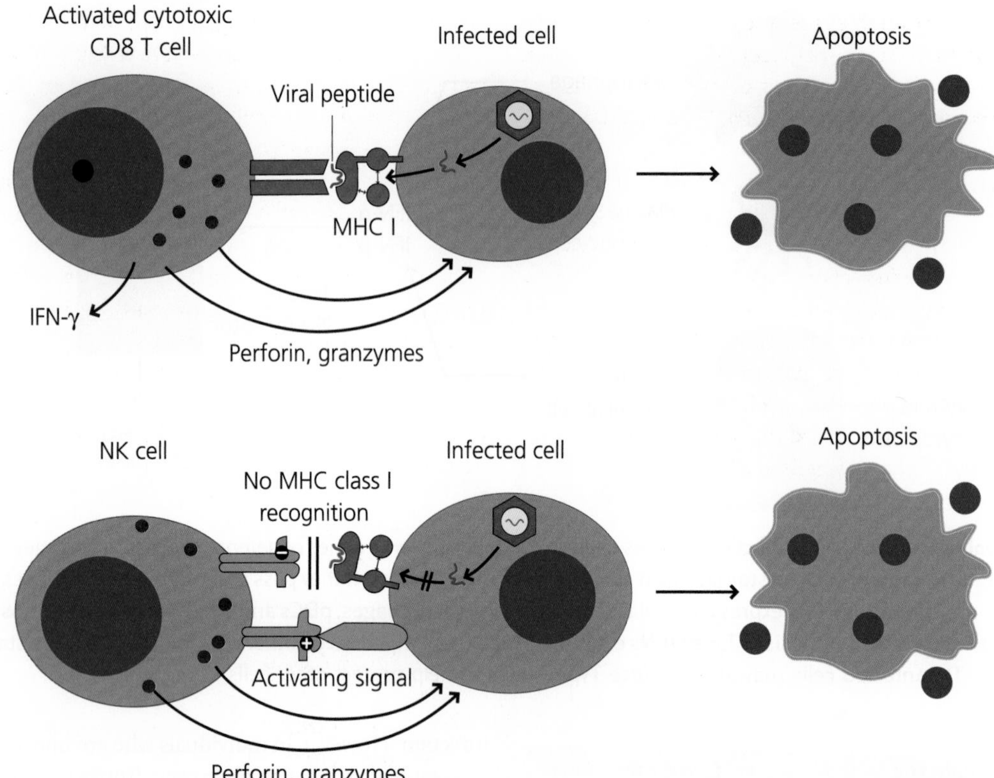

Fig. 2.34 Recovery from viral infection. In most cases of viral infection an adaptive response is needed to bring about recovery. This response involves both CD4 and CD8 T cells. CD8 T cells are often essential and generate two main effector mechanisms. Activated CD8 T cells can kill infected cells, and if this happens before new, infectious virus has been made, the spread of infection can be stopped (Section 1.4.5.2). Activated CD8 T cells can also secrete IFN-γ, which is a weak anti-viral agent but more importantly activates macrophages which are potent secretors of Type I IFNs. CD4 T cells may provide help in the activation of CD8 T cells and also help B cells to make antibodies – these may play some role in recovery, but are crucial in resistance to re-infection. In some cases, NK cells may also kill virally infected cells, aiding recovery from infection (Section 1.4.4).

and thus causing the characteristic flaccid paralysis. This cycle of events is, however, very rare. In most cases of polio infection the CNS is not involved and infection leads to transient diarrhoea or is asymptomatic.

Some more recent studies also suggest an additional or alternative pathogenesis. Mice cannot normally be infected with polio virus because they do not express the required PVR. Transgenic mice have been made that express the human PVR widely in their bodies and these mice can now be infected with polio. After infection of a transgenic mouse by injection of virus into one limb, the virus spreads to the CNS and causes paralysis. The first limb to be affected by paralysis is, however, the one that was infected. If the nerves connecting that limb to the CNS are cut, despite the virus having access to the blood, no paralysis occurs. This suggests strongly that, in this model, the virus actually travels up the nerves to reach the CNS, similarly to the rabies virus. We do not, however, know if this happens in human infections.

Case Study 2.5: Poliomyelitis

Clinical: A 12-year-old boy from an American family that does not believe in immunization goes swimming in a pool used by many other children. A few days later he develops a headache and mild neck stiffness. A few days after this he notices that his right arm is weak and over the next few days the weakness increases until he cannot move his arm at all.

Explanation: The swimming pool was contaminated by faecal material from other children, one of whom was excreting virulent polio virus. The patient ingested the virus which, after infecting his intestinal epithelium, travelled to the CNS where it induced the death of anterior horn motor neurons, leading to flaccid paralysis. Motor neurons cannot be regenerated and the paralysis is permanent.

What is apparent from this case study is that there remains much to be learned about the pathogenesis of polio infection and that unravelling the pathogenesis of an infectious disease can be very difficult if there are no suitable animal models.

Prevention There is no treatment for paralytic polio once it has developed. Polio can, however, be prevented very effectively by vaccination. Two kinds of vaccine are used: the live attenuated vaccine (Sabin vaccine, noted above), which is now used less commonly, and an inactivated, killed vaccine (Salk vaccine). Immunization with both vaccines leads to the synthesis of antibodies that prevent the virus binding to epithelial cells. The advantages of the Sabin vaccine is that it can be given orally (often on a sugar lump), and that it is very similar to the natural infection, leading to long-lived intestinal anti-polio IgA secretion. The disadvantage is that, being a live virus, it is subject to mutation that can lead to recovery of virulence and immunized recipients may excrete virulent virus in their faeces as in the case study above. This also happened fairly recently in Haiti and the Dominican Republic, and led to a small outbreak of paralytic polio. The Salk vaccine does not carry this risk. However, the disadvantage of the Salk vaccine is that it has to be given by repeated injections and that the antibody response is mainly IgG rather than IgA. Both vaccines do, however, very effectively prevent infection. Vaccination against polio is on of the most successful immunizations and world-wide eradication of polio is an attainable goal. One of the main reasons for this is that there are no natural non-human hosts that could act as vectors for the polio virus. See Figure 2.35.

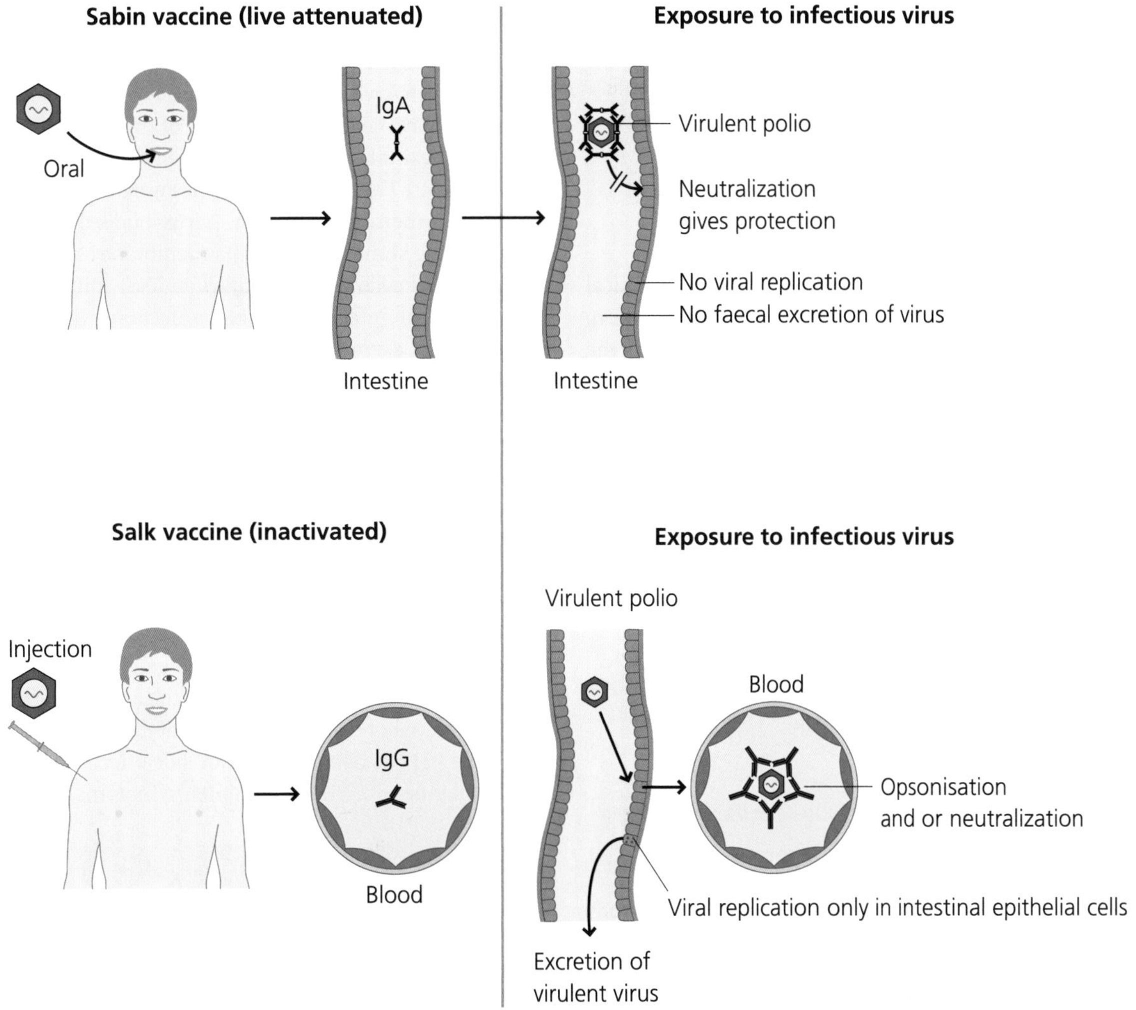

Fig. 2.35 Immunity to polio virus infection. Polio vaccines consist of either attenuated (weakened) live virus (Sabin) or inactivated virus (Salk). The live, Sabin vaccine is given orally (often on a sugar lump). The virus infects intestinal epithelial cells, mimicking the natural infection, and induces a local IgA response. This prevents virulent virus from attaching to the epithelial cells and replicating, and virulent virus is not excreted. There is, however, a small risk that the vaccine virus will revert to virulence. The Salk vaccine is given by injection and induces an antibody response that is mainly IgG. This prevents the spread of the virus beyond the intestine by opsonization and neutralization, but does not prevent infection of intestinal epithelial cells and the subsequent excretion of live, virulent virus. In practice both are very effective vaccines at both individual and population levels.

Case Study 2.6: Influenza

Clinical: One cold winter in the Northern Hemisphere you are in contact with an individual suffering from influenza. For a couple of days you remain symptom free. You then start to feel unwell, and develop a cough and a sore throat. You feel cold and shiver violently. Your muscles ache. You take your temperature and find you have a marked fever. You lose your appetite and feel sleepy. Your symptoms increase in severity for a few days, but you begin to recover and in time are back to normal. Despite the flu outbreak still raging, you do not get infected again and when the same strain of influenza appears next year you do not get symptomatic infection. The year after, however, in another flu epidemic, you are once again infected and get similar symptoms. How can we explain these phenomena?

Explanation: You were infected with a virulent influenza virus, which although only infecting respiratory epithelial cells, caused systemic symptoms. You generated a cell-mediated immune response that cleared the viral infection and which resulted in memory for the particular strain of virus that had infected you and prevented you becoming re-infected with the same strain of virus the following year. However, because of the specificity of adaptive immunity, this immunological memory could not protect you when a new strain appeared.

Q2.19. Polio virus infects humans by the oral route. However, most humans infected with polio virus do not develop paralysis. What might determine whether an infected individual does develop paralysis?

2.4.3.2 Viruses that Cause Acute Inflammation

Many viruses, that infect locally or systemically, cause clinical disease in which the main symptoms are of inflammation. This applies to most respiratory tract infections (e.g. the common cold and influenza) and to systemic infections (e.g. measles and smallpox). The pathology of the inflammation is not, however, the typical acute inflammation seen with pyogenic bacteria. It is characterized by accumulation of lymphocytes and macrophages. This signifies the involvement of a T cell-mediated adaptive immune response, and the inflammation is actually a by-product of the immune response that is responsible for eliminating the virus.

Q2.20. Why are colds and influenza more common in the winter months?

Q2.21. What might be the consequences of the early stages of influenza being asymptomatic?

Pathogenesis Infection and viral replication. In case study 2.6, the flu-infected individual you were in contact with transmitted the virus to you. Typically this could occur through droplets when they sneezed near you or perhaps by direct contact (e.g. shaking your hand after they blew their nose). The virus adhered to epithelial cells in the upper respiratory tract, adhesion being mediated by viral HA that binds to sialic acid on cell surface molecules. The virus was endocytosed and the low pH in the endocytic vesicles caused a conformational change in the HA, enabling it to fuse with proteins in the membrane of the vesicle. The viral contents could now enter the cytoplasm of the cell and the viral RNA (in eight separate segments) was transported to the nucleus. The viral RNA was copied as mRNA which codes for a variety of proteins. Some of these are concerned with replication of the RNA, others with direct synthesis of proteins that will be incorporated into new virus particles. The new virus particles were assembled under the plasma membrane and the virus budded off the cell, incorporating some of the plasma membrane as its envelope. The epithelial cells were not killed (at least initially) and the early stages of infection – when infectious virus is being released – are asymptomatic. See Figure 2.36.

Symptomatology. All is not quiet, however. The infected cells release cytokines such as Type I IFNs, which alter the expression of adhesion molecules on local endothelial cells, permitting the recruitment of monocytes and the accumulation of inflammatory macrophages. These cells are stimulated to secrete more cytokines (e.g. IL-1 and TNF-α) and other molecules such as prostaglandins. Locally these molecules initiate an inflammatory response that, via neuronal stimulation, causes the sore throat and cough. Some molecules also enter the bloodstream, and cause the generalized (systemic) symptoms such as fever, loss of appetite and headache. (It is relevant that when cancer-bearing patients were first treated with Type I IFNs, one of the major side-effects reported was the development of systemic flu-like symptoms.)

Q2.22. Why do you feel cold and shiver while you are developing a fever?

Q2.23. What might be the evolutionary function of fever in defence against infection?

Recovery from primary infection. Local secretion of IFNs may serve to slow down the infection, but recovery depends on an adaptive immune response. Influenza virus antigens (or maybe infected cells that have become apoptotic) are endocytosed by DCs and transported to local lymph nodes, where they activate both CD4 T cells and CD8 T cells. These activated

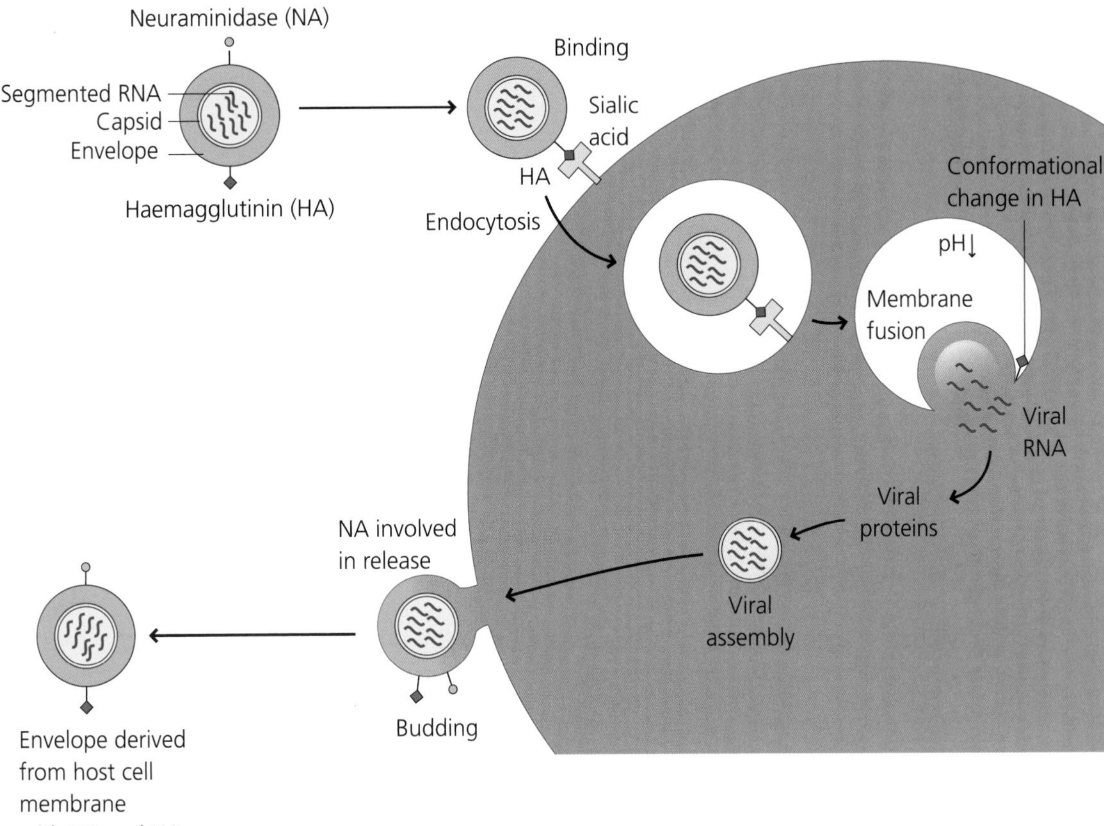

Fig. 2.36 Life cycle of the influenza virus. HA on the viral envelope binds to sialic acid on the host cell. The virus is internalized by endocytosis and the low endosomal pH (increased acidity) alters the HA so that it can mediate fusion of the envelope with the endosomal membrane. The viral RNA (eight discrete segments) is released, and induces synthesis of early proteins, involved in RNA replication, and late proteins that are assembled into new viral particles. The virus buds from the plasma membrane of the infected cell, incorporating some of the membrane as its envelope. The viral NA is involved in releasing the virus from the cell. Different forms of HA and NA give different strains of influenza viruses their characteristic names (H1N1, H5N1, etc.).

T cells migrate back to the site of infection because they can recognize adhesion molecules expressed on the inflamed endothelium. We know from adoptive transfer studies in mice that CD8 T cells are essential for recovery from influenza. How do they mediate recovery?

Activated CD8 T cells can develop into CTLs that recognize viral peptides expressed on MHC class I molecules of infected cells and induce apoptosis in these cells (e.g. Fig. 2.34). When influenza (and other viruses) infect a cell, they disassemble their structure and there is a period, known as the eclipse phase, during which no infectious virus is present in the infected cell. If the infected cell can be killed during the eclipse phase, infectious virus cannot be assembled and the spread of infection is stopped. In other words, the cell is sacrificed to prevent infection of other cells. Thus, the patient recovers, and the local and systemic symptoms disappear. Activated CD8 T cells also secrete IFN-γ, which can increase expression of MHC class I molecules on cells, further increasing their susceptibility to killing. See Figure 2.37.

Resistance to re-infection. Despite being in repeated contact with infected individuals, you do not become infected again in the same flu season. This resistance is primarily due to your having made antibodies earlier. During the initial infection, activated CD4 T cells provided help for B cell activation which then developed into plasma cells which secreted antibodies specific for molecules expressed on the viral surface, mainly HA and neuraminidase (NA). You are resistant to re-infection because these antibodies prevent the virus binding to epithelial cells (IgA may be particularly important in preventing binding). We know this because of an unusual but crucially important property of flu viruses. Different strains of virus express different forms of HA and NA, giving rise to the terminology for the infamous H5N1 (bird flu) and H1N1 (swine flu) strains. If someone is infected with, for example, a H1N1 strain, and then comes in contact with a different strain that expresses either H1 or N1, they are protected. If, however, they meet, say, H2N2 they become infected. See Figure 2.38.

Epidemics and pandemics. Epidemics are outbreaks of an infectious disease that infect many individuals in a limited area (e.g. country or region). Pandemics are world-wide epidemics.

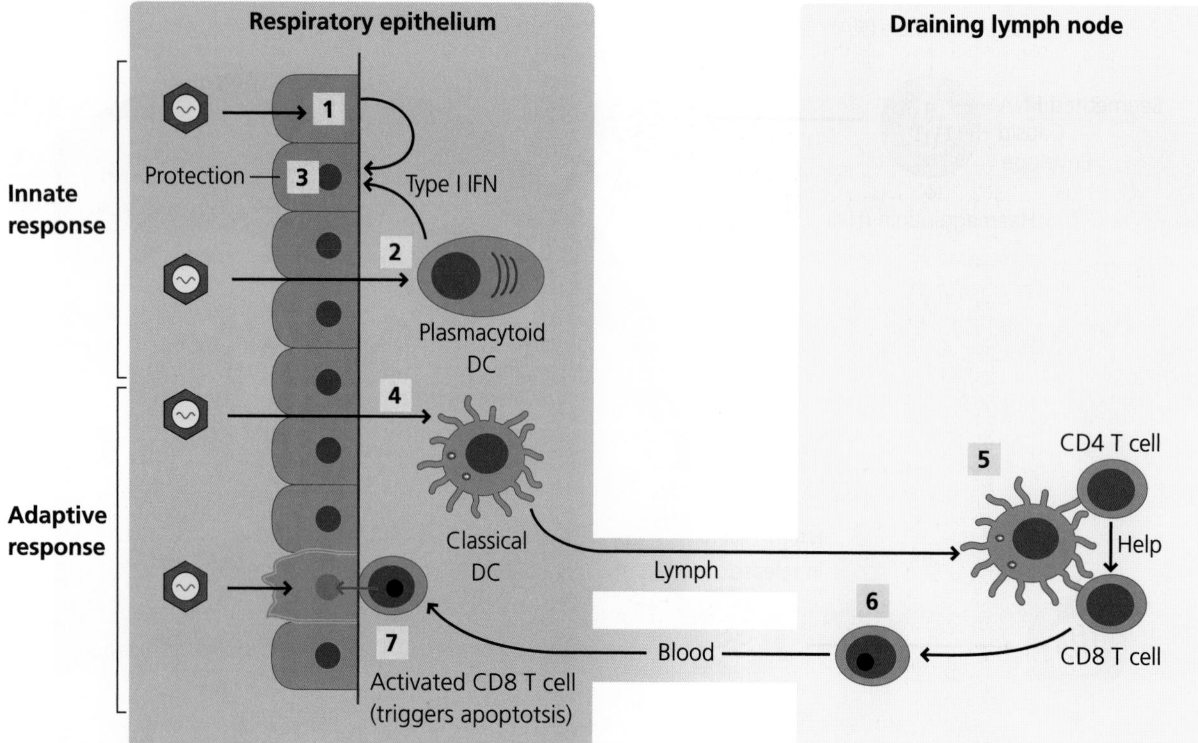

Fig. 2.37 Immunity to influenza I: recovery. (1) Influenza virus infects and replicates in respiratory epithelial cells and is released by budding. (2) Viral PAMPs and or cytokines released by the epithelial cells stimulate pDCs, which make large amounts of Type I IFNs. (3) Type I IFNs, which can also be produced by the infected epithelial cells, induce resistance to viral infection in other cells. (4) PAMPs also activate classical DCs which acquire viral antigens either from viral debris or by taking up apoptotic infected cells. (5) These DCs migrate to lymph nodes where they activate CD4 and CD8 T cells. (6) The activated CD8 T cells migrate to the inflamed epithelium (7) and are able to kill infected cells, hopefully before new virus is assembled. They may also secrete IFN-γ, which is weakly anti-viral, but also activates macrophages which may then secrete Type I IFNs.

- Influenza epidemics: Most viruses are subject to mutation (HIV massively, polio very little). Influenza undergoes frequent point mutations in its HA and NA genes. This is known as antigenic drift; the changes are small but some of these mutations make pre-existing antibodies ineffective. Individuals immune to the original virus are susceptible to the mutant strain. These mutant viruses are the cause of the epidemics which occur every few years.
- Influenza pandemics: The flu virus genome is unusual in that it consists of eight separate segments of RNA. It is possible for a single cell to be infected with two different flu viruses at the same time. Let us hypothesize that a pig in China is infected with a mammalian flu virus and a bird flu virus at the same time. A new virus may be assembled in which most of the gene segments are from the mammalian virus but the HA segment comes from the bird virus. This is known as antigenic shift. No humans will have been exposed to this virus and no-one will have anti-HA antibodies. There will be nothing to limit the spread of the virus in the human population, this is how pandemics occur. This is bad enough, but, for healthy adults, flu is not a life-threatening problem, most deaths occur in the very young and the very old. The 1918, Spanish flu and the H5N1 bird flu are, however, exceptions; the mortality in humans infected with H5N1 is around 50% but it is healthy adults who are most at risk. One hypothesis to explain the very high mortality of the H5N1 is that this virus is unusually effective at stimulating the innate immune system; the result is release of vast amounts of cytokines and it may be these that cause the mortality.

Q2.24. If it were proven that a massive production of cytokines in response to H5N1 infection causes mortality, what new therapeutic approaches might be developed?

2.4.3.3 Viruses that Cause Chronic Inflammation

In some cases of viral infection, the inflammation that results is directly due to damage to the infected cells which, in turn, release molecules that activate the inflammatory response. In others it is more complex, with the inflammation reflecting an immune response against virus-infected cells, for instance in viral hepatitis. In hepatitis B, viral infection of liver cells stimulates a strong CD8 T cell-mediated (CTL) response. The T cells kill infected cells and abort the infection; however, if many liver cells are infected, the degree of killing may be so great that the liver cannot function and liver failure ensues. If the virus is not successfully eliminated, the immune

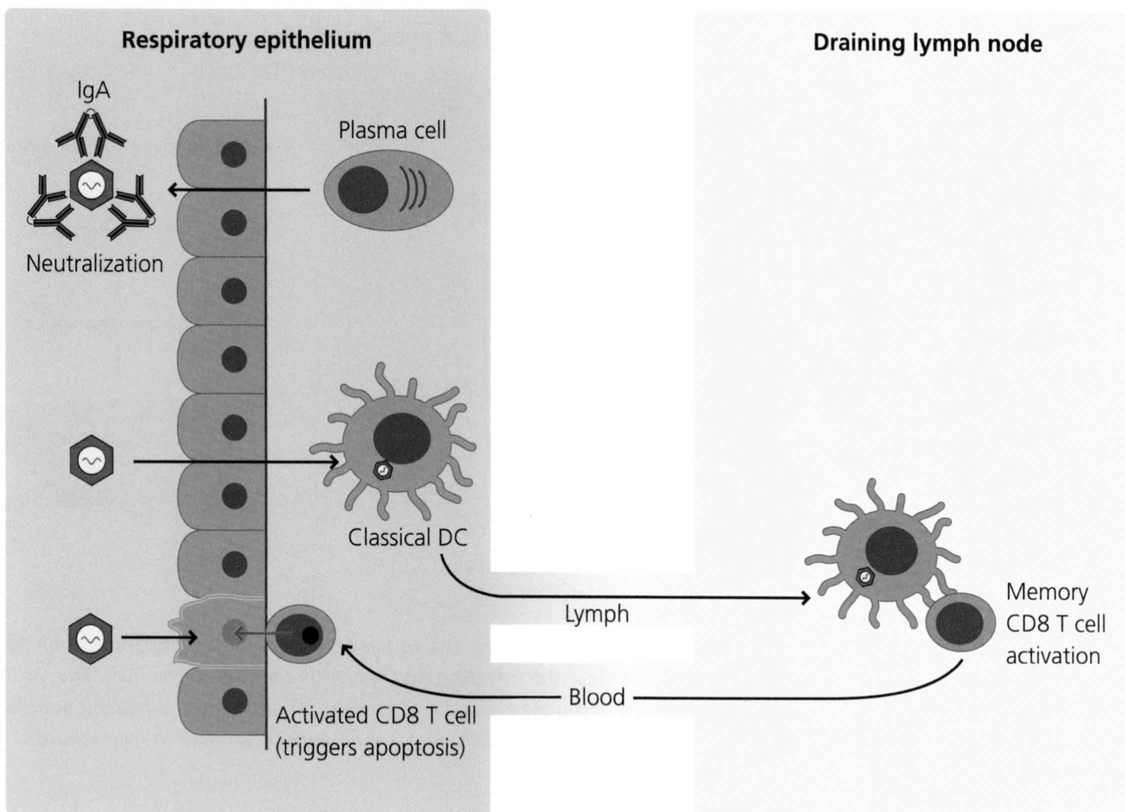

Fig. 2.38 Immunity to influenza II: resistance to re-infection. Antibodies, particularly of the IgA class and specific for HA, prevent the virus binding to epithelial cells. Additionally, but later, DCs may acquire viral antigens, travel to lymph nodes and activate CD8 memory T cells. The latter may become activated effector cells that may kill epithelial cells that are infected by any viruses that escape neutralization, thus preventing further release of live virus.

response may continue at a low level for months or years. Every inflammatory response is accompanied by a healing response involving the laying down of collagen by fibroblasts and over time this fibrosis can impair organ function. Thus, in hepatitis C, patients may be infected for years without knowing it and it is only when the degree of fibrosis is so severe that it inhibits liver function that the infection becomes clinically apparent. This is known as cirrhosis, and hepatitis C is one of the major causes of this condition. See Case Study 2.7.

Pathogenesis A remarkable feature of infection with hepatitis C is that in over 95% of cases the virus is not cleared, but persists as a chronic infection. This chronic infection leads to an ongoing immune response, which inevitably leads to chronic inflammation of the liver and collateral

Case Study 2.7: Hepatitis C

Clinical: A 67-year-old man visits his doctor because he has noticed that the white part of his eye has developed a yellow tint. His alcohol consumption is very low, he is not an intravenous drug use and he has none of the risk factors for hepatitis B infection. Liver function tests suggest moderate to severe liver damage. He had surgery for carcinoma of the colon 17 years ago, but there is no evidence of tumour recurrence or metastasis. Ultrasound examination of his liver shows marked nodularity and a liver biopsy shows advanced cirrhosis. A blood test shows the presence of anti-hepatitis C antibodies and hepatitis C RNA is detected in his blood. He receives a liver transplant and 5 years later remains well.

Explanation: His surgery for the carcinoma involved blood transfusion and this happened before routine screening for hepatitis C was instituted. The virus in the transfused blood infected his hepatocytes and induced an anti-viral immune response. This included the activation of CD8 T cells specific for the virus that killed some of his hepatocytes, but this was not sufficient to clear the infection. An ongoing immune response to the virus induced the cirrhosis.

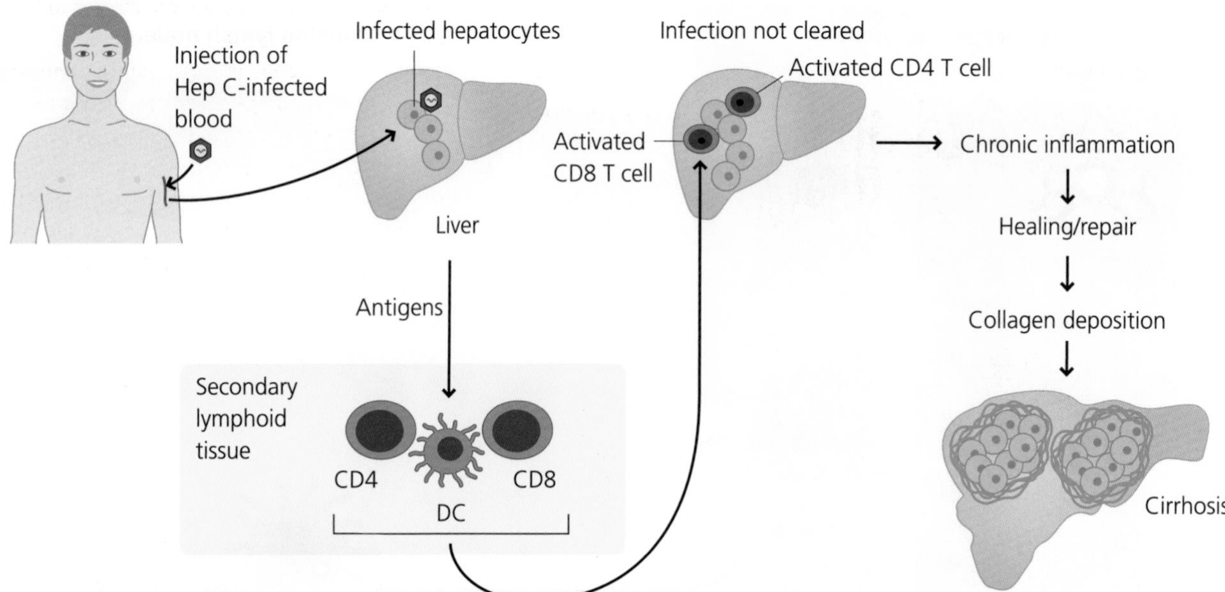

Fig. 2.39 Hepatitis C. Hepatitis C is usually transmitted by infected blood. The virus infects hepatocytes and induces an adaptive immune response. In many cases this response does not clear the infection and the virus persists in the liver. The virus continues to stimulate the immune response, which itself stimulates, as part of chronic inflammation, a healing and repair response. This leads to excessive deposition of collagen (cirrhosis), which over a long period can lead to liver failure.

damage. In some cases this does not appear to cause clinical disease, but in others disease may manifest itself after many years. In some patients carcinoma of the liver develops, while in others, as in Case Study 2.7, liver failure develops as a result of cirrhosis. In cirrhosis there is excessive laying down of connective tissue in the liver. This happens because a healing/repair response is an integral part of a chronic inflammatory process. In many sites this would not cause disease, but in liver the collagen is laid down around small groups of regenerating hepatocytes throughout the liver, essentially isolating them and preventing them functioning normally, leading to liver failure. The cirrhotic process may be slowed down by drugs, but once liver failure has occurred, the only treatment available is a transplant. See Figure 2.39.

2.4.3.4 Viruses that Cause Tumours

In animals, many viruses have been shown experimentally to cause tumours. In humans there is a very close association between many tumours and viruses such as Epstein–Barr virus in Burkitt's lymphoma, a tumour of B lymphocytes and nasopharyngeal carcinoma; hepatitis B virus and liver cancer; and a human herpes virus in Kaposi's sarcoma, a complication of AIDS. In only one case has a virus been shown directly to cause a human tumour (i.e. fulfilling Koch's postulates): injecting papilloma virus into volunteers' skin induced papilloma development at the site of injection (skin papillomas are benign tumours, warts). See Figure 2.40.

Carcinoma of the uterine cervix is a major killer of women. It has been known for a long time that there is a strong correlation between the number of sexual partners a women has and the risk of developing cervical cancer. It was then found that in most cases of cervical cancer, there was evidence for previous infection with human papilloma virus (HPV but not the same strain that causes warts). It was not possible to show that this is more than a strong association, but the association was so strong that it was thought justifiable to develop vaccines against the virus. These vaccines are being used widely to immunize girls before they become sexually active, and trials have shown that the vaccines both prevent infection with the virus and the development of the early stages of cancer. See Figure 2.41.

Q2.25. Does the finding that vaccination against papilloma virus prevents the development of cervical carcinoma prove a causal relationship between the two?

It has been known for a long time that patients who have received a kidney transplant have a high incidence of malignant tumours – around a 100-fold increase. All these patients are receiving immunosuppressive treatment and it has been suggested that the increase in tumour incidence is due a lack of immune surveillance, permitting newly formed mutant cells to develop into tumours (Chapter 7). If this were the case we might predict that the frequencies with which different tumours appeared in transplant recipients would reflect their frequencies in normal individuals. In fact most of the tumours seen arise from lymphoid cells, but skin cancers are an exception; the risk of developing skin cancer in some studies is increased 250-fold. However, in such cases it is often possible to isolate viral nucleic acid from the tumour, particularly from herpes and papilloma viruses. Similarly, patients infected with HIV (see below) have a much increased risk of

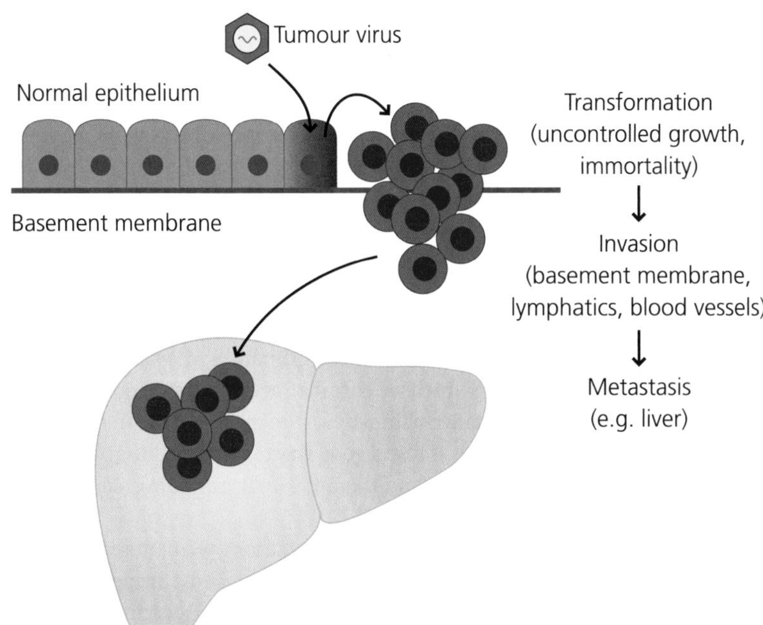

Fig. 2.40 Virally induced tumours. If a tumour causing virus (an oncogenic virus) infects a cell it may insert its own genes or induce abnormal activation or suppression of host genes that regulate growth of the cells (oncogenes and tumour suppressor genes respectively). This results in a clone of cells becoming able to divide indefinitely (immortality) and independently of external growth factors, and lack control of proliferation through cell–cell contacts. A tumour starting to develop may also induce angiogenesis, helping to support its growth. Tumour cells may invade into normal tissues, and escape into blood or lymph to seed more distant sites, leading to metastasis.

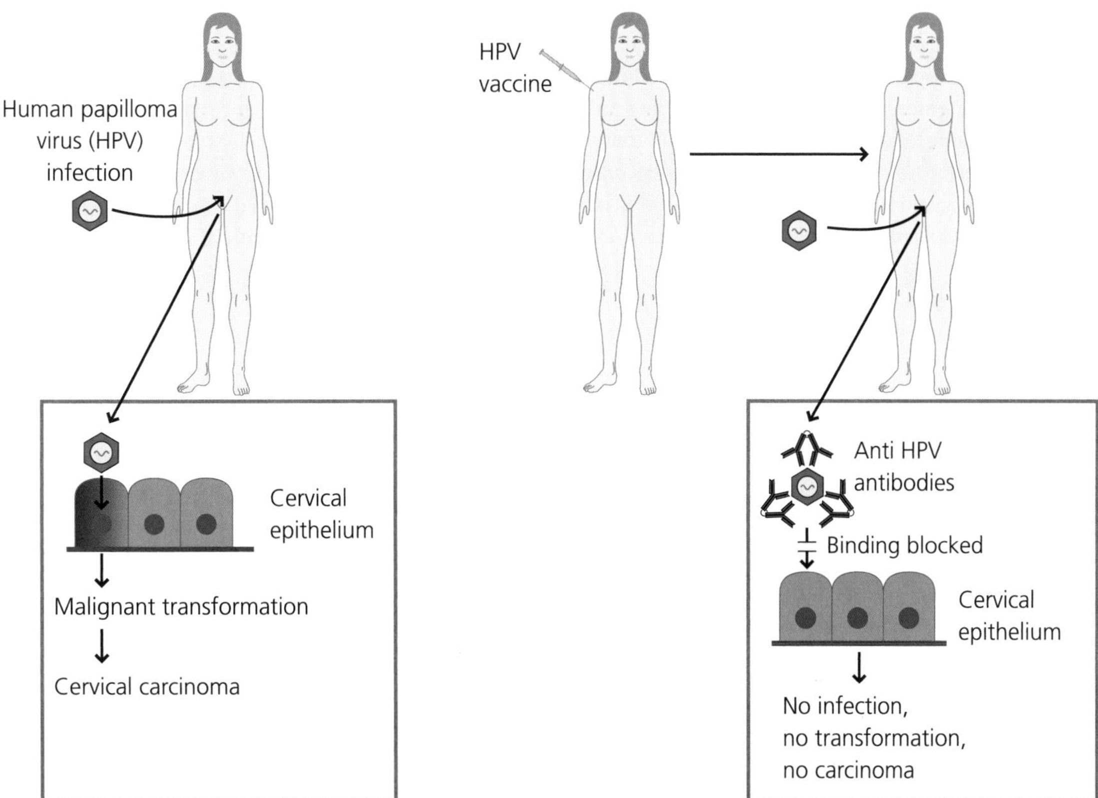

Fig. 2.41 Prevention of cervical carcinoma by vaccination. Carcinoma of the uterine cervix is very strongly associated with genital infection with sexually transmitted strains of HPV. Young women are immunized (prophylactically, before they have become infected) with an anti HPV virus vaccine which induces an immune response, mainly antibody, that prevents the virus infecting cervical epithelial cells. This is proving very effective in preventing subsequent tumour development. It must be emphasized that the immunization prevents infection: it is not an immunization against the tumour itself. Since the virus is typically transmitted through sexual intercourse there may be some value in vaccinating young men as well.

developing Kaposi's sarcoma and HPV nucleic acid can also be isolated from these tumours. These observations suggest that the increased incidence of tumours seen in immunosuppressed patients could represent a failure to clear viral infections rather than a primary failure of immunosurveillance against newly formed tumour cells, although this remains somewhat controversial.

2.4.3.5 Viruses that Cause Immunosuppression

As part of their evasion strategies, many viruses can induce more or less severe immunosuppression. This may lead not just to increased survival of the infecting virus, but also to infection with other viruses. See Case Study 2.8.

Pathogenesis HIV-1, the causative virus of AIDS is a retrovirus. It infects human cells that express CD4 and a particular chemokine receptor, CCR5 or CXCR4. Different viral strains generally use one or the other of the latter to infect, and are known as R5 and X4 strains, respectively. Human macrophages, DCs and activated CD4 T cells express CD4 and CCR5, while many cell types, including CD4 T cells, express CXCR4. The importance of CCR5 for HIV infection is shown by individuals with a mutation in the gene for this receptor: if they are homozygous for this mutation they cannot express CCR5 and are therefore resistant to infection by R5 strains of HIV. After binding to the cell, the contents of the virus, including the viral single-stranded RNA, enter the cell. The enzyme reverse transcriptase copies the RNA into double-stranded DNA which enters the nucleus and becomes integrated into the genome of the cell. The DNA can remain dormant for long periods of time but it eventually becomes activated and replicates, and RNA is synthesized, exported into the cytoplasm, and packaged into new virions. These are then released from the cell by budding and can infect other host cells, or can be transmitted to other individuals. See Figure 2.42.

Evasion Mechanisms Why can the immune system not eliminate the HIV virus? As with many viruses, HIV has evolved several different mechanisms for evading the immune response. The infected individual does in fact make vigorous CD4 and CD8 T cell-mediated and cytotoxic responses against the virus, as well as antibody responses, and these persist for very long times but are ultimately ineffective. It is crucial to remember that the presence and nature of evasion mechanisms, some of which we consider below, is very good evidence for the importance of those evasion mechanisms in resistance to the virus. This is also a good way to start thinking about how to design vaccines and therapies that might overcome these evasion mechanisms.

Elimination of CD4 T cells. In the long term, the destruction of CD4 T cells can be viewed as the most important HIV evasion strategy, since this prevents the induction of effective CTL and antibody responses. CD4 T cell elimination is also clinically crucial: it is this that permits the onset of opportunistic infections and it is these infections, not HIV itself, that kill the patient. Several distinct mechanisms contribute to the loss of CD4 T cells; the half-life of these, when newly infected, is only 1–2 days. Infected CD4 T cells increase the expression of molecules such as Fas (a death receptor) and Fas ligand, as well as TRAIL (TNF-related apoptosis-inducing ligand; another death receptor). Expression of anti-apoptotic molecules such as Bcl-2 is decreased so they become more susceptible to apoptosis. Infected CD4 T cells may also become targets for HIV-specific cytotoxic T cells. It is important to realize that the great bulk of CD4 T cells that die are not infected by HIV. The mechanisms causing the death of uninfected cells are poorly

Case Study 2.8: HIV

Clinical: A 35-year-old man has unprotected vaginal sex with a prostitute. About 2 weeks later he feels unwell, and has a mild fever and joint pains. He goes to his doctor but does not mention the unprotected sex. His doctor suggests he takes paracetamol and the man recovers over the next few days. He remains well for around 7 years, but then develops a fever and shortness of breath. His wife notices that he has a purple swelling on his back. He then goes back to his doctor who finds that he has signs of a mild pneumonia and generalized swelling of his lymph nodes. The purple swelling has the appearance of a vascular tumour. Blood tests reveal that the numbers of circulating CD4 T cell are very low and that he has antibodies to HIV. His sputum is positive for *Pneumocystis jirovecii*. He is persuaded to tell his wife and although she is completely asymptomatic, she is found to have anti-HIV antibodies. They are both put on HAART, (highly active anti-retroviral therapy in which three or four anti-HIV drugs are given together to reduce the chances of the virus becoming resistant; to do so it would require three or four simultaneous mutations. Three years later the wife remains HIV-infected but is well, while the husband develops metastases from his tumour and dies.

Explanation: The man acquired HIV during his one episode of unprotected sex (this is unusual since the chances of infection in these circumstances are less than 1: 100). The virus infected his macrophages and CD4 T cells and proliferated in these cells. This infection caused his fever. Had his blood been tested at this time is would have shown a very high level of free HIV. In time his CD4 T cell numbers started to decline and he became infected with *Pneumocystis carinii*, a yeast-like fungus, which is harmless to normal individuals. However, in his case he could not activate the adaptive immune mechanisms needed to control the organism and thus developed pneumonia. The purple swelling on his back is Kaposi's sarcoma. This tumour is very strongly associated with a herpes virus (HHV-8). Sarcomas are malignant tumours of connective tissue cells.

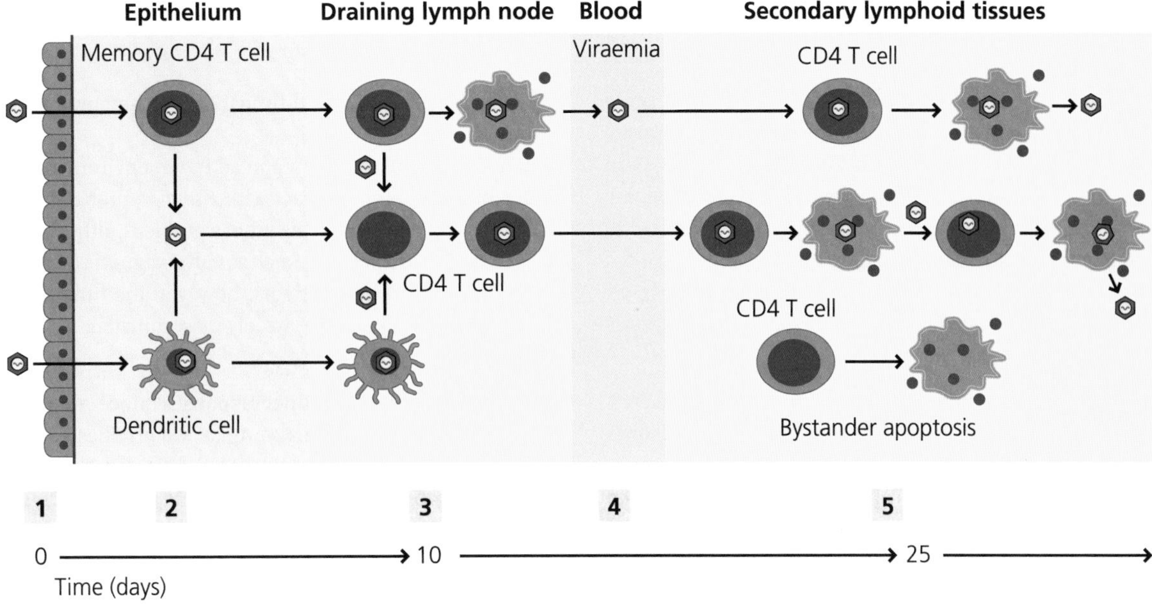

Fig. 2.42 Viral events in human immunodeficiency virus infection. (1) Viral infection usually occurs via mucosal surfaces. (2) Often a single viral particle crosses the epithelium and infects submucosal memory (CCR5) CD4 T cells and DCs. (3) There is an eclipse phase of 5–10 days after infection when virus cannot be detected in the blood. During this phase virus has entered the draining lymph nodes as either free particles or transported by CD4 T cells or DCs. The virus replicates in the node, killing many CD4 T cells, forming viral reservoirs and destroying the architecture of the node. (4) The virus now disseminates widely via the blood (a viraemia can be detected) to secondary lymphoid organs. (5) Here, it continues to replicate, destroying many CD4 T cells directly and inducing bystander apoptosis in many other non-infected CD4 T cells. Eventually CD4 T cell numbers decline to a level where opportunistic infections can occur.

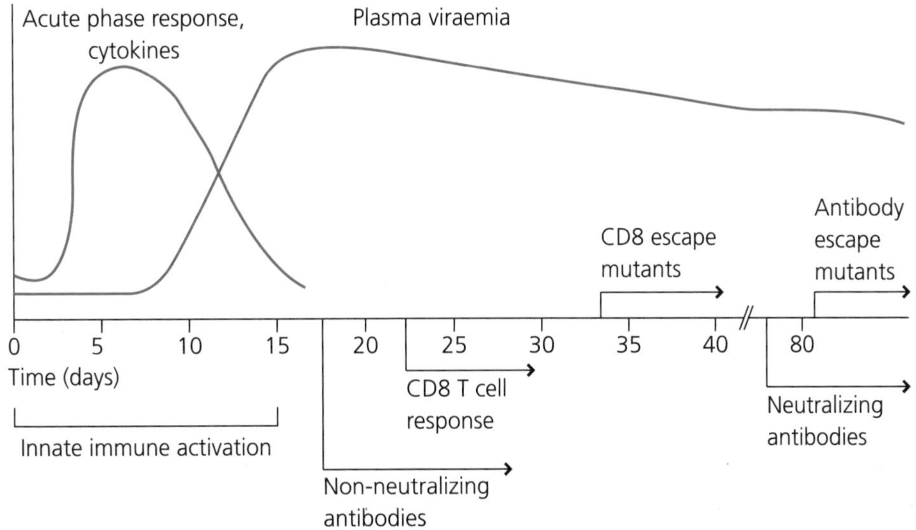

Fig. 2.43 Immunological events in human immunodeficiency virus infection. Early after infection, innate immune activation leads to the appearance of acute phase reactants and pro-inflammatory cytokines in the blood. The first adaptive immune response leads to the appearance of non-neutralizing antibodies which do not select for escape mutant virus. Soon after, a CD8 T cell response starts, which does select for escape mutants. Repeated cycles of CD8 activation occur in response to appearance of many different escape mutants. It is only after around 70–80 days that neutralizing antibodies appear, again selecting for escape mutants. The processes of immune selection of escape mutants continues for years, until CD4 T cell numbers have declined sufficiently to permit opportunistic infection to occur, with the onset of AIDS. At this point the viraemia increases substantially (not shown).

Case Study 2.9: Fungal Infection after Bone Marrow Transplantation

Clinical: A 6-year-old child was diagnosed with acute leukaemia. He was given chemotherapy to kill the leukaemic cells and, because this also kills bone marrow stem cells, he was given a bone marrow transplant from his sister. He was put on immunosuppressive therapy to prevent the transplanted cells being rejected. Six weeks after the transplant he developed a fever. Cultures of his blood showed the presence of the fungus *Candida*. Despite active anti-fungal chemotherapy, he died 2 weeks later.

Explanation: *Candida* is widespread in the environment so we are probably all in contact with *Candida* probably on a daily basis. For most of us this is not a problem, although for some it can cause infections such as vulvo-vaginitis (thrush). In immunosuppressed patients, however, the fungus can enter the body and become widely disseminated.

understood but include killing by Fas ligand-expressing CD4 T cells that are themselves infected. If an infected CD4 T cell does survive it can live for long periods, but once it is activated, HIV replication is initiated and the T cell dies rapidly. If the activated T cell is specific for an infecting microbe, resistance to that organism will disappear. CD4 T cell elimination is a long-term (years) strategy for viral evasion of immunity and HIV survives effectively even when there are large numbers of CD4 T cells available. Shorter-term strategies used by HIV include, amongst many others, those below.

Antigenic variation. Copying RNA to DNA by reverse transcriptase is very error prone. This leads to the introduction of numerous mutations in the virus that appear on a daily or weekly time scale. From the beginning there is a vigorous CD8 T cell response to the virus which is effective in killing virus-infected cells. If, however, the virus mutates so that the peptide target for CD8 T cells is no longer recognized, that variant will have a selective advantage and will come to predominate. This will of course induce a new population of CD8 T cells able to recognize the mutant peptide, but fresh mutations will render these T cells ineffective. These variant HIV virions are known as escape mutants. It is possible to identify successive waves of CD8 cytotoxic T cells specific for these mutants during the course of an infection.

Antigenic variation may also enable HIV to evade antibody-dependent destruction. Because antibodies are also generated in responses against HIV, they will select for HIV variants that escape neutralization. This can be shown by growing HIV *in vitro* in the presence of a neutralizing antibody, variants emerge that are resistant to the antibody. Additionally, antibodies to HIV may be made against epitopes that are not accessible in the infectious virus, reflecting conformational differences in viral surface proteins at different stage of the life cycle.

Other mechanisms. HIV uses many other strategies to evade immunity. For example, the viral protein Nef causes decreased expression of MHC class I molecules, but not of non-classical MHC molecules (HLA-C and -E) that inhibit NK cell function. Another viral protein, Vif, inactivates a generic mechanism of defence against retroviruses by inactivating a key RNA-editing enzyme (APOBEC-3G) that otherwise introduces lethal mutations in the viral genome. Other viral mechanisms interfere with complement activity and also recruit complement inhibitory proteins to the viral envelope, secondary antibody responses are impaired because of destruction of B cell follicles, and so forth. HIV is perhaps not unusual in the number of different mechanisms that it possesses to evade immune responses since many other viruses possess a similar array. Again we emphasize that the maintenance of genes important in different mechanisms underlying immune evasion highlights the importance of these mechanisms in defence. See Figure 2.43.

Q2.26. How might the discovery of such viral immune evasion mechanisms affect our thinking about vaccine design?

2.4.4 Pathogenesis of Yeast and Fungal Infections

Fungi are divided into those that grow as single cells (yeasts) and those that grow as multi-cellular organisms (moulds). Fungal infections are very common and usually are a nuisance rather than a cause of severe disease. Thus, the yeast *Candida albicans* commonly causes superficial infections (e.g. vulvo-vaginitis or thrush) and *Tinea fungi* cause skin infections (ring-worm). In immunosuppressed patients, however, fungal infections are a major cause of illness and death. In AIDS, many patients die of systemic infection with fungi; *Pneumocystis, Candida, Aspergillus* and *Cryptococcus* are important examples. Again, the increased risk of fungal infection in immunosuppressed patients is strong evidence that the adaptive immune response is continually active in normal individuals, so these infections are normally subclinical and do not cause disease. See Case Study 2.9.

2.4.5 Pathogenesis of Parasitic Infections

As noted in Section 2.2.2.4, parasites comprise a hugely diverse range of organisms, ranging from single-celled protozoa (e.g. malaria) to very large metazoan worms that inhabit the intestine. Some (rare) parasites give rise to short-lived infections, but the majority have very long or permanent relationships with their hosts. Many parasites have evolved to be able to exist in a balanced relationship with their hosts so that the immune responses generated by the host do not eliminate the parasite; this is also the case for many bacteria and viruses. Thus, humans have co-evolved with intestinal worms and in developing countries a large proportion of the population still carry such worms.

Case Study 2.10: Malaria

Clinical: A young child in the Gambia becomes unwell. She develops a high fever that peaks at irregular intervals of around 48–72 h. After about a week her condition improves, but then she develops another bout of fever episodes. This pattern repeats itself for several months. She is then disease-free for several months, but then becomes unwell again and the illness, again consisting of repeated bouts of fever, is repeated. This pattern goes on for several years, but the frequency of the bouts tends to decrease. By the time she is in early adulthood she only develops fever at very infrequent intervals. She is, however, involved in a road accident and her spleen is ruptured, necessitating its removal. When she returns into the community she finds that she is again stricken by repeated episodes of fever and that they do not diminish.

Explanation: The child was infected with malaria and made an adaptive immune response that permitted her eventually to recover from the first infection (below), but she did not become immune to further infections with different variants of the same parasite. Over time her immunity increased so that eventually she became immune to almost all variants. This immunity required her having an intact spleen.

Q2.27. Until very recently, essentially all humans co-existed from neonatal stages with intestinal parasites. These parasites generate T_h2-biased immune responses in their hosts. In the developed world we no longer co-exist with these parasites. How might this change have affected our overall immune responsiveness?

In this book we cannot explore the huge diversity of parasite interactions with the immune system. We will just use one clinical example that highlights the complexity of the interactions and the difficulty of generating effective vaccines against parasitic infections. See Case Study 2.10.

Pathogenesis In Case Study 2.10 a mosquito had acquired *Plasmodium falciparum*, the pathogen causing the most severe form of malaria, from an infected individual (there is no animal host for *Plasmodium falciparum*). The parasite reproduced in the mosquito and migrated into its salivary glands. The child was bitten by this mosquito and the parasites entered her bloodstream. The pathogen invaded hepatocytes, reproduced in these cells and was then released into the bloodstream where it invaded red blood cells (RBC). It divided asexually in the RBC, eventually causing them to burst, releasing large numbers of parasites into the blood and these parasites invaded new red cells. The release of molecules, largely unidentified, from infected red cells when they burst, stimulated the fever, probably acting by stimulating macrophages or monocytes to release cytokines such as TNF-α and IL-1.

Immune responses were initiated to antigens expressed by both the liver and red cell forms of the parasite. The responses to the liver form occurred too slowly to prevent the release of parasites into the blood. Antibodies can, however, be made against molecules expressed by the red cell form of the parasite. Why do these antibodies not prevent further infection of red cells? The answer reveals a very important strategy used by several parasites to evade immune responses – antigenic variation. The proteins expressed by the blood form are encoded by a large number of different genes in the parasite genome, but any individual parasite only expresses a small number of these genes. During asexual reproduction different members of the gene pool are selected for expression. This means that although a red cell may have been infected by a single parasite, many of the parasites released from the red cell are antigenically different from the infecting parasite, and are thus not susceptible to targeting by the antibody and can invade new red cells. Over time, however, antibodies are made to all the antigens that can be expressed by the parasites that originally invaded the boy and the infection dies away. If, however, he is bitten by a mosquito that carries parasites that contain genes coding for antigen variants not present in the original parasites, he will not be protected and a new round of infections starts. The number of antigenic variants that the strain of parasite carries is not, however, unlimited, and after time, if she survives, she will have made antibodies to all the variants that exist in that strain, and is thus fully immune.

It is a clinical observation in humans, and an experimental observation in animals, that immunity to malaria depends on having an intact spleen. Why this should be is not known. You can see that all of this makes designing vaccines to prevent malaria very difficult indeed. A straightforward approach to preventing transmission is to prevent mosquito bites in the first place, mosquito nets are highly effective. But what potential strategies for vaccination might we consider, at least in principle? These include the following (Figure 2.44):

- A vaccine made against the mosquito stage of the parasite might induce antibodies that would be ingested by the mosquito when it bites a vaccinated individual and which could therefore prevent multiplication of the parasite in the mosquito.
- A vaccine that induced antibodies to the molecule(s) that parasites use to bind to hepatocytes could prevent invasion of liver cells.
- A vaccine that stimulated the production of CD8 T cells that recognized malarial peptides (expressed on MHC class I molecules) might bring about the killing of the infected liver cells before infectious parasites were produced.

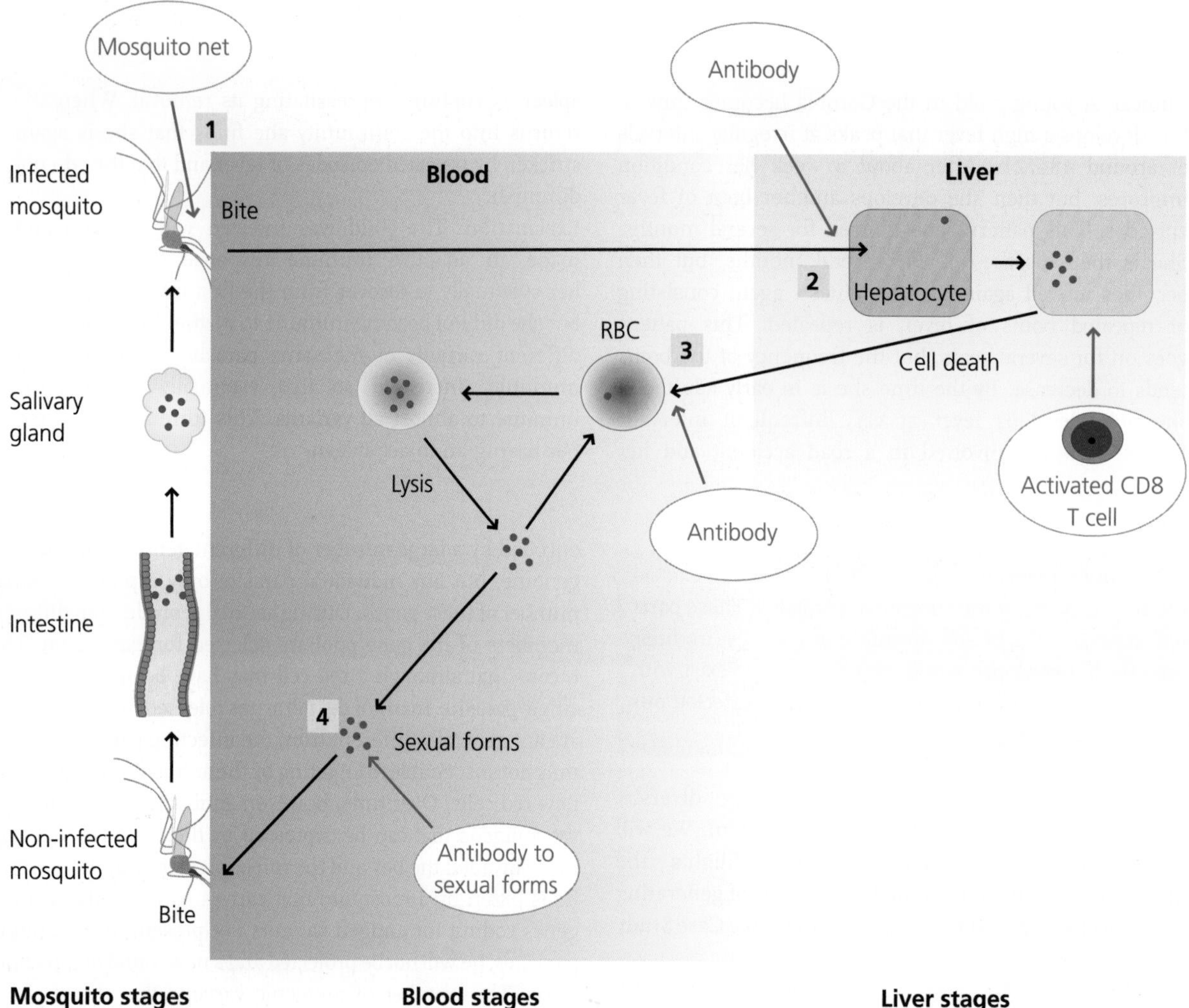

Fig. 2.44 Malaria parasite life cycle. The *Plasmodium* protozoan is acquired by an Anopheline mosquito from an infected person. It reproduces in the mosquito intestine and migrates to the salivary glands. (1) When a new subject is bitten, the parasite enters the blood and infects and reproduces in hepatocytes. (2) It is released from the hepatocytes and (3) infects RBCs, where it again reproduces and is released to infect other RBCs when the infected RBCs lyse. This cycle may recur many times and, should a mosquito bite the infected individual, the parasite starts the cycle over again. (4) The parasite changes form and divides asexually or sexually at different stages of its life cycle. Some points at which potential preventative measures might be introduced, to control transmission or induce immunity by vaccination, are indicated (green text).

- Finally (although there may be still other options to consider), a vaccine against an invariant region on the molecule used by the parasite to bind to erythrocytes might prevent infection of these cells and inhibit an important stage of the viral life cycle.

Q2.28. Can you think of any examples when it might be to a pathogen's benefit to kill its host?

2.5 Immunity and Vaccines

In this section we discuss different ways in which it is possible to provide immunity against infectious diseases, focussing particularly on vaccination. To start with, it is important to appreciate that immunity against infectious agents, may be *passive*, usually through transfer of antibodies, or *active*, through exposure to an infectious agent or antigen(s) from it, to stimulate a protective immune response. It may also be *natural*, such as through maternal transfer of antibodies to the foetus and neonate, or generated after recovery from infection, or *artificial* such as following intentional immunization with a synthetic antigen(s). Vaccination refers to the induction of active immunity by the artificial administration of antigen(s). Below we focus on prophylactic vaccination, which is designed to induce a state of immunity in individuals before they have become infected (as opposed to therapeutic vaccination for individuals with disease). Let us briefly consider forms of passive immunity before focussing on active forms of immunity and vaccines.

2.5.1 Passive Forms of Immunity

Passive immunity can be transferred naturally from the mother to the foetus or the neonate. As one example, IgG can cross the human placenta. Thus, the foetus receives antibodies from the mother and this gives protection until the infant's own immune system can be activated. Another example are the high levels of IgA in colostrum, the early milk, which protect the infant against intestinal infections. Many studies show that breast-fed children have a much lower incidence of diarrhoeal disease than those given artificial milk from a bottle. This type of immunity is of course not long-lasting since it will disappear as the antibodies are catabolized. (Hence, many primary immunodeficiencies, leading to infection, do not become evident until several months after birth as the maternal antibodies are removed.)

Passive immunity can also be transferred artificially. For example, if a patient is at risk of infection – say they have a wound containing much dead tissue and they have not been vaccinated against tetanus – they can be injected with antibodies against the tetanus toxin which have been raised in horses. The intravenous administration of pooled antibodies from normal human donors (intravenous immunoglobulin, IVIG therapy) to treat patients with primary antibody deficiencies is another example.

2.5.2 Active Forms of Immunity

It is a firm clinical observation that individuals who have recovered from a natural infection are usually very resistant to re-infection and that this type of resistance is often stronger than that achieved by vaccination. This is the form of active immunity that occurs naturally in immunocompetent individuals and is the focus of much of this book (pathological settings are, however, considered in Chapter 7).

The artificial induction of active immunity is what we usually understand by vaccination. Since its development by Jenner in the eighteenth century this measure has, on its own, prevented many millions of cases of infectious disease and death. This form of immunization is active because it involves activation of the adaptive immune system. Let us first consider some of the desirable properties of a vaccine before considering the different ways of inducing active immunity by vaccination.

2.5.2.1 What Makes a Good Vaccine?

What then are the properties needed by a good vaccine? Some of these are scientific, while others reflect economic or social considerations, and include the following:

i) It must be effective in generating protective immunity. This of course is the while primary concern. Some vaccines are essentially 100% protective in normal individuals. Good examples are vaccinia virus for smallpox and tetanus toxoid for tetanus. Other vaccines, such as the killed bacteria that were used to immunize against typhoid, or living bacteria such as BCG used against tuberculosis, give only partial protection for reasons that are not entirely clear.
ii) It must be safe. Vaccines consisting of inactivated whole organisms, such as those used in influenza, or where only parts of organisms are used, such as tetanus toxoid or the hepatitis B vaccine, are generally safe. Vaccines consisting of attenuated (weakened) live organisms are generally safe in normal individuals, but can give rise to disastrous infections if immunocompromised individuals accidentally receive the vaccine. For example T cell-deficient individuals who have been vaccinated with vaccinia or BCG before they have been diagnosed are likely to develop widespread, potentially fatal infections. Additionally, there is always the possibility that an attenuated living organism may revert to virulence as we described earlier in relation to the Sabin polio vaccine. All vaccines have potential side effects. Many of these are very minor, typically local pain at the site of injection or short-lived fever and malaise. Indeed, given what we know about the need to activate the innate immune system in order to stimulate a strong adaptive response, you would not be surprised if all vaccines needed to initiate local damage or inflammation to be successful. Thus, many vaccines are not given alone, but in association with an adjuvant (Chapter 4).
iii) A good vaccine needs to be effective in public health terms. This means that it needs to be cost-effective (the developing world cannot afford expensive vaccines). It also needs to be deliverable. Many vaccines need to be kept cold or they lose their efficacy and this provision of a cold chain is problematical in many developing countries. The vaccine must also be acceptable; oral vaccines cause less suffering than those given by injection. Ideally the vaccine should also be one shot: there is a real problem in getting parents to bring their children back for boosters, particularly if, as is often the case in developing countries, parents have to travel long distances.

2.5.2.2 How Best to Vaccinate?

Infection with Virulent Organisms Prior to vaccination, variolation – infection of an individual with smallpox from a mild case – was effective in preventing later infection. The downside is that in some cases variolation resulted in full-blown smallpox, with its consequences of scarring and possible death. For reasons that are unclear, many viral infections are less severe in young children than in adults. Chickenpox is an example and it is not uncommon for parents to put their young children in contact with an infected child in order to transmit the infection to them (chickenpox parties). In this way, recovery from a relative mild childhood disease that can be much more serious in adults can give life-long protection against the disease-causing agent provided it does not change in form or mutate.

Infection with Attenuated Organisms The original attenuated organism was the cowpox virus used by Jenner (of course he did not know it was a virus). This member of the pox virus family shares some antigens with the smallpox virus, thus enabling cross-protection. (By the way, the currently used vaccinia virus is not cowpox; no-one knows where vaccinia originated.) In general, such low-virulence relatives of pathogens are rare. Pasteur, however, found that pathogenic bacteria and viruses can be altered to generate strains that retain immunogenicity, but which are much less virulent than the parental microbe. This is termed attenuation and it can be done in a variety of ways, e.g. by subculturing the microorganism repeatedly *in vitro*. When this was first carried out there was no understanding of the basis of the observed reduced virulence and there was always a worry that the microbe could revert to full virulence. More recently, as the molecular basis of virulence has been identified, it has been possible to generate microbes in which the genes responsible for virulence are crippled in ways that make it highly improbable that they will be able to revert. A good example is modified virus Ankara (MVA). This is a variant of vaccinia virus that was developed by multiple passages in chicken fibroblasts. It has lost about 10% of the viral genome and can only undergo very limited replication in mammalian cells. It still, however, generates a strong anti-viral response in humans.

Such microbes have another major potential advantage in vaccinology. It is possible to insert completely unrelated genes into the microbe and these genes will be expressed when the subject is vaccinated. Hence, it is possible to introduce antigens from another infectious agent into a genetically modified virus. Thus, if (and this is a big if) the proteins have been identified in a pathogen that can stimulate a protective immune response, then the inherent strong immunogenicity of the vector (e.g. MVA) ensures a strong response to the inserted genes. The use of such vectors is the subject of intense research at present, but to date none are in general clinical use.

Subunit Vaccines It is not always necessary to use a whole organism for vaccination. An example is the vaccine used for protection against hepatitis B. This virus is very difficult to grow in culture, preventing its production in sufficient quantities for vaccination. However, as the complete structure and genome of the virus has been described, it is possible to make recombinant viral proteins in large quantities. One of these, a viral envelope protein, stimulates a strong protective antibody response after three injections. The protein is not, however, given alone – it is absorbed to the adjuvant alum which stimulates immunity by activating the innate immune system (Chapter 4).

Q2.29. What would be the outcome if the protein was injected without an adjuvant?

Conjugate Vaccines Many pathogenic bacteria (e.g. *Streptococcus pneumoniae, Haemophilus influenzae* and *Neisseria meningitidis*) have carbohydrate capsules. These prevent phagocytosis and are important virulence factors. Antibodies against the capsule acts as opsonins and facilitate recovery from infection. The difficulty is that anti-carbohydrate antibodies are T cell-independent, and immunizing with capsular polysaccharide primarily induces only a T-independent IgM response with very little memory (Section 2.4.2.2). To overcome this, vaccines have been developed in which the carbohydrate is covalently linked to a protein carrier. The protein is processed and presented to CD4 T cells, and because the protein is linked to the carbohydrate, B cells specific for epitopes on the carbohydrate can now receive helper signals from the T cells, make high levels of IgG and long-lasting memory responses are generated. These vaccines are proving highly effective against meningitis caused by *Haemophilus influenzae* Type B, a major cause of meningitis in young children, and other vaccines are under development.

DNA Vaccines Immunity to the influenza virus depends largely on pre-formed antibody to the viral HA. In the virus the HA gene exists as RNA, but complementary DNA coding for HA can be made and incorporated into a plasmid. If this DNA is injected into a mouse, HA is expressed, the mouse makes anti-HA antibodies and is thus protected against infection with the virus. Why should the expression of a harmless protein lead to antibody synthesis? The answer probably lies in the nature of the injected DNA. As well as containing the HA DNA the plasmid contains other sequences derived from bacterial DNA. Some of these express the non-methylated CpG motif that is a strong agonist for TLR9 (see Chapter 4). Thus the DNA is able to stimulate the innate system effectively – it has a built-in adjuvant.

DNA vaccines have several potential advantages over conventional vaccines. DNA is a very stable molecule. Think of the DNA recovered from ancient organisms. Researchers commonly dry DNA onto a filter paper and send it through the post; at the other end it is dissolved in buffer and remains intact. Thus, there is no need for a cold chain. Additionally, compared to the costs of synthesizing proteins, making DNA is very cheap. Clinical trials of DNA vaccines have in fact started: the US National Institutes of Health has started a small-scale trial of an anti-influenza H5N1 vaccine, using the H5 gene. There are, however, some concerns that the DNA may integrate into the host genome, but to date there is no evidence that this does occur.

2.5.2.3 Future Prospects

It is chastening to realize that essentially all the successful vaccines we have today work by stimulating antibody synthesis and that there are none that are fully successful for infectious diseases where T cells, rather than antibodies, are the mediators of defence. It is also chastening to realize that, in general, recovery from an actual infection give much better protection against re-infection than do our current vaccines.

HIV, tuberculosis, leprosy, cholera, typhoid fever and malaria are examples of infectious diseases where, despite intensive research, vaccines are currently unavailable or only partially effective. In the case of tuberculosis, for example, BCG is effective in preventing tuberculosis in the developed world but almost completely ineffective in the developing world, and we have little understanding of why this is the case.

Q2.30. Why are HIV and malaria somewhat similar, in terms of the challenge they represent to designing a vaccine against them?

The conjugate vaccines represent the only really new developments in effective vaccine design in recent years. Why should this be? Well, from an evolutionary viewpoint, pathogens have been selected by the immune system, over thousands of years, to be able to evade or avoid from their point of view the deleterious effects of immune responses against them. We, in making vaccines, are struggling to overcome this huge evolutionary disadvantage. Researchers are trying every available avenue but to date results are, to say the least, disappointing.

A central question in designing a vaccine is to understand what is required for a protective immune response. Is it antibodies, cytotoxic T cells, or helper T cells that can activate macrophages for example? In many cases, such as tetanus, the answer is very clear; antibodies in this case. However, for other pathogens the picture is not clear; for HIV is the best bet antibodies or cytotoxic T cells? We do not yet know. In some cases animal models give insight, but for infections such as HIV and malaria, which are difficult to model in animals, we are still uncertain. It is important to realize that in some infections, generating the wrong response may actually be harmful. In Dengue and yellow fever, patients with antibodies against the virus may suffer more severe disease; because the virus lives preferentially in macrophages, antibodies may simply target the virus efficiently to its preferred host cells.

There is also much we do not understand about effective vaccine design. Take for instance HIV. Researchers have devised a number of different candidate vaccines, and for some of them trials in volunteers in the developed world have given encouraging results in terms of antibody synthesis and/or generation of cytotoxic T cells. However, when trials were extended to the developing world, the responses generated were much weaker, resulting in several trials being aborted. The reasons for these differences are very unclear. Until we understand how immune responses are regulated in humans in different environments and of different ages, vaccine design will remain largely guesswork.

Learning Outcomes

By the end of this chapter you should be able to understand, explain and discuss the following topics – (the relevant sections of the chapter are indicated). You should understand some of the evidence from human and animal studies supporting what we know about these topics. You should have some idea of the areas where our understanding is incomplete. You may be able to suggest ways in which our understanding could be advanced.

- Pathogens and infectious disease (Section 2.1)
  - What is a pathogen and what are the main types of pathogen?
  - What makes a pathogen pathogenic and why are non-pathogens non-pathogenic?
  - What are commensal organisms and why are they important?
  - What is an opportunistic infection and what is a subclinical infection? Why may subclinical infections be important?
  - What is the difference between an infection and an infectious disease?
- What types of cell are involved in the immune elimination of different types of infectious agent (Section 2.2)?
  - Which cells of innate immunity are involved in the major types of infection and what are their roles?
  - How do different types of T cell and different types of antibody help in defence against different types of infectious agents?
- How do different cells and molecules integrate their activities in defence against different types of infectious agent (Section 2.3)?
  - How, in general, do different types of pathogen cause disease?
  - Give some examples of different types of virus and the mechanisms of host defence used against them
  - Give some examples of different types of bacteria and the mechanisms of host defence used against them
  - Give some examples of different types of yeasts and parasites and the mechanisms of host defence used against them
- What is a vaccine and how do different types of vaccines induce protective immunity against infection (Section 2.4)
  - Give examples of current successful vaccines – how do they work?
  - Give some examples of infectious diseases where vaccines are urgently needed. Why do we not have vaccines for these infections?
  - What are the potential dangers associated with different types of vaccine?
- INTEGRATIVE: How does the biology of an infection relate to the types of defence needed for its elimination, and how should this inform vaccine design?

3
Functional Anatomy of the Immune System

3.1
Introduction

All plant and animal cells are bounded by barriers that provide protection from the outside world and thus help to prevent infection. In the case of multi-cellular organisms there is a new problem: they can potentially be infected anywhere in the organism, so how can they ensure that the mechanisms needed for defence against infection are available wherever it occurs? One way is to have the defence mechanisms located throughout the entire body at all times, but this would be inherently inefficient. Another solution has evolved.

Most multicellular animals – certainly those that are more complex – possess mechanisms for moving cells and molecules around the body. Thus these can be used for bringing defence-related cells and molecules to the sites of infection and damage so that the infectious agents can be eliminated and tissues can be repaired. These recruited mechanisms are found in one form or another in all complex organisms. We know these as innate defences or innate immunity. However complex organisms first arose about half a billion years ago, during the Cambrian explosion and, starting with the earliest fish, it is possible to trace the evolution of specialized tissues and organs that reflect the emergence of an entirely new form of immunity. This is a adaptive immunity, which is mediated by lymphocytes.

In this chapter we examine the anatomical and physiological basis of innate and, adaptive immunity, and show how this relates functionally to defence against infection. We first consider the anatomical, physiological and chemical barriers that provide the first line of defence against infection, and which need to breached before infection can occur (Section 3.2). Infection can occur at any site in the body, and we next discuss the changes induced locally by infection that result in inflammation. Inflammation allows the cells and molecules of innate immunity to be targeted to infected sites, enabling the immediate and later recruitment of effector mechanisms that may help to eliminate the infectious agent and, afterwards, to heal the damage that has occurred. We show how this local response also drives the production of molecular messages that induce physiological responses in other tissues and organs at more distant sites (Section 3.3).

We then turn to the tissues and organs involved in adaptive immunity, the lymphoid tissues (Section 3.4). We describe how lymphocytes and other cells continually traffic though these organs, how lymphoid tissues become modified when infection occurs, and how lymphocytes respond in these sites to infectious agents. The lymphocytes then generate new effector mechanisms that can help to eliminate infection and later provide "memory", a feature unique to adaptive immunity that provides increased protection if the same pathogen is encountered in the future.

We next turn to where these immune cells are produced, examining the specialized tissues and organs that contain the haematopoietic stem cells (HSCs) and later progenitors that differentiate into the cells of innate immunity and lymphocytes (Section 3.5). We discuss some of the factors that control the development of these tissues, and their pathological induction at other sites in disease. Finally, we touch on how bone marrow stem cells can be used to treat immunodeficiencies and how, potentially, they can be genetically modified to develop treatments for immunological and other defects (Section 3.6).

By the end of this chapter you will appreciate that a full understanding of immunity to infection requires an understanding both of the patho-physiology of infection, and of the anatomy and physiology of the tissues involved in responses to infection. You will have a basic understanding of the changes that occur in these tissues during infection that permit elimination of the infecting microbes.

3.2
Natural Barriers

The first line of defence against infection consists of barriers, which may be structural, chemical or biological. Examples are keratin in the outermost layer of the skin, acid in the stomach and commensal bacteria in the intestine, respectively. These natural (as opposed to induced) defences are present in all normal individuals before any infection occurs and they normally need to be breached by any microbe that is going to cause disease. Their importance

Exploring Immunology: Concepts and Evidence, First Edition. Gordon MacPherson and Jon Austyn.

is shown where any of these barriers is defective, leading to a greatly increased risk of infection at these sites and beyond.

The skin is the most obvious barrier, but it is in fact the smallest of the surfaces in contact with the external environment, in contrast the surface areas of the lungs and the gastrointestinal tract are respectively 50 and 150 times that of the skin. That these huge areas of tissues can be monitored effectively for infection is perhaps surprising and it explains why inflammation is needed to target defence mechanisms to specific areas when required (below). See Figure 3.1.

3.2.1 Skin

The skin has a major protective function, not only against infection but also against mechanical trauma. Thus, the outer layers are hard and difficult for microbes to penetrate because of the presence of much keratin, and this provides a barrier to infection. The importance of this is shown by diseases such as eczema where the skin is damaged: bacterial (e.g. staphylococcal) and fungal infections are common in the damaged areas. Similarly, where the skin is breached by a wound or a burn, bacteria can get into subcutaneous tissues which can be fertile grounds for them to thrive. Patients with serious burns are at severe risk of dying from massive infection, and also from the side effects of the inflammatory mediators that the immune system produces to try to counter the infection. Apart from this mechanical barrier function, skin secretions, such as sebum produced by the sebaceous glands, contain anti-microbial chemicals. The relative paucity of sebaceous glands in some skin areas, such as the feet, may relate to the increased incidence of fungal infections in these areas (e.g. athlete's foot). See Figure 3.2.

Q3.1. Can you think of some other examples that illustrate the barrier role of skin?

3.2.2 Mucosal Surfaces

Mucosae are epithelial tissues that in general contain goblet cells, the source of mucus. The main mucosae are the gastrointestinal tract, the respiratory tract and the urogenital tract, but the eye is also included. Given that the main functions of mucosae are transport (absorption and secretion) they could not function if they were keratinized. They are also often only one cell thick. Thus, pathogens have a much easier task if they are to penetrate mucosae. Mucosae do, however, have their own defence mechanisms. Some mechanisms are common to most mucosae. All mucosae need lubrication and in most cases this is mediated by mucus or, in the case of the eye, by tears. Mucus itself is thought to provide a mechanical barrier to infection and may carry anti-microbial molecules (e.g. defensins: see Chapter 4). However, children suffering from

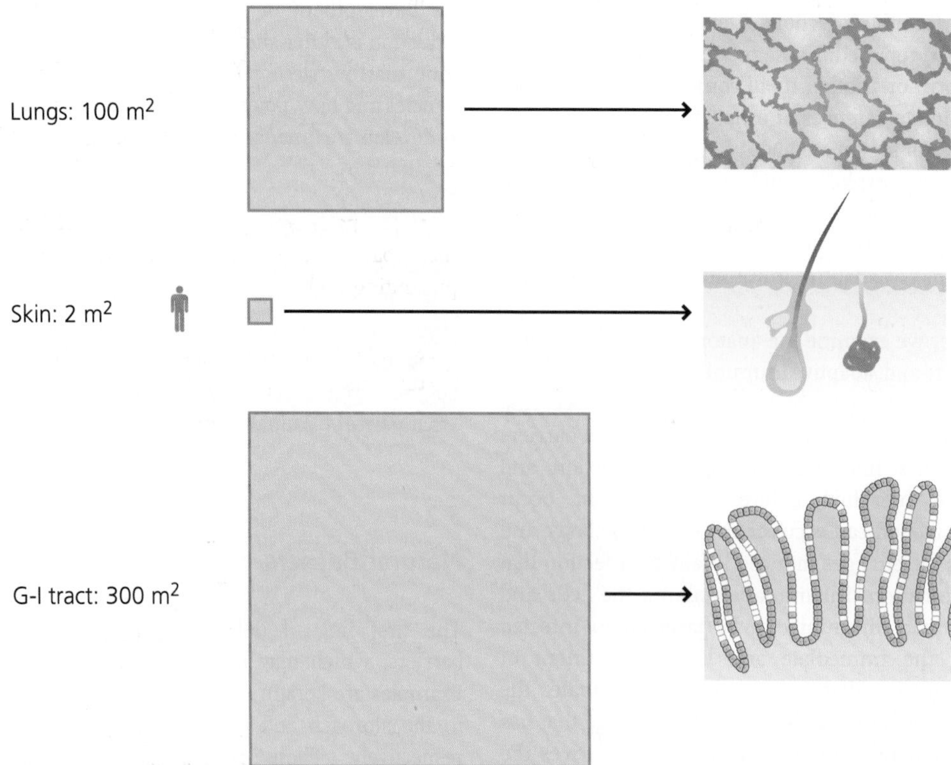

Fig. 3.1 Surface areas of skin and mucosal tissues. Mucosal tissues have many times the surface area of the skin. In adult humans the respiratory tract has a surface area of roughly the size of a tennis court and the gastro-intestinal (G-I) tract of a football pitch. These surfaces are also much thinner, often only one cell thick. It is not surprising that most infections start at mucosal surfaces.

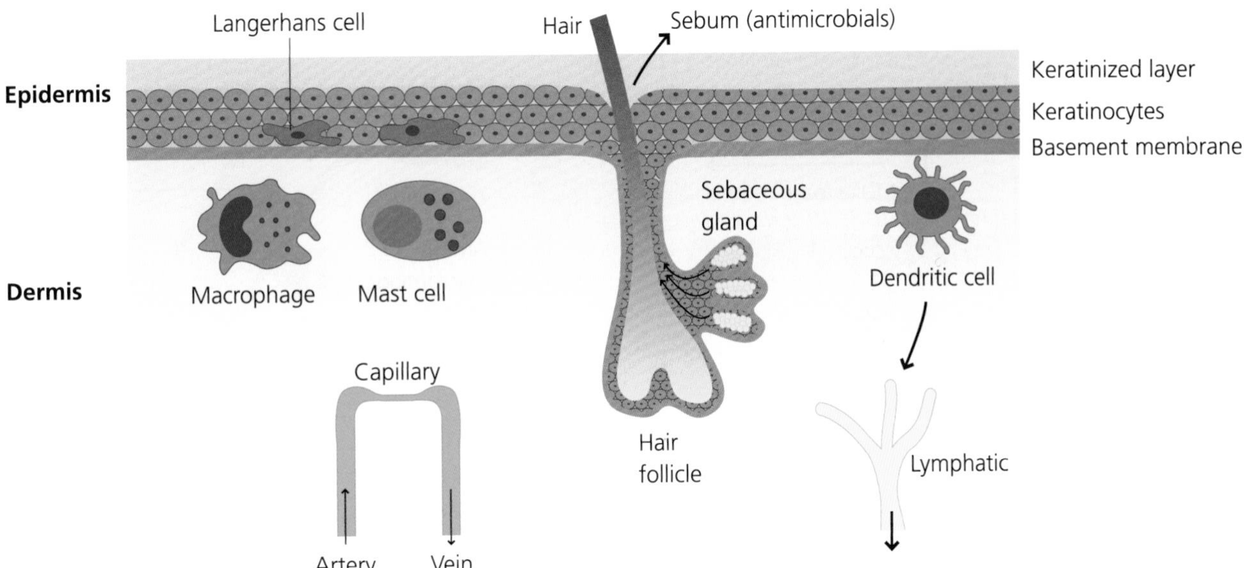

Fig. 3.2 Structure of skin showing defence barriers. The epidermis varies in thickness in different parts of the body, but is always several cells thick and the outer layers are always keratinized, forming a mechanical barrier to infection. Sebaceous glands discharge their contents, including anti-microbial molecules, into the hair follicle and thence onto the surface of the skin. Both the epidermis and the dermis contain classical dendritic cells (DCs), Langerhans cells and dermal dendritic cells, respectively involved in the initiation of adaptive immunity. The dermis also contains macrophages and mast cells that can sense infectious agents that have breached the outermost layers.

cystic fibrosis, where the mucus in the respiratory tract is abnormally thick, have an increased incidence of bacterial lung infections because microbes cannot be efficiently expelled from the lungs by coughing. Tears wash away pathogens but also contain lysozyme, a molecule capable of breaking down some bacterial cell walls.

3.2.2.1 Respiratory Tract

The epithelial cells of much of the respiratory tract are ciliated, and these cilia beat coherently to move secretions and particles towards the external world. In chronic cigarette smokers the cilia have disappeared and the epithelial cells are flattened (squamous metaplasia). Thus secretions containing bacteria pool in the lower parts of the tract and provide first-class growth environments for the bacteria. Lower down the respiratory tract, in the alveoli, are large numbers of macrophages, capable of phagocytosing and killing a range of microbes. Some of the lubricant surfactant proteins secreted into the lungs can also help to eliminate microbes. See Figure 3.3.

Q3.2. What other mechanisms might protect the respiratory tract against infection?

3.2.2.2 Gastro-Intestinal Tract

The gastro-intestinal tract possesses several kinds of barrier. Many microbes can be killed by acid. Hydrochloric acid is secreted into the stomach and its primary function is to aid digestion, but it can also kill many microbes. Hence, patients suffering from a deficit in stomach acid may be more susceptible to intestinal infections with bacteria such as Salmonella. Bile is secreted from the gall bladder into the upper small intestine. The main function of bile is also to aid digestion, but bile salts are also capable of killing many bacteria. Thus, culture media designed to isolate enteric bacteria from complex mixtures are formulated to contain bile salts because only the resistant intestinal bacteria can grow in these conditions. Bile salts can also destroy the envelopes of viruses, perhaps explaining why viruses that infect the intestine do not possess a lipid envelope. Peristalsis serves to drive out pathogens in the faeces (this may, however, be to the pathogen's advantage in spreading infection) but many bacteria are able to bind to intestinal epithelial cells, preventing their expulsion. See Figure 3.4.

Q3.3. The avian H5N1 virus is an enveloped virus. In birds, viral transmission is via the oro-faecal route. How might the virus be able to survive in the bird intestine?

3.2.2.3 Urogenital Tract

Bacterial cystitis, inflammation of the bladder, is very common in women, but very rare in men. Might this relate to the relative lengths of the ureters in the two genders – much longer in men – and to the anatomical juxtaposition of the urethra and anus in women? One survey of men and women travelling to work sampled their hands for faecal bacteria: more than 20% of those sampled carried faecal bacteria, and the incidence was higher in women than in men. Motile bacteria, such as *Escherichia coli*, also have much further to swim in men and tidal urine flow is more likely to wash them out. Not surprisingly, conditions leading to urinary

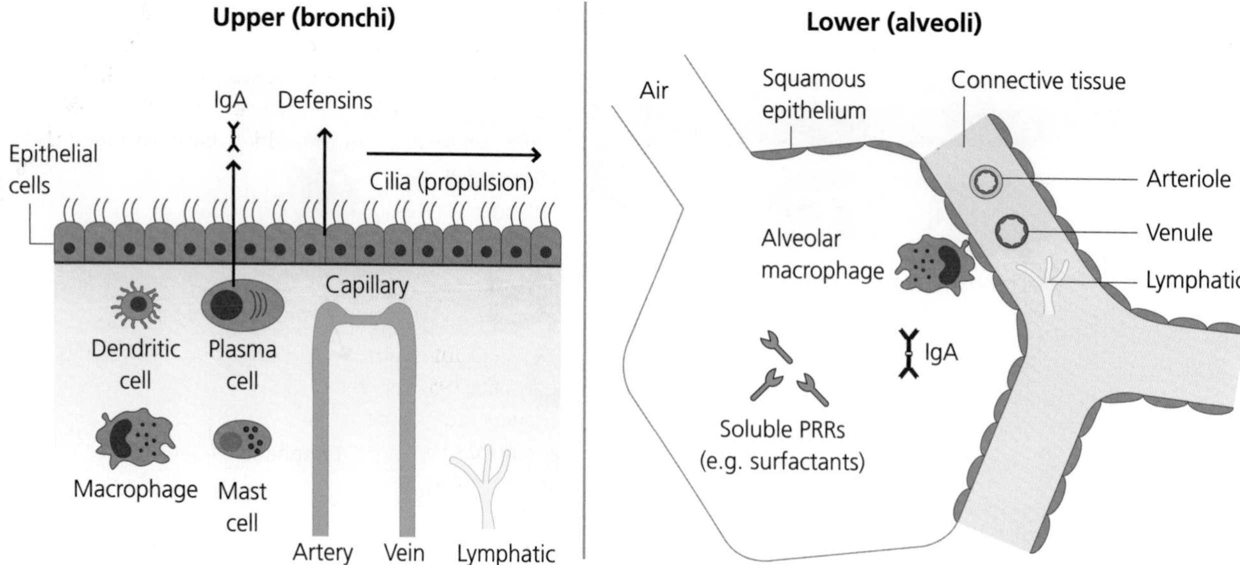

Fig. 3.3 Mucosal tissues I: respiratory tract. The upper respiratory tract (trachea and bronchi) is lined by ciliated epithelial cells – the cilia beat coherently to move secretions towards the mouth. The alveoli, where gas exchange occurs, are lined by flat, squamous epithelium. They contain alveolar macrophages that may have a defence function, but also clear inhaled particles. Other cells in the alveoli secrete molecules such as surfactants which are lubricants but may also have anti-microbial effects, acting as pattern-recognition receptors (PRRs), see Chapter 1. Some IgA may be produced as natural antibodies before adaptive immunity is triggered and may help to provide an extra layer of defence against infection of the epithelia.

obstruction often result in infections, illustrating the importance of this flushing effect of urine for basic defence.

3.2.2.4 Commensal Flora

Another important barrier to infection by pathogens is the commensal bacteria that coat many mucosal surfaces. Commensals are present in huge numbers in the large intestine. As we noted in the Introduction there may be about 10 times more bacteria in the human colon than all the cells that make up an average adult human body. Commensals also coat many other epithelial surfaces including the skin, the respiratory tract, and the urogenital tract of women. In most of us these bacteria do no harm and probably act to prevent pathogenic bacteria from causing disease: they form an effective

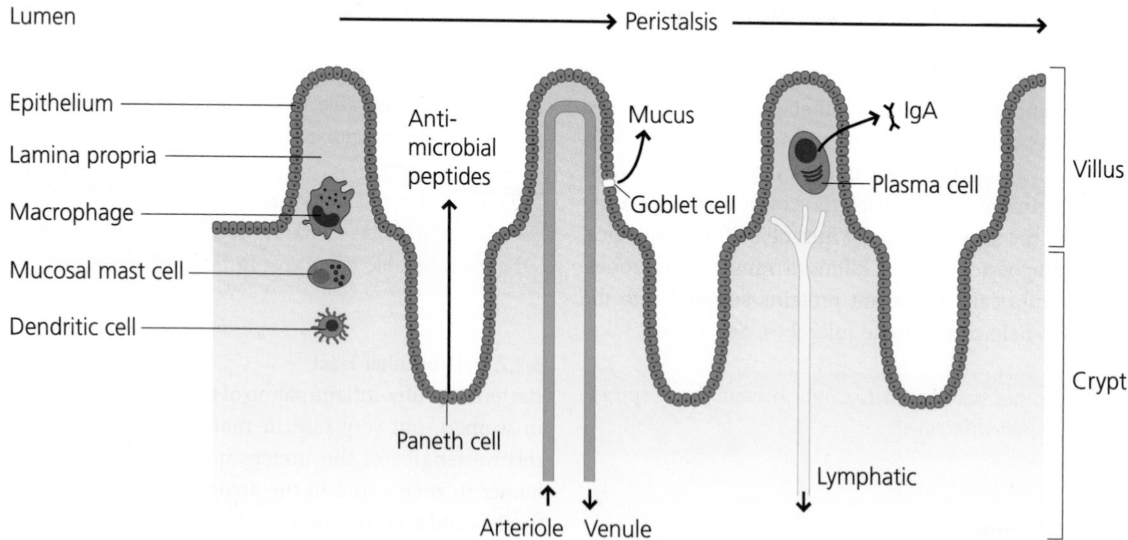

Fig. 3.4 Mucosal tissues II: intestine. The intestine is lined by a single layer of epithelial cells. In the small intestine villi and crypts are present. The large intestine has a smooth surface, but contains many deep tubular glands. Within the epithelium, goblet cells secrete mucus that is a lubricant, but also has mechanical barrier functions. In the crypts of the small intestine, specialized Paneth cells secrete a variety of anti-microbial peptides such as defensins. The lamina propria of the intestine contains macrophages and mucosal mast cells, classical DCs, usually eosinophils (not shown), and plasma cells that typically secrete IgA that is transported into the lumen. Peristalsis serves to propel the intestinal contents towards the anus.

barrier to colonization by pathogens. In some instances of large bowel surgery, patients used to be given large doses of oral antibiotics to get rid of most of the intestinal bacteria, so as to reduce the chance of bacteria infecting the peritoneal cavity. In such cases ingestion of the bacterium *Staphylococcus aureus* sometimes led to a serious condition known as enterocolitis in which parts of the wall of the large intestine were destroyed (necrosis), often leading to perforation and peritonitis that could be fatal. Commensals can also help to make the local environment more resistant to infection. For example, lactobacilli in the vagina produce acid that is thought to reduce the risk of bacterial or fungal infection. Indeed there is evidence to suggest that vaginal douching, by removing the lactobacilli, leads to a rise in pH, creating conditions in which other pathogens can thrive.

Q3.4. A technician working in a bacterial laboratory developed a severe upper respiratory tract infection and was treated with a broad-spectrum antibiotic. On his return to work he set up some large-scale cultures of a normally harmless bacterium, Haemophilus influenzae. He developed a very severe respiratory tract infection with the Haemophilus influenzae and nearly died. Why? What might have been done to prevent the infection?

3.3 Functional Anatomy of Innate Immunity

The natural barriers, outlined above, are absolutely crucial as the first lines of host defence against infection. It is, however, almost a defining feature of a pathogen that it can avoid or evade these barriers. The next major line of defence is the innate immune system, and a phenomenon that is absolutely central for effective functioning of innate immunity (as well as adaptive immunity) is acute inflammation. Here, we will discuss the major principles of inflammation in relation to the structure and function of the tissues in which inflammation occurs. We also briefly introduce some of the cells and molecules that are involved, and which are discussed in more detail in Chapter 4.

3.3.1 Features of Inflammation

What is acute inflammation? Inflammation was defined by Menkin in the 1950s as "the complex vascular, lymphatic and local tissue reaction elicited in higher animals by the presence of micro-organisms or of non-viable irritants". Inflammation is a process, not a state, and inflamed tissues are undergoing continual change. The cardinal features were recognized by the Greeks: redness and heat, reflecting dilatation of small blood vessels and increased blood flow; swelling, reflecting the increased accumulation of excess extravascular fluid, and which is called oedema; local pain, caused by increased tissue tension and the release of chemicals that stimulate pain nerve fibres; and, sometimes, loss of function. Acute inflammation is generally thought of as inflammation that occurs rapidly within minutes, hours or a few days of the initial stimulus and which subsequently resolves. In contrast, chronic inflammation has a time course of weeks, months or years. Chronic inflammation generally results from the persistence of the initiating inflammatory stimulus. It is important to realize that acute inflammation in a tissue may not be clinically apparent. In dealing with non-pathogenic microbes exactly the same defence mechanisms may operate that are used to deal with pathogens, but the local changes in infected tissues may be so minor as to be unapparent. These are called subclinical infections. See Figure 3.5.

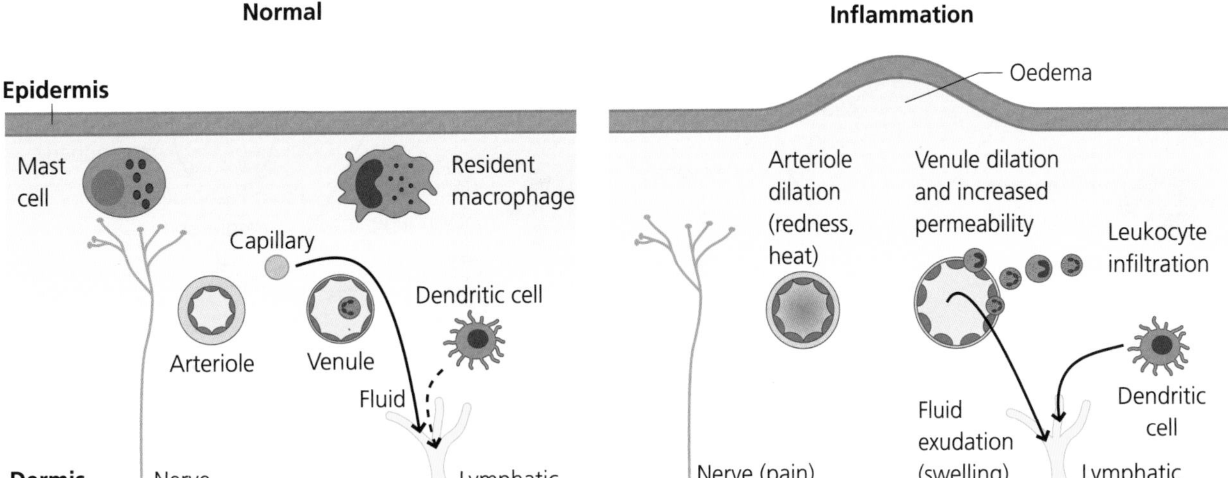

Fig. 3.5 Acute inflammation in the skin. Acute inflammation induces vascular changes: dilatation of arterioles and venules and increased permeability of venules. This increases overall blood flow to the area, but also slows the flow in venules, increasing the probability of leukocytes attaching to the endothelium. Changes in the expression of venule adhesion molecules permit the recruitment of neutrophils and monocytes into the inflamed tissue. The increased permeability, due to gaps forming between endothelial cells, permits exudation of water and solutes including macromolecules such as antibodies and complement. The water causes swelling – oedema, one of the main features of acute inflammation; another is often pain.

3.3.2 Initiation of Local Inflammation

Inflammation has two distinct but overlapping and closely related functions in defence against infection. (i) By increasing the permeability of blood vessels, inflammation permits macromolecules involved in defence and healing to enter extravascular tissues. (ii) By increasing the adhesiveness of the blood vessel endothelium, it enables leukocytes to be recruited from the blood into the inflamed area. Thus, the requisite molecular and cellular effector mechanisms can be rapidly targeted to the site of infection. If, however, inflammation is to be initiated, the presence of tissue damage has to be recognized. Given the number of ways in which tissue damage can be caused, by trauma as well as infection, it is not surprising that the mechanisms that recognize damage are complex and as yet only partially understood. See Figure 3.6.

3.3.2.1 Responses to Trauma

In the case of mechanical trauma there are several pathways involved in the recognition of tissue damage. For example, trauma induces mechanical damage in mast cells that reside in connective tissues. In turn, they degranulate, releasing histamine and other mediators of inflammation. Trauma in almost all cases also induces haemorrhage (bleeding into the tissues). This eventually results in the stimulation of the healing reaction (Section 3.3.6). Cells that have died by necrosis break down their plasma membranes and it is suggested that they may release molecules called damage-associated molecular patterns (DAMPs). DAMPs, which are discussed a little more in Chapter 4, are poorly understood, but may include a very wide range of molecules that signal to the immune system that cellular or tissue damage has occurred.

3.3.2.2 Responses to Infection

Mechanical trauma may lead to entry of pathogens, as for instance in an infected wound. However, many pathogens can breach the natural barriers in the absence of trauma and it is a necessary property of many pathogens that they can do this (e.g. following inhalation or ingestion). For example, influenza virus can infect and damage the epithelia lining the airways, despite the presence of mucus. In fact this often leads to secondary infection of the damaged tissues by bacteria since these barriers have now been disrupted. Similarly Salmonella can evade stomach acid, bile and mucus to invade the intestinal epithelium (in contrast to the commensal bacteria which do not invade). On the other hand, some viruses and other pathogens can be injected directly into the bloodstream through the bite of an insect or mammal, such as those that cause malaria, plague and rabies. In addition, the larval forms of schistosomes can even burrow through the skin from infected water and ultimately spread through the bloodstream to infect other organs such as the liver (see Chapter 2).

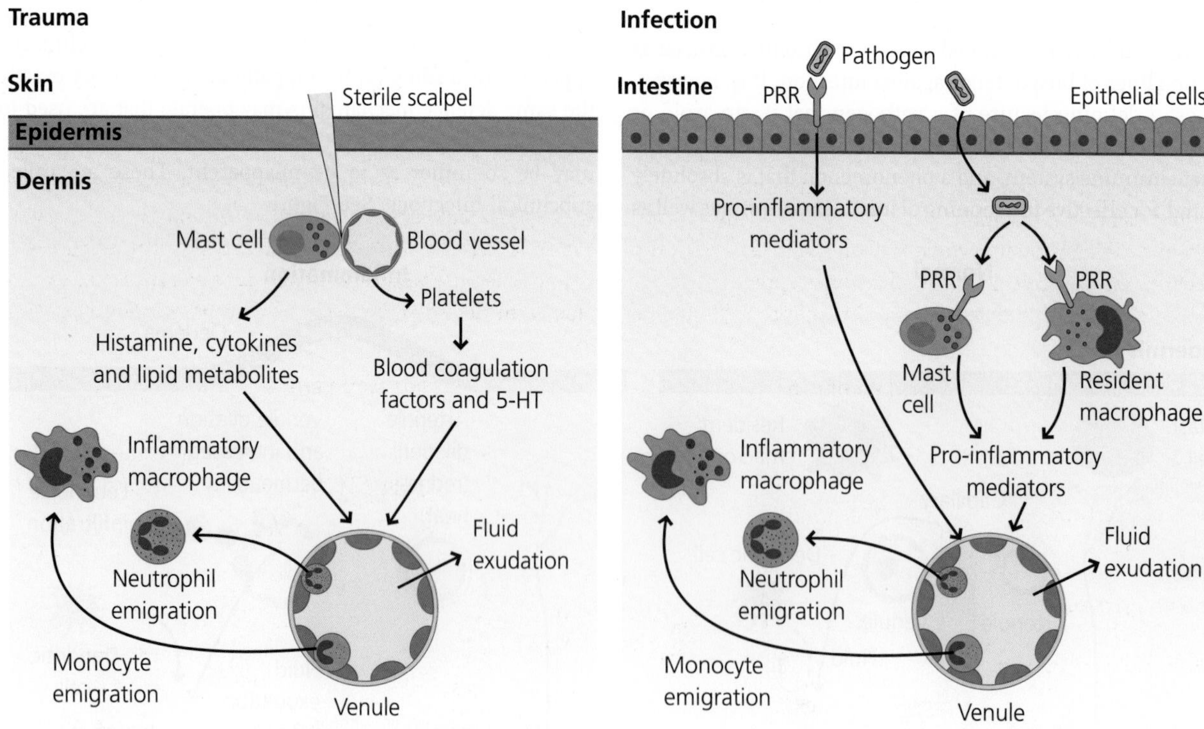

Fig. 3.6 Initiation of acute inflammation. Acute inflammation can be induced by sterile trauma or infection. Sterile trauma involving damage to blood vessels activates blood platelets and the coagulation cascade. It may also damage mast cells. All of these can release mediators (some examples are shown) that induce the vascular changes typical of acute inflammation. In infections, alarm cells such as epithelia, macrophages and mast cells can recognize pathogens via pattern recognition receptors (PRRs) which stimulate the release of inflammatory mediators. The skin and gut are illustrated, but these principles apply to epithelial tissues in general.

Given the ability of pathogens to evade or avoid barriers, you will understand that it is essential for mechanisms to be present in all parts of the body, in all tissues and organs, that are able to detect and respond to any breach in natural defences which represents danger and to signal alarm to other components of the immune system.

The recognition that infection has occurred in tissues requires the presence of cells that can sense that this has occurred. These cells include the epithelial cells lining anatomical barriers which are usually the first point of attack of a pathogen. Underlying these surfaces, and in essentially all tissues, are tissue-resident (or at least very long-lived) immune cells that may be specialized to sense danger. These particularly include resident macrophages and mast cells. The general feature shared by these cells in peripheral tissues is that they have receptors called pattern recognition receptors (PRRs) that are able to recognize molecules, which are present only in or on pathogens, called pathogen-associated molecular patterns (PAMPs) (Section 1.2.3.1). PAMPs act as agonists for different types of PRRs and stimulate resident cells to secrete molecules that act on blood vessels to cause local inflammation (inflammatory mediators).

Crucially in early defence, resident macrophages can rapidly detect the presence of pathogens and quickly initiate inflammatory responses and innate immunity. For example they can be triggered to synthesize and secrete pro-inflammatory cytokines such as interleukin (IL)-1, IL-6 and tumour necrosis factor (TNF)-α. Mast cells are also present in all loose connective tissues such as those underlying mucosal epithelia. They contain cytokines and other pre-formed inflammatory mediators in their granules. Following stimulation (e.g. by trauma, as above, as well as infection) they release these mediators, and synthesize and secrete others that contribute to the inflammatory response. These responses are discussed in more detail in Chapter 4.

For now, the important principle to understand is that, after infection, the dynamic anatomy of innate responses involves induced changes in the structure and function of peripheral tissues at the local site (local inflammatory responses) so that effector cells and molecules can be recruited to eliminate that infection. Other changes, in more distant tissues (systemic inflammatory responses) are also needed, so that the availability of these cells and molecules can be increased.

Q3.5. How might we show that mast cells are important in the initiation of acute inflammation following infection with a pyogenic bacterium such as Staphylococcus aureus?

3.3.3 Local Inflammatory Responses

The effector cells and molecules that are needed to mount effective innate immunity are not normally present in tissues; they cannot cross non-inflamed endothelium and are restricted to the blood. Blood is primarily a distribution system, and in defence its function is to send cells and molecules to sites where they can mediate their effects. How can cells and molecules be directed to the sites where they are needed? Arterioles are involved – nervous impulses from the area of inflammation trigger an axon reflex that leads to arteriolar dilatation, and the increased blood flow delivers more leukocytes to the inflamed area. However, in acute inflammation, changes in the endothelial cells of specialized blood vessels, called post-capillary venules, are central and crucial. Post-capillary venules are thin-walled vessels lacking the layers of muscle and connective tissue found in larger vessels. Hence, it is easier for things to get across them, from the blood into the tissues, but to do so the endothelial cells that line them first need to be changed or activated. Endothelial cells in venules can be activated by a variety of mediators (discussed in more detail in Chapter 4). Venule dilatation causes slowing of blood flow, giving leukocytes increased opportunity to interact with the endothelium. Increases in permeability permit egress of macromolecules while changes in adhesion molecule expression permit leukocyte binding and recruitment. See Figure 3.7.

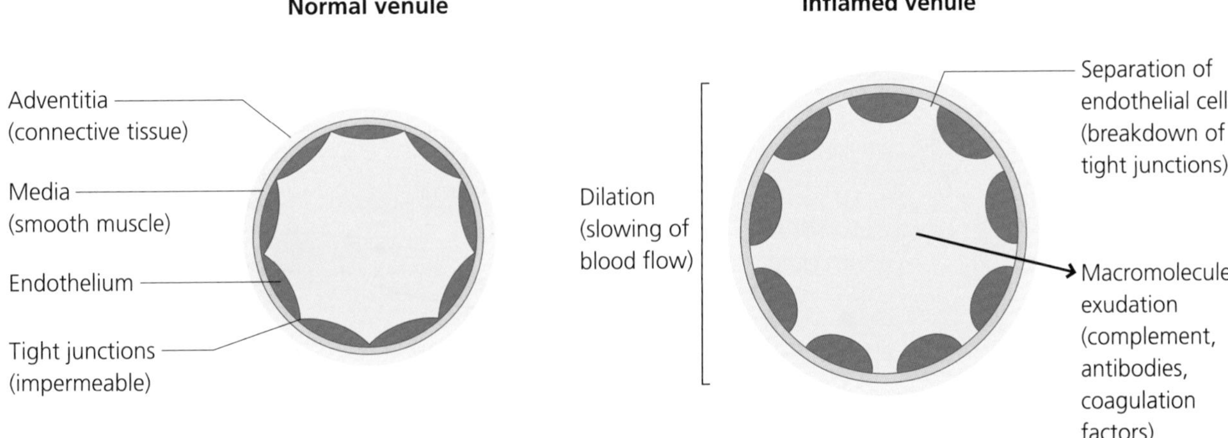

Fig. 3.7 Vascular changes in acute inflammation. Inflammatory mediators such as histamine act on venule smooth muscle to cause dilatation, leading to slowing of blood flow. Mediators also act on endothelial cells, leading to the breakdown of tight junctions – forming gaps that permit the exudation of blood plasma containing macromolecules such as antibodies, complement and coagulation factors. Changes in adhesion molecule expression also lead to leukocyte recruitment (see Figure 3.8).

3.3.3.1 Recruitment of Soluble Effector Molecules

A major alteration in the physiology of infected tissues is that the permeability of venules in the infected area is increased. Molecules such as TNF-α (secreted by macrophages and mast cells) and histamine (released from mast cells) stimulate the breakdown of tight junctions between endothelial cells, causing the formation of physical gaps. The increased permeability permits large amounts of plasma to enter the tissues, causing the characteristic swelling (oedema) of acute inflammation. The plasma contains proteins normally confined to the blood and these include large molecules such as complement components and (usually much later) antibodies which can now enter the extravascular space. Two main properties of these molecules are to further stimulate acute inflammation, mainly via the actions of breakdown components of activated complement proteins on mast cells, and to act as opsonins, coating pathogens to facilitate their uptake by the neutrophils and additional macrophages that are recruited to these sites (below).

> **Q3.6.** Can you suggest some situations in which oedema may not be beneficial to the host?

3.3.3.2 Recruitment of Cellular Effectors

The increased permeability seen in acute inflammation is not, however, sufficient to target leukocytes to the inflamed area. The only way cells can identify the site where they need to go is via molecules expressed on endothelium. Hence, changes in molecules expressed on the endothelium of venules permit the migration of blood leukocytes into extravascular tissues. (Molecules secreted into the blood would be of no use – they would just be washed away.) Inflammation changes the adhesiveness of endothelial cells for leukocytes in selective ways, ensuring that migration is regulated.

Leukocyte–Endothelium Interactions The first stage in leukocyte recruitment requires that the appropriate cells are targeted to the inflamed tissue. (Remember that each leukocyte has its own specialized functions and that, in general, these can only deal effectively with certain types of infection; Section 1.4.2). Three particularly important and distinct molecular recognition systems interact to direct specific types of circulating leucocytes to the appropriate tissue. Selectins on leukocytes bind carbohydrate ligands on endothelial cells to permit loose adhesion to the endothelium and vice versa. This loose adhesion then permits the leukocytes to bind to chemokines that are themselves bound to heavily glycosylated molecules on endothelial cells. This binding subsequently stimulates tight adhesion mediated via integrins. There are different selectins and integrins, and many different chemokines and chemokine receptors. Hence, different permutations and ombinations of these molecules permit the delivery of different leukocytes to different sites in a highly selective manner. The crucial outcome of this system is that it permits the recruitment of host defence cells to sites of infection quickly and appropriately – to the right place at the right time. Thus, in acute inflammation, neutrophils and monocytes are delivered. In chronic inflammation, monocytes and (later) activated lymphocytes are delivered. In parasitic and allergic responses, eosinophils and/or basophils arrive at the appropriate sites. In other words, these molecules act as an address for leukocytes and the complementary molecules on endothelial cells are in fact sometimes called vascular addressins. This can be described as a post code system, or ZIP code system if you happen to live in the United States. See Figure 3.8 and Box 3.1.

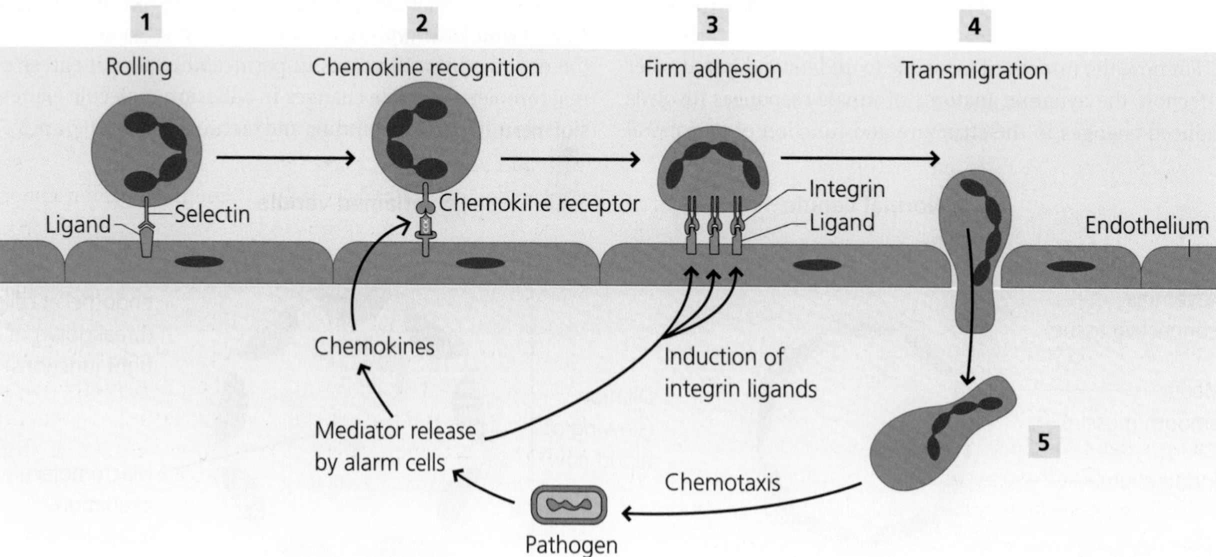

Fig. 3.8 Leukocyte recruitment in acute inflammation. (1) Leukocytes, particularly neutrophils, use selectins to form loose adhesions to carbohydrate ligands on venule endothelium and they roll along the endothelium. (2) This permits them to interact with chemokines bound to endothelial surface molecules. (3) The chemokines stimulate neutrophil integrins to increase their affinity, permitting strong adhesion to their ligands on the endothelium. (4) Neutrophils migrate to junctions between endothelial cells and cross the endothelium into extravascular tissues. (5) Chemotactic factors then draw the neutrophils to the site of infection. The underlying alarm cells are responsible for producing the chemokines and inducing the endothelial ligands (vascular addressins) in this process.

Box 3.1 Leukocyte Adhesion Deficiency

Rarely, young children present with an increased incidence of pyogenic bacterial infections. This may have multiple causes, such as a lack of neutrophils, or neutrophils that cannot kill bacteria. In some of these children, however, there are increased numbers of neutrophils in the blood, and the neutrophils are fully able to phagocytose and kill bacteria *in vitro*. However, there is little or no pus formation at the sites of infection. In many of these children, examination of neutrophils shows that expression of one of the adhesion molecules, an integrin (LFA-1; CD11a/CD18), is defective. The result is that although the cells can identify areas of inflamed endothelium and can roll along them by adhering to selectins, thus enabling them to sample endothelial-bound chemokines that can signal to the neutrophil, they cannot form the firm adhesions crucial for emigration. This is a direct demonstration of the importance of integrins for neutrophil migration from the blood. This condition is known as leukocyte adhesion deficiency (LAD) I. Other types of LAD also exist. For example LAD II occurs because of an abnormality in fucose metabolism, resulting in defective expression of selectin molecules. Another form, LAD III, occurs because of a mutation in the gene encoding RAC2 which is involved in the organization of the actin cytoskeleton, crucial for neutrophil migration.

In experimental models, if neutrophils (or other leukocytes) are treated *in vitro* with pertussis (whooping cough) toxin and injected intravenously into a normal recipient, they are unable to migrate to inflamed tissues. By intravital microscopy they can be seen to roll along the endothelium but do not form tight adhesions. This occurs because pertussis toxin blocks G-protein-coupled receptor (GPCR) signalling. Chemokine receptors on the neutrophil recognize chemokines bound to endothelial cells and, via G-proteins, induce the activation of integrin molecules on the neutrophil. Activation of integrins is essential for tight adhesion and subsequent extravasation.

These examples, although very rare or experimental, illustrate how the different steps involved in neutrophil migration into inflamed tissues have been identified and analyzed. In the former clinical cases they also demonstrate the importance of neutrophil migration in defence against infection by extracellular bacteria in peripheral sites.

Q3.7. How might children with LAD be treated?

The principle of selective delivery of leucocytes to appropriate sites is absolutely crucial to understanding defence against infection, and is also crucial for thinking about potential therapies in non-infectious situations such as autoimmune diseases, transplant reactions and cancer (Chapter 7). Thus, modulation of leukocyte migration is being tested in clinical trials for treatment of autoimmune diseases and could possibly be used to prevent transplant rejection or even to stimulate tumour destruction.

Q3.8. How might we attempt to identify a novel molecule involved in neutrophil adhesion to endothelial cells?

Leukocyte Function in Inflammation Virtually all tissues and organs can be stimulated by infection to generate signals that may act locally to trigger secretion of defence molecules. For example most epithelial cells can be stimulated to secrete defensins, which have antimicrobial activity (see Chapter 4). Importantly, infection can also stimulate the recruitment of cells and molecules (above) that cooperate to halt, or at least slow down, the development of the infection.

In acute inflammation stimulated by infection the major, initial, cellular effect is the recruitment of neutrophils into the site of infection. These cells are normally present in large numbers in the bloodstream and, once they have been recruited to inflammatory sites, they can efficiently phagocytose and kill some bacteria before dying. Stored neutrophils can also be released rapidly from bone marrow (below). Partly because of their rapid mobilization, neutrophils are particularly important in defence against pyogenic (extracellular) bacteria, which divide rapidly in extracellular sites, and some fungi (see Chapters 2 and 4). A little later, but in smaller numbers, blood monocytes are recruited into the acutely inflamed tissues where they differentiate into macrophages with additional, specialized functions compared to the resident cells that helped to trigger the initial inflammatory response. Their main function in pyogenic infection, however, is probably to regulate healing, repair and tissue remodelling. In other infections, such as those with mycobacteria, which induce chronic inflammation, recruited macrophages can be activated to become potent anti-microbial cells. These classically activated macrophages are able to kill intracellular bacteria such as Mycobacterium tuberculosis, but are also capable of causing much tissue damage because of their excessive secretory activity. See Figure 3.9.

3.3.4 Effects of Inflammatory Mediators on Distant Tissues: The Acute-Phase Response and Systemic Inflammatory Responses

The cytokines produced by tissue-resident macrophages and mast cells in response to infection do not always just act locally.

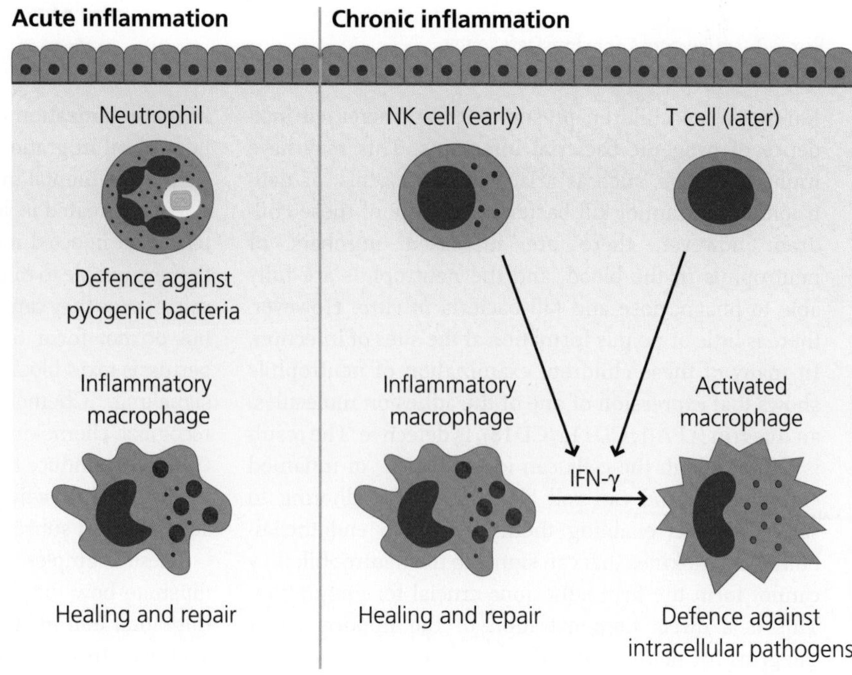

Fig. 3.9 Leukocyte function in inflammation. In acute inflammation caused by pyogenic bacteria, neutrophils are the most important bactericidal cells. Recruited (inflammatory) macrophages are important in removing the damaged tissues and regulating the healing response. In chronic inflammation, recruited (activated) macrophages are crucial for killing intracellular bacteria such as mycobacteria following their activation interferon (IFN)-γ from NK cells or T cells. Macrophages also both cause tissue damage and are responsible for regulating healing and repair.

If they are produced in large enough amounts they have effects on anatomically distant sites. For example, TNF-α can signal to the bone marrow to increase the output of neutrophils, IL-1 can signal to the hypothalamus to increase temperature (fever), and IL-6 can signal to the liver to induce the synthesis of a large number of defence molecules at high levels, including opsonins and other agents. The latter is called the acute-phase response. See Figure 3.10.

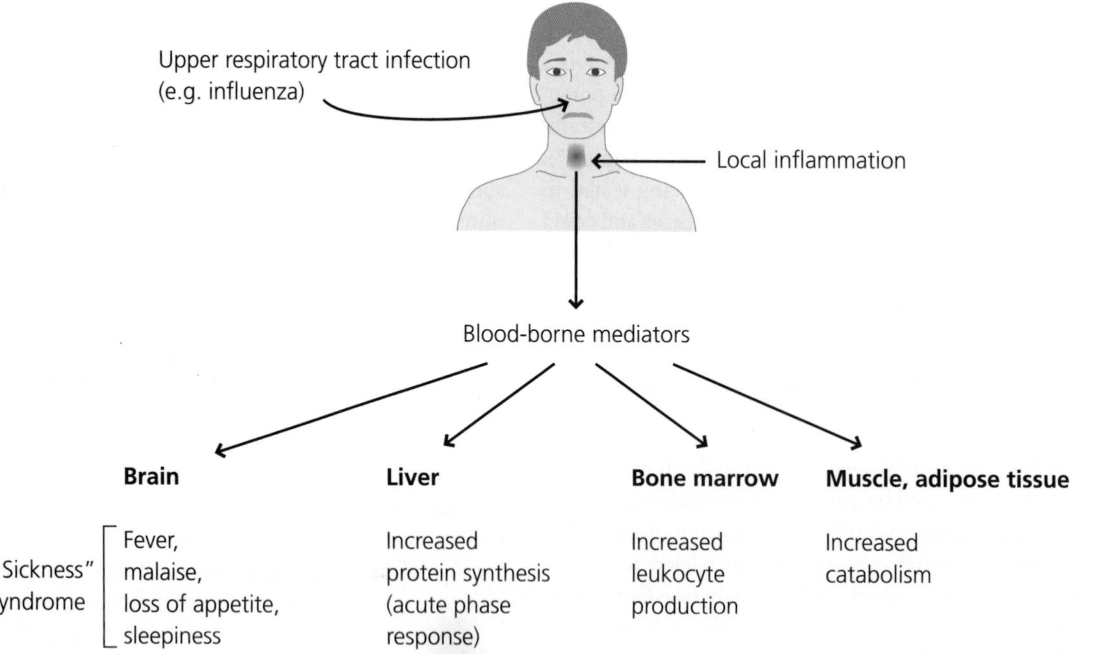

Fig. 3.10 Systemic effects of inflammation. Cells in inflamed tissues release pro-inflammatory mediators into the blood. These can act on distant organs such as the brain to cause fever, malaise and loss of appetite, and on the liver to increase synthesis of defence-related proteins such as complement components. They also act on the bone marrow to cause release of stored leukocytes and to increase the production of leukocytes from stem cells. They act on muscle and adipose tissue to increase catabolism, generating energy. The latter is crucial; to increase body temperature by only 1 °C requires as much energy as an adult walking 35–40 km.

3.3.5
Initiation of Adaptive Immunity

The final important principle to emphasize is that many of the changes that occur in inflammation are absolutely crucial to the initiation and regulation of adaptive immune responses against infectious agents. The ways in which T cells differentiate after activation is very much determined by signals generated by the innate immune system. In fact, generally speaking, it is likely that adaptive immune responses could not be triggered at all without innate responses occurring. This topic is discussed in a little more detail in Section 3.4.

3.3.6
Regulation of Inflammation, Healing and Repair

3.3.6.1 Regulation of Inflammation

What is it that stops the inflammatory process? Of primary importance, of course, is the removal of the agent that initiated the inflammation. We see the importance this in situations where the agent cannot be removed: it may be resistant microbes such as Mycobacteria, indestructible particles such as silica or a self component as in autoimmune disease (Chapter 7). In all these cases the inflammation is prolonged, becoming chronic. Cessation of inflammation however, is also an active process. Lipid mediators such as lipoxins and resolvins inhibit the recruitment and activation of neutrophils, and may direct macrophages to areas of cell death, enabling them to clear the debris. Cytokines such as IL-10 may later modulate the functions of macrophages, for example by stimulating them to secrete transforming growth factor (TGF)-β that promotes healing.

3.3.6.2 Healing and Repair in Acute Inflammation

Acute inflammation inevitably causes some tissue damage, and it is a crucial part of all inflammation that it stimulates the processes of healing and repair. In the simplest forms of damage, say a cut with a sterile scalpel during a surgical procedure, blood platelets adhere to the edges of the cut vessels, forming a platelet plug. Fibrin formed by blood coagulation stabilizes this plug and forms the first layer of repair. Blood platelets degranulate as they adhere, releasing factors such as platelet-derived growth factor (PDGF), which are crucial for later stages of repair. These and other factors cause changes in the expression of adhesion molecules by local venules, causing monocytes to adhere and cross into the inflamed area. See Figure 3.11.

The recruited monocytes can now differentiate into inflammatory" macrophages See Chapter 1. These macrophages have crucial functions in repair including the following:

i) They ingest and digest damaged cells, clearing up the mess.
ii) They secrete enzymes such as collagenase and elastase which help to degrade damaged connective tissue in the inflamed area.

1 Blood coagulation

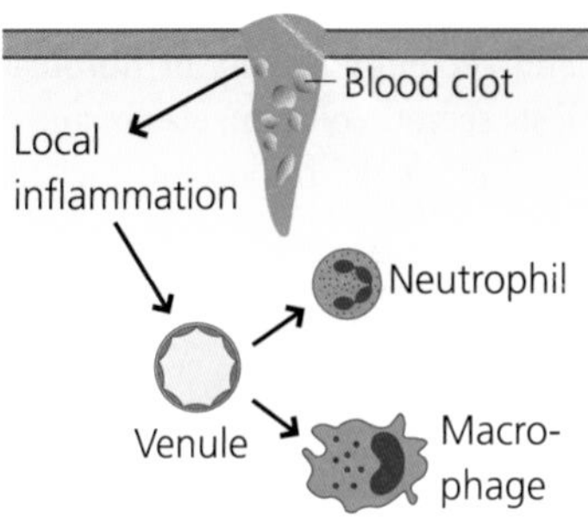

2 Re-epithelialization

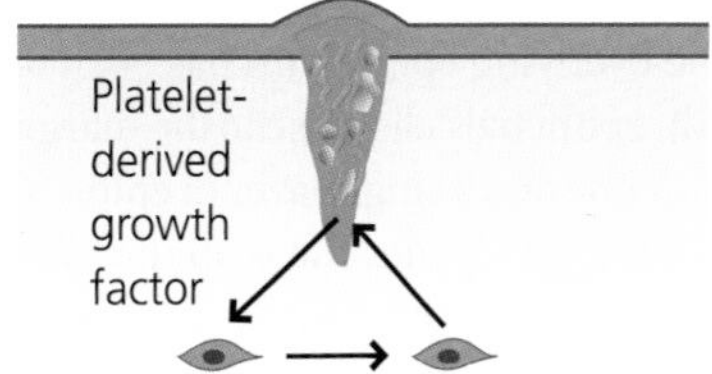

3 Angiogenesis

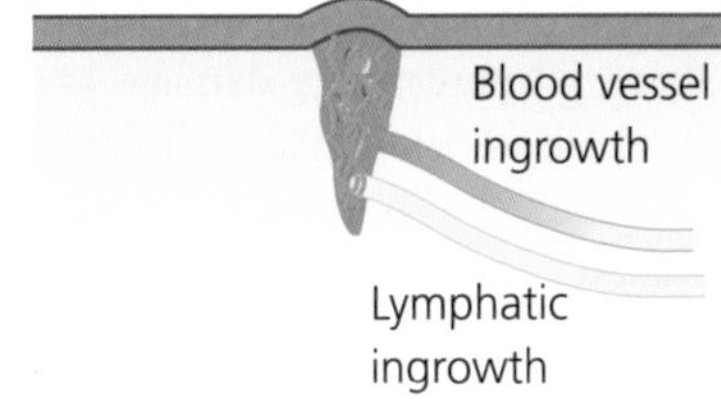

4 Scar formation

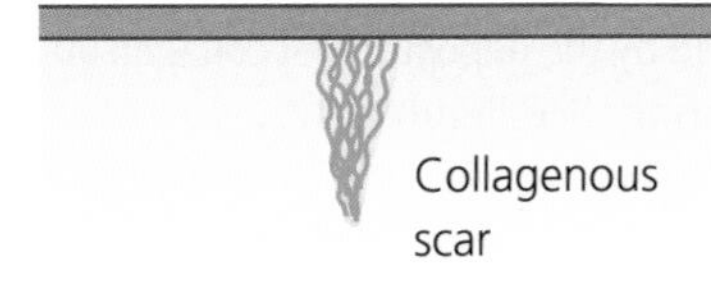

Fig. 3.11 Healing of a sterile wound. (1) A blood clot forms at the site of the wound. Local inflammation stimulates the activity of macrophages to release growth factors for fibroblasts, as do activated platelets in the clot. (2) These factors induce the recruitment and division of the fibroblasts and stimulate them to make collagen and other connective tissue molecules. (3) Other factors stimulate the in-growth of blood and lymphatic vessels (angiogenesis). (4) Epithelial cells divide to cover the wound and eventually the site of the wound is represented by a small, collagenous scar.

iii) They also secrete factors such as platelet-derived growth factor (PDGF) that stimulate the recruitment, proliferation and secretory activity of fibroblasts. In turn the fibroblasts secrete collagen, elastin and other connective tissue components important for connective tissue repair.

At the same time other factors stimulate the in-growth of blood and lymphatic vessels.

At this stage the healing tissue becomes very vascular with lots of fragile vessels (we all know how profuse bleeding can be if a wound scab is removed too early). This healing tissue is known as granulation tissue, not to be confused with the granulomas found in some forms of chronic inflammation. These processes occur in the first few days after the injury. If the overlying epithelium has been damaged (as in a skin wound), epithelial cells close to the margins of the wound start to proliferate and a single layer of epithelial cells migrates over the wound (and under the scab in a skin wound). Over time the structure of the healing tissue changes: macrophages migrate away from the area, the number of blood vessels decreases, and the amount of collagen laid down by fibroblasts increases. By 2–3 weeks after the injury the wound is represented by a relatively avascular, collagenous scar.

This repair process occurs in sterile wounds and is sometimes called "healing by first intention". If, however, the wound is infected, or if the inflammation is caused by infection, the processes described above are more or less perturbed. Inevitably there is more tissue damage and, although the processes of repair are taking place all the time, their ability to bring about healing is diminished. The result is that healing is delayed and often much more collagen is laid down, leading to increased scarring.

In some cases, as for example following lobar pneumonia caused by Streptococcus pneumoniae (Chapter 2), restoration of lung structure and function may be complete. This is known as resolution. However, this is not the case in most instances of clinical acute inflammation since healing is accompanied by the deposition of collagen and this may often lead to scarring. See Figure 3.12.

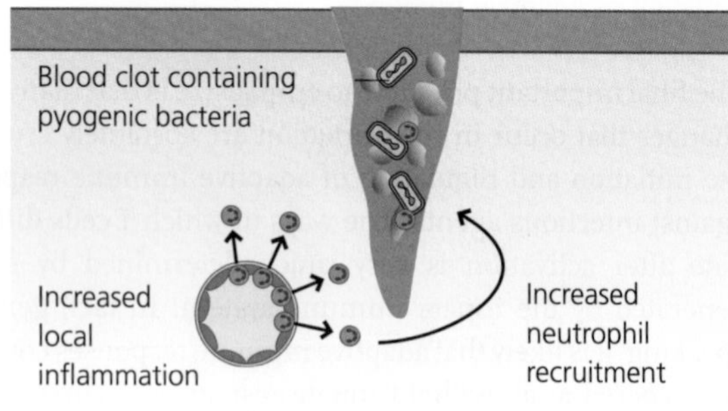

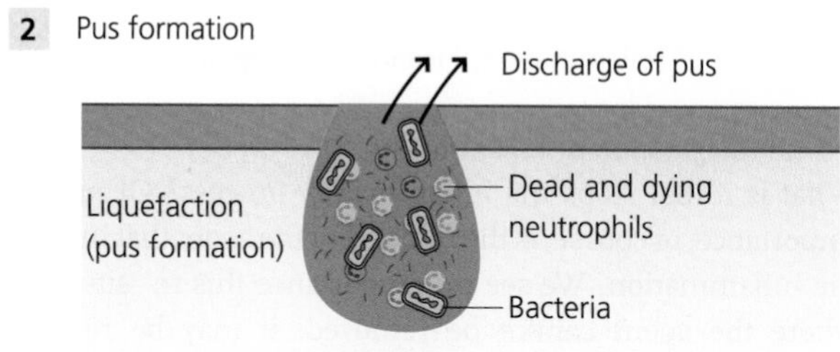

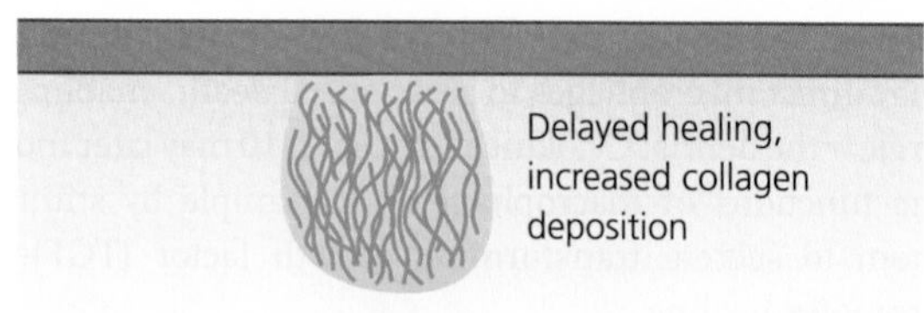

Fig. 3.12 Healing of an infected wound. If a wound is infected by pyogenic bacteria there is increased local inflammation and increased recruitment of neutrophils. (2) This may lead to increased tissue liquefaction and the formation of pus. (3) The healing process is delayed, there is increased collagen formation and increased scarring. (*cf.* Figure 3.11.)

Q3.9. Why may lobar pneumonia resolve with little or no permanent tissue destruction, whereas a pyogenic skin infection resulting in an abscess leads to permanent scarring?

3.3.6.3 Repair in Chronic Inflammation

In chronic inflammation there is a persistent inflammatory stimulus, which stimulates persistent activation of healing and repair. In these conditions the repair response may be the actual cause of the clinical disease. Thus, ingested alcohol damages and kills liver cells (hepatocytes). Hepatocytes have a remarkable capacity to regenerate: if 90% of a rat's liver is removed surgically, the remaining 10% regenerates to restore completely the original liver mass. However, the damage to the hepatocytes causes inflammation and this stimulates a healing response, leading to collagen deposition. If the stimulus is persistent, as in chronic alcoholic subjects, the healing process carries on indefinitely. This leads to the deposition of large amounts of collagen, surrounding and enveloping the regenerating hepatocytes, essentially strangling the liver. This process is known as cirrhosis. Importantly the same process can be caused by the immune response against certain viruses, such as hepatitis B and C (Chapter 2). See Figure 3.13.

Q3.10. If 90% of a rat's liver is removed surgically, the remaining hepatocytes start dividing and continue until the liver has regained its original mass. They then stop dividing. What mechanisms might enable the liver to regenerate to its previous mass?

1 Long-term liver damage (e.g. hepatitis C, alcohol)

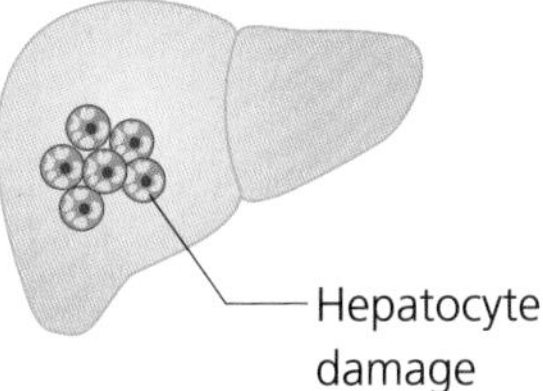

2 Chronic inflammation

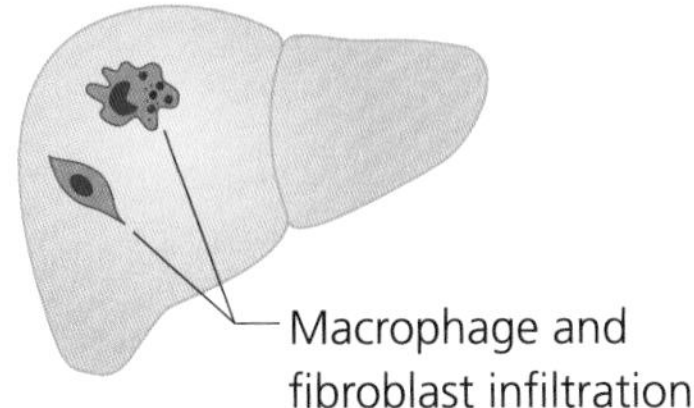

3 Healing response

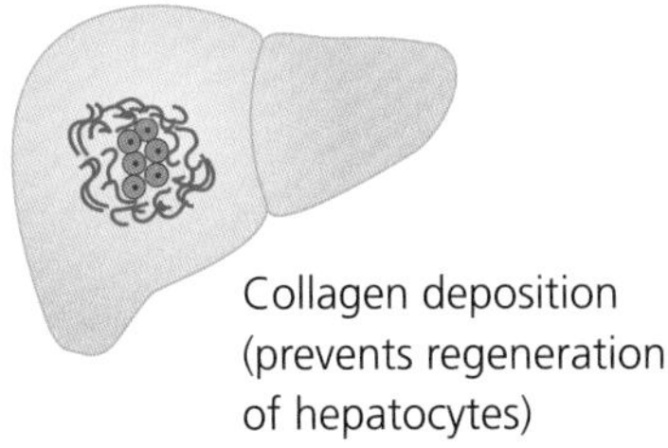

4 Liver failure

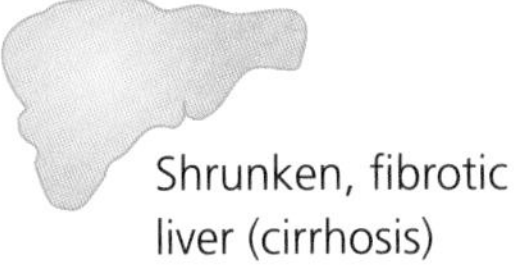

Fig. 3.13 Disease caused by an excessive healing response – cirrhosis. If the liver suffers long-term damage (e.g. from chronic infection with hepatitis viruses or excessive alcohol consumption), there is a continual healing response with large-scale deposition of collagen. The collagen surrounds and encloses groups of regenerating hepatocytes, preventing them from restoring normal cell numbers and function. The end result is a shrunken, fibrotic liver that cannot function effectively, leading to liver failure.

3.4 Functional Anatomy of Adaptive Immunity

All multi-cellular organisms, and even prokaryotic organisms such as bacteria, have evolved innate (or innate-like) immune mechanisms. However, the evolution of an entirely new type of immunity, adaptive immunity, which is mediated by lymphocytes, required the development of specialized tissues. These include particularly the secondary lymphoid tissues, the main function of which is to act as sites where adaptive immune responses are initiated and regulated. These tissues include the lymph nodes, the spleen and specialized tissues associated with mucosal sites. The latter are collectively termed mucosal-associated lymphoid tissues (MALTs) and they include Peyer's patches in the small intestine. See Figure 3.14.

3.4.1 Why Do We Need Secondary Lymphoid Tissues?

Why is it that adaptive immunity needs specialized secondary lymphoid tissues? Part of the answer lies in the nature of antigen recognition by lymphocytes. In innate immunity, the receptors involved (PRRs) can recognize a large range of pathogen-associated molecules and have a very broad range of specificities. Moreover, multiple PRRs are expressed on all cells of the innate immune system and they stimulate rapid cellular responses. Thus, the populations of innate immune cells that are present in tissues under resting conditions, or which are recruited in response to infection, are sufficient to mount a fast, large and (often) effective response. In stark contrast, the proportion of lymphocytes able to recognize a particular antigen is vanishingly small and only around $1:10^5$ to $1:10^6$ is specific for any given antigen. Thus, something like 99.99% of lymphocytes cannot recognize a particular antigen. Yet infection can strike anywhere. The reasons why we have evolved lymphocytes with such exquisite specificity is discussed elsewhere (Chapters 5 and 6), but it is not difficult to see that this very low frequency of antigen-reactive lymphocytes makes for very real logistical problems. Given that there are so few antigen-specific lymphocytes, how do they get to meet their specific antigens?

The first principle is that the adaptive immune system uses an antigen-focusing mechanism to concentrate antigens into sites where they can encounter lymphocytes. Vertebrates have evolved lymphatic systems. Almost all tissues and organs have lymphatic vessels that collect fluids from extravascular tissues and drain them into lymph nodes. Thus, a single lymph node receives fluids from a large volume of peripheral tissue. Should this happen to drain a site of infection, it will contain molecules derived from the pathogen or even whole pathogens. A lymph node is thus ideally placed to monitor peripheral tissues for infection. Not all potential sites of infection have lymphatic drainage, but other mechanisms have evolved to monitor these sites. Thus, the blood is monitored by the spleen and the intestine by lymphoid tissues in the walls of the organs; in the small intestine these are called Peyer's patches. Other types of MALT include the tonsils and adenoids in the back of the mouth and nose (nasal-associated lymphoid tissues; NALTs), the appendix, and tissues associated with the respiratory tract (bronchus-associated lymphoid tissues; BALTs). See Figure 3.15.

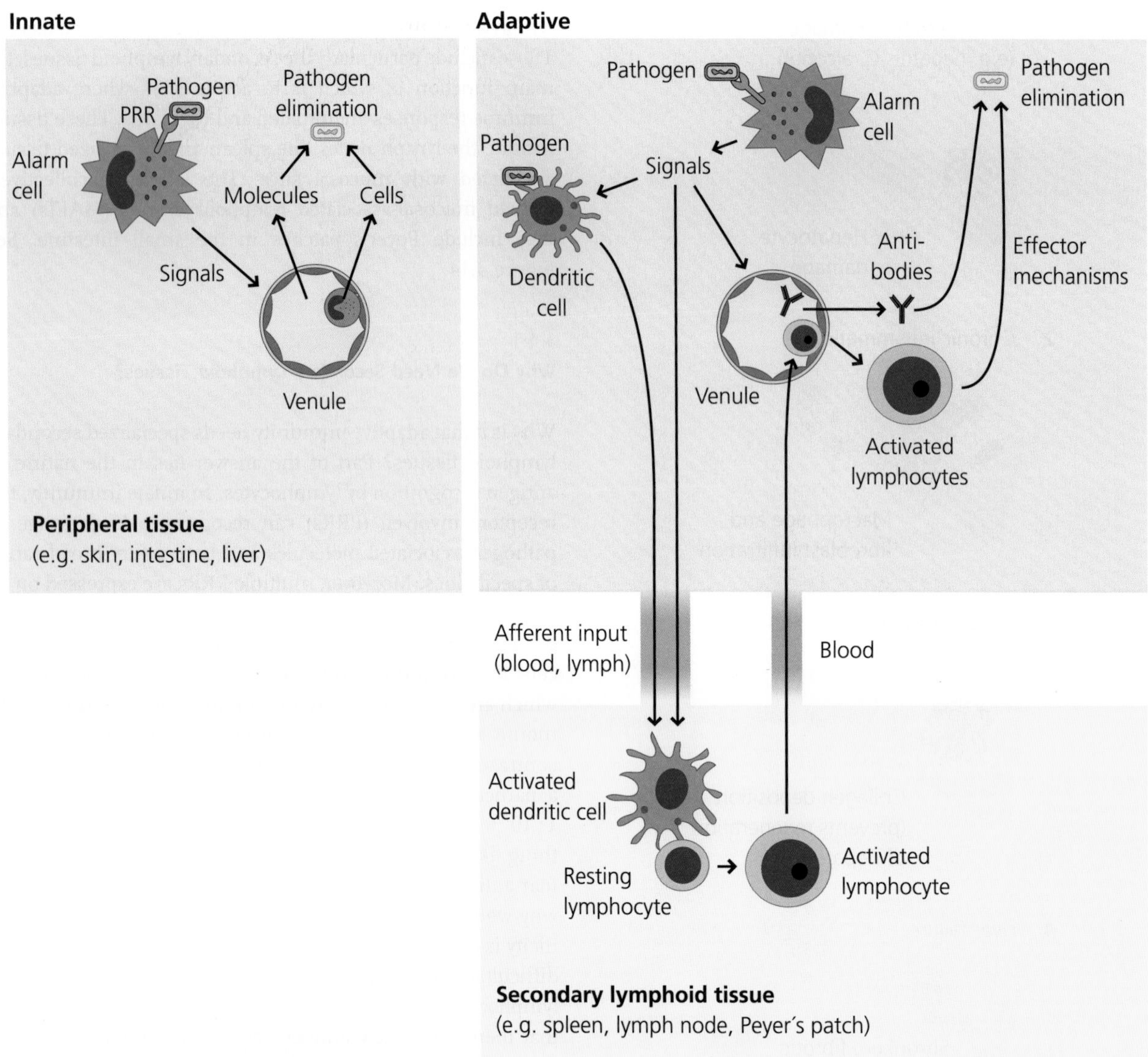

Fig. 3.14 Anatomical organization of innate and adaptive immune systems. Alarm cells of the innate system are widely dispersed throughout the body. Their local activation by infection serves to recruit effector cells and molecules to the site of infection. The adaptive system is based largely in secondary lymphoid tissues that are relatively distant from the site of initial infection. Antigens from the infecting organisms (typically carried by DCs) and signals from the innate system are transported into the secondary lymphoid organs to initiate and regulate the adaptive response. Effector mechanisms such as antibodies and activated T cells can then be transported or migrate back to the inflamed site of infection to mediate elimination of the microbe.

There is still, however, the problem of the very low frequency of antigen-specific lymphocytes. It would not be possible to have enough lymphocytes resident in each lymph node or Peyer's patch, or indeed the spleen, to cope with all potential pathogens. However, if you examine a histological section of a lymph node it is absolutely stuffed with small lymphocytes. These lymphocytes are not resident in the node; they are transient residents, spending only a few hours in the node before moving out again. This means that each node (and other secondary lymphoid organs) is monitored by huge numbers of lymphocytes. Studies in sheep, for example, have revealed that a single lymph node in this species is traversed by around 10^7 lymphocytes each hour. It is only because of this huge migration that sufficient antigen-specific lymphocytes can monitor secondary lymphoid tissues and ensure that the very rare antigen-specific lymphocytes are able to meet the antigens for which they are specific. Having met their antigen, these rare lymphocytes must be activated, stimulated to proliferate, so as to reach a critical mass that is able to deal with the infectious agent, and instructed as to which effector mechanisms they need to generate. Hence, adaptive immune responses take much longer than innate responses to become fully induced on a first exposure to an infectious agent. (Subsequent memory responses are, however, much faster.)

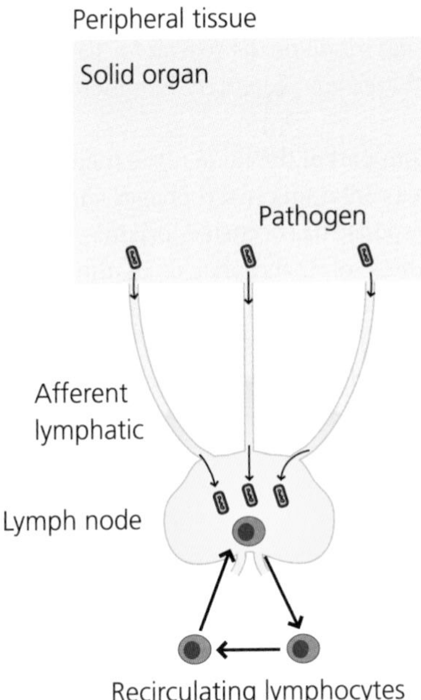

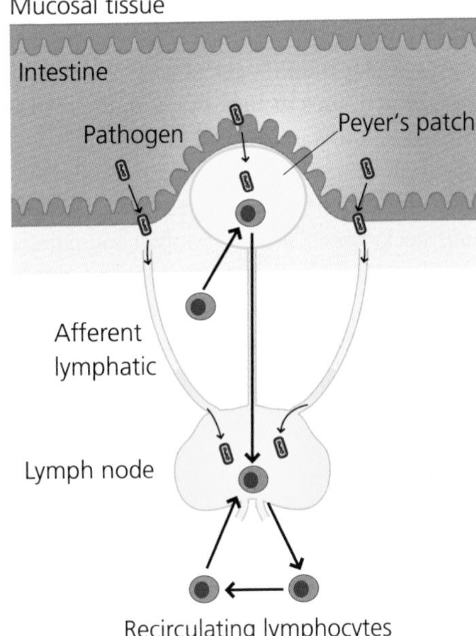

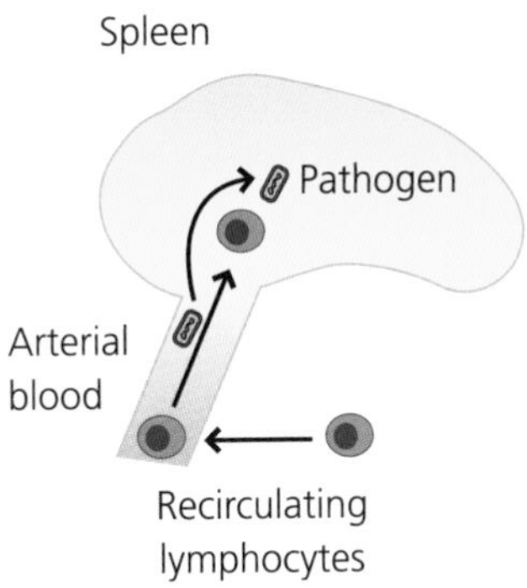

Fig. 3.15 Tissue monitoring in adaptive immunity. Most peripheral tissues and solid organs are monitored by lymph nodes which are connected to these sites by afferent lymphatics. The blood is monitored by the spleen. The intestine and some other mucosal tissues have secondary lymphoid organs embedded in their walls that monitor the lumen; in the small intestine these are Peyer's patches. Theses tissues also drain to lymph nodes. Naïve lymphocytes continually migrate to, and recirculate between all these secondary lymphoid organs, monitoring them for the presence of microbial and other antigens.

3.4.2
Structure and Function of Secondary Lymphoid Organs

Understanding how secondary lymphoid organs work in immune responses is very dependent on knowing how their structures are related to their complex functions, and all secondary lymphoid organs show basic similarities in their structure. We are, however, still struggling to understand fully the detailed structure–function relationships in these highly organized organs. In this subsection, to start with, we will use lymph nodes as generic examples of secondary lymphoid tissues, emphasizing features that are common to all such tissues; the functions of lymph nodes are perhaps the best understood of all secondary lymphoid tissues. We will then briefly discuss the specializations seen in other forms of secondary lymphoid tissues, taking Peyer's patches (a type of MALT) and the spleen as examples. These specializations include (i) the localization of these tissues, (ii) the peripheral tissues they monitor and the modes by which antigens are delivered to them, and (iii) the specialized compartments or distinctive cell subsets that they contain.

We will first examine the function of secondary lymphoid tissues in the steady state and will then discuss the changes that occur during active immune responses. It is important to realize that when we talk about the steady state, although there are no overt immune responses occurring, antigens (self-derived components and harmless non-self antigens such as those from food and commensal bacteria) are continually interacting with the immune system to induce tolerance or regulation.

3.4.2.1 Lymph Node Structure and Function in the Steady State

Lymph nodes continuously monitor peripheral tissues in case infection occurs. These tissues include the skin, the epithelia of the mucosal tissues, and vascularized organs such as the heart and kidney. Lymph nodes are small, kidney-shaped or ovoid organs often embedded in fat. (These are the glands you might feel in your neck, which become swollen and painful if you have a throat infection. If you have ever looked carefully while preparing a leg of lamb for cooking, you may have seen the greyish popliteal lymph node embedded in white fat deep inside the leg.)

Histological examination of lymph nodes shows that they have a complex structure (See Figure 3.16). They are contained in a fibrous capsule. This capsule is pierced by several small afferent lymphatic vessels that empty into the subcapsular sinus, the space between the capsule and the cellular body of the node. The subcapsular sinus contains specialized macrophages. It also serves to prevent macromolecules diffusing directly into the body of the node. The body of the node contains three main, anatomically distinct areas. The outer part, the cortex, contains spherical follicles that are the B cell areas Between the follicles are areas containing macrophages and migrating dendritic cells (DCs; below). Below the cortex is the paracortex which is the T cell area.

Q3.11. What might be the reason for having separate T cell and B cell areas in secondary lymphoid organs?

The bottom part of the node is the medulla, which contains wide sinuses containing macrophages and, in nodes where an immune response has occurred, plasma cells. The presence of large numbers of macrophages in the medulla illustrates another major function of lymph nodes: they act as filters for a variety of particles (including bacteria) and prevent, or at least slow down, the entry of these into the blood. The sinuses in the medulla then come together to empty into the efferent lymphatic at the hilum of the node, usually as a single vessel. Efferent lymph leaving in this vessel may pass through other nodes via afferent and efferent lymphatics in turn (they often form chains), but eventually is collected into larger lymphatic vessels and ultimately drain back into the venous system.

Q3.12. Why might the examination of the thoracic cavity of a long-term smoker illustrate the filtration function of lymph nodes?

The blood supply of the node consists of an artery which enters at the hilum and branches to form capillaries. In the

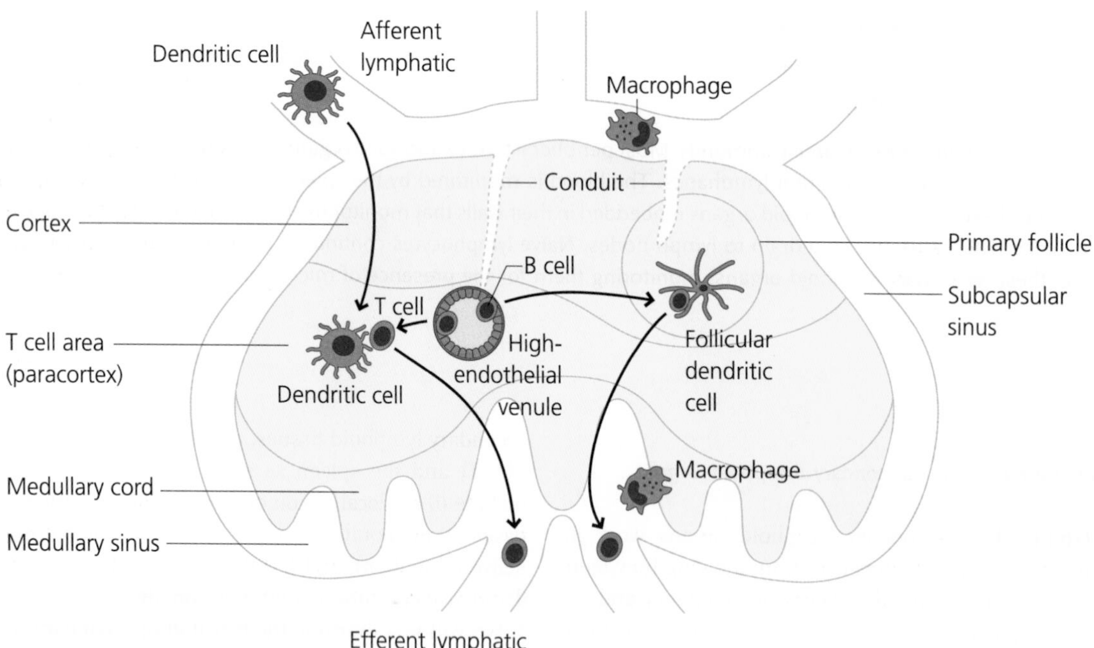

Fig. 3.16 Lymph node structure. Lymph nodes receive lymph from peripheral tissues through several afferent lymphatics that drain into the subcapsular sinus. Cells and molecules in the lymph are delivered to different specialized areas of the node. Cortical B cell follicles contain recirculating B cells and resident follicular dendritic cells (FDCs). The paracortex contains recirculating T cells and classical DCs, some of which have entered from the lymph. This area also contains high endothelial venules (HEV) which are the sites where recirculating T and B cells enter from the blood. The medulla contains sinuses that are lined with macrophages which have a filtering function. Lymph leaves the node via the efferent lymphatic(s) and is eventually drained back into the blood.

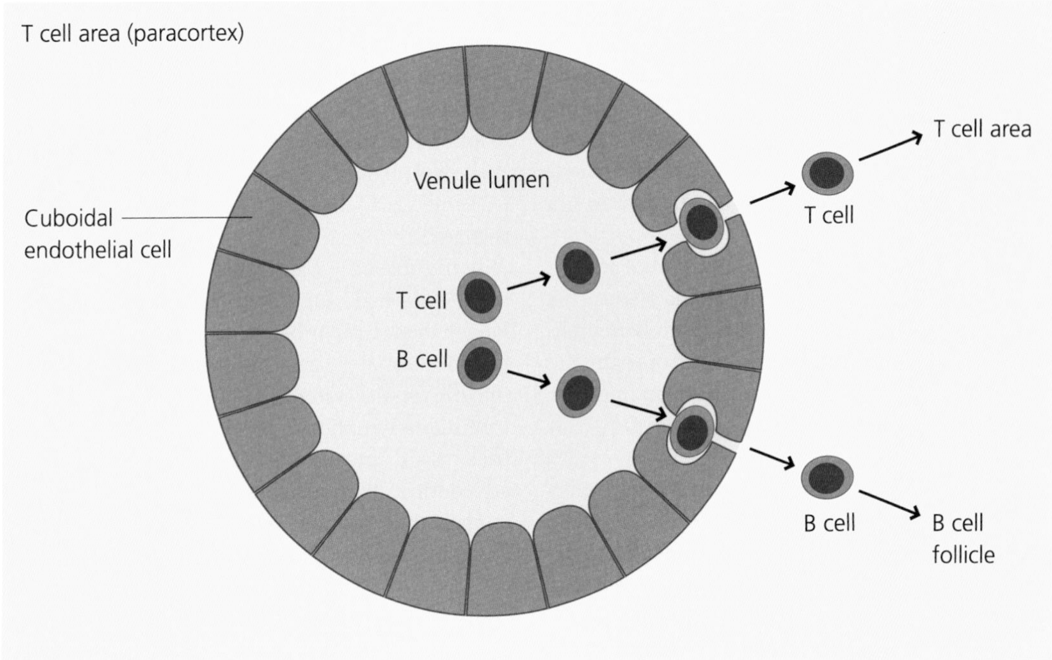

Fig. 3.17 High-endothelial venules. These post-capillary venules have cuboidal endothelial cells that express adhesion molecules complementary to lymphocyte surface molecules, permitting lymphocytes to attach and migrate between the endothelial cells into the node. HEVs are present in lymph nodes and mucosal secondary lymphoid tissues such as Peyer's patches, but not the spleen. HEVs are very efficient, extracting around 30% of lymphocytes that enter a node from the arterial blood during each passage through it.

paracortex the capillaries join to form post-capillary venules with an unusual structure. The endothelial cells of these venules are much more cuboidal than typical endothelial cells and the vessels are known as high endothelial venules (HEVs). Lymphocytes can be seen adhering to HEVs and also present within the endothelial layer; electron microscopy shows the lymphocytes between the cuboidal endothelial cells, apparently crossing from the blood into the body of the node itself. HEVs then join to form a vein that exits the node at the hilum (see Figure 3.17).

There is another structure in nodes, only recently recognized, that is involved in the transport of molecules across the node. This is the fibroblastic reticular network. Tubular structures called conduits, surrounded by fibroblastic cells and not by endothelial cells, run from the floor of the subcapsular sinus through the T cell areas and terminate at HEVs. Similar conduits also run from the subcapsular sinus into the B cell follicles. Possible functions of these conduits are discussed below.

Functionally, secondary lymphoid tissues can be thought of as dynamic structures that have evolved to enable the meeting of different streams of cells and molecules that need to interact in the induction of adaptive immune responses. Lymph nodes have two such streams. The first of these is the afferent lymph that drains peripheral tissues. The afferent lymph carries both cells and molecules that play crucial roles in inducing tolerance (Section 1.6.1) in the steady state. Following infection in the peripheral tissues, afferent lymph then transports the cells and molecules needed to induce an active immune response. In this context the most important cells in afferent lymph are probably DCs. These cells monitor peripheral tissues continually and migrate to draining lymph nodes, carrying antigens that can be recognized by specific T cells. DCs also transmit information about the state of the peripheral tissue; whether it is inflamed or infected and, if so, what type of inflammation or infection is present. They can do this because they respond, directly or indirectly to signals generated by PAMPs or DAMPs binding to their PRRs. Only lymph nodes have a supply of afferent lymph; Peyer's patches and the spleen do not, because these are designed to monitor infections in other sites that lack lymphatics.

The second type of cellular input into the nodes are the lymphocytes derived from the blood. HEVs in the T cell areas (paracortex) of the node express molecules on the luminal side of their endothelial cells for which naïve, resting lymphocytes have receptors (this is the basis of lymph node post code for lymphocytes, as well as some DCs). Lymphocytes that bind to HEVs via these receptors migrate into the extravascular tissues of the node in a remarkably efficient process. Both T and B cells enter the node via HEVs in T cell areas, but whereas T cells eventually migrate directly from the T cell area into efferent lymph, B cells first pass through the follicles and thus take longer to traverse the node. HEVs are also present in Peyer's patches but are not present in the spleen.

The exit routes from the node are the blood and the efferent lymph. Examination of efferent lymph from steady state nodes shows that it contains very large numbers of lymphocytes – both T and B cells, mainly as naïve, resting cells. In contrast to afferent lymph, DCs are very rare in efferent lymph. Careful analysis has shown that at least 95% of DCs that enter the node are filtered out and do not leave; they die in the node.

DCs in the node are not all derived from afferent lymph. Some enter nodes as precursors directly from the blood. This has been shown by isolating DC precursors from bone marrow and injecting them into the blood; the precursors enter the T cell areas of the node from the blood and develop into DCs (Box 3.3). A subset of classical, and plasmacytoid, DCs may enter the node in this way (Section 1.4.6).

To summarize, in the steady state, afferent lymph brings a continual stream of DCs and soluble molecules into the node. The DCs migrate into T cell areas of the node and die therein by apoptosis; in this sense they can be viewed as a conveyor belt for antigen delivery. In the T cell areas these DCs are interrogated by naïve T cells for the presence of specific antigen. If the T cell does not recognize antigen it leaves the T cell area and the node via the efferent lymph, re-enters the blood, and migrates to another secondary lymphoid organ. A very similar process occurs with naïve B cells. This process is known as lymphocyte recirculation (see Box 3.2).

3.4.2.2 Lymphocyte Recirculation

As we have seen (above), most lymphocytes enter nodes from the blood; the same is true for Peyer's patches and the spleen (below). In the steady state, these lymphocytes are small, resting, mainly naïve cells (i.e. they have not recognized antigen). Their entry appears to occur largely at random: naïve lymphocytes do not appear to be specialized to enter a particular type of secondary lymphoid tissue such as the lymph nodes, Peyer's patches or spleen. (This situation is somewhat changed, however, once they have been activated.)

In the absence of antigenic stimulation naïve lymphocytes reside in a node for a short time (a few hours; B cells stay longer than T cells) before leaving the node in the efferent lymph, ultimately entering the thoracic duct which empties into the venous system and thus they return to the blood. This continuous lymphocyte recirculation from blood to lymph and back again means that all secondary lymphoid tissues are continually surveyed by lymphocytes for the presence of antigen that they can recognize (sometimes called cognate antigen). See Figure 3.18.

3.4.2.3 T Cell Responses in Lymph Nodes

If infection occurs in a peripheral tissue, the draining lymph nodes undergo substantial changes – as becomes evident by the swollen glands that, for example, may be felt in the neck during severe throat infections. This swelling reflects inflammation in the nodes and is due to both oedema and cell accumulation.

T Cell Recognition of Antigens Before T cells can acquire their specialized effector functions, they first need to be activated.

Box 3.2 Discovery of Lymphocyte Recirculation in Rats

It is possible to insert a cannula (tube) into the thoracic lymph duct (the main vessel carrying lymph from peripheral tissues into the blood) and thus to collect lymph. It had been known for many years that thoracic duct lymph, obtained from a variety of mammalian species, contained very large numbers of small lymphocytes. The origins and fates (and functions) of these cells were not, however, understood. It was thought, for instance, that there could be continual production of lymphocytes somewhere in the peripheral tissues, and that these cells entered blood from lymph and then soon died. To explore these problems, it was shown in the 1960s that if the thoracic ducts of rats were cannulated for a period of days and all the exiting lymphocytes were collected and discarded, the numbers of lymphocytes present in the lymph declined progressively. If, however, the thoracic duct lymphocytes that were collected were re-infused intravenously into the cannulated rat, the decline in thoracic duct lymphocyte numbers was much reduced. This showed that lymphocytes were not being produced continually, but rather were undergoing a process of repeated rounds of migration from blood into lymph. It was for these reasons that this process was termed lymphocyte recirculation.

To determine the anatomical basis of recirculation, rat thoracic duct lymphocytes were collected, labelled with a radioactive RNA precursor that became incorporated into the cells, and then re-infused into the veins of other rats. At intervals, lymph nodes from the recipient rats were prepared for histology (i.e. fixed, embedded in wax, finely sectioned and simply stained to reveal basic tissue structure). Then the lymph nodes sections were overlaid with photographic emulsion; this enabled the localization of the radioactive cells to be detected because the radioactive emissions caused silver grains to be deposited when the emulsion was developed; this technique is called autoradiography. It was found that, soon after injection, the lymphocytes appeared to attach to specialized regions of the lymph nodes (the HEVs) and then to cross into the substance of the node. It was therefore inferred that lymphocytes migrated from blood via HEVs into lymph nodes and subsequently exited in the efferent lymph; they were then collected into the thoracic duct and re-entered the blood to repeat the cycle.

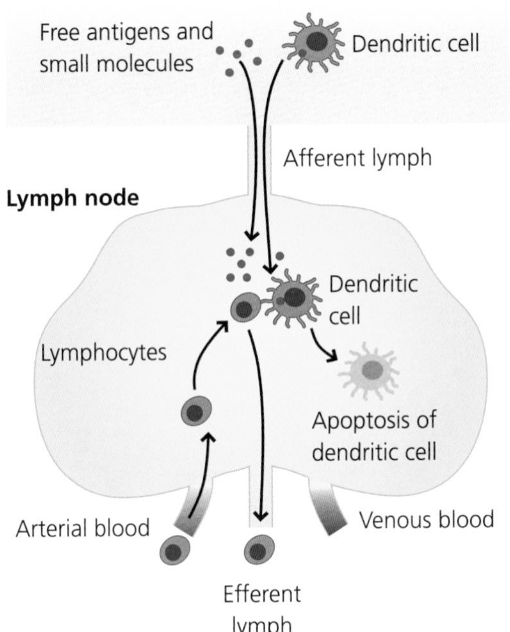

Fig. 3.18 Cellular and molecular streams in a lymph node (steady state). Afferent lymph brings soluble molecules and migratory DCs from peripheral tissues into the node. The DCs enter the T cell paracortex and molecules are delivered to different areas of the nodes. Naïve B and T cells enter the paracortex from the blood via HEVs (not shown). T cells interact with DCs in the paracortex and then leave via efferent lymph. B cells migrate to the follicle before they too enter efferent lymph and thence back to the blood (not shown). The continuous passage of lymphocytes into and out of lymph nodes and other secondary lymphoid tissues is known as lymphocyte recirculation.

This is introduced in Chapter 1 and discussed in more detail in Chapter 5, but for now some key points are as follows:

i) T cell activation first requires that T cell receptors (TCRs) the peptide–Major Histocompatibility Complex (MHC) complexes for which they are specific.
ii) T cells must additionally receive co-stimulatory signals to become fully activated; most evidence indicates that DCs are crucial for delivering these signals to T cells.
iii) CD4 T cells (and probably CD8 T cells) need to receive further signals to instruct them to produce a response that is most appropriate to the infectious agents they need to help eliminate, and to ensure they reach the correct site.

These stages in T cell activation and polarization of functions largely or exclusively occur within the secondary lymphoid tissues.

How do antigens from microbes reach the lymph node so they can be recognized by the rare antigen-specific T cells that are recirculating through it? If we consider DCs to be essential in this process (below) here are two main possibilities. (i) DCs in peripheral sites are continually sampling their environment. If there is a microbial infection, DCs can internalize the microbes, or antigens derived from the microbes, before migrating to nodes in afferent lymph. In the node they enter the T cell areas where they can interact with these lymphocytes. (ii) There is another population of DCs that are resident in the nodes and which migrate directly there from the blood. Small macromolecules less than 70 kDa, perhaps including soluble antigens, can enter the substance of the node from the afferent lymph through the conduits. These resident DCs attach themselves to the conduit and appear to extend processes into gaps between the cells forming the conduit, where they may be able to sample small antigens. Thus, T cells may be able to recognize their specific antigens because they can potentially interact with two different populations of antigen-sampling DCs, which can then initiate or drive the activation of the T cell. See Figure 3.19 and Box 3.3.

Q3.13. DCs also migrate from normal, steady-state tissues in the absence of infection, although at a lower rate. Why might this be important?

Lymph Node Shut Down and Lymphocyte Recruitment Infection in peripheral sites leads to significant changes in the physiology of the draining lymph nodes. Efferent lymphatics can be cannulated easily in large animals such as sheep or cows; this is very difficult in small animals such as mice and rats. If the efferent lymph from a stimulated node in a sheep is collected there is a found to be a very rapid decrease in lymphocyte exit from the node. This is called shut down and is perhaps due to Type I IFN secreted by plasmacytoid DCs (pDCs). These cells represent a distinct, specialised population of DCs whose functions in immune responses are not fully understood. They are called plasmacytoid because they are shown by electron microscopy to contain very large amounts of rough endoplasmic reticulum. Importantly, they rapidly secrete very large amounts of Type I IFN when they are stimulated via PRRs, and are thought to play important roles in regulating T cell activation and in resistance to viral infection. Experimentally, lymph node shut down can be stimulated by direct injection of Type I IFN into peripheral tissues, so it is tempting to speculate that pDCs may be primarily responsible for this phenomenon *in vivo*.

The next stage is lymphocyte recruitment. If lymphocytes recognize their cognate antigen they are retained in the node and activated. Thus antigen-specific lymphocytes can accumulate in a stimulated node and, even though the frequency of

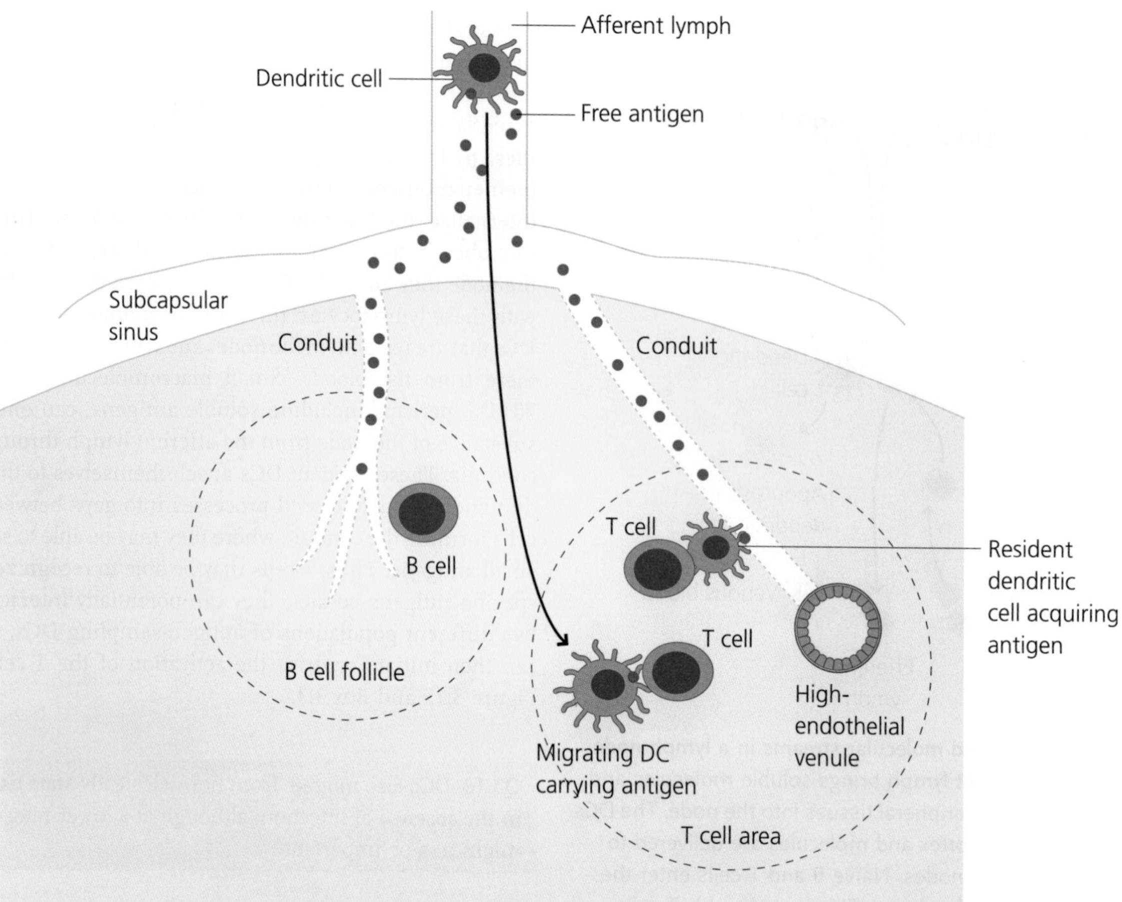

Fig. 3.19 Antigen delivery to lymph nodes. Antigens in peripheral tissues may be acquired by classical DCs and be transported by them to T cell areas in the node. Soluble antigens enter via afferent lymph and some enter connective tissue conduits which penetrate both B cell follicles and T cell areas. Resident DCs in the T cell areas can acquire antigens from the conduits. The functions of these different forms of antigen delivery are not yet fully understood (see Box 3.3).

such antigen-specific is very low, this recruitment means that the specific response can be amplified rapidly in the antigen-stimulated node. Lymphocyte recruitment has also been shown to occur in the spleen and applies to both T and B cells. See Figure 3.20 and Box 3.4.

T Cell Polarization in Secondary Lymphoid Tissues Once activated, naive T cells can differentiate down several different pathways (e.g. T_h0, T_h1, T_h2, T_h17 or T_{reg}; Section 1.4.5.1). What determines which pathway they adopt? In the absence of danger signals such T cells become T_{reg} or are induced to die by apoptosis and their absence results in antigen-specific tolerance. If, however, they are activated in a node draining inflamed tissue they may adopt any of the other pathways. This differentiation happens in the lymph node and the inflammation is in the peripheral tissue. The only direct connection between the two compartments is the afferent lymph and thus the information that determines the T cell's fate must be carried in the lymph. This information is transmitted both by cells and molecules that enter the node via afferent lymphatics.

During infection and inflammation in peripheral sites, neutrophils are recruited to the tissue, as are newly formed macrophages and DCs, both derived from blood monocytes, and these will express different surface molecules and patterns of secretion depending on the nature of the inflammation they encounter. In some types of inflammation eosinophils and basophils also enter the tissue. All these cell types can subsequently leave the tissue via afferent lymph and migrate to the draining node. Additionally, there will be many different bioactive molecules produced in the inflamed tissue, and lymph will collect all these molecules and transport them to the node. These molecules may include cytokines, chemokines, lipid mediators and other molecules, such as activated complement components. We know that all this novel information will arrive at the node and may be integrated to regulate T cell differentiation. What we do not yet know is the relative importance of these different sources of information or how this information is actually integrated to determine the CD4 T cell differentiation pathway. See Figure 3.21.

Box 3.3 How DC Activate T cells in Lymph Nodes of Mice

Antigen Transport To Lymph Nodes

If a protein antigen which happened to contain lipopolysaccharide (LPS)) is injected subcutaneously into the tip of a mouse ear this can prime antigen-specific T cells and when the mouse it is later challenged with the same antigen it shows a delayed-type hypersensitivity (DTH) response (Chapter 7) if antigen is injected as above into a naïve mouse, two pools of antigen enter lymph nodes; an initial soluble pool that drains rapidly from the site of injection, and a cell-associated pool carried by DCs that later migrate from the site. The soluble antigen enters the node very rapidly, within minutes to hours; and, if it is small enough (below 70 kDa) and labelled with a fluorochrome to enable visualization, it may seen in the conduits. DCs adjacent to the conduits can acquire small amounts of this soluble antigen, apparently by inserting processes into the conduit. DCs carrying larger amounts of antigen do not appear until 24 h or longer after injection. Which of these sets of DCs is important in initiating active immunity?

By surgically removing the tip of the ear from anaesthetized mice at different intervals after injection of the antigen it is possible to dissect the events that occur subsequently. If the tip of the ear is removed a few hours after injection, the soluble antigen arrives in the node but the migrating DCs carrying large amounts of antigen have not yet arrived. However the resident, conduit-associated DC do acquire small amounts of antigen. If antigen-specific transgenic T cells (all of the same specificity; see below) are adoptively transferred into the mice, they become activated and proliferate. However, if these mice are injected with the same antigen again, they do not display a DTH response. Hence, it appears that while the rapidly entering pool of antigen can partially activate T cells, full activation and polarization (e.g. to T_h1 cells which drive a Th1 response) requires their interaction with the later-arriving antigen-bearing DC.

Interactions of DCs and T Cells in the lymph Node

How do we know that the migrating DC actually interact with antigen-specific T cells and that this interaction is important for T cell activation? In other studies, first small numbers of TCR-transgenic CD4 T cells that had been labelled with one fluorescent probe were transferred to mice. This is because the frequency of antigen-specific T cells is very low, but their numbers can be artificially increased by transferring transgenic T cells of single antigen specificity enabling them to be more readily visualized; this approach is informally termed spiking. These mice were also injected subcutaneously with DCs which were labelled with a different probe and had been treated (pulsed) with the antigen for which the T cells were specific. At intervals after injecting the DCs, the draining lymph node was removed and sections of the node were examined by fluorescence microscopy. These experiments showed that some antigen-bearing DCs reached the node, and within a few hours had bound antigen-specific T cells which rapidly expressed activation markers. This did not happen if the DCs were not antigen-pulsed. This is strong evidence for the role of DCs in the activation of naïve CD4 T cells.

Q3.14. The investigators noticed that at later time points, there were fewer labelled DCs present remaining in the node if the DCs carried the antigen than if they did not. What might be the significance of this observation?

Real-time examination of these interactions can also be made in living animals: in an anaesthetized mouse the relevant lymph node can be exteriorized and examined using a multi-photon fluorescent microscope, which permits visualization of structures relatively deep in the node. Thus, the interaction of lymphocytes and DCs or other cells can be studied over a period of time and the complexity of these interactions observed.

Q3.15. If labelled chemokines are injected subcutaneously into a mouse, within 30 min the chemokines are expressed on the luminal surface of HEVs. How might this phenomenon aid immune responses.

Q3.16. How might we be able to start sorting out the relative importance of different peripheral influences on T cell differentiation.

3.4.2.4 Fates of Activated T Cells

Having been activated in the T cell area, CD4 T cells have several different potential fates:

i) They may remain in the T cell area and interact with CD8 T cells to help in their activation (e.g. to become cytotoxic T lymphocytes (CTLs)).
ii) They may interact with B cells in the margins of the T cell area and then migrate into the B cell follicle to provide help in B cell activation (below).
iii) They may leave the node in efferent lymph, enter the blood and migrate to inflamed tissues. Here, depending on how they have become polarized, they may, for example, help to regulate macrophage function or to recruit different types of granulocytes to these sites. Unlike naïve T cells, activated T cells, to a certain extent at least, can migrate preferentially to different peripheral tissues. Most activated T cells express adhesion molecules for inflamed tissues, enabling them to migrate to

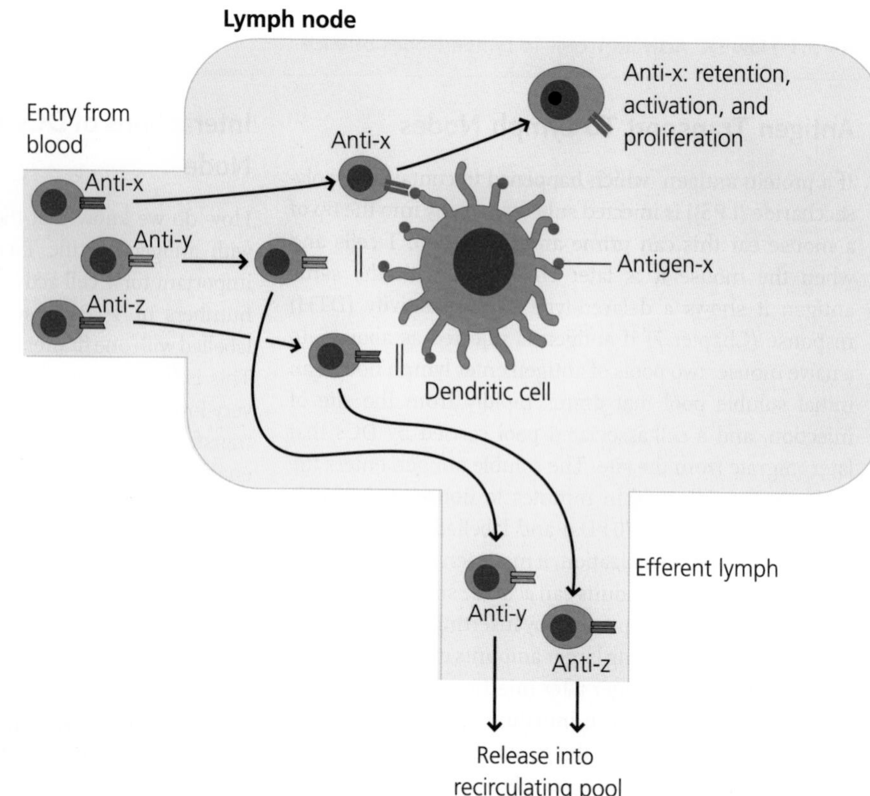

Fig. 3.20 Lymphocyte recruitment. Naïve B and T cells are continually entering lymph nodes from the blood. If a B or T cell does not recognize antigen in the node it leaves again after a few hours. If, however, antigen is recognized, the lymphocyte is retained in the node and may become activated. This recruitment of lymphocytes from the recirculating pool represents a method of rapidly increasing the numbers of antigen-specific lymphocytes at the secondary lymphoid tissue where they are needed to make a response. After activation, the clones of antigen-specific lymphocytes are expanded rapidly by proliferation (clonal expansion).

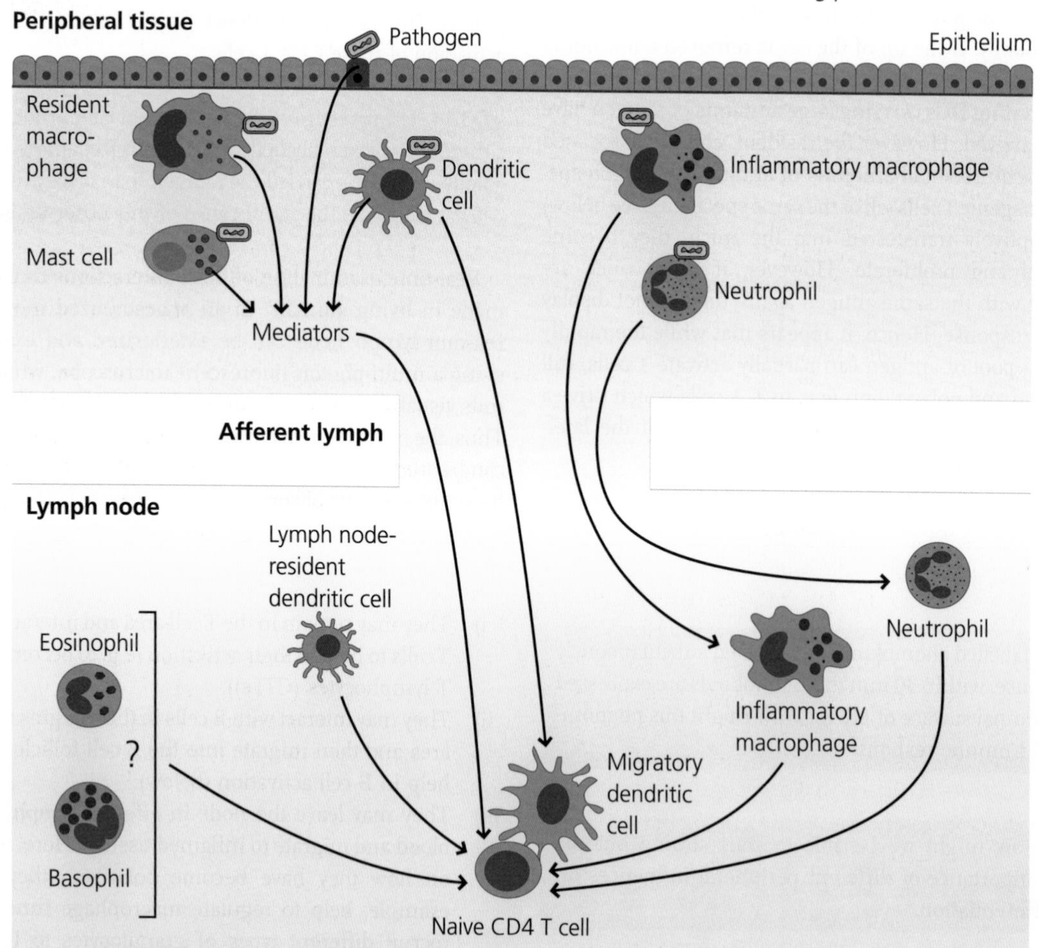

Fig. 3.21 Influences on naïve CD4 T cell differentiation. A naïve T cell, when activated, can adopt one of several distinct differentiation pathways, and which pathway is adopted depends on external signals received by the T cell in the node. Signals, which could be cell-bound or soluble, may derive from a number of different sources. Migrating DCs are a major source of such signals, but the roles of other potential sources are not well-understood; some of these potential sources are illustrated.

Box 3.4 Recruitment and Activation of Antigen-Specific T Cells

The approach of spiking normal, wild-type mice with fluorescently labelled transgenic T cells, to enable these cells to be visualized and traced during immune responses, is introduced in Box 3.3. Using this approach, further information on the kinetics of recruitment and activation of T cells can be obtained, as in the following examples.

Lymphocyte Recruitment

Groups of mice can be injected with the antigen for which the labelled T cells are specific, and the node draining the injection site can be removed at intervals after injection. Within a few hours labelled T cells are seen to start to accumulate in the lymph node that drains that tissue bed. Thus, antigen-specific lymphocytes are recruited into the nodes that have received the antigen. This recruitment is also seen for B cells and means that large numbers of antigen-specific lymphocytes can be directed rapidly to where they are needed. This rapid delivery, and retention, of antigen-specific lymphocytes in antigen-stimulated secondary lymphoid tissues is the central function of lymphocyte recirculation; it also happens in the spleen for lymphocytes specific for blood-borne antigens.

Lymphocyte Activation

At later time points, if they recognize their specific antigen, lymphocytes become activated, express novel molecules on their surfaces and start to divide. Two markers expressed rapidly by activated T cells are CD69 and CD25 (one chain of the high-affinity IL-2 receptor). If antigen is given at a particular site, the place where lymphocytes first show signs of activation can be determined. Thus, if antigen is given subcutaneously, the first site where T cells expressing CD69 and CD25 are found is the draining lymph node. Activated T cells are not found in any other lymph nodes at these early time points. Soon after expressing these activation markers, lymphocytes start to divide. One way of assessing division is by the technique of dye dilution. Each time a fluorescently labelled lymphocyte divides, the amount of fluorescent dye the daughter cells contain is halved. By extracting lymphocytes from secondary lymphoid organs and measuring their fluorescence by flow cytometry, cell division can be detected and quantified. Thus, if antigen is injected subcutaneously, lymphocytes are first activated in the draining lymph node, whereas intravenously injected antigen activates lymphocytes in the spleen. These and other similar experiments have provided good evidence that naïve lymphocytes are first activated in the secondary lymphoid tissue that drains the site of antigen deposition, and not in other lymphoid tissues or within the peripheral tissue itself.

sites of infection. T cells may also be programmed during their activation to migrate to particular types of tissues. For example, T cells activated in lymph nodes draining mucosal tissues may home preferentially to mucosal tissues, while T cells activated by antigens in the skin may home to the skin. This specialization may depend in part on the origin of the DCs that activated them – the DCs themselves having been programmed by tissue-specific signals. See Figure 3.22.

3.4.2.5 B Cell Responses in Lymph Nodes

B lymphocytes are the precursors of antibody-forming plasma cells and memory B cells. As for T cells, naïve B cells first need to be activated before they can develop further. In many instances, such as in antibody responses against protein antigens, these B cells need to receive help from the T cells themselves, particularly from activated CD4 T cells. In such T-dependent (TD) responses CD4 T cells control, the type of antibodies that the B cells will make when they subsequently develop into plasma cells (Section 1.4.5.1). Here, we will focus on TD responses; more details of these, and T-independent (TI) B cell responses, are discussed in Chapter 6.

The activation of B cells, like that of T cells, occurs within secondary lymphoid tissues. Unlike T cells, however, B cells have the option of changing their B cell receptors in the course of an immune response, so they can produce higher affinity antibodies of a different class for example (Section 1.5.3.2). The initiation of a B cell response, and these later events in the course of it, occur in different anatomical compartments of lymph nodes and of other secondary lymphoid tissues. Moreover, if B cells develop into plasma cells, these may either remain in the node or travel back to the bone marrow (from where the B cells were originally produced) and survive for long periods as antibody-secreting cells. Let us now look at some of these processes in a little more detail, in relation to how and where they occur. See Figure 3.23.

Antigen Recognition by B Cells B cells enter lymph nodes through the same HEVs as T cells and then migrate into the B cell follicles. How do they recognize antigen so they can be activated; where does it come from? Native (unprocessed) protein antigens present in the periphery will get to lymph nodes via afferent lymph (or to the spleen directly from blood and to Peyer's patches from the intestinal lumen; see later). Some antigen diffuses freely; this is clearly shown for lymph. There is evidence that macrophages in the subcapsular space can trap antigens in lymph nodes and somehow enable its subsequent transfer to B cells in follicles. In addition, more recent studies have revealed the existence of conduits (Section 3.4.2.1) that may deliver small soluble antigens

Box 3.5 Localization of Memory T Cells

To examine the characteristics of memory T cells, adoptive transfer experiments were used. T cells were isolated from secondary lymphoid organs of immunized mice and transferred into normal, genetically identical mice. Challenge of the transferred mice with the specific antigen elicited a memory response. It was then possible to examine the T cells from the original mice by flow cytometry to identify subsets expressing different patterns of surface molecule expression. These subsets could then be separated by cell sorting and the sorted cells transferred into normal mice. By challenging the transferred mice with the antigen, the surface phenotype of the memory cells could be determined. It was thus shown that at short intervals after immunization, the memory cells showed characteristics of activation – they were large cells, expressed activation markers and did not express lymph node homing receptors. They tended to migrate to inflamed endothelium. These cells were called **effector memory** cells. At longer periods after immunization, however, the memory cells were much more like naïve T cells – they were small, did not express activation markers, did express lymph node homing receptors and migrated to lymph nodes. These were called **central memory** cells.

directly to the follicles directly. It is also possible that DCs play a role. These cells, which we think of primarily as T cell activators, can in fact retain intracellular protein antigens in native form for relatively long periods (unlike macrophages, which degrade proteins very rapidly). DCs migrate to the T cell areas and may be able to release antigen for recognition by B cells which are migrating through the T cell area from HEVs. This could serve as an efficient way of delivering antigen to B cells.

The B cell follicles in a non-immunized node contain two main cell types: small, recirculating, naïve B cells (above) and a very distinctive non-lymphocytic cell, the follicular DC (FDC). FDCs are a quite different cell type from the DCs that express (processed) antigen as peptide–MHC complexes and which trigger T cell activation, and they are also quite distinct from pDCs (above). FDCs are not bone marrow-derived, they are very long-lived, and their primary function appears to be to

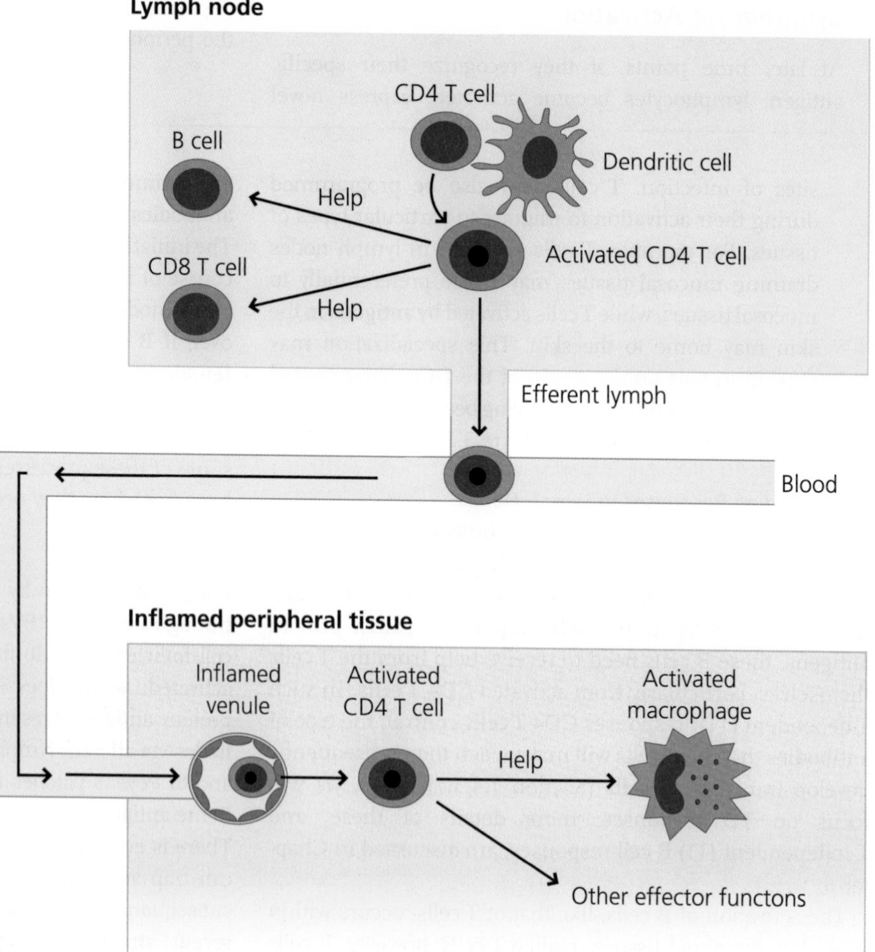

Fig. 3.22 Migration of activated CD4 T cells. Some activated CD4 T cells migrate to B cell follicles and help B cells generate antibody responses. Others remain in the T cell area and give help to CD8 T cells. Some effector CD4 T cells leave the node in efferent lymph and enter the blood from where they migrate to peripheral sites of inflammation. Here, depending on how they have differentiated, they may activate cells such as macrophages or help to recruit the different types of granulocyte needed to deal with different types of infection.

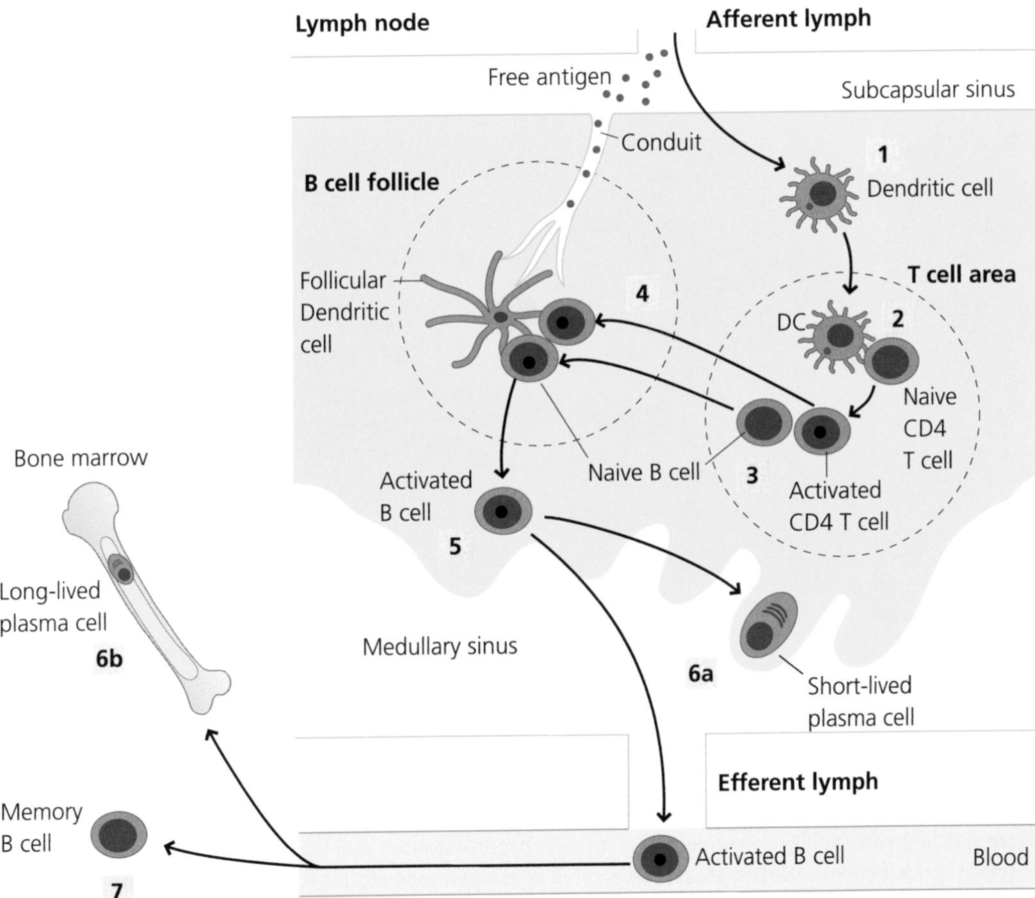

Fig. 3.23 T cell-dependent B cell activation. (1) Following antigen presentation by DCs in the T cell area, (2) activated CD4 T cells migrate to the edge of B cell follicles and (3) interact with antigen-specific B cells. (4) The B T cells migrate into the follicles, and the B cell starts to divide, forming a germinal centre. (5) Activated B lymphoblasts leave the germinal centre – some become plasma cells, secreting antibody either in the medulla of the node itself (these tend to be short-lived cells) (6a) or after migration to other sites such as spleen and bone marrow (bone marrow plasma cells may be very long-lived) (6b). Other activated B cells do not become plasma cells, but differentiate to become memory cells.

retain unprocessed (native) antigen on their surfaces in complexes with antibodies and complement. What is the purpose of this particular source of antigen (in contrast to the antigen we mentioned above that is needed for B cell activation)? There are two main purposes: (i) it is involved in selecting B cells with higher-affinity receptors that have been generated during the germinal centre reaction that we describe below, and (ii) it serves as a reservoir and appears to be able to stimulate memory B cells continually, making a crucial contribution to long-term B cell memory described later.

Germinal Centre Reaction In B cell responses to protein antigens, help from CD4 T cells is almost always required. The T cells are activated by DCs in the T cell area as described above. The activated T cells then move to the edge of the follicle where they interact with B cells. The B cells, probably accompanied by the T cells, cells then move into the follicle. Some of the activated B cells rapidly leave the follicle and become plasma cells; often in the medulla of the node. Other B cells in the follicle start to divide very rapidly. The B cell follicles enlarge and develop a novel region containing many dividing B cells. The follicle is now called a secondary follicle, and the area of B cell proliferation is called the germinal centre (some B cells are dividing every 6 h, remarkably fast for a mammalian cell) See Figure 3.24.

The germinal centre has two distinct areas – the dark and light zones. These changes reflect very important events in the differentiation of B cells and in antibody synthesis. It is in the dark zone that the B cells are proliferating – those in the light zone are not (or are proliferating to a much lesser extent areas). In the light zone the heavy chain constant region of the immunoglobulin molecule can be changed. Thus, a B cell may switch from expressing IgM to expressing IgG or other isotypes. In addition, those B cell variable region genes undergo very rapid mutation. In the latter case, the antigen bound to the surface of the FDC, noted above, selects the B cells with higher-affinity receptors. As a consequence, both the quality and quantity of subsequent immune responses to the same antigen are considerably changed. These events are discussed in more detail in the Chapter 6.

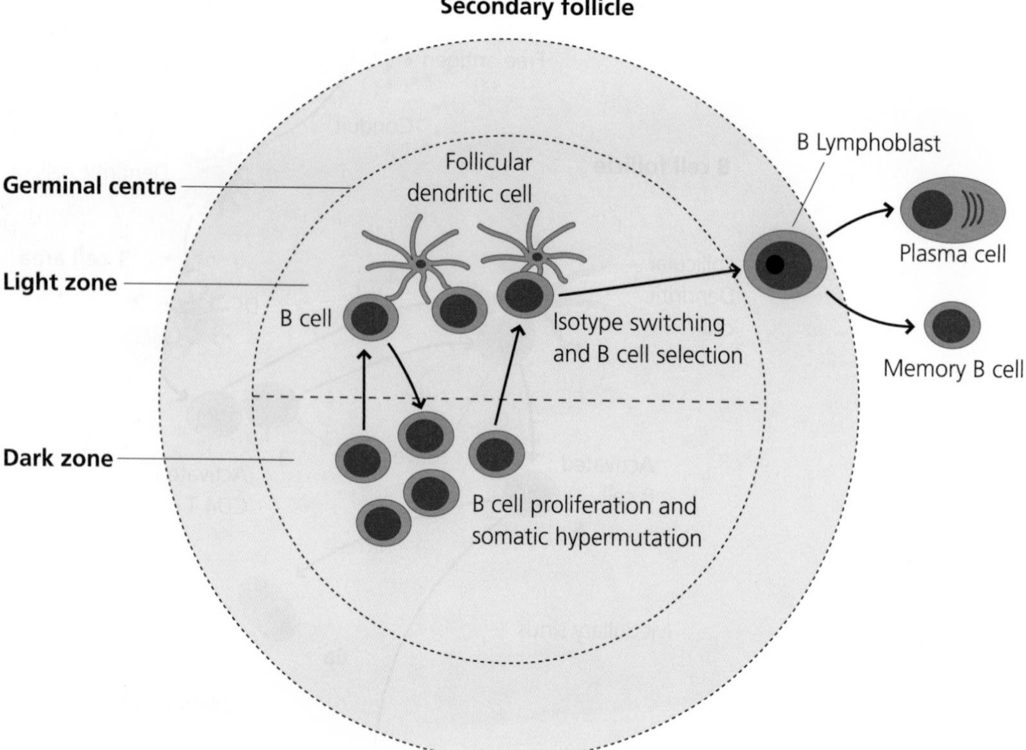

Fig. 3.24 Germinal centres. Germinal centres form in B cell follicles during TD antibody responses. They consist of dark areas where B cell are dividing very rapidly and undergoing somatic hypermutation, and light areas where B cell undergo isotype switching. B cells migrate between the two areas. FDCs serve as antigen repositories for the differentiating B cells to help select those of higher affinity for the antigen.

The end result of B cell activation is the generation of antibody-forming plasma cells and memory B cells. In the lymph node, plasma cells are found in the medulla, not in the follicles or germinal centres. However, these are not the only sites where antibodies are made. Plasma cells derived from lymph nodes also migrate to the bone marrow where they can live, secreting antibody, for long periods (many years in some examples). This is a most important source of antibody in long-term protection against infection.

3.4.2.6 Development of Memory Responses

One of the most important characteristics of an adaptive immune response is that administration of the immunizing antigen on subsequent occasions induces a faster, larger and often qualitatively different response. This is immunological memory. How is this brought about? The answer is that not all lymphocytes activated in a primary response become terminal effector cells. In the case of T cells, there is a rapid expansion of antigen-specific cells in the primary response. For CD8 T cells in a response to a virus, antigen-specific cells can represent 50% or more of the total CD8 T cell pool at the height of the response in some cases. These very large numbers of antigen-specific T cells decrease rapidly once the virus has been cleared, but even when the response has completely died down there are still increased numbers of antigen-specific T cell clones present. These are the memory T cells.

Memory B cells also develop after a primary immunization. When B cells are activated, a proportion of B lymphoblasts develop into plasma cells, secreting large amounts of antibody. These are terminally-differentiated cells and remain as antibody-secreting cells for the rest of their lives. Some B blasts, however, lose markers of activation and come to resemble naïve B cells. They may, however, have switched the antibody isotype they express. It was thus shown that long-term memory B cells for IgG responses are small, recirculating B cells that express surface IgG. Some memory B cells do not, however, seem to recirculate. For example, in individuals vaccinated against smallpox, the spleen (below) contains memory B cells for many years, but these are not found in significant numbers in the blood.

3.4.2.7 Peyer's Patches

The intestine, lungs and other mucosal tissues have lymphatics leading to draining lymph nodes, which in general function similarly to somatic nodes (those that drain non-mucosal tissues). In addition, the intestinal and respiratory tracts, for example, have local secondary lymphoid organs actually embedded in their walls, collectively known as MALTs. In the small intestine these are known as Peyer's patches. These latter organs function to monitor the intestinal lumen for antigens. See Figure 3.25.

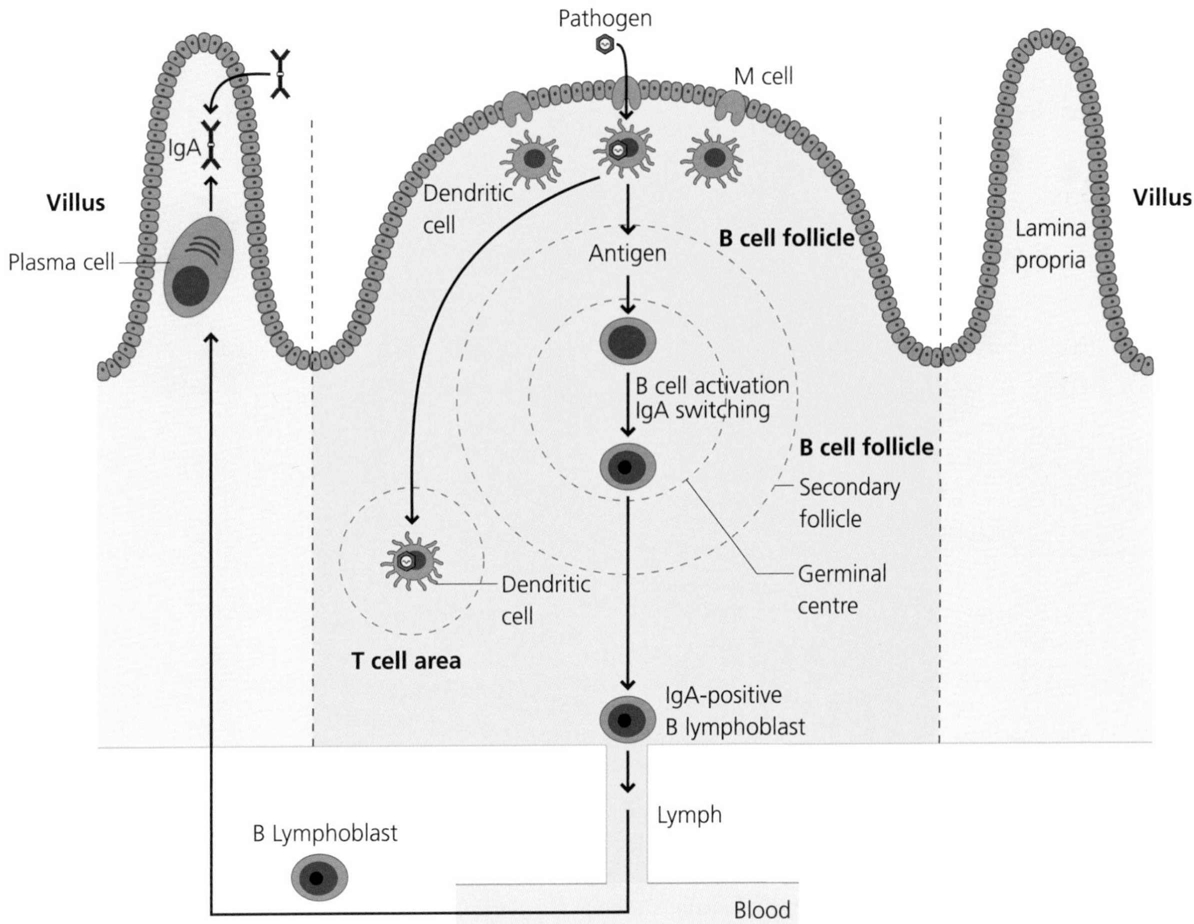

Fig. 3.25 Peyer's patches. Peyer's patches are secondary lymphoid organs present in the wall of the small intestine, related structures are present in some other mucosal tissues. They do not possess afferent lymphatics. Instead, antigen is transported from the intestinal lumen by specialized epithelial cells called M cells. DCs and other cells lie under the M cells and acquire antigens from them. Peyer's patches contain B and T cell areas where these cells can be activated. The activated lymphocytes then leave the patches via efferent lymph and pass through mesenteric lymph nodes before eventually entering the blood. B cells activated in Peyer's patches usually switch to producing IgA and migrate to many mucosal tissues. Peyer's patches are particularly involved in IgA responses to intestinal antigens.

Peyer's patches and their equivalents in other mucosal tissues, but unlike any other secondary lymphoid tissues, possess a specialized type of epithelial cell, the M cell. Rather than being an absorptive or secretory cell, as is the case for other intestinal epithelial cells, M cells have few microvilli and are specialized to transport molecules and particles from the gut lumen, and to release them into a pocket under the M cell, where they may be acquired by DCs. In some cases (e.g. reovirus and Salmonella) pathogens can also use M cells as a portal to bypass the epithelial cell layer. Peyer's patch T cells are able to switch B cells to express the heavy chain of IgA. Following activation these B lymphoblasts enter the blood and selectively migrate to other mucosal tissues. Here they then develop into plasma cells secreting IgA, which is transported across the epithelium and mediates protection against mucosal infections. This has given rise to the concept of the common mucosal immune system, so important in breast milk giving protection to intestinal infections to new-born infants.

3.4.2.8 Spleen Structure and Function

The spleen has many functions. It acts as a filter for blood-borne particles and is the site of destruction of worn-out erythrocytes. However, in immunity, its primary function is the generation of immune responses against blood-borne antigens. In contrast to lymph nodes, the spleen does not have any afferent lymphatic input. It does, however, have some similarities to lymph nodes in its functional organization: it has T and B cell areas, and the red pulp (site of erythrocyte destruction) contains many macrophages and plasma cells. Blood-born antigens give rise to the same types of immune responses as are seen with lymph nodes in terms of T cell polarization and Ig isotype synthesis. The spleen, however, also has an additional specialized region, the marginal zone. This area contains two types of specialized macrophages, whose functions are still poorly understood, and a distinct type of specialized, non-recirculating B lymphocyte, the marginal zone B cell. This type of B cell is specialized to respond in a T cell-independent manner to polymeric carbohydrate

Spleen

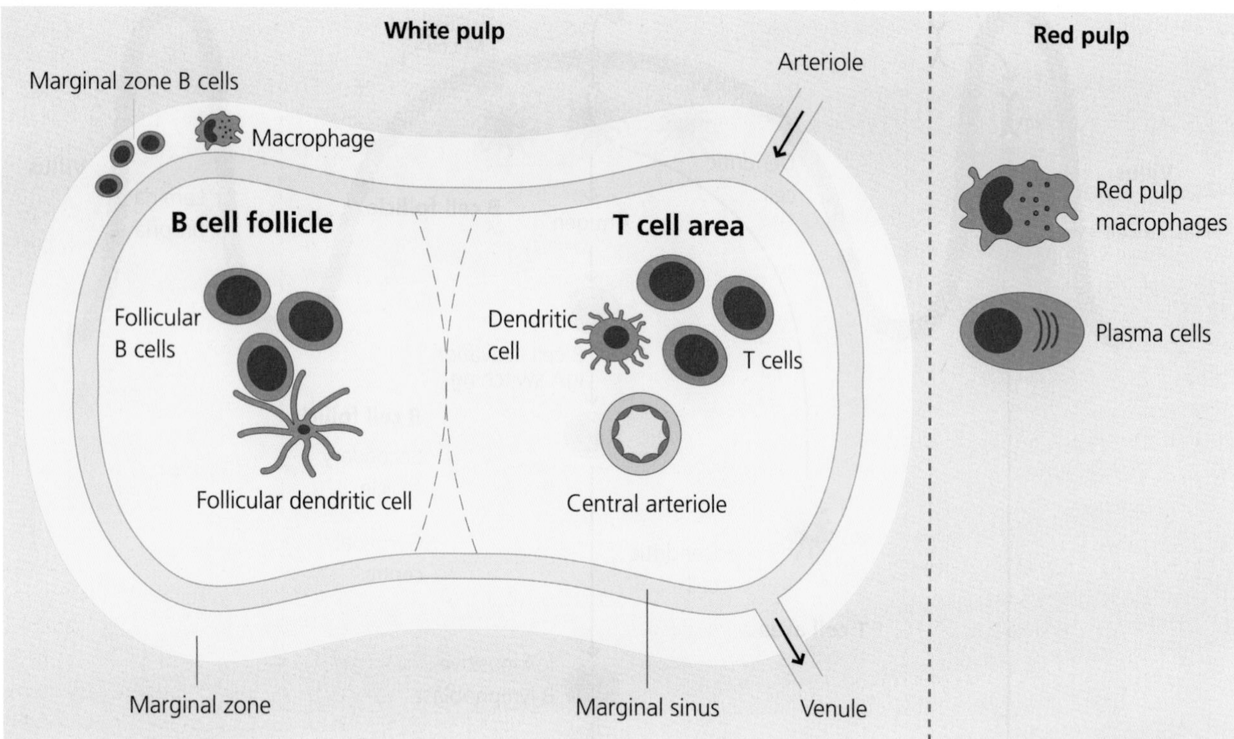

Fig. 3.26 Spleen. The spleen is a secondary lymphoid organ monitoring blood-borne antigens. The white pulp is the lymphoid area, containing T and B cell areas as in other tissues. It also possesses a marginal zone that contains specialized macrophages and resident, non-recirculating B cells that respond to TI antigens. The red pulp of the spleen contains plasma cells, but also macrophages that are involved in housekeeping functions such as the removal of old RBCs. In some circumstances the spleen may become a primary haematopoietic organ (Section 3.5.1).

antigens such as those present in some bacterial capsules (e.g. Streptococcus pneumoniae). Children lacking a spleen are much more susceptible to infection with these bacteria than their normal counterparts. See Figure 3.26.

Q3.17. As splenectomized children grow older it appears that their susceptibility to infection with encapsulated bacteria decreases. Why might this happen?

3.5 Development of Blood Cells and Organs of Immunity

In this section we briefly outline the origins of the different organs and tissues, and the leukocytes, that contribute to host defence. The development of all blood cells, including lymphocytes, occurs through a process called haematopoiesis. If lymphocyte development is considered on its own, this is called lymphopoiesis. In contrast the development of other cells, such as monocytes and macrophages, granulocytes, is called myelopoiesis. Below we will focus on stem cells that can generate all the leukocyte lineages. In later chapters we will discuss the development of the cells of innate and adaptive immunity in more detail (Chapter 4, and Chapters 5 and 6, respectively). Here we will also see that we can distinguish between primary, secondary and, in fact, tertiary lymphoid tissues, and will consider how secondary lymphoid tissues are formed in a little more detail.

3.5.1 Primary Lymphoid Tissues and Haematopoiesis

3.5.1.1 Haematopoietic and Primary Lymphoid Organs

Where and how do cells of the immune system develop? This question is not just an academic exercise. Defects in the development of immune cells can lead to important if rare diseases, and additionally throw light on the mechanisms involved in development. Similarly, immunodeficiency diseases, although also rare, help us understand the functions of the different cells and molecules that contribute to immunity.

In the adult mammal most cells of the immune system are formed, by haematopoiesis, in the bone marrow; in the foetus this takes place in the spleen and liver, and at still earlier stages in other sites. Under abnormal circumstances, such as in severe anaemia or if the marrow space becomes filled with connective tissue (myelofibrosis), haematopoiesis can occur outside the bone marrow, for example in the spleen, liver and lymph nodes.

With regard to lymphopoiesis, specialized tissues exist where lymphocytes undergo their development. In all mammals the thymus is the organ where T cells undergo most of their development and from where they are released as mature lymphocytes. In humans and mice, B cells undergo most of their development in the bone marrow; in chickens, however, there is a specialized tissue called the Bursa of Fabricius, where B cells undergo an analogous process. In the sheep, some B cells undergo much of their development in ileal Peyer.s patches (the ileum is the terminal part of the small intestine), in contrast to humans and mice where Peyer's patches are solely secondary lymphoid organs. In rabbits, much B cell development occurs in the appendix.

Haematopoietic and lymphopoietic organs are not only composed of the haematopoietic cells themselves. Non-haematopoietic stromal cells such as epithelial cells or fibroblasts, as well as bone marrow-resident macrophages, provide important signals that are involved in the regulation of haematopoiesis.

3.5.1.2 Stem Cells and Haematopoiesis

It is obvious that the fusion of only two cell types, a sperm and an egg, gives rise to a fertilized egg that can produce all the different lineages of cells in the body, some perhaps remaining to be discovered. The fertilized egg, at an early stage of development, develops into a blastocyst. It was discovered that blastocysts contain embryonic (ES) stem cells that can potentially give rise to any cell type of the body and which are therefore termed totipotent. Techniques were then developed that allowed these ES cells to be maintained indefinitely in culture, where they could be manipulated. This laid the foundations for transgenesis (Fig 2.7) and the development of genetically modified animals, and offers promise for the generation of specialized tissues that might be used to replace defective tissues and organs for the treatment of disease.

Stem cells are the precursor cells for other cell types. They have the capacity for self-renewal, hence maintaining their population as a whole. Unlike other cells, however, they can give rise to cells of other lineages. In the bone marrow of adults there is a haematopoietic stem cell (HSC) that can give rise to all blood cells, including those of the immune system. It cannot, however, generate cells of other lineages and is therefore termed multipotent or pluripotent as opposed to totipotent: it gives rise to multiple cell lineages, but not to all. Unlike ES cells, HSCs cannot yet be maintained indefinitely in culture. When one of these pluriotent stem cells divides, the division is usually asymmetrical. One daughter cell remains as a stem cell while the other starts off down one of the differentiation pathways. In other words, one of the daughter cells starts to become committed to a different cell lineage. When, eventually, a cell is produced that cannot develop into any other cell type (at least, under normal circumstances) it is said to be terminally differentiated. See Figure 3.27 and Box 3.6.

It is convenient, although perhaps not entirely accurate, to think of HSCs as initially being able to give rise to just two different types of stem cells that are more committed in their potential. These are the common myeloid progenitors (CMPs) and the common lymphoid progenitors (CLPs). As their names imply, the former progenitors (or precursors) can be

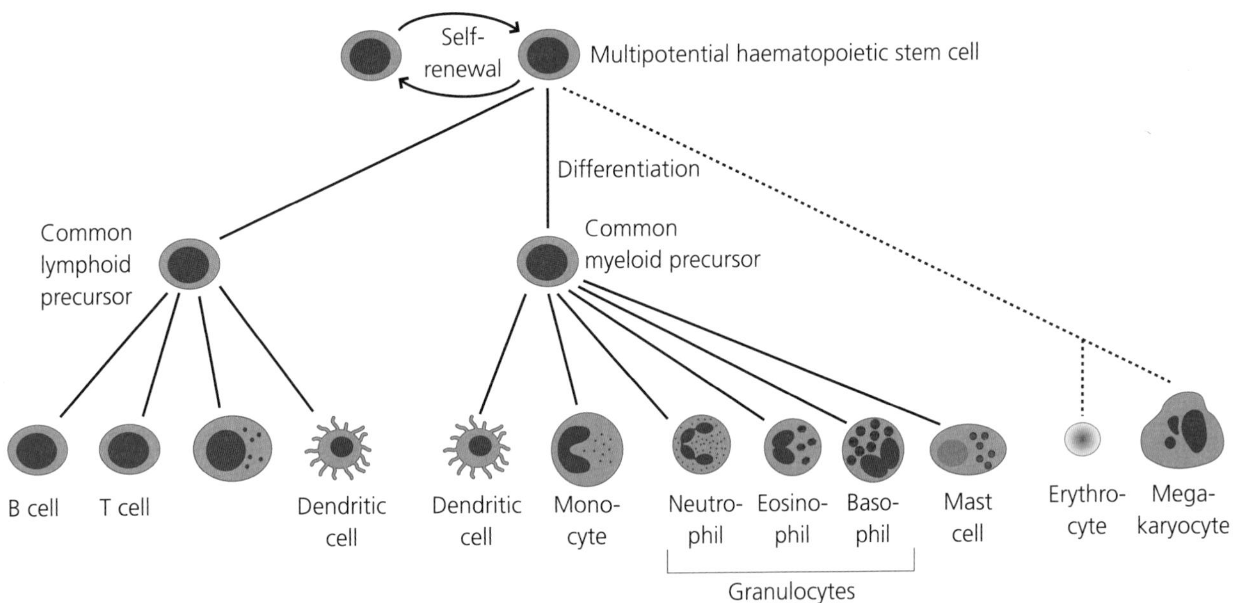

Fig. 3.27 Haematopoiesis. In adult mammals, the bone marrow is the major haematopoietic organ. Multipotent (pluripotent) stem cells divide to give another stem cell and a differentiating precursor cell which becomes committed to one of several lineages. Common lymphoid precursors (CLP) give rise to B and T cells, as well as NK cells and some DCs. The common myeloid precursor (CMP) gives rise to all the other leukocytes including monocytes and macrophages, different types of granulocytes, other DCs but not megakaryocytes (the source of platelets) and erythrocytes. For simplicity, mast cells are shown as being myeloid cells, although they may develop from a distinct progenitor.

Box 3.6 Identification of Multi-Potent Haematopoietic Stem Cells

How do we know there is a multi-potent stem cell? As noted in the text, ionizing irradiation kills rapidly dividing cells, including bone marrow cells. If a mouse is irradiated to destroy its own bone marrow stem cells, all its leukocytes can be replenished by giving it bone marrow from a normal donor. To trace the fates of the transferred cells and to ensure that the leukocytes did arise from the donor, the donor and recipient must differ in some form of genetic marker. In early experiments, karyotypic markers such as T6 were used; these are naturally occurring alterations in chromosome structure that are present in all cells of the body. Allelic variants of cell surface proteins can also be used: For example CD45 is a commonly used molecule expressed by all leukocytes that has two allelic variants. These studies clearly demonstrated that the bone marrow contains cells that can give rise to all haematopoietic cell lineages, since all the cells that were produced carried the donor marker.

Does this mean that there is a multi-potential stem cell? No. In the transplanted population of bone marrow there could be different stem cells for all the different lineages. We need to show that a single cell has the potential to give rise to all cell types. This means that we need to generate genetic markers that are induced in one cell only but which can be identified in all the progeny of that cell. Mild irradiation of cells causes damage to their chromosomes but may not kill the cells or prevent them dividing. The pattern of chromosomal damage is random and thus each individual cell has a unique pattern of damage. When lethally irradiated mice were reconstituted using lightly irradiated bone marrow cells, in some cases the same pattern of chromosomal damage was seen in some cells of all the different lineages examined. This can only mean that a single cell was able to give rise to all these lineages (i.e. it is multi-potential).

thought of as being able to give rise to all myeloid cells, while the latter can give rise to all the lymphoid cells. As these cells are more committed than the HSCs (they give rise to cells of one lineage but not the other) they can be termed multipotent stem cells to distinguish them from the pluripotent HSCs. One reason that this simple model may not be entirely accurate is that some lineages (perhaps mast cells in particular) may originate from neither the CMP or CLP, while others (e.g. DCs) can develop from both but may have a distinct precursor, the common dendritic cell precursor.

After the CMP and CLP there are precursor cells that are still more committed and which can give rise to a smaller number of cell types but these are of course still multi-potent. Multipotent stem cells were first identified by injecting bone marrow cells into the circulation of irradiated mice. (Rapidly dividing cell populations such as those in bone marrow are particularly sensitive to irradiation; this is why, for example, bone marrow transplants are needed to treat cancer patients who may have been treated with high doses of irradiation that were actually intended primarily to kill the tumour). When the spleens of these reconstituted mice were later examined, they were found to have bumps on their surfaces. These turned out to contain blood cells that were surprisingly of different types, for example granulocytes, macrophages, megakaryocytes (that give rise to platelets) and erythrocytes. The number of bumps was directly proportional to the number of bone marrow cells injected, indicating that each arose from a single cell (i.e. it was clonal). The cells that gave rise to these bumps were termed colony-forming units of the spleen (CFU-S).

Further evidence for multi-potent stem cells also came from *in vitro* studies in which bone marrow cells were cultured with different growth factors or cytokines as we now also know them. By using very dispersed cells in semi-solid culture media such as agar it was possible to constrain the movement of individual cells and to see which of them proliferated to form a colony. It was found that, depending on the particular growth factors and concentrations of these that were used, colonies of mixed or single cell types could be produced. For example, under different circumstances, some contained both granulocytes and macrophages, while others contained one or the other. This led to the concept of a colony-forming unit (CFU) that reflected whether a cell was, for example, committed to producing one or more than one type of cell. For example, those that generated mixed cultures of granulocytes and macrophages were called CFU-GM. As each individual growth factor was characterized it was called a colony-stimulating factor (CSF). These are of course cytokines, but in many instances this terminology has stuck. For example the cytokine that promotes differentiation of both granulocytes and macrophages is hence termed GM-CSF. This is further discussed in Chapter 4, where we also point out that the production of myeloid cells, and other leukocytes, is under tight feedback control in ways that are still not fully understood.

Q3.18. How, experimentally, might we test the hypothesis that the production of monocytes is under feedback control during normal, steady-state conditions?

3.5.1.3 Thymus

The earliest precursors of T cells are formed during haematopoiesis in the bone marrow. These early T cell progenitors then enter the thymus where they eventually develop into mature T cells that are released into the periphery and which recirculate between secondary lymphoid tissues or, for some specialized populations, populate specific tissues. Early T cell progenitors are not fully committed to the T cell lineage and, at this stage, these cells can be driven into development of other lineages under experimental conditions. Nevertheless, the thymus can be thought of as the central organ where these cells, partly under instruction by specialized stromal cells and cytokines that are produced by them, progressively become committed and undergo the remainder of their development into mature lymphocytes.

Q3.19. Why might it be that all animals that have evolved T lymphocytes have also evolved a thymus in which these cells develop?

How do we know that the thymus is the site of T cell development? Evidence comes from both clinical and experimental studies. A paediatrician observed that some very rare children were unusually susceptible to a variety of viral, bacterial and fungal infections. These are infections that require T cell-mediated immunity for recovery and these children had very few if any peripheral T cells. It was found that such children had a congenital absence of the thymus. Thus, the thymus appeared critical for T cell development. This condition is known as DiGeorge syndrome. However, there are many other abnormalities in tissues and organs of these children, so this is not definitive evidence. Researchers then surgically removed the thymi from new-born mice (neonatal thymectomy). When these mice grew into adults they were defective in their ability to mount cell-mediated immune responses such as DTH and to make antibodies to protein antigens. It was later found that these mice did not possess significant numbers of peripheral T cells (which we no know are essential for many B cell responses) but that their B cell numbers were close to normal. Thus, it was concluded that the thymus was essential for T cell development. See Figure 3.28.

Q3.20. A researcher in Australia removed thymi from foetal sheep while they were still in the uterus. When these sheep were born they showed normal cell-mediated immunity. The worker concluded that the thymus was not important in the development of lymphocytes involved in cell-mediated immunity. Why was he wrong to make this conclusion?

3.5.2
Development of Secondary Lymphoid Organs

Humans possess 400–600 lymph nodes, around 60 Peyer's patches and one spleen. As for any organ, it is of interest to understand how the development of primary and secondary lymphoid tissues is controlled. It was a complete surprise when some members of the TNF–TNF receptor family were found to play a crucial role in development of these tissues. The genes

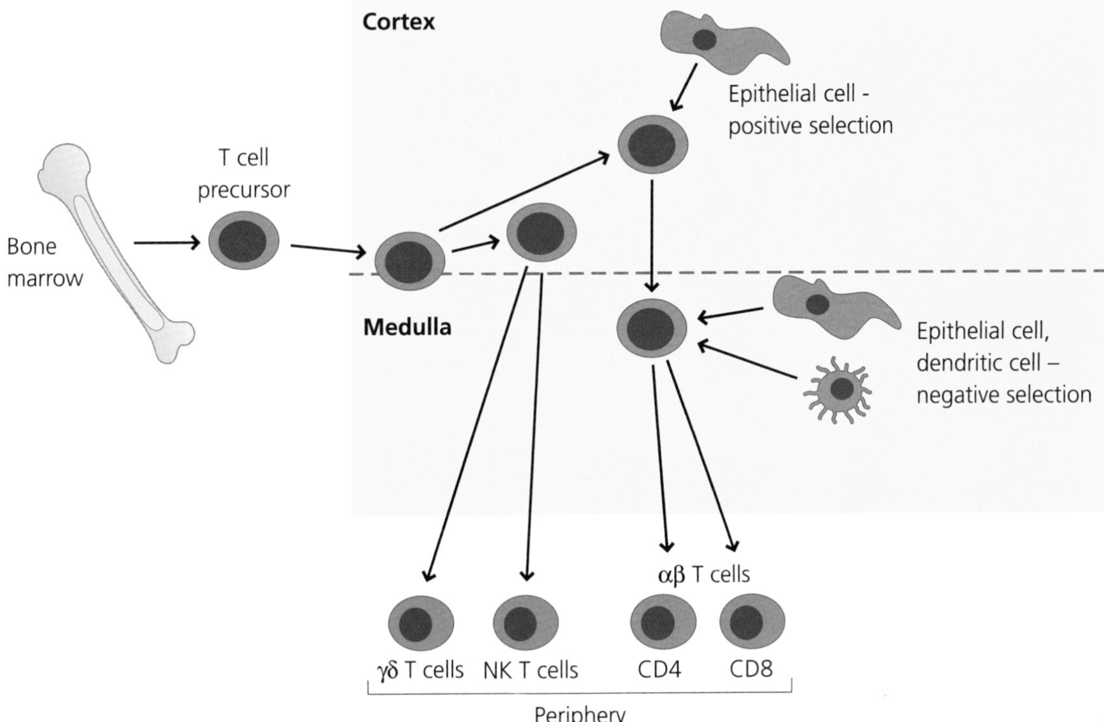

Fig. 3.28 Thymus. The thymus is the organ in which T cells differentiate. Early T cell precursors enter from the bone marrow and divide to populate the thymic cortex with thymocytes. If a thymocyte recognizes MHC molecules on cortical epithelial cells it survives (positive selection); otherwise it undergoes apoptosis. Surviving thymocytes enter the medulla and those that recognize self peptides on medullary epithelial cells (MECs) or DCs apoptose (negative selection). The surviving thymocytes leave the thymus and become peripheral T cells. Positive and negative selection respectively ensure that TCRs recognize antigens appropriately (peptides bound to MHC molecules, rather than either alone) and are generally unable to recognize normal cells (i.e. are not autoreactive).

involved include TNF and lymphotoxin (which is of different types), and their receptors that can bind members of each type of cytokine. These mice may lack peripheral or mesenteric lymph nodes. They may also lack or have reduced numbers of Peyer's patches and typically the organization of the spleen is disrupted. Typically, they also lack FDCs in any lymphoid tissues that may be present. One of the main uses of these mice is that they may enable analysis of the functions of different sets of secondary lymphoid organs. It is thus possible, for instance, to ask if Peyer's patches and mesenteric lymph nodes have different roles in responses to intestinal antigens.

Current theories suggest that lymph node development depends upon the interaction of stromal lymphoid tissue organizer cells and haematopoietic inducer cells, involving members of the TNF-TNF receptor family of molecules. Recently a cell population has also been identified in the lymph nodes of foetal mice which, if injected into the skin or mesentery of another mouse, induces *de novo* lymph node formation. Further analysis of these cells will provide insights into lymph node development. This finding may also relate to the well-recognized development of tertiary lymphoid structures in sites of chronic inflammation (below) See Box 3.7.

3.5.3
Tertiary Lymphoid Organs

In some chronic inflammatory conditions, cells recruited into tissues or organs can form structures very similar to secondary lymphoid tissues. Typically these conditions are associated with autoimmune diseases such as rheumatoid arthritis or Hashimoto's thyroiditis (Chapter 7). As these structures resemble secondary lymphoid tissues, but they appear in peripheral sites, they are termed tertiary lymphoid organs. Thus, T cell areas containing DCs and HEVs can be found, as well as follicles containing B cells and FDCs. Whether they have any defence function is unknown – it is quite possible that they represent an aberrant activity of organizer/inducer cells. For example, it is known that one type of lymphotoxin is needed for the expression of adhesion molecules on lymph node HEVs, so perhaps aberrant secretion of this molecule is involved in the development of tertiary lymphoid tissues.

Q3.21. What might be the functional significance of having tissues resembling secondary lymphoid tissues present in an inflamed organ?

3.6
Stem Cell and Gene Therapy

The underlying concept of stem cell therapy is that, if a particular lineage(s) of cells is deficient or defective, stem cells that are delivered to the appropriate anatomical compartment may be induced to differentiate into competent end cells. As we suggested earlier (Section 3.5.1.2) there is much interest in using ES cells for therapy as they can be induced to differentiate into a large variety of different end cells, theoretically of any lineage, depending on the conditioning that is used.

Another form of stem cell therapy that very is widely and successfully used is of course bone marrow transplantation. This is being used to treat a variety of immunodeficiencies of both the innate and adaptive systems. Thus, chronic granulomatous disease (CGD), in which neutrophils have a genetic defect, can be cured by transplanting bone marrow cells from a normal

Box 3.7 Lymphoid Tissues in Other Species

The evolution of adaptive immunity is accompanied by the development of specialized secondary lymphoid organs in which immune responses are initiated. Fish first evolved during the Cambrian explosion, initially as jawless fish. Then jawed fishes appeared now and, with them, some of the typical features of adaptive immunity that are now present in mammals. These evolved into the cartilaginous and then bony species of fish. In the jawless fishes there are accumulations of myeloid cells in the intestinal connective tissues. These may be the sites where cells resembling lymphocytes are activated, but this is far from clear. As well as possessing a spleen and thymus, as do higher vertebrates, cartilaginous fish additionally possess three other organs which all contain lymphocytes and plasma cells: epigonal organs, a Leydig's organ and an intestinal spiral valve. In bony fishes the anterior kidney is also an important immune organ. In an informative experiment cells from rainbow trout thymuses, spleens and anterior kidneys were incubated with concanavalin A (Con A) or lipopolysaccharide (LPS) typical mitogens for T cells and B cells, respectively. The thymic cells only responded to Con A, the spleen cells responded to both, while the anterior kidney cells responded only to LPS. This suggests that, as in mammals, the thymus is where T cells develop (the thymus contains some mature T cells) and the spleen is a site where T and B cells are both present. Perhaps analogous to the Bursa of Fabricius in birds, the anterior kidney may be a site where B cells develop. In birds, the arrangement of secondary lymphoid organs is more like that of mammals. As well as a spleen that is functionally similar to that of mammals, birds possess some encapsulated lymph nodes. Additionally, however, birds possess an Harderian gland that is associated with the eyes and is responsible for some IgA synthesis.

individual, and SCID (severe combined immunodeficiency), in which both T and B cells are absent, can be similarly treated.

Gene therapy refers to the insertion of genes into somatic cells with a view to curing genetic diseases. The principle is very straightforward. For example, if a child has a mutant NADPH oxidase gene this may lead to defective killing of bacteria by neutrophils, causing CGD. The current treatment for CGD is bone marrow transplantation. This is, however, a dangerous therapy (Chapter 7). It is theoretically possible to take some of the child's own bone marrow cells, isolate HSCs and insert into these cells a normal gene coding for NADPH oxidase. These stem cells following transfer back to the child are hoped to establish themselves in the marrow and to generate normal neutrophils. It should be clear that this principle could be applied to many other diseases where mutations had led to defective leukocytes. At present, however, there are very few examples of successful gene therapy and some trials have revealed dangerous side effects. For example, the inserted gene can lead to dysregulation of gene expression, leading to the induction of leukaemia, for instance. Gene therapy is still in its infancy; its potential is however enormous and its development will lead to the treatment or cure of many important diseases.

Learning Outcomes

By the end of this chapter you should be able to understand, explain and discuss the following topics, the relevant sections of the chapter are indicated. You should understand some of the evidence from human and animal studies supporting what we know about these topics. You should have some idea of the areas where our understanding is incomplete. You may be able to suggest ways in which our understanding could be advanced.

- Natural barriers (Section 3.2)
  - What are the major, pre-formed barriers against infection?
  - How can these barriers be breached by pathogens and what happens subsequently?
- Functional anatomy of innate immunity (Section 3.3)
  - What are the characteristic local and systemic features of acute and chronic inflammation?
  - How are inflammatory responses triggered? Give some examples of cells and soluble molecules involved in the initiation of inflammation.
  - Why is inflammation important in defence against infection?
  - How do leukocytes and soluble effector molecules get to inflammatory sites?
  - What are some mechanisms that cause tissue damage in inflammation?
  - How can tissue be damage repaired
  - What are the effects of inflammatory responses on distant tissues? Why are these important?
- Functional anatomy of adaptive immunity (Section 3.4)
  - Why do we need secondary lymphoid tissues?
  - What are the similarities and differences in structure and function between lymph nodes, spleen and Peyer's patches?
  - What are the main compartments of secondary lymphoid organs, what are their functions and why are they compartmentalized?
  - What changes occur in the structure and physiology of secondary lymphoid organs during peripheral inflammation and infection?
- Development of cells and organs of the immune system (Section 3.5)
  - What is a stem cell?
  - What is haematopoiesis and where does it occur?
  - What is lymphopoiesis and where does it occur?
  - How do secondary lymphoid tissues develop?
  - What is a tertiary lymphoid tissue; give some examples?
- Stem cell and gene therapy (Section 3.6)
  - How can our knowledge of stem cells be exploited for therapy? (Section 3.6.1.4?)
- GENERAL: How have natural and experimental defects shed light on the structures and functions of different tissues and organs in immunity?
- INTEGRATIVE: How do differences in the structure and function of different types of tissue relate to their roles in innate versus adaptive immunity?

Further Study Questions

Qa. What types of constitutive versus inducible defences might the natural barriers have? (Section 3.2)
Hint These barriers are external (skin), or situated at topologically external sites lined by epithelia (e.g. lungs, gut, tract). Epithelial cells would be in a good position to secrete some basic defence molecules constitutively, and perhaps to secrete more of these, or new ones, if infection occurs. Perhaps think also about how much the secretion of mucus, or the acidity of these sites, might be controlled.

Qb. To what extent is it possible to identify different types of inflammation and their purposes? (Section 3.3)
Hints It might be helpful to start with the settings of sterile trauma and infection in general. Then we might consider different types of viruses, microbes or larger parasites – and the sorts of tissues they might infect. This then brings us on to inflammatory responses in different organs. We know, for example, that those in the brain can be very different to those in other non-lymphoid tissues such as skin. What about responses in different types of lymphoid tissues?

Qc. How much do we know about the regulation of adaptive immune responses in lymphoid tissues by events that occur in peripheral tissues? (Section 3.4)
Hints We find the conduits fascinating structures and are sure we have much to learn about how much and what type of information they might be able to transmit directly into the core of a lymph node, for example. It also appears that leukocytes recruited to inflammatory sites might then traffic to such tissues (some could be recruited directly from the blood). Perhaps start by considering neutrophil swarms or if there is anything known about granulocytes such as basophils in lymph nodes.

Qd. To what extent can different secondary lymphoid organs be considered also to function as primary lymphoid tissues, or vice versa? (Section 3.5)
Hints Perhaps start by thinking about different settings. Are we talking about the normal steady state in the absence of infection or after an adaptive immune response has occurred? What about if an organ is removed (e.g. a ruptured spleen in a road traffic accident) – might the functions of other lymphoid organs be affected? What about different species or different stages of maturity (humans versus mice, adults versus neonates)?

Qe. What are the indications and complications of stem cell therapy (and/or the potentials for gene therapy) particularly for the treatment of immune-related diseases? (Section 3.6)
Hints We need to define what sort of stem cell therapy we mean – transplantation of bone marrow (e.g. HSCs) or of organs grown from ES cells? The complications of the first are in fact very well known, those of the latter rather more theoretical at present. How about gene therapy? What sort of immune-related diseases can we think of where replacement of a defective gene would be advantageous? What are the potential risks of messing about with a person's DNA? (Do we want to get into ethics as well?)

4
Innate Immunity

4.1
Introduction

In a world governed by natural selection it is inevitable that there will be competition for resources. The available resources include all living organisms, as a source of food or as hosts for reproduction, distribution, shelter or other reasons. In several cases (e.g. infections with pathogenic Staphylococcus and Streptococcus), these bacteria are more like saprophytes. The molecules that they release and that cause tissue damage – we call them toxins – are used to kills cells and digest tissues in order to provide food for the bacteria. It is in the interests of hosts to defend themselves against those who would use them as resources. In most cases it is also in the interests of the users not to wipe out their hosts for then they would then lose their resources. It is thus not surprising that probably all living organisms have evolved mechanisms to defend themselves against the attentions of invaders.

Innate immunity is the only form of immunity available to the vast majority of organisms and in these organisms it provides effective defence against infection. Innate immunity appears very early in evolution and many of its components can be traced back to the most primitive organisms, with new types of components being progressively added or modified during evolution. As most of the components of innate immunity are pre-formed, defence mechanisms can be activated very quickly after infection. In higher vertebrates, innate immunity can provide rapid defence against some types of infection, although these species have come to rely on adaptive immunity for maximal protection. Innate immunity is, however, crucial for the activation and regulation of adaptive immunity, and adaptive immunity, in turn, utilizes innate mechanisms as effector mechanisms for the elimination of microbes.

In this chapter we examine the different components of innate immunity and their contributions to host defence (Section 4.2). We start by considering the recognition of different types of infectious agents by innate immune cells and the ways in which these cells can direct the responses most appropriate for their elimination. This includes the secretion of specialized molecules (cytokines) that enable communication between different cells and which act to modify their functions. We then discuss the specialized cells resident in all tissues that function to detect infection and trigger inflammation (Section 4.3). This enables recruitment of the effector cells and molecules that are needed for elimination of the microbe. We explain how these recruited cells and molecules provide defence against different types of infectious agent, using studies of human defects or genetically modified mice as evidence for their importance, and how they can help to trigger adaptive immunity (Section 4.4).

In the penultimate section we discuss the origins of the cells of immunity: where they come from and how their production is regulated (Section 4.5). Finally, we show how our knowledge of innate immunity is beginning to inform the design of more effective vaccines, in particular by using components of infectious agents or other biomolecules and even inorganic molecules as adjuvants to boost immune responses against pathogens (Section 4.6).

By the end of this chapter you will comprehend the importance of innate immunity in eliminating infections, triggering inflammation and responses in more distant organs, and in initiating (and later contributing to) adaptive immunity

4.2
Induction of Innate Immunity

4.2.1
The Concepts: Pattern Recognition and Danger

A pivotal moment in the development of immunology, and in our thinking about immunity, came about some two decades ago when it was postulated that the detection of infection might be mediated by receptors of the innate immune system, rather than by the antigen receptors of T cells and B cells. It was suggested that, following pathogen sensing, the innate immune system instructed the adaptive immune system to respond if the antigen was of microbial origin. At

Exploring Immunology: Concepts and Evidence, First Edition. Gordon MacPherson and Jon Austyn.

the time these innate receptors were unidentified but, for a variety of reasons, it was suggested that they needed to be germline-encoded and that they would recognize conserved features of microbes termed pathogen-associated molecular patterns (PAMPs). The receptors themselves have become known as pattern recognition receptors (PRRs). This concept was put on a firm experimental basis when the germline-encoded Toll-like receptors (TLRs) were discovered. Initially "Toll" was discovered in flies as a molecule involved in development, but it was then found that this receptor also plays a role in defence. TLRs were then discovered in other species, including humans, and we now know that these and other, functionally related, types of receptors have the potential to discriminate between different components of pathogens.

Q4.1. Is it in any way surprising that the Toll family of molecules can be involved in functions as diverse as development and defence against infection?

Then, in the early 1990s, the danger hypothesis was put forward. It was suggested that rather than the immune system being concerned with what is self and what is not (self–non-self discrimination), it was actually discriminating between what represents danger and what does not. It was proposed that anything that can cause tissue damage could stimulate an immune response. Pathogens, because of the damage they cause to host tissues, would represent one obvious source of potential danger. However, this hypothesis also implied that other forms of tissue damage (e.g. sterile surgical trauma) might also trigger immune responses. More recently attention has been turning to danger (or damage)-associated molecular patterns; DAMPs. These are produced within or released from damaged cells and tissues, and can be recognized by the innate immune system whether or not they result from microbial infection. The difficulty we currently face in immunology is that while we have an increasing understanding of the PRRs that sense PAMPs, those that might recognize DAMPs are rather more elusive, although we are beginning to gain some insights into their potential nature (Section 4.2.2.4).

Generally speaking, a crucial outcome of recognition of both PAMPs and (so far as we understand them) DAMPs by the innate immune system is the initiation of inflammation. However, there are multiple forms of inflammation that have different functions in homoeostasis in the body (Section 3.3.2). The inflammation caused by sterile injury, for example, may be different, with different functions, from that caused by microbial infection. In the first case, it may primarily enable repair and healing of the damaged tissues, while in the latter it may principally enable the recruitment of soluble and cellular effectors that can help to eliminate the infectious agent (and contribute to healing, later). We ourselves subscribe to the view that innate immunity is indeed concerned with the recognition of danger – in the most general sense – and find it convenient to describe the outcome of recognition as being the production of alarm signals that for example trigger the different forms of inflammation.

4.2.2 Pattern Recognition Receptors (PRRs)

PRRs are now generally thought of as receptors that recognize microbially-derived PAMPs. So what are these molecules? In general, they are not proteins; there are few if any generic differences between bacterial, viral and vertebrate proteins, although there are some exceptions (below). However, there are clear generic differences between other classes of macromolecule expressed in vertebrates and pathogens. One of the best characterized PAMPs is lipopolysaccharide (LPS), also known as endotoxin, a complex of lipid and carbohydrate that is present in the cell wall of Gram-negative bacteria. Another is lipoteichoic acid, present in the cell walls of Gram-positive bacteria. These molecules are integral components of these different types of bacteria and are therefore not dispensable, microbes cannot easily evolve so that they do not express these molecules. Other PAMPs include bacterial and fungal carbohydrates with external mannose residues, glycolipids, and a few unique proteins such as flagellin, a major structural component of bacterial flagellae. See Figure 4.1.

Pathogens can inhabit extracellular tissues, may be internalized by endocytosis or phagocytosis into the endosomal pathway, and some can gain direct access to the cytoplasm during infection of cells. If pathogens are not to escape detection by the immune system, there is a need for sensors at the cell surface and in all of these compartments. These sensors are the PRRs and they exist in a variety of molecular forms. Many are cell-associated, although a number of others are extracellular soluble molecules, which are sometimes termed pattern recognition molecules (PRMs). PRRs are widely expressed throughout the immune system, including the tissue-resident cells of innate immunity that initially sense infection and even by lymphocytes. See Figure 4.2.

4.2.2.1 Cell-Associated PRRs that Promote Uptake

In the case of phagocytes there are structurally different types of PRR at the cell surface that can also promote internalization of what is recognized, through receptor-mediated endocytosis or phagocytosis. These include some C-type lectin receptors such as the macrophage mannose receptor, which recognizes mannose residues on bacteria. Other members of this family are dectin-1 and dectin-2 that recognize complex carbohydrates (glucans) found in fungi, and which lead to their uptake and also signal to the nucleus to change gene expression (below). Another type of PRR is the family of scavenger receptors, which recognize a wide range of pathogen-associated ligands. One such receptor is CD36, which plays an important role in the recognition, uptake and clearance of apoptotic cells by macrophages for example. Other scavenger

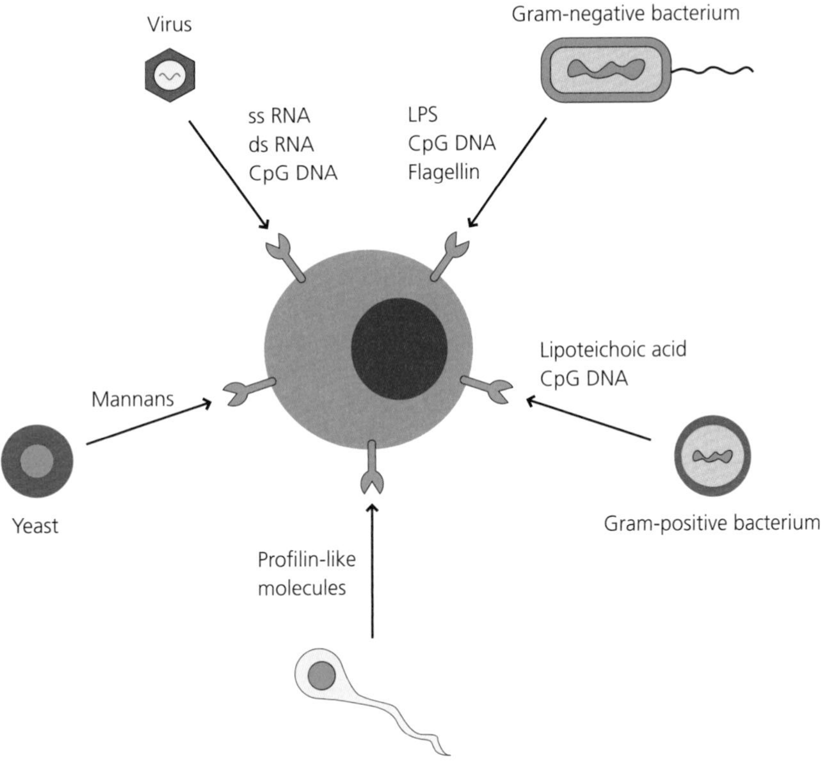

Fig. 4.1 Pathogen-associated molecular patterns. Many microbes express molecules that are not expressed by mammals. These include unique carbohydrate and lipid structures, as well as nucleic acids such as single-stranded (ss) or double-stranded (ds) RNA, or DNA containing CpG motifs and a few proteins. Mammals have evolved PRRs that recognize broadly-shared features of PAMPs. These receptors are encoded in the germline. A few examples of the type of molecule that can be a PAMP are indicated.

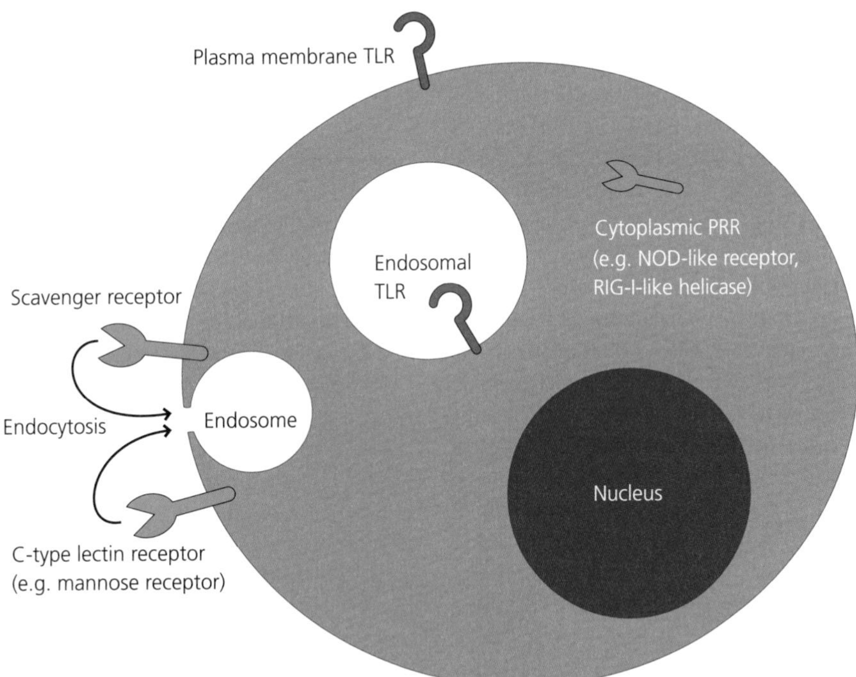

Fig. 4.2 Cellular locations of pattern recognition receptors. PRRs are located where they are most likely to be able to interact with different types of PAMPs and often where there is little possibility of their meeting host-derived molecules that are cross-reactive with PAMPs. Thus, TLRs may be expressed on the plasma membrane or in endosomes, while other types of PRRs are present within the cytoplasm. Some PRRs can also promote uptake of infectious agents.

receptors also recognize other forms of altered self components such as modified low-density lipoprotein, very important in the pathogenesis of atherosclerosis. Hence the concept of a PRR may need to be extended to include these types of ligands or agonists, as well as those that are microbially derived (below.) See Figure 4.3.

4.2.2.2 Cell-Associated PRRs that Signal to the Nucleus

A large number of other PRRs are critically important to host defence because their ligation induces profound changes in gene expression in the cells the express them; this capacity is crucial for inducible defence. For example, in the case of macrophages, PRR signalling can trigger changes in the expression of hundreds of different genes. These encode molecules such as cytokines and chemokines, anti-microbial peptides, as well as molecules involved in coagulation and tissue repair; we will focus particularly on the former. Some cytokines have direct defence functions, whereas others initiate the local and systemic features of inflammation, and regulate the quality of adaptive immune responses. They include particularly the Type I interferons (IFNs) and important pro-inflammatory cytokines (below).

The signalling PRRs that change gene expression belong to structurally or functionally different families and they have distinct cellular localizations. As indicated earlier, the first set of PRRs to be thoroughly characterized were the TLRs. A subset of TLRs is mainly expressed on the plasma membrane of cells. In general these recognize bacterial components, some acting singly, others acting in pairs. Bacterial components recognized include lipoproteins (TLR2 paired with TLR1 or TLR6), lipoteichoic acid found in the cell wall of Gram-positive bacteria (TLR2 and TLR6) and bacterial flagellin (TLR5). Some surface-expressed TLRs also recognize parasitic molecules; for example TLR11 recognizes another protein, a profilin-like molecule expressed by Toxoplasma gondii. In some cases, TLRs do not directly recognize microbial components although they signal in response to them; for example is TLR4 for which LPS is an agonist (below); TLR4 can also be translocated to endosomes. See Box 4.1

Another set of TLRs is primarily or exclusively expressed in the endosomal system where many viruses and bacteria can be digested after uptake. These TLRs are conventionally viewed as being able to recognize or respond to internalized

Box 4.1 Identification of TLR4 and its Role in Lipopolysaccharide Responsiveness

LPS at high concentrations can cause shock and death Section 4.4.6. Different strains of mice differ in their susceptibility to LPS-mediated endotoxin shock. One strain, C3H/HeJ, is resistant to LPS. Another closely related strain, C3H/HeN, is not resistant. This resistance versus sensitivity is controlled by a single gene versus which was called *Lps*. It was originally thought that this gene would encode CD14, known to be part of the LPS receptor, but no association of CD14 with *Lps* could be shown. Different groups attempted to identify the *Lps* gene by positional cloning. In positional cloning the approximate position of the gene is identified by its linkage with markers whose chromosomal location has been identified. Overlapping pieces of DNA are then isolated, starting from the known marker, until the gene of interest can be identified. In the case of *Lps* a DNA region was sequenced and found to contain only one fully functional gene, this was *Tlr4*, which encoded the TLR4 molecule. The *Tlr4* gene was among several that had been identified in humans because of similarities to the insect *Toll* gene, which controlled susceptibility to fungal infections. It was found that the C3H/HeJ mice had a mutation in *Tlr4*, not present in the closely related LPS-sensitive strain C3H/HeN. It is interesting that the non-responding allele, if transfected into responsive macrophages as a single copy, totally suppresses responsiveness to LPS; thus the mutant protein acts as a dominant suppressor.

TLR4 is not, however, a direct receptor for LPS. Instead the picture is much more complicated and CD14 appears to be the molecule that actually binds LPS. This is supported by results from three approaches: transfection, knock-out mice and direct binding studies. LPS is not, however, free when it binds to CD14. There is a plasma protein, LPS-binding protein (LBP), which is found in association with LPS *in vivo* and appears to extract single LPS molecules from micelles (it has also been suggested that LPS-binding protein may act to prevent the toxic actions of LPS). Another molecule that appears to be crucially involved is MD2, a small protein that binds to the extracellular domain of TLR4 and is needed for signalling via TLR4. Is it believed that LPS bound to LBP, is captured by CD14 and transferred to MD2, which then interacts with CD14, facilitating signalling. Other molecules have also been implicated in TLR4 signalling and these differ in different cell types e.g. macrophages or dendritic cells (DCs) versus B cells.

Many other TLRs were identified by genetic approaches, but as for almost all receptors, understanding the structure of a receptor often does not help in identifying its physiological ligand or agonist. Other approaches have been used to identify TLR agonists. Thus, the TLR5 gene was transfected into a tissue culture cell line and different bacterial preparations were tested for their ability to signal through TLR5. TLR5 did not respond to any of the known PAMPs, but did respond to bacterial culture supernatants. Mass spectrometry was used to identify the active molecule and it was shown to be flagellin; the major protein found in bacterial flagellae.

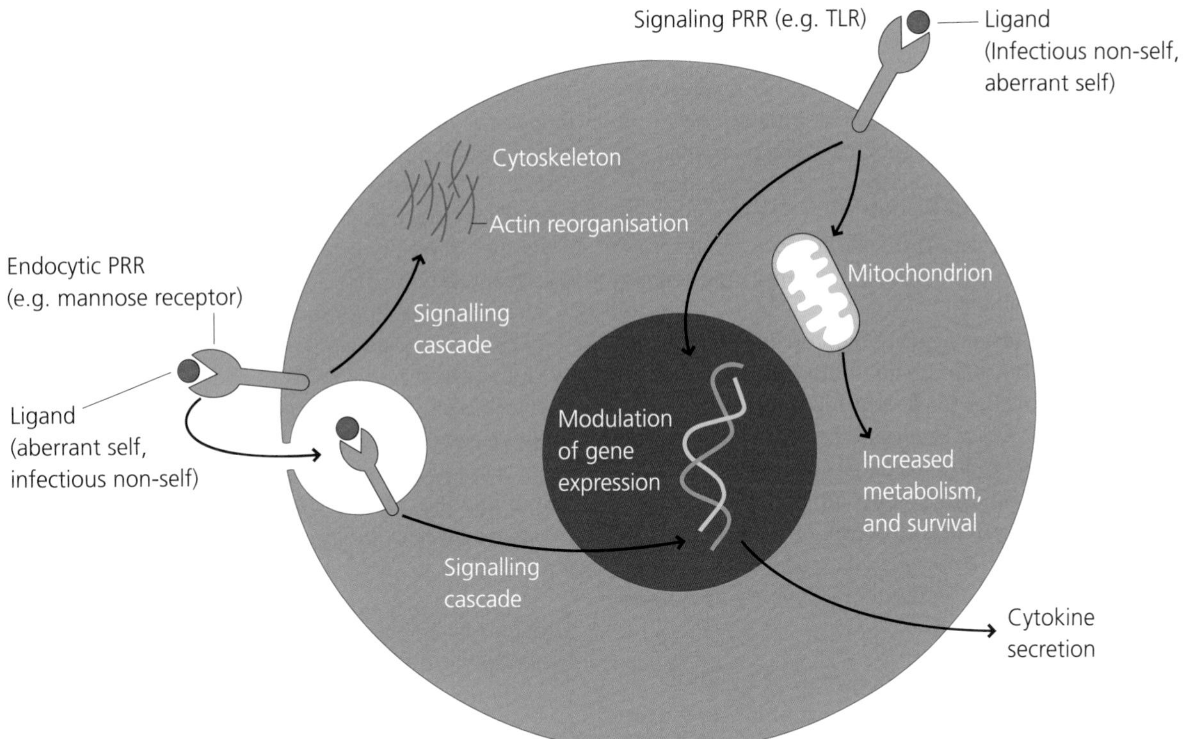

Fig. 4.3 Functions of pattern recognition receptors. Most PRRs have signalling properties while some, such as the mannose receptor, are also endocytic receptors promoting uptake (e.g. phagocytosis). Agonist binding to PRRs activates intracellular signalling cascades of different types that modulate cellular functions. Some of their effects can be broadly divided into metabolic, cytoskeletal and changes in gene expression leading, for example, to cytokine secretion. Most of the cytokines secreted are pro-inflammatory.

viral or bacterial nucleic acids, including double-stranded RNA (TLR3), single-stranded RNA (TLR7/8), and viral and bacterial DNA with characteristic features called CpG motifs in which the nucleotides are methylated (TLR9). In fact, it is possible that these particular TLRs may not actually discriminate between self and non-self nucleic acids but instead they may sense these components in an abnormal location. Thus, host cell DNA and RNA is normally confined to the nucleus or cytoplasm; if it is not, there is something wrong.

All viruses and some bacteria can also enter the cytoplasm of cells when they have infected cells. To cope with these, PRRs are expressed within the cytoplasm. However, these are not members of the TLR family. They include a family of so-called Nod-like receptors (NLRs) which recognize components of intracellular bacteria such as muramyl dipeptide, and a family of RIG-I-like helicases (RLHs) that apparently recognize double-stranded RNA produced during viral replication. The former include NOD1, NOD2 and Nalp3 or cryopyrin (Section 4.2.2.4), and the latter RIG-I and MDA5.

It is now absolutely clear that all the PRRs we have noted above can initiate cellular responses to components of infectious agents, PAMPs. There is, however, increasing evidence that some of these receptors may also recognize modified host components. Thus, some TLRs can apparently also recognize host components such as heat shock proteins, fibrinogen, and fragments of heparan sulphate and hyaluronic acid (TLR2, TLR4) which are all associated with signs of stress or damage to the host. If so, the concept of a PRR needs to be extended to include components of both microbial PAMPs and "stressed self" as agonists. It is not yet clear to what extent the latter can also be viewed as representing DAMPs (Section 4.2.2.4).

4.2.2.3 Structure and Function of TLRs

Structures Revealed by Crystallography All TLRs have a common structural organization with an extracellular recognition domain composed of leucine-rich repeats (LLRs), a transmembrane domain that anchors them to cell surface or endosomal membranes, and an intracellular signalling domain. The latter is called a TIR domain as it is shared by TLRs and the interleukin (IL)-1 receptor. The extracellular domains of TLRs are in the shape of a horseshoe and it appears that many TLRs function as dimers. In some cases, PAMPs (or altered self components) can bind directly to TLRs and therefore they act as true ligands for these receptors. In other cases, the interaction is indirect and involves the ligand binding to other molecules which then interact with the TLR. Binding of ligands or agonists is thought to induce dimer readjustment that can subsequently initiate signalling. Interestingly, it also appears that TLRs can use different molecular mechanisms to bind their respective ligands or agonists (e.g. in the convex or the concave region of the horseshoe-shaped extracellular LRR domain). Moreover, in some cases, the ligand or agonist has to be chaperoned to the TLR by another molecule(s). See Figure 4.4.

Our increased understanding of the structure and function of TLRs has led to much interest in the development of drugs that inhibit their function for therapeutic purposes. For example, antagonists of TLR4 are being designed as mimetics of the lipid A component of LPS for potential therapy of septic shock in which there is massive production of tumour necrosis factor (TNF)-α as a consequence of ligation of TLR4 on blood monocytes by LPS (Section 4.4.6). Others are being developed as microbially derived adjuvants for use in vaccination (Section 4.6.2).

Signalling via TLRs What is the outcome of the PRRs recognizing the presence of microbial components? In the case of TLRs we can generalize and say that particularly important outcomes are the secretion of anti-microbial peptides and of pro-inflammatory cytokines that include IL-1, IL-6 and TNF-α (below). Activated TLRs bind adapter molecules which trigger signalling cascades within the cytoplasm of the cells. These result in the activation and nuclear translocation of transcription factors that induce gene expression. For most TLRs the adapter molecule is Myd88. While it might seem that genetically modified mice in which the Myd88 gene is deleted would be informative in terms of dissecting TLR signalling, this adapter is also used by the IL-1 receptor and, in fact, many mouse knock-out phenotypes are actually due to defective signalling through this receptor rather than the TLRs. One of the most important transcription factors for the synthesis of pro-inflammatory cytokines is the nuclear factor NF-κB; another is AP-1, which is induced by triggering of the mitogen-activated protein (MAP) kinase pathways. See Figure 4.5.

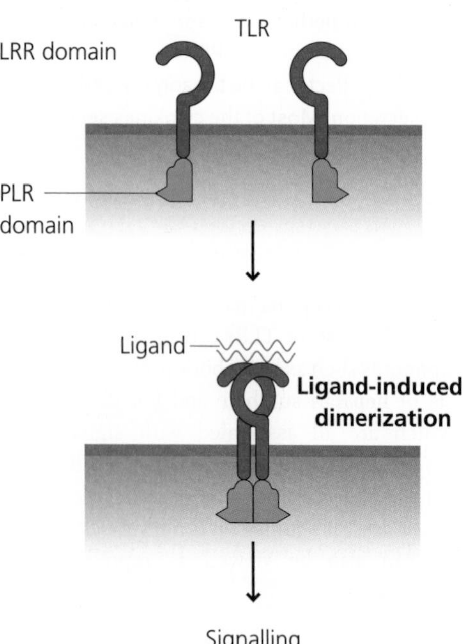

Fig. 4.4 General structure of Toll-like receptors. TLRs have similar general structures, but the ways in which they interact with their agonists varies greatly. TLRs have a horseshoe-shaped LRR domain connected to a globular TIR domain that is involved in signalling. In some cases, the agonists interact directly with the LRRs. In others cases, one or more accessory molecules may be involved in binding to LLRs and the triggering of signalling can involve several distinct molecular interactions. In general, the activation of signalling requires the dimerization of two TLR molecules.

> Q4.2. How might the relative roles of TLR versus IL-1 receptor signalling be dissected experimentally?

It is very important to appreciate that signalling through TLRs does not trigger the release of a single bolus of cytokines, rather it is a carefully regulated process. For example, different pro-inflammatory cytokines are produced by the responding cells at different times after stimulation and probably with different kinetics in different cell types. If TLR stimulation is sustained, the cells eventually become refractory to TLR stimulation; they stop producing cytokines, presumably to limit the host damage that might be otherwise be caused. In contrast, the capacity to produce anti-microbial peptides usually remains.

TLR3, which binds viral double-stranded RNA, is unusual in that it signals via an adaptor called TRIF instead of Myd88 (and TLR4 can in fact signal via both). Again we can generalize and say that one particularly important outcome of signalling through this pathway is the secretion of Type I IFNs. TLR3 and some RLHs signalling, leads to activation of IFN-regulated factors (IRFs), which are transcription factors that induce the expression of Type I IFN genes. The signalling pathways involved in all these responses are very complex, but understanding how they work is crucially important for the development of therapies (above). It is also very important to stress that while many components of these pathways are shared between different pathways and different cells, there are many that are cell-specific or pathway-restricted See Box 4.2.

Consequences of Defects in TLR Signalling TLRs are crucial for many aspects of vertebrate host defence against infection. This is reflected by the diversity of different viral components that can interfere with signalling at different points. Viruses can sequester or inhibit the upstream adaptor proteins thus preventing the initiation of TLR signalling. Some examples include vaccinia virus which sequesters multiple adaptors,

Box 4.2 Investigating Toll-Like Receptor Signalling

Intracellular signalling by TLRs is triggered when an agonist binds directly or indirectly to the receptor. As for any other signalling pathway, this initiates cascades of protein interactions in the cytoplasm that ultimately lead to altered activities of transcription factors in the nucleus. These cascades involve enzymes – kinases or phosphatases, respectively – that add or remove phosphates from proteins at specific sites, such as tyrosine residues in particular motifs of the protein, in order to activate or deactivate other components in the pathway. Hence, if a known signalling component becomes phosphorylated in a cell that has been stimulated with a TLR agonist (or any other receptor ligand), this strongly suggests that it is involved in the pathway.

One experimental approach to study signalling pathways is immunoblotting or Western blotting. In this, cell extracts are electrophoresed on a gel and the proteins in the gel are subsequently transferred to a membrane. The membrane can then be probed with an antibody specific for the component of interest. Binding of this primary antibody can be detected by using a labelled secondary antibody – this could be radio-labelled (e.g. ^{32}P) or light-emitting (e.g. luminol) – and its binding identified using radiosensitive or light-sensitive films, respectively. As primary antibodies can be generated that are specific for the phosphorylated and dephosphorylated forms of signalling components, the relative intensity of the bands that are detected in Western blots provides a semi-quantitative measure of the extent of activation or deactivation of that component. Thus, at different times after stimulation of cells with a TLR agonist, phosphorylation of specific signalling molecules can be assessed. In addition, if cells are fractionated into different components – such as cell membranes, endosomes and nuclei – similar techniques can be used to determine the cellular localization of the active components with time after stimulation. See Figure 4.6.

An alternative, and complementary, approach is immunoprecipitation in which the component of interest is physically separated from a solubilized extract of stimulated cells (e.g. by using antibodies coupled to beads). If that component is associated with another, such as an adaptor protein that has bound to the phosphorylated protein, this will be co-precipitated and can subsequently be characterized. If the proteins in the stimulated cell were labelled with a radioisotope (e.g. by metabolically labelling them with a ^{35}S-labelled amino acid) the co-precipitated proteins will also be labelled. If the isolated proteins are electrophoresed on a gel under non-reducing or reducing conditions they will appear as either one heavy band or as two lighter bands, thus enabling the molecular weight of the second component to be estimated. It is also sometimes possible to cut out the bands and sequence part of the protein, after which gene databases can be searched using the predicted DNA sequence of the protein, and the gene encoding it can be cloned. See Figure 4.7.

A combination of the above techniques could be used to demonstrate that TLR4 is coupled indirectly both to TRIF (through the TRAM adaptor) and Myd88 (through TIRAP), and that it signals through both pathways; in contrast TLR3 is coupled directly only to TRIF and other TLRs directly only to Myd88.

The localization of TLR4 and its respective partners can be visualized by combining the use of fluorescent probes with microscopy. For example macrophages were transfected with green fluorescent protein-labelled TRAM or TIRAP and stimulated with LPS. Using markers of different cellular compartments (e.g. lysosomal enzymes), it was thus possible to trace the movement of these molecules with time after stimulation. These and other techniques (including the use of nuclear reporters for IFN-responsive elements and NF-κB which are activated by the respective pathways) have revealed that TLR4 signals through Myd88 when at the cell surface, but through TRIF when it is in endosomes.

These types of study are crucially important for furthering our understanding of disease processes, such as how different TLR mutations in humans can lead to susceptibility to infection and how they are involved in immunity. They are also central to new therapeutic approaches. For example, the identification of a novel signalling component in a pathway can enable the design of small molecule inhibitors that can be used to inhibit cell responses. Thus tyrosine kinase inhibitors are now widely used for the treatment of some myeloid or lymphocytic leukemias.

and hepatitis C virus which uses one of its proteins (ND5A) to bind Myd88 and a protease (NS3-4A) to cleave TRIF. Other viruses can inhibit the downstream transcription factors IRF3 or IRF7 by degrading or sequestering them, or by competing for binding to the promoter sequences of the DNA. Other microbes, such as certain protozoa, can also prevent or reduce TLR signalling.

Evidence for the role of TLRs in defence against infection comes from studies of patients with genetic defects in the respective receptors, the adaptors or other signalling components, and from gene knock-out mice in which the corresponding genes have been selectively targeted. Thus, exceedingly rare individuals have been identified with defects in TLR3 or a crucial component in its signalling

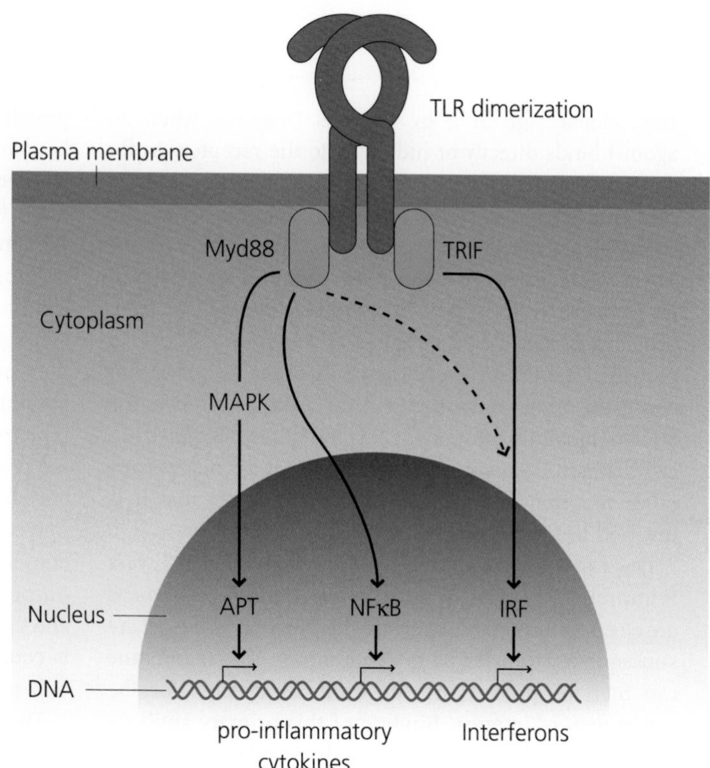

Fig. 4.5 Signalling pathways from Toll-like receptors. Following agonist binding and dimerization, different adapter molecules bind to TLRs. The key ones are Myd88, which is used by most TLRs, and TRIF, which is only used by one, TLR3; TLR4 can use either depending on its location. The canonical Myd88 pathway leads to NF-κB activation, as well as activation of the MAP kinase pathways which activate the AP-1 transcription factor; typically this results in activation of genes for pro-inflammatory cytokines. In contrast, the TRIF pathway typically leads to activation of IRFs that act on "IFN-responsive elements" in the DNA (not shown) and the activation of genes for Type I IFN production. Cross-talk between these pathways can occur (e.g. stimulation of the Myd88 pathway can also lead to IFN production). Each pathway involves multiple signalling components that are not shown for simplicity.

pathway. These patients present with a herpes simplex virus (HSV)-related encephalitis (inflammation of the brain), thus demonstrating the importance of this pathway, at least in the central nervous system (CNS) in defence against this class of virus. Other individuals with defects in Type I IFN production also suffer from increased viral infections.

Q4.3. Why might the effects of defective TLR3 signalling be seen only with herpes viruses and only in the CNS?

Rare patients with genetic defects in Myd88 or IRAK-4 which are involved in TLR signalling have also been identified.

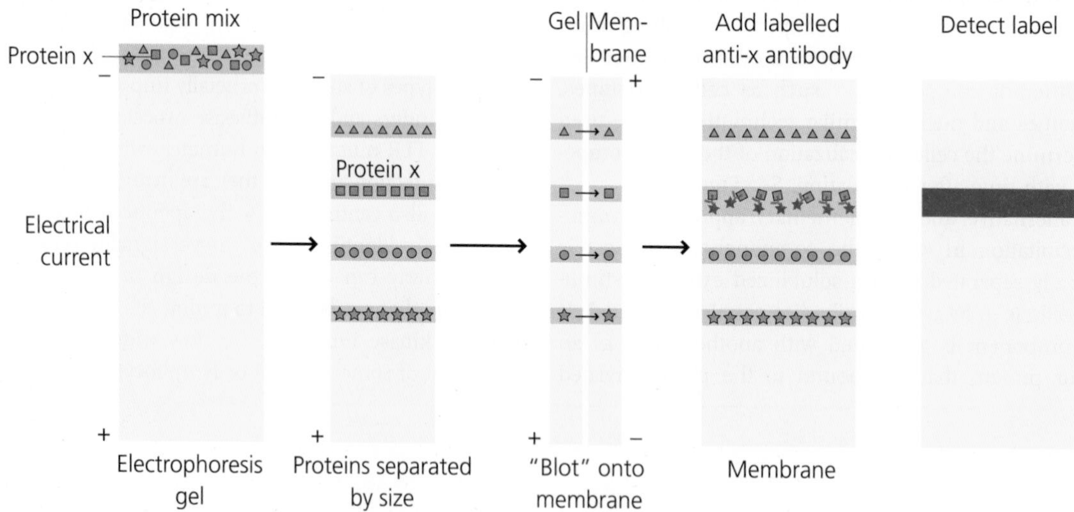

Fig. 4.6 Western blotting. This technique permits identification of specific proteins in a cell lysate. Proteins are dissociated by typically treating with a detergent, sodium dodecylsulfate (SDS). The mix is then electrophoresed on a polyacrylamide gel in which the rate of a protein's migration when a charge is applied is proportional to its molecular size. This process is called SDS–polyacrylamide gel electrophoresis (PAGE). Antibodies cannot bind to proteins in the gel so typically a nitrocellulose membrane is apposed to the gel, and the proteins are made to migrate into the membrane, again by applying a current; this process is often called "blotting". Once in the membrane, the proteins can be detected by using antibodies labelled with enzymatic or other labels.

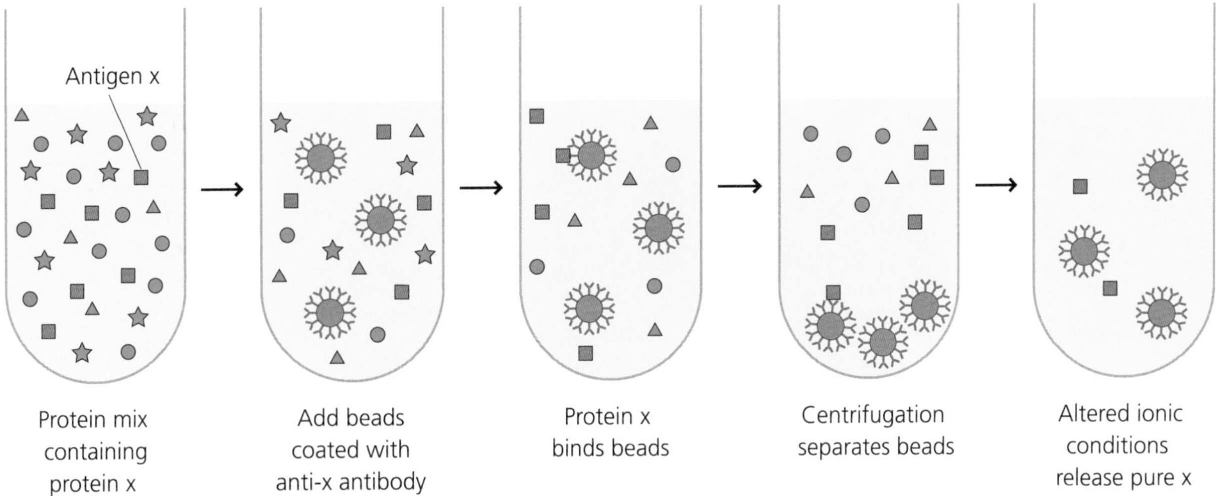

Fig. 4.7 Inflammasomes. TLR agonists can lead to the transcription and translation of the IL-1β precursor, pro-IL-1β. Inflammasomes are molecular complexes in the cytoplasm that contain multimerized NLRs, such as Nalp3, associated with a proteolytic enzyme, caspase-1. In activated inflammasomes, caspase-1 processes pro-IL-1 into the mature, functional form of IL-1, which can then be secreted. Hence, particularly in DCs, two stages are needed for production of IL-1 (IL-18 is similarly processed), but how these pathways interact is not well understood; these pathways and requirements may differ in other cell types such as macrophages.

These individuals suffered in early life from recurrent pyogenic infections, particularly invasive pneumococcal disease. Surprisingly, however, these infections were not apparent in later life. In contrast, knock-out mice in which the gene for TLR4 was targeted were susceptible to a very wide range of microbial infections for life. (Note, however, the earlier point that IL-1 receptor also signals through Myd88.) This clearly demonstrates that we must be very cautious in extrapolating from studies in mice to the human situation.

Q4.4. Why might recurrent pyogenic disease be seen only in childhood in people with defective Myd88 signalling?

4.2.2.4 Inflammasomes and Autoinflammatory Diseases

The original concept of a PRR was that of a receptor for a conserved microbial component, a PAMP. As noted earlier, however, there is increasing evidence that some TLRs may also recognize self components that have become altered or modified through cellular damage or stress, although to what extent this does occur or is relevant *in vivo* is still not clear. What is, however, clear is that the NLRs can sense both PAMPs such as bacterial peptidoglycans (above) and non-microbial signs of danger including metabolic stress.

Some, possibly all, NLRs – including Nalp3 (also called cryopyrin and by other names) – can act as platforms for the assembly of molecular complexes called inflammasomes which are involved in the activation of specialized proteases – the inflammatory caspases. An example is caspase-1 that processes the precursor form of IL-1β into the active cytokine; the same is also true for a related cytokine IL-18. Thus, a PRR such as a TLR may induce translation of pro-IL-1, but the active molecule cannot be secreted until it is processed by an inflammatory caspase; we still have much to learn about the interactions of these pathways. See Figure 4.8.

Inflammasomes can be activated by a variety of non-microbial danger signals such as those associated with cell damage perhaps, for example, by sensing abnormal levels of intracellular potassium ions or extracellular ATP, as well as reactive oxygen intermediates (ROIs). Large particles such as monosodium urate crystals, formed from uric acid as an end-product of purine metabolism, as well as silica and asbestos can apparently activate the Nalp3 inflammasome directly or indirectly. Currently, there is much interest in the roles of the Nalp3 inflammasomes in the actions of aluminium-based adjuvants (Section 4.6.1).

IL-1β, produced after inflammasome activation, has multiple biological activities. Crucially, it functions as an endogenous pyrogen in stimulating fever through its actions on the hypothalamus (this term must not be confused with pyogenic, something that induces pus; cf. pyogenic bacteria. For some time it has been known that rare individuals suffer from unexplained and recurrent episodes of fever often accompanied by severe inflammation. These conditions are often familial and known as hereditary periodic fevers. These conditions are examples of autoinflammatory diseases which are known or suspected to result from mutations in components of inflammasomes such as Nalp3 and other, potentially related, signalling pathways such as the TNF receptor and the IL-1 receptor antagonist. Crucially, most of these mutations result in gain of function, rather

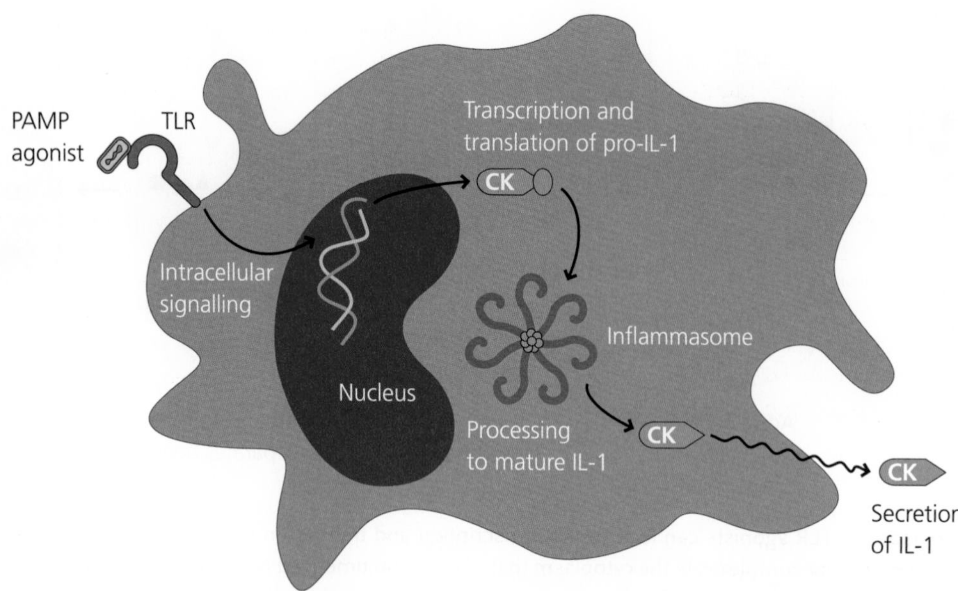

Fig. 4.8 Immunoprecipitation. This technique is used to isolate specific molecules from a complex mix in solution. Beads coated with antibodies specific for a particular molecule are added to the mix and will bind the specific molecule. Centrifugation can be used to isolate the beads that have bound the molecule; alternatively, magnetic beads can be used and can be captured in a magnetic field. By altering the ionic conditions, the specific molecule can be released from the beads and can be further characterized.

than being mutations that lead to deficiencies of the components involved. Mutations in the IL-1 receptor antagonist illustrate the importance of this molecule in regulating inflammation. These diseases involve only innate immunity and the adaptive response does not appear to play any part in their pathology. Although rare, these diseases give important insights into the regulation of inflammatory processes.

Q4.5. The autoinflammatory diseases are not associated with an increased risk of infections. Why might this be?

4.2.2.5 Cellular Distribution of PRRs

If PRRs are to act as alarm molecules, on which cells should they be expressed? Clearly the closer they are to the sites of infection, the earlier they will be activated. It is thus not surprising that some epithelial cells, such as those in the intestine, express PRRs. The PRRs expressed by epithelial cells are not uptake receptors; epithelial cells have no known defence role in phagocytosing pathogens. Interestingly, in the intestine, the expression of these PRRs is polarized to the basal, rather than luminal, surface of the cells and this presumably enables them to discriminate between infectious microbes that have penetrated the epithelium and the commensals situated above. Stimulation of these receptors results in secretion of pro-inflammatory cytokines into the underlying connective tissues (and of antimicrobial peptides into the lumen).

Connective tissues contain resident cells of innate immunity that also express PRRs – primarily macrophages and mast cells – that are ideally placed to detect infection. Macrophages are phagocytic cells that express high levels of certain C-type lectins and scavenger receptors, and they also express the PRRs associated with signalling to the nucleus. In contrast, mast cells may only express the latter. There is, in addition, another cell type present in connective tissues which expresses PRRs, the DC. These cells are, however, primarily thought of as being concerned with the activation and regulation of adaptive immunity, and we discuss them in Chapter 5. Endothelial cells also express some TLRs, and mice that have been engineered to express TLR4 exclusively on endothelial cells can recruit neutrophils to inflammatory sites as efficiently as normal mice and can clear a peritoneal infection with Gram-negative bacteria at least as effectively as normal mice. This indicates that endothelial cells may sometimes play a very significant, direct role in host defence, in addition to being regulated by other cells in the vicinity.

To summarize, in all tissues there are cells that can respond quickly to the presence of a pathogen, and the end result of these responses is the activation of innate immune effector mechanisms (and usually also adaptive immunity). How the activity of these different cells and signals is coordinated is far from clear. For instance, what is the relative contribution of signals generated by epithelial cells and macrophages to innate immune activation? What is the relative importance of macrophages and mast cells in connective tissues? Do these differ in different types of infection, and so on?

Q4.6. How might we be able to determine the relative contributions of TLR activation in intestinal epithelial cells and macrophages to innate immunity?

4.2.3 Cytokines in Innate Immunity

Signalling through PRRs such as TLRs alters the properties of the cells expressing the PRR, but crucially also enables these cells to modulate the properties of other cells through their secretion of cytokines. Here, we will discuss the roles of some of these cytokines in host defence and inflammation.

4.2.3.1 Interferons (IFNs) and Anti-Viral Resistance

As we have seen (above), Type I IFNs can be produced when viruses are detected in endosomal or cytoplasmic compartments by different PRRs. This occurs, for example, when phagocytes internalize viruses or when viruses infect cells. Type I IFNs are of two main types, α and β (there are others). Both IFN-α (which is of multiple types, and IFN-β bind to a single dimeric receptor. This receptor signals through the JAK–STAT pathway, common to many cytokine signalling pathways, and additionally signals through an IRF (above). Transcription of IFN-stimulated genes leads to the coordinated transcription of a wide variety of genes that confer anti-viral resistance on the cell. Their actions include degradation of viral RNA, suppression of protein translation, inhibition of viral transcription and trafficking, and RNA editing which effectively scrambles the genomes of retroviruses (APOBEC3G). They also induce the synthesis of RLHs, further enhancing the detection of viral presence. Type I IFNs are different from IFN-γ (also termed Type II IFN) that signals through an entirely different dimeric receptor (IFNG1 and 2, although also linked to the JAK–STAT pathway) and which has relatively weak anti-viral activity, but other crucial functions in host defence such as macrophage activation. See Figure 4.9.

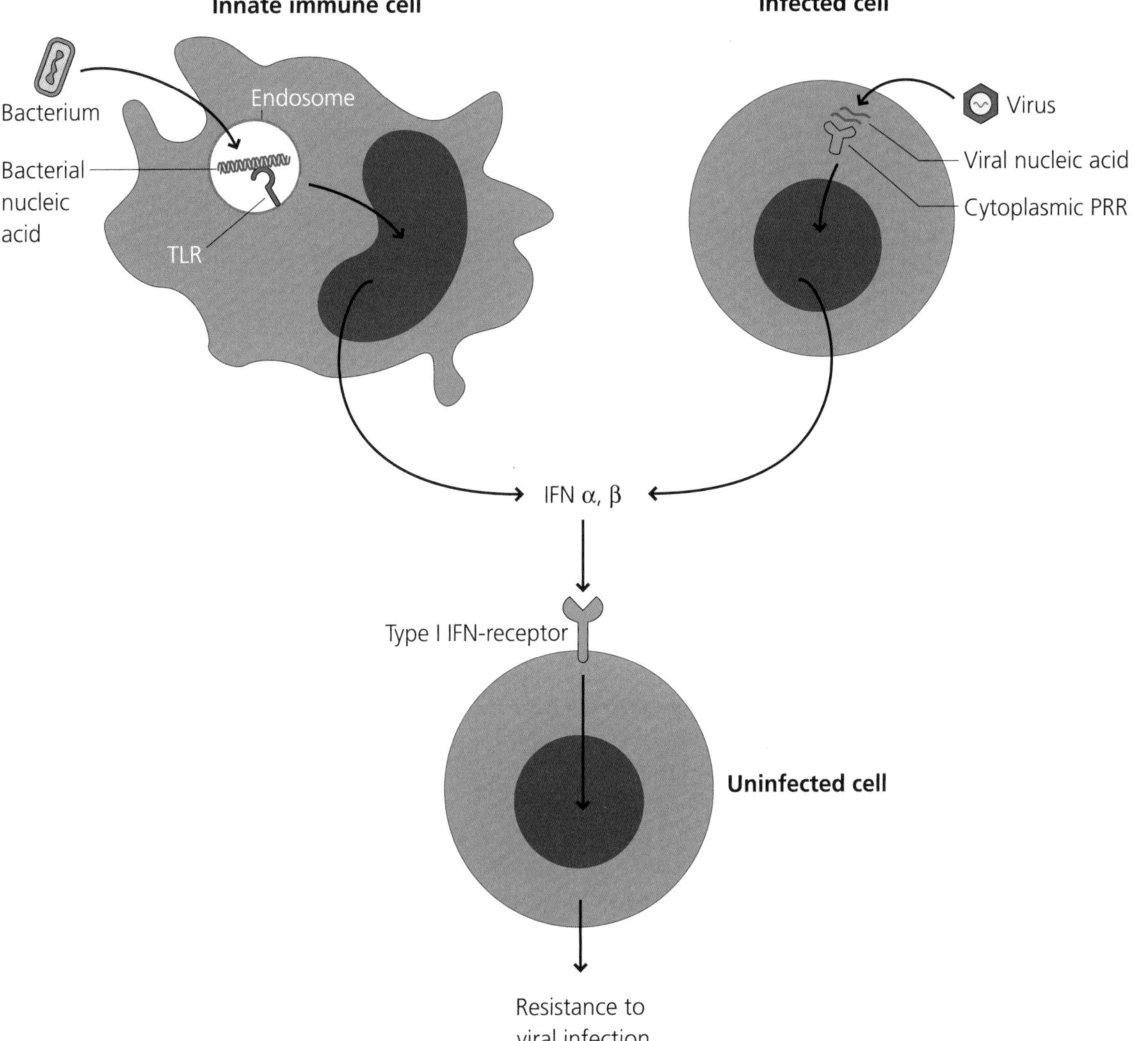

Fig. 4.9 Type I interferons. Type I IFNs (e.g. α and β) are potent anti-viral molecules. They are secreted by cells containing viruses or other microbes for example, after recognition of viral nucleic acids by cytoplasmic PRRs. They are also secreted by cells of the innate immune system such as macrophages and plasmacytoid dendritic cells (pDCs) after PRR ligation. Type I IFNs act on other cells through a single plasma membrane receptor to induce synthesis of a variety of proteins that collectively act to inhibit viral replication.

Box 4.3 Analysis of Cytokine Production by Immune Cells

Cytokines are molecules that have central roles in mediating communication during immune responses. It is thus crucial in both experimental and clinical situations to be able to measure cytokine production by different cells during these responses. The commonest approach to measure bulk production of cytokines by cell populations is the enzyme-linked immunosorbent assay (ELISA). In one form of this assay, an antibody specific for the cytokine is used as the detection reagent. This antibody is bound to the bottom of wells in a plastic tray; this is the capture antibody. For example, to measure IL-12 production in a macrophage activation assay, the wells are coated with an anti-IL-12 monoclonal antibody. Supernatants from the macrophage cultures are collected and dilutions of these are added to the wells. Any IL-12 will bind to the capture antibody. Binding of IL-12 can then be detected by adding an anti-IL-12 antibody specific for a different epitope on IL-12 that binds quantitatively to the IL-12. Binding of this second antibody is detected by adding a third antibody, specific for the particular immunoglobulin of the second antibody which is labelled with an enzyme such as horseradish peroxidase. The amount of the third antibody bound is assessed by using the enzyme to generate a coloured reaction product which can be measured in a spectrophotometer. By comparing the binding from the supernatant with the binding of a standard preparation of IL-12, the amount of IL-12 in the supernatant can be estimated. See Figure 4.10.

ELISA assays have proved invaluable in assessing cytokine production and in many other ways, but one of their disadvantages is that they can only measure secretion by a whole population of cells. Using this technique with a mixed population of cells it is not directly possible to determine the surface molecules (phenotype) expressed by the cytokine-secreting cell, which cells are secreting the cytokine, or whether an individual cell is secreting more than one cytokine. This can, however, be achieved by intracellular cytokine staining. Antibodies cannot penetrate intact cell membranes but, if the cells are permeabilized, anti-cytokine antibodies can enter. Cytokine secretion is usually blocked by using a drug such as Brefeldin A that prevents protein exit from the Golgi apparatus, thus increasing the amount present within the cell. By using two antibodies against different cytokines, labelled with different fluorescent dyes, and analyzing the cells by flow cytometry, non-, single- and double-secretors can be quantitated. By using monoclonal antibodies to cell surface markers, the phenotype of the secreting cells can be assessed. See Figure 4.11.

4.2.3.2 Pro-Inflammatory Cytokines

These molecules can be secreted by resident cells such as macrophages, mast cells, and in some cases epithelial cells (e.g. keratinocytes) and even endothelial cells. Their production is largely dependent on the transcription factor NF-κB. As noted earlier, typical cytokines in this group include IL-1, IL-6 and TNF-α.

The IL-1 family of cytokines includes IL-1α and -1β, and IL-18. The cytoplasmic domains of the IL-1 receptor molecules are very similar to those of TLRs and they also act via Myd88. They stimulate a variety of pro-inflammatory responses such as leukotriene and prostaglandin secretion, nitric oxide secretion, and increasd of adhesion molecule expression. They act on the hypothalamus to cause fever and on the liver to stimulate the increase in protein synthesis seen in the acute-phase response. The activity of IL-1 is in part regulated by natural antagonists; the IL-1 receptor antagonist is a "decoy" molecule that is now used therapeutically in a variety of arthritic conditions.

IL-6 signals through the JAK–STAT pathway. This cytokine has multiple effects on inflammation, immune responses, the bone marrow and the nervous and endocrine systems. The use of IL-6 gene knock-out mice has demonstrated its central roles in stimulation of the acute-phase response, induction of fever and mucosal IgA synthesis. Two other cytokines, IL-12 and IL-23, are closely related to IL-6 and also signal through the JAK–STAT pathway. As we shall see (below), IL-12 is probably crucial for acute responses against intracellular pathogens, because it can trigger the production of IFN-γ by natural killer (NK) cells, leading to macrophage activation. In contrast, IL-23 may play a role in the induction of chronic inflammation and the formation of granulomas.

TNF-α was initially discovered in two separate ways: as a molecule that could stimulate the destruction of some tumours in mice, and as the molecule causing the wasting syndrome (cachexia) seen in chronic infections and late-stage malignant tumours. Neither of these relate to the important roles of TNF-α in defence against infection. TNF-α is produced primarily by macrophages and mast cells, but many other cell types can be stimulated to secrete TNF-α. TNF-α signals through two TNF receptors. Such signalling can lead to activation of the cell, with cytokine secretion, or in some cases to apoptosis of the cell (as is seen with some tumours). At low concentrations TNF-α acts locally to increase expression of adhesion molecules on leukocytes, and to stimulate a variety of cell types to secrete pro-inflammatory molecules. At high concentrations TNF-α enters the blood and stimulates many cell types to make excess IL-1 and IL-6. It also stimulates endothelial cells to synthesize nitric oxide, causing venous dilatation. Together these effects can induce potentially fatal cardiovascular shock. One of the main agents inducing TNF secretion is LPS from Gram-negative bacteria. The importance of TNF-α in defence against intracellular bacteria is illustrated by rheumatoid arthritis patients treated with an anti-TNF monoclonal antibody – this can lead to re-emergence of active tuberculosis.

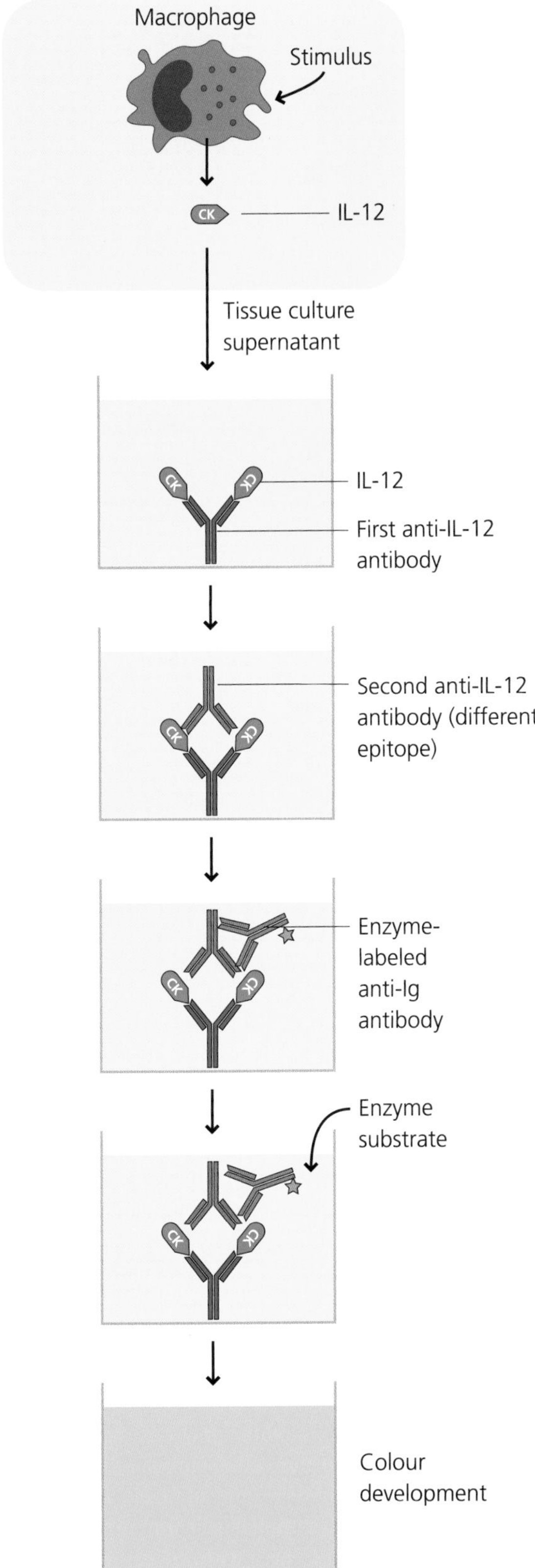

Fig. 4.10 Analysis of cytokine secretion I: enzyme-linked immunosorbent assays. Cytokines in solution can be assayed by using ELISAs, as can any other molecule for which antibodies recognizing different epitopes are available. In a multi-stage procedure as illustrated for IL-12, a specific antibody can be attached to a surface to capture the cytokine, and a second antibody to a different epitope on the cytokine can be added to detect the bound cytokine. Binding of the second antibody can be detected by adding a third, enzyme-labelled, anti-Ig antibody. Adding an enzyme substrate leads to production of a coloured reaction product than can be measured quantitatively. By comparing this with a standard preparation of the cytokine, the concentration can be measured. This particular type of ELISA is known as a sandwich ELISA (referring to the two different anti-IL12 antibodies that are used); it is also called an antigen-capture assay.

4.2.3.3 Anti-Inflammatory Cytokines

Inflammatory responses are central to defence against infection but can, and do, cause tissue damage. It is thus crucial that inflammatory responses are regulated to minimize such damage. One of the most important ways in which inflammation is regulated is via the secretion of counter-acting, anti-inflammatory cytokines. These include the following examples:

- **IL-10.** This cytokine is secreted by many cell types including macrophages (particularly alternatively activated macrophages) and T_h2 T cells in mice (in humans both T_h1 and T_h2 cells secrete IL-10). IL-10 acts on macrophages, neutrophils and other leukocytes, largely by inhibiting the actions of NF-κB and thus preventing the synthesis of pro-inflammatory cytokines and preventing biasing of CD4 T cell activation to T_h1; in fact, it promotes T_h2 biasing. The importance of IL-10 in regulating inflammation is illustrated by IL-10 knock-out mice, which develop a severe inflammatory bowel disease.
- **Transforming growth factor (TGF)-β.** TGF-β can be secreted by many haematopoietic cells, such as macrophages, lymphocytes and NK cells, but also by some other cells present in the CNS and kidney, for example. TGF-β acts via a receptor expressed on many cell types. Downstream signalling involves interactions with a family of transcription factors called Smads. TGF-β inhibits macrophage activation, cytokine synthesis by macrophages and lymphocytes, and can block the effects of pro-inflammatory cytokines on endothelial cells and neutrophils. It also stimulates angiogenesis and activates fibroblasts, thus having an important role in the induction of healing and repair.

4.3 Tissue-Resident Cells of Innate Immunity

In peripheral tissues, both resident immune cells, and component cells of the tissues, are crucial for innate defence against infection. See Figure 4.12.

4.3.1 Epithelial Cells

How does activation of the alarm system lead to the generation of effector mechanisms that can kill or limit the growth

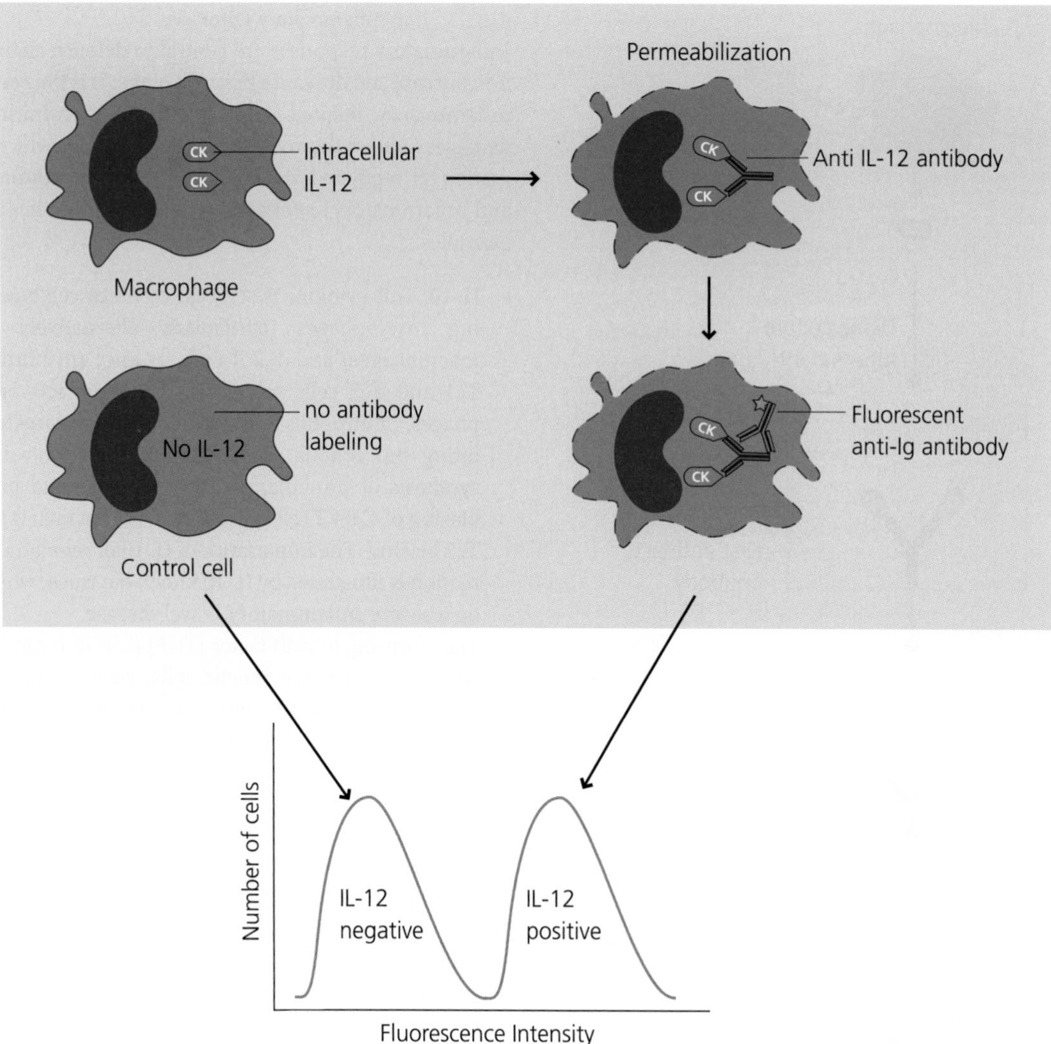

Fig. 4.11 Analysis of cytokine secretion II: intracellular flow cytometry. It can be important to know which individual cells in a population are secreting a particular molecule. To determine this, cells in suspension are permeabilized, allowing antibodies to enter. Antibody binding is detected by fluorescent labelling and the cells are analyzed in a flow cytometer. This permits the identification of specific cell types by their expression of particular cell surface molecules (phenotypic markers), and determining whether that cell has synthesized the cytokine in question. This technique is illustrated for a mixed population of macrophages in which only a subset is secreting IL-12; analysis of phenotypic markers, which can be done simultaneously, is not shown.

and spread of an infectious agent? Some effects are very direct. For example activation of epithelial cells stimulates the release of small proteins with direct anti-microbial effects; these include the defensins and other families of molecules. Many of these molecules appear to work by inserting pores into the microbial surface, causing osmotic lysis. It has proved difficult to define the role of defensins in immunity because there are so many of them that their genes cannot all be knocked out in transgenic mice. One example is provided by Paneth cells in the crypts of the small intestine that synthesize defensins following TLR stimulation. To become microbicidal these defensins need to be activated by an enzyme called matrilysin and, if the gene encoding this enzyme is knocked out in mice, they become more susceptible to intestinal Salmonella infection. Epithelial cells are not the only cells that make defensins. For example, these molecules are also stored in neutrophils and are released on recognition of agonists (below).

Another set of molecules, with functions that are closely related to defensins, are the cathelicidins, which are also small, anti-bacterial peptides. One of these molecules is released by urinary tract epithelial cells into urine following bacterial contact. By using mice with the cathelicidin gene (Camp) knocked out it was shown that epithelial cell-derived cathelicidin plays an important role in protection against infection with Escherichia coli. Additionally, in humans, cathelicidin-resistant Escherichia coli strains caused more severe urinary tract infections. Thus, cathelicidin seems to be a key factor in mucosal immunity in the urinary tract. See Figure 4.13.

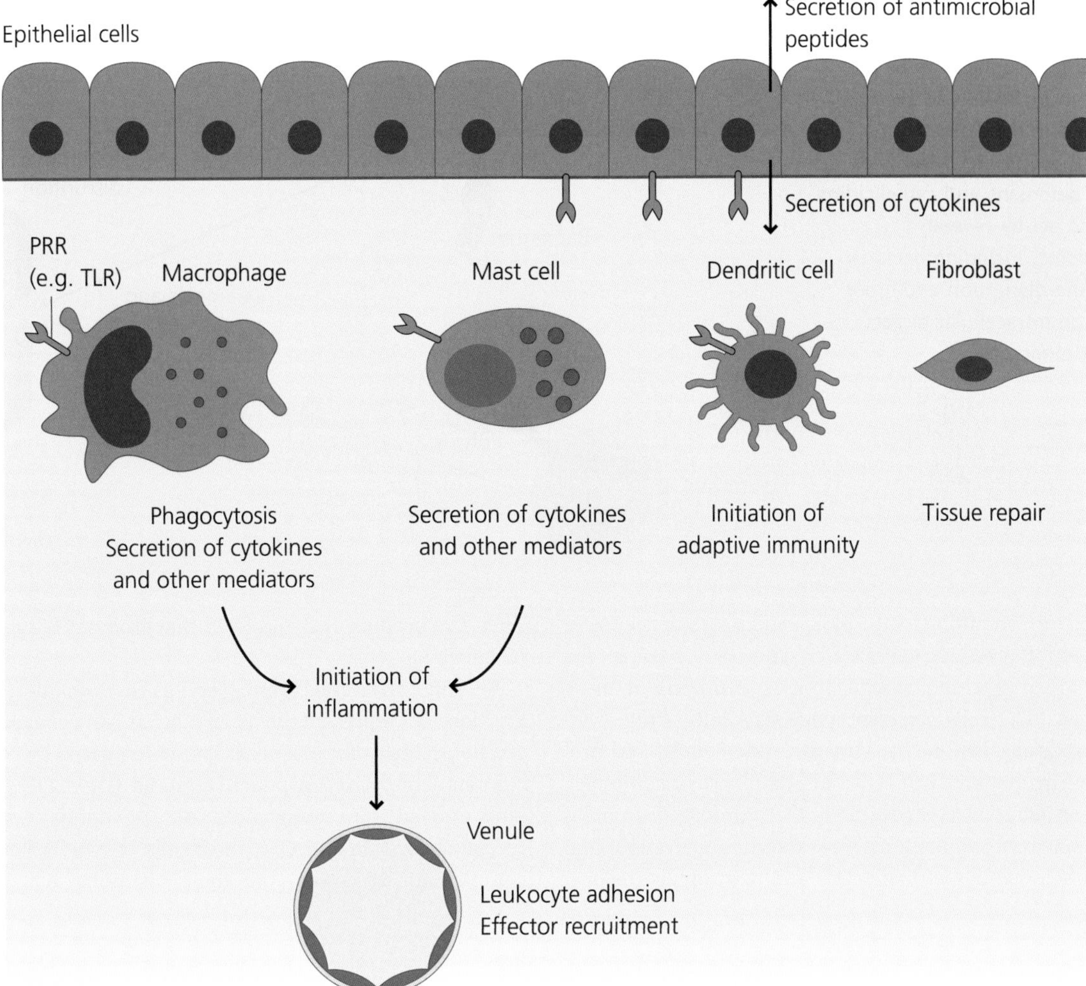

Fig. 4.12 Tissue-resident cells of innate immunity. Tissue-resident cells can detect the presence of a microbe or its products. Many cells in peripheral tissues such as macrophages and mast cells in connective tissues, and epithelial cells in mucosal tissues, express PRRs that recognize PAMPs from infectious agents. PRR agonists trigger different cellular responses, often including cytokine secretion that triggers inflammation and initiates innate immune responses. DCs, responding to the PAMPs or the signals delivered by other resident cells, become activated and start to trigger adaptive immune responses. Other tissue-resident cells, such as fibroblasts, are involved in repair of tissues damaged by the microbe.

4.3.2 Mast Cells

All connective tissues contain mast cells that play crucial roles in both defence and immunopathology; if you suffer from hay fever, blame your mast cells. Mast cells are very similar in their properties and functions to blood basophils, one of the granulocyte lineages. Mast cells are bone marrow-derived cells, but it is not clear how they relate to other haematopoietic cells in terms of their origins. They may develop independently of both the common myeloid and lymphoid precursors (CMPs and CLPs). See Figure 4.14.

What do mast cells do? Unlike many other immune cells, mast cells can act immediately upon stimulation. Their granules store mediators such as histamine and other inflammatory mediators which can be released very rapidly when they degranulate. Mast cells can even be made to degranulate by mechanical forces. Try marking your skin by dragging a blunt point over the inside of your arm. What do you observe? You should see a series of changes along the course of the mark; first a local reddening, then a reddening that spreads laterally from the scratch. Finally, in some of you, after a minute or two, there will a raised paler area along the scratch (oedema). This is the classical triple response described in the 1920s. It is due to mechanical damage to mast cells causing degranulation and the release of histamine.

Q4.7. How could we show that the triple response is caused by histamine?

Apart from these stored mediators, mast cells can rapidly, but later, synthesize other inflammatory mediators. Particularly important are the eicosanoids – products of arachidonic acid metabolism, such as prostaglandins and leukotrienes. These molecules are also closely involved in acute inflammation. Evidence for this comes from the efficacy in the treatment of

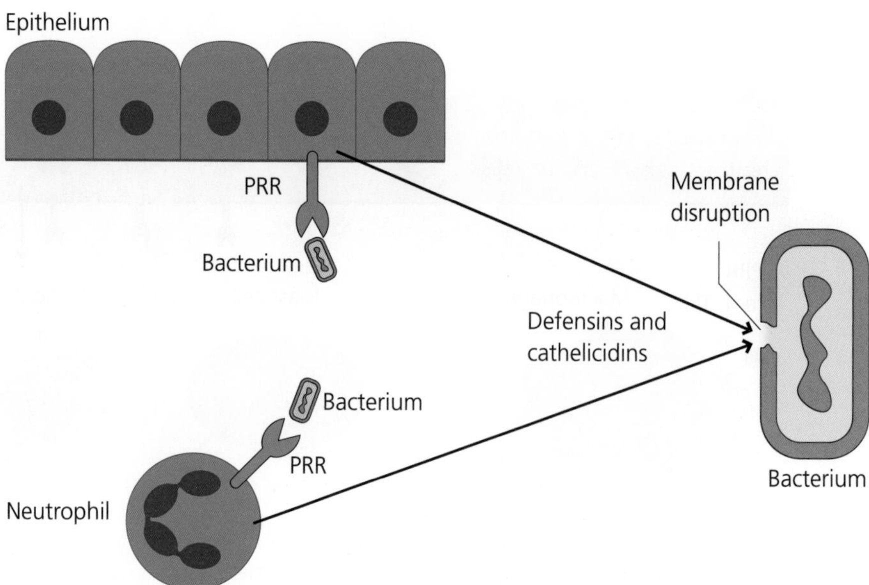

Fig. 4.13 Anti-microbial peptides. Cells such as mucosal epithelial cells and neutrophils can be stimulated by PAMPs to secrete peptides that have direct anti-microbial properties. Molecules such as defensins and cathelicidins appear to act by several mechanisms, including bacterial membrane disruption and other actions on intracellular targets within the microbe.

inflammatory conditions with drugs such as indomethacin which inhibit cyclo-oxygenases, enzymes involved in the breakdown of arachidonic acid and the synthesis of these mediators. Mast cells can also synthesize and secrete cytokines, especially TNF-α that activates endothelium and promotes inflammation; IL-3, IL-5 and granulocyte-macrophage colony stimulating factor (GM-CSF) that stimulate eosinophil production from the bone marrow and their activation; and chemokines that attract monocytes and neutrophils (below). They also produce IL-4 and IL-13 that promote T_h2 responses (Chapter 5).

Although mast cells are susceptible to mechanical damage this does not represent the usual mechanism of triggering. In acute inflammation mast cells can be triggered by two small complement-derived peptides, C3a and C5a, as well as proteins derived from neutrophils and eosinophils. Apart from receptors for C3a and C5a, mast cells possess several other complement receptors, but their functions are not well-

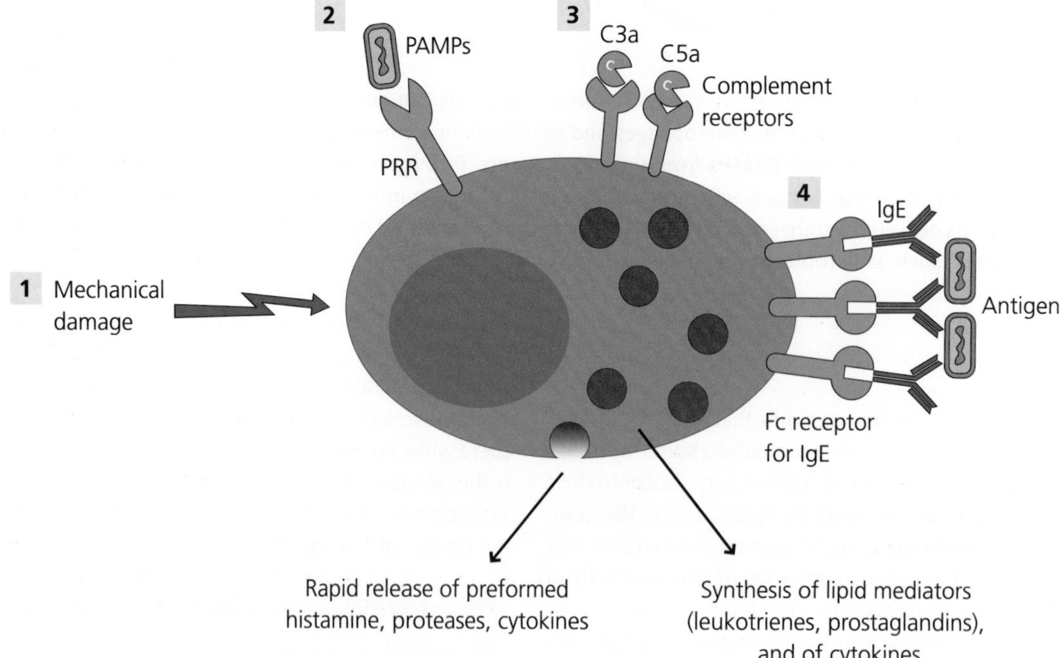

Fig. 4.14 Mast cell functions. Mast cells are resident in connective tissues throughout the body. The can be activated by (1) mechanical damage, (2) by PAMPs acting on their PRRs, (3) through small molecules such as the C3a and C5a complement fragments, (4) and via cross-linking of IgE bound to their specific FcRs. Activation results in very rapid degranulation with release of histamine and some other pre-formed, stored mediators including some cytokines, and the later synthesis and release of lipid mediators such as leukotrienes and prostaglandins, and of many cytokines and chemokines.

defined. However, the mast cell triggering most familiar to us is that stimulated by pollens and other allergens (Section 1.6.4.2). This is not innate immunity, as it depends on IgE, but is another example which shows that in mammals innate and adaptive immunity cannot be separated easily.

Q4.8. How might mast cells be able to "sense" mechanical damage?

4.3.3 Macrophages

In evolutionary terms, macrophages represent one of the most ancient components of innate immunity and one of their most obvious functions is phagocytosis of particles, including microbes. This was recognized by Metchnikoff in the nineteenth century when he saw cells in starfish engulfing foreign particles and suggested that phagocytosis might be important for defence, part of the work for which he was awarded a Nobel Prize; these cells, called amoebocytes, may be the phylogenetic ancestors of macrophages. Macrophages are phagocytic; they can engulf microbes and then release molecules into the phagosome, where the microbe can be killed and digested.

4.3.3.1 Resident Macrophages

Resident macrophages are present in all loose connective tissues underlying epithelia and within solid organs. These cells are specialized for their own particular anatomical niches. For example, there are macrophages in liver (Kupffer cells), kidney (mesangial cells), lung (alveolar macrophages), bone (osteoclasts) and the central nervous system (microglial cells). These may share some common functions, but in other respects each is adapted specifically to its own environment. For example, all macrophages can probably help to reshape tissues during normal development by removing apoptotic cells and regulating blood vessel development (angiogenesis), and they also help to regulate wound healing after trauma. As examples of their tissue specializations, brain microglia and macrophages in the intestinal lamina propria are very unresponsive to a variety of stimuli compared to peritoneal macrophages. Those in the intestine, and perhaps in other mucosal tissues, may play an important role in maintaining the barrier function of the epithelial layer and so help to limit inflammation and contribute to repair if damage occurs. The importance of macrophages in normal tissue functions (homeostasis) and development is shown for example by mice where macrophage development is defective resulting in major defects in bone formation. It is worth noting that these macrophage deficiencies are all only partial; complete macrophage deficiency is probably incompatible with life.

As discussed later (Section 4.4.4), tissue damage and inflammation leads to the recruitment of new populations of macrophages into tissues from their precursors the blood monocytes. In acute inflammatory responses monocytes are recruited after the initial influx of neutrophils. These recruited monocytes can develop into macrophages, called inflammatory macrophages, which have properties and functions that differ from resident macrophages. Under other circumstances they may acquire very different functions. It is convenient to think of resident macrophages as alarm cells that help to trigger inflammation if infection occurs, but their relative importance as effector cells in relation to newly recruited macrophages is unclear. See Figure 4.15.

Q4.9. If Kupffer cells are isolated from a mouse's liver, and tested for their ability to synthesize reactive oxygen intermediates, they are very inefficient compared to recently recruited macrophages. Why might Kupffer cells be adapted in this way?

4.3.3.2 Phagocytosis

Endocytosis refers to uptake of molecules and particles where the substances are delivered into the endosomal system. Phagocytosis refers to the uptake of particles into the same system. Phagocytosis is probably always receptor-mediated, although in some cases the receptor has not been identified. Under normal circumstances phagocytosis is a property of a limited range of cells such as macrophages and neutrophils which are particularly important in defence. The ability to phagocytose can, however, be induced in cells that are normally non-phagocytic. Salmonella bacteria, when they adhere to intestinal epithelial cells, inject some of their own proteins into the cell. This makes the epithelial cell able to phagocytose the bacteria (Section 2.4.2.4). Not all forms of phagocytosis are morphologically the same. In many cases the phagocyte extends processes around the particle, which is then drawn into the body of the cell. In some cases, such as in macrophages phagocytosing via the CR1 complement receptor, the particle sinks into the cell. Legionella bacteria (the cause of Legionnaire's disease) induce phagocytosis in which the plasma membrane of the macrophage coils around the bacterium. The significance of these different patterns remains largely unclear.

The molecular basis of phagocytosis is becoming better understood. The binding of a particle to a receptor initiates signalling which leads to activation of the cell's actin cytoskeleton. This involves the activation of the Rho family of small GTPases. For Fc receptors (FcRs) these include Rac and Cdc42. These initiate polymerization of actin resulting in the pushing out of membrane extensions around the particle. This form of phagocytosis depends on sequential interactions between the FcR and antibody; "zippering". This was shown most directly by using particles where only part of the particle was coated with antibody and in these cases there was only partial internalization, thus phagocytosis is not "triggered" by a single ligand-receptor interaction. For phagocytosis via complement receptors it is Rho that is involved. See Figure 4.16.

4.3.3.3 Phagocytic Receptors on Macrophages

Perhaps it is not surprising that we are still discovering new phagocytic receptors expressed by macrophages. These are professional phagocytes and they need to be able to deal with a huge range of microbes as well as self components such as apoptotic cells. When a receptor is discovered, it can be tested

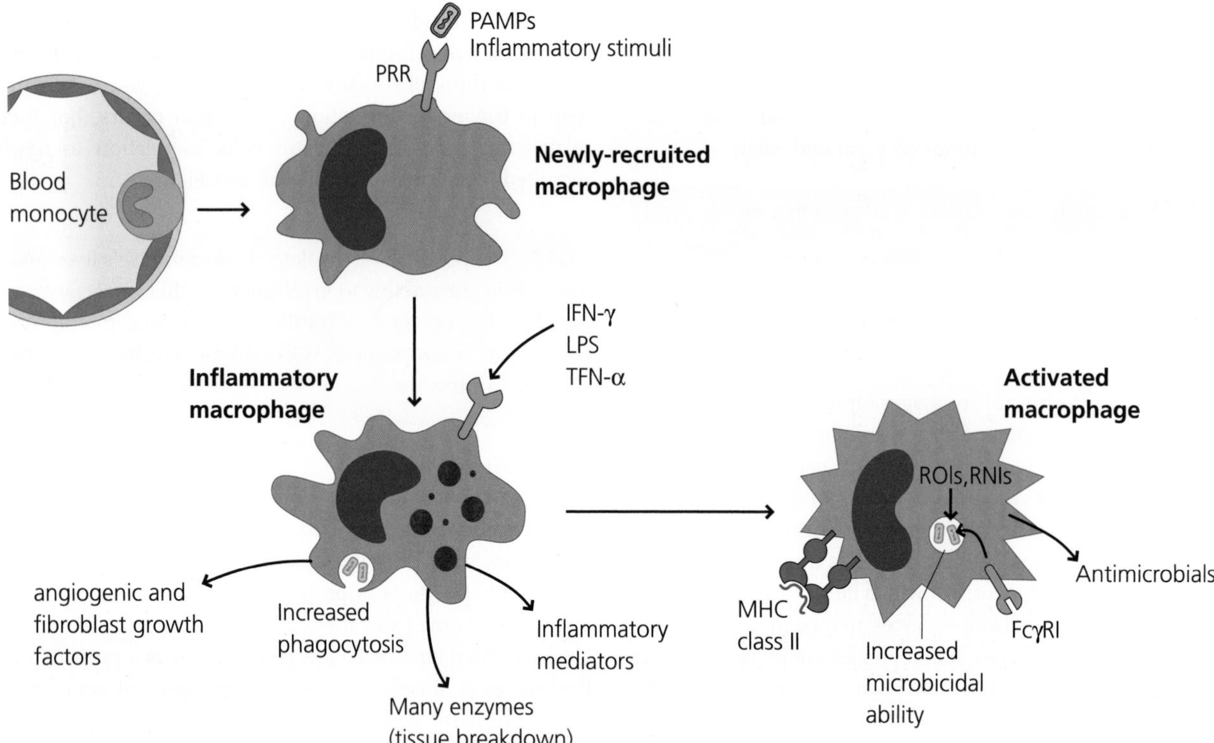

Fig. 4.15 Macrophage activation. Blood monocytes are recruited into inflamed tissues where they can differentiate into macrophages. Stimulation via PRRs or other receptors leads to changes in their properties that include increased phagocytic ability, secretion of enzymes such as elastase and collagenase, and of angiogenic and fibroblast growth factors. These cells are now known as inflammatory macrophages. If these cells are primed by cytokines such as IFN-γ (the most important) or TNF-α, and stimulated via PRRs e.g. by LPS, they can acquire potent anti-microbial properties. These fully activated macrophages produce many reactive oxygen and nitrogen species (ROIs, RNIs), and acquire specific FcRs (e.g. FCγR1) to target opsonized microbes into their microbicidal machinery. The most important early source of IFN-γ is probably NK cells. Activated macrophages have increased expression of MHC class II molecules, which enables later recognition by activated CD4 T cells and delivery of IFN-γ.

for its ability to bind potential ligands, and of course many may be identified. This does not, however, tell us the real function of these receptors, or whether they have evolved to defend against infection, to mediate homeostatic functions or both.

Resident macrophages play crucial roles in remodelling tissues during development; for example they are crucial for removing cells that have died through apoptosis. Apoptosis is one mechanism of cell death, the other being necrosis. The significance of apoptosis is that it is immunologically silent, whereas necrosis (e.g. resulting from trauma) is generally associated with inflammation, often leading to tissue damage and scarring. Macrophages have a diversity of receptors that allow them to internalize apoptotic cells, and cell fragments called apoptotic bodies, without generating an inflammatory response.

As a general rule it appears that some of the apoptotic receptors, such as CD36 (a scavenger receptor noted earlier), may inhibit inflammatory responses such as those that can be generated through TLRs. In contrast, other receptors that are involved in phagocytosis by macrophages, such as some of the FcRs, can trigger potent anti microbial responses in the cells and also induce inflammation. As these are antibody-dependent mechanisms they may, however, be more strictly related to adaptive immunity. It is difficult to assess the importance of these different receptors in human infectious disease, but work with animal models, particularly gene knock-out mice, suggests that they may play roles in the initial stages of infection. Thus mice lacking one type of scavenger receptor are more susceptible to infection with Gram-positive bacteria See Box 4.4.

Q4.10. How might we attempt to discover a novel phagocytic macrophage receptor?

4.3.4
Endothelial Cells and Inflammation

Endothelial cells are the gateway between the blood and sites of inflammation or infection. It is the properties of endothelial cells that determine the recruitment of the molecules and cells crucial for defence against infection. We describe the phenomenology of the above events in Chapter 3. Here, we discuss the underlying cellular and molecular mechanisms. One important question is how infection, which in many cases is limited to the epithelium, induces the changes in

Box 4.4 Studying Apoptosis and Necrosis in the Immune System

Cells can die in two ways. Apoptosis represents a quiet death. It does not normally lead to inflammation and apoptotic cells are rapidly removed *in vivo*, largely by macrophages. Apoptosis is a physiological process involved in tissue organization, (for example in the destruction of embryonic webs, leading to the formation of spaces between our fingers). Apoptosis is also induced in cells by cytotoxic T cells and NK cells. Necrosis, on the other, hand is a pathological process; necrotic cells release their contents into their environment and often stimulate inflammation. If an apoptotic cell is not cleared by phagocytosis in a few hours it may undergo necrosis. Experimentally, complement-mediated lysis of cells causes immediate necrosis. The extent of necrosis can be measured in a flow cytometer by using a fluorescent dye, typically propidium iodide (PI), which enters through the disrupted membranes of necrotic cells and binds to the DNA in the nucleus.

Apoptosis can be assessed in several ways. (1) One of the characteristic features of apoptosis is the early breakdown of DNA. This can be assessed in bulk cell populations by electrophoresis. DNA is broken down into fragments of multiples of 180 base pairs (representing one DNA nucleosome) and following electrophoresis is seen as a laddering pattern. (2) The breakdown of DNA can also be visualized by the TUNEL (terminal deoxynucleotidyl transferase dUTP nick end-labelling) assay. Apoptotic DNA has multiple free 3′-ends. These can be detected by using the enzyme terminal deoxynucleotidyl transferase (the enzyme that is important in generating T and B cell receptor (TCR and BCR) diversity) to add a modified nucleotide to these ends. The modified nucleotide can then be detected by using standard immunocytochemistry techniques. (3) Another feature of apoptotic cells is that phosphatidylserine, usually a component of the inner leaflet of the plasma membrane, translocates to the outer leaflet early in the apoptotic process. Annexin V is a protein that binds specifically to phosphatidylserine and Annexin V binding can be detected by antibodies; this is very useful in flow cytometric assays in conjunction with PI staining. Thus, cells undergoing apoptosis can be detected in several different ways. See Figure 4.17.

It is important in cellular immunology to be able to detect cells that have been killed by cytotoxic T cells or NK cells. Early assays were based on the release of radioactive chromium. Chromium passes though the plasma membrane and binds to intracellular macromolecules. These are released when the plasma membrane disintegrates in necrosis. Thus although cytotoxic cells induce apoptosis, by 6–8 h after killing their targets become necrotic and release chromium. More recently assays have been developed that do not use chromium. The JAM assay uses the release of labelled DNA fragments while other assays measure release of enzymes such as lactic dehydrogenase. Why "JAM" assay? It's "just another method".

endothelial cells that can permit the egress of macromolecules and cells from venules, when these venules are relatively distant from the site of the actual infection. Thus, for instance, in gonoccocal infection of the uterine (Fallopian) tube, the bacteria attach to the luminal surface of the epithelium and do not enter the connective tissues, but the venules displaying changes lie deep in the connective tissues. The answer is of course that resident alarm cells such as epithelial cells, macrophages and mast cells express PRRs that respond to microbial PAMPs by secreting cytokines and chemokines that diffuse to the endothelial cells and modulate their functions.

4.3.4.1 Leukocyte Transendothelial Migration See Figure 4.18

Different types of molecule regulate leukocyte migration from the blood into tissues (Section 1.5.2). Neutrophils constitutively express selectin ligands, but resting endothelial cells do not express the corresponding selectins "Post" or "Zip" code. If, however, endothelial cells are treated with histamine, TNF-α or IL-1 they increase expression of P- and E-selectins, permitting the rolling interaction. Firm adhesion of leukocytes to endothelial cells is then, mediated by integrin molecules. One way in which this was shown was by using antibodies that bound to leukocyte integrins: when injected intravenously these antibodies did not interfere with leukocyte rolling, but did prevent firm adhesion. Integrins can exist in two conformational states that differ in affinity for their ligands. Resting neutrophils express the low affinity form. Activation of neutrophil integrins to the high-affinity form requires signalling via a G-protein-coupled chemokine receptor (GPCR) on the neutrophil surface. How do neutrophils find a chemokine? Chemokines secreted by epithelial or tissue-resident innate cells in response to infection diffuse to the endothelium. They are taken up by the endothelial cells and transported onto the luminal surface where they bind to highly glycosylated molecules. The rolling activity of the neutrophils allows them to interact with the bound chemokines, leading to increased integrin affinity and firm adhesion and flattening of the cell. See Figure 4.18.

Q4.11. L-selectin is the molecule used by lymphocytes to adhere loosely to endothelial cells in high endothelial venules of lymph nodes. Neutrophils and monocytes also express L-selectin, but do not emigrate into lymph nodes under normal conditions. Why might this be?

The tightly adherent leukocytes then migrate between the intercellular junctions of the endothelium and cross into the extravascular tissues. A variety of other molecules expressed

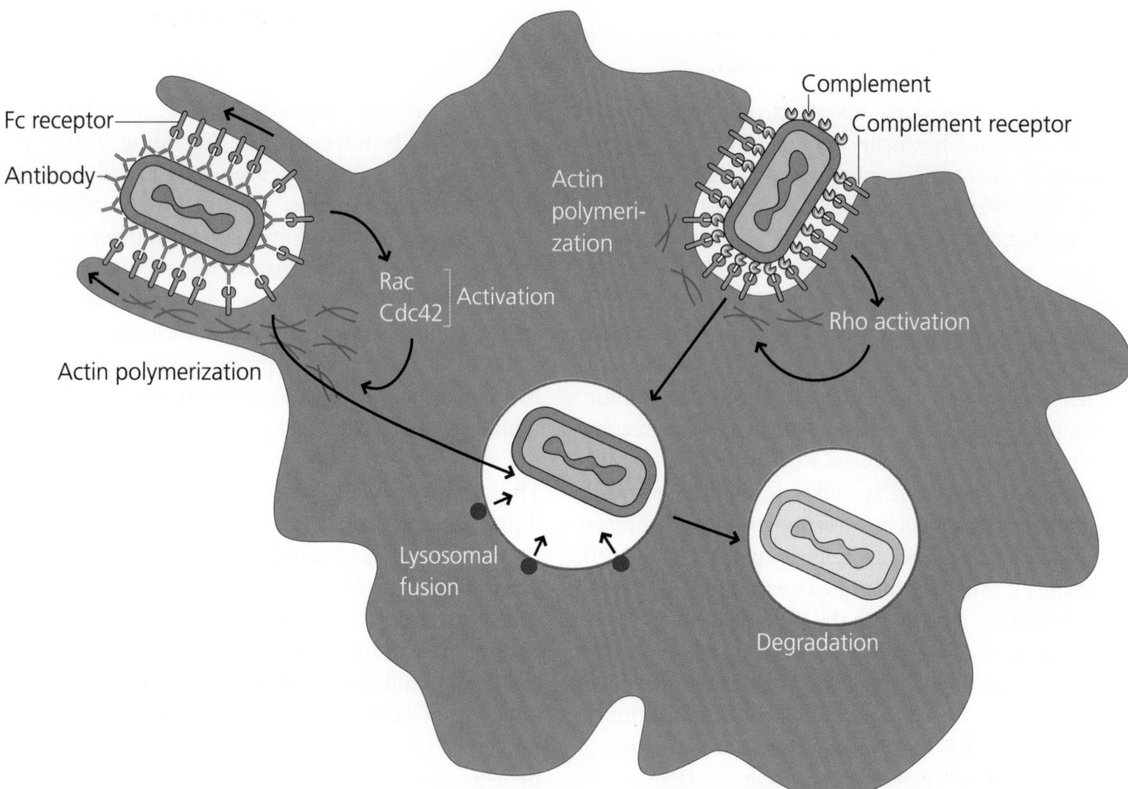

Fig. 4.16 Phagocytosis by macrophages. Particles can be internalized by macrophages in several ways. The figure illustrates two of the most important mechanisms. Antibody-coated particles bind to FcRs and this stimulates actin-dependent extension of cytoplasmic processes that engulf the particle by a zippering mechanism. In contrast, particles coated with complement C3b or iC3b appear to "sink" into the cell, but again this is actin-dependent. The activation of the actin cytoskeleton involves signalling through the Rho family GTPases, examples of which are shown. In both cases the vacuole containing the particle, the phagosome, fuses with lysosomes, which discharge their contained enzymes into the phagosome.

on both the endothelial cells and the leukocytes mediate this transmigration. Leukocytes usually cross between the junctions of the endothelial cells but, in some cases, actually appear to migrate through the endothelial cells. Having crossed the endothelium the cells may have to penetrate the basement membrane to reach the infecting microbes, and it is likely that proteases secreted by the leukocytes are involved in breakdown of the membrane and/or the connective tissue. As we have outlined elsewhere (Section 3.3.3.2), the permutation and combination of different molecules that is used to recruit different populations of leukocytes from the blood into different sites can be thought of as a post code principle. See Figure 4.18.

Q4.12. Can you think of some situations where understanding the "post code" principle may lead to innovations in the treatment of disease?

4.3.4.2 Regulation of Leukocyte Recruitment by Endothelial Cells

Further studies have revealed that endothelial cells are more than just passive players in inflammatory responses; they have intrinsic abilities to control leukocyte recruitment themselves. One type of endothelial activation can be triggered by histamine, which stimulates endothelial cells to synthesize nitric oxide. This acts on smooth muscle to cause dilatation of small arteries, leading to increased blood flow. Histamine also induces the breakdown of tight junctions between endothelial cells in venules, leading to leakage of fluid into the tissue, and promotes the binding of neutrophils by stimulating the expression of P-selectin. This typically occurs over a few minutes and then shuts off. In contrast, pro-inflammatory cytokines such as IL-1 and TNF-α cause another type of activation that leads to different and more sustained responses. Interestingly, one feature of this type of response is that the endothelial cells spontaneously change their ability to recruit different leukocytes, initially neutrophils, later, monocytes and, if an adaptive response has occurred, effector T cells as well. The nature and timing of these responses may relate to the different timing of the release of histamine (very rapid) and cytokines (slower, requiring protein synthesis). Together with the fact that different cell types can secrete different pro-inflammatory cytokines with different kinetics, and that endothelial cells apparently have some autonomy in

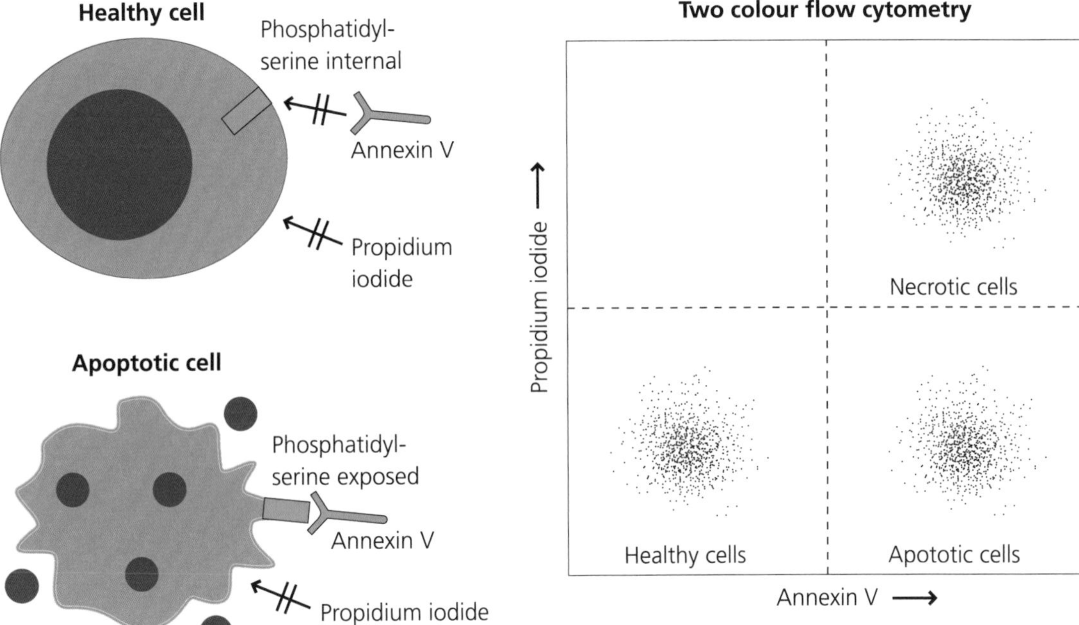

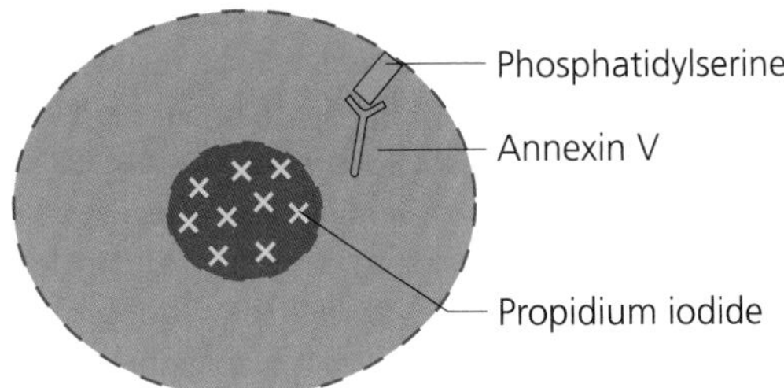

Fig. 4.17 Assessment of apoptosis and necrosis. Apoptotic cells "flip" phosphatidylserine from the inner to the outer surface of the plasma membrane, where it can be detected by the binding of fluorescent Annexin V. Their plasma membranes are, however, still intact and this prevents the entry of molecules such as propidium iodide (PI), which binds DNA. In necrotic cells, however, PI can enter the cell and access and bind DNA. Using two-colour flow cytometry, apoptotic cells (Annexin V-positive) can be distinguished from necrotic cells (double-positive for Annexin V and PI) and live cells (negative for both). The figure shows a dot-plot as generated by two-colour flow cytometry.

controlling the different types of cells they can recruit, you can see that leukocyte recruitment is a highly complex and tightly regulated process, and that our current understanding of this process is likely to be very incomplete.

4.4 Recruited Effectors of Innate Immunity

Effectors that are recruited a little later into the innate immune response include both soluble molecules such as complement, and cells such as neutrophils and monocytes. Efficient recruitment of these components of innate immunity is essential to bring about recovery from infection and, later, effective healing and repair of the damaged tissues. We will first discuss some examples of different classes of soluble PRRs before considering complement in more detail.

Q4.13. Where do all the molecules that enter inflamed tissues go to?

4.4.1 Recruited Molecules

Some PRRs are not cell-associated, rather they are secreted. Amongst these are the collectins and ficolins that recognize microbial carbohydrates but not those of the host, and some

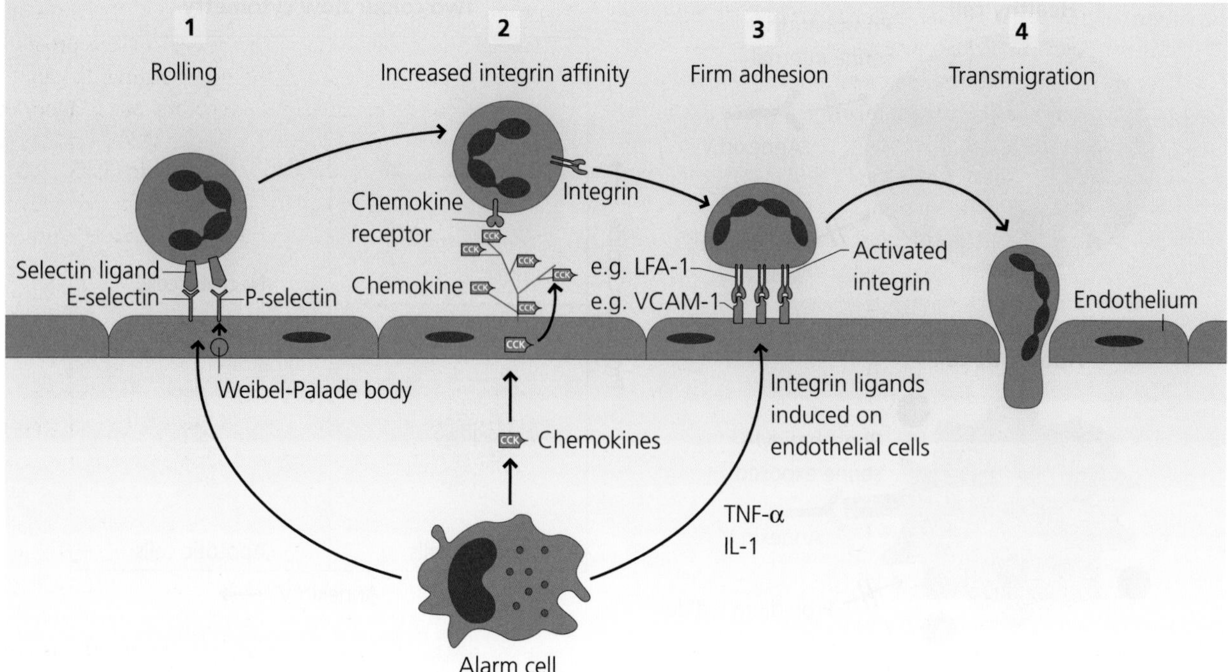

Fig. 4.18 Endothelial cells and leukocyte emigration. Leukocyte emigration from the blood into the tissues (extravasation) requires a highly regulated set of molecular interactions, shown here for neutrophils. (1) Inflammatory mediators secreted by alarm cells in response to PAMPs, and perhaps by direct action of PAMPs on endothelial cells, stimulate endothelial cells to express E- and P-selectins on their luminal surface; the latter is stored in specialised granules (Webel-Palade bodies) in endothelial cells, and is released on activation of the endothelial cell. Neutrophils have complementary ligands that make low-affinity interactions with the selectins, allowing them to roll along the endothelium. (2) Chemokines, secreted by alarm cells, are translocated across the endothelium and bind to glycosylated molecules on the luminal surface. Chemokine receptors on the neutrophils recognize the corresponding chemokines and signal to integrin molecules, increasing their affinity for their ligands. (3) New expression of these corresponding ligands (e.g. stimulated by alarm signals) enables the neutrophils to adhere tightly and to flatten on the endothelium. (4) Other molecules such as CD31 then enable the neutrophils to migrate between, or through, the endothelial cells into the underlying connective tissue.

pentraxins that recognize phospholipids of microbes and apoptotic cells. As we shall see (below) different members of these families can activate complement, thus helping to opsonize their bound microbes, promoting their uptake by phagocytes and contributing to the initiation of acute inflammation.

Another group of secreted PRRs that are important in defence are some of the surfactants. Epithelial cells in the alveoli of the lungs secrete surfactants into the alveolar spaces. Some of these have no known defence role, being involved in lubrication of the respiratory tract, but surfactant proteins SP-A and SP-D are thought to play roles in defence by activating complement or by aggregating bacteria. Polymorphisms in the SP-A and SP-D genes are associated with an increased risk of respiratory syncytial virus infection in children.

There are potentially important interactions between the coagulation and kinin systems and inflammatory responses. Factor XII (Hageman factor) is a protein that can be activated by contact with negatively-charged surfaces such as collagen in extravascular tissues. This initiates the coagulation cascade but can also initiate the kinin cascade, leading to the formation of bradykinin which causes increased vascular permeability, and is also direct a very potent stimulator of pain nerve fibres. There are also direct interactions between the coagulation system and the complement system (Section 4.4.2). Activated Hageman factor can also cleave complement C1, leading to activation of the classical complement pathway. Later coagulation factors – activated Factors X and XI – can cleave complement C3 and C5, leading to the production of C3a and C5a. The fibrinolytic protease, plasmin, can also cleave C3 and C5 directly. The roles of these different activation mechanisms in defence against infection are, however, difficult to unravel.

4.4.2 Complement System

Complement is crucially important in inflammation and defence against microbes (Section 1.5.5.2) It also has important roles in clearing damaged or altered host cells and macromolecules. Here, we will discuss the nature of complement, its activation and regulation, and its functions. What is complement? Some crucial things to understand about complement are the following.

i) It represents in integrated system of nearly 30 soluble proteins present abundantly in blood, but not present in large amounts in normal extracellular tissues. Inflammatory macrophages can produce many, if not all, of these components and this local secretion may increase the availability of complement at inflammatory sites. In addition there are at least eight complement receptors expressed by different cells and three membrane-bound complement regulators expressed on nearly all host cells.
ii) Many complement components form a cascade, meaning that activation of one component leads to activation of more components at the next stage; activation of complement is based in part on a series of enzymatic (proteolytic) cleavages.
iii) Complement is activated on surfaces, normally microbial or damaged host surfaces; it does, however, have the potential to bind to and attack normal host cells.
iv) Complement activation must be highly regulated; If complement components are activated, but do not bind to a microbe, they may attach to host cells which must therefore protect themselves from attack. (the role of the three membrane-bound complement regulators mentioned above). Otherwise, they are rapidly inactivated.
v) Complement has at least four main functional roles in defence: it induces acute inflammation; it opsonizes microbes for phagocytosis; it kills some microbes by inducing osmotic/colloidal lysis; it helps to remove damaged host material (such as apoptotic cells). It also helps to solubilize or remove antibody–antigen complexes from circulation by breaking up the lattices . Complement also helps to regulate the activation of B cells in adaptive responses (Chapter 6).

4.4.2.1 Complement Activation and Functions

The complement cascade can be activated by three pathways. These are known as the lectin, classical and alternative pathways of complement activation. Apart from the components involved in their initial activation (below), the lectin and classical pathways are identical. These two cascades sequentially involve components C1–C9, with the sole exception that C4 is activated before C3. Note also that cleavage of complement components generates small and large subunits that are designated "a" and "b". The alternative pathway involves different proteins, including Factors B, D, H, I and properdin. See Figure 4.19

Initial activation of the lectin and classical pathways leads to the formation of a molecular complex of C4b with C2a (C4b2a), while activation of the alternative pathway produces a complex of C3b with Bb (C3bBb). The crucial thing about these complexes is that they are C3 convertases which can all cleave and activate C3. Hence, all three activation pathways come together at C3 and thereafter the rest of the pathway, involving C5, C6, C7, C8 and C9, is the same. Remember, however, that successive activation of the early components occurs on surfaces, typically the surfaces of microbes, and that the number of components is amplified at several steps. The central functions of complement in innate host defence are mediated by different complement components as follows (Figure 4.19):

i) Some components can act as opsonins. C3b, covalently attached to target surfaces, is cleaved by the protease Factor I to form iC3b. Both these fragments can be recognized by complement receptors on phagocytes (CR1 for C3b, and CR3 and CR4 for iC3b) leading to phagocytosis of the target. (As an aside iC3b slowly breaks down further to C3d, for which the receptor is CR2. CR2 is not involved in phagocytosis, but in enhancement of B cell responses (Chapter 6).
ii) The small, diffusible components C3a, C4a and C5a are called anaphylatoxins. These are involved in the inflammatory functions of complement. C3a, for example, is a potent stimulator of acute inflammation. For example it activates mast cells to release their mediators and acts on endothelial cells to stimulate leukocyte adhesion. C5a can also activate mast cells, and is additionally a potent activator and chemoattractant for neutrophils.
iii) C5b initiates the assembly of the late complement components (C5b–C9) into a non-specific pore, called the membrane attack complex (MAC), that inserts into the lipid bilayer of cells and permits free ingress and egress of water and electrolytes, leading to osmotic lysis of some microbes or virus-infected host cells.
iv) Complement is important for solubilizing immune complexes and removing them from the circulation. In the first case, immune complexes of IgG antibodies bound to their antigens are solubilized when complement is activated and C3b binds to them, because their lattice structure is disrupted. In the latter case, erythrocytes have CR1 receptors that can bind to C3b or C4b attached to immune complexes. The erythrocytes transport these complexes to the liver where they are effectively captured and degraded by liver macrophages, Kupffer cells, and the erythrocytes are released.

We will now say a few more words about the different activation pathways, and then discuss some examples of how they are regulated, describe some consequences of defects in complement components, and note how some pathogens can subvert the complement system.

Lectin Pathway Activation Mannose-binding lectin (MBL) is a low-abundance, soluble PRR molecule belonging to the collectin family. It is secreted by hepatocytes in larger amounts during the inflammatory acute-phase response. It binds to specific types of sugar clusters that are found on many bacteria and fungi (yeasts), and some viruses, but not on host cells. MBL is associated with two proteases, MBL-associated serine protease (MASP)-1 and -2 that become activated when it binds. These cleave the next two complement components to form a C3 convertase and thereby trigger the remainder of the pathway. In this way the lectin pathway can mediate opsonization and the initiation of acute inflammation. The ficolins also associate with the protease MASP-2, and can bind to targets and activate complement by the lectin pathway. The molecular specificity of ficolins is not yet established, but they can certainly bind to many bacteria.

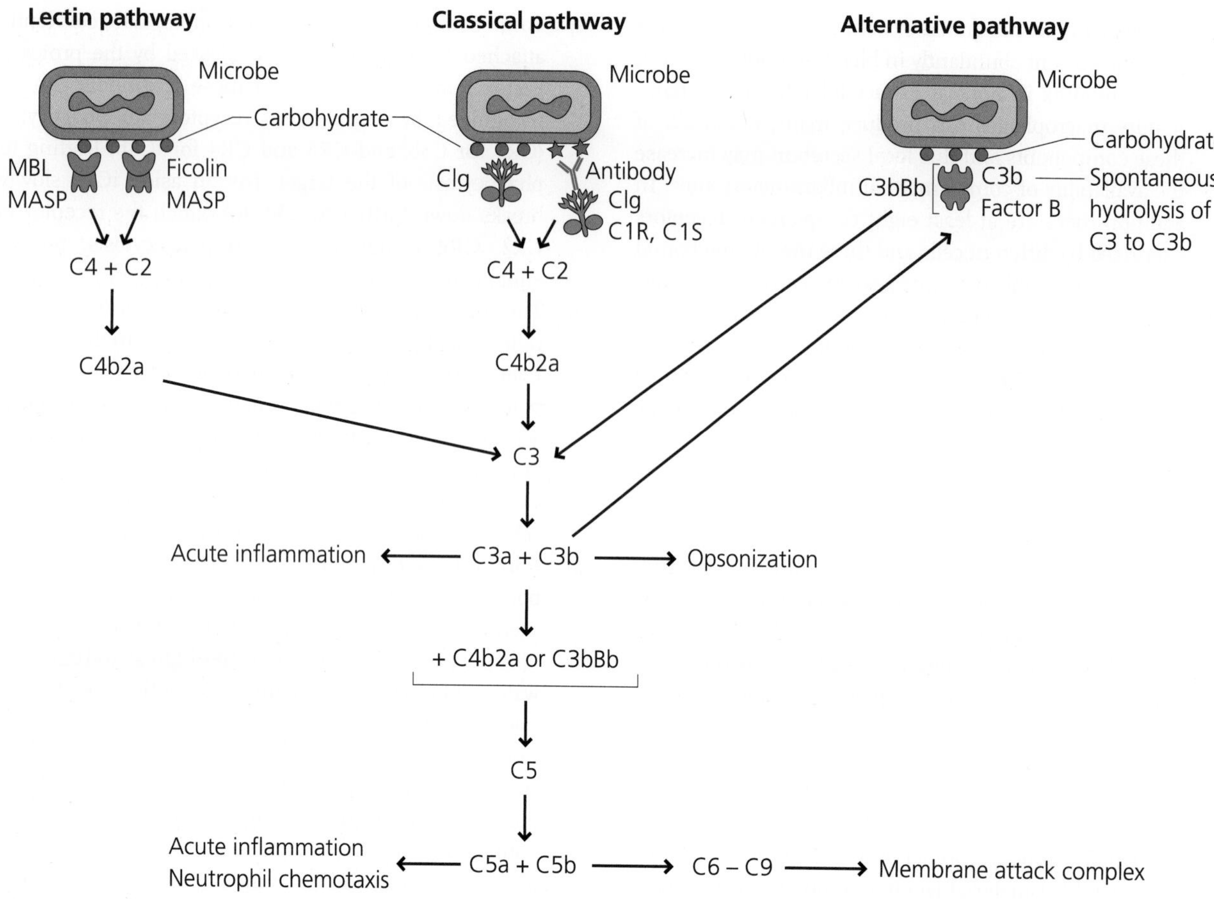

Fig. 4.19 Complement activation. The three pathways of complement activation converge with the formation of C3 convertases. In the lectin pathway, MBL or ficolin bind to bacterial carbohydrates and in association with the MASP proteases, activate C4 and C2 to form the C3 convertase C4b2a. In the classical pathway, C1q can bind to antibodies bound to a surface and in association with C1r and C1s, again activates C4 and C2 to form C4b2a. C1q may also bind to bacterial carbohydrates and initiate the cascade. In the alternative pathway, C3 is being continually broken down, but further activation is normally blocked by inhibitory factors. If C3 deposits on microbial surfaces however, inhibition is not effective and C3b combines with Factor B to form the C3 convertase C3bBb. C3b in combination with either C4a2a or C3bBb then creates a C5 convertase which cleaves C5 and activation continues to form the membrane attack complex (MAC). Each early stage generates a large (b) fragment, which is surface-associated (e.g. on a microbe), and a small (a) fragment, which diffuses away. There is also enormous amplification of key components (e.g. C3b, C5b) as the cascade progresses.

Classical Pathway Activation The first component of the classical pathway, C1q, is also associated with two proteases, (C1r and C1s). These are very homologous to the MASPs and also become activated when C1q binds to a target, leading to the formation of a C3 convertase. C1q can bind directly to many types of target: Gram-negative bacteria, some viruses, particles from damaged cells (mitochondria, chromatin) and altered host proteins (e.g. amyloids and antibodies that are already bound to their antigen). For historic reasons, classical pathway activation by antibody–antigen complexes has been very extensively studied, but it is now realized that direct binding of C1q by bacteria may be important in initiating complement activation. IgM and some subclasses of IgG are the major antibody classes that can activate the classical pathway. IgG can diffuse into normal extravascular tissues, whereas IgM can only enter these sites in significant amounts when inflammation has induced increased venule permeability. C1q is much more abundant in blood than MBL or the ficolins, so the classical pathway is quantitatively the major activation pathway.

Alternative Pathway Activation As we have seen above, activation of the lectin and classical pathways leads to cleavage of C3 and generation of C3b that attaches to cell surfaces. In addition C3 undergoes very slow spontaneous hydrolysis and generates a C3b-like molecule called C3(H_2O). This molecule can bind Factor B, which is cleaved by Factor D to form C3 (H_2O)Bb, the other C3 convertase. This, of course, cleaves more C3 molecules, forming genuine C3b, which can bind randomly to nearby surfaces. Once bound, the C3b molecule can form another complex with Factor B, generating more C3bBb, and thus amplifying C3b deposition. Alternatively,

C3b can bind the regulatory protein Factor H and then be cleaved by Factor I to form iC3b, anopsonin.

4.4.2.2 **Regulation of Complement Activation**

If C3b can deposit randomly on cell surfaces, why do we not continually lyse our own cells? The answer is that host cells are protected from destruction because they possess cell surface inhibitors that rapidly inactivate deposited C3b. This prevents formation of C3bBb, thus blocking further activation. The inhibitors are decay accelerating factor (DAF; CD55) and membrane cofactor protein (MCP; CD46). Another major inhibitor of complement activation is Factor H, an abundant soluble C3b-binding protein that can bind selectively to host cell surfaces to protect them from complement attack. Another protector of host cells is protectin (CD59) which prevents assembly of the MAC, and so prevents lysis of the cell.

4.4.2.3 **Complement Deficiencies**

Natural genetic mutations (polymorphisms and deficiencies) in almost all of the complement components have been identified in the human population and these have provided much evidence as to the roles of complement in defence. We shall use some examples to illustrate this.

Early Complement Component Defects People with loss of function mutations in C1q, C2 and C4 (involved in the classical and MBL pathways) have an increased risk of inflammatory and autoimmune disorders such as systemic lupus erythematosus and other immune complex diseases. These are discussed in Chapter 7, and they highlight the importance of complement in removing circulating immune complexes and debris from damaged or dying cells. In contrast, mutations in the early alternative pathway components (Factors B and D and properdin) lead to increased susceptibility to infection with encapsulated bacteria such as Staphylococcus, Streptococcus pneumoniae and Neisseria meningitidis. Deficiencies in MBL are common and are associated with slightly increased susceptibility to bacterial or yeast infections, especially in infancy.

Central Complement Defects Deficiencies of C3 are particularly serious as they give rise to severe, recurrent, pyogenic infections. It is an essential feature of all enzymatic cascade systems that active components can be inactivated, otherwise there would be nothing to stop the cascade continuing to activate all the downstream components. As a different example, if there were no inhibitors of activated coagulation factors our whole blood system might clot. This feature is also true of complement. Cleavage of C3b to the inactive iC3b is carried out by the plasma protease, Factor I. Factor I works only when a cofactor protein, such as Factor H, has bound to C3b. Deficiencies in Factor I or Factor H cause massive depletion of C3, because C3b is not broken down and continues to form the C3 convertase C3bBb until there is no Factor B or C3 left. Deficient individuals thus also have increased susceptibility to pyogenic infections. Factor H deficiency is also associated with severe kidney damage. An animal model, the Norwegian Yorkshire pig, dies soon after birth from renal failure associated with massive deposition of C3 and later complement components in the glomeruli.

Late Complement Defects While formation of the MAC is one of the best known outcomes of complement activation to many people, it appears relatively unimportant in human defence against infection. Individuals with defects in the later complement components (C5–C8) have increased susceptibility to Neisseria infection but not to other bacteria. People with C9 deficiency, the final component essential for complement-mediated lysis, are generally symptom-free, but have an increased risk of infections with Neisseria (meningitis and gonorrhoea). This suggests that formation of the MAC is of limited use in defence against other types of bacterial infection, at least in humans.

4.4.2.4 **Subversion of Complement Activation by Pathogens**

The importance of complement activation in immunity to infection is shown by the number of pathogens that can capture host molecules able to interfere with complement activation, particularly C3 cleavage. Many bacteria and parasites capture host Factor H on their surface. Some viruses, such as human immunodeficiency virus (HIV) and vaccinia, incorporate inhibitory proteins into their membranes as they bud from infected cells, and poxviruses (e.g. vaccinia, smallpox) encode proteins similar to MCP and DAF. Other microbes inhibit assembly of the MAC.

4.4.3 Neutrophils

Neutrophils, also known as neutrophilic polymorphonuclear leukocytes (PMNs), are the immediate effectors of anti-bacterial defence. These cells belong to the granulocyte series; other members are eosinophils and basophils. Neutrophils are phylogenetically much more recent than macrophages, apparently only evolving in more advanced vertebrates such as amphibians. Their most well-recognized function is in anti-bacterial defence and they carry a potent array of anti-bacterial defence mechanisms. They may also modulate adaptive immune responses when some of them subsequently migrate into lymph nodes (Chapter 3). Thus in lymph nodes of mice infected with Toxoplasma "swarms" of neutrophils appear around the sites of infection. See Figure 4.20.

Neutrophils are present in the blood in large numbers and make up around 70% of all blood leukocytes in normal humans. Neutrophils are very short-lived cells, their average life span is 2–3 days, but their production in the bone marrow can also be increased very quickly during infection, possibly through the action of G-CSF and or GM-CSF. In addition there are stores of mature PMNs held in the bone marrow that can be released very rapidly in times of need. Thus, neutrophils can be recruited very rapidly to sites of infection.

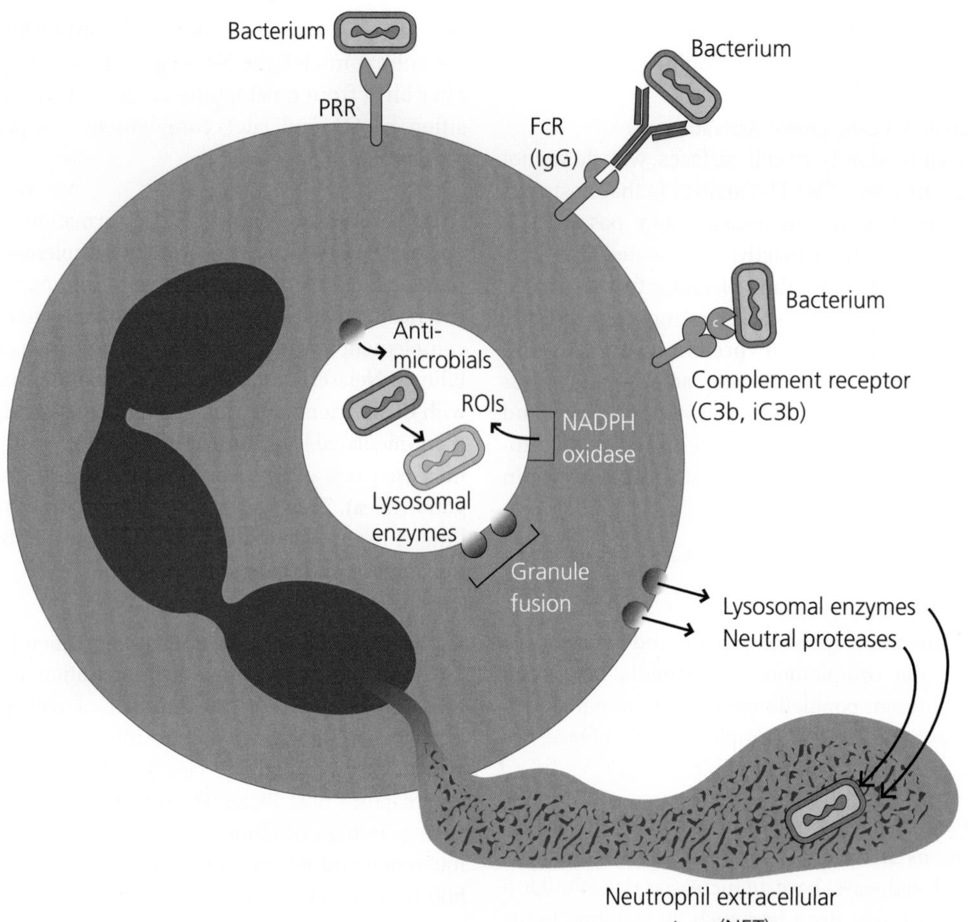

Fig. 4.20 Anti-microbial functions of neutrophils. Neutrophils can phagocytose opsonized microbes through their FcRs and complement receptors, and also express other receptors (e.g. PRRs) that promote phagocytosis. They possess several intracellular antimicrobial mechanisms including ROIs (for which NADPH oxidase is crucial), lysosomal enzymes such as acid proteases, and antimicrobial proteins and peptides, all of which can be released into the phagosome. Neutrophils can also form neutrophil extracellular traps (NETs) that contain chromatin and which may bind bacteria. These, perhaps together with secretion of other enzymes such as neutral proteases, may enable neutrophils to kill microbes extracellularly.

To carry out their functions in destroying bacteria, neutrophils first need to be delivered to the site of infection. We have already discussed migration through inflamed endothelium (Section 4.3.4.1). However, once in these tissues how does the neutrophil find the pathogen? The answer is chemotaxis or directed cell migration. Neutrophils have receptors for chemotactic factors and can distinguish differences in concentration of these factors over the diameter of the cell. Thus, they can sense the direction of the concentration gradient and activate their cytoskeleton so that they migrate towards regions of higher concentration. Chemotactic factors for PMNs also include molecules secreted by other tissue cells such as macrophages and mast cells. Some of these are proteins (e.g. chemokines), others are lipids such as some leukotrienes. Other chemotactic factors include breakdown products of complement such as C5a (above). That these chemotactic factors are important in host defence is suggested by the fact that Streptococcus pyogenes, one of the most important pyogenic bacteria, secretes an enzyme that breaks down chemokines and another that breaks down C5a. It is a general principle that microbes will not retain genes which do not give them an evolutionary selective advantage and we can thus hypothesize that the ability to destroy these two classes of chemotactic factor gives a selective advantage to the bacteria. See Figure 4.21.

Bacteria themselves release molecules that are chemoattractants for neutrophils. These are often formylated tripeptides such as f-Met-Leu-Phe (fMLP). In time-lapse films of cultures of neutrophils with bacteria, the neutrophils are clearly "chasing" the bacteria and eventually phagocytose them. All these chemoattractants bind to G-coupled receptors on the phagocyte membrane and activate the actin cytoskeleton. As neutrophils can detect differences in chemoattractant concentration between the front and the rear of the cell, they can organize the cytoskeleton so that a pseudopod is pushed out from the front of the cell, enabling it to move forward.

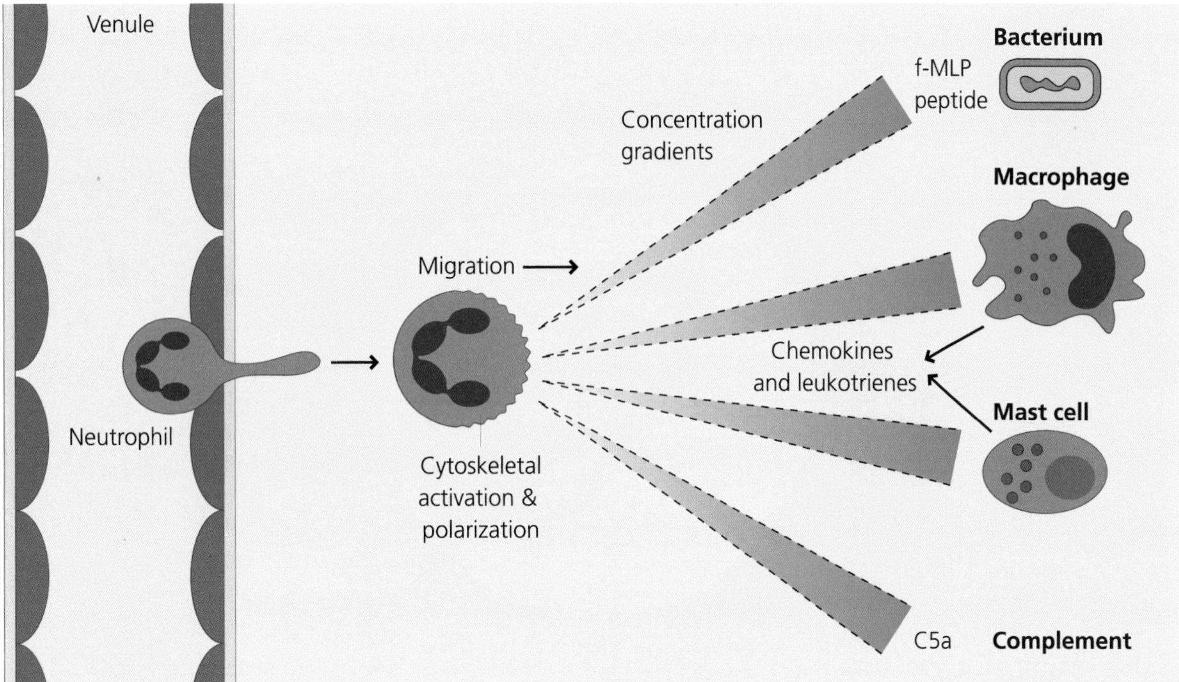

Fig. 4.21 Neutrophil chemotaxis. Neutrophils that have entered inflamed connective tissues from blood move towards the site of infection by migrating up a chemical concentration gradient, a process termed chemotaxis. Chemotactic factors are recognized by receptors on the neutrophil surface, that typically signal through G-proteins and activate the cytoskeleton. The neutrophil can sense concentration differences over the length of the cell and this permits polarization of the cell, with pseudopodia being extended from the part of the cell in contact with the highest concentration. Important chemotactic factors for neutrophils include formylated bacterial peptides such as f-MLP, specific types of chemokines and leukotrienes released from macrophages and mast cells, and complement component C5a.

Q4.14. Suppose we have identified a novel small protein that is secreted by intestinal epithelial cells when they are cultured with *Salmonella* bacteria. How could we determine if this protein is chemotactic for neutrophils?

4.4.3.1 Bacterial Killing by Neutrophils

Having reached the site of infection, neutrophils need to phagocytose the bacteria in order to kill them. Many pyogenic bacteria are resistant to phagocytosis, for example some have capsules made of polysaccharides or proteins for which PMNs do not have receptors. If PMNs are cultured *in vitro* with such bacteria the bacteria do not bind to the PMN surface. PMNs do, however, have receptors for complement components (particularly C3b and iC3b; CR1, CR3) and for some classes of antibody (particularly IgG). These components can opsonize microbes and promote phagocytosis. Apparently George Bernard Shaw was also aware of this phenomenon when he wrote, in his play The Doctor's Dilemma, "The phagocytes won't eat the microbes, unless the microbes are nicely buttered for them".

Having phagocytosed the bacteria, the PMN now needs to kill it. Unlike macrophages, neutrophils constitutively posses a diversity of mechanisms that are able to kill bacteria effectively. They contain all the components of the killing mechanisms in their granules and these can be rapidly assembled or released when the neutrophil phagocytoses a microbe. Neutrophils contain two main types of granules: secondary (specific) granules which are released first and primary (azurophilic) granules, which are released later. The secondary granules contain all the components necessary for the generation of reactive oxygen intermediates (ROIs). See Figure 4.22.

In the generation of ROIs, molecules assemble on the phagosome surface as soon as it is formed to form a multicomponent enzyme complex, NADPH oxidase. This produces superoxide anion (O_2^-) which is itself toxic to bacteria, but can be further modified by the enzyme superoxide dismutase to produce hydrogen peroxide, a more potent anti-bacterial agent. Myeloperoxidase, another component of neutrophil granules, can convert hydrogen peroxide into hydroxyl radicals and hypochlorous acid, both of which are highly toxic to many bacteria. The importance of hydrogen peroxide and later ROIs is suggested by the possession by many pathogenic bacteria of the enzyme catalase, which breaks down hydrogen peroxide. Nevertheless, Streptococcus pyogenes is catalase-negative but highly pathogenic, suggesting that catalase is not essential for pathogenicity. However, is the primary function of NADPH oxidase actually to produce directly anti-microbial ROIs? Some recent evidence suggests that this enzyme is

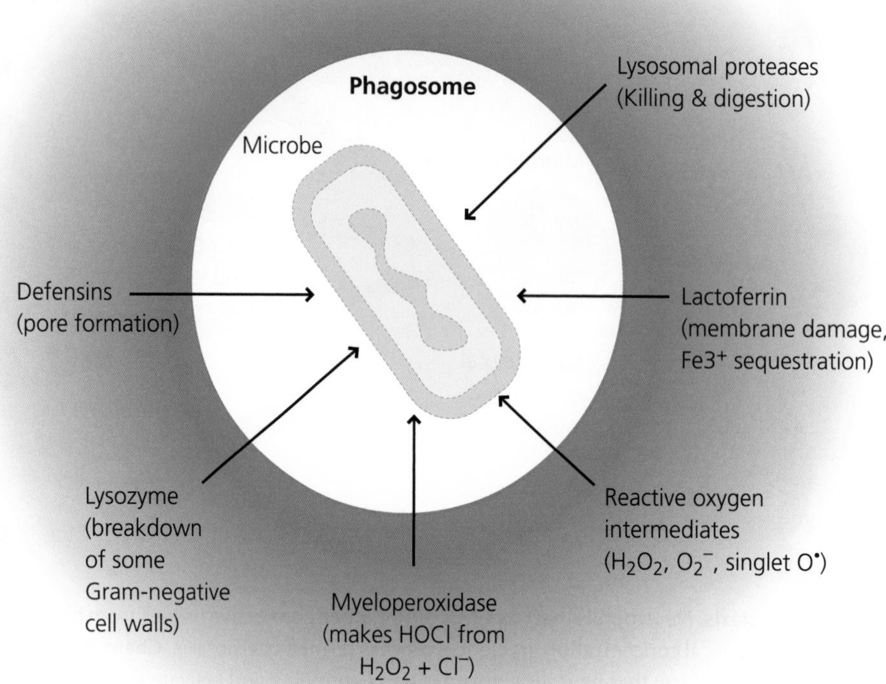

Fig. 4.22 Phagosomal killing by neutrophils. Bacteria in neutrophil phagosomes are subject to several different killing mechanisms. ROIs such as hydrogen peroxide (H_2O_2), superoxide anion (O_2^-) and singlet oxygen ($O^•$) can be directly toxic, but may also act to generate pH conditions appropriate for the actions of antimicrobial enzymes such as acid proteases. H_2O_2 can interact with chloride ions (Cl^-) in the presence of myeloperoxidase to generate hypochlorous acid (HOCl). Proteins such as lactoferrin can sequestrate iron (Fe^{3+})and make it unavailable for microbial metabolism. Lysosomal proteases break down dead bacteria but may also have more direct roles in killing. Anti-bacterial peptides such as defensins cause membrane disruption and lysozyme can break down the cells of some Gram-negative bacteria.

instead be needed to regulate the acidity of the endosomes, and that this is important for the function of cathepsin G and elastase-mediated killing. Thus, the direct effects of ROIs on microbes may not be as important as once thought.

Q4.15. If neutrophils are treated *in vitro* with low concentrations of ethanol, their ability to generate ROIs is reduced. Might this observation have clinical significance?

At the same time, secondary granules and lysosomes fuse with the phagosome, and a variety of anti-microbial agents and proteases are released into the phagosome. These include: lactoferrin and transcobalamin II, that sequester iron from essential microbial enzymes, preventing their function; lysozyme that can digest cell walls of certain bacteria; and cathelicidins (above). The cathelicidins may directly kill bacteria, after which the lysosomal enzymes may be important for digesting the dead bacteria. (Figure 4.22).

Hence, neutrophils can certainly kill bacteria intracellularly. There are suggestions that they may also be able to trap and kill bacteria extracellularly. This came from the discovery that neutrophils, when they reach an infected site, can undergo a remarkable change in which their nuclei apparently release sheets of loose chromatin into the extracellular tissues. These structures appear to be able to trap bacteria and they have hence been termed neutrophil extracellular traps (NETs). Furthermore, it has also been suggested that NETs can associate with microbicidal molecules released from the neutrophil granules. For example, the primary granules of neutrophils (released later) contain a bacterial permeability-inducing protein as well as defensins and lysozyme, any or all of which could potentially help to kill bacteria extracellularly. Hard evidence for the importance of NETs is, however, so far lacking.

Not surprisingly, pyogenic bacteria have evolved multiple ways of countering at least some of the effector mechanisms of neutrophils indeed this ability is needed for them to be pathogenic. Some of their counter-defences include the production of toxins such as the leucocidins that are secreted by

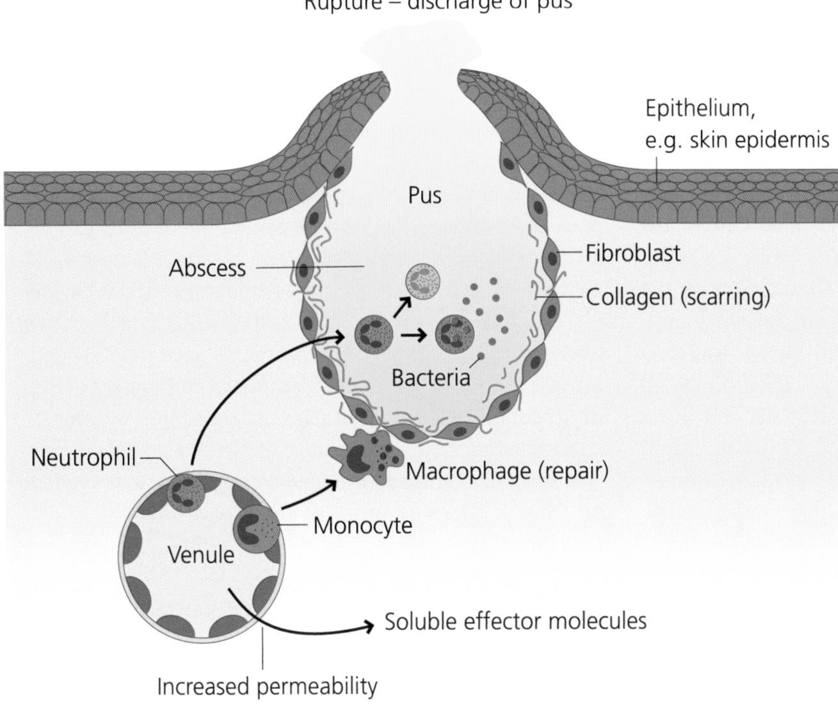

Fig. 4.23 Structure of an abscess. If pyogenic bacteria such as *Staphylococcus aureus* enter connective tissue they initiate acute inflammation, resulting in oedema and neutrophil recruitment. The bacteria can, however, resist and kill the neutrophils, and enzymes released by the bacteria and neutrophils cause liquefaction of the tissue, forming pus. *Staphylococcus aureus* also secretes coagulase, which precipitates fibrin, forming a barrier around the site of infection (not shown). At the same time, a healing reaction is starting to lay down collagen, secreted by fibroblasts, at the margins of the site of infection. The structure thus formed is an abscess. As the abscess develops, it erodes towards a surface, and will eventually rupture the surface and discharge its contents including live bacteria. This can be very serious if it discharges, for example, into the peritoneal cavity or a blood vessel, resulting in septicaemia and sometimes septic shock.

the bacteria and which can kill PMNs. Furthermore, we have already mentioned the role of bacterial capsules as possessed by Streptococcus pneumoniae in preventing phagocytosis. Staphylococcus aureus expresses a protein, called Protein A, on its surface. This binds to the Fc region of IgG and blocks binding of the antibody-coated bacteria to the FcRs of the neutrophil so they cannot be internalized by this route. That pyogenic bacteria have evolved these mechanisms is strong evidence for the crucial role of neutrophils in resistance to pyogenic infections. As yet, we do not know if or how bacteria might evade the extracellular trapping or killing mechanisms that are suggested by the discovery of NETs.

Q4.16. Can you suggest some other ways in which bacteria might have evolved to avoid being killed by neutrophils?

The ability of bacteria to resist killing by neutrophils, and indeed to kill neutrophils, has an important clinical consequence, the formation of pus. Pus is the fluid contents of abscesses, and consists of dead leukocytes and tissue cells, degraded connective tissues, and living and dead bacteria. Liquefaction is largely brought about by enzymes released by neutrophils. See Figure 4.23.

4.4.3.2 Neutrophil Defects

How do we know that neutrophils are important in host defence? The best evidence is clinical. Any neutrophil defect can lead to a greatly increased incidence of infections with bacteria that are associated with acute inflammation, the pus-forming, pyogenic species such as "Staph. and Strep.". Importantly, individuals with these PMN disorders do not suffer from an increased incidence of infection with viruses or intracellular pathogens such as mycobacteria. There are several different types of neutrophil deficiencies which may be due to defects in production, defects in migration or defects in their ability to kill bacteria.

Quantitative Defects Anything that reduces the production of blood cells in the bone marrow will affect neutrophil production and, because neutrophils have very short life spans, they are among the first blood cells whose numbers decrease (platelets are a close second). Thus, ionizing irradiation, which kills dividing stem cells and other progenitor cells, or high doses of chemotherapeutic drugs designed to act on tumour cells, will affect PMN numbers before those of other blood cells, leading to an increased risk of pyogenic infection.

Q4.17. Can you suggest two ways in which people affected by the radiation leakage at Chernobyl might present clinically?

Defects in Migration Rarely children present with a skin ulcer which is clearly infected, but there is very little pus. Cultures

Case Study 3.1: Chronic Granulomatous Disease

Case history. A 4-year-old boy is investigated because of repeated episodes of ear infections and pneumonia. On examination he is underweight for his age. He has a yellowish exudate coming from his right ear. His body shows scars representing healed abscesses. One of his brothers and two uncles had suffered from a similar condition. Laboratory tests showed an increase in the numbers of blood neutrophils. Levels of IgG and complement were normal. As a neutrophil defect was suspected, his neutrophils were tested for their ability to generate a respiratory burst. His neutrophils and those from a control subject were incubated with a non-fluorescent molecule that becomes fluorescent when oxidized by ROIs. The cells were stimulated with the chemotactic agent f-Met-Leu-Phe to give them an activation stimulus. The control cells became highly fluorescent, the patient's cells did not.

Explanation. Increased susceptibility to pyogenic infection can have many causes. Defects in antibody synthesis and complement were excluded by the blood tests, as were quantitative defects in neutrophil production. Neutrophils stimulated by chemotactic factors undergo a respiratory burst – resulting in the production of ROIs, which are powerful oxidizing agents. The generation of ROIs can be assessed in a number of ways, one of which is described above. The inability to generate a respiratory burst is typical of CGD. In this case the clinical history suggested that only males were affected. This is commonly seen in CGD and represents an X-linked defect: a mutated component of the multi-subunit NADPH oxidase is coded for on the X chromosome. Many mutations in NADPH oxidase may lead to CGD but not all mutated components are encoded on the X chromosome so they may be inherited autosomally.

Q4.18. How might patients with CGD be treated? What might be the problems with these approaches?

from the ulcer reveal a pyogenic infection, but there are no neutrophils in the exudate. His or her blood neutrophil count is, however, markedly raised and they can kill bacteria effectively. The defect in this case may result from the inability of the neutrophils to reach the site of infection. This type of defect is called leukocyte adhesion deficiency (LAD) and can have a variety of molecular causes Box 3.1.

Defects in Killing Bacteria Occasionally children who present with an increased incidence of pyogenic infections possess neutrophils that can migrate and phagocytose quite normally, but these cells are unable to kill the bacteria they have phagocytosed. Such children may be suffering from chronic granulomatous disease (CGD) See Case Study 3.1.

4.4.4 Recruited Macrophages and Inflammatory Homeostasis

Any inflammatory process involves tissue disruption. Bacteria secrete toxins that degrade connective tissue and kill cells; neutrophils enter, die and release enzymes such as collagenase and elastase. In the most extreme cases this can lead to liquefaction and pus formation. It is crucial that this damage is, so far as is possible, limited and repaired. Central to these processes are blood monocytes that are recruited into sites of acute inflammation and which may develop into inflammatory or elicited macrophages. For example, elicited macrophages can be studied following injection of an irritant into the peritoneal cavity of mice. This stimulates the recruitment of monocytes, and the resulting macrophages can be isolated easily by washing out the cavity and studied in culture.

Although the elicited peritoneal macrophages that are studied experimentally may not be totally representative of those recruited to inflamed connective tissues, they suggest that such cells may have at least four main functions (See Figure 4.24):

i) They phagocytose the debris and dying cells: the receptors used for this are not fully identified. They then migrate away from the area, often via lymph, and might modulate later adaptive responses in the node (Section 3.4.2.3).
ii) Elicited macrophages also secrete many proteases and other enzymes that are crucial in tissue remodelling, such as collagenase and elastase; these enzymes are involved in the remodelling of connective tissue that is central to repair.
iii) These macrophages secrete factors such as platelet-derived growth factor (PDGF; yes, it does also come from macrophages!) that are involved in stimulating the recruitment and secretory activities of fibroblasts that are crucial for tissue repair. Macrophage-derived factors also regulate angiogenesis by stimulating endothelial cells, leading to the in-growth of blood and lymphatic vessels.
iv) Elicited macrophages secrete many proteins normally found in the blood, including most of the complement components. These may be important in extravascular sites, adding to the soluble effectors that can enter inflamed tissues directly from the blood.

The induction of inflammatory macrophages is due, at least in part, to their response to microbial stimuli such as LPS (i.e. via PRR recognition). If, however, these macrophages have first been primed by IFN-γ, they respond to microbial stimuli by developing into activated macrophages with potent anti-microbial properties. Such cells are for example able to kill

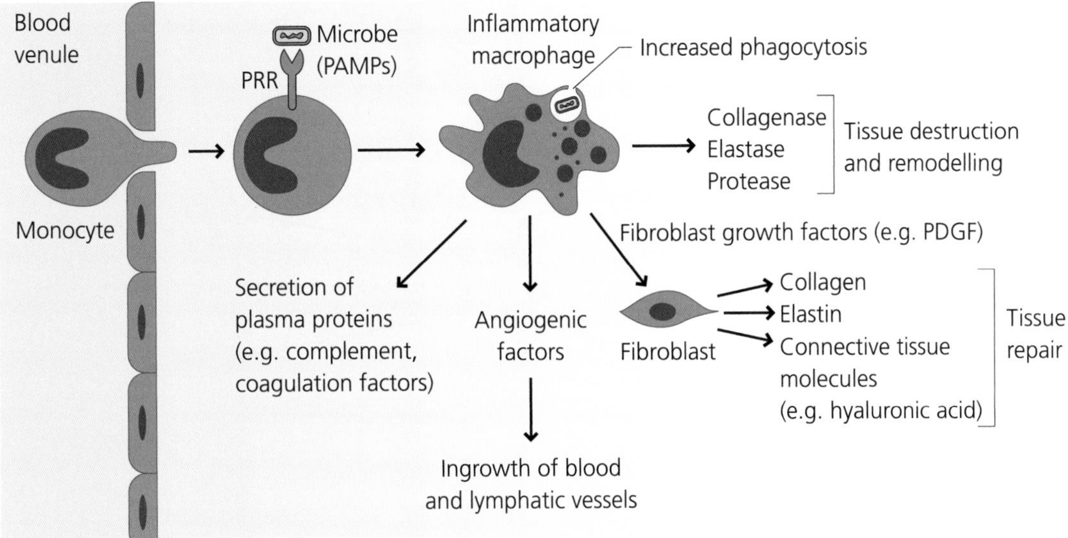

Fig. 4.24 Functions of inflammatory macrophages. Monocytes recruited into inflamed tissues can become macrophages that develop properties distinct from those of resident macrophages. These inflammatory macrophages are highly phagocytic. If acted on by PAMPs they are active secretors of catabolic enzymes that help to remodel tissues, release growth factors for blood vessels and fibroblasts, and secrete some plasma proteins such as complement components and coagulation factors. The latter help to increase the concentrations of those present at inflammatory sites, over and above the proteins that are recruited from the blood. The normal functions of inflammatory macrophages include assisting in healing and repair, but they can also be activated by IFN-γ to become potent anti-microbial cells.

intracellular pathogens such as Mycobacterium tuberculosis which otherwise could live quite happily and replicate within these cells. Where does this IFN-γ come from? Soon after infection, a major source may be NK cells that reside in some tissues and which can also be recruited to inflammatory sites (below). Macrophages that have phagocytosed pathogens, as well as DCs, are stimulated (via TLRs) to secrete the pro-inflammatory cytokines noted earlier and another cytokine, IL-12. This binds to IL-12 receptors on NK cells which are, in turn, stimulated to produce high levels of IFN-γ. Later, once an adaptive response has been triggered, activated T cells can also produce IFN-γ. A circuit is thus set up in which macrophages can become fully activated before an adaptive response has had time to kick in, and this circuit is important in early defence against such facultative intracellular parasites. The later secretion of IFN-γ by activated CD4 and CD8 T cells is probably crucial to recovery from many intracellular infections.

Q4.19. Some workers have claimed that peritoneal macrophages can secrete IFN-γ. Others, however, have suggested that this secretion may come from a small population of contaminating NK cells. What sort of experiments could we do to help sort out this controversy?

4.4.5 Natural Killer (NK) Cells

In the early 1970s, researchers wondered if patients with cancers could make immune responses against their tumours. They reasoned that if this was the case, patients might have lymphocytes in their blood that would kill the tumour cells in culture. Yes, they found that such lymphocytes did exist, but being careful workers, they set up controls using blood lymphocytes from normal (non-tumour-bearing) individuals. They were surprised to find that cells from these individuals were also able to kill the same tumour cells. As these cells were present without prior immunization, they were termed natural killer (NK) cells. This phenomenon was so far out of line with thinking at that time that many workers refused to accept NK cells as more than an artefact. Today, however, NK cells are regarded as being central to innate immunity and immune regulation.

4.4.5.1 NK Cell Subsets

NK cells are present in the blood, the liver and the secondary lymphoid tissues, particularly the spleen and the MALT; they are also present in the endometrium and decidua in pregnancy. Recently, it has become apparent that NK cells actually comprise different subsets and that they probably have different functions in immunity. Two subsets can be identified, in part, according to whether or not they express CD16, the Fc receptor FcγRIII. The majority of NK cells in blood express CD16, whereas those in spleen do not, while a distinct subset appears to reside in MALTs.

NK cells are bone marrow-derived cells. Adoptive transfer experiments have shown that NK cells arise from the common lymphoid precursor (CLP), which also gives rise to T and B lymphocytes, and some DCs. However, their development is completed in the bone marrow, unlike T cells and NKT cells which complete their development in

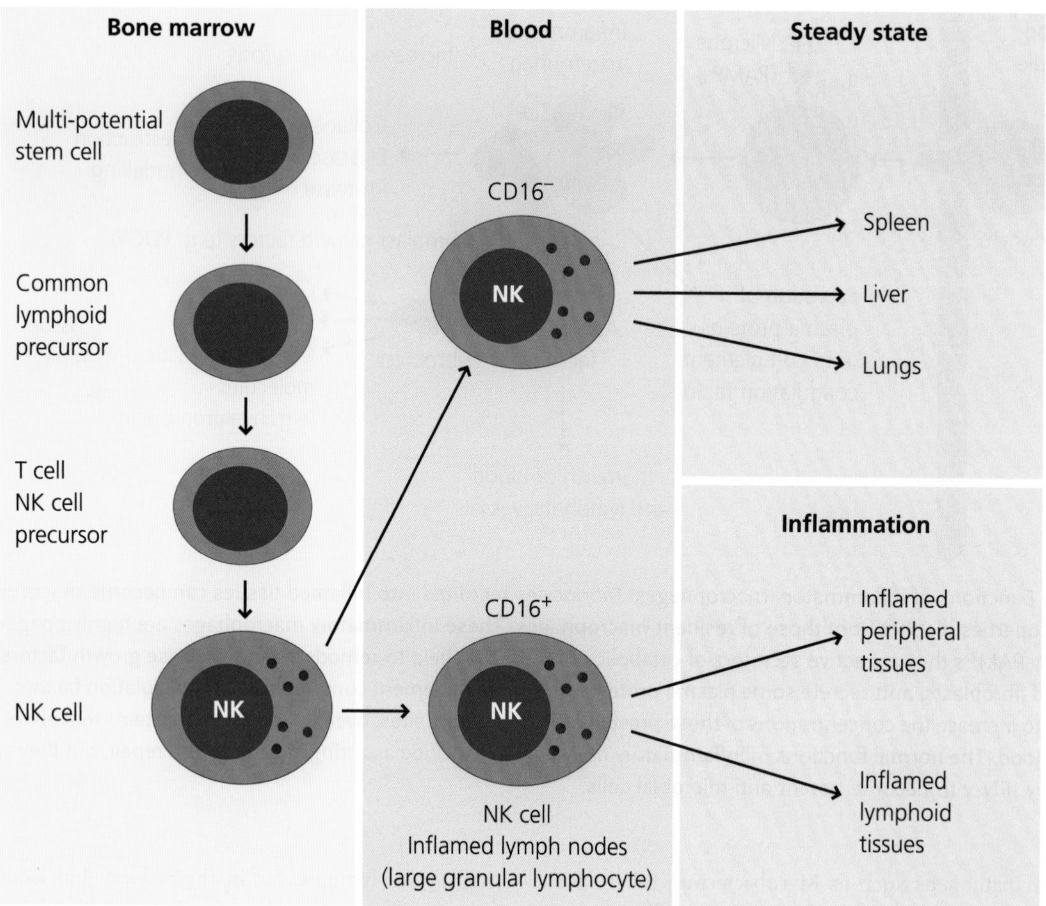

Fig. 4.25 Natural killer cell life history. NK cells develop from CLP in the bone marrow. They are released into the blood as lymphocyte-like cells, with conspicuous cytoplasmic granules (large granular lymphocytes). In the steady state a CD16$^-$ subset of NK cells may enter the liver, lungs and spleen or return to the marrow. During inflammation another CD16$^+$ subset may be recruited to inflamed tissues and their draining lymph nodes. NK cells are not long-lived, most being replaced from precursors in days or weeks.

the thymus. Little is known about the detailed regulation of NK cell development and NK cell homeostasis, although there is good evidence that IL-15 is important for their generation from CLPs e.g. certain severe combined immunodeficiencies; SCIDs (Chapter 5). See Figure 4.25.

The predominant CD16$^+$ subset of NK cells in blood look rather like lymphocytes. They are large mononuclear cells with round or oval nuclei, but they differ from most other large lymphocytes in that they have conspicuous granules in their cytoplasm; these cells had long been recognized in the blood and were called large granular lymphocytes. This observation gives important clues as to NK cell function since we now know the granules store preformed cytotoxic machinery. Hence, these cells are poised to kill, unlike cytotoxic T cells, which have to synthesize the molecules involved from scratch (Section 1.4.5.2). These cells express the relevant chemokine receptors that enable them to home to peripheral sites of infection (e.g. the IL-8 receptor and the fractalkine receptor that is also expressed by a subset of "inflammatory monocytes"). Here they may kill antibody-opsonized target cells by antibody-dependent cell-mediated cytotoxicity (ADCC) when their FcR is ligated. Killing involves the exocytosis of granule contents, including perforin and granzymes, onto the target cell; they also express Fas ligand, important in other forms of cellular cytotoxicity, such as in defence against virally infected cells. See Figure 4.26.

In contrast, the CD16$^-$ subset of NK cells in lymphoid tissues appears to be the major producer of NK cell-derived cytokines and chemokines. Two of the most important cytokines NK cells secrete are IFN-γ and TNF-α. However, NK cells may first need to be activated by cytokines such as the Type I IFNs, probably derived particularly from plasmacytoid DCs (pDCs), and IL-12 that is produced by macrophages and DCs following TLR stimulation. The IFN-γ that NK cells secrete may play an important role in early, innate defence against infections such as tuberculosis where macrophage activation is critical in mediating recovery. Thus, there is much cross-regulation between NK cells, macrophages and DCs in innate immunity against infection.

What is the actual role of NK cells in defence against infection? Human NK cell deficiencies are very rare, indeed it has been argued that this rarity actually demonstrates their crucial importance. However, very rare deficiencies have been identified and the characteristic feature that has led to their

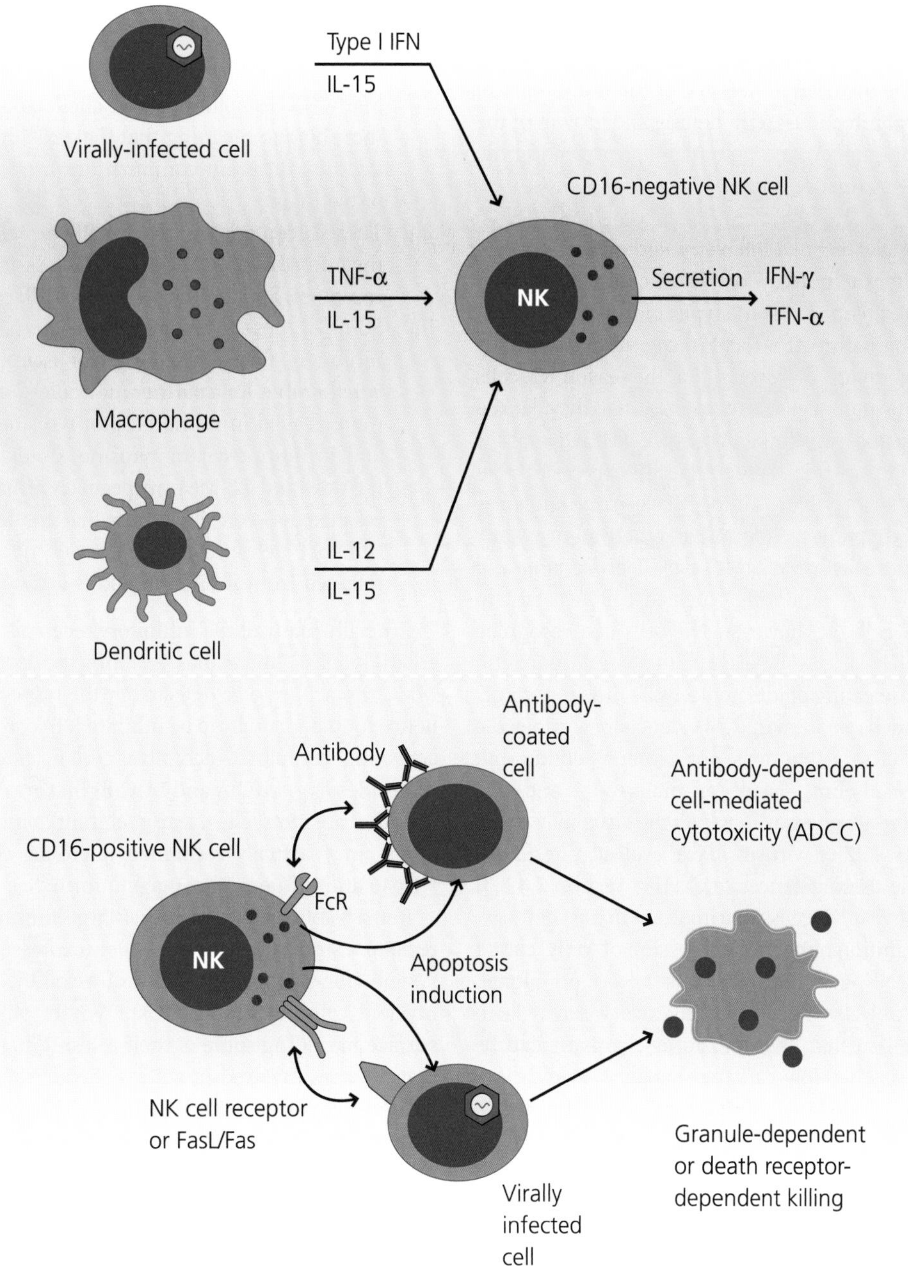

Fig. 4.26 Functions of natural killer cell subsets. NK cells comprise two main subsets based in part on expression of CD16, a specific FcR for IgG. CD16$^-$ NK cells are active secretory cells. They themselves can be activated by cytokines secreted by different cells including macrophages (IL-12, IL-15, TNF-α), DCs (IL-12, IL-15), activated T cells (IL-2, not shown) and virally infected cells (Type I IFNs). These activated NK cells can the secrete cytokines including IFN-γ and TNF-α. CD16$^+$ NK cells appear to be primarily cytotoxic cells. They may kill virally infected cells through granule-dependent or death-inducing receptor-dependent mechanisms and they may also mediate ADCC when their FcRs are ligated. ADCC can be induced experimentally in culture, using antibody-coated target cells, but its *in vivo* significance remains uncertain.

identification is increased susceptibility to viral infection, particularly herpes viruses. Why mainly herpes viruses? It is not clear. In animals, no selective deficiency of NK cells has been established. Young mice, which show a relative NK cell deficiency, have increased susceptibility to murine cytomegalovirus (CMV) infection and this can be prevented by giving adult NK cells to young mice. However, the relative importance of the different subsets of NK cells in these apparent defensive roles remains unclear.

Q4.20. How might we assess whether NK cell defence against herpes virus infection is due to their cytotoxic activity or their ability to secrete cytokines?

CD16$^-$ NK cells are also present in the uterus, although some of their characteristics are different from those in lymphoid tissues and they may represent a distinct subset. These cells may be involved in defence of the non-pregnant uterus against viral infections, remodelling and growth of the decidua during pregnancy, and potentially in protecting the developing foetus against rejection.

4.4.5.2 NK Cell Activating and Inhibitory Receptors

For a long time after their discovery it was a complete mystery how NK cells were able to kill some types of tumour cells but not others, and the nature of NK cell recognition was totally obscure. It was eventually discovered that the reason NK cells could kill some tumour cells *in vitro* was that they lacked surface major histocompatibility complex (MHC) class I expression. Since it was this apparent lack of expression that rendered the cells susceptible to killing, this became known as the missing self hypothesis for NK cell activation. Many viruses can interfere with the MHC class I processing and presentation pathways, rendering infected cell resistant to killing by CD8 T cells. If, however, NK cells can recognize and kill cells which lack MHC class I, this would permit a second means of detecting of infected cells. Is this important? Some viruses, such as human CMV, encode a molecule similar to MHC class I that does not present peptide, but which inhibits NK killing – a decoy molecule – and this presumably represents one way in which the virus may evade NK cell defences. Other viruses have evolved alternative strategies to evade these defences (e.g. HIV; Section 2.4.3.5) so we can surmise that NK cells are important in host defence.

Thus, the recognition of normal expression of MHC class I molecules by a NK cell normally acts as an off signal. However, a NK cell also needs to recognize when it is in contact with a cell so that its cytotoxic mechanisms can be activated appropriately. How can this happen? Many more years of research revealed that there are also activating receptors on NK cells that recognize ligands on target cells. These can activate and deploy the cytotoxic machinery, unless an inhibitory signal is also received which represents the normal situation in the absence of infection. For convenience these different types of inhibitory and activating receptors can collectively be termed NK receptors. They are completely different from the TCRs and BCRs that lymphocytes use for antigen recognition.

Structurally, NK receptors can be grouped mainly into those that are members of the immunoglobulin or C-type lectin superfamilies. The terminology of the different types of NK receptors is immensely complex and can be very confusing. For example, many members of the respective structural families are encoded by large multi-gene complexes termed the leukocyte receptor complex (LRC) and the NK cell complex (NKC) respectively. The LCR encodes a group of molecules that include the killer cell inhibitory-like receptors (KIRs), while the NKC encodes another group of killer lectin-like receptors (KLRs). Moreover, in humans many NK receptors structurally belong to the immunoglobulin superfamily, while those with analogous functions in mice are members of the C-type lectin receptors. Despite this complexity there are some straightforward general principles that explain how these receptors work–

i) Many NK receptors exhibit allelic and haplotypic polymorphisms, meaning that they exist in different forms and that different combinations are found in different individuals.
ii) These receptors are not clonally expressed on NK cells, meaning that one NK cell can possess one combination of receptors, and another NK cell a different combination.
iii) Some of these receptors can exist in both activating and inhibitory forms, differing, for example, according to whether they (or a partner molecule) possess cytoplasmic motifs called immunoreceptor tyrosine-based activation and immunoreceptor tyrosine-based inhibition motifs (ITAMS and ITIMS), respectively (Chapter 5). It is the balance between the signals that are delivered to an NK cell that determines whether or not that NK cell is released from inhibition and is activated.

Crucially many of the inhibitory receptors recognize different MHC class I molecules, enabling the normal expression of MHC molecules to be detected by the respective NK cells and hence for their activity to be inhibited. In contrast, some of the activating receptors recognize certain non-classical MHC molecules (e.g. MIC-A and MIC-B) that are only expressed by cells that are stressed – clearly relevant to infection – whereas others may actually recognize microbial components (e.g. NKp46 and influenza haemagglutinin). See Figure 4.27.

Remarkably, it turns out that large numbers of inhibitory receptors similar to KIRs are widely expressed in the immune system (for example they are expressed by different types of myeloid cells, including DCs). We are only scratching the surface in finding some of their roles. It may, for example, be that they are involved in setting the activation thresholds of different leukocyte populations, but this finding hints at an entirely novel level of innate (and potentially adaptive) immunity that we have yet to appreciate.

4.4.6 Systemic Effects of Innate Immune Activation

The effects of a local infection are often not confined to the actual tissues infected. We are all aware that when we have a cold or flu we feel generally unwell (e.g. fever, headache, muscle pains, loss of appetite, etc.). This reflects the distant (systemic) effects of cytokines released at the site of infection. It is relevant that when patients with tumours were being treated experimentally with Type I IFNs, they developed flu-like symptoms, and one of the effects of viral infection is to cause release of Type I IFNs. A major systemic effect of inflammation is the induction of fever, although it is not entirely clear why we have evolved the generation of fever as part of defence against infection. (That it might be beneficial is suggested by an early treatment for patients with syphilis who were infected with malaria to induce fevers in an attempt to kill the causative bacteria.) Fever can be caused by the release of

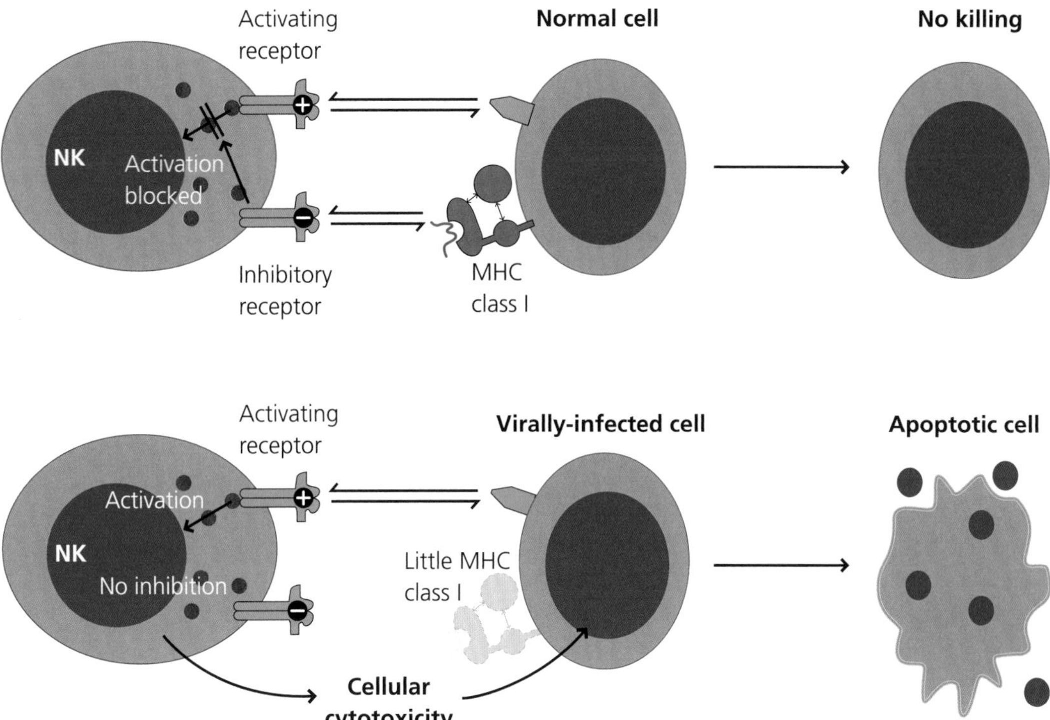

Fig. 4.27 Natural killer cell recognition and activation. NK cells possess granules that contain the pre-formed apparatus (perforin and granzymes) needed to kill cells they recognize. They express several activating receptors which, on recognizing a ligand on the target cell, activate the killing mechanisms. NK cells, however, also express inhibitory receptors which, if they recognize MHC class I on the target cell, block the activation of the cytotoxic mechanisms. Thus, NK cells kill target cells on which MHC class I expression is decreased e.g. by viral infection or some tumour cells.

mediators such as IL-1 and TNF-α. These cytokines act on the hypothalamus and reset the temperature regulation centre, so the body reacts as if it was cold and starts to generate heat. This is why we shiver, because muscle contractions generate heat. They also stimulate metabolism of fat and muscle tissue, possibly to provide a source of energy. If this stimulation is excessive, it can lead to tissue wasting (cachexia), which is also seen in patients with severe infections or advanced cancers. The cytokine responsible for cachexia is TNF-α, which is more toxic on a mole per mole basis than hydrogen cyanide.

Q4.21. How experimentally might it be possible to dissect the respective roles of IL-1 and TNF-α in systemic inflammatory responses?

Other systemic effects of inflammation include those of IL-6, particularly on the liver: many proteins are secreted in increased amounts and some are only secreted in these situations. This constitutes the acute-phase response and many of the proteins involved in the response are thought to have roles in defence. These include many complement components together with collectins (e.g. MBL), ficollins and pentraxins (Section 4.4.1). As we have seen, some of these can act as opsonins and also activate the complement pathway. However, for many of these proteins we do not understand their roles in infection and the reasons for their increased synthesis remain a puzzle. See Figure 4.28.

Q4.22. Can you suggest two ways in which the functions of acute-phase proteins such as serum amyloid A could be explored?

Pathological Effects of Systemic Inflammation Clearly the effects of systemic inflammation (above) can be unpleasant and may become serious. Under other circumstances, however, they can be frankly fatal. Blood monocytes are highly reactive to TLR stimulation. Normally there are very few if any PAMPs present in blood, even during an infection. If however, dead or dying bacteria enter the bloodstream (septicaemia), PAMPs released by the bacteria can activate the monocytes with potentially disastrous consequences leading to septic shock. Septic shock can have other consequences. Decreased organ perfusion can lead to organ failure (e.g. kidney failure). The major problem is that the low blood pressure (hypotension) is refractory to treatment; all the usual drugs used to raise it are ineffective. In animal models, if antibody to TNF-α is given early (before hypotension has set in), the hypotension can be prevented. This treatment is, however, ineffective in human septic shock, probably because by the time it is given, the TNF-α has done its work. Most septic shock is caused by LPS from Gram-negative bacteria via TLR4, but it can also be caused by lipoteichoic acid from Gram-positive bacteria though TLR2 and 6. See Figure 4.29 and Section 2.4.2.5.

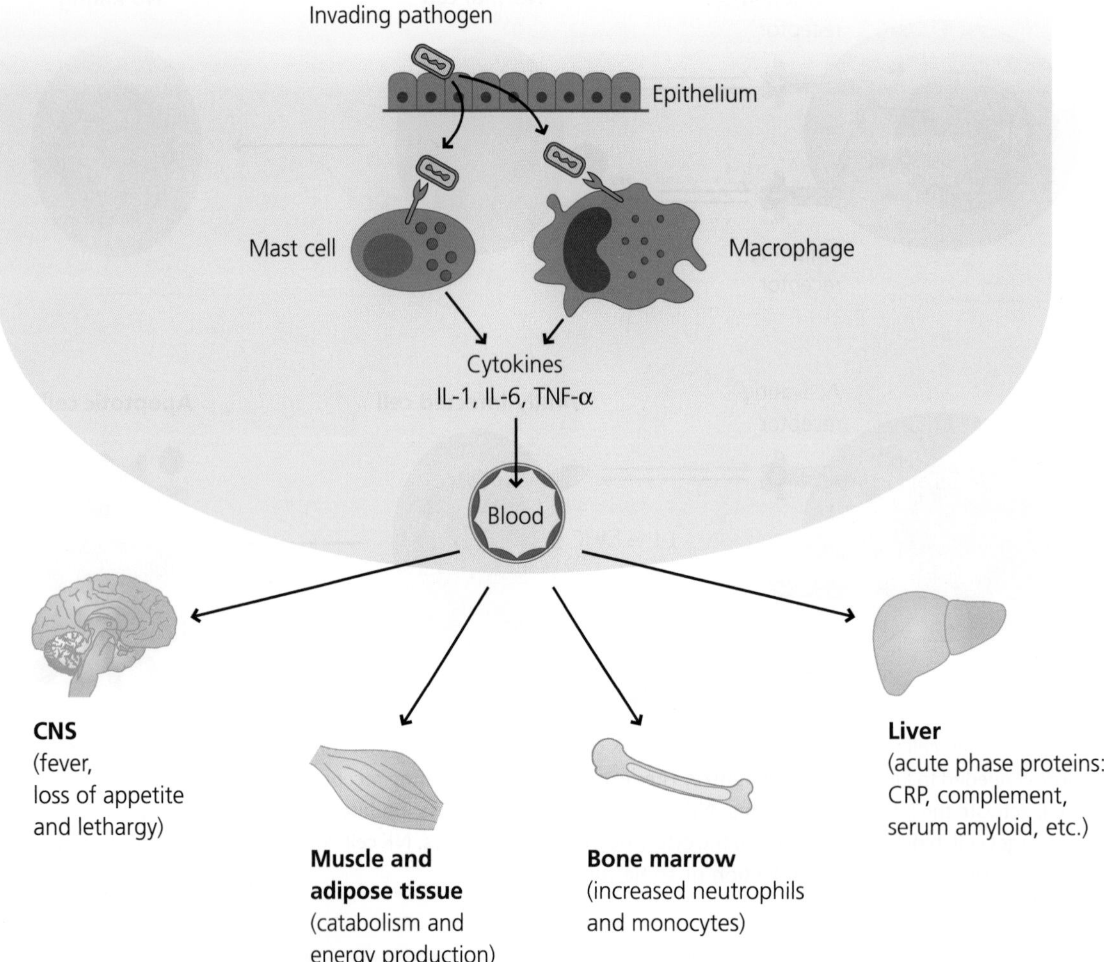

Fig. 4.28 Systemic effects of inflammation. Cytokines secreted by cells in inflamed tissues can enter the blood and affect distant tissues. These tissues include the liver where the acute-phase response involves the increased synthesis of several proteins, some of which such as complement components have well-defined roles in defence, whereas the functions of others are still obscure. Effects on the bone marrow include increased production and release of myeloid cells, particularly neutrophils and monocytes. Effects on the CNS include the induction of fever, loss of appetite and lethargy; on muscle and adipose these effects include increased catabolism to gererte energy.

4.4.7
Innate immunity and the Induction of Adaptive Responses

Innate immune activation is essential for the initiation of adaptive immunity, and the quantity and quality of innate responses contribute significantly to the different types of adaptive immune response. Why is this? Naïve lymphocytes need to be informed whether any antigen they recognize is harmful in which case they need to make an active response, or harmless (e.g. self or food antigens) in which case they need to be switched off – tolerized or turned into regulatory cells. Thus, the response of the naïve lymphocyte should reflect conditions in peripheral tissues – whether or not there is inflammation and whether or not infection has occurred. That is, lymphocytes need to integrate the antigenic information they receive with the context within which that information was received and make an appropriate response (Section 1.4.5.1).

We discuss in more detail elsewhere (e.g. Chapters 2 and 5) how naïve T cells have the potential to differentiate down a variety of pathways with very different outcomes once they are activated. Naive T cells are activated in secondary lymphoid organs, often distant from the site of infection, yet must differentiate in a way that reflects conditions in the peripheral tissues (Chapter 3). DCs are the prime candidates for the transmitters of information about peripheral conditions to T cells, DCs make tight interactions with the T cells they are activating, facilitating the transfer of soluble or membrane bound signals. We should not, however, forget that there are other potential means of information transfer. Afferent lymph contains other cells, particularly during inflammation; these include monocytes or macrophages, neutrophils, and, in some circumstances, eosinophils and basophils. Additionally, lymph collects the extracellular fluids in peripheral tissues, and these may contain cytokines,

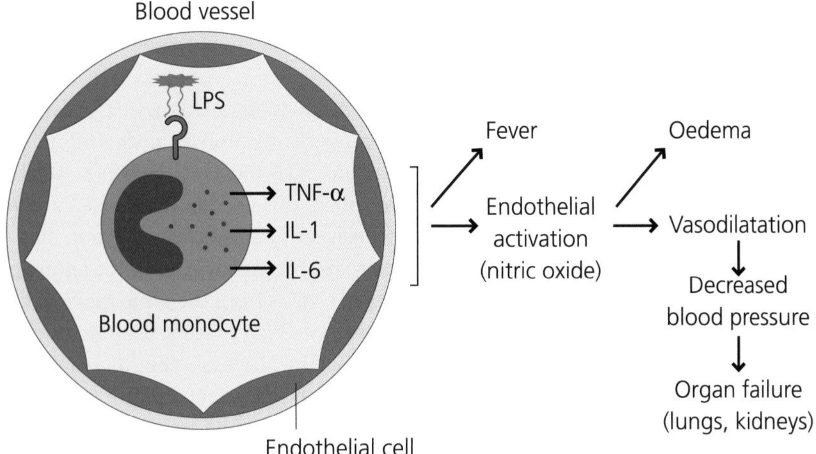

Fig. 4.29 Septic shock. If LPS from Gram-negative bacteria or lipoteichoic acid from Gram-positive bacteria enters the bloodstream it can activate monocytes via TLR4 or TLR2/6 respectively. This can result in the release of pro-inflammatory mediators such as IL-1, IL-6 and TNF-α. In high concentration these cytokines can have potentially fatal effects. As well as causing fever, they can act on blood vessels to induce nitric oxide, and cause venous dilatation and increased permeability. This may lead to decreased venous return to the heart, resulting in severe hypotension (low blood pressure) and failure of organs such as the kidney and lungs.

chemokines and other mediators, such as prostaglandins and leukotrienes. The roles of these different cell types in determining T cell differentiation is not well understood and the roles of the soluble factors is quite unknown. For immune responses initiated in other secondary organs such as spleen and Peyer's patches, similar influences are likely to operate, but our understanding of their nature and roles is just as limited See Box 4.5.

4.5 Haematopoiesis and Myeloid Cells

As we outline in Section 1.4.1, all blood cells originate from multi-potent haematopoietic stem cells (HSCs) that are normally situated in the bone marrow of adult mice and humans. Here, we very briefly discuss how haematopoiesis

Box 4.5 Innate Immunity in Other Species

We suggested at the start of this chapter that the evolution of defence mechanisms is essential for survival in a world governed by natural selection and that it is thus not surprising that all organisms have evolved mechanisms to defend themselves against the attentions of invaders. Even prokaryotes such as bacteria can be infected by viruses called bacteriophages. Bacteria use restriction enzymes to cut the nucleic acids of bacteriophages (and which have become crucially important to molecular biology). Plants are of course also subject to infection, often the bane of farmers and gardeners. Plants do, however, possess complex innate defence mechanisms. Plants use PRRs to recognize PAMPs on pathogens; these include bacterial flagellin, LPS and peptidogylcans. The nature of the PRRs involved is not fully understood but some of them contain the LRRs and other molecular features found in NLRs. The outcomes of signalling via these receptors can in some ways be similar to those found in insects and mammals, including the generation of a respiratory burst and nitric oxide synthesis. There are of course plant-restricted outcomes, such as strengthening of the cell wall.

In the animal kingdom, innate immunity is well developed in invertebrates. To take TLRs as an example, these were first discovered in insects, as we noted earlier. TLRs are widely expressed throughout invertebrate groups and in some their organization appears more complex than in vertebrates. Sequencing of the sea urchin genome has revealed that these invertebrates express 222 molecules resembling TLRs, and a further 203 that resemble NLRs and which are primarily expressed in the gut. It is important to realize that in all non-vertebrates, innate immunity is the only form of immunity available and that it has thus evolved in different ways to meet the needs of different species. In vertebrates, however, the evolution of adaptive immunity has given innate immunity an extra role. As well as forming first-line defence if barriers have been breached, it is the innate immune system that informs and regulates the activity of the adaptive system. Pathogens have evolved multiple ways of interfering with innate immunity in ways that weaken the ability of adaptive immunity to mount effective protection.

and the production of leukocytes can be regulated. We focus particularly on the role of cytokines in this process in relation to stem cells and leukocytes, with emphasis on myeloid cells. (Development of lymphocytes is discussed in Chapters 5 and 6.)

4.5.1 Regulation of Haematopoiesis

Haematopoiesis is regulated, at least in part, though the contacts of the developing cells with stromal and other cells, and through the actions of cytokines (below). Collectively, these control induction and repression of gene expression in stem cells and other precursor cells, leading to the development of leukocytes with specialized phenotypes. Cytokines, for example, bind to their receptors and initiate signalling cascades that lead to the expression of transcription factors that regulate gene expression. Some of the new genes that are expressed can lead to inhibitors of the inducer, or regulate other signalling pathways in a positive or negative manner. Hence, the process is immensely complicated and very difficult to analyse. For example there may be very good evidence that a particular transcription factor is essential for the development of a given cell lineage. Knocking it out in a transgenic mouse, however, can lead to highly unexpected effects on other cell lineages, including their failure to develop, simply because it may also be needed by these for a short time at one small stage of development. An example is PU.1 which is known to be required for development of the CMP. Surprisingly, genetically engineered mice lacking this transcription factor are found to lack T cells from the very earliest stages and, also surprisingly, these precursors develop instead into monocytes or DCs. Thus, the development of HSCs and other precursors into different lineages is controlled by a complex set of influences that we have yet to fully understand.

4.5.2 Stem Cells and Cytokines

How stem cell numbers are regulated is largely unknown. Several cytokines have been shown to influence stem cell properties *in vitro* and to have effects on them if they are administered to mice. Hence, there are growth factors such as IL-3 and Flt3 that can support stem cells, and stimulate their division and differentiation, but how these factors act in the steady state is still a matter for conjecture. The difficulty is that, in general, mice lacking the ability to make these cytokines or their receptors do not show marked, or even any, defects in haematopoiesis under steady-state conditions. One exception is stem cell factor (SCF; c-Kit ligand) and its receptor, c-Kit. Mice lacking either of these molecules display severe anaemia and die before birth, and bone marrow cells from these mice are unable to repopulate leukocytes in irradiated mice. It thus appears that, apart perhaps from SCF, no cytokines are indispensible for maintaining stem cell function under steady state conditions.

4.5.3 Cytokines and Leukocytes

The production of leukocytes is also tightly controlled. In the absence of infection the numbers of different blood leukocytes remain very constant. However, relatively small changes in numbers can be vital diagnostic aids (e.g. indicating that an otherwise unsuspected infection has occurred). If there is increased demand, such as for neutrophils in pyogenic infections, output from the bone marrow can, however, be rapidly increased. This reflects accurate homeostatic control of their production. In part the increased output of neutrophils reflects the fact that mature neutrophils are stored in the marrow and can be readily mobilized in times of need. After this, however, continued output needs increased recruitment of stem cells into the required lineage and increased division of precursor cells within that lineage. Experimentally, if leukocyte numbers are reduced, they rapidly return to normal levels, often after an overshoot. Most evidence suggests that the production of leukocytes is regulated by cytokines – indeed some cytokines were first identified because of their ability to stimulate the growth of particular leukocytes *in vitro* such as the colony-stimulating factors (CSFs; Chapter 3). So, *in vitro*, GM-CSF drives production of both granulocytes and macrophages, and it now seems clear that, in turn, G-CSF and M-CSF, respectively, control differentiation into one lineage or the other. The roles of these cytokines *in vivo* are not, however, fully elucidated. See Figure 4.30.

Q4.23. In pyogenic infections, the numbers of circulating neutrophils are greatly increased. How could we design an experiment to test the hypothesis that this increase is related to the presence of a blood-borne growth factor?

Many cytokines that stimulate leukocyte development *in vitro* do not appear to be essential for leukocyte production in vivo. Thus, G-CSF knock-out mice show a 70–80% reduction in neutrophil numbers, but neutrophils are not completely absent. Mice with the GM-CSF gene knocked out show few abnormalities in steady-state maintenance of leukocytes, but do show marked lung pathology, possibly because of defects in macrophage-mediated clearance of lung surfactant. Mice deficient in M-CSF or its receptor show marked abnormalities of some macrophages, particularly osteoclasts (macrophage-derived cells essential for bone remodelling) leading to multiple bone abnormalities. What seems clear is that, despite the role of cytokines in determining lineage commitment in vitro, there is no single cytokine that is solely responsible for the regulation of leukocyte production in the steady state or under conditions of increased demand. This does not mean that individual cytokines cannot be used therapeutically. Thus, G-CSF is used to mobilize stem cells from the bone marrow into blood for transplantation (Section 3.6) and GM-CSF is used to stimulate leukocyte production after irradiation. In addition, GM-CSF and IL-4 are used to generate DCs from human blood monocytes *in vitro*.

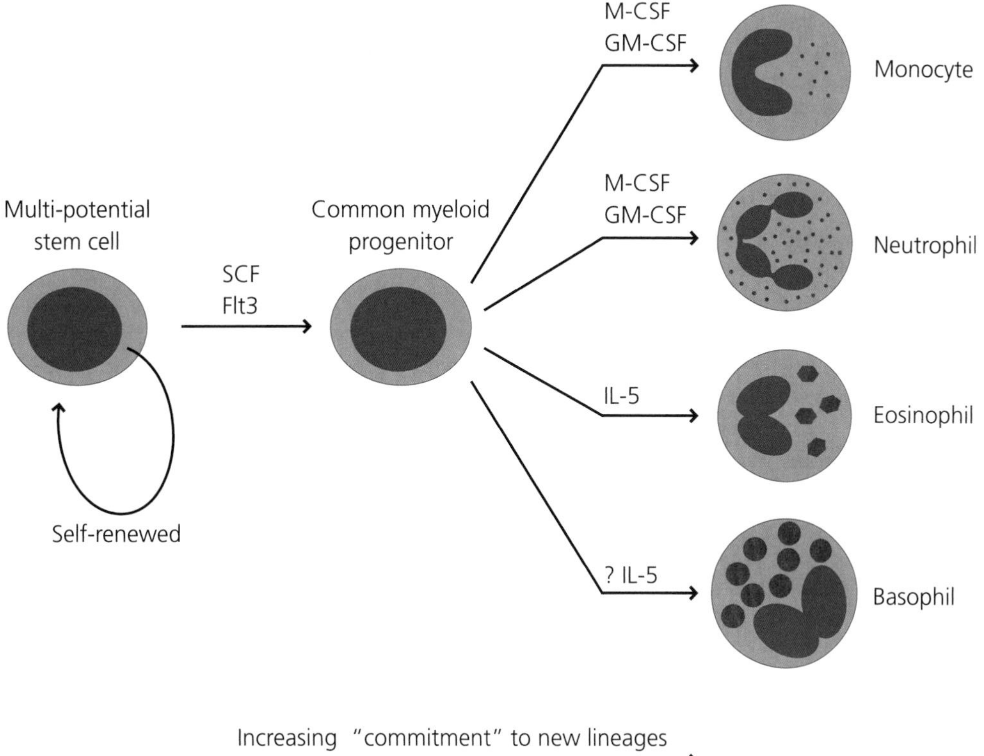

Fig. 4.30 Cytokines in myelopoiesis. Bone marrow multi-potential stem cells are acted on by SCF and Flt3, and probably other factors, to become CMPs. The CMP responds to other growth factors by dividing to form the different myeloid cell types. Some that are known or thought to be essential for production of monocytes and different types of granulocytes are shown. While it is clear that the production of the different myeloid cells is under feedback control, the details of the control mechanisms are still poorly understood.

If the roles of cytokines in regulating leukocyte production are not fully understood, how the production of these cytokines is regulated is even less clear. We do know which cells secrete the cytokines. Often these are stromal cells in the bone marrow, but epithelial cells, endothelial cells, activated macrophages and activated lymphocytes can also be sources. At present, however, we have little evidence as to how the production of these cytokines is regulated, and thus how the production of leukocytes is regulated under steady state and perturbed conditions. This contrasts with our understanding of red cell production, where we know that measurement of blood oxygen levels in the kidney regulates erythropoietin production by juxtaglomerular cells, and that erythropoietin is the main cytokine stimulating red cell production in the marrow.

4.6 Vaccines and Adjuvants

Under steady-state conditions the adaptive immune system ignores, or makes tolerogenic or regulatory responses, to the myriad of harmless antigens that continually bombard us. A successful vaccine needs to overcome this tolerogenic state and alert the immune system to make an active response. It must also inform the immune system about the quality of the adaptive response needed to give protection. This is why vaccine adjuvants are needed. All vaccines that are designed to provide protection against infectious diseases must contain an antigen(s), a component of the causative agent of disease, and an adjuvant. The antigen determines the specificity of the adaptive response, while the adjuvant initiates and partly regulates the innate response. Roughly speaking we can consider two different types of adjuvant. One is strictly an inorganic chemical, alum being the best known example (below). The other may be biological or at least a large organic molecule, in the form of a viral vector or DNA for example (below). The former may cause cellular damage and the production of DAMPs, while the latter may contain inbuilt adjuvants in the form of PAMPs, both of which can lead to inflammatory responses and help the induction of adaptive immunity.

4.6.1 Current Adjuvants

Tetanus is a totally preventable disease. In developed countries everyone is likely have been immunized against tetanus. Vaccination involves the injection of a modified tetanus toxin (toxoid) that has lost its toxicity but retains its antigenicity. Most of us remember that at the site of injection there was a short-lived painful swelling (inflammation). Why should injection of a simple protein do this? If the vaccine were just protein, this would not happen. There would be no inflammation and probably no protective immune response. The vaccine, however, contains the adjuvant alum which is a poorly characterized precipitated aluminium salt that causes local inflammation after infection. Alum has been used for decades as a safe and effective adjuvant in billions of people, although associated with transient side effects such as we have noted. One major problem with alum is that it is very good at stimulating effective antibody responses, but it does not trigger effective cytotoxic T cell responses, which might be essential for generating effective vaccines against many infectious diseases and tumours Section 1.6.6. Only recently have we begun to get a better idea of how alum works. We hope that the insights gained from understanding alum will allow us to develop more effective vaccines in the future.

How does alum work? For many years it was believed that its principle function was to provide a long-lasting depot of antigen at the site of injection. Certainly the site of alum deposition becomes infiltrated with inflammatory cells, it takes a long time to be cleared (if ever), and if the residue is dissected from an experimental animal such as a mouse and is injected into another animal, it provokes another immune response. Although there is still controversy, one possibility is that small particles of alum induce damage in the cells that internalize it and that the subsequent DAMPs that are produced trigger an inflammatory response. These may also result in the activation and migration of DC from the site of injection to lymph nodes where they initiate the adaptive response. For example, some workers have suggested that alum triggers the production of uric acid as a metabolite of purine metabolism in the cells. Conflicting evidence further suggests that uric acid acts as a DAMP and activates inflammasomes, leading to caspase processing of IL-1β, for example, and secretion of this pro-inflammatory cytokine. See Figure 4.31.

Apart from alum, there are very few adjuvants that have been licensed for human use, and most of those used in animal experiments cannot be used in humans. Those that are

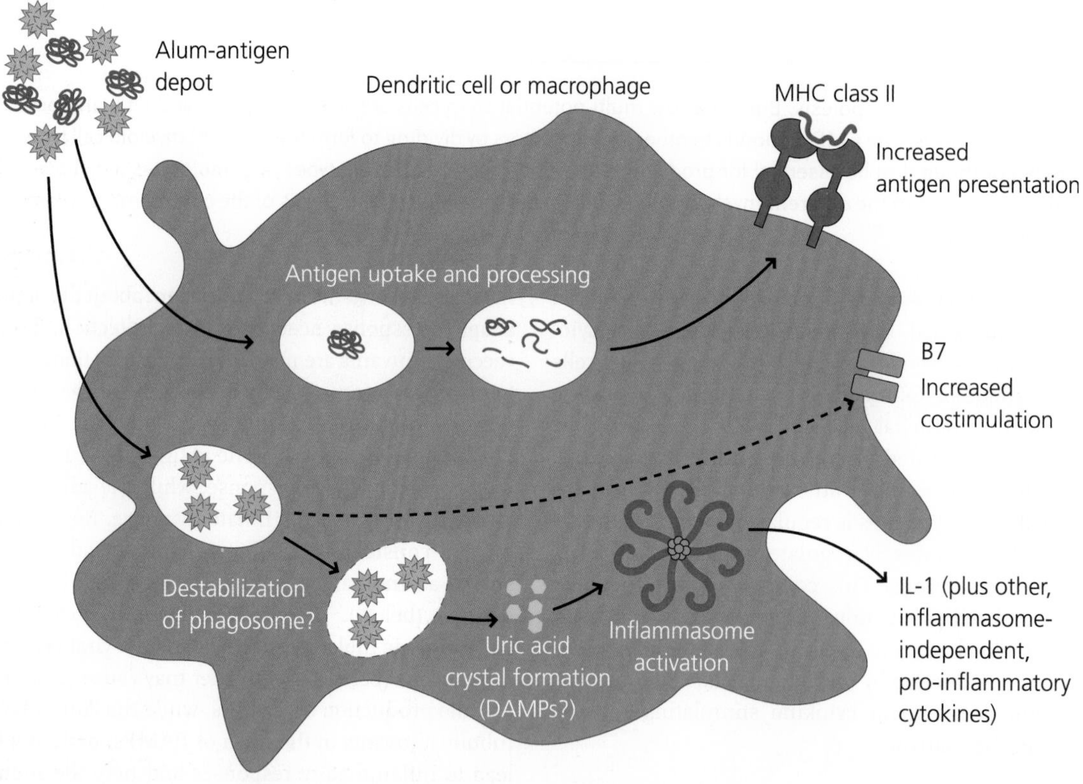

Fig. 4.31 Possible mechanisms of alum adjuvanticity. Vaccines containing alum adjuvant stimulate very good, protective antibody responses against the antigen(s) they contain (e.g. tetanus toxoid). Alum may act as an adjuvant in several ways. (i) It may provide a "depot" of antigen at the site of injection, increasing the amount available to cells such as DCs. (ii) Through unknown routes it may lead to increased costimulatory molecule expression on DCs. Both would increase antigen presentation to, and activation of antigen-specific T cells that could then help antibody responses to be triggered. (iii) It may be phagocytosed (e.g. by macrophages) and lead to damage or destabilization of endosomes. This may lead to activation of inflammasomes, perhaps through production of uric acid as a DAMP, and synthesis of pro-inflammatory cytokines (e.g. IL-1) to enhance innate immune responses and, in turn, adaptive immunity. We are gaining increased insights into these (and other) different mechanisms, but still have far to go, including understanding why alum does not efficiently stimulate other types of immunity (e.g. cytotoxic T lymphocytes) that may be more effective for vaccination against virally-infected cells and tumours.

most effective in experimental animals cause quite severe inflammation. For example, Freund's complete adjuvant consists of a suspension of killed mycobacteria thus expressing PAMPs, in an oil-in-water emulsion which is believed to create a depot. When people have been injected accidentally with Freund's complete adjuvant they are likely to develop a granuloma at the site of injection that is very resistant to treatment. Nevertheless a derivative of LPS has recently been licenced as an adjuvant for clinical use in Europe and the United States.

4.6.2 New Adjuvants

What of the future? There is an active search for effective, safe adjuvants for use in humans. We end with a few brief examples where there are reasons for optimism.

In animals, cholera toxin is a very potent adjuvant for both local mucosal (IgA) and systemic (IgG) antibody responses. In humans, however, cholera toxin is a very powerful stimulant of diarrhoea: it is said that 1 μg of toxin will cause 1 litre of diarrhoea (who was the volunteer?). Researchers are now modifying the toxin to determine if they can reduce or abolish the toxicity while maintaining the adjuvant effect for mucosal vaccines.

DNA vaccines have their own inbuilt adjuvants; for example they contain the unmethylated CpG DNA sequences that are ligands for TLR9 (Section 2.5.2.2). At present it seems that effective DNA vaccines would need very large amounts of DNA. Attempts are being made, however, to use DNA vaccines may incorporate extra innate immune stimuli into the DNA construct, or to include sequences in the vaccine that will selectively stimulate cells such as DCs.

Another approach is to insert vaccine antigen genes into viruses or bacteria that have been genetically modified to reduce or abolish their pathogenicity. Modified virus Ankara (MVA) is a modified vaccinia virus that is being used in many trials as a vaccine while others are using modified Salmonella or Bacillus Calmette-Guérin (BCG) bacteria. These vectors carry inbuilt innate system stimulants and may obviate the need for external adjuvants.

Learning Outcomes

By the end of this chapter you should be able to understand, explain and discuss the following topics and questions, the relevant sections of the chapter are indicated. You should understand some of the evidence from human and animal studies supporting what we know about these topics. You should have some idea of the areas where our understanding is incomplete. You may be able to suggest ways in which our understanding could be advanced.

- Induction of innate immunity (Section 4.2)
  - What are PRRs and where are they found?
  - What are PAMPs and DAMPs?
  - What are the general structures and functions of different TLRs?
  - How, in general terms, do TLRs work as signalling molecules?
  - What are inflammasomes?
  - Give some key examples of the roles of cytokines in mediating local and systemic inflammatory effects.
- Tissue-resident cells of innate immunity (Section 4.3)
  - Which tissue-resident cells are involved in initial sensing of infection?
  - How do tissue-resident cells trigger inflammation?
  - What are the main features of inflammation and why are they important?
- Recruited effectors of innate immunity (Section 4.4)
  - How are leukocytes recruited to inflammatory sites? What are the key molecular players?
  - What are the key similarities and differences between the effector mechanisms of macrophages and neutrophils
  - How can the functions of macrophages be modulated?
  - What types of neutrophil defects are known?
  - What is complement and why is it important for host defence?
  - How is complement activated, what are its main effects and how can host cells resist complement damage?
  - What are NK cells and what do they do?
  - How is the cytotoxic activity of NK cells controlled?
- Development of innate myeloid and lymphoid cells (Section 4.5)
  - How is the development of the cells of innate immunity regulated
- Vaccines and adjuvants (Section 4.6)
  - What are adjuvants and why are they needed for vaccines?
  - What types of molecules are involved in adjuvanticity?
- GENERAL: How have natural and experimental defects in innate immunity helped us understand the mechanisms and functions of innate immune response in defence against infection?
- INTEGRATIVE: How do secondary lymphoid tissues integrate the signals generated by activation of the innate immune system and how do these signals regulate adaptive immune responses?

It should be very clear that understanding how to modulate immune responses – and perhaps particularly an understanding of innate recognition pathways – is a central goal of immunotherapeutic research, with applications in vaccine design, and, as we see in Chapter 7, the treatment of immune-mediated diseases, prevention of transplant rejection and immunotherapy of tumours.

Q4.24. How likely is it that adjuvants will be developed for use in vaccination with absolutely no side effects at all?

Further Study Questions

Qa. To what extent can altered, damaged or stressed self components act as agonists for PRRs? (Section 4.2)

Hints We do know that some types of PRRs, such as the scavenger receptors, can bind modified self components such as oxidized low-density lipoprotein. Others, such as some of the TLRs, are also thought to respond to components of the host, including heat-shock proteins that are produced by stressed cells (e.g. after infection). We do, however, need to be cautious because some apparent responses might actually be due to traces of contaminants in the preparations used, LPS being a prime example.

Qb. How much do we understand about the regulation of inflammatory responses? (Section 4.3)

Hints Inflammation is dangerous and needs to be carefully controlled. Do we want to consider local or inflammatory responses, or regulation at the tissue, cell or molecular level? There is, for example, increasing understanding of the nervous and endocrine control of inflammation (e.g. the hypothalamus–pituitary–adrenal axis). How about specific types of cells making anti-inflammatory response to counter-balance it? What about the kinetics of secretion of different types of molecules, including cytokines, that might dampen it down?

Qc. To what extent can we define the different stages of macrophage activation, the function of different NK cell receptors and their ligands, or the roles of acute-phase reactants in defence? (Section 4.4)

Hints Yes, we apologise (a little) – this is three questions in one. There is in fact a great deal known about each of these. However, for macrophages, are these really defined stages or is there more of a continuum? For NK cells, are these receptors expressed differently by different subsets, and what types of responses might they control? For acute-phase reactants, in some cases the answer is very clear, but in others it is probably true to say we still have absolutely no idea of why these molecules are produced.

Qd. How much do we know about the control of production of different types of immune cells? (Section 4.5)

Hints The best place to start is probably to choose one cell lineage that takes your fancy. (For us this would probably be DCs; we also find mast cells interesting because their origins are still relatively obscure.) Where are the precursors of these cells first produced? What types of stromal cells might they contact there, what types of cytokines might be produced to help them develop and what types of transcription factors might regulate development? Do they get into the blood as precursors, or at a mature stage? Where do they go? How do these populations change if infection or inflammation occurs?

Qe. To what extent is it possible to define different types of DAMPs and their receptors?

Hints This is a minefield at present. Perhaps start by defining what we mean by a DAMP. Is it an "altered self" component or something produced by cells or tissues in response to damage, and to what extent might it overlap with a PAMP? Then perhaps find out about how much we do or do not yet know about the mechanism(s) of alum as an adjuvant, how inflammasomes are activated, and such like. But please tread carefully!

5
T Cell-Mediated Immunity

5.1
Introduction

An entirely new form of immunity, adaptive immunity, first evolved around the time of the Cambrian explosion. The first true lymphocytes appeared when fish developed jaws around 450 million years ago and adaptive immunity exists in all jawed vertebrates. Adaptive immunity has evolved in the presence of innate immunity, and depends in large part on the utilization of innate components for its activation and effector functions. In mice and men, innate immunity has now become largely subservient to the adaptive form in defence against infection. Why an additional form of immunity evolved is unknown, although presumably it could reflect an "arms race" between infectious agents and their hosts. Nevertheless, the ability of lymphocytes to recognize infectious agents may have been facilitated by mobile DNA elements ("jumping genes") which are present in complex organisms. Stable insertion of such elements may have contributed to the evolution of a recognition system based on rearranging gene segments to generate multiple different antigen receptors. This system is used by T and B lymphocytes in vertebrates, and ultimately enables "anticipatory" recognition of antigens expressed by infectious agents.

In this chapter, we examine immune responses that are primarily dependent on T lymphocytes. We start by noting that different populations of T cells have evolved, before focusing mainly on one conventional type, the αβ T cell (Section 5.1). We introduce classical major histocompatibility complex (MHC) molecules, which co-evolved with T lymphocytes and that enable them to recognize cells containing infectious agents, and describe how these molecules work (Section 5.2). (We also note that some related types of non-classical MHC molecules can enable non-conventional T cells to recognize and respond to infection.) We next examine the responses of conventional T cells, where they are activated and how they are triggered (Section 5.3), and what is known of how they can subsequently develop specialized functions that are needed to eliminate different types of infectious agents (Section 5.4). After this we discuss the role of the thymus in T cell development, including the generation of a huge repertoire of antigen receptors by random DNA changes – which is essential for the anticipatory recognition of infectious agents (Section 5.5). We also note how non-conventional T cells also develop within in the same organ. Finally we turn to therapy and briefly describe how T cells can be used in vaccines against infectious disease and potentially against malignant tumours (cancers) (Section 5.6).

By the end of this chapter you should have acquired a fresh understanding of T cell-mediated immunity and its central role in host defence.

5.1.1
T Cell Populations

The primary division of T cells into different populations is in terms of the molecular nature of their antigen receptors. T cell receptors (TCRs) are composed of two chains, both of which contribute to the antigen recognition site. The best understood and most studied are the conventional αβ T cells, where α and β refers to the two chains of their antigen receptors. Later we outline some features of γδ T cells and invariant NKT (iNKT) cells (not to be confused with natural killer (NK) cells; Section 1.4.5). It is important to appreciate that the development of all these different populations of lymphoid cells occurs in the thymus, which is relatively shielded from the rest of the body, and that they are generated before any infection occurs. Their subsequent effector responses are, however, generally dependent on the prior sensing of infectious agents, particularly by components of innate immunity. See Figure 5.1.

5.1.2
How Do Conventional T Cells Recognize Antigens?

The αβ T lymphocytes have evolved primarily to interact with cells containing antigens derived from infectious agents. In contrast, B cells can directly recognize antigens that are soluble or form part of an infectious agent. Many pathogens that have infected mammals spend much of their life inside host cells; these include all viruses, many bacteria and some protozoa. To function effectively, the adaptive immune system needs to detect and deal with such infections. This poses

Exploring Immunology: Concepts and Evidence, First Edition. Gordon MacPherson and Jon Austyn.

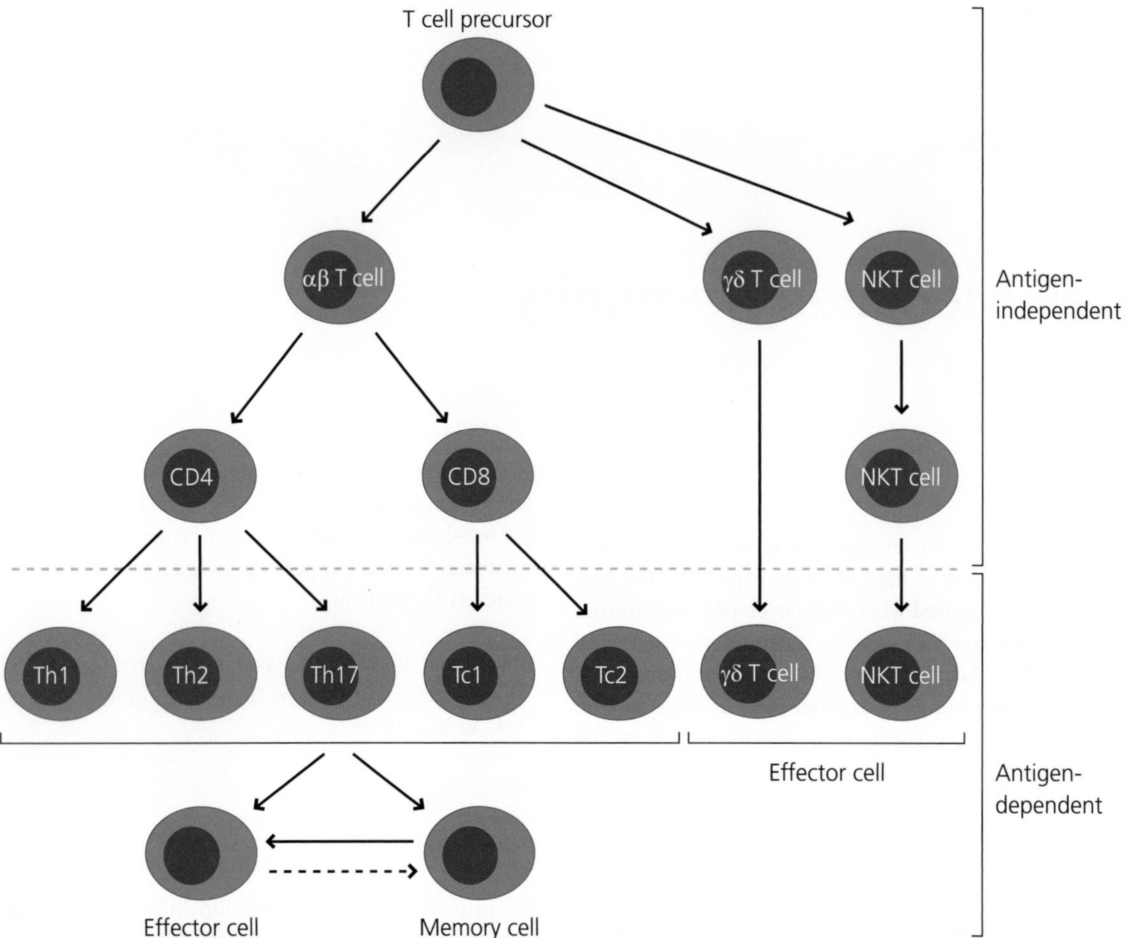

Fig. 5.1 T cell populations. The same T cell precursor can give rise to several T cell subsets with different functions. Conventional αβ T cells develop into CD4 and CD8 populations in the absence of foreign antigens. When they become activated by antigen they can develop further into different effector subsets (e.g. T_h1, T_h2 and perhaps T_c1, T_c2 cells, etc.) or into memory T cells. Some can also develop into regulatory T cells (not shown). Non-conventional T cell populations include γδ T cells and iNKT cells. (There are also non-conventional αβ cells that express neither CD4 nor CD8, and different types of NKT cells; not shown.)

two problems for T cells – how can they recognize that they are interacting with another cell, and how can they detect that that cell is infected? The adaptive immune system has evolved a sophisticated mechanism to limit T cell activation to situations where the T cell is in contact with another cell. This is the primary function of the classical MHC molecules. These MHC molecules act as receptors for small peptides that are generated within different cellular compartments. In turn, the TCRs of conventional T cells recognize parts of bound peptides and the MHC molecules to which they are bound. This means that a single T cell shows dual antigenic specificity; both for the peptide bound to the MHC molecule and for the MHC molecule itself. The specificity of a T cell for a particular MHC molecule is known as its MHC restriction. We now know that this is because each chain of any given αβ TCR has three particular regions that differ most from those of any other TCRs (i.e. they are hypervariable) and which determine complementarity between the receptor and its ligand; hence they are termed complementarity-determining regions (CDR1–3 for each chain, respectively). For each, CDR1 and 2 primarily interact with exposed parts of the MHC molecule while CDR3 makes contact with the bound peptide. (Antibodies also exhibit similar hypervariable regions, but in these cases they only contact the antigen itself; Chapter 6.) See Figure 5.2.

5.1.3
Subpopulations of Conventional T Cells

Conventional αβ T cells are divided into two main subpopulations: CD4 and CD8 T cells. The CD4 molecule binds to MHC class II molecules, whereas CD8 binds to MHC class I. This binding generates signals that contribute to T cell activation and is crucial in determining the different functions of the subsets. MHC class I is expressed on almost all nucleated cells. Expression of MHC class II is, however, restricted to a limited number of cells involved in CD4 T cell development, activation or effector functions. These include thymic epithelial cells, dendritic cells (DCs), B cells and some macrophages. See Figure 5.3.

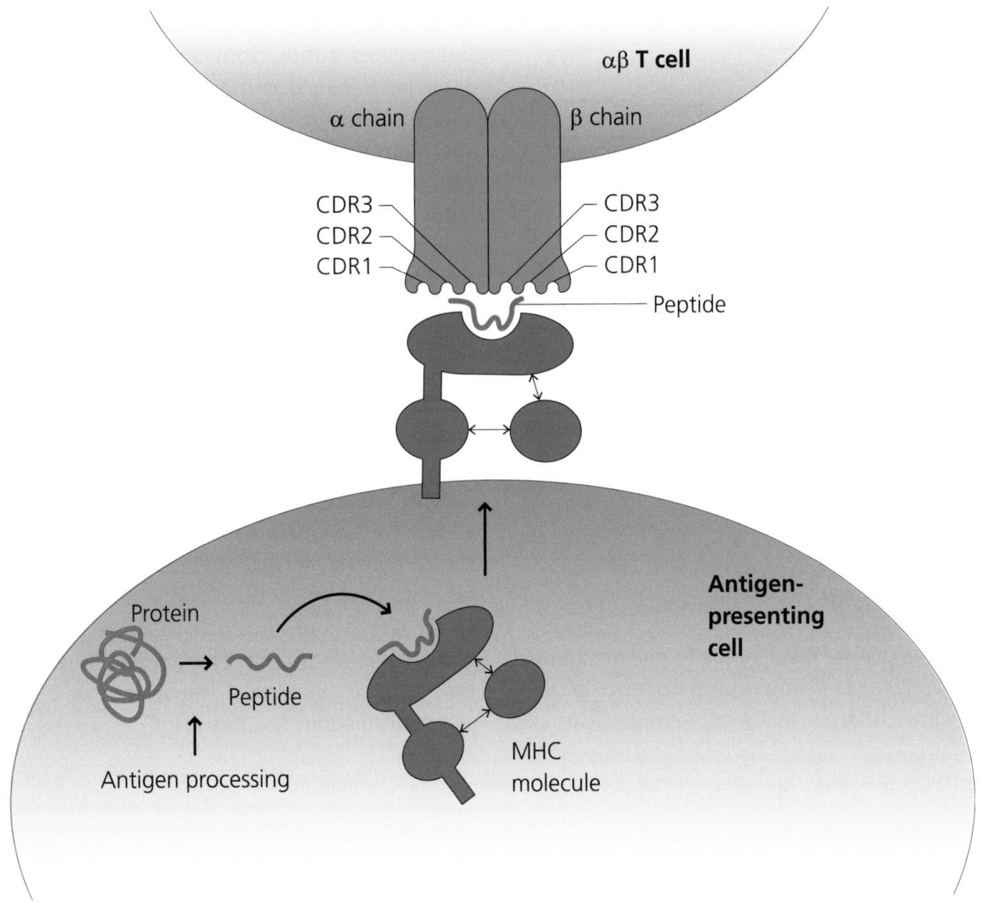

Fig. 5.2 Interactions of T cell receptors with peptide–MHC complexes. αβ TCRs recognize peptides bound to MHC molecules. These peptides are generated from proteins within cells by mechanisms collectively known as antigen processing and bind to MHC molecules before they are exported to the plasma membrane. The TCR α and β chains each contain three hypervariable CDRs (CDR1–3). The CDR3 regions interact primarily with the peptide while the others interact predominantly with the MHC molecule itself.

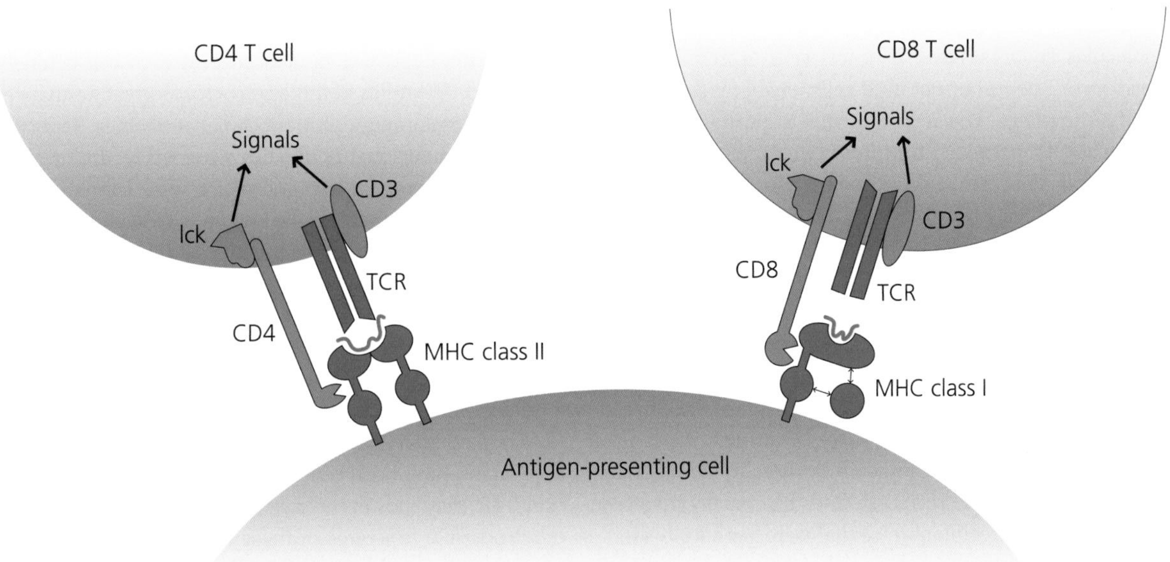

Fig. 5.3 Interactions of CD4 and CD8 with MHC molecules. CD4 and CD8 molecules are first expressed on thymocytes during T cell development, and which one of these a T cell eventually expresses determines whether it will interact with MHC class I or II. CD4 and CD8 bind to invariant parts of MHC class II and class I respectively. The TCR is associated with CD3, a multimolecular complex (not shown). CD4 and CD8 are constitutively associated with the tyrosine kinase, Lck, which phosphorylates specific regions ITAM motifs of CD3 (not shown) in the early stages of T cell activation.

5.1.4 Effector Functions of T Cell Populations

In this chapter we focus particularly on conventional CD4 and CD8 T cells. In general, CD4 T cells are the cells that regulate adaptive immune responses. For example, they can instruct B cells to develop into plasma cells that make different types of antibodies, recruit different types of granulocytes to specific sites of infection and regulate how macrophages function. They are also primarily responsible for many clinically important diseases or conditions, including allergies and autoimmune diseases, much forms of transplant rejection, and much of our inability to reject tumours (Chapter 7). CD8 T cells are generally thought to have less of a role in immune regulation, but are certainly crucial effector cells in both immunity to infection and in a number of important immune-mediated diseases.

5.1.5 T Cell-Mediated Immunity Against Infection

What is the evidence that we need T cells in defence against infectious disease? There is strong clinical evidence for this. For example, some children are born without thymuses (DiGeorge syndrome). These children and others with T cell deficiencies suffer from an increased incidence of viral infections. If such an individual is accidentally infected with vaccinia virus in smallpox vaccination, rather than the normal short-lived self-healing lesion, the infection spreads and may destroy large areas of the skin. Similarly, children who were genetically T cell-deficient (i.e. they suffered from a primary immunodeficiency disease) and who were immunized with Bacillus Calmette-Guérin (BCG) during vaccination programmes became infected and developed a widespread, potentially fatal infection. In acquired T cell deficiencies, such as acquired immunodeficiency syndrome (AIDS), a similar increase in susceptibility to tuberculosis is also seen. In addition, AIDS patients may become infected with strains of mycobacteria such as Mycobacterium avium, or with Pneumocystis jirovecii and cytomegalovirus (CMV), which are harmless to normal individuals. This can only mean that in normal individuals, the adaptive immune system, particularly the T cell system, is continually acting to recognize and eliminate such microbes without there being any clinical signs. There is also abundant experimental evidence for the crucial importance of T cells in defence against infection. See Box 5.1

Box 5.1 Analysis of T Cell Functions by Adoptive Transfer

Early studies showed that antibodies were made by bone marrow-derived cells (B cells) and that responses such as delayed-type hypersensitivity (DTH; Section 1.6.4.2) were mediated by thymus-derived cells (T cells). T cells were, however, able to mediate a variety of functions: as well as DTH, they could help B cells to make antibodies, could develop cytotoxic properties and could bring about allograft rejection. How can we determine if these functions are mediated by different subsets of T cells? It was shown that molecules such as Thy-1 in mice were expressed by all peripheral T cells, but not by B cells, enabling these to be used as markers. The next step was to purify these cells and to ask if a particular function is mediated by a defined population. This can be done in vitro using assays such as those which measure cytotoxicity, but in many cases experiments involved transfer of purified subsets into recipient mice, for example, that acted as in vivo culture vessels. This is termed adoptive trasnfer. In these studies, the recipients were usually either naturally T cell deficient (as in nude mice which lack a thymus) or had been irradiated (naïve lymphocytes are unusual in that they are particularly sensitive to ionizing radiation during interphase). Obviously it was also important to be able to distinguish between the transferred cells and those of the recipient, and this usually involved using donor and recipient mice that were genetically identical except for one polymorphic cell surface marker, such as CD45, which exists in different allelic forms in mice.

Initially, cell populations were purified by techniques such as complement-mediated lysis or by sticking cells to antibody-coated plates (panning) to deplete a subpopulation; this type of experimental approach is termed negative selection (not to be confused with the process of the same name that occurs in the thymus; below). The problem with these approaches is that the efficiency of the separation technique cannot be estimated. For example, how many surface molecules does a cell need to express if it is to be killed by complement? With the introduction of fluorescence-activated cell sorting (FACS), subpopulations could be separated (sorted) with a high degree of purity and, importantly, the sensitivity of the technique could be estimated because the degree of fluorescence of each cell could be measured. The utility of this approach was greatly enhanced with the development of monoclonal antibodies. Thus, for example, monoclonal antibodies were made that defined two T cell subpopulations, and which were later shown to bind to molecules that became known as CD4 and CD8. The sorted subsets could then be transferred into T cell-depleted animals and their functions studied. It was thus shown that CD4 T cells were the helper cells for DTH responses and antibody synthesis, while CD8 T cells were the precursors of cytotoxic cells. Other immunodeficient mice are now available as recipients such as recombinase-activating gene (RAG) gene knock-outs which have no T or B cells. These approaches have been central to understanding T cell functions. (It may, however, have biased our appreciation of other possible functions of these cells because our understanding is inevitably limited by the assays we can use!)

Q5.3. Might there be any problems in using fluorescently labelled lymphocytes for long-term adoptive transfer experiments?

Q5.1. How might we determine the life span of T cells in the mouse?

5.2 Major Histocompatibility Complex (MHC) and Antigen Presentation

5.2.1 Classical MHC Molecules

It is not possible to understand how T cells work without understanding the role of the MHC. The function of conventional MHC molecules is to act as peptide-binding receptors that are recognized by antigen receptors on T cells. The MHC was discovered from early transplantation studies. It was known long ago that transplants between different members of the same species (allografts) were destroyed (rejected) very rapidly (Section 1.6.5). Early experiments led to the identification of genetic regions that controlled the rejection of such transplants. These were termed histocompatibility loci. One locus in particular controlled the most rapid rejection of these transplants and was thus termed the major histocompatibility locus. This locus was subsequently found to include a large complex of genes and became known as the MHC. In mice, the MHC is known as H-2 (because historically it was the second locus to be identified from transplant studies). In humans, it is known as HLA (short for human leukocyte antigens which were originally identified using panels of antibodies).

The MHC contains two different sets of genes that code for the classical MHC peptide-binding molecules. These are the MHC class I and class II loci and both contain several discrete structural genes. In humans, the MHC class I region contains three structural genes: HLA-A, -B and -C. In mice, the equivalents are H-2K and H-2D. The human MHC class II genes are called HLA-DP, -DQ and -DR, and in mouse they are called I-A and I-E. See Figure 5.4.

MHC moleucles are codominantly. Thus, any individual is likely to express six different MHC class I molecules, each binding a different set of peptides. Another most important feature of the classical MHC loci is their extreme polymorphism. Thus, within human and mouse populations there are often more than 100 variants (alleles) of some structural MHC genes. Many other genes have multiple allelic variants in populations, but usually these variants are rare. The MHC is different in that most alleles are present with significant frequencies. The importance of this polymorphism is that it makes it difficult for a pathogen to escape recognition by a T cell both in an infected individual and in a population. The result is that if a pathogen mutates so a peptide from it can no longer bind to any given MHC molecule, there are five other MHC molecules that could potentially bind other peptides from the pathogen. At a population level, the vast numbers of MHC alleles available means that even if any pathogen mutates so that no peptides

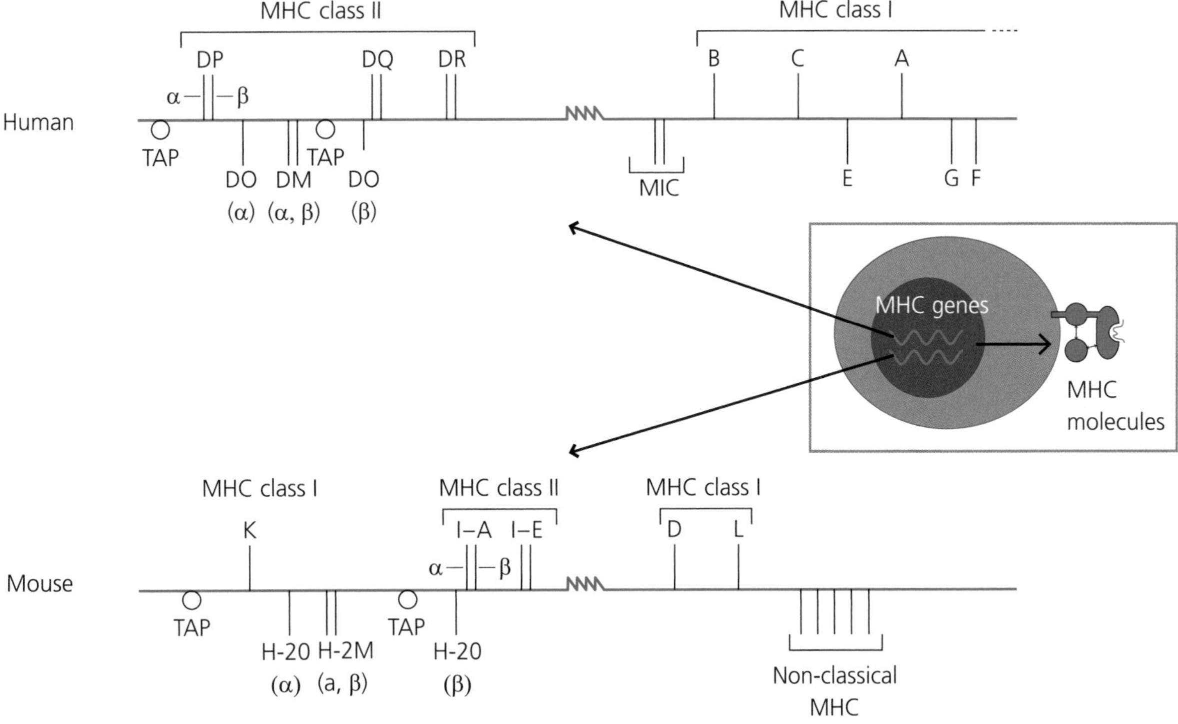

Fig. 5.4 Genetic organization of human and mouse MHC. The MHC consists of a large stretch of DNA, in humans called HLA and in mice H-2. Several distinct structural genes encode the α chain of classical MHC class I molecules (which associate with β_2-microglobulin, encoded outside of the MHC), and the α and β chains of class II molecules. Many other molecules are also encoded in the MHC region. Some of these are involved in peptide delivery to, or loading of, classical MHC class I or class II molecules; these, respectively, include TAP, and in humans HLA-DM and -DO (which are also dimers; A, B); the mouse equivalents are H-2M and H-2O. There are also non-classical MHC molecules, such as MIC(-A and -B), and HLA-E, with specialized functions.

Box 5.2 Histocompatibility and Self–Non-Self Discrimination in Other Species

It is crucial for all complex organisms that cells know their place, and recognition by a cell of where it is depends on molecules expressed on the plasma membrane of the cell. In jawed vertebrates T lymphocytes have evolved to interact with other cells and this interaction depends on them recognizing a particular MHC molecule. T cells during their development are selected so that only those capable of recognizing the MHC molecules expressed in the thymus are allowed to survive. This restriction by the MHC serves at least two functions. It means that T cells that assemble TCRs unable to interact with the particular MHC molecules that are expressed by their host are eliminated, so they will not crowd out T cells that can react. In addition the requirement for MHC recognition means that T cells can only be activated if they are interacting with MHC-expressing cells. Thus, the MHC is involved in limiting interactions between T cells and host cells.

Mechanisms for limiting cell–cell interactions occur in many other circumstances. Within mammals, cell–cell recognition can be cell-type specific. Thus, if suspensions of cells from different tissues are mixed in culture, mixed aggregates form at first, but rapidly the aggregates come to contain only one of the cell types. Amongst invertebrates, sponges show very strong histocompatibility: if mixed suspensions of cells from two sponges, even from the same species, are mixed, the aggregates that form contain cells from only one of the donors. This mechanism represents allogeneic recognition and depends on a highly polymorphic gene system that does not involve MHC molecules.

Many plants are bisexual – they make both pollen and ova. It is important that self fertilization does not occur in these plants and this is also prevented by a polymorphic histocompatibility system. In this case it is a self–self incompatibility system: self–self recognition results in the prevention of pollen meeting the ovum by a variety of different mechanisms including destruction of pollen tubes and induction of apoptosis. Three different incompatibility systems exist in plants and the molecules used are quite different and the mechanisms used also differ.

What this emphasizes is the importance of cell–cell recognition in all complex organisms. However, we must not forget that even bacteria can make population-based decisions in a coherent manner – in this case they use secreted molecules for quorum sensing.

can be recognized by T cells of a particular individual, it is highly unlikely that any other individual will express the same set of alleles and thus most individuals within the population will remain resistant. The importance of this polymorphism is illustrated by African cheetahs. Cheetahs in captive colonies are susceptible to widespread infections. These animals suffered a reproductive bottle-neck at some time in their evolution, resulting in them having a very limited MHC diversity and this may explain their susceptibility. See Box 5.2

Q5.2. How many different MHC class II molecules is any individual human likely to express?

5.2.1.1 How Do Peptides Bind to Classical MHC Molecules?

The amino acid sequence of MHC molecules was determined by biochemical techniques but this did not reveal how they interacted with peptides. Many theories existed until X-ray crystallography showed the three-dimensional structure of a MHC class I molecule. This revealed that the outer part of the MHC molecule contained a groove into which a single peptide could be bound. A similar structure was identified for MHC class II molecules.

MHC molecules consist of two chains. In MHC class I molecules the heavy chain is a trans-membrane protein consisting of three domains and it wholly contains the peptide-binding groove. The heavy chain is associated non-covalently with a smaller molecule, β_2-microglobulin, which represents the fourth domain, but this is encoded outside of the MHC. MHC class II molecules are formed from two chains, α and β, both of which are trans-membrane molecules and both of which contribute to the peptide-binding site. Structurally, the domains of MHC molecules are members of the immunoglobulin superfamily.

It was long known that peptides added to cells in culture could enable recognition by specific T cells, but the nature of the peptides that were naturally bound to MHC molecules was unknown. To resolve this, MHC molecules were isolated from cells and, by altering the ionic conditions, peptides were eluted from the MHC molecules and their characteristics determined by chemical or physical approaches (e.g. sequencing or mass spectrometry). These experiments illustrated several important features of MHC peptide interactions. Thus, peptides bound to MHC class I molecules tend to be 8–10 amino acids in length (typically 9), whereas those bound to MHC class II molecules can be considerably longer. The reason for this is that in MHC class I molecules the peptide-binding groove is closed at both ends, constraining the length of peptide that can be accommodated. In MHC class II molecules the groove is more open, permitting peptides of greater lengths to bind. See Figure 5.5.

Peptides eluted from a particular MHC molecule are very heterogeneous in overall composition. However, it was found that, for MHC class I molecules, amino acids of similar types were present at particular positions in the eluted peptide (2 or more positions in each peptide) suggesting that these amino acids might be involved in anchoring the peptides to the MHC molecule. In fact, pockets in the peptide groove were subsequently identified from X-ray crystallographic studies.

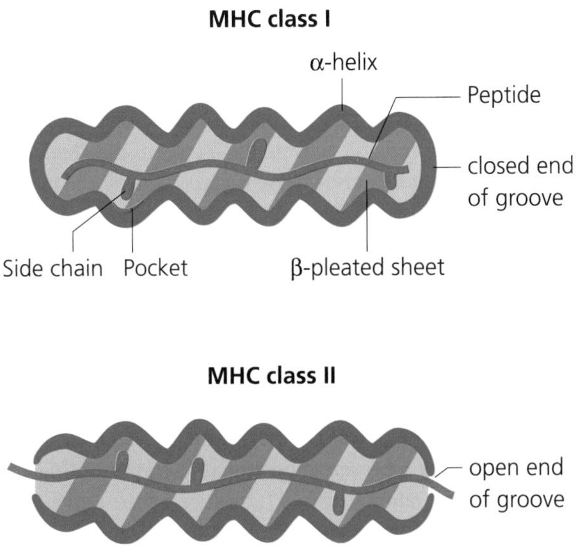

MHC class I
α-helix
Peptide
closed end of groove
Side chain
Pocket
β-pleated sheet

MHC class II
open end of groove

Fig. 5.5 Interactions of peptides with MHC molecules. The peptide-binding grooves of MHC molecules consist of two stretches of α-helix which overlie a β-pleated sheet base. Most of the differences (polymorphisms) between MHC molecules are found in the groove and hence determine precisely which peptides can be bound by different MHC alleles. Class I molecules have grooves closed at both ends, restricting the size of peptides that can bind. They also contain pockets into which side chains from the peptide amino acids can bind non-covalently as anchor residues to hold the peptide in place. Class II molecules have more open ends, allowing peptides of greater and more varied lengths to bind.

Binding of peptides to MHC class II also involves anchor residues although these tend to be more dispersed along the groove. Hence specific residues of the respective MHC grooves are crucial for high-affinity binding of peptides. Comparison of different MHC alleles also confirmed that most of the polymorphic residues are situated in the walls and base of the groove, thus determining which peptides can bind to any given MHC molecule. What is equally important is that because only relatively small number of residues are involved in binding to MHC class I molecules, for example, the remaining residues can be highly variable. Hence, a single MHC molecule can bind a multitude of different peptides, probably numbering many tens of thousands, making them highly promiscuous receptors.

Clearly, MHC molecules are capable of binding peptides from infectious agents. But what do they bind in the absence of infection? When MHC-bound peptides were identified and sequenced, many of them were found to derive from normal cell-associated proteins. MHC class I-associated peptides are mainly derived from intracellular proteins (nuclear and cytoplasmic), whereas MHC class II peptides come from external proteins (such as those present in the culture media of cells) and also from proteins from the plasma membrane and endocytic compartments of the cell itself. What this suggests, and indeed what we now know to be the case, is that peptides from infectious agents in the cytoplasm (such as viruses) are likely to bind primarily to MHC class I molecules, while those from endosomal compartments (such as phagocytosed bacteria) will mostly bind to MHC class II molecules. See Figure 5.6.

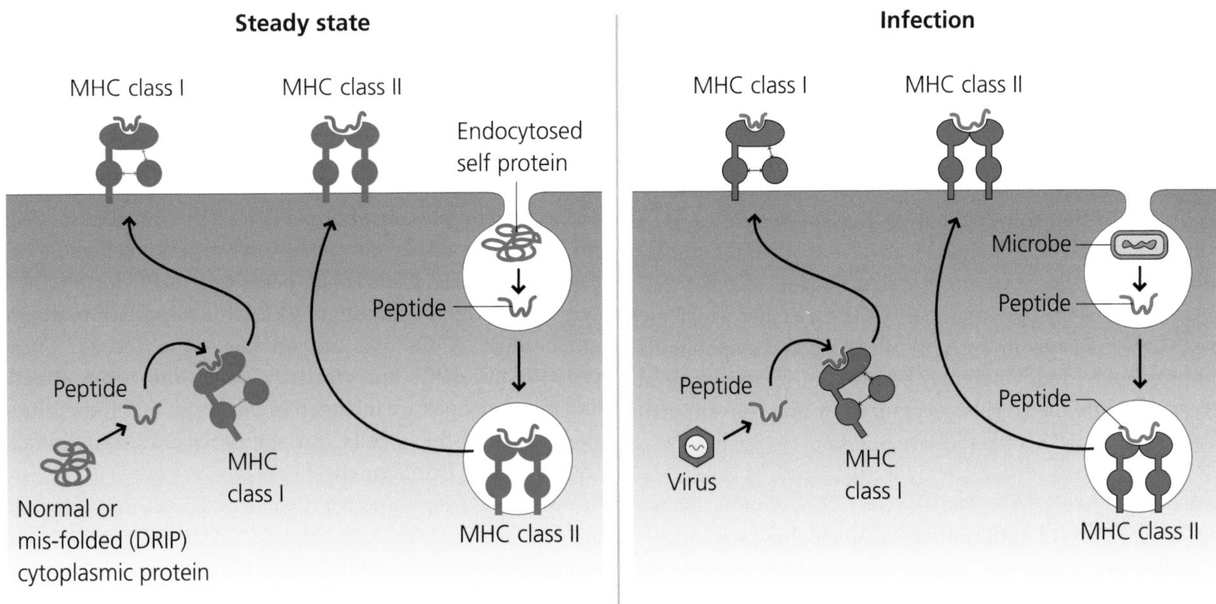

Fig. 5.6 Origins of peptides bound to MHC molecules. Under steady-state conditions, peptides bound to MHC class I molecules are mostly derived from abnormally-folded cytosolic proteins (defective ribosomal products; DRIPs). In contrast, peptides bound to class II molecules may originate from extracellular proteins that have been endocytosed or from plasma or endosome membranes. During infection peptides from viruses or microbes that have gained access to the cytoplasm, or which have been endocytosed, may become loaded onto either MHC class I or class II molecules, respectively.

5.2.1.2 Cellular Distribution of Classical MHC Molecules

In humans, apart from some central nervous system (CNS) neurons, all nucleated cells express MHC class I; erythrocytes, which lack nuclei, do not. MHC class II molecules show a much more restricted distribution. Under steady-state (resting) conditions, the cells in rodents that express MHC class II are DCs, B lymphocytes, thymic epithelial cells and epithelial cells in the small intestine. In humans a wider range of cells, including macrophages and activated T cells, may be MHC class II-positive. However, during active immune responses interferon (IFN)-γ can stimulate expression of MHC class II on many different cell types and can additionally increase the expression levels of MHC class I molecules.

Q5.4. Why might it be that some CNS neurons do not express MHC class I?

Q5.5. In epithelial cells, MHC class II molecules are present, not at the cell surface, but in intracellular vesicles. What might be the significance of this finding?

5.2.2 Non-Classical MHC Molecules

Other molecules are encoded within the MHC region that are structurally similar to those of the classical MHC, but which differ functionally. These are non-classical MHC molecules. They mediate a diverse range of functions. For example, HLA-E (Qa-1 is similar in mice) is a non-polymorphic molecule. It binds leader sequence peptides from newly synthesized MHC class I molecules; these sequences target the MHC molecule to the rough endoplasmic reticulum (RER)). Signal sequences bound to HLA-E can be recognized by NK cells and are important in inhibiting NK cell killing. Two other non-classical MHC molecules include human MIC-A and MIC-B (which do not appear to bind peptides) that become expressed by injured or stressed cells, and may then be recognized by non-conventional γδ T cells (Section 5.4.5) as well as NK cells. Other molecules are also structurally similar to MHC molecules, but are encoded outside of the MHC region. Perhaps the best known are CD1 molecules which are able to bind lipid-containing molecules such as phospholipids and glycolipids, and which may be involved in anti-bacterial defence because they can be recognized by non-conventional T cells such as iNKT cells (Section 5.2.8.2).

5.2.3 The Cellular Basis of Antigen Processing and Presentation

It was known from very early studies that antibodies could bind directly to protein antigens in their natural three-dimensional confirmation but not when the protein had been denatured (i.e. if its tertiary three-dimensional structure was destroyed). However, when the activation of T cells was first studied it was clear that their antigenic requirements were very different. If T cells were removed from an immunized animal and stimulated in vitro with the antigen alone they did not respond. If, however, as well as antigen, other cell types such as macrophages were added to the cultures, the T cells became activated. It was soon found that the macrophages could be pre-cultured with the antigen and still activate the T cells in the absence of free antigen. It was also shown that macrophages could present denatured proteins, but needed to be metabolically active to present native or denatured proteins. Metabolically inactivated macrophages could, however, still activate T cells if peptide fragments from the antigen were added. The concept thus arose that antigens needed to be "presented" to T cells by other cells, rather than the T cells being able to recognize them directly. The antigen also needed to be processed, before it could be recognized. These are the origins of the terms antigen processing, antigen presentation and antigen-presenting cells (APCs). Now of course, as we introduce above and further discuss below, we have a clear mechanistic explanation for these early findings, but the terms have stuck.

What are APCs? Historically this term was first used for any cell that could activate helper (CD4) T cells. In contrast, studies of killing by CD8 cytotoxic T cells generally used cells called target cells. Hence a distinction came about between the concept of an APC (as a cell that expressed MHC class II molecules for CD4 T cell recognition) and a target cell (as a cell that expressed MHC class I molecules for CD8 T cell recognition). The later discovery of the crucial roles of DCs in activating naïve CD4 T cells led to further muddying of the waters. Cells that could activate naïve T cells were called professional APCs. Unfortunately there is still a general tendency to limit the use of these terms to CD4 T cell activation, although we now know that DCs are also often essential for CD8 T cell activation.

Even though we now have a clearer understanding of the molecular basis of peptide generation and loading onto MHC molecules, and of T cell activation by APCs, it is still convenient to use the terms antigen processing and presentation. Perhaps, however, the term APC should be used for any cell that expresses MHC molecules and which can be recognized by any conventional T cell. Thus, a cell expressing MHC class I can become an APC for a CD8 T cell and a cell expressing MHC class II can become an APC for a CD4 T cell. As mentioned, amongst APCs it is important to distinguish those APCs that can activate naïve T cells. These professional APCs are primarily DCs. Sometimes B cells and macrophages are included in this term, but there is little evidence that these cells can activate (or at least initially trigger) naïve T cells in vivo.

5.2.4 Antigen Processing and MHC Class I Presentation

CD8 T cells were first characterized as cells that can kill other cells after they have been activated, and we now know that MHC class I molecules are crucial for recognition. How was this discovered? Doherty and Zinkernagel found that mice infected with lymphocytic choriomeningitis virus (LCMV) developed CD8 T cells that could kill virally infected mouse target cells in culture. However, for killing, it was essential that the target cells shared at least one MHC class I molecule with the infected mouse. This was the origin of the concept of MHC restriction which led to the award of a Nobel Prize to Doherty and Zinkernagel. Subsequent experiments by many groups showed that CD8 T cells recognized short peptides derived from cytoplasmic proteins and that these peptides were bound to MHC class I molecules. Generally speaking, peptides derived from the cytoplasm, whether or not they originate from self proteins or infectious agents (e.g. viruses or microbes that have infected cells), are termed endogenous antigens. In general, endogenous peptides are presented to CD8 T cells by classical MHC class I molecules. See Figure 5.7 and Box 5.3.

Q5.6. How might the antigen-processing mechanisms distinguish between normal self proteins and those derived from pathogens?

5.2.4.1 The Proteasome

Endogenous peptides bind to MHC class molecules, but how are the peptides generated? Here we have an example of the adaptive immune system using a pre-existing mechanism from another system. Proteasomes are large, multi-protein complexes which degrade proteins into peptides and are part of normal cellular housekeeping, degrading proteins into short peptides for subsequent recycling of their constituent

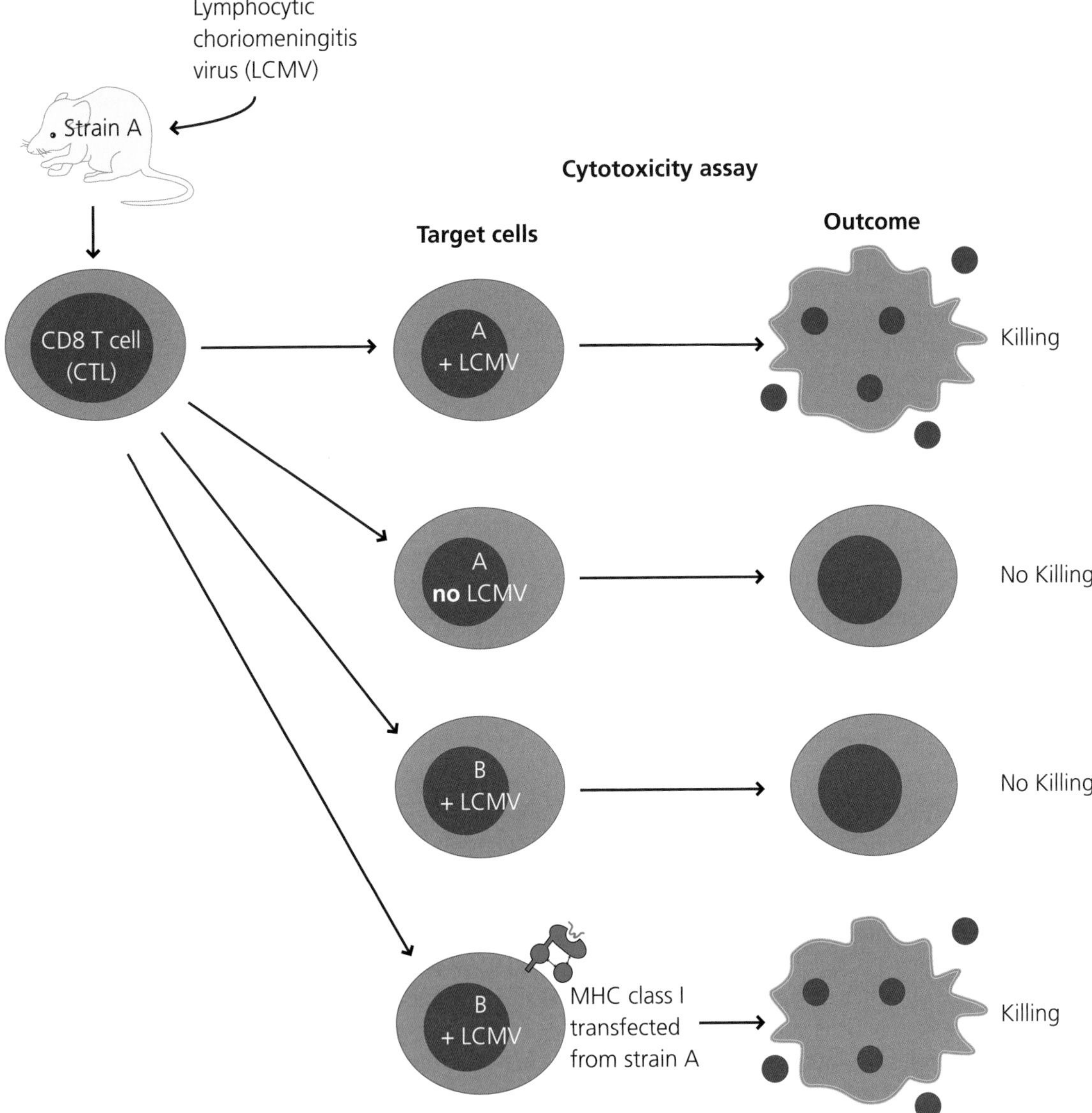

Fig. 5.7 MHC restriction of CD8 cytotoxic T cells. Mice infected with many viruses generate $CD8^+$ CTLs. These can be isolated and are able to kill cells in culture that are infected with the same virus. Doherty and Zinkernagel, using a particular virus, LCMV, found that only cells of the original mouse strain could be killed (e.g. A), while cells from other strains (e.g. B) could not even if they were infected with the same virus. Eventually they discovered that, if infected cells were to be killed, they needed to express the same MHC class I molecules as the mouse that was initially infected. This became known as MHC restriction. This principle applies to both CD8 and CD4 T cells: the former are restricted to MHC class I molecules, the latter to MHC class II.

Box 5.3 Cytotoxic T Cell Recognition of Intracellular Proteins

How do cytotoxic T cells recognize the cells they kill? It was originally supposed that the antigens recognized by CD8 cytotoxic T cells were whole proteins expressed on the cell surface; thus, for viruses these would be molecules that became exposed on the cell surface during viral assembly. Researchers working with influenza virus wanted to know which part of the virus was actually recognized by cytotoxic T cells. The advantage of influenza is that there are four major proteins present in the virus and different strains of virus express different variants of these proteins. Two of these proteins (haemagglutinin and neuraminidase – the H and N we are all too familiar with, e.g. H5N1, H1N1) are expressed on the viral surface, and on the surface of infected cells during viral assembly. The others, nucleoprotein and the matrix protein, are internal and are never expressed on the infected cell's surface membrane.

Mice were infected with one strain of the virus and the resulting activated T cells were tested for cytotoxicity on target cells infected with different strains, each of these strains only sharing one of the major protein variants with the infecting virus. It was a big surprise that, for killing in some mouse strains, the only molecule that needed to be shared between the virus variants used to infect the mice and the target cells was the nucleoprotein, which is never expressed on the surface of an infected cell. Therefore, there must be a mechanism for expressing antigen derived from cytoplasmic proteins on the cell surface. See Figure 5.8.

What is it that is being recognized? To answer this the researchers made a series of overlapping peptides that covered the whole nucleoprotein molecule and incubated uninfected target cells with these peptides. They were able to identify short peptides that would sensitize the target cells for killing. This showed that somehow the cytotoxic T cells were recognizing peptides on the target cell surface and that this recognition was MHC class I-dependent. How the peptide and the MHC related to each other was not known until the crystal structure of MHC class I was solved, and it became clear that MHC molecules bound short peptides and that the T cell recognized the peptide–MHC complex with a single receptor.

Q5.7. If peptides are needed to get MHC class I molecules to the cell surface, how can MHC class I molecules bind peptides that are simply added to the cultures?

amino acids. Many proteins are targeted to proteasomes after they have been modified with multiple copies of another small protein, ubiquitin. Proteasomes consist of a stack of ring-like protein complexes, rather like a stack of American doughnuts. These form two smaller chambers at each end with a larger, central chamber with multiple proteolytic activities. Proteins are unfolded (by mechanisms we do not fully understand) and passed into the antechamber, then into the central chamber where they are cleaved, and peptides emerge at the other end. In response to IFN-γ the structure of the housekeeping proteasome is changed to form the immunoproteasome, which has different proteolytic activities. These tend to generate peptides that are better targeted to the MHC as they have C-termini that are good anchors for MHC class I molecules in one pocket of the groove. How do we know that proteasomes are important in processing for MHC class I? Lactacystin is a bacterial product that is a selective inhibitor of proteasomal activity and cells treated with lactacystin show inhibition of processing of some, but not all proteins. Hence, other cytoplasmic proteolytic mechanisms must exist. An important example is the ER-resident aminopeptidase-1 (ERAP1) enzyme that can cleave that N-terminal region of peptides, sometimes generating ends that fit well into an other pocket of the groove. See Figure 5.9.

5.2.4.2 The TAP Transporter

MHC class I-binding peptides are generated in the cytoplasm, but the MHC molecules themselves are synthesized in the RER, separated from the cytoplasm by a membrane. How can the peptides get across this membrane? They are transported by a bi-molecular complex, TAP, which is expressed in the RER and transport is ATP-dependent. TAP is selective in the sizes of peptides it can transport and for peptides that have particular types of terminal amino acids the fit better into MHC class I grooves. The TAP transporter belongs to a family of molecules specialized to transport a variety of small molecules across membranes (ABC transporters), so yet again the adaptive system has adapted a pre-existing system for its own ends.

5.2.4.3 The Peptide-Loading Complex and Peptide Editing

Once the peptide has been transported into the RER it may bind to a MHC class I molecule. Newly formed MHC class I molecules are attached to TAP by a complex including a molecule called tapasin to form the peptide-loading complex. Other molecules associated with this complex act as chaperones to control the retention, stability and structure of the MHC class I molecules. Tapasin may also increase the probability of the MHC molecule meeting transported peptides because the spectrum of peptide–MHC complexes expressed in tapasin gene knock-out mice and wild-type mice is different. Further modification of the peptide may also occur via enzymes present in the RER such as ERAP1. If in the end a stable peptide–MHC complex is formed, it is transported out of the RER through the Golgi apparatus to the cell surface. It is important to appreciate that peptide editing occurs throughout the MHC class I peptide-loading pathway, from immunoproteasomes, through TAP and tapasin, to help bias the spectrum of peptides towards those that are best able to bind to MHC

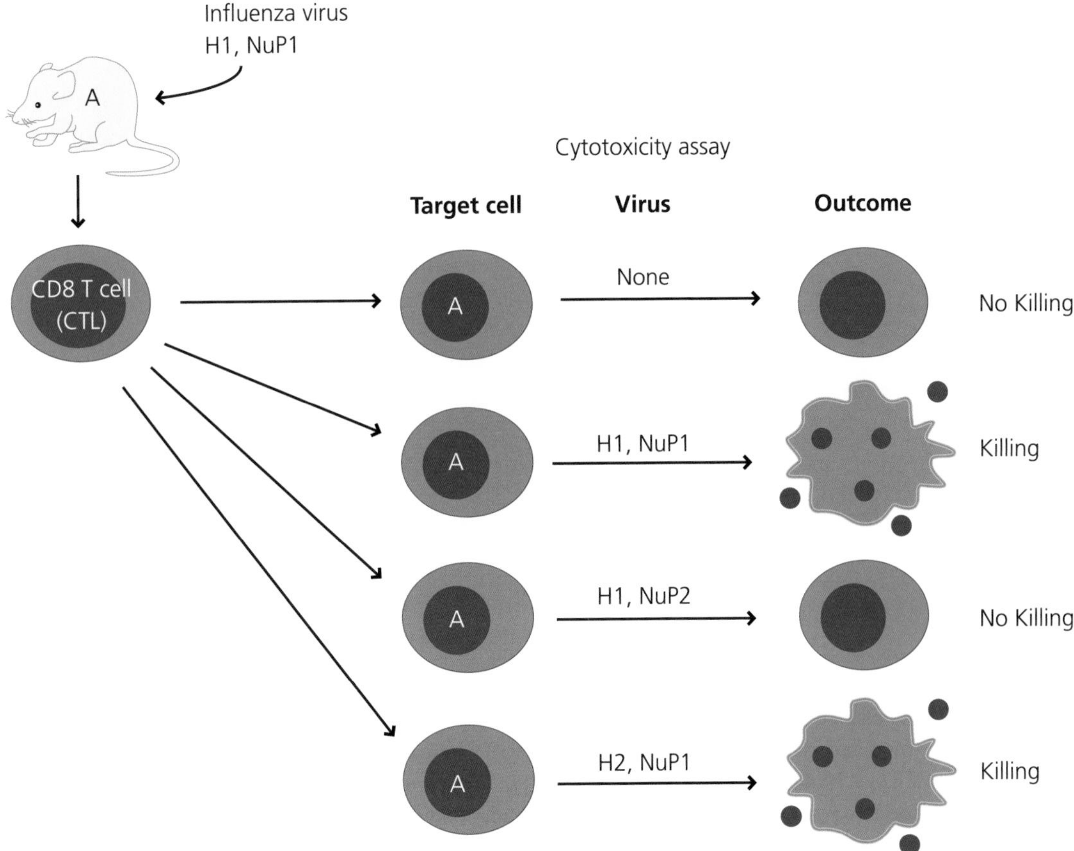

Fig. 5.8 Cellular localization of proteins recognized by CD8 cytotoxic T cells. Different strains of influenza virus express different forms of a relatively small number of proteins. One of these is the nucleoprotein (NuP) that is never expressed on the surface of infected cells. Mice were infected with one strain of the virus and their CTLs were tested for the ability to kill infected target cells in culture. The target cells were infected with strains of virus that shared different variant proteins with the original virus used to infect the mice. In some cases it was found that in order to render the target cells sensitive to killing, the only protein that needed to be identical in the two viruses was NuP. At the time this was a real puzzle. However, we now understand that the NuP protein is degraded intracellularly and peptides from it are about to MHC class I molecules and transported to the cell surface where these complexes can be recognized by CTL.

class I molecules, sometimes by mechanisms which are still unclear. Peptide editing also occurs at different stages of the MHC class II pathway (below).

5.2.5 Antigen Processing and MHC Class II Presentation

Around the time that Zinkernagel and Doherty were investigating immune responses to viruses (above), other groups were examining the genetic basis of the immune response to proteins in mice. To simplify the experiments they used synthetic polymers of small numbers of amino acids. When different strains of mice were immunized with these artificial antigens they could be classified as high or low responders in terms of the antibody responses they made on secondary stimulation. Again, classical breeding experiments showed that a small number of genes controlling responsiveness were coded for in the MHC. These were originally called immune response (IR) genes and their products therefore became known as IR-associated (Ia) antigens. They are now known to be classical MHC class II antigens. We might explain these early findings at the level of the APC, on the basis of whether or not the synthetic polypeptides were able to bind to the particular alleles of MHC class II molecules. However, we could also explain these results at the level of the T cell. Since thymic MHC class II molecules control the repertoires of T cells produced in different strains of mice (Section 5.5.2), they could determine whether or not the T cells that develop can recognize these peptide–MHC complexes.

5.2.5.1 Endosomal Degradation of Foreign Antigens

Peptides that bind to MHC class II molecules are generated in the endosomal pathway, in contrast to those that bind to MHC class I which originate in the cytosol. Most cells have mechanisms for taking up and degrading extracellular proteins in the endosomal pathway, and the adaptive immune system uses these mechanisms for the generation of MHC class II-binding peptides. Proteins are taken up by fluid-phase or receptor-mediated endocytosis, or by phagocytosis, and degraded by lysosomal proteases in the progressively-acidified vesicles of the endosomal pathway.

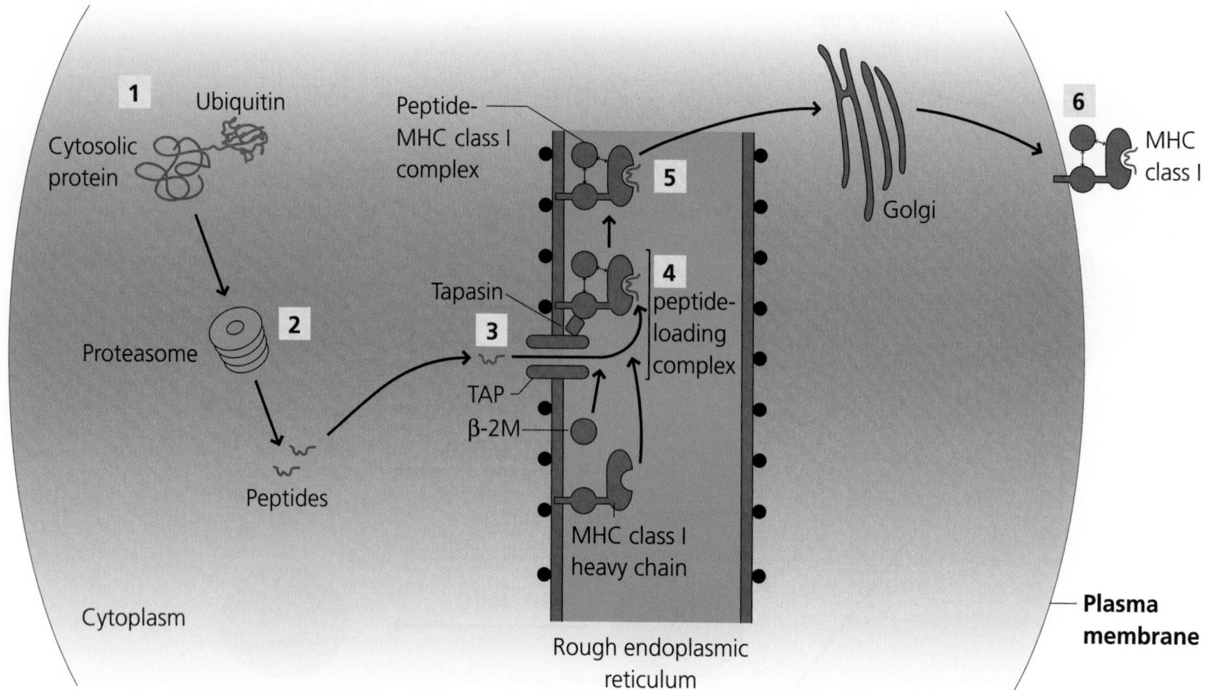

Fig. 5.9 The peptide–major histocompatibility complex class I pathway. (1) Cytosolic proteins, often attached to ubiquitin, (2) are targeted to proteasomes (modified as immunoproteasome; no shown) where they degraded. (3) Peptides of appropriate sizes are transported into the RER by ATP-dependent TAP transporters. (4) These are associated, via tapasin, to newly assembled MHC class I molecules, with associated β_2-microglobulin, to form the peptide-loading complex. Suitable peptides can bind to the MHC molecules, (5) which are then released and (6) transported through the Golgi apparatus to the plasma membrane. (The peptide-loading complex also contains chaperone proteins; not shown.)

The importance of lysosomal proteases was shown in experiments in which protease inhibitors were added to cultures of APCs with whole proteins and it was found that MHC class II presentation was inhibited. Lysosomal proteases act at low pH, and if the endosomal pH was raised by adding weak bases such as chloroquine, again presentation was inhibited. DCs and perhaps B cells do show some specialization in terms of the proteases used and their regulation, but the principles are very much the same whatever the cell type involved. Generally speaking, peptides that originate from the endosomal pathway, whether or not they derive from self proteins or infectious agents such as phagocytosed bacteria, are known as exogenous antigens. Generally, exogenous antigens bind to classical MHC class II molecules and are presented to CD4 T cells.

Q5.8. Chloroquine, added to cultures of APCs and protein, strongly inhibits antigen presentation. Are there possible reasons that might account for this other than inhibition of proteolysis? How might we attempt to set up control experiments to exclude the other possible reasons?

5.2.5.2 Peptide Loading in the MHC Class II Pathway

Exogenous proteins are broken down in the endocytic pathway, but how do they become bound to MHC class II molecules? MHC class II molecules are synthesized in the RER; however, unlike many proteins, MHC class II molecules do not travel directly to the plasma membrane. They need to interact with peptides and these peptides are present in the endosomal pathway. MHC class II molecules are therefore targeted to the endocytic pathway and this is a function of the **invariant chain** with which they are initially associated. The invariant chain is synthesized in the RER and forms a complex with newly synthesized MHC class II molecules. A targeting sequence in the cytoplasmic tail of the invariant chain directs the complex via the Golgi apparatus into the endosomal pathway, where it can sample peptides in different parts of the pathway. See Figure 5.10.

Another crucial function of the invariant chain is to block the peptide-binding grooves of MHC class II molecules. This is important because it ensures that peptides delivered into the ER for binding to MHC class I molecules (above) are unable to bind to newly synthesized class II molecules, and hence ensures these pathways are functionally discrete. However, once the class II molecules are delivered to the maturing endosomes, proteases start to cleave the invariant chain at specific sites until a single portion (CLIP) remains and continues to block the peptide-binding groove. Clearly peptides from foreign antigens cannot bind to class II molecules until this is released. Sometimes this can happen spontaneously. In other cases, however, CLIP release requires a different, MHC class II-related molecule – HLA-DM in humans or H-2M in mice – that is restricted to the endosomal pathway. In fact, this class II-related molecule (which is only present in endosomes) probably plays a more general catalytic role in peptide selection for binding to the groove, perhaps by maintaining a more open conformation of the peptide-binding

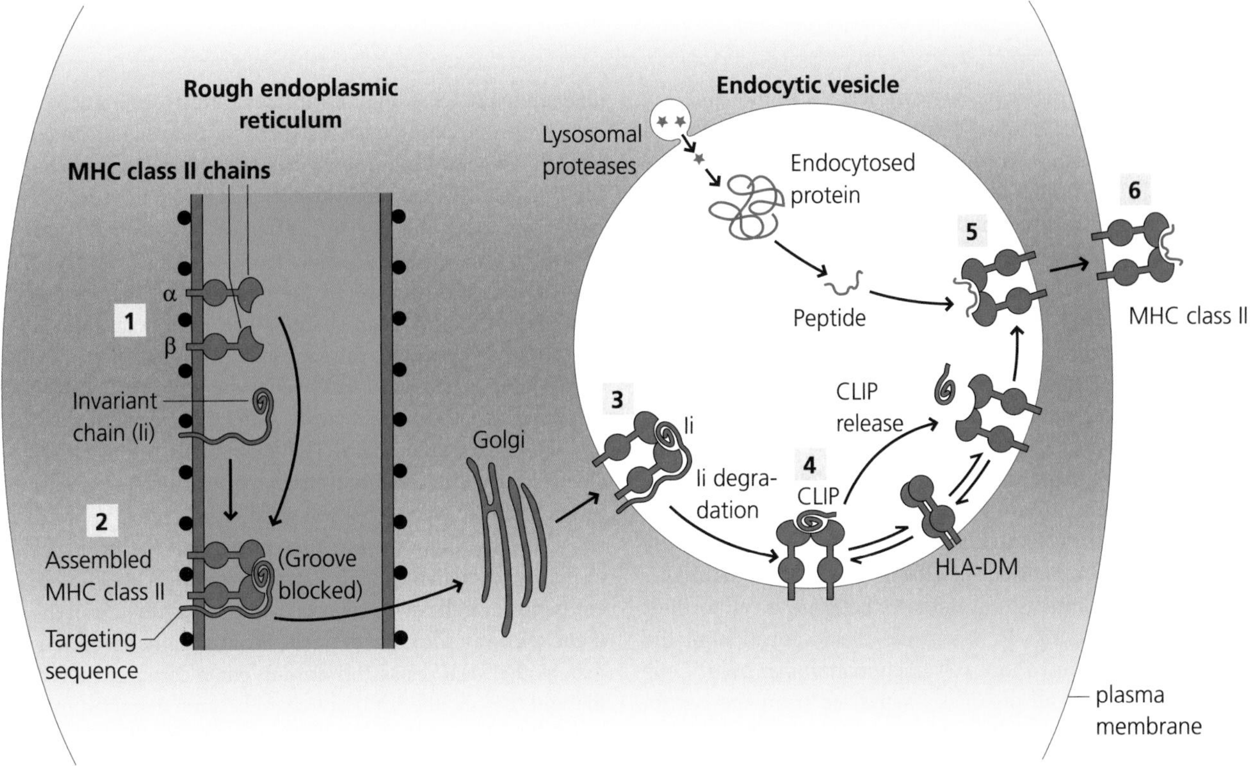

Fig. 5.10 The peptide–major histocompatibility complex class II processing. Proteins in the endocytic pathway are broken down to peptides by lysosomal proteases. Meanwhile, (1) MHC class II α, β and invariant chains are synthesized and (2) assembled in the RER. (3) The invariant chain both blocks the peptide-binding groove in the MHC molecule and targets the complex, through the Golgi, to the endocytic pathway. (4) Here, the invariant chain is digested in a step-wise manner to leave a small fragment, CLIP, which blocks the groove. (5) CLIP then either dissociates spontaneously or dissociation is facilitated by a MHC class II-related molecule, human HLA-DM (or mouse H-2M), which additionally promotes peptide exchange until a high-affinity peptide binds into the groove. (6) The peptide–MHC complex can then be transported to the plasma membrane.

groove until a high-affinity peptide binds. The peptide-bearing MHC class II molecule can now be transported to the plasma membrane where it can potentially be recognized by CD4 T cells. In some cells, particularly B cells and some DCs, the function of HLA-DM or H-2M is regulated by another class II-related molecule, HLA-DO in humans or H-2O in mice. See Box 5.4

Q5.9. Why might it be important to have mechanisms that increase the likelihood of peptides binding with high affinity to class II molecules in the acidic conditions of the endosomal pathway?

Q5.10. How might we attempt to show that the cytoplasmic tail of the invariant chain was responsible for targeting MHC class II molecules to the endosomal system?

5.2.6 Cross-Presentation

Most viruses are very selective in terms of the cell type(s) they can infect (tropism, Section 2.2.2.1). We know from depletion and adoptive transfer experiments that CD8 T cells are crucial for recovery from many viral infections, and also that CD8 T cells recognize MHC class I-peptide complexes derived from cytoplasmic proteins. We also know that, generally speaking, DCs are the principle, if not the only APCs that activate naïve T cells effectively *in vivo* (Section 5.4.1). This raises a real problem. How can a DC present antigens from all the different viruses that might infect a given species? It might be that DCs can be infected by all viruses; however, given viral tropism, this would imply that DCs need to express all the surface molecules used by viruses as receptors which is clearly very unlikely. At least one subset of DCs has, however, evolved a different solution to this problem. These cells can deliver exogenous peptides derived from the endosomal pathway (which would normally be delivered to class II molecules) onto MHC class I molecules instead. This process is known as cross presentation. This means that DCs can activate CD8 T cells to viruses that do not actively infect the DCs. Although several mechanisms have been proposed to explain cross-presentation, this is still an area of controversy. It has, for example, been suggested that, for this to occur, phagocytosed antigens need to be delivered into specialized endosomes and that soluble antigens, internalized by receptor-mediated endocytosis, might be targeted into one pathway or the other depending on the actual receptor involved. In some cases whole proteins, and in others

Box 5.4 Characterization of Major Histocompatibility Complex Class II Biosynthesis

MHC class II molecules are synthesized in the RER, but appear on the cell surface carrying peptides that are derived from proteins broken down in the endosomal system. How can this be studied? To understand this we need to investigate the life history of molecules within a cell – which sites they occupy, how they get there, how long they stay there. Several complementary approaches have been used. Pulse-chase labelling allows us to follow a cohort of proteins though a cell. Cells are grown in medium containing a radioactive amino acid, say [^{35}S]-methionine. This will become incorporated into any protein that is being synthesized at that time. If the radioactive residue is only present for a short time (i.e. the cells are pulsed with it) the cohort of proteins synthesized at this point in time will be labelled. If the cells are then cultured in the absence of the label for further periods (i.e. the label is chased) the cohort of labelled proteins will migrate though the cell to their final destinations. To determine the localization of the labelled protein cohort, cell fractionation can used. Cells are disrupted and organelles separated from the cytosol using their physico-chemical properties (e.g. density, size or charge). To isolate the labelled proteins of interest from fractionated cells, immunoprecipitation (Chapter 4, Fig 4.7) can be used. All the proteins are solubilized (e.g. by detergent) and beads coated with an antibody specific for the protein of interest are added to the mixture. The particular protein will bind to the beads and can then be separated by centrifugation (immunoprecipitation, sometimes called pull-down). In altered ionic conditions the protein will separate from the antibody and can be electrophoresed and characterized on a gel. Alternatively, to determine the location of molecules within a cell, microscopy is used. Fluorescence and electron microscopy can be used to localize molecules within cells by detecting the binding of antibodies labelled with fluorescent dyes or gold particles, respectively, in different organelles and compartments.

An example of the use of these approaches lies in the discovery of the invariant chain. When newly synthesized MHC class II molecules were immunoprecipitated, in addition to the two known (α and β) chains of the MHC class II molecule, another chain with different electrophoretic mobility was pulled down. The mobility of the α and β chains is variable in different strains of mice, reflecting their different allelic forms, but the third chain always showed the same mobility and was therefore called the invariant chain. By precipitating MHC class II molecules after different periods of chase it was possible to follow their biosynthetic pathway through the cell. What became clear was that, as the MHC class II–invariant chain complex moved thought the cell, the invariant chain became degraded in a step-wise manner. X-ray crystallography showed that newly formed invariant chain bound to the MHC class II molecule and that part of the invariant chain (CLIP) bound to and blocked the peptide-binding groove of the MHC molecule.

Q5.12. How might we investigate different molecules that may be associated with MHC class I molecules at different stages in their life cycle within a cell, such as in the RER?

peptides appear to be transported from the endocytic pathway into the cytosol. See Figure 5.11.

Q5.11. How might we determine if cross-presentation requires TAP for the delivery of peptides to the RER?

5.2.7 Subversion of Antigen Processing by Pathogens

If a pathogen carries genes that enable it to avoid, evade or otherwise subvert a particular component of immunity, this is suggestive evidence of the importance of that component in resistance to infection. No pathogen would carry these extra genes if they did not give it a selective advantage. Many pathogenic viruses can interfere with both MHC class I and II processing pathways. As just one example, human CMV carries more than 100 genes that are not essential for its growth in vitro and which could thus be involved in different mechanisms of immune subversion. In relation to MHC class I processing this virus encodes proteins which can, for example, inhibit peptide transport by TAP (US6), cause MHC class I proteins to be retained in the RER (US3) or induce newly synthesized MHC class I molecules to be translocated back into the cytoplasm and be degraded by proteasomes (US2 and US11). Of course if an infected cell down-regulates expression of MHC class I it becomes a potential target for NK cells (Section 1.4.4). Human CMV gets around this by additionally encoding a MHC class I-like decoy protein that is expressed on the surface of the infected cell and inhibits NK cell activation. Do not think, however, that human CMV only uses MHC class I subversion: it can also interfere with MHC class II processing and presentation, with the activity of cytokines and chemokines, and with complement and antibody. Many other viruses carry a similar range of immune subversion genes.

5.2.8 Presentation of Lipid Antigens by CD1 Molecules

Microbial antigens are not limited to simple proteins. They may contain carbohydrates (e.g. glycopeptides) or lipids (e.g. glycolipids). Whereas antigenic peptides and some glycopeptides are presented by classical MHC molecules to

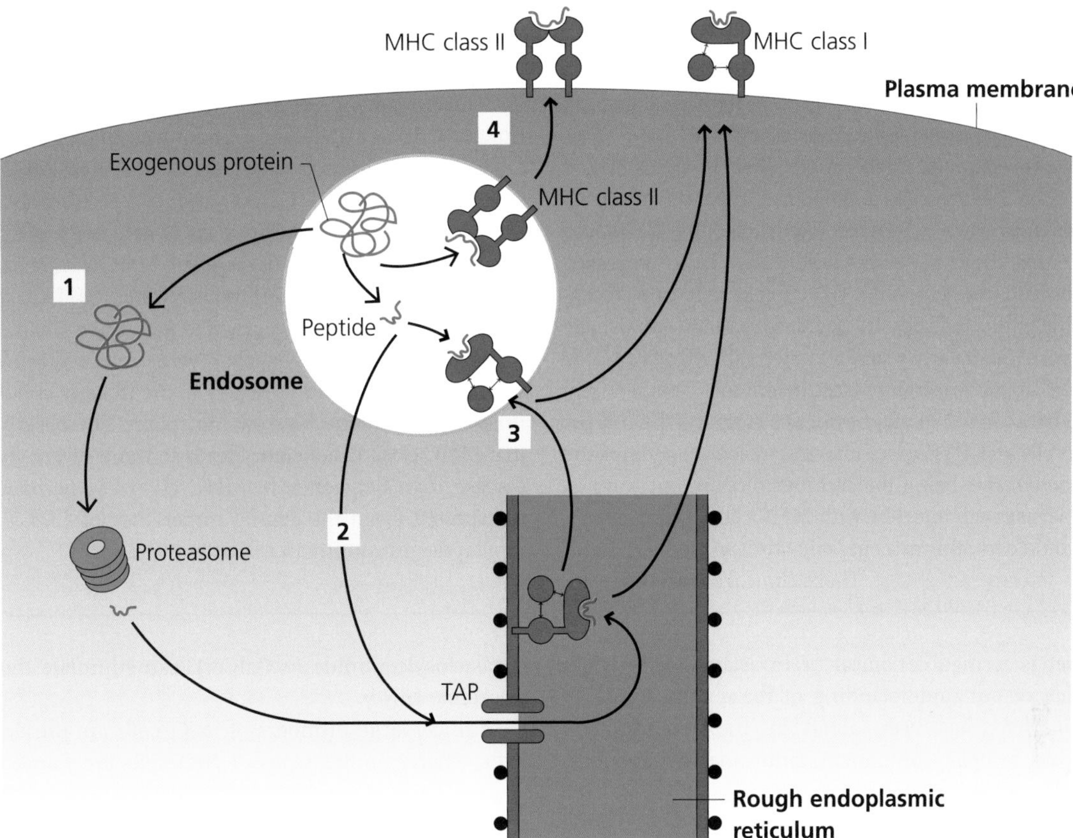

Fig. 5.11 Cross-presentation. In some DCs, peptides from endocytosed proteins can be displayed on MHC class I molecules. This is known as cross-presentation. The mechanisms are still not fully understood. (1) Proteins or (2) peptides may exit the endocytic pathway into the cytosol. (3) Alternatively, endosomes may acquire MHC class I, possibly from the RER, permitting direct peptide binding. (4) Cross-presentation enables peptides derived from microbes in peripheral tissues to be presented on MHC class I molecules to CD8 T cells, without the need for the DCs to be actively infected. Such DCs can also present exogenous peptides on MHC class II, additionally permitting CD4 T cell activation through the normal route.

conventional T cells, lipid-containing antigens can be presented by a different set of molecules, the CD1 molecules, to certain types of non-conventional T cells.

5.2.8.1 CD1 molecules

CD1 molecules are MHC-like in structure and function, but they are encoded outside the MHC. Structurally, in terms of their domains and their association with β_2-microglobulin, they are very similar to classical MHC class I molecules. The key difference is, however, that CD1 molecules possess hydrophobic antigen-binding grooves that can accommodate lipid-containing molecules such as glycolipids. Functionally, however, CD1 molecules resemble MHC class II molecules, in that they acquire their lipid-containing antigens from endosomal compartments. These compartments contain not just proteases but also lipases, meaning that both protein- and lipid-containing molecules from phagocytosed microbes can be degraded. Hence, CD1 molecules can acquire lipid-containing molecules from endosomes. However, we still have much to learn about the function of CD1 molecules and their roles in immunity.

In humans, five genes encode different groups of CD1 molecules (CD1a–e) divided into two groups; CD1e is thought to play a regulatory role in lipid loading. Some of these molecules recycle between the plasma membrane and early or late endosomes, or between different endosomes. This applies particularly for the group 1 molecules (CD1a–c) which can subsequently present microbial lipids, such as those from mycobacteria, to some γδ T cells as well some poorly understood non-conventional αβ T cells. The natural ligands for the group 2 (CD1d) molecules are largely unknown. What is becoming clear, however, is that CD1d molecules, which can for example be expressed by DCs or macrophages, present these lipid antigens to a specialized subset of non-conventional iNKT cells (below). The same appears to be the case in mice, which in fact have only two genes for CD1, both related to human CD1d. See Figure 5.12.

5.2.8.2 Invariant NKT (iNKT) Cells

Some lymphocytes, called NKT cells, express αβ TCRs, CD4 and a molecule called NK1.1 which was first identified on NK cells (Section 4.4.5). One subset of these is unusual in that it expresses a TCR that is relatively invariant: these contain canonical arrangements of $V_\alpha 24$–$J_\alpha 18$ with $V_\beta 11$ in humans and $V_\alpha 14$–$J_\alpha 18$ with $V_\beta 8$ in mouse. Hence,

Box 5.5 Bare Lymphocyte Syndrome

Some, very rare, patients who present with an increased incidence of infection are found to have very reduced or absent expression of MHC molecules. In cases where MHC class I expression is defective (Type 1 syndrome), patients may have a mutant TAP molecule. We would predict that these patients would suffer from increased viral infection because their CD8 T cells could not recognize virally infected cells. In fact, these patients are usually asymptomatic in early years, but later suffer from chronic bacterial upper respiratory tract infections. This is a real puzzle because much other evidence supports the role for CD8 T cells and MHC class I in viral infection, perhaps the most persuasive being the number of different ways in which viruses can interfere with MHC class I expression – they would carry this extra genetic burden if it did not give them a selective advantage. These clinical observations are difficult to explain but serve yet again to illustrate how much we do not understand about immunology.

Patients with absence of MHC class II expression (Type 2 syndrome) or both MHC class I and II (Type 3) are also very rare. In these cases it MHC class II deficiency is usually due to a defect in genes that regulate MHC class II expression particularly the class II transactivator (CIITA, a member of the NOD-like receptor (NLR) family; Chapter 4). These patients generally show severe combined immunodeficiency (SCID) and are likely to die in early childhood unless given a bone marrow transplant. The observation that MHC class II deficiency leads to more severe clinical disease than deficiency in MHC class I is perhaps not unexpected given the central importance of CD4 T cells in adaptive immunity to infection.

this subset is termed canonical or invariant NKT (iNKT) cells. Much of our understanding of these cells, which is still incomplete, has been aided by the serendipitous finding that a lipid component from marine sponges, α-galactosylceramide (α-GalCer), can stimulate these cells to secrete IFN-γ.

In mice, large numbers of iNKT cells are present in the liver, whereas other types of NKT cells are found in bone

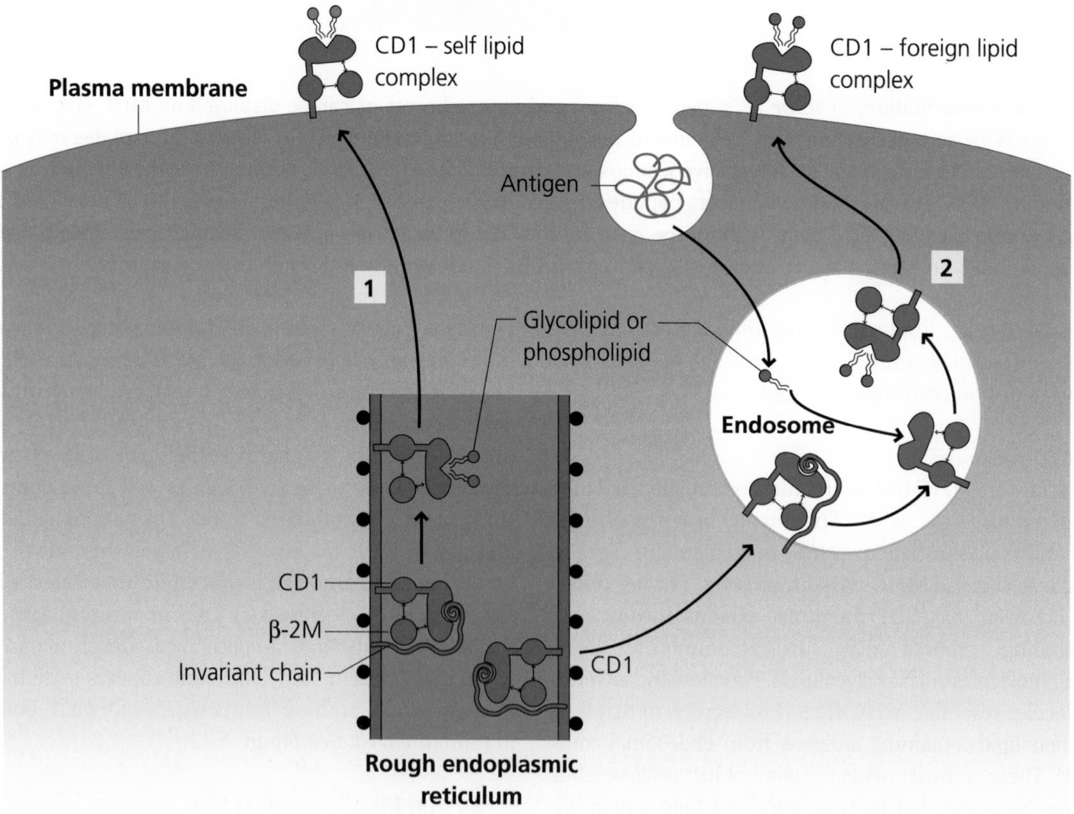

Fig. 5.12 The lipid–CD1 pathway. CD1 molecules are structurally similar to classical MHC class I molecules, but possess a hydrophobic groove, permitting lipid-containing molecules to bind. Unlike classical MHC class I molecules, CD1 molecules may associate with the invariant chain and are targeted to the endosomal pathway. They may acquire lipid-containing molecules at different stages of their life history, including the RER (1) and the endosome (2). Different CD1 molecules traffic through different endosomes. The natural ligands of CD1 molecules may include glycolipids derived from mycobacterial cell walls and phospholipids. Glycolipids bound to CD1 molecules can be presented to non-conventional T cells such as iNKT cells in the case of human CD1d.

marrow, spleen, lymph nodes or gut epithelium. iNKT cells resemble activated or effector cells, and rapidly secrete IFN-γ and interleukin (IL)-4 when they are stimulated in vitro with, for example, α-GalCer. The real function of iNKT cells (and indeed of other types of NKT cells) in host defence is, however, unclear. The most direct evidence comes from studies of knock-out mice lacking the $J_{\alpha}18$ gene, in which iNKT cells do not develop. These mice are unable to clear infections by a specific type of microbe, the α-proteobacteria, which are Gram-negative organisms that do not express lipopolysaccharide (LPS). Some CD1d-presented glycoslyceramides from these bacteria can activate iNKT cells in vitro. It is an intriguing hypothesis that iNKT cells might have evolved to combat microbes that cannot act as TLR agonists or that evade TLR recognition. However, other studies in NKT- or CD1-deficient mice have suggested iNKT cells may also be involved in defence against mycobacteria and malarial parasites in the liver.

Q5.13. How might we attempt to identify a natural ligand for a CD1 molecule?

5.3 T Lymphocyte Activation

5.3.1 Anatomical Basis of T Cell Responses

Having introduced T cells, the MHC and APCs, we will now briefly revisit the activation of T cells in vivo, discussed in more detail in Section 3.4.2.3. In contrast to innate immune responses, which are initiated in peripheral tissues, all (primary) T cell responses are initiated in secondary lymphoid organs. In brief, naïve $\alpha\beta$ T cells, after leaving the thymus, patrol the secondary lymphoid organs, probably quite randomly. When they encounter appropriate molecules on high endothelial venules (HEVs) in lymph nodes or Peyer's patches, or on sinusoids in the spleen, they adhere and migrate into the organ. This migration is directed by adhesion molecules expressed on the lymphocyte, e.g. L-selectin, and the endothelium, e.g. peripheral node addressin (PNAd). They spend a few hours in the T cell areas of the organs, and then leave and migrate, directly (spleen) or indirectly via lymph (nodes and Peyer's patches) into the blood. At the same time, DCs are migrating from peripheral sites of infection to the T cell areas of secondary lymphoid organs.

When the T cells arrive in T cell areas they form short-term adhesions to the DCs. If their antigen receptors do not bind to any combinations of MHC and peptide on the DCs surface with sufficient affinity, they release themselves from the DCs and move on. In this way they may interrogate several DCs before they leave the organ, enter blood and then migrate to other secondary lymphoid tissues. This pattern of lymphocyte recirculation may go on for very long periods. If, however, a T cell does recognize a peptide–MHC complex with sufficient affinity, it makes a longer-lasting interaction with a DC. These interactions can by visualized by the technique of intra-vital microscopy (see Box 3.3) in which cellular events occurring in living organs and tissues can be observed microscopically over periods of hours. They may ultimately result in the T cell becoming activated.

Most activated T cells leave the secondary lymphoid tissue (although follicular T helper cells migrate to the B cell follicles) and travel back to peripheral tissues. In general, activated T cells migrate to inflamed tissues in a non-antigen-specific manner. In some cases, however, their migration may be directed to particular types of tissue. Thus, CD4 T cells activated in mucosal lymphoid tissues tend to migrate back to mucosae; they express the integrin $\alpha_4\beta_7$ which recognizes the mucosal addressin cell adhesion molecule (MadCAM) on mucosal endothelium. Similarly, T cells activated in skin-draining nodes tend to migrate back to the skin; they express different receptors for the vascular addressins (Section 3.3.3.2) in this site. Having been activated the T cell can now recognize any cell that expresses its cognate peptide–MHC combination (not just DCs) and carry out its function.

Q5.14. Endothelial cells express MHC class I molecules. Might this have a role in directing the migration of antigen-specific CD8 T cells, in addition to the usual post code signals (Section 3.3.3.2)?

Before they can function in immunity, T cells first need to be activated and then instructed as to which effector functions that should acquire, depending on the type of infection that has occurred. This is the function of both the TCR and other molecules on the T cell surface that generate additional activation signals, and of cytokines which contribute significantly to regulating T cell differentiation. It is important to realize that in the steady state, in the absence of inflammation or other forms of danger, T cells that recognize antigen usually die (leading to tolerance) or in the case of some CD4 T cells, may become regulatory cells. It is only in the presence of adequate costimulation that can they become fully activated effector cells.

5.3.2 Molecular Requirements for T Cell Activation

The antigenic specificity of T cell activation is determined by the interaction of the TCR and the peptide–MHC complex. What might be the consequence if this was the only requirement for T cell activation? Remember that most MHC molecules have bound self peptides and that all DCs will be expressing peptides from normal cellular components. In the case of those migrating from the gut, they will also express peptides from food proteins or commensal bacteria. Would there not be a real chance of generating autoimmune diseases or food hypersensitivity if the only requirement for T cell activation was peptide–MHC recognition? It is almost certainly a good thing that T cell activation is rather more complex. It is now clear that a naïve or resting T cell requires at least two sets of signals to become activated. The first set of signals is delivered by peptide–MHC through the TCR and its associated CD3 complex. This is sometimes called Signal 1. The second set of signals is known as costimulation and can involve several different molecules on the APC interacting with complementary molecules on the T cell. This is generally referred to as Signal 2. Other cell-associated molecules and secreted factors (cytokines) determine how an activated T cell will differentiate, e.g. whether it will become polarized to T_h1, T_h2 or T_h17 (Signal 3), and where it will migrate (the latter could perhaps be termed Signal 4). See Figure 5.13 and Box 5.6.

5.3.3 Molecular Anatomy of T Cell Activation

When T cells were examined using fluorescence microscopy during the activation process, it was seen that each formed

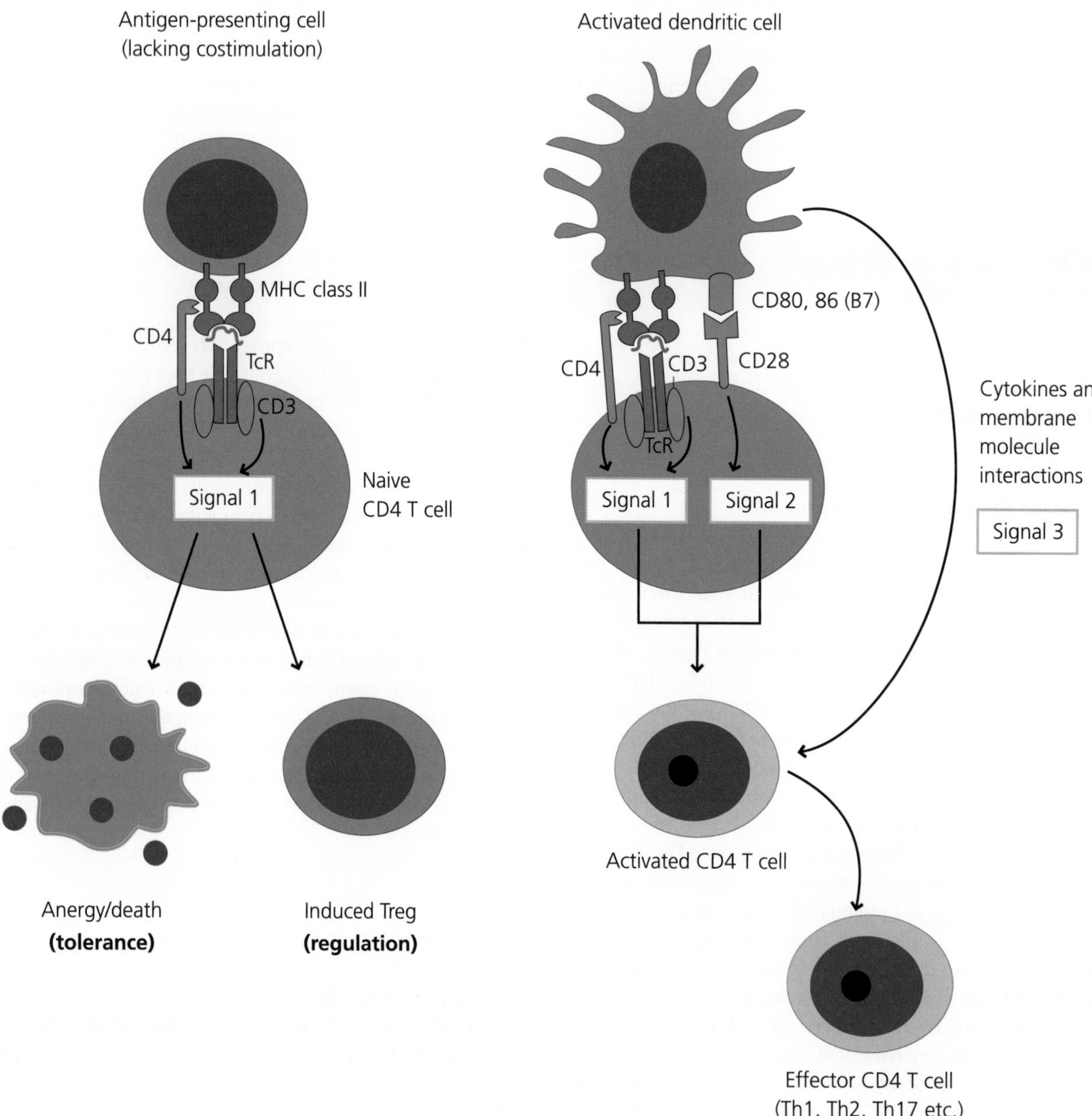

Fig. 5.13 Costimulation of CD4 T cells. Naive CD4 T cells may become unresponsive (anergic), die (apoptose) or differentiate into regulatory T cells (T_{reg}) if they recognize a peptide–MHC complex on an APC in the absence of further stimuli (Signal 1 alone). This can lead to antigen-specific tolerance. If however the T cell additionally receives positive signals through CD28, from B7 family members CD80 and/or CD86, it will become full activated (Signal 1 plus Signal 2). (In contrast, negative signals delivered by molecules such as CTLA-4 or PD1 can inhibit T cell activation; not shown.) Activated (mature) DCs are the usual source of such costimulatory signals for initial, naïve CD4 T cell activation. Other signals regulate the type of effector response that develops, such as into T_h1 versus T_h2 cells (Signal 3). The T cells may also be instructed as to where they should migrate (Signal 4: not shown)

Box 5.6 Characterization of the T Cell Receptors

The molecular nature of the TCR was controversial for many years – was it an immunoglobulin or an immunoglobulin-like molecule, or something completely novel? The answer came from quite different approaches.

In one approach researchers hypothesized that the TCR would be unique to a single clone of T cells, and that monoclonal antibodies might be generated that were specific for its unique antigen-binding site. To test this, large numbers of T cells all expressing the same TCR were needed. This was achieved by using T cell hybridomas: single T cells of known specificity are fused with T cell lymphoma (tumour) cells in a procedure similar to that used in making monoclonal antibodies with B cells. These T cell hybridomas can be grown as clones and they divide spontaneously, allowing the production of very large cell numbers. Panels of monoclonal antibodies were made against a single hybridoma and screened for antibodies that bound only to the hybridoma they had used to immunize. They found such antibodies, which were termed clonotypic, and were able to use these to isolate the molecule to which they bound and to gain structural information. This showed that the molecules were immunoglobulin-like, but were not antibodies.

In another approach, researchers hypothesized that the TCR would be formed by a genetic rearrangement similar to that already known to occur in B cells, that it would not be expressed in B cells, but that the vast majority of proteins would be identical in B and T cells. They therefore devised strategies to identify such genes. One example is subtraction hybridization in which cDNA derived from T cell mRNA is hybridized (bound) to B cell mRNA and cDNA that does not bind to B cell mRNA, but which does bind to T cell mRNA, was examined for rearrangement. Two groups used this kind of approach to identify TCR genes in humans and mice, respectively. When another group sequenced a human TCR chain identified by clonotypic antibodies, the sequence was identical to that predicted from the genetic analysis. Thus, two quite distinct approaches contributed the discovery of the TCR. It was some time later that the crystal structure of the TCR was solved and it became clear how it interacted with the peptide–MHC complex.

very tight connections, called the immunological synapse, with APCs. This synapse has a characteristic structure consisting of concentric rings in which different molecules are arranged in different areas of the synapse. In the central area are the TCRs, the associated CD4 or CD8 molecules interacting with MHC, and the costimulatory molecule CD28 interacting with CD80 and CD86. Outside this central area are the adhesion molecules, LFA-1 on the T cell interacting with intercellular cell adhesion molecule (ICAM-1) on the APC, for example. Importantly, CD45, a molecule involved in inhibition of activation, is a very large molecule and is excluded from the central area of the synapse, as are other large molecules such as CD43. The picture is complicated because of the identification of lipid rafts that contain focal collections of molecules that may coalesce to form the synapse. The synapse is not a fixed collection of molecules. Molecules such as the TCR and MHC molecules are continually entering and leaving, presumably to facilitate the meeting of TCR with its cognate peptide–MHC. See Figure 5.14.

Q5.15. How might we attempt to estimate the concentration of a particular secreted molecule in the immunological synapse?

Q5.16. Might the immunological synapse have roles in infection other than being involved in T cell activation or function

5.3.3.1 Role of the TCR and Other Molecules in T Cell Activation

The $\alpha\beta$ chains of the TCRs of conventional T cells are associated with several other proteins in the cell membrane – collectively called CD3 – the main function of which is signal transduction. CD3 consists of six polypeptide chains (one γ, one δ, two ε and two ζ chains). TCR binding to peptide–MHC ultimately leads to phosphorylation of immunoreceptor activation motifs (ITAMs) in the cytoplasmic portions of the CD3 chains. Subsequent binding of intracellular signalling components to these ITAMs then triggers the biochemical cascade that leads to the changes in cellular functions that are recognized as the early stages of T cell activation.

Many signalling receptors undergo dimerization as an essential component of signalling. This is probably not possible for TCRs as the frequency of cognate peptide–MHC complexes is too low. Rather it seems that the concentration of TCRs and MHC molecules in the synapse permits enough cognate interactions to occur to add together and generate the required overall strength of signalling. For naïve T cells this is estimated to require hundreds of cognate interactions, but for memory/activated T cells as few as one to ten interactions may suffice. In addition to the TCR and CD3, CD4 and CD8 also play roles in T cell activation. Their function is to bind directly to MHC molecules (MHC class II and I, respectively). CD4 and CD8 are associated with an intracellular tyrosine kinase, Lck, which plays a very important role in initiating intracellular signalling. The interactions of CD4 or CD8 with MHC molecules, and LFA-1 with ICAM, as well as other interactions between adhesion molecules, may also

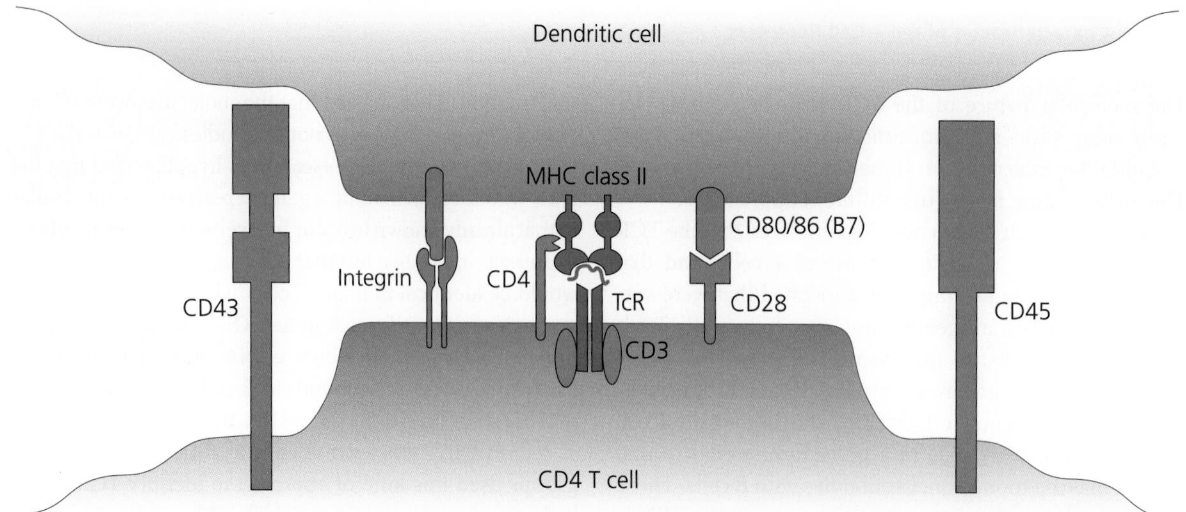

Fig. 5.14 The immunological synapse. When a CD4 T cell interacts with a DC the cytoskeleton of both cells reorganizes to form a tight cell–cell interaction zone. This is the immunological synapse in which different sets of molecules are arranged in concentric rings. Molecules involved in T cell activation – the TCR and MHC, and costimulatory molecules – are concentrated in the central areas and are surrounded by adhesion molecules. Some large molecules, such as CD45 which has inhibitory functions, and CD43, which may serve as a barrier molecule preventing close contact between cells, are excluded from the synapse. (Such synapses also form when other cells of adaptive immunity interact, e.g. CD4 T cells with B cells and CD8 T cells (CTLs) with target cells; not shown.

influence the number of TCR–MHC interactions needed for activation to be triggered. It is important that T cell activation is tightly regulated. CD45 is a protein expressed in different isoforms on all leukocytes. In T cells CD45 acts as a protein tyrosine phosphatase responsible in part for keeping T cell activation in check by regulating phosphorylation by Lck and other kinases.

5.3.3.2 Intracellular Signalling in T Cell Activation

Signal 1: CD4 T Cell Recognition of Peptide–MHC Complexes How does the generation of a signal via the TCR and CD3 lead to changes in gene expression? In essence the ITAM motifs are phosphorylated by Lck, associated with CD4 or CD8. This permits binding and activation of other molecules such as ZAP70 which phosphorylates a large, membrane-associated scaffold protein called LAT. This binds a number of adapters and signalling molecules, leading to many downstream events including activation of essential transcription factors. The main outcomes include changes in the cytoskeleton, leading to tighter interactions between the T cell and the APC, and the initiation of cytokine secretion – particularly IL-2 – the central growth factor for T cells in the early stages of their activation. See Figure 5.15.

The changes induced in the cytoskeleton are primarily in actin polymerization. A central molecule in this chain is Wiskott– Aldrich syndrome protein (WASP), which is expressed in haematopoietic cells. In Wiskott–Aldrich syndrome, the WASP protein is defective and patients are susceptible to bacterial, viral and fungal infections. Defects in several cell types may contribute to this disease, but T cell activation is clearly affected. The avidity of the T cell–APC interaction is also increased by signalling from the TCR which leads to increased affinity of LFA-1–ICAM adhesion.

IL-2 secretion is regulated by the activity of several transcription factors. Different pathways downstream of LAT generate these factors. A phospholipase enzyme hydrolyses a membrane phospholipid to generate two lipid mediators, diacylglycerol (DAG) and inositol triphosphate (IP_3). These stimulate separate pathways. DAG stimulates two pathways that ultimately lead to activation of the transcription factors AP-1 and NF-κB. IP_3 stimulates waves of calcium fluxes in the cell, and the calcium activates the protein phosphatase, calcineurin, which dephosphorylates and activates the transcription factor NF-AT. Activation of IL-2 gene transcription requires simultaneous occupancy of its gene promoter region by different transcription factors; this means that IL-2 is synthesized only when there is strong signalling from the TCR. This array of signals is not, however, sufficient to induce IL-2 secretion. Instead there is a requirement for further signals generated by costimulation (Signal 2).

> **Q5.17.** Why does robust activation of IL-2 synthesis require strong signalling?

Signal 2: Costimulation of CD4 T cells The most important costimulatory system involves two molecules on the APC, CD80 and CD86 (collectively known as B7), which bind to a functionally important molecule that is constitutively expressed by naïve T cells, CD28. Signalling through CD28 is thought to lead to more robust T cell activation by amplifying existing pathways that are initially triggered by TCR ligation and probably by inducing additional pathways. For example, the cytoplasmic tail of CD28 may become phosphorylated by

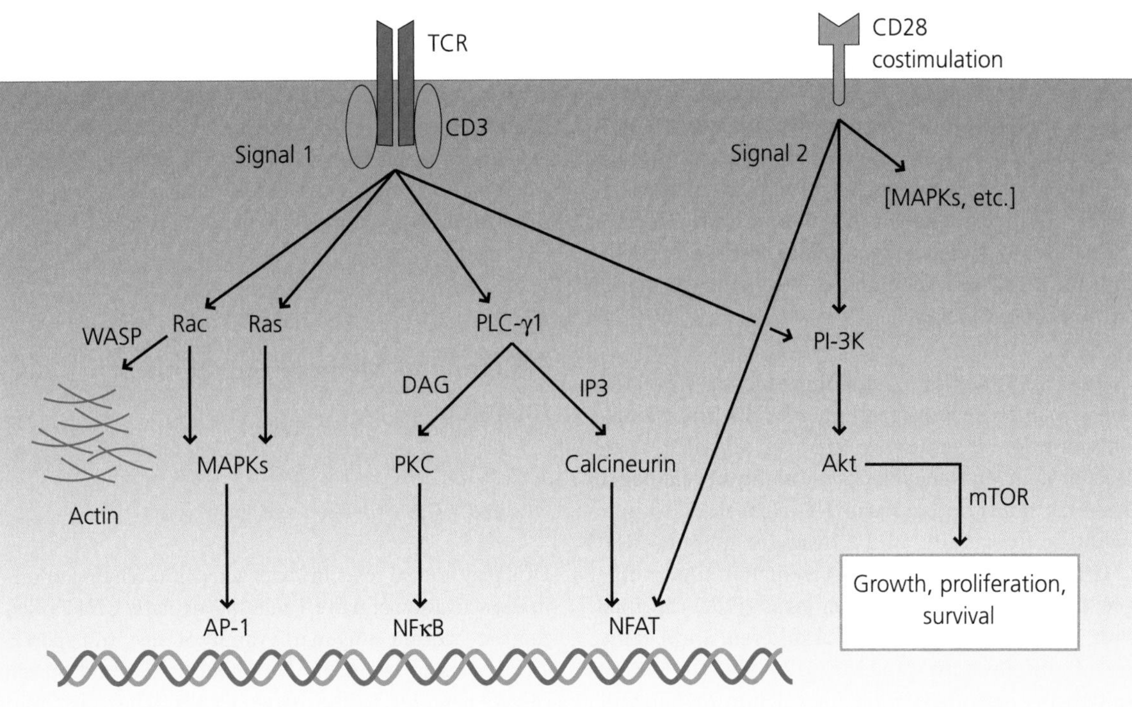

Fig. 5.15 Intracellular signalling pathways for T cell activation. A highly over-simplified scheme of some major intracellular signalling pathways involved in T cells. Signal 1. Ligation of peptide–MHC complexes (and co-recognition by CD4 or CD8; not shown) initiates signalling via the CD3 complex which activates multiple downstream pathways. Several of these lead to the activation and nuclear translocation of transcription factors AP-1, NF-κB and NF-AT, the latter two via pathways involving DAG and IP_3 respectively. Rac is involved in activation of the actin cytoskeleton (which also involves WASP). Signal 2. Additional signalling from CD28 increases activation of key transcription factors and the PI3 kinase pathway (involving mTOR) which regulates T cell growth, proliferation and survival. See the text for further details.

TCR-activated kinases and subsequently activates the phosphatidylinositol 3-kinase (PI3) kinase/Akt pathway. This pathway involves the "mammalian target of rapamycin" (mTOR) which regulates protein synthesis during the cell cycle. Overall, signalling through CD28 fully activates the IL-2 gene, leading to robust and sustained cytokine secretion. CD28 thus mediates the delivery of positive costimulatory signals to the T cell. In addition, signalling through CD28 is involved in feedback loops that can amplify function. For example, they lead to induction of CD40 ligand on the activated CD4 T cell that can then interact with CD40 on DCs to stimulate the expression of other costimulatory molecules (see CD8 T cells, below). CD40 and CD40 ligand are members of the tumour necrosis factor (TNF)–TNF receptor family.

T cell responses eventually need to be turned off. Activated T cells increasingly express a molecule called CTLA-4 (another member of the immunoglobulin superfamily) which also binds B7 and competes with CD28. In contrast to CD28, CTLA-4 mediates the delivery of negative costimulatory signals to the T cell, inhibiting further responses, as do several other receptors such as PD-1 which binds another B7 family member PD-1 ligand, B7H1. The importance of CTLA-4 in negatively regulating T cell responses is underlined by CTLA-4 gene knock-out mice which exhibit uncontrolled lymphocyte proliferation and by the observation that CTLA-4 variants may be present in some autoimmune diseases. Patients and mice with defects in Fas or Fas ligand, involved in lymphocyte apoptosis, may show similar lymphoproliferative, autoimmune conditions.

5.3.3.3 Defects, Evasion and Manipulation of T Cell Signalling

How do we know that the molecular interactions described above are important? Evidence comes from several sources including human and mouse immunodeficiencies, the actions of immunosuppressive drugs, and the strategies used by pathogens to interfere with signalling.

Immunodeficiencies Mutations in T cell signalling molecules can lead to human immunodeficiencies. One example is WASP (noted above) which is crucial for both T and B cell immunity and defects lead to recurrent bacterial, viral and fungal infections. Another is defects in ZAP 70 which lead to an absence of circulating CD8 T cells and unresponsive CD4 T cells, causing one form of severe combined immunodeficiency (SCID).

Immunosuppressive Agents Glucocorticoids used for immunosuppression have many effects but these include inhibition of NF-κ-B activation. Cyclosporin and tracrolimus

(FK506) are both calcineurin inhibitors that are used as immunosuppressants to prevent transplant rejection. These drugs inhibit NF-AT that is normally activated after TCR binding to peptide–MHC complexes. Rapamycin (sirolimus), another drug used to prevent rejection, inhibits mTOR, which is activated during costimulation through CD28. The inhibitory and immunosuppressive effects of these agents demonstrate the importance of these respective pathways for T cell activation and of T cells for mediating allograft rejection; Chapter 7.

Microbial Evasion Strategies A vast number of pathogens can interfere with intracellular signalling in T cells. For example, Helicobacter pylori, the causative organism of gastric ulcers, encodes a protein VacA that blocks the ability of calcineurin to activate the transcription factor NF-AT. (VacA also interferes with the functioning of the invariant chain of MHC class II in antigen processing and presentation.) As another example, human immunodeficiency virus (HIV) can interfere with T cell activation in several different ways. Thus, binding of the envelope protein gp120 to CD4 inhibits protein tyrosine phosphorylation and calcium mobilization after TCR ligation, and NEF can bind to several signalling proteins including Lck and PI3-kinase. Herpes simplex virus (HSV), after infecting a cell, blocks the downstream signalling events that follow LAT activation. It has been suggested that some viruses, including HIV and the respiratory syncytial virus, can also interfere with the formation of the immunological synapse and thus inhibit T cell activation. That these pathogens encode genes capable of causing such interference is strong evidence for the importance of these signalling pathways and T cell activation for defence against infection.

Finally, we will note that some pathogens can stimulate robust polyclonal T cell activation, rather than inhibition of function. Pyogenic bacteria such as Staphylococcus aureus, for example, secrete toxins called exotoxins. These can bind to shared regions of different TCR β chains and to conserved regions of MHC class II molecules and act as superantigens, stimulating considerable CD4 T cell activation and the production of high levels of cytokines (a cytokine storm) that can lead to shock and even death. To what extent this represents a microbial subversion or evasion mechanism, or is simply an unfortunate coincidence for the host, is not entirely clear.

5.4 Effector and Memory Functions of T Cells in Infection

5.4.1 DCs and T Cell Activation and Polarization

DCs play crucial roles in T cell activation. Their central role is in the activation of naïve T cells, particularly CD4 T cells. To do this they acquire antigen in a variety of ways, transport it to the T cell areas of secondary lymphoid tissues and present processed peptides to the naive T cells. They also transduce information from the periphery that regulates T cell differentiation. Several types of DC with different functions and properties have been identified. See Figure 5.16 and Box 5.9.

5.4.1.1 Classical DCs

Classical DCs are formed from bone marrow precursors. They may arise from both common lymphoid progenitors (CLPs) and common myeloid progenitors (CMPs). DCs originating from both lineages can migrate from blood into peripheral tissues where they spend a short time (a few days) before migrating via afferent lymph to lymph nodes. In peripheral tissues DCs are specialized to acquire antigens and hence express a variety of endocytic receptors; these antigens can subsequently be expressed as peptide–MHC complexes for T cell recognition. They also express pattern recognition

Box 5.7 Dendritic Cells and the Initiation of T Cell Responses

A technique has been developed in mice that permits DC to be killed selectively. Diphtheria toxin (DTX) kills human cells because it can bind to a particular molecule, the DTX receptor, on the cell surface. Mouse cells do not express the DTX receptor and are therefore normally resistant to the toxin. It is, however, possible to generate transgenic mice that express the human receptor for DTX. If the mice are fed DTX (e.g. in their drinking water) any cells expressing DTX will be killed. Hence, if DTX could be expressed specifically in DC, it would be possible to kill them selectively and to assess the subsequent capacity of these mice to mount different types of immune responses. To have a particular protein expressed in a cell it has to be under the control of a specific promoter – a stretch of DNA that lies upstream of the gene and controls transcription of the gene. CD11c is a protein expressed (almost) exclusively on DCs. Hence, genetically engineered mice were created in which the human DTX receptor gene was placed under the control of the CD11c promoter. Thus, the gene was expressed almost exclusively in DCs and therefore only DCs expressed the DTX receptor. When researchers injected these mice with DTX the DCs were indeed killed. When they tried to immunize these mice after killing the DCs, they could not generate T cell-dependent (TD) responses. Very rare natural deficiencies of DC (often including other cell types such as monocytes) have recently been identified in humans. These deficiencies lead to increase susceptibility to infection, and in some cases, autoimmune disease.

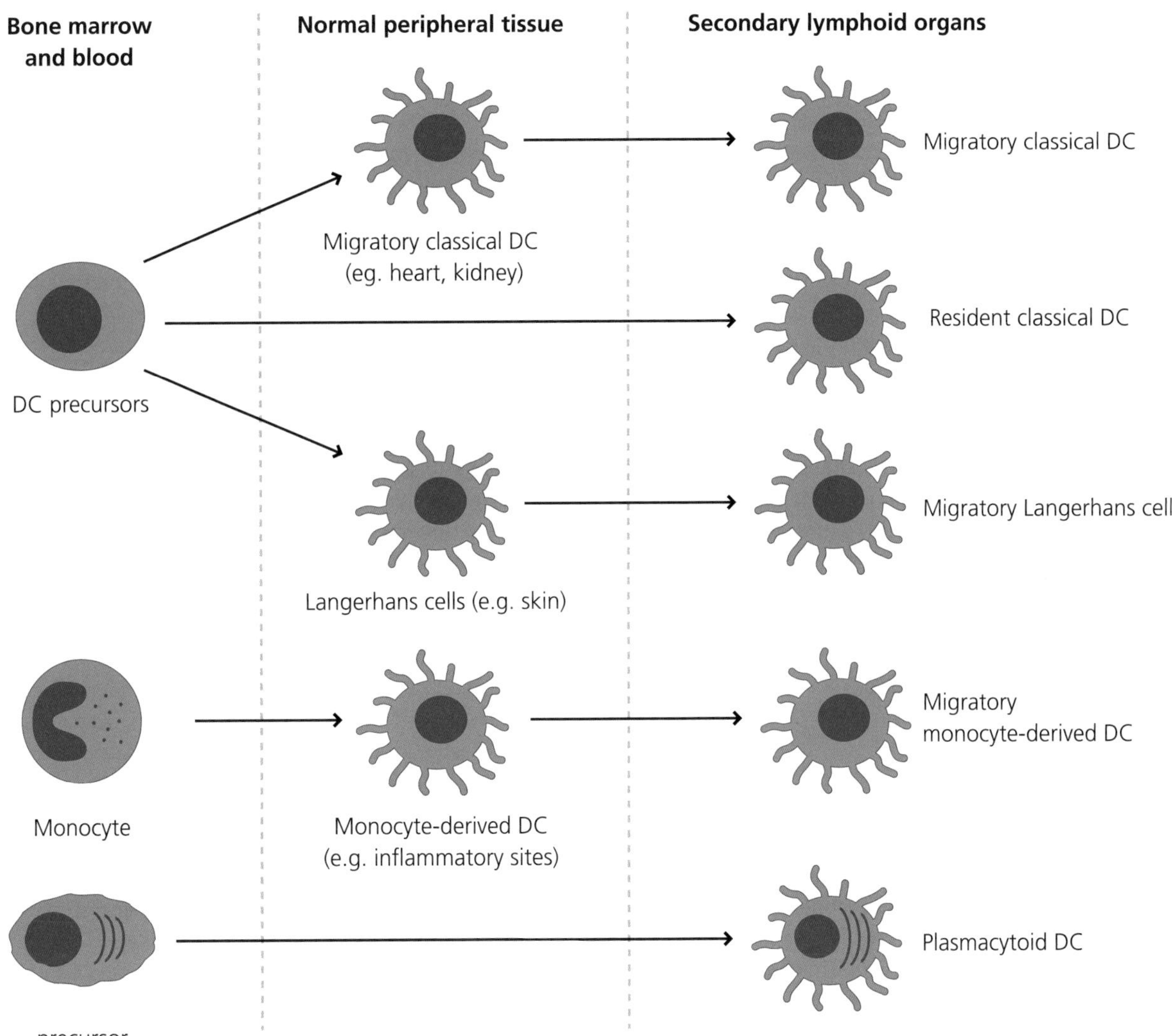

Fig. 5.16 Dendritic cell populations. Classical DCs and plasmacytoid DCs arise from different precursor cells. Some classical DC precursors enter peripheral tissues and then migrate to secondary lymphoid tissues; a similar pathway is followed by specialized Langerhans cells of skin. Other classical precursors enter lymph nodes directly from the blood and comprise a separate lymph node-resident DC population. DCs can also develop from monocytes that are recruited to sites of inflammation, and may also migrate to secondary lymphoid tissues. Plasmacytoid DCs are a distinct population that can migrate directly from the blood into secondary lymphoid tissues (or to peripheral tissues and tumours in disease settings; not shown). These cells must not be confused with follicular DCs (FDCs), which are entirely unrelated in origin or in function (not shown).

receptors (PRRs) that, directly or indirectly, lead to up-regulation of costimulatory molecules, such as CD86, needed for T cell activation. More recently another, DC-restricted precursor has also been identified. The latter precursors enter blood and some migrate directly to T cell areas of secondary lymphoid tissues to become resident DC in lymph nodes, for example; their functions are not completely understood.

5.4.1.2 Langerhans Cells

Langerhans are specialized DCs that are present in the epidermis of the skin, and related cells are present in other stratified squamous epithelia such as the vagina. They differ from other DCs in that they represent a very long-lived, self-renewing population. Their physiological functions in immune responses are still uncertain. If the skin is inflamed, they rapidly migrate to the draining lymph nodes, as do other types of DC, but there is some evidence they may play more important roles in triggering cellular responses than antibody responses.

5.4.1.3 Monocyte-Derived DCs

In addition to developing from multiple types of DC precursors (above), DCs can also differentiate from an important subset of circulating inflammatory monocytes. During inflammation, monocytes are recruited to the inflamed tissue where they may develop into DCs or macrophages depending on environmental cues that are not fully understood. Hence, DCs in larger numbers can be released from inflamed tissues and migrate to lymph nodes. These DCs are activated (they express high levels of costimulatory molecules) and are able to induce full activation of naïve T cells. Under certain circumstances it is possible that some of these monocyte-derived DCs may also become potent secretors of TNF-α and nitric oxide which is formed from inducible nitric oxide synthase (iNOS); these are therefore known as TNF- and iNOS-producing (TIP) DCs.

The discovery that DCs could develop from monocytes paved the way for the generation of large numbers of DCs in culture that could be used for therapeutic purposes (e.g. as vaccines for cancer; Sections 5.6.2 and 7.6.5). Typically monocytes are isolated from human peripheral blood and cultured for several days in the presence of cytokines, particularly granulocyte-macrophage colony-stimulating factor (GM-CSF) and IL-4 (the latter suppresses their differentiation into macrophages). These monocyte-derived DCs can then be harvested, and used for experimental and clinical studies. DCs can also be generated using similar techniques from bone marrow cells and particularly for studies in mice. To what extent these artificially generated populations of DCs reflect the natural populations remains unclear, but they have revolutionized this field because DCs can only be isolated from tissues in comparatively low numbers.

5.4.1.4 Plasmacytoid DCs

Plasmacytoid DCs (pDCs) are normally present in secondary lymphoid organs, although they can also be detected in some peripheral tissues such as skin in pathological conditions and in certain types of tumour. Their most prominent property is perhaps their ability to secrete very large amounts of Type I IFNs in response to viral infection. They may also regulate T cell activation by secretion of cytokines such as IL-6 and TNF-α, but their roles as direct APCs are still unclear and controversial.

Q5.18. Why might we need so many different types of DC?

5.4.2 Effector Functions of CD4 T Cells

5.4.2.1 CD4 T Cells in the Steady State

Under steady-state conditions classical DCs are continually migrating from peripheral tissues to lymph nodes. These DCs have high levels of surface MHC class II but low levels of costimulatory molecules. The antigens they present to T cells are self or harmless foreign (such as food) proteins. If a naïve CD4 T cell recognizes cognate peptide–MHC complexes with sufficient avidity to generate signalling through the TCR, but receives low levels of costimulation, it may undergo partial activation but then become unresponsive (anergy) or die. This is one of the mechanisms of tolerance. Under other conditions, which are not fully understood, but which may include the influence of transforming growth factor (TGF)-β, the T cell may become activated but then develop the features of a regulatory T cell, able to kill DCs or otherwise prevent the activation of other T cells (Section 5.5.5).

5.4.2.2 CD4 T Cells in the Non-Steady State

If there is inflammation in peripheral tissues, DCs will become fully activated and able to initiate the development of effector function in other cells. There is not, however, a single pathway for T cell effector function generation. At present we know of T_h0, T_h1, T_h2 T_h17 and T_{fh} cell pathways, and other polarized subsets have been suggested. Activated DCs express high levels of costimulatory molecules, are capable of driving CD4 T cells to full activation and can be programmed to induce these different types of response in these T cells.

There are, however, other potential influences on CD4 T cell differentiation that are not directly DC-associated and which are still poorly understood. We will not attempt to discuss this in detail, but will try to give some idea of the complexity (see also Section 3.4.2.3). When a pathogen infects a peripheral tissue (e.g. the intestine) many cells and systems will be perturbed (Section 4.2). Resident cells – epithelial cells, DCs, resident macrophages, mast cells and stromal cells such as fibroblasts – may all be stimulated to secrete cytokines or other mediators. The inflammation, depending on its nature, will recruit neutrophils and monocytes that may differentiate into inflammatory macrophages or new DCs, or in other situations will recruit eosinophils and basophils. Many, if not all, these cells are capable of migrating to the draining nodes. The lymph, draining into lymph nodes, will contain cytokines and other mediators. It has also been shown that mast cells can release mediator-containing particles into the lymph. All these may arrive at the lymph node and can potentially influence the T cell during its activation. We are starting to understand some of the most important influences on T cell differentiation through the use of knock-out mice and in vitro experiments, but there are still many areas of uncertainty. For example, T cells can also express Toll-like receptors (TLRs; Section 4.2.2). Hypothetically at least these could signal to the T cell, in response to agonists carried in the lymph, for example, and start to bias the cell to towards a particular differentiation pathway before the cell receives full instructions from migratory DCs or other cell types.

Below, we discuss some different possible fates of an activated CD4 T cell, and what we know and do not know about the regulation of this differentiation.

5.4.2.3 T_h0 Cells

When a CD4 T cell is initially activated a large set of changes in gene expression are initiated. Very early, the T cell starts to secrete IL-2 and express the IL-2 receptor (IL-2 is a T cell growth factor and this permits autocrine stimulation). Activated CD4 T cells at this stage are sometimes called T_h0 cells. When populations of newly activated T cells are examined they are found to be secreting a number of different cytokines (e.g. IFN-γ and IL-4) that, individually, are associated with polarized responses. Although it is often assumed that T_h0 is an intermediate stage in CD4 T cell differentiation, this may not actually be the case. When human CD4 T cells specific for tetanus toxoid were cloned from immunized individuals

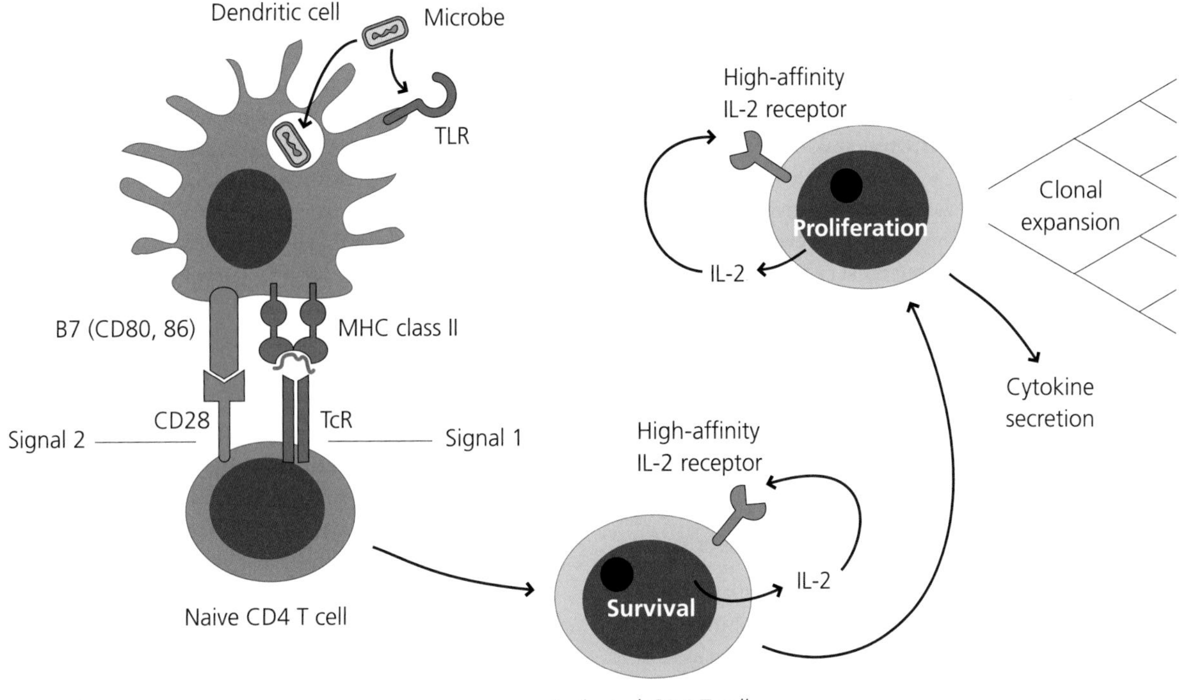

Fig. 5.17 Induction of T_h0 CD4 T cells. DCs acquire microbial antigens and present them, as peptide–MHC II complexes, to CD4 T cells. In response to signalling through TLRs, for example, they also up-regulate CD80 and CD86 (B7) which provide costimulation for naïve T cell activation. These signals stimulate the T cell to express the high-affinity IL-2 receptor (CD25) and also to secrete IL-2. The IL-2 acts in an autocrine manner to induce anti-apoptotic mechanisms and to stimulate proliferation. The result is increased survival and clonal expansion of the activated T cell, which can secrete both T_h1 and T_h2 cytokines. Some CD4 T cells may remain as T_h0 cells without further differentiation.

they often showed a T_h0 pattern of cytokine secretion, even though the last immunization had happened a long time previously. Thus, T_h0 might be an end stage for CD4 T cells. This may make sense: if responses were always fully polarized to T_h1 or T_h2 a pathogen might evolve means to subvert these responses. Perhaps it is better for the immune system to generate a broad panel of effector responses so that a more rapid polarized response can be generated depending on the infection that occurs. See Figure 5.17.

5.4.2.4 T_h1 and T_h2 cells

When CD4 T cell clones from mice were repeatedly re-stimulated in vitro with APCs in the presence of different cytokines, the T cells became polarized in terms of the cytokines they secreted. T cells cultured in the presence of IL-12 became secretors of IFN-γ (T_h1 cells). In contrast T cells cultured with IL-4 became polarized to secrete IL-4, IL-5 and IL-13 (T_h2 cells). It then became clear that naïve CD4 T cells can adopt several other differentiation pathways following activation. These include T_h17 cells and regulatory T cells (T_{reg}), and a distinct subset of T_{fh} cells is also found in vivo. This may not, however, be the final list and apparently-new subsets continue to be described. What is important is that each CD4 T cell subset is induced by different patterns of cytokines, expresses a different range of signalling molecules, including subset-specific transcription factors, secretes a particular patterns of cytokines, and expresses distinct chemokine receptors that permit it to migrate to specific sites in the body. Two critical parameters that control T cell polarization include the cytokines to which they are exposed during activation, and the dose of antigen they sense.

Effects of Cytokines on T Cell Polarization The actions of cytokines on activated naïve T cells play a very important role in determining their subsequent differentiation pathways, although plasma membrane interactions between cells may also contribute significantly. From in vitro and in vivo experiments, largely carried out in mice, it has been shown that the critical cytokines for T_h1 induction are IL-12, IL-18 and IFN-γ, and for T_h2, IL-4. (The respective cytokines that polarize T_h17 cells are not fully understood but include IL-6, TGF-β and IL-23 being needed for survival; for T_{reg}, TGF-β and IL-2; and for T_{fh}, IL-6 and IL-21.) What is also clear is that this differentiation is accompanied by the activation of master transcription factors that are restricted to a particular subset. Thus T_h1 cells express T-bet while T_h2 express GATA-3. (In contrast T_h17 cells express RORγt, many T_{reg} express FoxP3, and T_{fh} express Bcl-6.) Generally speaking, if the genes for these transcription factors are transfected into T cells they develop the properties of the cells from which the factors were obtained, providing evidence of their key roles in polarization. The roles the different polarizing cytokines in human defence against infection are difficult to define accurately but, to give one example, children with defects in IL-12 or its receptor have increased susceptibility to infections with mycobacteria and Salmonella.

Where do the critical polarizing cytokines come from and how are they induced? It is clear that the most important source of IL-12 may be the DCs themselves, and perhaps macrophages. The source of IL-4 is, however, still controversial. DCs do not secrete IL-4, but mast cells and basophils can secrete this cytokine: other cellular sources suggested include NK T cells. It may be that DCs are required initially to activate a naive T cell but that IL-4 secreted locally by a ell such as a basophil then drives Th2 differentiation. See Figure 5.18.

Q5.19. How might we begin to identify accurately the cell type which is the crucial secretor of IL-4 in the initiation of T_h2 polarization?

Effects of Antigen Dose on CD4 T Cell Differentiation How does the dose of antigen administered to an animal affect CD4 T cell activation and differentiation? This is not just an academic question since it will influence how we design and administer vaccines. Early experiments showed that if mice were treated with a low dose of bacterial flagellin, they made DTH responses that were biased towards T_h1. Higher doses, however, biased the response towards antibody synthesis, possibly a T_h2 response. More recently, CD4 T cells with transgenic TCRs specific for an ovalbumin peptide have been used in experiments assessing in vitro activation. DCs presenting very low or very high concentrations of the peptide biased the response towards T_h2, whereas intermediate doses biased towards T_h1. In a real infection, BALB/c mice infected with an intermediate dose of the protozoan parasite Leishmania major generate a T_h2-biased response and cannot control the infection (below). If, however, BALB/c mice are infected with a very low dose of the parasite, they generate a protective T_h1 response. Antigen given orally can also have differential effects. Intermediate or high oral doses of ovalbumin lead to systemic hyporesponsiveness to the antigen known as oral tolerance. Very low doses, however, induce an active T_h1 response. See Box 5.8

These observations show that there is much we do not understand about the regulation of CD4 T cell polarization. We also need to consider what happens in a real infection, in terms of the amounts of antigen reaching naïve CD4 T cells. A real infection will never mimic the administration of a single bolus of an antigen. At the start of the infection only very small amounts of antigen will reach the naïve T cells, but as the microbe multiplies, the amounts of antigen will increase, probably exponentially. Perhaps the most informative model might be to use a real infection, but to adoptively transfer transgenic T cells specific for a microbial peptide and analyze their activation at different times after infection. In the real world however, things will be very

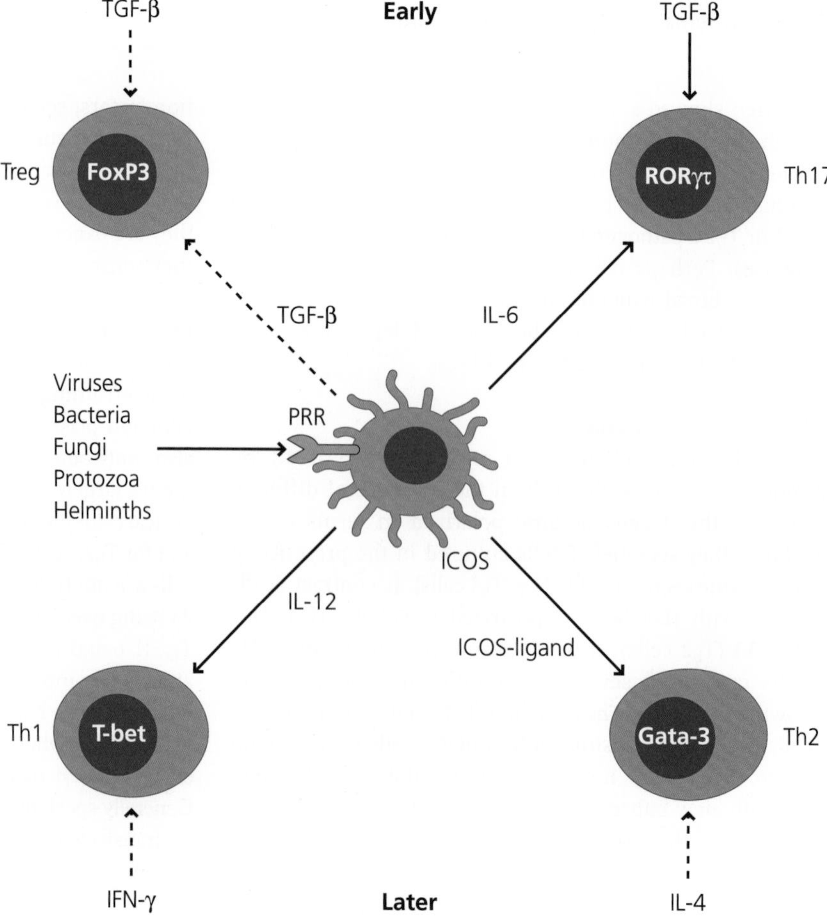

Fig. 5.18 Dendritic cell regulation of CD4 T cell differentiation. DCs are important regulators of both activation and differentiation of CD4 T cells. Different cytokines, and perhaps cell surface molecules such as ICOS expressed by DCs induce distinct patterns of gene expression in responding naïve T cells; some key examples are shown. These stimuli initiate signalling pathways that activate transcription factors that are signatures of a particular differentiation pathway; these are shown in the nuclei of the T cells. Contributions from other cell types are also likely to be very important. For example, IFN-γ contributes to T_h1 activation and may originate from NK cells; IL-4 is crucial for T_h2 differentiation and may come from mast cells, basophils or iNKT cells; and TGF-β, important for both T_{reg} and T_h17 cells, can be produced by many cell types, including DCs.

different; many different T cell clones of different avidities will respond and contribute to the overall response.

Functions of T_h1 and T_h2 cells The effector functions of activated CD4 T cells primarily seem to be required for recruiting or regulating the activity of other cells (some can also acquire the capacity to kill other cells, for example by expressing Fas ligand; below). Hence, for antibody synthesis to protein antigens, CD4 T cell help is essential Section 1.4.5.3. Both T_h1 and T_h2 cells can provide help for B cells to make antibodies. The antibodies made by B cells given T_h1 help are restricted to certain IgG isotypes, in particular those isotypes that can bind strongly to activating Fc receptors (FcRs) on neutrophils and macrophages. These isotypes are thus important opsonins for protection against pyogenic bacteria. The other two major functions of activated T_h1 CD4 T cells are in macrophage activation and in providing help in the activation of CD8 T cells. CD4 T cells can provide signals to other cells via direct cell–cell interaction (CD40 ligand on an activated CD4 T cell can interact with CD40 on B cells or DCs to give them activation signals). CD4 T cells also secrete cytokines that act locally on other cells. IFN-γ is the major macrophage-activating factor, inducing gene expression in macrophages that increase their microbicidal activity towards a variety of intracellular pathogens such as mycobacteria and Leishmania. This activity depends on the synthesis of mediators such as hydrogen peroxide and nitric oxide. Patients that cannot make or respond to IFN-γ are highly susceptible to mycobacterial infection, often with species that are harmless to normal subjects. In parallel, mice in which the gene for IFN-γ or its receptor has been knocked out are highly susceptible to similar infections. See Figure 5.19.

Q5.20. What might be the importance of IFN-γ stimulating MHC class II expression on epithelial cells?

T_h2 CD4 T cells also provide help for B cells, but in this case the isotypes secreted are not primarily opsonins. Instead, the T_h2-associated IgE isotype plays a role in immunity to multicellular parasites and IgA acts as a mucosal barrier to infection. T_h2 T cells can also induce alternative macrophage activation (e.g. through secretion of IL-4) and these cells are perhaps more involved in to healing and repair than to defence. This is discussed more fully in Section 6.2.4. See Figure 5.20.

5.4.2.5 T_h17 cells

T_h17 cells were identified as a subset of activated CD4 T cells that secretes IL-17. These cells also differ from T_h1 and T_h2 cells in their requirements for cytokines in their activation and in their use of a specific transcription factor, strongly suggesting that T_h17 cells should be viewed as a distinct differentiation pathway. T_h17 cells were identified as a subset of activated CD4 T cells that secretes IL-17. These cells differ from T_h1 and T_h2 cells in their requirements for cytokines in their activation and in their use of a specific transcription factor, strongly suggesting that T_h17 cells should be viewed as a distinct differentiation pathway. There may in fact be more than one type of T_h17 cell and different cytokines may be required for their generation in humans and in mice. Most evidence suggests that TGF-b, IL-6 and IL-23 are important in their generation and

Box 5.8 Evidence for Polarized Responses in Mice and Humans

The original T_h1/T_h2 paradigm derived from studies in two different strains of mice. Leishmania major is a protozoan parasite that infects humans and mice. In the mouse strain C57BL/6 subcutaneous infection with Leishmania major causes a local self-limiting lesion that heals, leaving the mouse resistant to further infection. In BALB/c mice, however, similar inoculation gives rise to a spreading infection that will eventually kill the mouse. Leishmania-specific CD4 T cells from infected C57BL/6 mice are potent secretors of IFN-γ, the major macrophage-activating factor, and activated macrophages are the major cell type able to kill the parasite. In contrast, CD4 T cells from infected BALB/c mice secrete IL-4, and IL-4 acts on macrophages to cause alternative activation; these macrophages are very poor killers of the parasite. To demonstrate the central role of the T cell-secreted cytokines, BALB/c mice were treated with neutralizing antibodies to IL-4 before infection; these mice survived and became resistant. Conversely, C57BL/6 mice treated with neutralizing antibody to IFN-γ became susceptible to infection and died. See Figure 5.21.

Definitive evidence for truly polarized T_h1/T_h2 responses has been much more difficult to obtain in humans. Perhaps the best evidence evidence comes from studies of leprosy. Most people infected with Mycobacterium leprae, which causes leprosy, recover without any clinical disease. A small proportion, however, do develop disease that shows a clinical spectrum from the tuberculoid (paucibacillary) form which shows small local lesions containing few bacteria, to the lepromatous (multibacillary) form with multiple large lesions containing large numbers of bacteria. The tuberculoid patients show strong cell-mediated immunity (DTH) to Mycobacterium leprae antigens and have little or no antibodies to Mycobacterium leprae. In contrast, lepromatous patients show weak cell-mediated immunity and have high antibody titres. When mRNA was extracted from tuberculoid and lepromatous lesions, it was found that tuberculoid lesions contained high levels of T_h1-related cytokine mRNA, particularly IFN-γ and IL-2, whereas lepromatous lesions contained high levels of mRNA for IL-4. Thus, in humans the disease spectrum to this intracellular bacterium can range from no disease, through strong T_h1 to strong T_h2 immunity. This clear polarization could reflect the very chronic nature (years) of this infection.

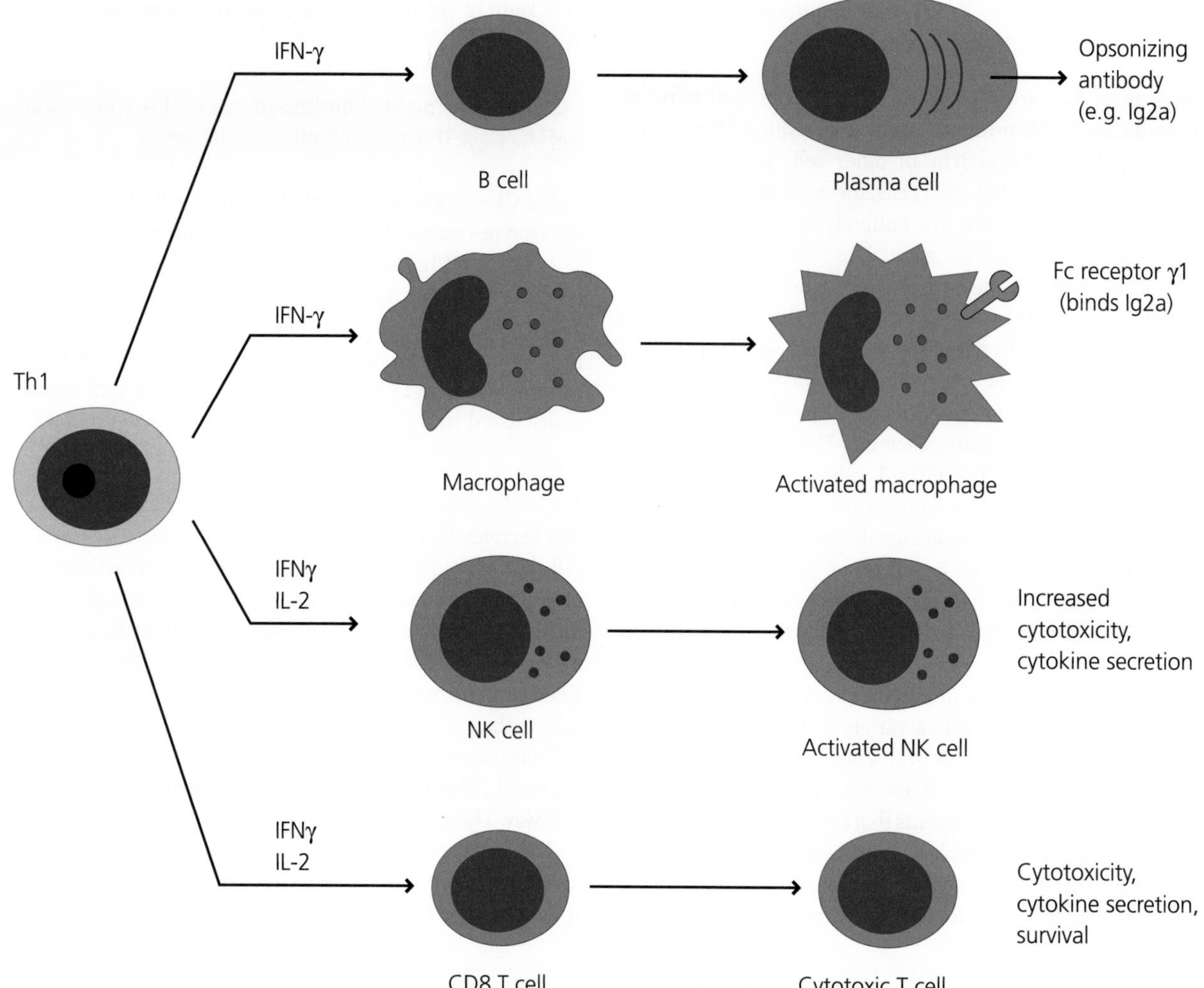

Fig. 5.19 Effector functions of T_h1 CD4 T cells. Two important cytokines secreted by T_h1 cells are IFN-γ and IL-2. These act on several different cell types. IFN-γ acts on B cells to promote secretion of opsonizing IgG antibodies and on macrophages to make them potent antimicrobial cells. IFN-r acts, with other cytokines, to increase NK cell cytotoxicity and on CD8 T cells with IL-2 to promote their activation, survival and proliferation. IgG2a is a major opsonizing antibody in mice.

maintenance. For T_h17 polarization in mice, STAT3 signalling is essential to activate the RORγt transcription factor. (In contrast, STAT1 and 4 signalling induce T-bet for T_h1 cells, and STAT6 signalling induces GATA3 for T_h2 cells.) IL-17 appears to play a role in defence against Gram-negative extracellular bacteria and fungi, particularly because it is involved in the recruitment of neutrophils. Patients have been identified with a defect in T_h17 cell generation and a mutation in the STAT3 gene. They have high levels of IgE, and suffer from increased susceptibility to infection with extracellular bacteria and fungi, closely resembling gene knock-out mice that are unable to generate T_h17 cells.

T_h17 cells were discovered some time after T_h1 and T_h2 cells were characterized. However there are suggestions that, in vivo, T_h17 cells may develop early in T cell responses that then, later, become polarized to T_h1 or T_h2. In the absence of infection, the default pathway may be T_{reg} (discussed in Section 5.5.5), which are polarized by TGF-β, but when infection occurs DCs may secrete IL-6 and the combination of these factors may thus initiate T_h17 responses. See Figure 5.22.

5.4.3 Effector Functions of CD8 T Cells

5.4.3.1 Activation of CD8 T Cells

Most CD8 T cell activation is dependent on signals from activated CD4 T cells (T cell help). In part this consists of extra stimulation from DCs that have themselves been stimulated by cognate interactions with CD4 T cells. One of the important initial interactions may be between CD40 on the DC and CD40 ligand on the newly activated CD4 T cell. This interaction may induce DCs to express new and additional costimulatory molecules that are needed for CD8 T cell activation. (Hence the activation threshold for CD8 T cells appears to be set higher than that for CD4 T cells.) Two important sets of such molecules appear to be 4-1BB/L and

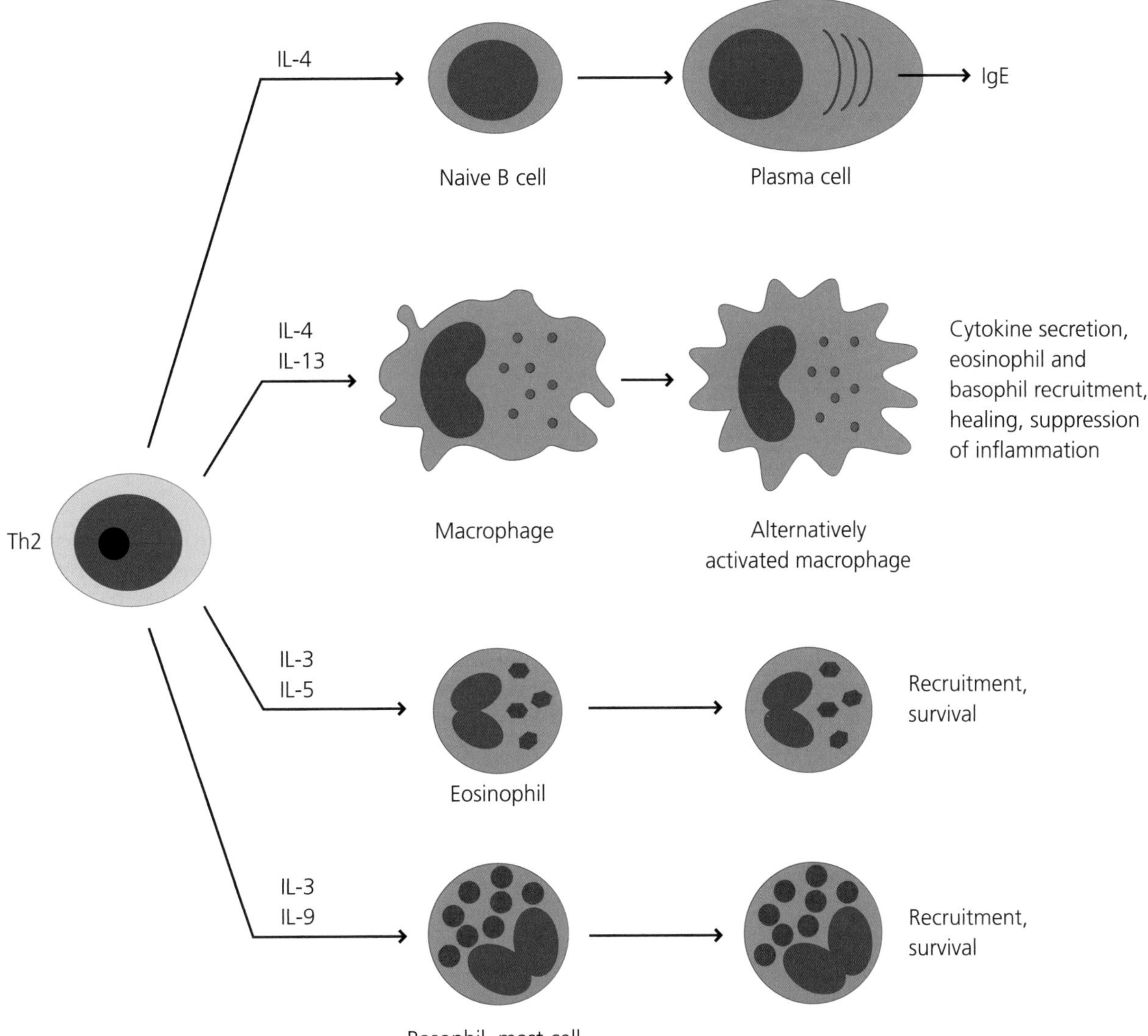

Fig. 5.20 Effector functions of T_h2 CD4 T cells. Four important cytokines that T_h2 cells secrete are IL-4, IL-5, IL-9 and IL-13. IL-4 promotes activated B cells to make barrier antibodies, including IgE. IL-4 and IL-13 act on macrophages to promote alternative activation – these cells have crucial functions in repair and healing. IL-5 or IL-9 (in concert with other cytokines and chemokines) act to promote esosinophil and basophil recruitment to, and survival in, inflamed tissues.

CD70/CD27, members of the TNF–TNF receptor family. IL-2 secreted from CD4 T cells, or from the CD8 T cell itself, may be another important signal involved in CD8 T cell activation or function. In the former case, for example, IL-2 secreted from activated CD4 T cells may drive clonal expansion of activated CD8 T cells in peripheral sites of infection. See Figure 5.23.

5.4.3.2 CD8 T Cell Cytotoxicity

An activated CD8 T cell, cytotoxic T lymphocyte (CTL), that recognizes a cell bearing an appropriate (cognate) peptide–MHC complex can adhere strongly to this target cell and can kill it. Unlike cell killing by complement, which involves direct osmotic lysis, CTLs induce the target cell to undergo apoptosis. When a CTL binds to a target cell, cytoskeletal activation directs cytoplasmic granules (which are secretory lysosomes) to the site of contact. At this site the immunological synapse forms. The secretory lysosomes of the CTL fuse with the plasma membrane and discharge their contents into the synaptic cleft. These components ultimately trigger target cell apoptosis. See Figure 5.24.

Granule-Dependent Mechanisms of Killing: Perform and Granzymes Perforin is a protein present in secretory lysosomes that is structurally similar to the C9 component of complement (although it most likely evolved independently because their gene structures are very different in terms of organization of exons and introns). Perforin monomers are secreted. They may be mtenalisim by the target cell and polymerize to form a non-specific pore (polyperforin), which may actually be formed in an endocytic vesicle. The pore then appears to enable other proteins to enter the target cell and to trigger apoptosis. The most important of these proteins are

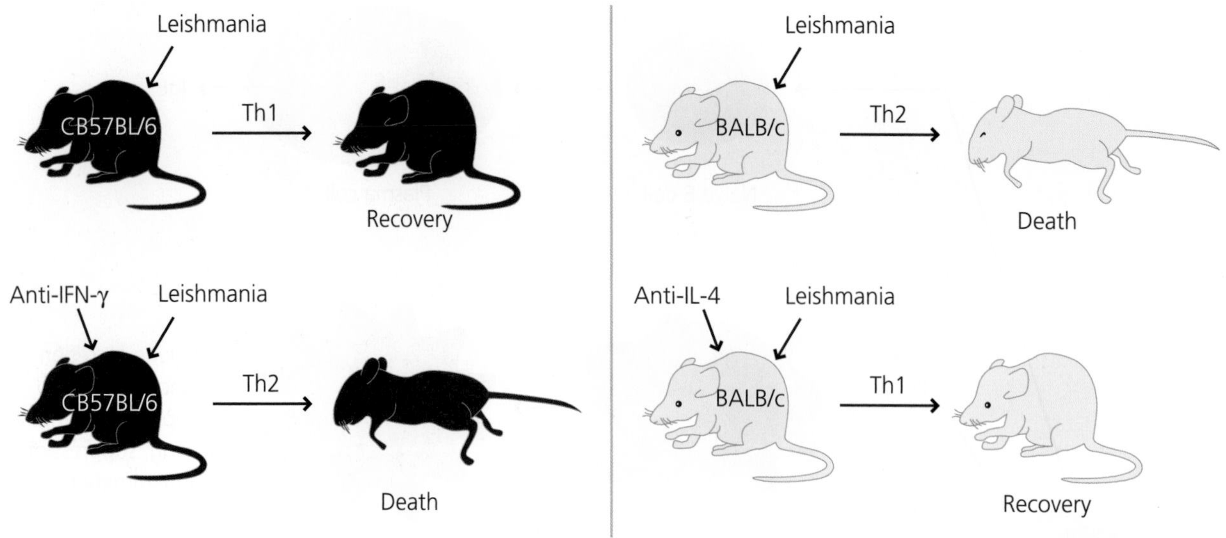

Fig. 5.21 CD4 T cell polarization in *Leishmania* infection in mice. Leishmania major is a human parasite that also infects mice. C57BL/6 mice infected intradermally develop a local lesion that heals spontaneously and leaves the mouse resistant to re-infection. This is due to a polarized T_h1 response. BALB/c mice, in contrast, develop a systemic infection and die. This is because their response is biased to T_h2. If, however, C57BL/6 mice are treated with a blocking antibody to IFN-γ, they make a T_h2 response and develop a fatal systemic infection. In contrast, if infected BALB/c mice are treated with a blocking antibody to IL-4, they make a T_h1 response, recover and are resistant to re-infection. These experiments show the crucial roles of IFN-γ and IL-4 in polarizing CD4 T cell responses to T_h1 and T_h2, respectively.

the granzymes, which are able to activate the apoptotic pathway in the target cell. CTLs from transgenic mice in which either the perforin or granzyme genes are knocked out are unable to kill target cells in vitro, suggesting a requirement for both in CTL cytotoxicity.

Q5.21. Why might perforin not polymerize in the CTL membrane and induce suicide?

Granule-Independent Mechanisms of Killing: Fas and Fas Ligand Perforin and granzymes are not the only way in which CTL can kill other cells. Many cells express Fas, a signalling molecule that is able to activate the apoptotic pathway. Activated CTLs express Fas ligand and following peptide–MHC recognition, by engaging Fas, can induce apoptosis in the target cell. What role Fas and Fas ligand play in defence is not clear: mice with gene defects in either do not generally die of infection, but rather they develop uncontrolled lymphocyte proliferation. Fas–Fas ligand interactions may instead be important in the regulation of T cell responses. It is crucial that populations of expanded antigen-specific T cells generated in response to infection are removed once the infection has been cleared, and Fas-induced apoptosis is probably particularly important for this.

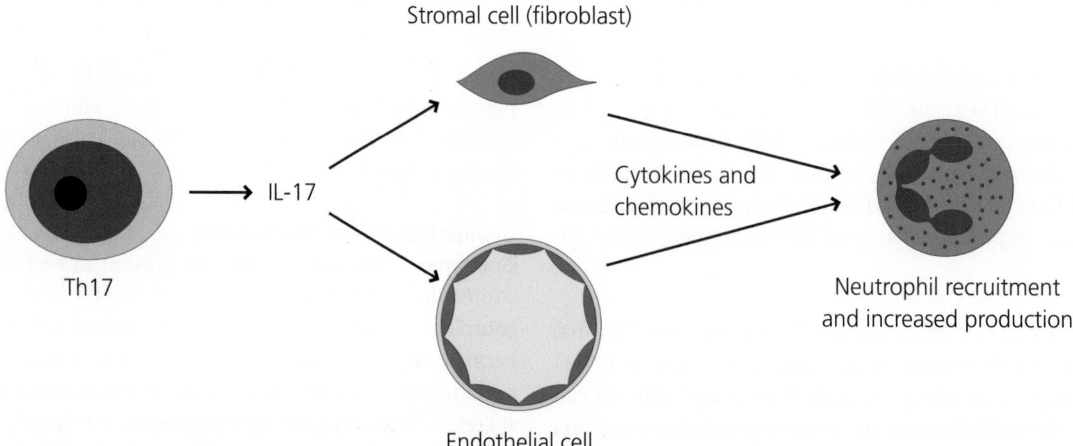

Fig. 5.22 Effector functions of T_h17 CD4 T cells. The major cytokine secreted by T_h17 T cells is IL-17. IL-17 acts on stromal cells and endothelial cells, inducing them to secrete growth factors, and cytokines and chemokines, which stimulate neutrophil production and recruitment to inflamed tissues.

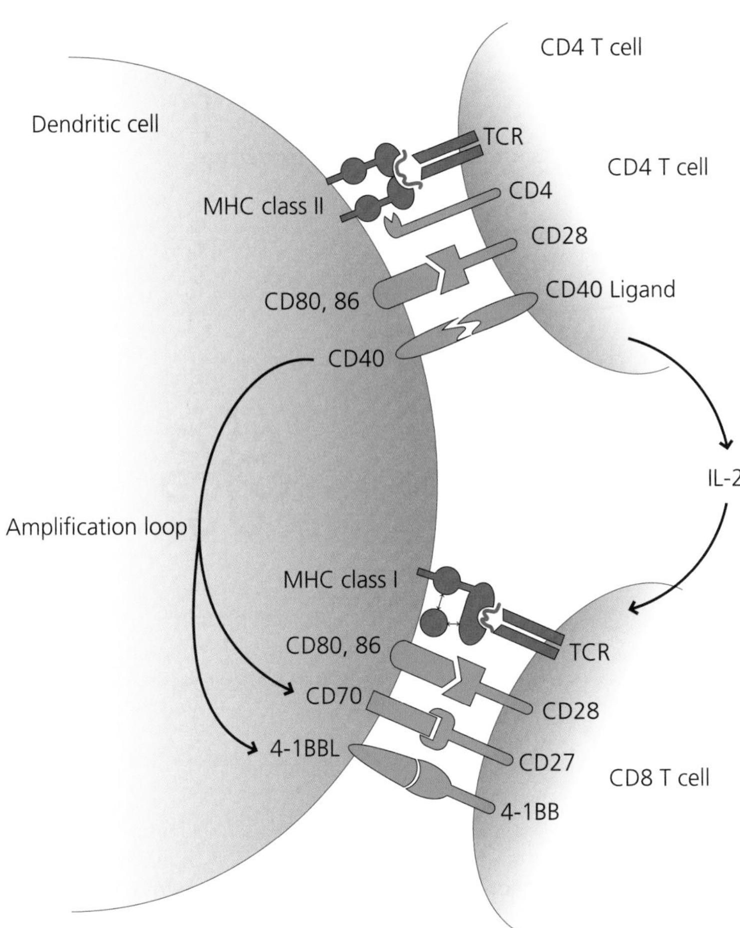

Fig. 5.23 Activation of CD8 T cells. In many cases, CD8 T cells need help from CD4 T cells to become activated. Evidence suggests that in some cases DCs first activate CD4 T cells, which then in turn signal back and super-activate the DCs. This may involve CD40–CD40 ligand interactions leading to increased expression of additional costimulatory molecules by the DCs (such as those shown). These additional signals are required for activation of CD8 T cells, which at the same time are recognizing peptide–MHC complexes on the DCs. IL-2 secreted by activated CD4 T cells may also lead to proliferation or expansion of CTLs, perhaps particularly in peripheral tissues.

Q5.22. What might be the benefits to the host of inducing apoptosis in target cells rather than straightforward lysis?

Granulysin Granulysin is a mystery molecule. It is stored in the secretory granules of activated human CD8 T cells and NK cells, and is secreted with perforin and granzymes. It is a potent anti-microbial agent, able to kill Gram-positive and Gram-negative bacteria, fungi and yeasts, and parasites such as the malaria parasite. It can also kill some tumour cells. Homologues of human granulysin have been found in the genomes of rats, pigs and cattle, but surprisingly not in mice. Experimentally, however, mice that are transgenic for human granulysin have been made and this will be an important tool for further investigations. Perhaps the most exciting aspect of granulysin to date is its anti-microbial activities. Small peptides with similar activities have been derived from the granulysin sequence and hold promise as novel chemotherapeutic agents.

5.4.3.3 Cytokine Secretion by CD8 T Cells

Activated CD8 T cells are potent secretors of cytokines. For example, their secretion of TNF-α can, under certain circumstances, induce apoptosis of cells that express the corresponding cytokine receptor. They also secrete IFN-γ, and this may also make a major contribution to their effectiveness in defence against viral and other intracellular infections. In addition to its macrophage-activating properties (but relatively weak anti-viral properties), IFN-γ may enhance the recognition of infected cells by activated CD8 T cells (e.g. CTLs) by increasing their expression of MHC class I molecules. It may also additionally enhance surveillance of other cell types by activated CD4 T cells by inducing the expression of class II molecules. It should be noted that CD8 T cells can be induced to differentiate along different pathways in a manner that is perhaps analogous to CD4 cells. Hence, T_c1, T_c2 and T_c17 CD8 T cells have been described, by analogy with the respective CD4 T cell subsets, but their respective roles in immunity to infection are unclear. See Box 5.9

Q5.23. How might one be able to assess the relative importance of CTL killing versus secretion of cytokines for eliminating infectious agents?

5.4.4 T Cell Memory

Most of us who have been vaccinated against an infection or who have recovered from an infection show long-lasting immunity, sometimes for a lifetime. This is termed immunological memory. For some infections, such as tetanus or

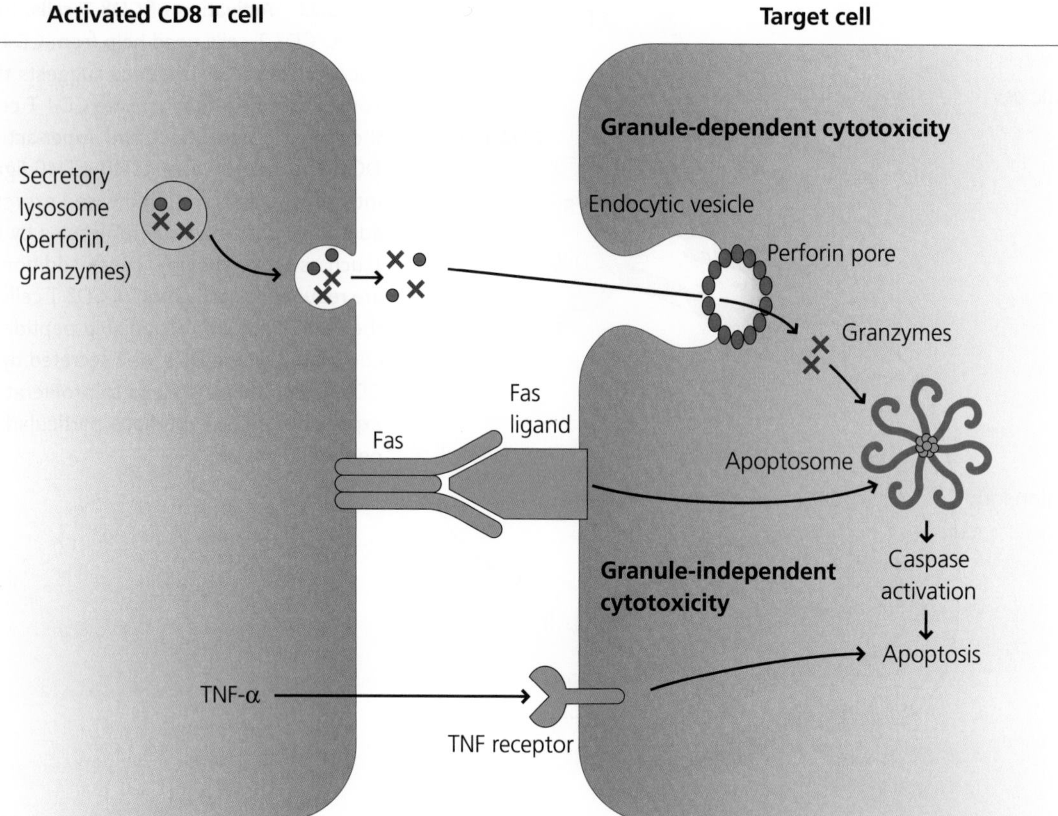

Fig. 5.24 Effector functions of CD8 T cells. Activated CD8 T cells can possess cytoplasmic granules that contain the cytotoxic machinery. Following recognition of peptide–MHC class I complex on a target cell, the T cell cytoskeleton is activated and the granules move to the immunological synapse. The granules discharge perforin and granzymes into the cleft. These molecules may be endocytosed by the target cell. Perforin may then polymerize to form a pore in the endocytic membrane. This allows entry of granzymes to the cytosol, which induce activation of the apoptosome, leading to caspase activation and apoptosis. Activated CD8 T cells can also express Fas ligand, which on binding to Fas on a target cell, may also induce apoptosis. Activated CD8 T cells can also secrete cytokines such as TNF-α. Binding of this cytokine to the respective cytokine receptor can in some cases also trigger apoptosis.

Box 5.9 How T Cell Responses Can Be Quantified: Major Histocompatibility Complex Tetramers

The outcome of B cell responses can be measured by the production of antibodies – this is a quantitative assay and represents a cornerstone of immunology. Measurement of T cell responses was, until relatively recently, much more difficult to assess quantitatively. In assays such as lymphocyte proliferation, DTH and T cell-mediated cytotoxicity, the activity of an entire T cell population is measured, and this cannot be related easily to the numbers or activities of individual specific T cells. One approach to measure individual T cell responses would theoretically be to make soluble peptide–MHC complexes that could be labelled and used as probes through binding to TCRs on the specific T cells. A major problem is, however, that the affinity of the TCR for peptide–MHC is very low compared to that of the B cell receptor (BCR) for antigen; the TCR dissociates very rapidly from peptide–MHC complexes. To overcome this, peptide–MHC tetramers, and more recently pentamers, have been constructed from recombinant proteins and peptides of choice. These contain four or five linked peptide–MHC complexes that are coupled to a fluorescent label. As they are multi valent, their binding strength (avidity) is much higher than the binding strength of a single molecule (affinity) because of receptor cooperativity. Hence, they bind with high avidity to any T cell with a complementary TCR and the fluorescent label enables these T cells to be visualized by flow cytometry. This approach has been very widely used. For example, the numbers of antigen-specific T cells in a viral infection can be directly measured. This has given truly remarkable results. For example, at the height of the response to certain viruses, such as Epstein–Barr virus (EBV), up to 50% of all circulating T cells in the blood have been shown to be specific for a single viral peptide–MHC complex. Tetramers can also be used, for example, to follow the specificity of the CD8 T cell response against HIV, revealing that a succession of waves of CD8 T cells specific for different peptides may be produced with time. See Figure 5.25.

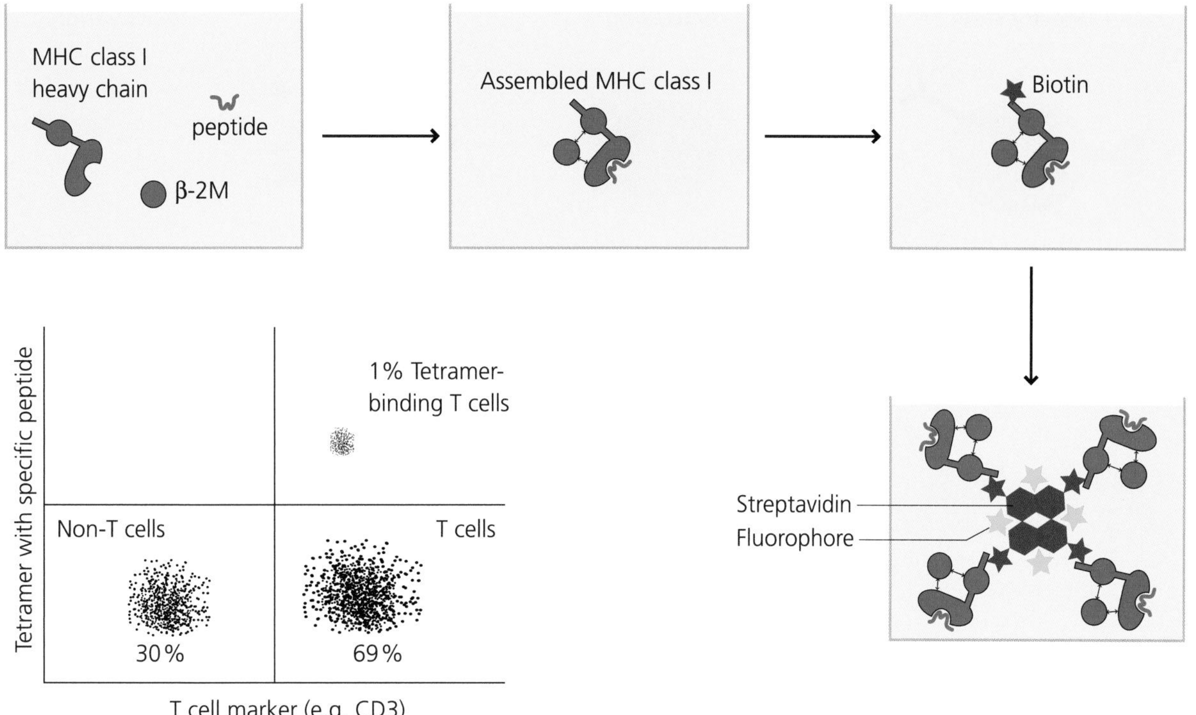

Fig. 5.25 Major histocompatibility complex tetramers. MHC tetramers (and pentamers) are used to quantitate the numbers of T cells that can recognize a particular peptide–MHC complex. The affinity of a TCR for its peptide–MHC complex is too low to permit stable binding in vitro. MHC tetramers are composed of four copies of a particular peptide–MHC complex and the avidity of the tetramer is sufficient for stable binding. Tetramers are made by refolding recombinant MHC class I (α and β_2-microglobulin) molecules in the presence of a specific peptide. Biotin is then bound enzymatically to the MHC. Fluorescent streptavidin, which has four binding sites for biotin, is added to complete the tetramer. The FACS contour plot shows an example in which a lymphocyte population has been labelled and around 1% of the cells have bound the tetramer.

influenza, protection is clearly mediated by antibodies, but in both these cases the protective antibodies are specific for proteins, and production is CD4 T cell-dependent. For other infections, such as tuberculosis, antibodies play little or no role in protective immunity and it is cell-mediated immunity, such as that mediated by CD8 T cells, that is important. While it is now clear that plasma cells can in some cases secrete antibodies for very considerable periods of time (Section 6.3.1.2), in other cases it would seem that T cell memory would be crucial to provide long-lasting immunity.

How is long-lasting T cell-mediated immunity to infection brought about? Several mechanisms that are not mutually exclusive have been suggested. It might be that some memory T cells are inherently long-lived in the absence of antigenic stimulation. In mice it is claimed that memory T cells can survive in the absence of antigen and even of the appropriate MHC molecules. In humans, memory T cells have been found in individuals vaccinated against smallpox more than 50 years previously and it seems very unlikely that viral antigens could survive so long in a host. However, in other circumstances, antigen persisting in the individual may be involved in the maintenance of memory. One mechanism is antigen storage on follicular DCs (FDCs). Thus following immunization with tetanus toxoid, native antigen is stored on FDCs note that (these are not the conventional DCs; Section 5.4.1), and it is suggested that B cells recognize, endocytose, process this antigen and present it to memory T cells (see Section 6.4). Another poorly understood mechanism is seen in tuberculosis. Patients who have recovered from tuberculosis retain living bacteria in their bodies and it is suggested that these may be giving low-level stimulation to memory T cells, which may protect against re-infection. We know that these interactions do occur because patients who become immunosuppressed (e.g. by HIV infection) are likely to suffer the re-emergence of clinical tuberculosis.

Experimentally, memory T cells differ in several important ways from naïve T cells. (i) Their activation requirements are different. Unlike naïve CD4 T cells, memory T cells can be activated by B cells, MHC class II-expressing macrophages and probably other MHC class II-expressing cells (e.g. those stimulated by IFN-γ). (ii) In many cases memory CD8 T cells do not need CD4 T cell help to be activated. (iii) Memory T cells show different migratory properties. Unlike naïve T cells some memory T cells can enter peripheral tissues efficiently and may additionally exhibit tissue tropism. Thus, CD4 memory T cells that were initially activated in mucosal tissue-associated secondary lymphoid tissues show a tendency to migrate back to mucosal tissues, while those activated in skin-draining nodes tend to migrate back to the skin. This is important in that it means that T cells activated in a particular tissue will patrol that tissue more effectively, since it is more likely to be re-infected with the same pathogen than other sites, and their frequency in the associated lymph nodes will be increased.

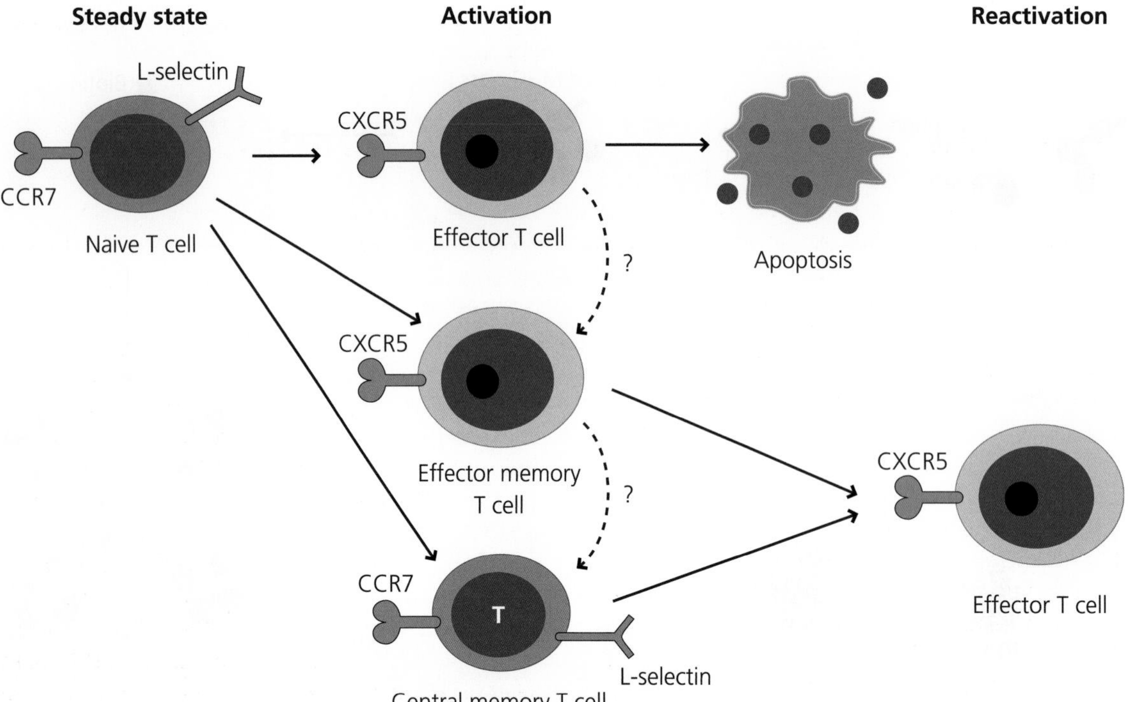

Fig. 5.26 Memory T cells. Following an adaptive immune response many effector cells apoptose but memory T cells persist; these are responsible for quicker, stronger responses following secondary exposure to the antigen and which can transfer immunity to a naïve animal. Some, effector memory T cells, are similar to effector T cells. For example, they express chemokine receptors such as CXCR5 that enable recruitment to inflamed tissues and they do not recirculate. Other central memory T cells resemble naïve T cells. For example, they express L-selectin and CCR7 and can recirculate. Both forms of memory T cell can be reactivated to become active effector cells.

It is generally thought that memory T cell may exist in two forms that can be distinguished according to the membrane molecules they express, such as different chemokine receptors. Central memory cells appear to migrate preferentially through secondary lymphoid tissues, whereas effector memory T cells may migrate preferentially through peripheral tissues. Whether these are in fact discrete populations or represent different ends of a continuum (e.g. with effector memory cells becoming central memory cells) is still uncertain. See Figure 5.26.

5.4.5 Gamma-delta (γδ) T cells

In humans and mice we tend to think of T cells as mainly expressing αβ TCRs. γδ T cells are often thought of as being less important. In humans and mice most of the recirculating T cells found in secondary lymphoid organs express the αβ TCR and γδ cells are found mainly in peripheral tissues such as the skin (in mice) and intestine. The same is not, however, true in many other mammalian species, in species such as cows and sheep γδ T cells make up a large proportion of the T cells in secondary lymphoid organs. In epithelia of mice γδ T cells may make up 50% of T cells (these are not present in humans). Most γδ T cells are of thymic origin, although some are found in athymic mice and may develop outside the thymus (e.g. in the gut). Their antigen receptors do not recognize classical MHC molecules and some appear to recognize pathogen-derived molecules directly without the need for presentation by another cell. Human γδ T cells can also recognize stress-induced MHC-related molecules such as MIC-A and MIC-B, and to do this some γδ T cells, especially those in epithelia, appear to use the cell-activating NKG2D receptor which is also expressed by NK cells. These T cells can also recognize other stress-related molecules such as heat-shock proteins.

In response to infection γδ T cells are activated very rapidly – much faster than αβ T cells and without appearing to need the stringent activation requirements of conventional T cells. They can therefore be viewed as representing a first line of defence and are thus more part of the innate immune system than the adaptive. They are recruited to sites of infection and stimulated to divide in those sites. They can generate effector mechanisms very similar to those of αβ T cell. For example, they can be cytolytic via perforin and can secrete pro-inflammatory cytokines such as IFN-γ. Human γδ T cells can also secrete the anti-microbial protein granulysin. Are γδ T cells important in immunity? In humans, many infections are accompanied by increases in the number and activation of γδ T cells. γδ T cells are able to kill some tumour cells in vitro, but their roles in tumour immunity in vivo are not defined. Human deficiencies have not been described but mice lacking γδ T cells are more susceptible to a variety of viral (e.g. HSV-1) and bacterial (e.g. Mycobacterium tuberculosis) diseases.

5.5 T Cell Development and Selection

The thymus is the site in which T cells develop; developing T cells in the thymus are called thymocytes (Figure 3.28). The thymus is first seeded by precursor cells from the bone marrow during foetal development and seeding continues throughout adult life. These cells are not fully committed to the T cell lineage, and can also give rise to B and NK cells experimentally (this does not happen in the thymus). They can also give rise to thymic DCs (this does happen in the thymus and may be important in the selection of T cells; below). These T cell precursors enter the thymus at the cortico-medullary junction and migrate towards the outer cortex. The progeny of these cells migrate back though the cortex and then enter the medulla, undergoing continuous development, some finally leaving the thymus as peripheral T cells. Of all the thymocytes formed by division in the cortex, only 1–3% will ever enter the peripheral T cell pool. Different stages of thymocyte development can be defined by patterns of surface marker expression on thymocytes, including pro-T cells and, later, pre-T cells. T cell development in the thymus is dependent on IL-7. As the thymocytes are developing the TCR is formed.

5.5.1 Generation of Diversity of Alpha-beta ($\alpha\beta$) TCRs

5.5.1.1 TCR Generation

The molecular processes involved in TCR and BCR generation are remarkably similar, involving rearrangement of gene segments. These mechanisms were first described in B cells by analysing the DNA coding for immunoglobulins in germline cells and B cells. The TCR β and γ chain gene segments resemble those of Ig heavy chains, being made up of V, D and J (and respective C) segments, while the α and δ cells resemble those of Ig light chains, being made up of only V and J (and respective C) segments. An unusual feature of TCR genes is that the δ gene region is contained within the α gene region on one chromosome (e.g. human 14), while the β and γ gene regions are separated on another (e.g. human 7). As for antibodies the V regions of conventional TCRs contain hypervariable regions – CDR1, CDR2 and CDR3. CDR1 and CDR2 are encoded within the α and β V region segments, whereas CDR3 is encoded by the joints between VDJ and VJ of the β and α chains, respectively. See Figure 5.27.

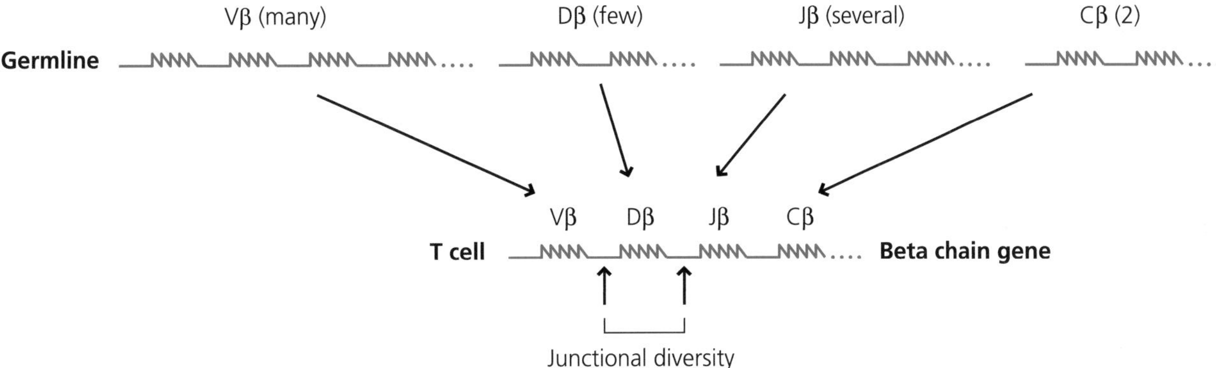

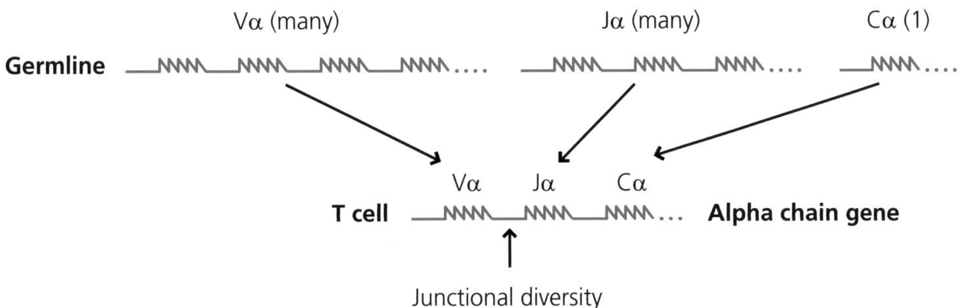

Fig. 5.27 T cell receptor generation. Complete TCR genes are only present in T cells. In germline DNA, separate gene segments encode parts of the V region or the C region. The β chain region contains numerous V_β segments, a limited number of D_β segments and multiple J_β segments. In developing T cells, one segment from each group is selected (largely randomly) and the selected segments are brought together to form the complete V region. The V_β region is paired to a C_β region segment. For α chains the principle is similar, and there are numerous V_α and J_α segments, but no D region segments. Additional diversity is generated by modification of the bases at the junctions between V, (D) and J segments.

5.5.1.2 Generation of the $\alpha\beta$ TCR

In the human α chain region there are around 70 V region, 60 J region and one C region gene segments. In the β chain region there are around 40 V, one D, 13 J and two C region segments. The early pro- and pre-T cells are described as double-negative since they do not express CD4 or CD8. In pro-T cells the β chain rearranges first, with D–J joining being followed by V–DJ joining, and this may produce a functional β chain. If this happens, the β chain combines with an invariant pre-Tα chain that is expressed on the cell surface and initiates constitutive signalling in the absence of any antigen. This occurs in pre-T cells and shuts off recombination of further β chains. This represents allelic exclusion, preventing T cells from expressing more than one β chain. These cells proliferate and, at this stage, rearrangement of variable α chains (V–J) starts, and the cell expresses both CD4 and CD8 (it is double-positive). Allelic exclusion of α chain expression is relatively incomplete. Thus a developing T cell may express one β chain paired with either of two α chains, potentially giving it two TCRs of different specificities, so other mechanisms generally ensure that one of these is down-regulated before the T cell leaves the thymus. The α chain(s) combine with the β chain and, together with CD3, are expressed at the cell surface. At this stage the cell is known as an immature T cell.

5.5.1.3 Mechanisms of Somatic Gene Rearrangement and Further Generation of Diversity of TCRs

Owing to the similarities between TCR and immunoglobulin genes (in terms of different gene segments being combined at analogous stages of development) it is perhaps not surprising that essentially the same mechanisms are used for somatic recombination in both. The endonucleases RAG-1 and RAG-2 play a crucial role in recombination. In their absence, TCR and immunoglobulin genes cannot undergo recombination and because production of a functional TCR or BCR is essential for survival of T cells and B cells, this results in SCID – a profound defect of adaptive immunity leading to early death if not treated. If the RAG genes are defective, rather than absent, this causes decreased numbers of lymphocytes and an increased pre-disposition to autoimmune diseases, known as the Omenn syndrome.

RAG (1, 2) is involved in the initial juxtaposition of pairs of D–J, V–DJ or V–J segments. The adjacent ends of these segments contain conserved heptamer and nonamer sequences separated by either 12 or 23 base pairs (corresponding to around one or two turns of the double helix). These are the recombination signal sequences (RSS-12 or RSS-23) to which RAG binds, bringing them together; after this other components join to form the recombinase-activating complex, which cuts and rejoins the gene segments. As a general (but not inviolate) rule, RSS-12 can only be paired with RSS-23. In the case of TCR gene segments, V region segments have RSS-23, D segments are bounded by RSS-12 and J segments also have RSS-12. This means that, unlike immunoglobulin genes, any given V region segment can pair either with a D or a J segment, and indeed multiple D–D joinings are also permitted. This considerably increases the diversity of TCRs that can be formed. This does not occur to anywhere near the same extent for immunoglobulin genes. Owing to of the nature of the enzymes involved in somatic recombination it is possible to generate palindromic sequences where the V(D)J joins occur – these encode P-regions in assembled TCR (and immunoglobulin) molecules, adding to diversity. Moreover, the enzyme terminal deoxynucleotidyl transferase (TdT) can add random bases to the joints, and these nucleotides contribute to N-regions. Essentially the same processes happen in the generation of immunoglobulin genes. See Figure 5.28.

5.5.2 Positive Selection and MHC Restriction of $\alpha\beta$ T Cells

T cells, as they develop, express TCRs that are formed randomly. These cells cannot predict the MHC that will be expressed in their host, and the thus the antigenic repertoire that developing T cells generate must be capable of recognizing all MHC molecules that may be expressed in the species. The great bulk of these MHC molecules (well over 90%) will not, however, be expressed in any one individual of that species. T cells that do not recognize host MHC molecules will not be able to function. To permit these non-functional T cells to enter the peripheral pool would mean that there would be a huge number of useless T cells around, competing with potentially functional T cells for migration, interaction with DCs and all other functions. To avoid this, only T cells that can recognize self MHC molecules expressed in the thymus are permitted to develop further. This is known as positive selection. See Figure 5.29.

Thymocytes which express a TCR and express both CD4 and CD8 (double-positives) migrate back across the cortex towards the medulla. As they migrate they interact with cortical epithelial cells (CECs), expressing self MHC and self peptides. This is mainly when and where positive selection occurs. If the thymocyte fails to recognize self peptide–MHC complexes expressed by the CECs, it undergoes apoptosis. If it recognizes self peptide–MHC with high affinity, it also dies. If, however, the cell recognizes peptide–MHC weakly, it survives, a crucial check point is passed and further development can occur.

The outcome of positive selection is a thymocyte with a TCR that is MHC-restricted. Importantly, if the TCR has recognized peptide bound to MHC class I the cell down-regulates expression of CD4 and becomes a single-positive CD8 T cell. Conversely a cell recognizing peptide bound to MHC class II becomes a CD4 T cell. Hence, both the MHC restriction and eventual function of the T cell is determined in the thymus. Interestingly some non-conventional T cells, such as iNKT cells (Section 5.2.8) are also positively selected during their development in the thymus, in this case, by CD1 molecules (CD1d in human), molecules that are expressed by thymocytes themselves.

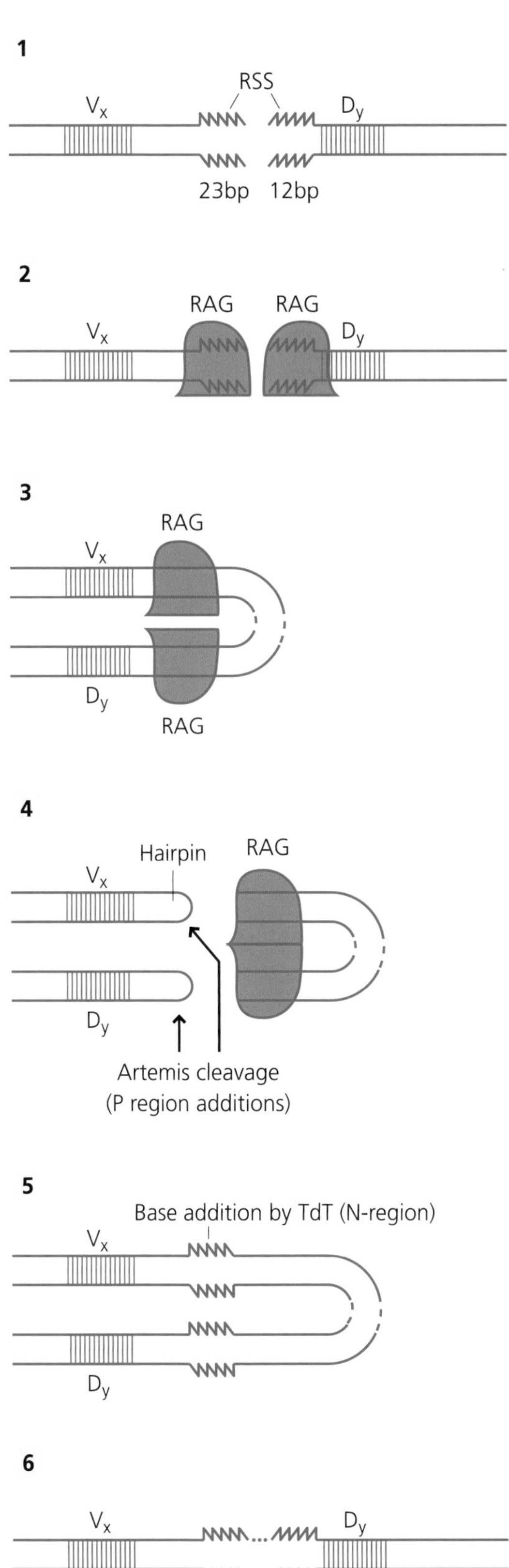

Fig. 5.28 Junctional diversity in antigen receptor gene formation. When antigen receptors are being formed in T or B cells, RAG molecules bind to recombination signal sequences (RSSs) that flank the V, D and J segments. Spacer segments of 12 or 23 base pairs in the RSSs ensure the correct joining of the different segments. The RAG molecules interact to cleave the flanking sequences of the gene segments, forming a hairpin between the coding sequences. This hairpin is cleaved by the enzyme Artemis, and subsequently repaired to form palindromic ends and P-regions. Additionally random nucleotides can be added to the free DNA ends by the enzyme TdT to form N-regions. The result is that additional bases can be added to the junctions between the segments, thus greatly increasing receptor diversity.

5.5.3 Negative Selection and Central Tolerance

After positive selection (above) a surviving single-positive αβ thymocyte now enters the medulla. Here, it interacts with both DCs (which may develop within the thymus itself) and specialized medullary epithelial cells (MECs, which are different from CECs). It is during this process that negative selection occurs. If the thymocyte recognizes self peptide–MHC complexes on either DCs or MECs with high affinity it undergoes apoptosis. If it does not recognize peptide–MHC it survives and leaves the thymus as a mature T cell with a TCR that is weakly self-reactive, but which may be able to recognize a foreign peptide–MHC with high avidity.

Why are DCs involved in negative selection? DCs are crucial cells for triggering T cell responses. Hence, it is essential that mature T cells do not recognize normal cell-specific components of DCs, and negative selection removes such T cells from the pool entering the periphery. DCs may also acquire self antigens that enter the thymus from the blood and it is even possible that some DCs enter the thymic medulla from the blood. (It is also possible that MECs can transfer self molecules or peptide–MHC complexes to DCs.)

It was a great surprise when researchers found that MECs expressed mRNA and proteins that had nothing to do with the thymus (e.g. coding for insulin), and it was soon shown that thymic mRNAs coded for other hormones, transcription factors, structural proteins, membrane proteins and secreted proteins that are normally made by highly specialized cells in other sites. This ectopic expression is in part regulated by the transcription factor AIRE, and humans lacking AIRE suffer from autoimmunity affecting many different organs, often involving the endocrine system. This condition is known as APECED (Autoimmune Polyendocrinopathy–Candidiasis–Ectodermal Dystrophy) and mice in which the AIRE gene has been knocked out develop a very similar autoimmune disease. Why should this mechanism for ectopic protein expression in the thymus have evolved? Presumably it is simply not possible for a representative sample of every self protein made in the body to access the thymus and be presented by DCs or other cells during negative selection; some indeed may be sequestered at specialized sites, such as the eye, testis or parts of the CNS, which tend to be shielded from the rest of the immune system.

The outcome of negative selection is the production of mature T cells that are generally not autoreactive or at least not with high avidity for available ligands. Negative selection cannot, however, be fully comprehensive. Not all self peptides

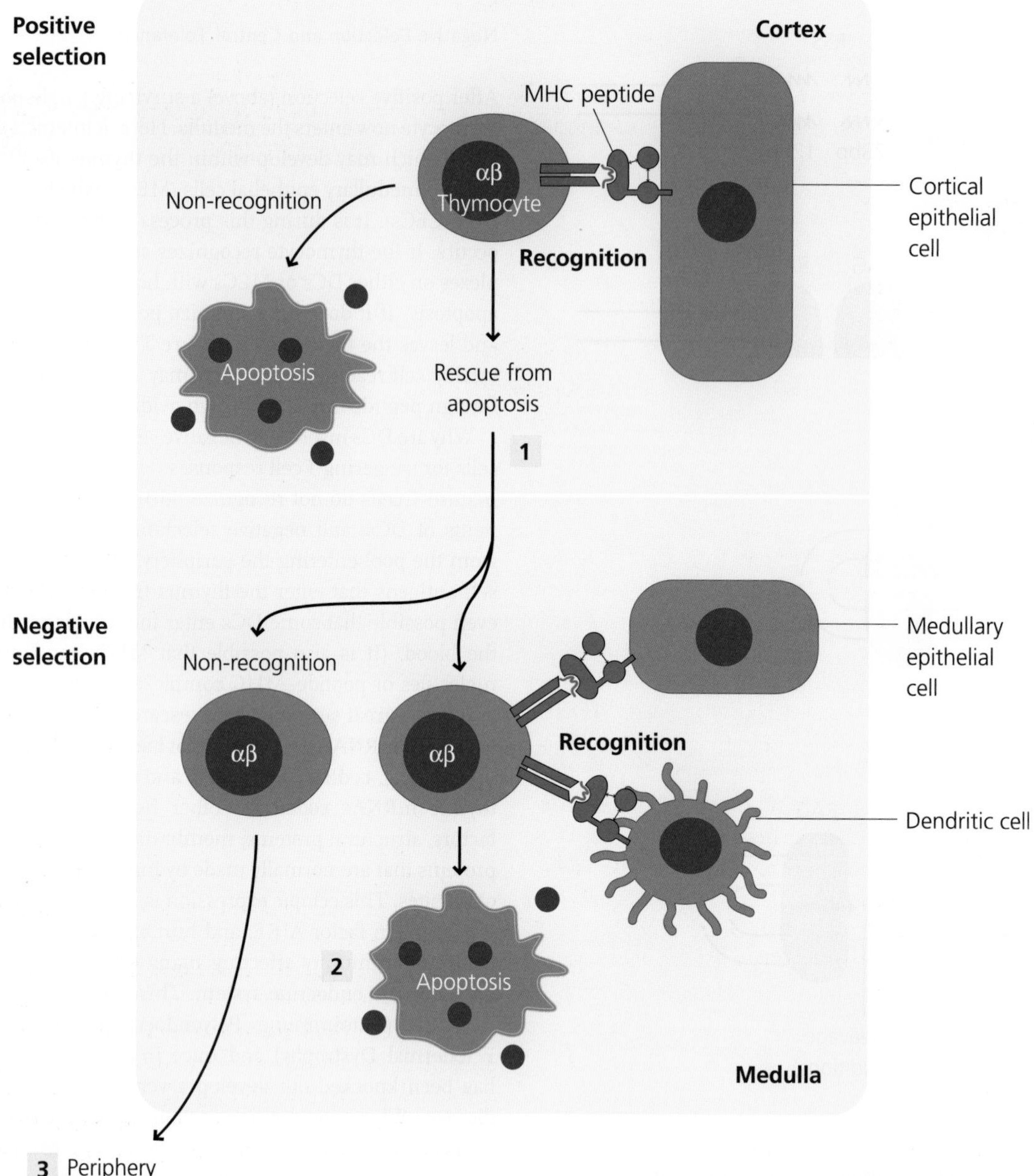

Fig. 5.29 Thymic selection of αβ T cells. The αβ TCRs generated in thymocytes can potentially recognize any MHC molecules expressed in a species. In the cortex, T cells that cannot bind with sufficient affinity to self MHC expressed on cortical epithelial cells undergo apoptosis while those that do recognize self MHC are rescued from apoptosis. This is called positive selection (1). Surviving thymocytes enter the medulla and interact with DCs and medullary epithelial cells that express AIRE. Those that now recognize self peptide–MHC complexes with medium or high affinity undergo apoptosis. This is called negative selection (2). After passing through both processes, T cells exit the thymus into the periphery. These cells are self MHC-restricted, very weakly or non-autoreactive, but can potentially recognize with high affinity foreign peptides bound to self MHC (3).

can be expressed in the thymus even with remarkable systems such as AIRE and potentially self-reactive T cells are released continually into the periphery. Some of the consequences of this are discussed in Chapter 7.

5.5.4 Generation of γδ TCRs

The TCRs of γδ T cells are generated in a very similar manner to those of αβ T cells, but there are some important differences. The numbers of γδ gene segments are less than those for αβ TCRs. According to one model, in developing pro-thymocytes, recombination of β, γ and δ chain gene segments occurs at around the same time. If a functional β chain is formed first, it pairs with pre-Tα and further development follows as described above. The subsequent rearrangement of the α chain deletes the δ locus (located between V_α and J_α segments), presumably reinforcing commitment to αβ lineage. If, however, a functional γ and a δ chain are formed before a β chain, the cell becomes committed to the γδ T cell lineage. See Figure 5.30.

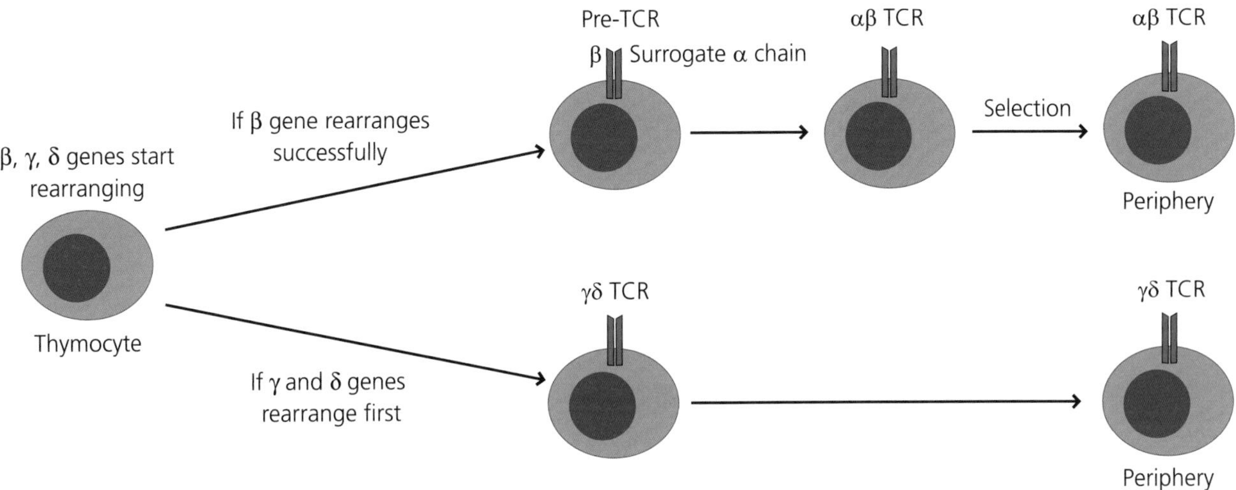

Fig. 5.30 Determination of αβ or γδ T cell receptor expression. According to one model which type of TCR is generated in a thymocyte depends on which gene(s) first rearrange successfully. If a β chain gene rearranges first, the β chain pairs with a surrogate α chain and γδ rearrangement is suppressed. If, however, both γ and δ genes rearrange successfully before the β chain, αβ rearrangement is suppressed and the γδ TCR is expressed. Other models to explain αβ versus γδ T cell development have also been proposed.

During early development (at least in the mouse), different waves of γδ T cells are formed. These cells express limited V region diversity (they do not express TdT) and these migrate to different sites. Cells expressing $V_\gamma 5$ and $C_\gamma 1$ migrate to the skin, forming dendritic epidermal T cells (these are not DCs and are not present in humans). Those that express $V_\gamma 6$ go to the epithelium of reproductive organs. Later, γδ T cells are formed continuously and these cells show much more V region diversity (they do express TdT) and they do not show particular tissue tropisms.

5.5.5 Peripheral Tolerance

Potentially self-reactive T cells are released continually from the thymus into the periphery. The clearest evidence for this is that people develop T cell-dependent autoimmune diseases (Section 7.1.1). If a T cell does not encounter antigen it cannot become activated, and many potential antigens (e.g. intracellular proteins) may not be released in sufficient quantities to initiate a response. This is termed immunological ignorance, and it is relevant to both T cells and B cells. There are, however, more active mechanisms available to prevent the development of active self-reactivity of peripheral T cells.

5.5.5.1 Regulatory T Cells

In the 1970s a number of researchers claimed that a subset of T cells could interact with other immune cells (B cells and CD8 T cells, in particular) to prevent their activation by activated CD4 T cells. These cells were called suppressor cells, but their existence was questioned. Things remained unclear until the 1980s when two groups working with rats and mice, respectively, first showed that a population of CD4 T cells existed in normal animals that could prevent naïve CD4 T cells from causing autoimmune disease. These were termed regulatory T cells (T_{reg}) and it is now clear that several other types of T_{reg} exist.

Natural Regulatory T Cells The T_{reg} that were originally described (above) are called natural T_{reg}. They are formed in the thymus and not in response to any external antigenic stimulus, and it is thus difficult to see how they could be specific for any exogenous antigen. These natural T_{reg} are characterized by the expression of the FoxP3 transcription factor. Is this gene important in immunity? New-born children that cannot express FoxP3 do not have detectable T_{reg} and develop a very severe, rapidly fatal autoimmune syndrome called– IPEX (Immune dysregulation, Polyendocrinopathy, Enteropathy X-linked). A mutant mouse strain (scurfy) develops a similar disease. It thus seems clear that the development of these T_{reg} in the thymus is essential to prevent anti-self T cells, released from the thymus, inducing autoimmunity. It is not fully clear how natural T_{reg} develop in the thymus, but it has been suggested that they represent cells that have been selected by medium level avidity for self peptide–MHC complexes.

Induced Regulatory T Cells Apart from the natural T_{reg} that are released from the thymus ready to act, other T_{reg} can be induced during on-going immune responses. Several types have been described. Some T_{reg} are induced under the influence of TGF-β and may express FoxP3. Other T_{reg}, such as T_r1 or T_h3 subsets, are also induced in the periphery under natural or experimental conditions and do not express FoxP3. Many of these T_{reg} populations have been generated in vitro and their in vivo roles are not fully understood. The generation and use of T_{reg} in therapy is, however, a crucial area for research (below).

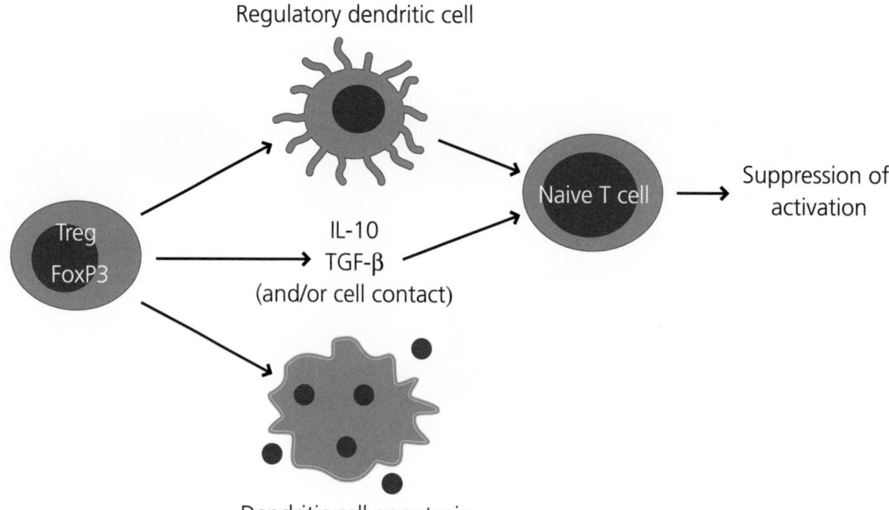

Fig. 5.31 Effector functions of regulatory T cells. T_{reg} can be natural (formed in the thymus) or induced during adaptive immune responses. The mechanisms by which T_{reg} suppress responses may include contact-dependent actions on DCs, which are then either killed or become unable to fully activate naïve T cells. Alternatively, or in addition, T_{reg} may secrete IL-10 and/or TGF-β, which may act on naïve T cells during their activation to suppress full activation. Receptor–ligand interactions, such as Notch and its ligands, may also be important for some functions of T_{reg} (not shown).

Function of Regulatory T Cells The function of T_{reg} in ongoing responses is thought to be to prevent collateral damage. This is partly due to their secretion of anti-inflammatory cytokines such as TGF-β and IL-10, but may also depend on cell–cell contact. They may also act on DCs to inhibit their ability to induce full activation in naïve T cells perhaps even by killing them. How in an ongoing response both effector T cells and T_{reg} are generated is not clear but they may represent sequential stages. See Figure 5.31.

5.5.5.2 DCs and Peripheral Self-Tolerance

We have noted above that DCs are continually migrating from peripheral tissues to lymph nodes, even in the absence of any innate immune activation. What is the function of this migration? These DCs will have acquired self or harmless foreign proteins in the periphery and, on reaching the nodes, will be interrogated by naïve T cells. The DCs will, however, not be expressing high levels of the costimulatory molecules needed to fully activate naïve CD4 T cells. They do, however, express high levels of MHC class II. It is suggested that these DCs can induce partial activation in these T cells, but that this partial activation leads to apoptosis of the T cell thus inducing tolerance to the peptide expressed by the DC.

Experimental evidence suggests that DCs can in fact contribute to peripheral tolerance. For example, an artificial antigen was made in which a peptide from ovalbumin (OVA) was attached to a monoclonal antibody specific for a molecule expressed by DCs. Normal mice were given small numbers of TCR transgenic CD4 T cells specific for the OVA peptide. When the antigen was injected subcutaneously, the transferred T cells accumulated in the draining node, showed markers of activation and started to divide, but rapidly disappeared and could not be detected anywhere in the mouse. It was found that the mice had also become unresponsive to the peptide. Thus, targeting an antigen to DCs in the absence of any innate immune activation can apparently lead to deletion of antigen-specific CD4 T cells. If the same experiment was carried out, but the mice were injected with an agonist monoclonal antibody to CD40, the transferred T cells became fully activated and the animals generated a DTH responses to ovalbumin. To what extent this contributes to peripheral tolerance in natural settings is, however, still largely unknown.

5.5.5.3 Oral and Nasal Tolerance

It has long been known that antigens delivered orally can induce systemic unresponsiveness, another form of tolerance. It is said that native Americans used to eat poison ivy leaves to protect them from its toxic effects (contact sensitivity; Section 7.1.1). Most proteins are digested in the intestine, but small amounts do enter the circulation and induce tolerance. In many cases this systemic tolerance is accompanied by a local IgA response against the protein. This is called split tolerance. More recently it has been found that peptides delivered nasally can also induce tolerance. If used clinically, this avoids the problems of digestion of the antigen when given orally. The mechanisms of oral and nasal tolerance are not fully understood. CD4 T cells are primarily involved, because oral tolerance can be induced in mice lacking CD8 T cells and B cells. Essentially, delivery of antigens without danger signals may induce anergy in CD4 T cells. In some cases a different kind of T_{reg} (T_h3) that secretes large amounts of TGF-β has been identified in oral tolerance. Whatever the mechanisms, the possibility of using oral tolerance to treat immune-related disease is currently being investigated Section 7.4.1.

5.5.5.4 Secondary Lymphoid Tissues and AIRE

As described above, the transcription factor AIRE was identified in the thymus and has a crucial role in the induction of self tolerance. Cells expressing AIRE have, however, been identified in lymph nodes, suggesting that negative selection of mature T cells may continue once they have reached the periphery (i.e. left the thymus).

5.6 Adoptive Cell Therapy

5.6.1 T Cell Vaccines

We have mentioned in other chapters (e.g. Section 2.5.2.3) that all the wholly successful vaccines appear to depend on antibodies for their efficacy. It is a major puzzle why it is so difficult to produce vaccines that work via induction of cytotoxic T cells or activated macrophages. In some cases, as in HIV, antigenic variation is a serious problem, but in others, such as tuberculosis, this is not the case. It is important to remember that these pathogens have evolved in the face of strong selection pressures from the immune system and it is an essential part of their pathogenicity that they have generated ways of evading or avoiding the immune response. Overcoming these barriers demands a thorough understanding of the biology of such pathogens; these topics are discussed in Chapter 2. To date, however, it is probably true to say that we do not yet have even a single vaccine that can induce a protective T cell response, that is mediated by T cells (e.g. CD8 T cells) rather than antibodies.

5.6.2 Immunotherapy

It is an attractive idea to use T cells to treat disease. The successful use of antibodies to do this is discussed in Section 6.6, but the use of T cells is much less advanced. The main areas under development for T cell therapy are the

Learning Outcomes

By the end of this chapter you should be able to understand, explain and discuss the following topics and questions; the relevant sections of the chapter are indicated. You should understand some of the evidence from human and animal studies supporting what we know about these topics. You should have some idea of the areas where our understanding is incomplete. You may be able to suggest ways in which our understanding could be advanced.

- T cell populations (Section 5.1)
  - What are the main types of T cells?
  - How do T and B cells differ in the ways they recognize antigens?
- The MHC (Section 5.2)
  - What are classical MHC molecules?
  - What are non-classical MHC molecules?
- Antigen processing and presentation (Section 5.3)
  - How are peptides generated from different cellular compartments and loaded onto classical MHC molecules??
  - What is cross-presentation and why is it important?
- The anatomical basis of T cell responses (Section 5.3)
  - Where do T cell responses occur in different stages of immune responses?
- T cell activation (Section 5.4)
  - What is Signal 1 and Signal 2 for T cell activation, and what happens if T cells receive the first alone or both?
- Effector and memory functions of T cells in infection (Section 5.4)
  - What is CD4 T cell polarization, why is it important in host defence against infection and how, generally, can this be regulated?
  - What roles do DCs play in host defence?
  - How are CD8 T cells activated and how can they kill infected cells?
  - What are some of the key cytokines produced by T cell subsets and their functions?
  - What are γδ T cells and what might they do in immunity to infection?
- T cell development and selection (Section 5.5)
  - How is the diversity of TCRs generated?
  - What is positive selection and how does it occur?
  - What is negative selection and how does it occur?
  - How can peripheral mechanisms contribute to T cell tolerance?
- Adoptive cell therapy (Section 5.6)
  - How might DCs and T cells be used for therapy of disease?
- GENERAL: What types of natural and experimental defect lead to deficiencies in T cell development and function, and what do these defects tell us about the roles of T cells in defence against infectious disease?
- INTEGRATIVE: What are the similarities and differences between cytotoxic T cells and NK cells in terms of their origins, life histories and functions in defence against infection?
- INTEGRATIVE: How does mucosal immunity differ from systemic immunity?

treatment of chronic infections and autoimmune diseases, prevention of allograft rejection, and therapy for malignant tumours. One approach is to take T cells from a patient, manipulate them in vitro and re-inject them, or to modulate the functions of T cells in vivo. Thus some investigators have taken T cells from tumour-bearing patients, isolated CD8 T cells specific for tumour antigens, activated and expanded them in culture, and re-injected them into the patient. However, this approach has had only limited success to date.

An alternative approach is to use DCs directly. In the treatment of malignant tumours, DCs, generated from a patient's own blood monocytes (Section 5.4.1.3), are made to express tumour antigens as peptide–MHC complexes that can potentially be recognized by antigen-specific T cells. This can be done by incubating the DCs with the antigens, by transferring tumour mRNA into the DCs or by fusing the DC with tumour cells. The DC are also given activation signals through combinations or cytokines or via PRRs since, to stimulate effective immunity, they need to express costimulatory molecules. The DCs are then injected into the patient in the hope that they will stimulate effector T cells, particularly CD8 cytotoxic cells. This approach is being explored in many clinical trials, but again with limited success so far.

5.6.2.1 Therapy of Infections and Immunopathological Diseases

Pathogens have evolved multiple ways of evading immune responses, autoimmune diseases are caused by inappropriately-biased immune responses, allografts are rejected by unwanted responses. In theory it should be possible to re-program T cell differentiation to generate the response or lack of response that is desired. In chronic infections such as HIV and tuberculosis, boosting the on-going but ineffective response is the desired outcome. This might be achieved by using similar methods to those being trialled in tumour therapy. In autoimmune disease and allograft rejection the aim is to alter the bias or to suppress the unwanted or inappropriate response. Again this could involve reprogramming DCs in vitro, or administering antigens in ways that induce tolerance or T_{reg}. That this approach is feasible is shown by the successful treatment of hay fever by giving sublingual extracts of timothy grass pollen. All these approaches are at present in their infancy but could hold huge potential for the future.

Further Study Questions

Qa. How much is known about the functions of non-classical MHC molecules, their ligands (if any), and their cell or tissue distributions?

Hint We have provided a few examples of non-classical MHC molecules, but many others have been identified in the human and mouse MHC region. Some have ligands such as peptides, but is antigen processing needed to generate these? Some are restricted to certain cell types or tissues, such as the placenta. For some, it is clear that in some cases they can be recognized by non-conventional lymphoid cells, but why?

Qb. To what extent do we understand the functions of different DC subsets in immunity?

Hint Perhaps start by defining what we mean by a subset. Multiple subsets of classical DC have been identified in mouse lymphoid tissues, for example, and we believe that some have specialized functions. What of pDCs in normal and pathological settings, and to what extent can they regulate lymphocyte responses? Do we want to consider FDCs as well?

Qc. How far can we define different polarized subsets of effector or memory T cells?

Hint The ability of CD4 T cells to adopt specialized functions (T_h1, T_h2, etc.) seems clear, at least in mice. However, additional subsets with different functions, such as T_h9 and T_h21, have also been postulated. In some cases CD8 T cells also seem to adopt specific functions (T_c1, T_c2, etc.) and this has even been suggested to be the case for iNKT cells. Might this also apply to other non-conventional T cells such as γδ T cells? To what extent is this apparent polarization retained or lost if the CD4 or CD8 T cells develop into central or effector memory cells?

Qd. To what extent might mechanisms of positive and/or negative selection apply to development of non-conventional lymphocytes?

Hint We know that, during development, follicular B cells and αβ T cells are subject to negative selection, and that the latter are also subject to positive selection. Some have postulated that positive selection also applies to B cells. It certainly applies to iNKT cells, but what about γδ T cells? If we wish to extend this to NK cells, it is clear these cells are selected to express receptors for specific MHC class I alleles, for example, but how? Does negative selection apply here as well?

Qe. What advances are being made in cell-based adoptive therapies for different immune-related diseases?

Hint Perhaps start by considering diseases in which one might wish to stimulate protective immunity or in which one might wish to suppress aberrant or unwanted responses. Then, ask which immune cell types might be best suited for one or the other. We have provided a few examples. However, new approaches are being tried even for these, such as transfection of tumour-specific TCRs into T cells, exposure of DCs to different types of stimuli to modulate their functions, and different ways of generating antigen-specific regulatory T cells. To what extent do we really want to repolarize, and redirect immune responses rather than simply turn them on or off?

6
Antibody-Mediated Immunity

6.1
Introduction

B lymphocytes play a crucial part in adaptive immunity. They give rise to plasma cells that secrete antibodies, as well as memory B cells. These are crucial for immediate and long-term defence against many infectious agents. B cells appear to have co-evolved with T cells. Antibodies are structurally very similar to T cell receptors (TCRs) and the processes that generate the enormous diversity of antibody specificities in B cells are very similar to those that occur in T cells. We do not fully understand how two different types of lymphocyte appeared in evolution, but it is clear that B and T cells do recognize antigens in very different ways. This evolution may have been facilitated by two entire genome replications that occurred before and during the Cambrian explosion some half a billion years ago. The duplicated sets of chromosomes now enabled a group of genes on one chromosome to diversify from their original forms and to assume new functions. This may have allowed the generation of closely related but different types of antigen receptor during evolution of the two main types of lymphocyte.

In this chapter we discuss the functions of different types of antibodies in host defence, and describe how B cells can make different types of response against infectious agents. We start by noting that there are different types of response against B cells that can develop into plasma cells that make different types of antibodies, sometimes against different classes of molecular structures (Section 6.1). We then explain how it is possible for antibodies to recognize so many different kinds of molecule, before considering the different types of antibody, where they are commonly found and how they help in host defence against infection (Section 6.2). We then describe how different types of B cell can recognize different types of infectious agents and produce antibodies against them in different ways – and how in many cases they need to collaborate with T cells in order to synthesize the most appropriate type of antibody. We describe the anatomy and physiology of B cell activation and antibody synthesis (Section 6.3), and show how long-term resistance to infection can occur through antibodies (Section 6.4). We explain how and where different types of B cell develop and why it is that antibodies do not normally recognize components of the host itself, a part of immunological tolerance (Section 6.5). Finally, we outline how antibodies can be produced and used as tools for the treatment of diseases (Section 6.6)

By the end of this chapter you will recognize the importance of antibodies in defence against infection and will understand how B cells become activated to differentiate into antibody-secreting plasma cells or memory cells, and how this activation is regulated.

6.1.1
B Cell Populations

B lymphocytes are the precursors of antibody-forming cells, known as plasma cells, and of memory B cells. In contrast to T cells that can develop into subsets (T_h1, T_h2, etc.) with very different functions, the major variation in the effector function of B cells lies in the biological properties of the different classes of antibody that they can secrete. In humans and mice, for example, these are the IgM, IgD, IgG, IgA and IgE isotypes. However, different subsets of CD4 T cells often control which class of antibody a plasma cell will secrete. B cells also differ from T cells in that they can change the structure of their antigen receptors, and thus of the antibodies they secrete, by somatic mutation after they have been appropriately stimulated by antigen.

Morphologically, newly produced, naïve B cells are indistinguishable from T cells. However, they do of course have a very different pattern of gene expression, and because they express different surface molecules they can be readily distinguished from T cells and other mononuclear cells. In mice and humans there appear to be three main populations of mature B cells:

i) Follicular B cells repeatedly migrate through the follicles that are present in all secondary lymphoid organs (Chapter 3). They are made continually in the bone

Exploring Immunology: Concepts and Evidence, First Edition. Gordon MacPherson and Jon Austyn.

marrow and have life spans of about 8 weeks. These B cells are responsible for making conventional antibody responses, typically against protein antigens. In general, these types of antibody response need help from CD4 T cells; these are called T-dependent (TD) responses.

ii) Marginal zone B cells are a specialized, sessile B cell population found almost exclusively in the spleen. They typically make antibodies against carbohydrate antigens and some other macromolecules with repetitive structures. These responses do not need T cell help in the way that those of follicular B cells do, and they are therefore called T-independent (TI) responses. (These B cells can, however, also make TD responses to some extent.)

iii) B-1 B cells are found mainly in the peritoneal and pleural cavities of mice. Some B-1 cells, which also express CD5, produce natural antibodies in the absence of overt antigenic stimulation. B-1 cells are best characterized in the mouse, although CD5 B cells are also present in similar locations in humans. Other B-1 cells can make TI responses.

Follicular B cells and marginal zone B cells are sometimes referred to as B-2 cells to distinguish them from B-1 cells (above). Because they need help from CD4 T cells, follicular B cells typically take longer to make antibody responses than the other types of B cells. In contrast, marginal zone B cells can rapidly make antibody responses to TI antigens, while some B-1 cells may continuously and spontaneously produce natural antibodies. See Figure 6.1.

6.1.2 How Do Antibodies Recognize Antigens?

B cells recognize antigens through their membrane-bound B cell receptors (BCRs). These receptors are secreted in a soluble form, as antibodies, after the B cells develop into plasma cells. What is the range of antigens against which B cells can make antibodies? Collectively, B cells can respond to a huge range of molecules including proteins, carbohydrates, nucleic acids, glycolipids or even small inorganic molecules such as dinitrophenol (DNP) which do not exist in nature. In contrast to conventional T cells which recognize linear peptides that can arise from any part of a protein, (sequential epitopes), bound of course to MHC molecules antibody molecules recognize small parts that are on exposed surfaces of a molecule called conformational epitopes. These epitopes need to possess a stable three-dimensional structure if antibodies are to be made against them. X-ray crystallography has enabled the different types of atomic interactions that can occur between antibodies and their corresponding antigens to be visualized, and has shown the variety of shapes that can be adopted by the antigen-binding sites of antibody molecules.

Q6.1. Why can we not make antibodies to gelatine? (denatured collagen)

6.2 Antibody Structure and Function

6.2.1 Immunoglobulins and Antibodies

The terms immunoglobulin and antibody are often used interchangeably. Strictly, however, immunoglobulin is the name given to a particular group of protein molecules in blood plasma or secretions with a size and electrophoretic mobility that characterise them as the γ-globulins. Antibodies are immunoglobulins for which an antigen has been identified (i.e. antibody is really a functional label). The term antigen was initially used to describe anything that triggered the production of antibodies. Now, however, this term is more generally used to describe anything that can be recognized by a lymphocyte, including T cells.

Immunoglobulins are made of different polypeptide chains: the heavy (H) and light (L) chains. Each immunoglobulins chains has a characteristic structure: they are built from a number of domains, which possess common features in their three-dimensional conformation. This general structure is known as the immunoglobulin domain. It is characterized by the immunoglobulin fold, a particular conformation of closely apposed β-pleated sheets that forms the structure of each domain. The immunoglobulin domain is not only found in antibodies. Many other molecules, often involved in recognition, including TCRs, CD4 and CD8, and the co stimulatory molecules involved in T cell responses (Chapter 5), are made up of immunoglobulin domains. Thus we can deduce that the immunoglobulin domain is evolutionarily ancient. The Thy-1 protein (CD90) is one of the most ancient; it contains a single immunoglobulin domain and is present in all vertebrates and even in squid, an invertebrate. All molecules containing immunoglobulin domains are members of the immunoglobulin superfamily. As happens so often the adaptive immune system has utilised and diversified a pre-existing structure for its own purposes.

6.2.2 Antibody Structures

The characteristic structure of antibodies was initially worked out using biochemical techniques. We now know that the classical Y-shaped antibody monomer is composed of two H chains and two smaller L chains that are disulphide linked to each other (Section 1.5.3.2). Each H and L chain has one portion, called the variable (V) region, that come together to form the binding site for antigen, so that each immunoglobulin monomer is bivalent. Each V region contains three stretches of amino acids that differ greatly in sequence between antibodies of different specificities, and are therefore known as hypervariable regions. In an antibody, when a H chain V region combines with a L chain V region, these hypervariable regions are brought together to form the

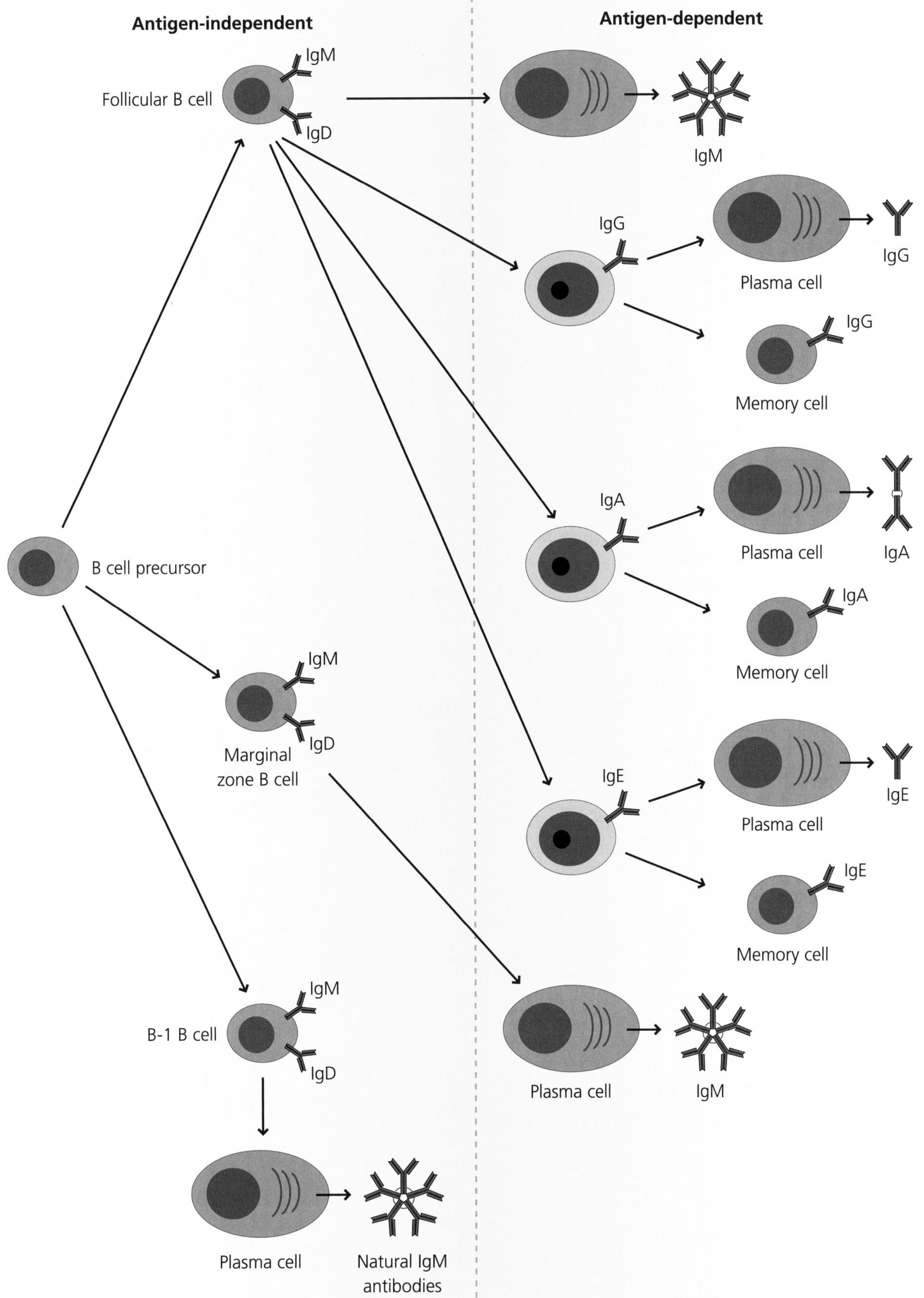

Fig. 6.1 B cell populations. The same B cell precursor can probably give rise to different B cell populations. All subsets initially express IgM and most express IgD. Activated follicular B cells can develop into plasma cells that secrete IgM and some IgD (not shown). These B cells can also switch the class of immunoglobulin they express to IgG, IgA or IgE and can develop into plasma cells or memory cells. Activated marginal zone B cells can become plasma cells that make IgM antibodies, typically to carbohydrate antigens, but they do not switch to most other immunoglobulin classes or become memory cells. Some B-1 B cells make natural IgM antibodies in the apparent absence of any antigenic stimulation and perhaps some IgA.

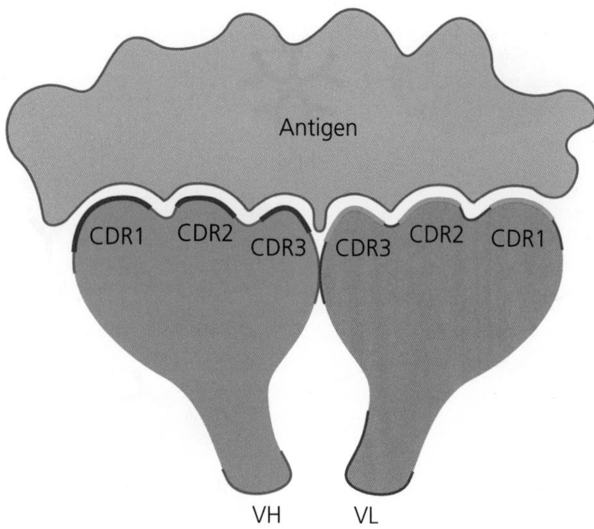

Fig. 6.2 Antigen-binding sites. When an immunoglobulin molecule folds into its tertiary structure, the six hypervariable regions (CDRs) – three each from the H and L chains – are brought together, they form the three-dimensional antigen-binding site, which can vary in structure from a flat surface as shown or a deep cleft (not shown).

antigen-binding site and thus determine the antigenic specificity. As the binding of antibody to antigen is controlled by complementarity in shape, the hypervariable regions are also called complementarity-determining regions (CDRs). They are numbered CDR1, CDR2 and CDR3 for each chain. Similar principles apply to TCRs Section 1.5.3.2. See Figure 6.2.

The remaining portions of the H and L chains are called constant (C) regions. For H chains in humans and mice, there are five main C regions, called μ, δ, γ, α and ε. These define the class (isotype) of antibodies produced – IgM, IgD, IgG, IgA and IgE, respectively. There can two or more forms of different classes of antibodies, particularly IgG and sometimes IgA, and these are called subclasses. For L chains there are two main types of C region, κ and λ, and the corresponding chains that contain them are named accordingly. The C regions of the H chain are entirely responsible for the different tissue localizations and effector functions of the different antibody classes; whether a κ or a λ C region is present in the L chain of an antibody does not seem to have any biological effect at all See Box 6.1.

6.2.3 Generation of Antibody Diversity

It was apparent from early days that any individual could make antibodies against any of a huge number of different antigens. Since these antibodies did not generally cross-react with other antigens it was argued there must be at least an equally large number of antibody genes. At the time it was very difficult to understand the genetic basis of this antibody diversity. Estimates suggested that humans or mice could make more than 10^7 antibodies. Did this require 10^7 genes? This was far beyond the number of genes that had been estimated to exist in the genome. In fact, we know this to be impossible because the human genome project has revealed that humans have only around 30 000 genes in total. Many theories were put forward to explain this paradox, and it was not until the mid-1970s that the answer was found.

Using molecular biological techniques (Southern blotting) it was shown that immunoglobulin genes were arranged differently in B cells compared to other cells types. It was found that segments of antibody genes that were widely separated in the germline DNA of other cells were closely linked in B cells. This suggested a mechanism in which the complete immunoglobulin gene present in a B cell was constructed by bringing together different segments of germline DNA. If there were multiple variants of each segment, a huge number of different antibody genes could be made by random combinatorial association of a relatively small number of segments. Further analysis by many groups showed that this hypothesis was correct.

It is now clear that complete immunoglobulin genes are constructed from different gene segments in humans and mice by rearrangement of DNA. The mechanism by which immunoglobulin gene (and TCR gene) rearrangement occurs in species such as humans and mice is termed somatic recombination because it occurs involves recombination of stretches of DNA in somatic cells rather than germ cells. Generally, this form of DNA rearrangement occurs during the early differentiation of B cells (and T cells) into mature lymphocytes (Section 1.5.3.2).

6.2.3.1 Rearrangement of Immunoglobulin Genes DNA

DNA coding for immunoglobulin genes exists as discrete groups of V, D and/or J, and C gene segments in the genome. A complete H chain gene is made by bringing together one segment from each of the V, D and J groups. The selection of an individual segment from a group is largely at random. The V region segments initially become associated with two of the H chain C region gene segments (μ and δ). Following B cell activation the cell may switch by associating a different C region (γ, α or ε) with its V region gene. The L chain V region genes are similar, but are made from only two segments, V and J. The V region gene segments will associate with either a κ or λ C region segment; κ and λ gene segments are present on different chromosomes. The number of gene segments in each group differs. Thus, in the human H chain region, there are around 40 V segments, 27 D segments and six J segments. In the human L chain regions there are around 40 (κ) or 30 (λ) V segments and five or four J segments. See Figure 6.4.

The order in which the different immunoglobulin gene segments rearranged could be established by examining populations of B cells at different stages of development. In some cases, examination of B cell tumours, in which the cells are arrested at a particular stage in development, was also very

Box 6.1 Antibody Structure: Biochemical Studies, Amino Acid Sequencing, Photoaffinity Labelling and X-Ray Crystallography

When IgG was digested with the enzyme papain, two different fragments were obtained. One of these could be crystallized and was called Fc. The other fragment could not be crystallized but retained the ability to bind antigen and was called Fab. Each Fab fragment could bind one molecule of antigen. This digestion produced two Fabs for every Fc fragment. If another protease, pepsin, was used to digest IgG, only one fragment was produced. This fragment could bind two antigen molecules and the fragment was called $F(ab')_2$. These observations suggested that IgG had a Y-shaped structure and was bivalent in terms of antigen binding. It was then found that, for IgG myeloma proteins (a myeloma is a tumour of the B cell lineage that secretes immunoglobulins), reduction of disulphide bonds released four chains from each molecule. Two were of a higher molecular weight and were called while the heavy chains while the two of a lower molecular weight were called light chains. See Figure 6.3.

One of the mainstays of understanding protein structure is determination of the amino acid sequence. When the H and L immunoglobulin chains from different myeloma proteins were isolated, sequenced and compared it was found that they differed in one region of the molecule and this was called the variable or V region. It was hypothesized that the V region contained the sites that actually bound antigen. The other regions of both H and L chains were much less variable between different myelomas and could be grouped into a small number of subsets with identical or very similar structures. These were called the constant or C regions (Figure 6.3). As more myeloma proteins were sequenced and compared (as well as Bence-Jones proteins, which are just L chains) it became clear that within the V region there were certain areas that varied greatly between different proteins, compared to other areas of the molecule. These were therefore originally termed hypervariable regions and are now more commonly known as CDRs (Figure 6.2).

It was not difficult to surmise that the amino acid sequence of the hypervariable regions, and hence their three-dimensional structures, determined the antigen-binding specificity of antibodies. This was confirmed by photoaffinity labelling. Antibodies were made (e.g. by immunizing mice or rabbits) against small molecules that were photoreactive (i.e. they would form covalent bonds with adjacent amino acids when exposed to light). When antibodies were incubated with their photoreactive antigens in the dark, and then were illuminated and subsequently digested with proteases, the antigen was found to be cross-linked to the hypervariable regions

The topography of antigen binding to antibody was eventually elucidated by other techniques such as electron microscopy, nuclear magnetic resonance and X-ray crystallography. In the small number of complexes where antibodies could actually be crystallised with their bound antigens, it was discovered that when the V regions of the H and L chains associated to form the three-dimensional antibody, the six hypervariable regions were brought together to form a cleft or surface where the antigen bound. Unlike the TCR, which always binds peptide-MHC complexes, antibodies can bind to a huge variety of different three-dimensional molecular shapes. This is reflected in the shape of the antigen-binding site. It can vary from a deep cleft into which small molecules can almost entirely fit, to a broad surface that can bind to parts of proteins in their native (folded) conformations. It is also possible to identify the individual atomic interactions that enable the antigen to bind non-covalently to the antibody.

informative. We now know that rearrangement of the H chain segments occurs first and this is followed by the L chain. For H chains, one D segment joins to one J segment and this rearranged DJ segment then joins to one V segment. In L chains, there is just V–J joining. For both H and L chains, the respective V gene segments encode both CDR1 and CDR2 while the junctions (from V to J, with or without D depending on the chain) encode CDR3. If all the V, D and J segments were used, they could generate around 6500 H chains and 200 L chains. Since any recombined H chain can in theory join with any L chain, this combinatorial process could generate around 1.3×10^6 antibody specificities. However, the actual potential diversity is much, much higher because additional mechanisms can introduce further variability (below).

Two important enzymes involved in joining the different V gene segments are the recombinase-activating gene (RAG) endonucleases RAG-1 and RAG-2. These proteins recognize recombination signal sequences (RSSs) (Section 5.5.1.3) that flank the gene segments, and these sequences are arranged so that the immunoglobulin gene segments can only join appropriately. So, for example, a H chain V segment is normally prevented from joining directly to a H chain J segment. Human patients who possess a defective RAG gene are unable to make either BCRs or TCRs, and the developing lymphocytes stop further differentiation and die. As these patients lack mature lymphocytes, they are highly susceptible to life-threatening infections; they suffer from severe combined immunodeficiency(SCID). That a RAG gene defect is sufficient to cause SCID

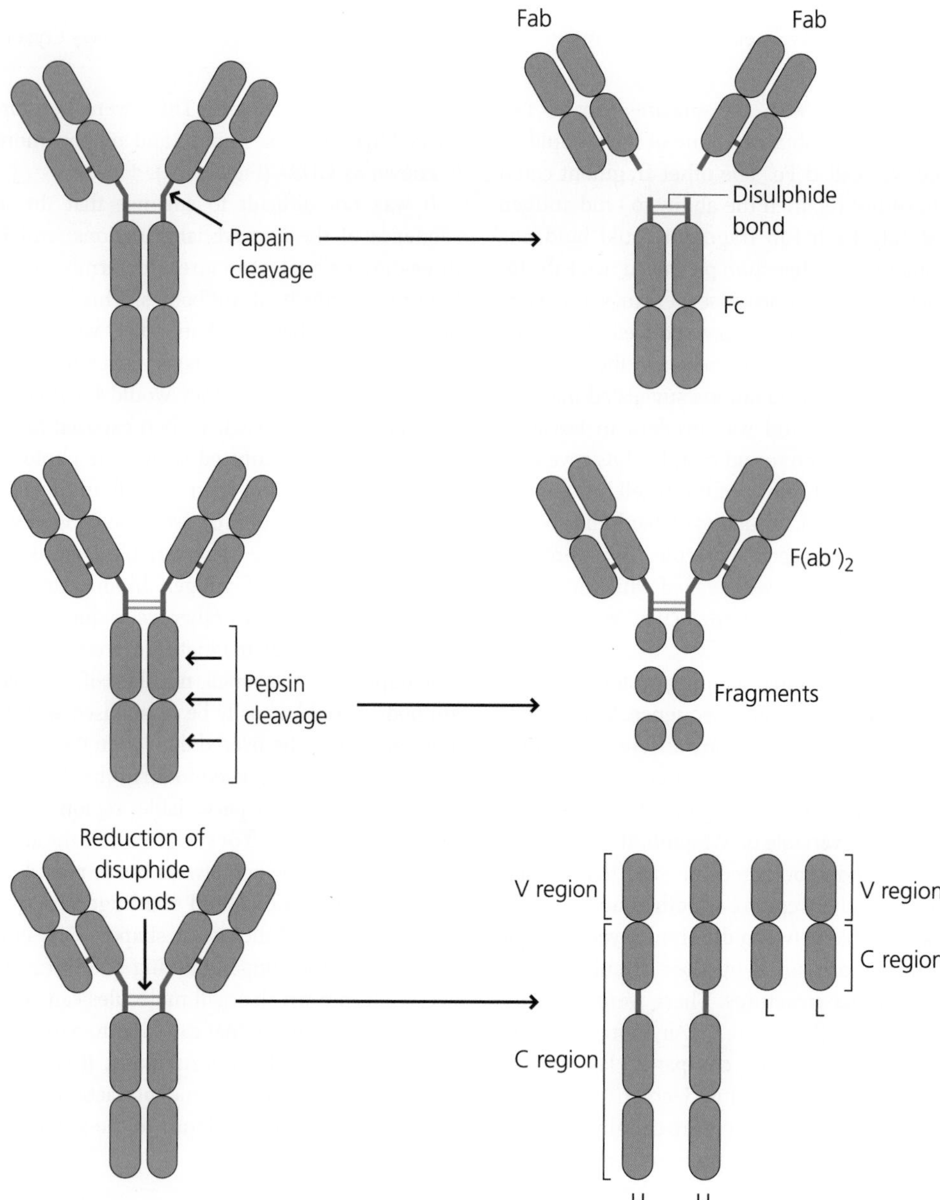

Fig. 6.3 Studies of immunoglobulin structure. When antibodies were digested with the protease papain, fragments of two sizes were obtained. One fragment could bind antigen and was called Fab; this fragment could not be crystallized. The other fragment did not bind antigen but could be crystallized and was called Fc. If however, immunoglobulin was digested with pepsin, a single large fragment was obtained that could bind two antigen molecules and was called $F(ab')_2$. When disulphide bonds were reduced, two complete H and L chains were released from each antibody monomer. The V and C regions of each chain are indicated.

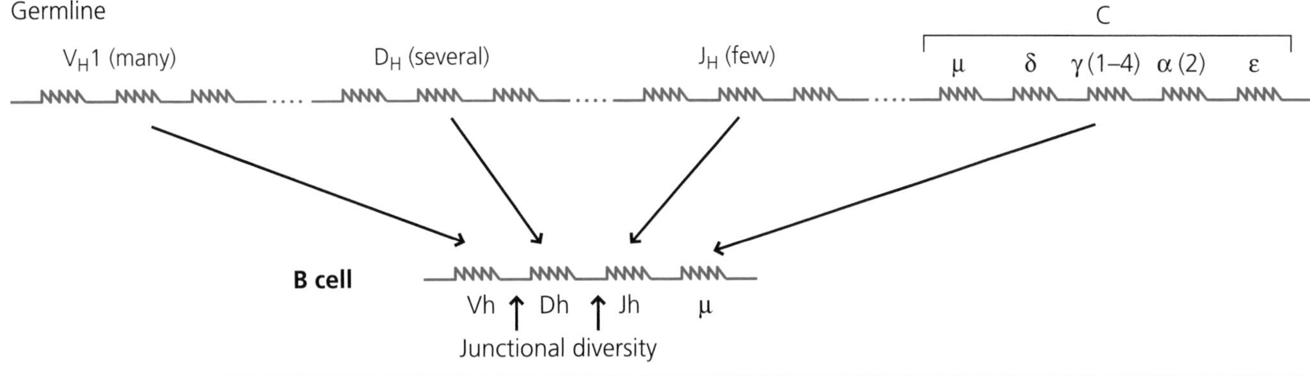

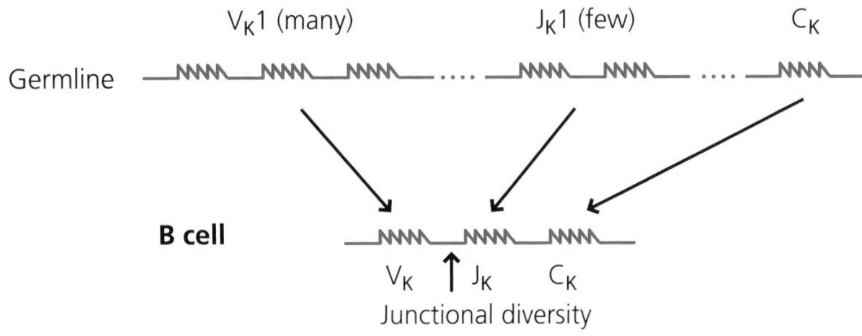

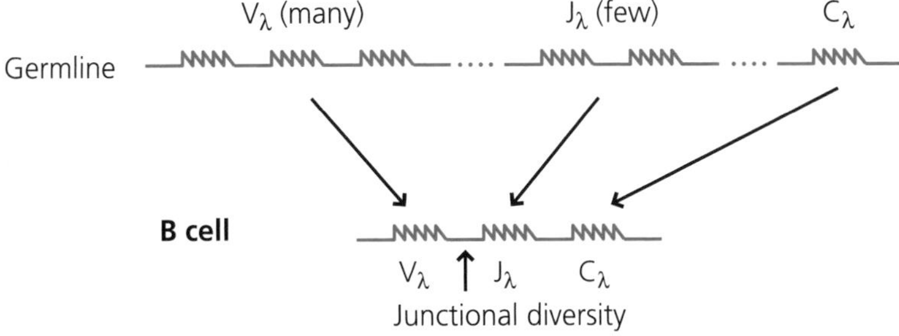

Fig. 6.4 Generation of immunoglobulin genes. Complete immunoglobulin genes are found only in B cells. In germline DNA, three separate loci on different chromosomes code for the H chain and the κ and λ L chains. At each locus there are multiple gene segments that code for parts of the V region (and C regions). The H chain region contains numerous V_H segments, a number of D_H segments and a small number of J_H segments. In B cells, one segment from each group is selected (largely randomly) and the selected segments are brought together to form the complete V region. The V region is paired initially with the μ (and δ; not shown) C region segments. For L chains the principle is similar, but there are no D region segments and the V region pairs with either κ or λ C region segments. During assembly of the complete H and L chain genes, additional bases can be inserted at the junctions between V, (D) and J segments, increasing junctional diversity (see Chapter 5, Figure 5.28 for details). The order of the C region genes is shown in a simplified form for humans.

has been confirmed in mice in which one of the RAG genes has been knocked out these mice have neither T nor B cells.

As mentioned, several other mechanisms contribute to diversity by varying the nucleotides present at the junctions between the gene segments. This junctional diversity is introduced by two different processes (Figure 5.28). First, an enzyme called Artemis acts after RAG during the joining process and may generate staggered breaks in the DNA. Depending on how this happens subsequent DNA replication at these sites can generate palindromic sequences, which are called P-regions. Not surprisingly, gene defects in Artemis are another cause of SCID. Second, another enzyme, terminal deoxynucleotidyl transferase (TdT), can insert additional nucleotides into the junctional sites and this gives rise to

Box 6.2 Generation of Antibody Diversity in Other Species

It is crucial for adaptive immunity that animals possess a diverse set of antigen receptors and that these are present in advance of infection – they are anticipatory. Different groups of vertebrates have, however, evolved somewhat different mechanisms to achieve this diversity. Humans and mice, as described above, generate their primary immunoglobulin diversity through somatic recombination of gene segments selected randomly from a pool of segments arranged linearly in the genome during B cell development in the bone marrow. This is not, however, true for all vertebrates.

In chickens, B cells are formed in the Bursa of Fabricius – a sac opening into the cloacae. In this species gene conversion, rather than somatic recombination, is the major mechanism involved in diversity generation. Chickens possess only one functional V_H and one V_L gene. Diversity is generated by inserting short sequences from other pseudo-genes (non-functional immunoglobulin genes) into the V regions of H and L chain genes.

In sheep and rabbits gene conversion is important, but somatic hypermutation (Section 6.3.3.8) also contributes to the primary repertoire. In the light of this it is intriguing that considerable B cell development and immunoglobulin diversification in these species (and in the chicken) occurs in gut-associated lymphoid tissue, including the appendix and Peyer's patches that of course are in contact with commensal bacteria after birth (or hatching). Some diversification occurs before this contact, but it seems likely that the bacteria drive further diversification.

In some cartilaginous fishes, such as sharks, there are substantial numbers of pre-existing cassettes each containing one V segment, one or more D segments, one J segment and a single set of C regions. In contrast, there is no evidence for immunoglobulin molecules (or TCRs) in the phylogenetically more ancient jawless fish such as lampreys and hagfish. Remarkably, however, it has more recently been demonstrated that these jawless fish do contain a, structurally, entirely different set of rearranging antigen receptors that have been termed variable lymphocyte receptors (VLRs). These appear to represent a dead-end in evolution, at least as far as jawed vertebrates are concerned.

so-called N-regions. The generation of P- and N-regions vastly increases the diversity that can be created during the construction of immunoglobulin genes and hence of the antibodies that can be formed. In humans, it is estimated that around 5×10^{13} different antibodies could potentially be generated; in contrast, the total number of B cells in an adult human is probably less than this by at least two orders of magnitude.

6.2.3.2 Allelic and Isotype Exclusion

All vertebrate somatic cells (i.e. other than the sperm and egg germ cells) are diploid; in other words, they possess a set of chromosomes from both parents. In theory it would be possible for the maternal and paternal immunoglobulin V gene segments of both the H and L chains to be rearranged and expressed in any given B cell, and to make and express several antibodies of different specificities. We know, however, that each B cell only makes one antibody specificity. The reason for this is, in part, that if there is a successful rearrangement on one chromosome, this inhibits rearrangement on the other chromosome, so either the maternal or the paternal allele can be expressed, but not both. This process is called allelic exclusion. In mammals, allelic exclusion only occurs for these immunoglobulin genes, the TCR β (but not α) genes, and the nasal olfactory receptors for the sense of smell. Moreover, because there are two different L chains loci (κ and λ) it is essential that if one rearranges successfully, the other is inhibited. This is called isotype exclusion. (It is of course also essential that rearrangement stops as soon as a functional immunoglobulin chain is produced from any locus, otherwise a B cell could keep producing an almost limitless number of antibody specificities during its life.)

Q6.2. If allelic and isotype exclusion did not occur, how many different antibodies could be generated in any given B cell?

6.2.4 Antibody Classes (Isotypes) and Their Properties

As we outlined above, antibodies can be grouped into five main immunoglobulin classes or isotypes by the characteristics of the C regions of their H chains. In some cases there are different versions of some of the isotypes called immunoglobulin subclasses. It is the C region that determines all the biological properties of an antibody, other than its antigen-binding specificity, and it is these properties that determine how and where antibodies work in defence against infection. See Figure 6.5.

6.2.4.1 IgM

All naïve B cells express surface IgM (they also co-express IgD of the same specificity; below). Secreted IgM is composed of five immunoglobulin units that form a pentamer joined by a polypeptide chain, the J chain. Thus, IgM is a very large molecule with 10 potential antigen-binding sites (in most cases an IgM molecule cannot bind 10 antigen molecules because of steric interference). Each binding site will have an

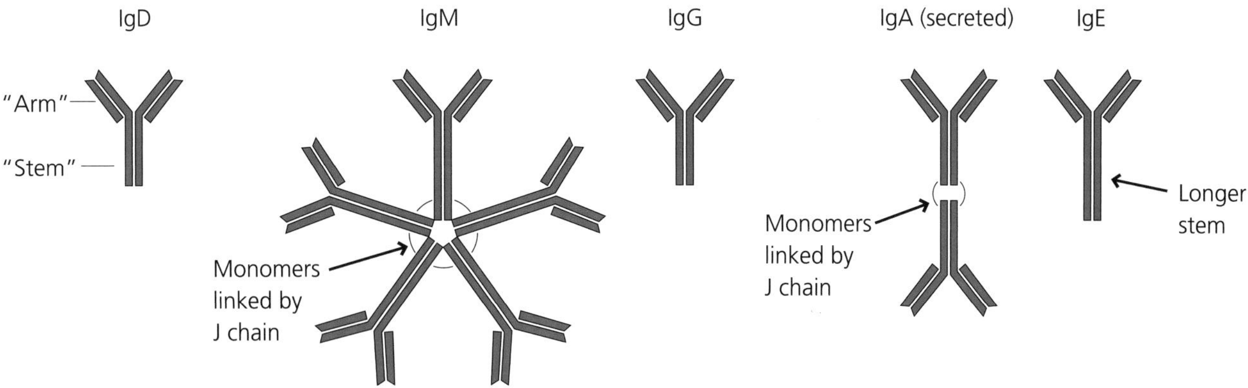

Fig. 6.5 Structures of different classes of antibodies. The five main classes of antibodies all have a similar monomeric structure. Some classes can also form multimers, such as IgM pentamers or IgA dimers in which the monomers are linked by another component called the J chain. However, the detailed structure of the monomers differs for each class (and subclass) of antibody. These differences lead to different properties for each antibody. For example, the arms of the monomer may be relatively rigid or flexible. The stem may be short or extended. The biological properties of antibodies are determined by the structure of the stem, enabling it to be recognized by different types of FcR, for example, or to activate complement. (Immunoglobulin subclasses for IgG and IgA are not shown.)

Box 6.3 Antibody Affinity and Avidity: Measurement of Rate Constants

The efficacy of an antibody in defence against infection is strongly dependent on its remaining bound to its antigen. The strength of the binding between a single antigen-binding site and a single antigenic determinant (epitope) is known as the affinity. Antibody affinity was first measured using binding to haptens. Haptens are small molecules (e.g. DNP). Haptens cannot stimulate antibody formation on their own but will do so if they are coupled to a protein molecule as a carrier. The affinity of hapten binding to antibody can be measured using the classical technique of equilibrium dialysis. In this technique an antibody is separated from a radiolabelled hapten by a semi-permeable dialysis membrane. This allows the hapten to cross the membrane and bind to the antibody, but not vice versa, and the concentrations of hapten on the two sides of the membrane can be measured when equilibrium is reached. The difference in concentration depends on the rate of association and dissociation of the hapten to and from the antibody, and the affinity can subsequently be calculated using a mathematical technique known as Scatchard analysis. While very useful for assessing the affinity of haptens, this approach cannot be used for measuring affinities of receptors of larger ligands, such as other proteins that cannot diffuse across a dialysis membrane.

The measurement of ligand-receptor binding has been greatly facilitated by using surface plasmon resonance (SPR). In this technique light is reflected off a gold-coated glass slide and this reflection generates SPR, which can be measured. If a molecule is bound to the glass this alters SPR quantitatively. If a solution containing a ligand for the bound molecule is allowed to flow over the slide, changes in SPR will be proportional to the amount of ligand that is bound. Thus, ligand–receptor binding can be measured quantitatively in real-time. These studies have revealed that antibodies bind to their antigens with variable but high affinity (dissociation constant, K_d from 10^{-6} to 10^{-10} M); in contrast other studies have demonstrated that the binding of a TCR to its peptide–MHC complex is much weaker (K_d from 10^{-4} to 10^{-7} M).

Secreted antibodies are always at least bivalent. The strength of the binding between a multivalent antibody and an antigen expressing multiple identical epitopes (e.g. on the surface of a bacterium) is known as the avidity. You might think that for IgG the avidity would be twice the affinity (two binding sites are involved). In fact measurement of the avidity of antigen–antibody interactions shows that the avidity may be a 100-fold or more higher than the affinity. This is because of receptor co-operativity. Because antibody–antigen interactions are non-covalent, receptors and epitopes are continually associating and dissociating. If the interaction is multivalent, the probability that all receptors will separate from their epitopes at any one time is very low compared to the probability of a single receptor epitope separating, thus it is much more likely that the receptor will remain bound. This phenomenon is biologically very important. It means that an antibodies such as IgM may have a low affinity, but a very high avidity (more so than IgG), and be equally or more effective in defence. See Figure 6.6.

Q6.3. What might be the biological relevance of the extremely different typical affinities of antibodies and TCRs for their respective antigens?

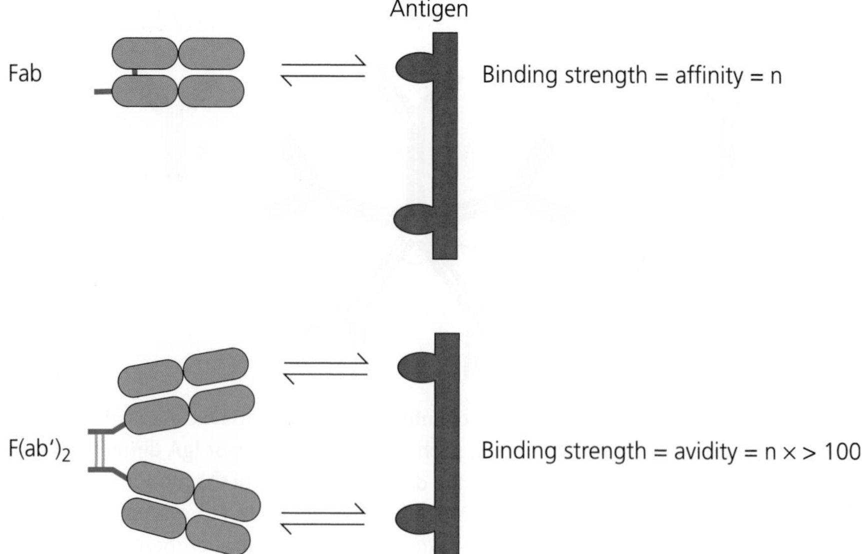

Fig. 6.6 Antibody affinity and avidity. The affinity of an antibody refers to the strength of binding of a single antigen-binding site to a single antigenic determinant (epitope). Antibody avidity refers to the overall strength of binding between all the antigen-binding sites and an antigen that expresses more than one identical epitope. The avidity of a divalent $F(ab')_2$ fragment is vastly greater than twice the affinity of a monomeric Fab fragment of the same antibody. Likewise, the avidity for a divalent IgG antibody may be 100 times its affinity, and for pentameric IgM 1000 times or more.

identical affinity for an antigen, but multi-valency dramatically increases the avidity of the IgM molecule for its antigen.

Under normal conditions, IgM is found only in the blood because it cannot cross the endothelial barrier. You will, however, remember that during acute inflammation, venule endothelial cells separate from each other and this allows IgM to enter extravascular inflamed tissues. How does IgM act in defence? Its main role is against pyogenic infections and, as we describe in Section 2.3.1.3, the main type of effector cell for these infections is the neutrophil. Phagocytes, such as human neutrophils, do not have Fc receptors (FcRs) for IgM, and IgM thus cannot act as a direct opsonin. IgM is, however, a very efficient complement activator. As secreted IgM is pentameric it has five Fc regions in very close proximity to each other, and this allows the C1q complement component to bind and to trigger the cascade (Section 4.4.2.1). Neutrophils have receptors for derivatives of the C3 complement component (C3b and iC3b), and thus bacteria coated with IgM and complement can be phagocytosed via complement receptors (Figure 2.10).

Natural Antibodies Natural antibodies are so-called because they are formed without any obvious or overt stimulation by foreign antigens. They are mainly IgM; however, some IgG, particularly IgG3 in mice and humans, and perhaps some IgA may also fall into this category. Natural antibodies are broadly reactive (i.e. they react with a range of pathogens with relatively low affinity). They may also be involved in autoimmune diseases such as systemic lupus erythematosus (SLE, Section 7.4.3). Natural antibodies do appear to have a role as first-line defence against infection. For example, they may be able to neutralize bacteria and viruses present in blood. Natural IgG antibodies may also be involved in the induction of some types of TI responses, by binding to blood-borne antigens and targeting them to cells, possibly macrophages, in the marginal zone in the spleen. Some natural antibodies may be autoreactive and it has been suggested that these may be involved in clearance of debris from self cells, and even that failure of this clearance may lead to autoimmunity but this is controversial.

Q6.4. How likely it is that natural antibodies are produced in the complete absence of any antigenic stimulation?

Natural antibodies are formed primarily by a subset of CD5 B-1 cells (called B-1a cells) in the peritoneal and pleural cavities, at least in mice. They are encoded by immunoglobulin genes that preferentially use certain V regions and, because little or no TdT is expressed by these cells, they have very limited N-region junctional variation; they also do not undergo somatic hypermutation (Section 6.3.3.8). Their formation is independent of overt antigenic stimulation and in germ-free mice they are present at similar concentrations to normal mice. Natural antibodies thus in some ways represent an innate immune mechanism, although their efficacy depends on their binding to antigens through a variable antigen-binding site, normally thought of as a characteristic of adaptive immune responses.

Q6.5. What might be the role of commensal organisms in stimulating the production of the intestinal IgA antibodies found in normal individuals?

6.2.4.2 IgD

For many years IgD has been a mystery molecule. It differs from all other immunoglobulin isotypes in that only trace amounts are normally found in the blood (some rare myelomas do, however, secrete large quantities of IgD). IgD is expressed on the surface of most naïve follicular B cells, together with IgM. This is because a rearranged V gene approximated to the μ and δ constant genes comprises a single transcriptional unit, and one or the other can be produced simply by alternative splicing of the mRNA that encodes them. Genetically engineered mice that lack IgD have lower than normal levels of B cells, affinity maturation (Section 6.3.3.8) of their antibody responses is delayed and they also produce lower levels of immunoglobulin isotypes such as IgE that are controlled by the cytokine IL-4 (Section 6.3.3.6). Recently it has been shown that secreted IgD can be produced by class switching, involving DNA rearrangement, that human basophils possess a receptor for IgD and that binding to this receptor can trigger basophil responses, at least in the upper respiratory tract. It therefore seems that IgD may play a role in defence against respiratory bacteria, and may have other functions in immunity and inflammation that are yet to be discovered.

6.2.4.3 IgG

IgG is the most abundant class of antibody in the blood. IgG is a monomeric molecule that comprises several different subclasses with different Fc regions (four in humans and mice). These subclasses have some differences in function such as the ability to activate complement and trigger inflammation, or to directly opsonize microbes for FcR binding (Figure, 1.31 and 2.10). Unlike IgM, most IgG synthesis is CD4 T cell-dependent. Which subclass is produced during adaptive immune responses depends largely on the type of CD4 T cell response that is involved (e.g. T_h1 versus T_h2; see later).

IgG is crucially important for defence against infection in early life, before the adaptive immune system is fully developed. In mice, rats and humans, IgG can cross the placenta via a specialized transport molecule, the neonatal FcR, making it available for protection of the late foetus. Because maternal IgG lasts for several months in the offspring's circulation and tissues it also provides defence for the new-born infant (as does IgA from the colostrum, early milk; below). It is for this reason that the severe infections that are seen in human primary immunodeficiencies such as SCID (above) often do not become apparent until several months after birth, when the maternal antibody has disappeared. See Figure 6.7.

6.2.4.4 IgA

IgA is the most important mucosal antibody. Most IgA is secreted by plasma cells in the connective tissues underlying mucosal epithelium in tissues such as the intestine, the respiratory tract, the uro-genital tract and the lactating mammary gland. IgA is secreted into the connective tissue, but is then actively transported (below) across the epithelial cells into the lumen overlying the epithelium, which is where it carries out its main protective functions. The most important action of IgA is to prevent pathogens and toxins binding to epithelial cells (Figure 2.24). Much IgA is secreted in these sites as a dimer, with the two monomers joined by a J chain (the same molecule that joins IgM monomers to form a pentamer). IgA in blood is mostly found as a monomer. The functions of this form are not entirely clear although it can activate the alternative pathway of complement, and some macrophages possess FcRs for IgA, suggesting it may be involved in opsonization of infectious agents. Two subclasses of IgA exist in humans, both being present in the blood and secretions.

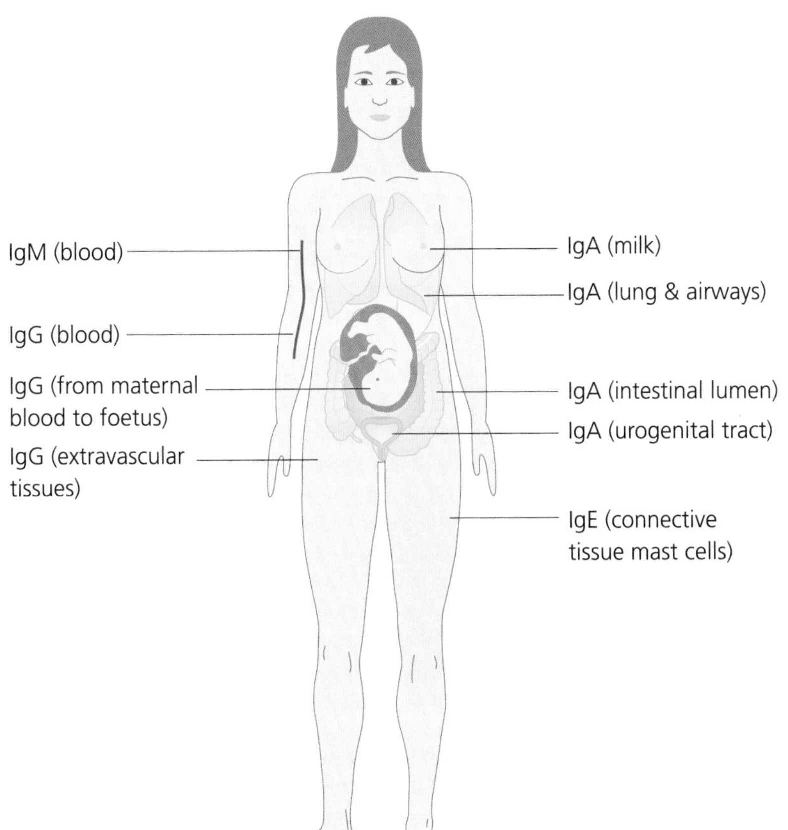

Fig. 6.7 Tissue distribution of antibodies. IgM is confined to the blood plasma unless there is acute inflammation, when it can pass between endothelial cells into extravascular spaces. IgG can cross endothelia into many extravascular spaces and, in addition, can cross the human placenta to give protection to the neonate. Monomeric IgA tends to be present in the blood, but dimeric IgA is transported onto mucosal surfaces such as the gastro-intestinal, respiratory and uro genital tracts, and the lactating breast. IgE is normally present at very low concentrations in blood but binds to mast cells, particularly those underlying the epithelia. IgD is usually only present in significant amounts on the surface of naïve B cells and is normally only present in the blood at exceedingly low levels; not shown.

One subclass (IgA_2) lacks a stretch of 13 amino acids present in the other and shows increased resistance to some bacterial proteases; this may increase its potency on mucosal surfaces. IgA deficiency is one of the commonest immunodeficiencies. In the developed world around 1: 500 people are found to be IgA-deficient. Most are asymptomatic, but some suffer from an increased incidence of respiratory and intestinal infections.

Mucosal Immunity Most infections enter the body through mucosal surfaces. It is thus not surprising that the adaptive immune system has evolved specializations to protect these surfaces. Immune responses to mucosal pathogens may be initiated in draining lymph nodes as for other tissues, but mucosal surfaces such as the intestine possess specialized secondary lymphoid organs in their walls (Section 3.4.1). In the small intestine these are the Peyer's patches Section 3.4.2.7.

The main function of the Peyer's patches and similar lymphoid aggregates appears to be the indeuctioin of IgA synthesis. Antigens from the gut lumen enter Peyer's patches via M cells and encounter dendritic cells (DCs) in the sub-epithelial areas (Figure 3.25). These DCs interact with recirculating naïve T cells and, unless there is very strong stimulation, induce T_h2 differentiation. This may be because the DCs are conditioned by their environment to secrete IL-10 and transforming growth factor (TGF)-β. The activated T cells interact with naïve recirculating B cells, activate them and induce isotype switching to IgA. This switch depends on the T cells secreting cytokines such as IL-4, IL-5 (in mice) and IL-10. The activated IgA lymphoblasts do not secrete IgA locally; rather they exit the lymphoid tissue in the lymph and subsequently enter the blood. These lymphoblasts express adhesion molecules (such as the $\alpha_4\beta_7$ integrin) that are complementary to vascular addressins expressed on mucosal venule endothelium. The most important of these addressins is the mucosal addressin cell adhesion molecule (MadCAM). The lymphoblasts can therefore migrate into mucosal connective tissues and mature into plasma cells secreting dimeric IgA.

Mucosal epithelial cells express a specialized receptor on their baso-lateral surfaces for dimeric IgA, as well as pentameric IgM, called the poly-immunoglobulin receptor. Binding of these antibodies to this receptor is dependent on the J chain present in multimeric IgA and IgM. Having bound the polymeric immunoglobulin, the immunoglobulin–receptor complex is endocytosed by the epithelial cells. Normally, endocytosed molecules are delivered to lysosomes and degraded. However, the vacuole containing the poly-immunoglobulin–receptor complex is treated differently. It is transported across the epithelial cell to the apical surface and its contents are released onto the mucosal surface and thus into the lumen. During transport the poly-immunoglobulin–receptor complex is cleaved, forming a complex which consists of the IgA plus part of the receptor; the latter cleaved fragment is known as the secretory piece. It is this complex that is released onto the mucosal surface, and the secretory piece additionally confers some resistance to proteolytic breakdown. See Figure 6.8.

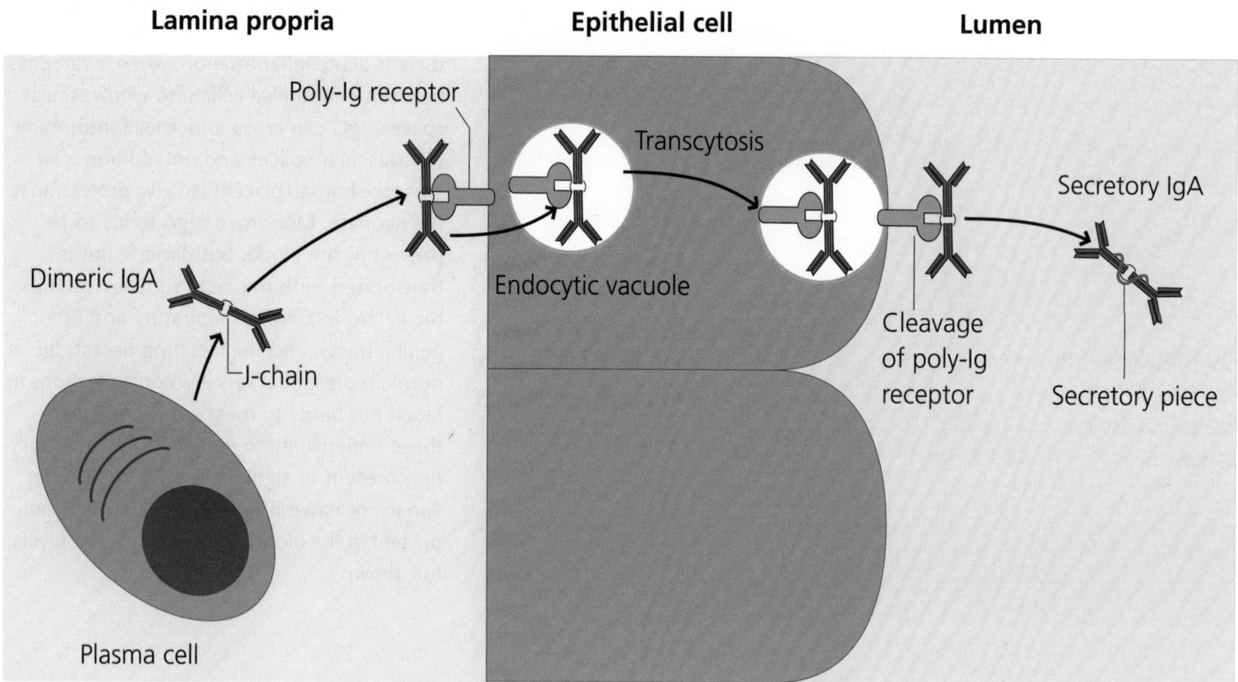

Fig. 6.8 Transport of IgA onto mucosal surfaces. Mucosal plasma cells secrete IgA as dimers, joined by the J chain. Dimeric IgA binds to poly-immunoglobulin receptors on the baso-lateral surfaces of mucosal epithelial cells and is endocytosed. These vacuoles are transported to the luminal surface of the epithelial cell where the poly-immunoglobulin receptor is cleaved. This releases the dimeric IgA, with a portion of the receptor known as the secretory piece, which renders the IgA less susceptible to proteolytic degradation.

Q6.6. If the poly-immunoglobulin receptor can bind both IgA and IgM, why might it be that IgA is the most abundant isotype in mucosal secretions?

The Common Mucosal Immune System The common mucosal immune system evolved so that immune responses initiated in one mucosal tissue can generate effector mechanisms, IgA in particular, that can operate in all mucosal sites. Some of the most dramatic examples of this come from studies on the protective effects of breast milk against intestinal infections. One study examined the incidence of diarrhoeal disease in young children in a large apartment block in Pakistan. The two groups of children being compared were those that were breast-fed and those that were not. The incidence of diarrhoeal disease was much lower in breast-fed children. This clinical observation is consistent with the concept of the common mucosal immune system: that immune responses induced in one mucosal site can lead to IgA (and activated T cells) being distributed to all mucosal sites. Thus, IgA-expressing B cells activated in the intestine against bacteria causing diarrhoea can migrate to the lactating breast, leading to IgA secretion into milk. These observations provide sufficient evidence to justify the encouragement of breast feeding in humans, particularly in the developing world. The common mucosal system can thus provide protection at all mucosal surfaces against an infection in just one site. See Figure 6.9.

Q6.7. The relationship between breast feeding and protection is a correlation, not an explanation. What other interpretations might be possible?

Q6.8. How might the concept of a common mucosal immune system be relevant for the generation of vaccines against mucosal infections?

6.2.4.5 IgE

IgE is a monomeric molecule, but binds strongly to mast cells and eosinophils because these cells express a high-affinity FcR for IgE. IgE production and eosinophil recruitment are often associated. In T_h2-biased responses B cells may switch to make IgE while eosinophils are being recruited to tissues in T_h2-mediated inflammatory responses. IgE probably has a role in immunity to parasites and may act via eosinophils by activating them and stimulating them to degranulate and to kill parasites by antibody-dependent cell-mediated cytotoxicity (ADCC; below), but the relative important of this in host defence is not clear. All sufferers from hay fever should have an interest in IgE because it is the antibody responsible for allergic responses. These include potentially fatal conditions such as the anaphylaxis caused by peanut allergy (Section 7.2.3).

6.2.5 Monoclonal Antibodies

If an animal is immunized with a foreign protein, it will make a very heterogeneous antibody response. Generally speaking, any larger antigen expresses many different epitopes and many different antibodies can be made to any single epitope. For example, the response to ovalbumin in a mouse may consist of hundreds or thousands of different antibodies, binding with different affinities to different epitopes of the protein. This is known as a polyclonal response, since many different clones of B cells have been activated and are making antibodies. Additionally, if different animals are immunized with the same antigen they will make a different set of antibodies, even if the animals are genetically identical, and the antigen is delivered at the same dose, by the same route and exactly under the same conditions. The outcome of this heterogeneous type of response is that it is difficult, if not impossible, to produce standardized antibodies by normal immunization. This is a real problem if antibodies are to be used therapeutically, or if they are to be used as reagents in clinical or experimental settings. In addition, even if large animals such as horses, sheep or cows are immunized, the amounts of antibody that can be produced are limited in relation to the amounts needed to treat human disease.

Q6.9. Why might genetically identical animals make different antibody responses to the same antigen?

Q6.10. What might be the limitations of using antibodies from another species to treat clinical disease (e.g. for passive immunization against tetanus)?

Many workers had thought about the possibility of producing antibodies of a single specificity in bulk (i.e. monoclonal as opposed to polyclonal antibodies). It might seem simple to take single antibody-forming cells or their precursors and to culture them *in vitro* to form clones of cells that would all produce the same antibody molecule. The difficulty is that, in general, antibody-forming cells quickly die in culture. In contrast, myelomas (tumours of antibody-forming cells; above) are immortal and will go on dividing indefinitely in culture. In the early 1970s, Kohler and Milstein fused antibody-forming cells, induced by immunizing a mouse with a given antigen, with myeloma cells that had lost their ability to secrete their own immunoglobulin. The resulting fused cells, called hybridomas, produce antibodies against the antigen used for immunization and are immortal. The hybridomas, were then tested (screened) to identify clones that synthesized the desired antibodies. Because the hybridoma contains only one functional H and L chain immunoglobulin gene, it makes antibodies of only a single specificity and is therefore monoclonal. Hybridomas continue to divide indefinitely and to produce the same antibodies, permitting the production of monoclonal

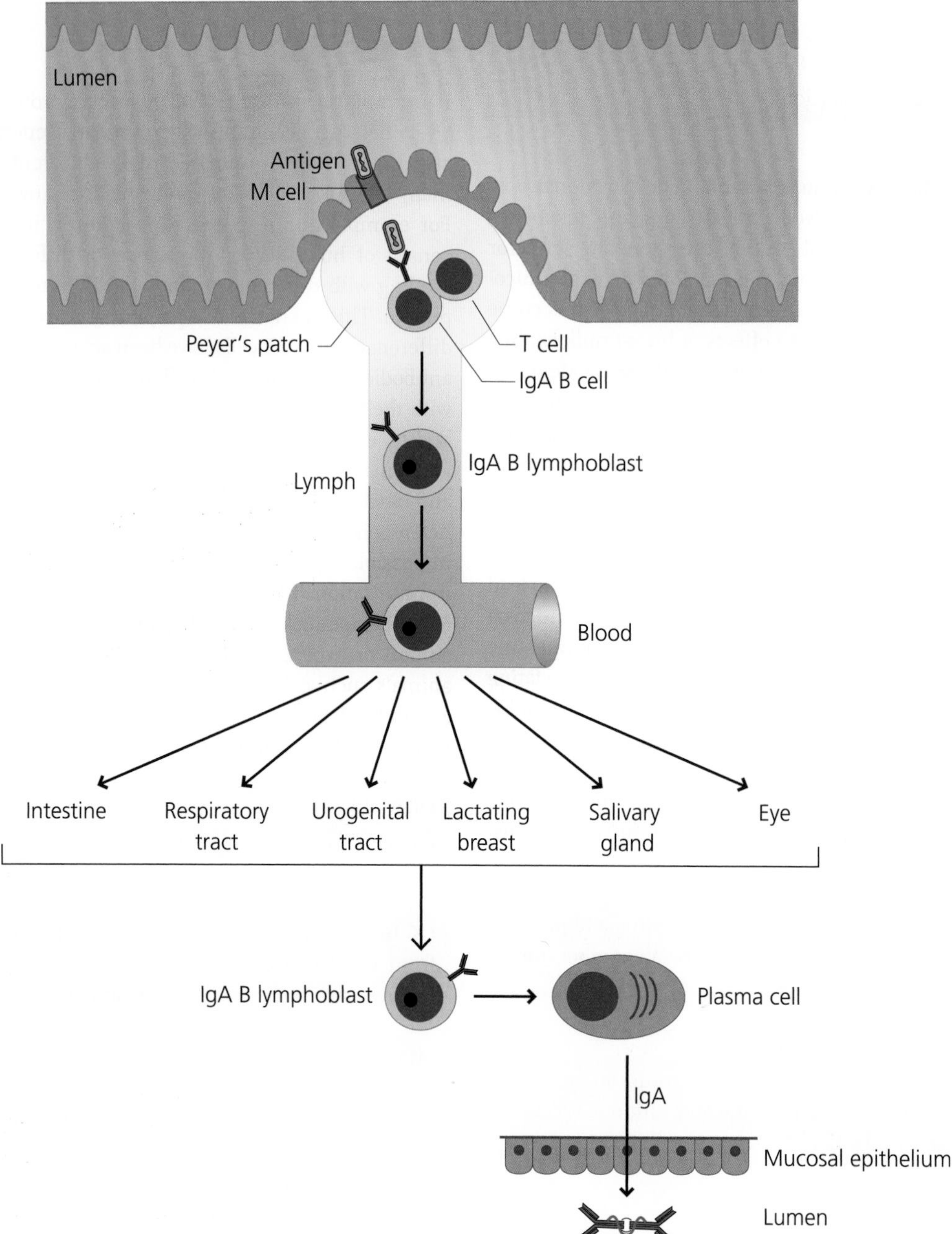

Fig. 6.9 The common mucosal immune system. Antigens from the gut lumen, for example, cross M cells into Peyer's patches (or similar secondary lymphoid organ in other mucosal sites) and, with T cell help, develop into B cells that express IgA. These cells then leave these secondary lymphoid tissues and enter the blood. They express homing receptors for adhesion molecules present on all mucosal venule endothelial cells and thus can migrate, relatively randomly, to any mucosal tissue. Within these tissues they mature into plasma cells secreting dimeric IgA which is transported onto the luminal surface of any mucosal tissue.

antibodies in large amounts sufficient for example in human therapy. See Figures 6.10 and 6.11.

Q6.11. In what ways might hybridomas might be screened for production of a monoclonal antibody against a given antigen?

The monoclonal antibody technique has revolutionized immunology, and gained the award of a Nobel Prize to Kohler and Milstein. In the context of immunological research, monoclonal antibodies can be made against molecules expressed at very low abundance (e.g. receptors on cell membranes), enabling the separation of cell subsets and the functional characterization of many cell surface molecules. In terms of therapy, standardized monoclonal antibodies can be grown on the industrial scales that are essential for clinical use; these are therefore called therapeutic antibodies (Section 6.6). In addition, highly specific monoclonal

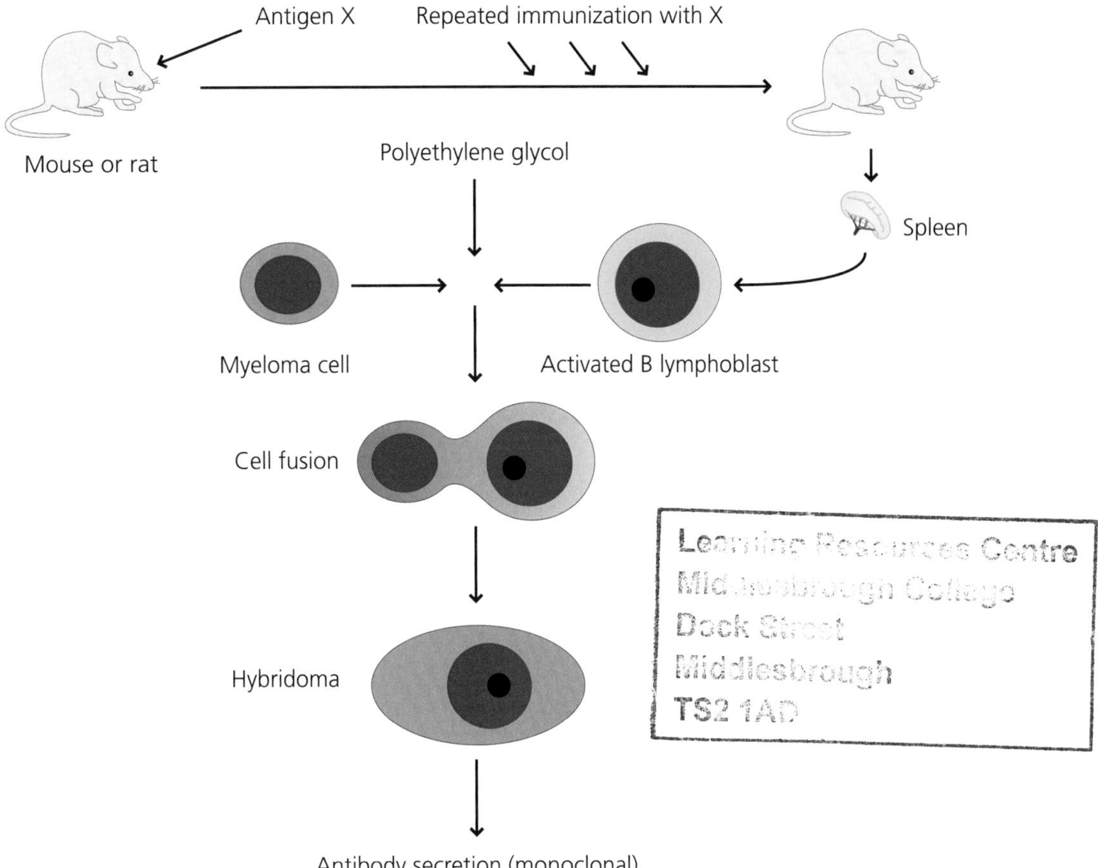

Fig. 6.10 Monoclonal antibody production. Mice (or another species such as rats) are immunized repeatedly with an antigen. Activated B cells (lymphoblasts) from the spleen are chemically fused (e.g. using polyethylene glycol) with non-antibody-secreting B tumour cells (myelomas), to form hybridomas. Each hybridoma secretes antibodies of a single specificity against the antigen used for immunization (i.e. they are monoclonal). The hybridomas are effectively immortal and will continue to proliferate indefinitely. They can be screened to select those of desired specificity (Figure 6.11).

antibodies can be produced as reagents for a whole range of sciences, not just immunology See Box 6.4.

Q6.12. In what areas other than immunology might monoclonal antibodies might be of use?

6.2.6 Antibody Functions in Defence Against Infection

In this section we will illustrate the ways in which antibodies work in the prevention of infection, in recovery from infection and in resistance to re-infection. There are only a small number of ways in which antibodies can function in defence. Some antibodies can act directly by binding to an antigen and blocking interaction with a host cell (neutralization). Antibodies may also activate complement leading to activation of different effector functions mediated by molecules produced during complement activation, including the induction of acute inflammation. Most, if not a the other functions of antibodies are mediated by binding of different isotypes of antibody to FcRs that are specific for those isotypes and which have different cellular distributions (Section 6.3.3.7).

6.2.6.1 Neutralization

Toxin Neutralization Bacterial exotoxins (Section 2.4.2.1) are secreted molecules that act on host cells. Exotoxins are mainly protein molecules and they need to bind to receptor molecules on the plasma membrane of a cell to damage or kill the cell. In such cases, the activity of antibodies is very straightforward. If they block the site on the toxin molecule that binds to the plasma membrane, the toxin cannot bind and is rendered ineffective. This is how vaccination against tetanus works (see Case Study 2.1) and similar mechanisms operate for cholera and diphtheria. In the case of cholera, anti-toxin IgA prevents the toxin binding to intestinal epithelial cells. See Figure 6.12.

Blocking Pathogen Binding In many cases, pathogens (all viruses, some bacteria and some protozoal parasites) need to bind to cells to exert their pathogenicity. As with anti-toxins, if an antibody binds to a molecule needed by the pathogen for binding to a cell, it can bring about protection by blocking binding. This is why patients who have recovered from influenza are resistant to infection with the same strain of the virus (see Case Study 2.6). In the case of cholera, IgA, as well as neutralizing the toxin, prevents the cholera bacterium from

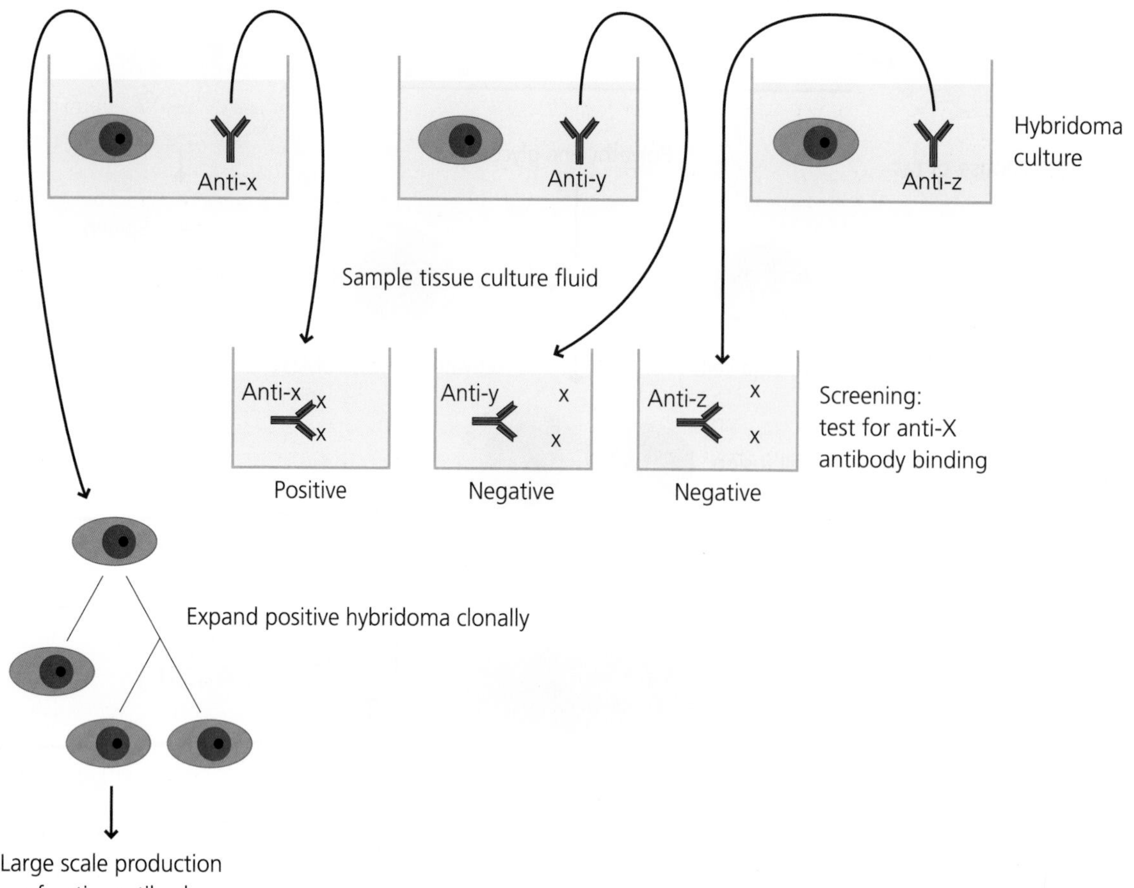

Fig. 6.11 Monoclonal antibody screening. The products of the cell fusion (See Figure 6.10) are divided into small wells of a culture vessel at limiting dilution, so most wells only contain one hybridoma cell. Non-fused B cells die rapidly and the culture medium is designed so that non-fused myeloma cells cannot survive. The fluid from individual cultures can then be tested (screened) for the presence of antibodies specific for a single epitope of the antigen (here, anti-X). Cells from selected cultures are grown as clones and expanded. These hybridomas can be grown on industrial scales and generate sufficient monoclonal antibodies to be used clinically.

binding to intestinal epithelia, permitting it to be eliminated in the faeces. See Figures 6.12 and 2.28.

6.2.6.2 Pathogen Opsonization

The delivery of pathogens to phagocytes by antibodies is a crucial part of resistance to many microbes, particularly extracellular pyogenic bacteria such as Staphylococcus and Streptococcus. IgM (with complement) and some isotypes of IgG are the most important opsonins. This reflects the expression of specific FcRs for these isotypes (and complement receptors) on neutrophils and macrophages. After uptake, these infectious agents can be killed intracellularly within the phagocyte. In some cases, however, pathogens can survive within phagocytes and in these infections opsonization may have no protective effect. Thus, Mycobacterium tuberculosis can survive inside neutrophils and non-activated macrophages, and all the antibodies do is to deliver the bacterium to its preferred target cells. It is also possible that antibodies may exacerbate infectious disease. Thus, in Dengue fever and yellow fever pre-existing antibodies to the respective flavivirus can be associated with increased disease severity. It has been suggested that the antibodies are serving to target the virus selectively to its preferred host cell, once again the macrophage; this is known as antibody-mediated enhancement of infection. See Figure 6.13.

> **Q6.13.** Monoclonal antibodies to CD4, CD8 and CD25 have been used to deplete T cell subsets *in vivo* in mice. Why might it be of relevance that other cells also express these molecules?

> **Q6.14.** Given that the human immunodeficiency virus (HIV) can live and replicate within macrophages, how should this make us think about vaccine design?

6.2.6.3 Induction of Acute Inflammation

Complement Activation Acute inflammation is central to defence against many pathogens, particularly pyogenic bacteria. While acute inflammation is primarily a manifestation of innate immunity, antibodies can play an important role. IgM and some isotypes of IgG are very potent complement

Box 6.4 Uses of Monoclonal Antibodies in Immunology

Monoclonal antibodies are very versatile, and are being used in a wide variety of molecular and cellular studies, as well as having therapeutic and diagnostic applications. These include some of the following.

Identification of Novel Molecules

Many important molecules (CD4 and CD8 are two examples) have been identified by making monoclonal antibodies against whole cells (lymphocytes in this case). A major advantage of the monoclonal antibody approach is that molecules present in very low copy numbers on a cell may be identified. This approach has been used for whole cells, for fractionated cells, or for complex mixtures of molecules such as in serum.

Antigen Isolation

If a monoclonal antibody is bound to beads, and the beads are mixed with a tissue homogenate containing specific antigen, the antigen will bind to the beads. Then by centrifugation, washing and then altering the ionic conditions, pure antigen can be isolated and characterized. Alternatively, if the beads are magnetic they may be separated by passing the suspension over a magnet Figure 4.7.

Antigen Detection

The binding of a monoclonal antibody to its specific antigen can be detected using direct labelling of the antibody, or more often by using a labelled secondary antibody that will bind to the monoclonal. The use of a secondary antibody permits amplification of the signal (more than one secondary antibody may bind to the monoclonal) and also means that a single secondary antibody can be used for many monoclonals antibodies. Labels are often fluorescent but can be radioactive, enzymes such as horseradish peroxidase, or, for electron microscopy, gold particles.

- **Immunocytochemistry**: By using monoclonals antibodies on histological tissue sections or on cells attached to a microscope slide, the localization of a particular antigen can be defined. Fluorescence microscopy can involve the use of several different monoclonals, each identified by a different fluorophore, to localize different molecules in cells or tissues on the same preparation.
- **Flow cytometry and Fluorescence-Activated Cell Sorting (FACS)**: The same principle is widely used to detect antigens expressed on the surface of cells in suspensions (e.g. collected from blood or released from a lymph node). In the flow cytometer, a stream of cells is passed very rapidly (up to 3000 or more cells per second) though a laser beam that excites the fluorescence on labelled cells. The fluorescence is measured quantitatively by a detector and the intensity of fluorescence on individual cells can be expressed graphically. Multiple different fluorescent labels can be used, and other parameters such as cell size and granularity are also assessed. Alternatively, a flow cytometric approach, called cell sorting, can be adopted to physically separate cells into those that do or do not express a given molecule (e.g. CD4 or CD8 in a mixed populations), enabling extremely pure populations of one or the other to be isolated (e.g. for further studies in culture or *in vivo*)
- **ELISPOT assays**: It is often important to know how many cells in a population are secreting a particular molecule (e.g. a cytokine). In the ELISPOT assay, cells are cultured on a plate to which an anti-cytokine antibody is bound. The cytokine secreted by the cells binds around the cell secreting it to the antibody and can be detected as a spot by using a labelled second antibody similarly to the ELISA assay Figure 4.10.

Functional Studies

Monoclonal antibodies to a receptor on a cell surface can block the binding of the ligand (as an antagonist) or in some cases can mimic the ligand and activate the receptor (as an agonist). Thus these monoclonal antibodies can be used to study receptor function. There are, for example, antagonist and agonist monoclonal antibodies against CD40, and these have proved important in understanding T and B cell activation. Monoclonal antibodies can also be used to deplete cells *in vivo*. Thus, anti-CD8 monoclonal antibodies have been used to deplete CD8 T cells. Beware, however, the possibility that more than one cell type may actually express the molecule used for depletion.

activators when bound to pathogens, and activated complement mediators such as C3a and C5a are crucial mediators of inflammation Section 4.4.2 (Figure 4.19). Acute inflammation serves to increase vascular permeability, permitting egress of macromolecules such complement and IgM, and to recruit the effector cells, primarily neutrophils that kill the bacteria (Sections 2.3.1.3 and 4.4.3). See Figure 6.14.

Sensitization of Mast Cells, Basophils and Eosinophils As noted (above) when IgE is produced it binds to high-affinity FcRs on mast cells that are resident in connective and mucosal tissues, or to similar receptors on basophils or eosinophils that have been recruited to inflammatory sites. Subsequent exposure to the same antigen can then lead to cross-linking of the FcRs, degranulation with the release of pre-formed mediators

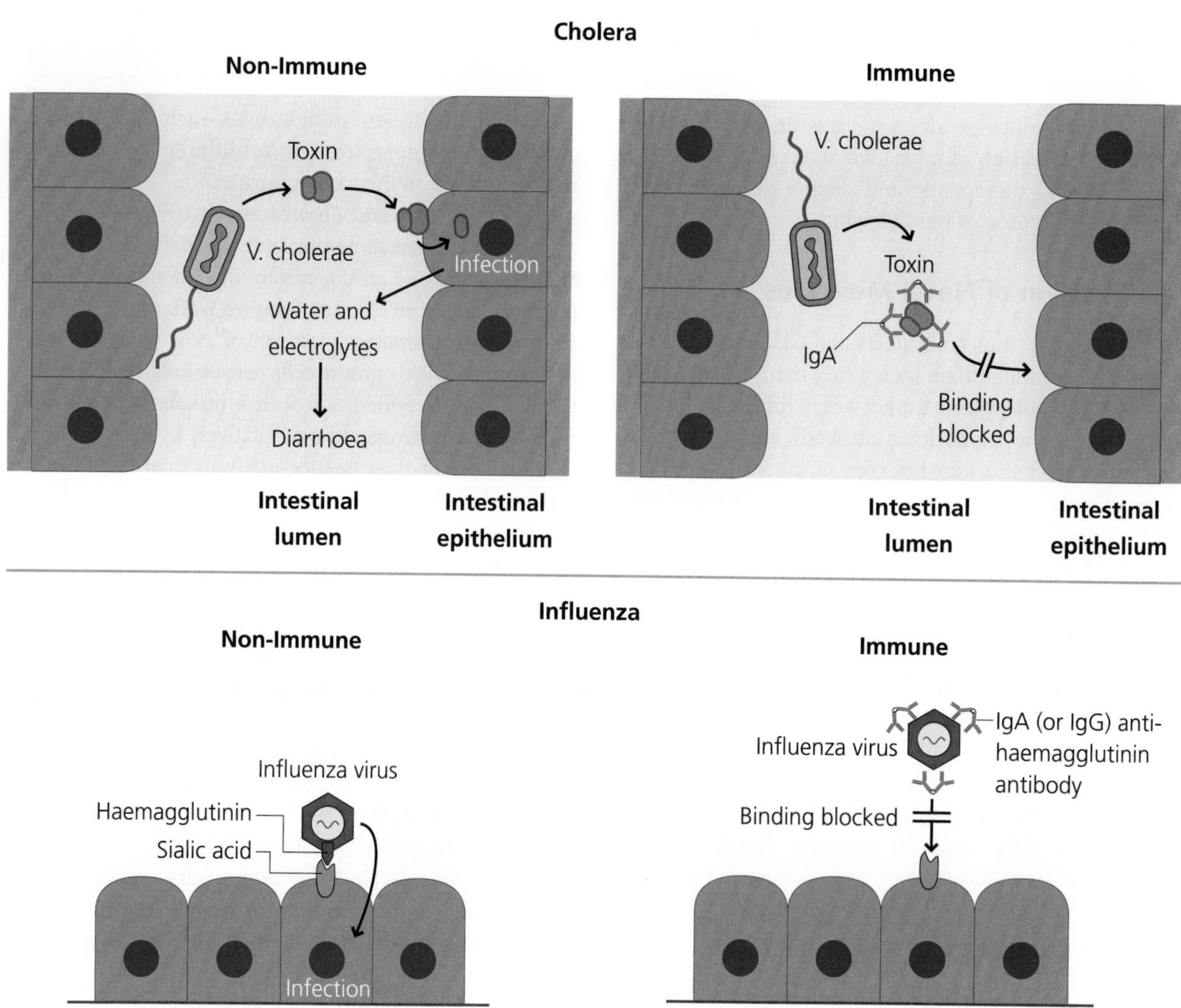

Fig. 6.12 Toxin and pathogen neutralization by antibodies. Bacterial exotoxins such as cholera toxin, and pathogens such as influenza virus, need first to attach to receptors on cells to damage or modulate cells, or to replicate. Antibodies that bind to the toxin receptor or the pathogen-binding site can block attachment and thus prevent toxin action or viral infection. This is known as neutralization.

of inflammation and pro-inflammatory cytokines, and later synthesis of other inflammatory mediators such as lipid mediators (Section 4.3.2).

> **Q6.15.** How might it come about that mast cells can be coated with IgE, yet serum levels of IgE are barely detectable?

6.2.6.4 Antibody-Dependent Cellular Cytotoxicity (ADCC)

ADCC is really a phenomenon looking for a function. Experimentally, if target cells are coated with antibodies, several different cell types can bind to the antibodies through their FcRs and kill the target cell. Cell types that can kill by ADCC include natural killer (NK) cells and monocytes (via IgG), as well as eosinophils (via IgE). Some evidence that suggests ADCC might be important in host defence against infection has come from experimental studies of herpes simplex virus (HSV) infection in mice. Thus, HSV infection is lethal for new-born and adult mice. Passive transfer of human anti-HSV antibodies into mice protects adults, but not new-borns against HSV because adult mice can mount effective ADCC, whereas new-born mice cannot. However, passive transfer of both anti-HSV antibody together with human mononuclear leukocytes (which would contain both monocytes and NK cells) conferred protection to new-born mice against HSV infection as well. It was hypothesised that HSV-infected host cells express viral proteins on their plasma membranes, that these are recognized by the anti-HSV antibodies, and that the human leukocytes can subsequently bind to and kill the infected cells, aborting the spread of infection. Some forms of parasites (e.g. immature schistosomula) can also be killed by eosinophils if the parasite is coated with IgG or IgE, at least *in vitro*, but the importance of this killing in host defence is unclear. See Figure 6.15.

> **Q6.16.** How might we determine if ADCC is actually carried out by NK cells, monocytes or indeed any other specific type of cell present in a mix of leukocytes?

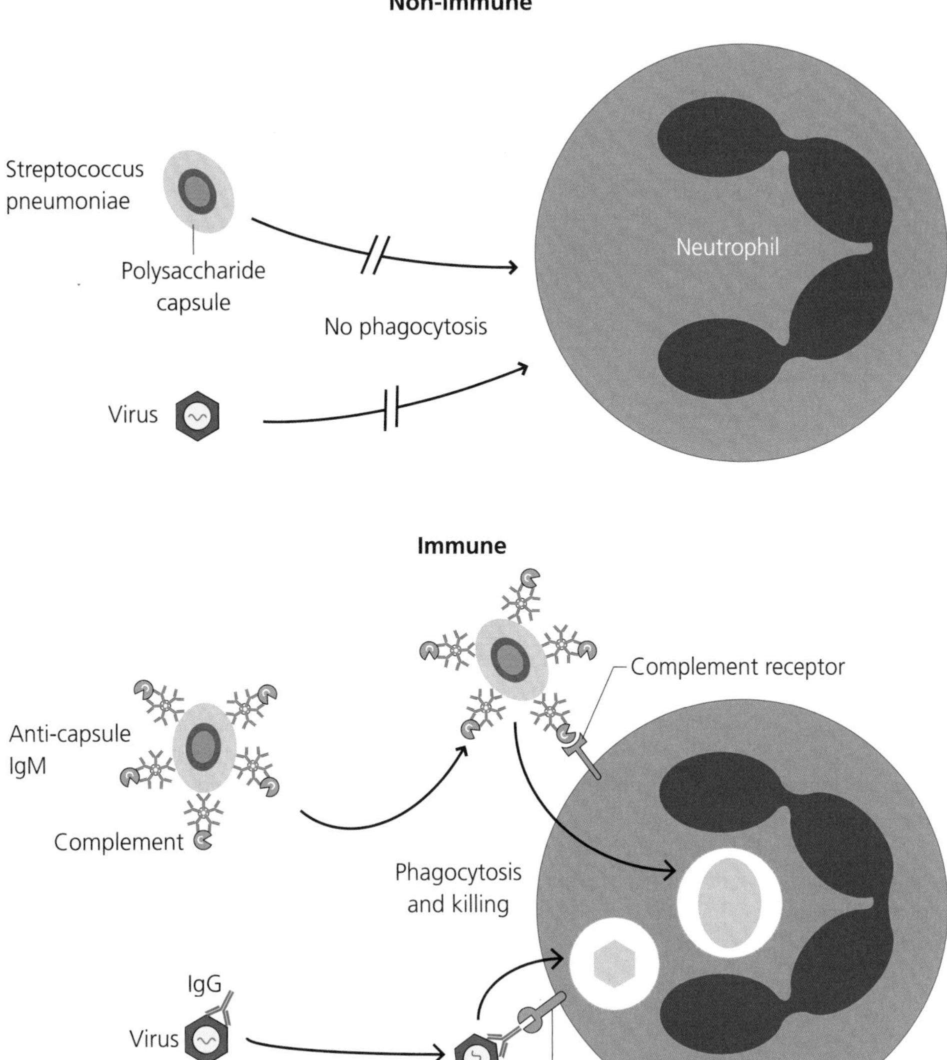

Fig. 6.13 Pathogen opsonization. Phagocytes (neutrophils and macrophages) can kill many phagocytosed pathogens intracellularly. Some bacteria avoid destruction by phagocytes because they do not express surface molecules that can be recognized by the phagocyte. If, however, the pathogen is coated with antibodies, it becomes opsonized and can be recognized either directly (through FcRs) or indirectly (through complement receptors) because the antibodies activate and bind complement. In this figure, *Streptococcus pneumoniae* is opsonized by anti-capsule IgM which binds and activates complement, permitting phagocytosis and killing of the bacterium. A virus is shown being opsonized by IgG.

6.2.6.5 Modulation of Adaptive Immunity

Pre-existing antibodies can modulate antibody responses to an antigen. Natural IgM antibodies may be involved in the delivery of antigens complexed with complement components to secondary lymphoid organs, especially the spleen. Here the binding of antigen to antigen-presenting cells (APCs) such as DCs and B cells may be enhanced via complement receptors. Pre-existing IgG, again with or without complement, may also enhance immune responses by increasing uptake of antigen by DCs or inhibit responses via inhibitory FcRs on B cells, (below). An example of antibodies modulating immune responses in humans is seen in the prevention of Rhesus disease (haemolytic disease of the new-born, Section 7.4.2.5). Foetal red blood cells (RBCs) enter the maternal circulation during childbirth and if the RBCs are Rhesus-positive and the mother Rhesus-negative, the mother may make anti-Rhesus IgG, which can cross the placenta during the next pregnancy and destroy foetal RBCs. If the mother is injected with anti-Rhesus antibodies immediately after delivery, she does make any anti-Rhesus antibodies and disease is prevented.

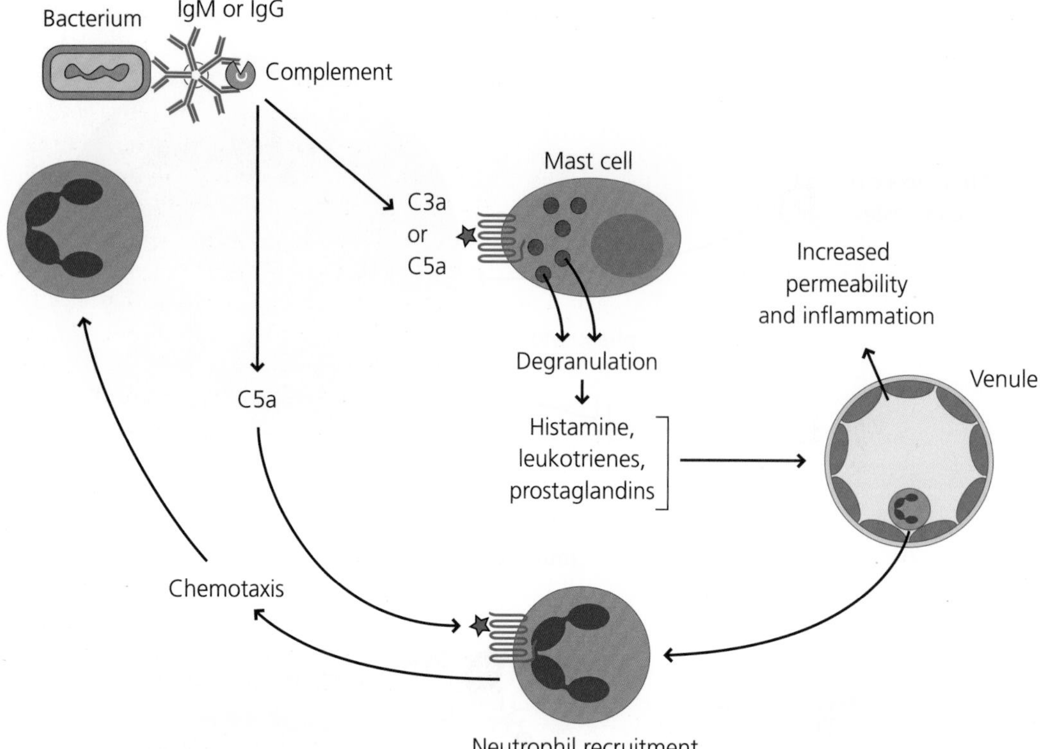

Fig. 6.14 Antibodies and the induction of acute inflammation. Antibodies IgM or IgG (not shown) antibodies bound to pathogens (immune complexes) may activate complement. Complement fragments C3a and C5a (which are called anaphylatoxins) can bind to receptors on mast cells and activate the cell. The mast cells release multiple pro-inflammatory mediators that act on local venules to increase permeability and recruit neutrophils and monocytes. C5a is also a powerful chemoattractant for neutrophils. (In some cases, uptake of immune complexes by phagocytes can also trigger the cells to secrete pro-inflammatory molecules; not shown.)

6.3 B Cell Responses

In this section we will discuss the triggering of different types of B cell responses. We start by considering responses at a tissue level, reiterating and expanding aspects that are introduced in Chapter 3. We next look at TI responses, and introduce some key aspects of B cell activation and regulation of antibody responses. Then we turn to TD responses, and the cellular and molecular bases of their regulation. We discuss TI responses before TD responses as they are conceptually simpler. However, we must emphasize that in many forms of defence against infection, TD responses are probably more

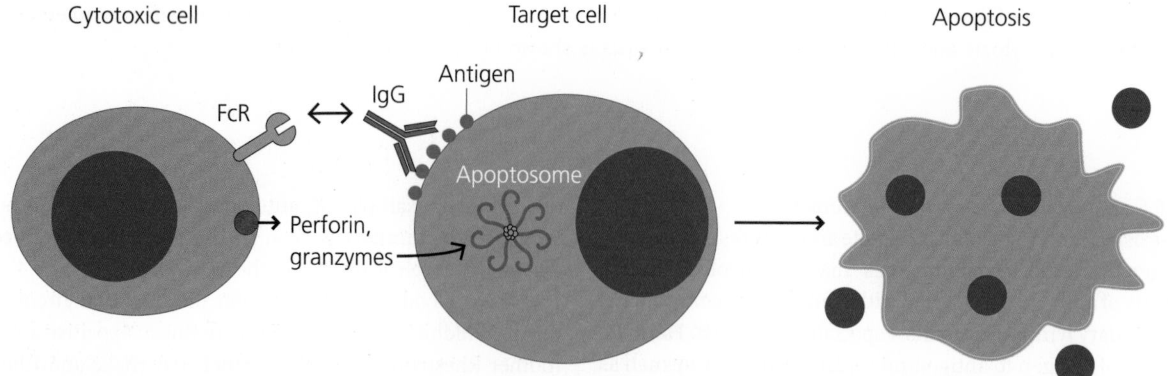

Fig. 6.15 Antibody-dependent cell-mediated cytotoxicity. Some cells with cytotoxic functions, such as NK cells, express FcRs for IgG. If a target cell is experimentally coated with IgG antibodies (e.g. that have been produced against a cell surface molecule of the target cell) it can be recognized through the FcRs of the cytotoxic cells. This stimulates the activation of cytotoxic mechanisms that include perforin and granzyme secretion, (the role of Fas-Fas-ligand interactions in ADCC is uncertain) that can induce apoptosis in the target cell. (Eosinophils may also use ADCC to kill some immature parasites, such as schistosomula, that are coated with IgE; not shown.)

important and that some of the cellular and molecular aspects that are introduced while discussing TI responses are also highly relevant to TD responses.

6.3.1 Anatomical Basis of B Cell Responses

Early studies in mutant T cell-deficient mice demonstrated that while antibody responses to protein antigens were generally dependent on helper T cells (i.e. they were TD responses), responses to other types of antigen could be made in the absence of T cells (i.e. they were TI responses). It then became clear, using mutant mice lacking certain B cell functions, that it was possible to divide TI antigens into two types: TI-1 and TI-2. The former could stimulate antibody responses in these mice, whereas the latter were defective. In general, high doses of TI-1 antigens act as mitogens, leading to the production of non-specific, polyclonal antibodies (below). However, TI-2 and TD antigens can both stimulate the production of antigen specific antibodies, generally at different sites.

6.3.1.1 Functional Anatomy of TI-2 Antibody Responses

TI-2 responses occur largely in the marginal zone of the spleen and depend on marginal zone B cells. However, many of the fine details of TI-2 responses *in vivo* remain obscure. Carbohydrate antigens, or others with highly repetitive structures, may be delivered to the marginal zone after they have bound complement (natural antibodies may play a part in this delivery) and TI-2 antigens injected into the bloodstream localize to marginal zone macrophages in the spleen (Section 1.4.5.3). The role of these macrophages in TI-2 responses is, however, unclear. They may for example bind the antigens in a form that can cross-link the BCR of marginal zone B cells. Another cell population in the marginal zone is the marginal zone metallophil. Again, this cell is thought to be a macrophage, but its functions in immune responses and potential roles in antibody responses are unclear.

6.3.1.2 Functional Anatomy of TD Antibody Responses

In the steady state, follicular B cells are continually migrating into secondary lymphoid tissues from blood. Here they spend variable times and, if no antigen is encountered, they leave to recirculate back to the same or other tissues. In lymph nodes and Peyer's patches, B cells enter through the same high endothelial venules (HEVs) used by T cells (Section 3.4.2.1) in the spleen they enter from the marginal sinus (Section 3.4.2.8). They then traverse part of the T cell area before entering the primary B cell follicle. Their migration into the follicle is perhaps guided in part by chemokines secreted by stromal cells of the tissue. More recent microscopic studies suggest, however, that naïve follicular B cells may use the processes of follicular DCs (FDCs) as guidance systems to enter the follicle. If they do not recognize antigen the B cells rapidly exit the organ, in the efferent lymph in the case of lymph nodes and Peyer's patches or directly into the blood from the spleen. If, however, they do recognize antigen a complex series of events is initiated.

Interactions Between T and B Cells During TD Antigen Responses Activation of both T cells and B cells is crucial for TD responses. CD4 T cells are initially activated in the T cell zones when they encounter DCs expressing the peptide–MHC complexes for which they are specific. Intra-vital microscopy of lymph nodes after the injection of antigen into peripheral tissues shows that the initial interactions between T and B cells in lymphoid tissues occur on the margins between the T cell area and the follicle, rather than in the follicles themselves. After this initial interaction, some B cells (possibly those with higher-affinity BCRs) develop locally into lymphoblasts. These cells do not enter follicles but migrate to medullary cords in lymph nodes and the red pulp in the spleen. Here, they secrete IgM, and some may switch to produce IgG or other isotypes. Other B cells that have recognized antigen, together with CD4 follicular helper (T_{fh}) T cells, migrate into the B cell follicle. It is not fully clear how T_{fh} cells relate to other T cell subsets, but they appear to be distinct (e.g. the transcription factor Bcl-6 seems crucial for their development). This migration into the follicle is followed by complex local structural and functional changes known as the germinal centre reaction. See Figure 6.16.

Germinal Centre Reaction In the steady state, secondary lymphoid tissues contain have follicles containing an abundance of B cells. During TD responses, however, these develop into secondary follicles where germinal centres develop (Section 3.4.2.5). Germinal centres are crucially important because these are the sites where: (i) B cells can class switch from producing one class of antibody to another by recombination of their V region gene with a new C region gene; (ii) mutations are introduced though somatic hypermutation into the V region of class-switched BCRs, which may lead to the generation of B cells with higher affinity receptors; and, crucially, (iii) memory B cells (bearing these higher affinity receptors) develop. See Figure 6.17.

Germinal centre B cells are one of the most rapidly dividing cell populations described in mammals, with a cell cycle time of as little as 6 h. This rapid proliferative rate facilitates somatic hypermutation, since mutations can only be introduced into the DNA encoding V regions during the cell cycle. (It may also be important for class switching, but B cells can also become committed to producing a different isotype outside the follicles.) Relatively few B cells enter each germinal centre, possible only one, but each dividing B cell produces multiple, and potentially thousands of, daughter cells. The germinal centre itself is divided into dark and light areas. The dark areas are the main sites of cell division and are where somatic hypermutation primarily occurs. The light areas are sites where B cell are residing before or after division and the place where B cells with higher affinity receptors are selected by antigen on the surface of FDCs (remember that FDCs are quite separate from classical DCs). In fact, some B cells leave the light areas, re-enter the dark areas and the process can occur in an iterative manner, potentially further increasing receptor affinity. The great majority of B cells dividing in the dark area die by apoptosis in the germinal centre. It contains

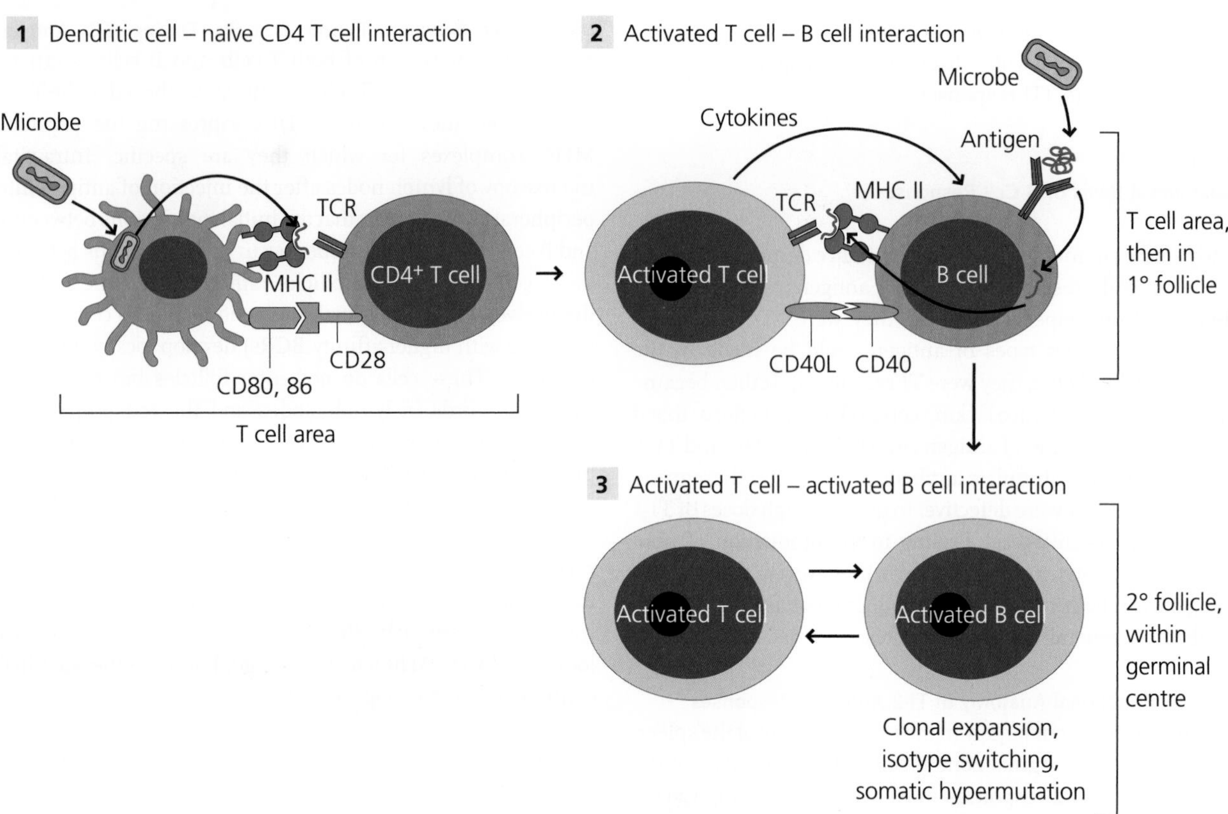

Fig. 6.16 Cellular interactions in T-dependent B cell activation. (1) Initially a CD4 T cell is activated by an activated DC that expresses antigen-derived peptides on MHC class II in the T cell area of a secondary lymphoid tissue. (2) An antigen-specific B cell endocytoses the antigen and presents peptides on its own MHC class II. The activated T cell recognizes the peptide–MHC complex on the B cell and delivers activation signals to the B cell. This happens on the margins of the T cell area, close to the B cell follicle. (3) Within the follicle, further interactions between B cells and activated T cells can lead to isotype switching and somatic hypermutation.

a population of macrophages called tingible body macrophages that contain the remnants of apoptotic B cells (originally called tingible bodies).

What determines whether or not germinal centre B cells live or die? One essential factor is the recognition by the B cell of antigen bound to FDCs. Antigen-binding to FDCs depends on the presence of antibodies (e.g. natural antibodies in the early phase of innate responses or antigen-specific IgG once class switching has occurred) and complement. Evidence for the complement dependency of this process originally came from the use of cobra venom factor, which depletes C3 and was found to prevent the persistence of injected antigen on the FDCs. During this phase of the response, immunoglobulin gene hypermutation has started in the dividing B cells. Those B cells that generate higher affinity receptors will have a selective advantage in conditions of low antigen concentration, and can potentially survive and become antibody-secreting plasma cells or memory cells.

Q6.17. What consequences may follow if the mutated V region generated during somatic hypermutation makes the antibody autoreactive?

Fates of Activated B Cells In the early phases of TD responses, some B cells are activated on the borders between T cell areas and the follicles (Section 3.4.2.5), and may become short-lived plasma cells. In the later phases of the B cell response, IgG, possibly secreted by these early plasma cells, can form complexes with antigen in the follicles and these complexes bind to the resident FDCs. B lymphoblasts in the germinal centre that recognize this cell-bound antigen, and which additionally receive signals from activated CD4 T cells (e.g. through CD40–CD40 ligand interactions), can be rescued from apoptosis and can have two potential fates. Some migrate to other tissues, such as the spleen and bone marrow, where they develop into plasma cells. Some plasma cells, particularly those in the bone marrow, are very long-lived. In mice these cells may survive for months or years, and in humans possibly for most of the life span of the individual. Other B lymphoblasts become memory cells which do not actively secrete antibodies but which can do so on restimulation (below). Some of these migrate to tissues such as the spleen where they may become resident memory cells, while others become small, recirculating memory B cells. See Figure 6.18.

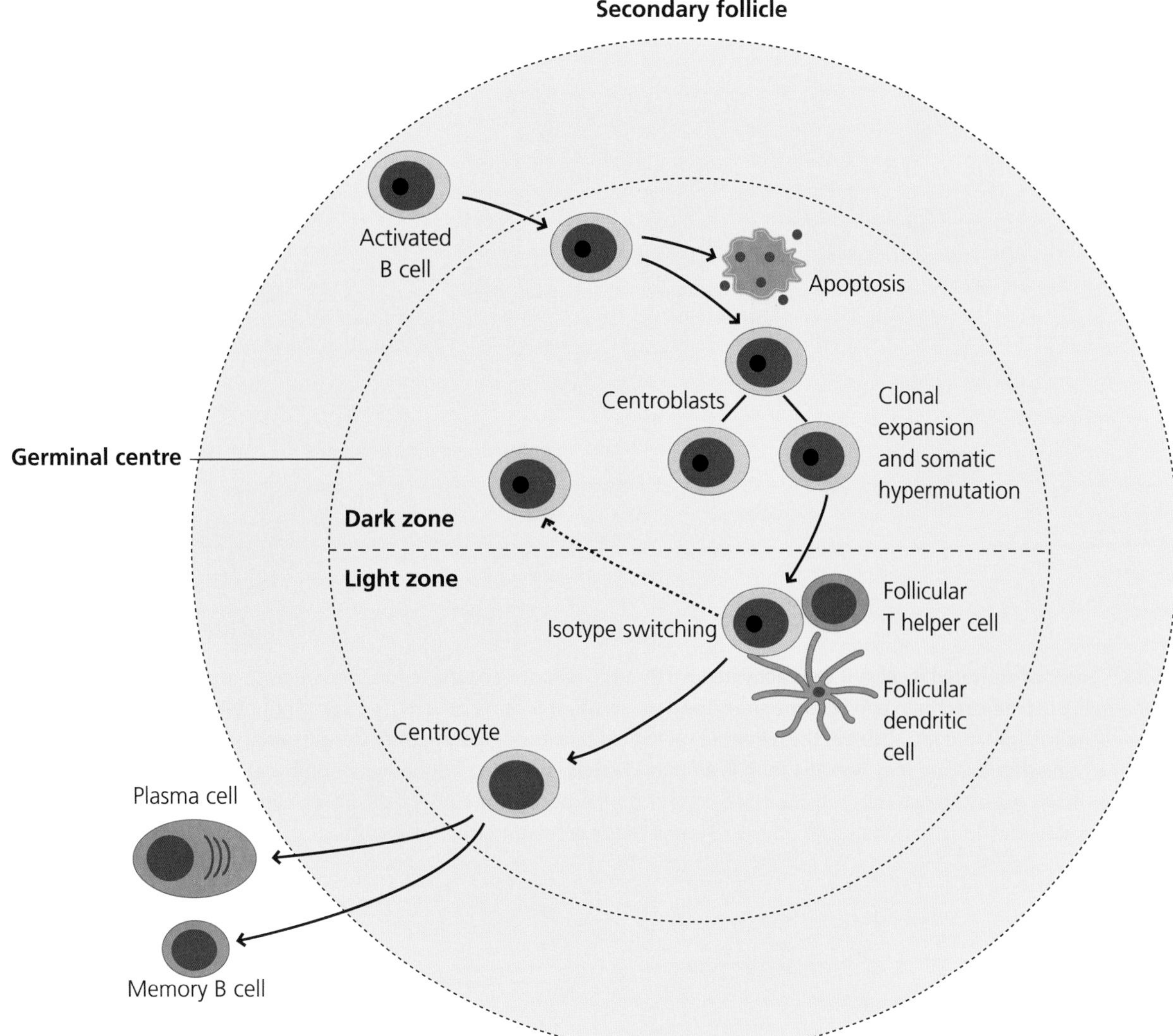

Fig. 6.17 Germinal centre reaction. Germinal centres form in B cell follicles during TD antibody responses. A very few activated B cells (possibly only one) enter each germinal centre. In the dark zone, the B cells (centroblasts) proliferate very rapidly to form clones that undergo somatic hypermutation. Many of these B cells apoptose. Surviving B cells enter the light zone as centrocytes where they interact with Follicular T helper cells (T_{fh}) and follicular dendritic cells (FDCs), undergoing further differentiation and possibly isotype switching. Again, many B cells apoptose here. Some B cells may then re-enter the dark zone, expand and undergo further mutation in an iterative process. Other B cells leave the germinal centre and may become plasma cells or memory cells.

6.3.2 T-Independent (TI) Antibody Responses

As noted above, antigens that can stimulate B cells in the absence of T cells are divided into two classes, TI-1 and TI-2. At high concentrations, some TI-1 antigens are mitogenic for B cells. That is, at high concentrations they stimulate B cells to divide in a non-antigen-specific manner, to develop into plasma cells and to secrete immunoglobulins that are not specific for the antigen. Lipopolysaccharide (LPS) is perhaps the most studied example of a TI-1 antigen, while DNA containing CpG sequences is another. These molecules are of course agonists for the respective Toll-like receptors TLR4 and TLR9 (Section 4.2.2.3). There is increasing evidence that both LPS and CpG can trigger proliferation of all types of B cell under experimental conditions, potentially by acting as agonists for the respective TLRs rather than by ligating BCRs. It should, however, be noted that at lower concentrations some TI-1 antigens, such as LPS, can stimulate the secretion of antigen-specific antibodies. The strength of this type of response may be because one part of the LPS molecule (the O-antigen portion) ligates the antigen-specific BCR, while the other (lipid A) acts as an agonist for TLR4. Whether all types of B cells or only specific subsets can respond in this way, particularly *in vivo*, is still not entirely clear. See Figure 6.19.

Q6.18. What might be the evolutionary advantage of a TI-1 response to LPS?

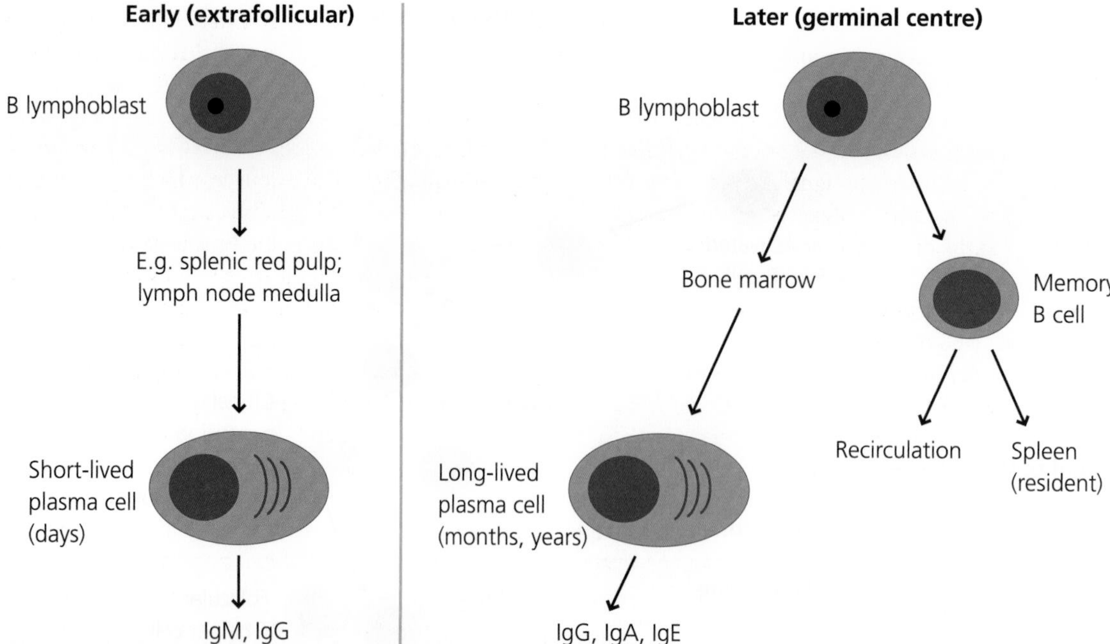

Fig. 6.18 Fates of activated B cells. B cells activated early in the response, outside follicles, migrate to areas such as the lymph node medulla or splenic red pulp and become short-lived (a few days) plasma cells that secrete antibodies of different isotypes perhaps especially IgM and IgG. Later in the response, activated B cells originating from the germinal centre can migrate to areas such as the bone marrow and may become long-lived plasma cells, secreting antibodies of different isotypes (IgG, IgA or IgE) for months or years. Other B cells activated late in the response become memory cells. Some of these are retained in tissues such as the spleen for long periods while others re-enter the recirculating pool of B cells.

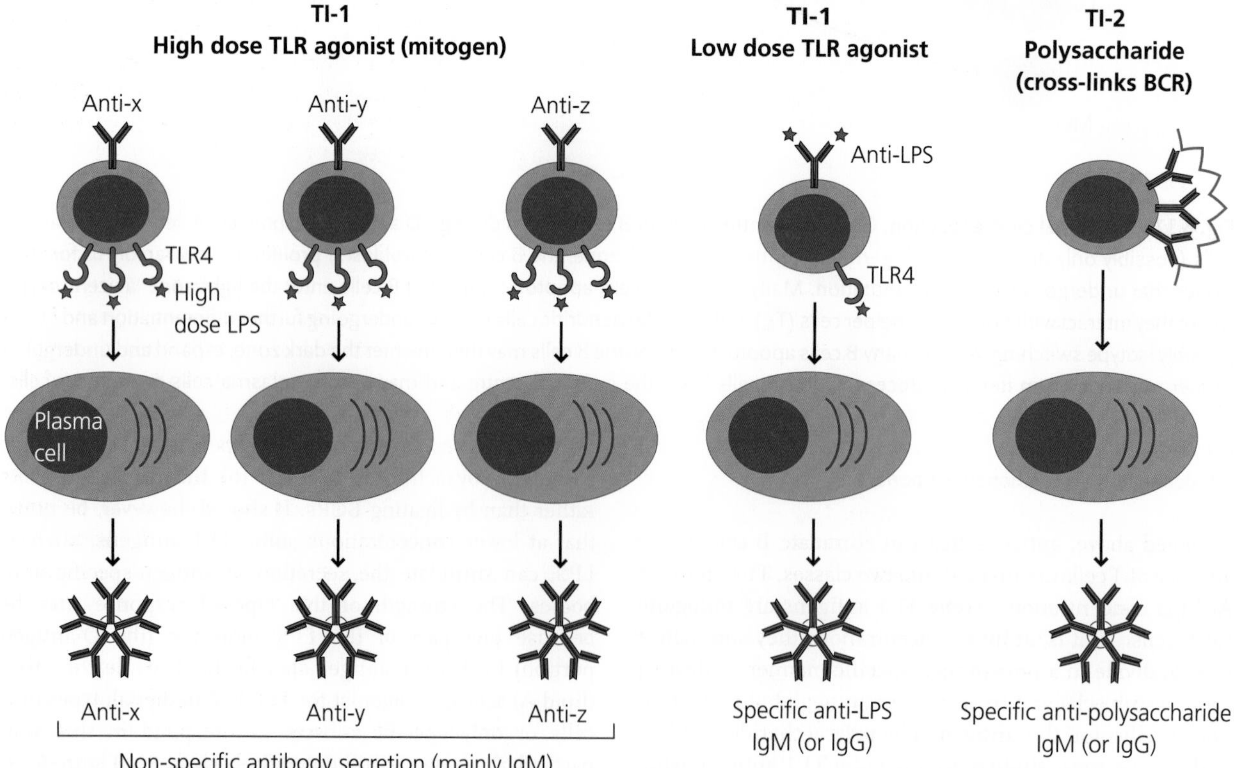

Fig. 6.19 T-independent antibody synthesis. TI antigens are of two types. TI-1 antigens are exemplified by LPS. LPS is mitogenic for B cells at high concentrations and stimulates non-specific immunoglobulin synthesis. This involves signalling via TLR4. At low concentrations, however, LPS induces specific anti-LPS antibody synthesis perhaps by signalling through both antigen-specific BCRs and TLRs. TI-2 antigens are usually polymeric carbohydrates that can cross-link surface immunoglobulin. Typically, the antibodies produced in TI responses are IgM, although some degree of switching to IgG3 and IgA can occur. TI responses show little if any memory or affinity maturation.

Conventional TI-2 antigens are often carbohydrates that have many similar, repeating antigenic determinants (e.g. the capsular polysaccharide on Streptococcus pneumoniae). It is thought that this type of antigen can cross-link several antibody molecules on the surface of the B cell and it is this cross-linking that activates the signalling pathways within the B cell. This form of activation can be mimicked experimentally by using antibodies to surface immunoglobulin on the B cell. As noted earlier, the marginal zone B cells seem to be specialized to produce this type of response, at least *in vivo*, because individuals who have had their spleen removed (e.g. because it was ruptured by trauma) become more susceptible to infection with encapsulated bacteria. Marginal zone B cells perhaps more closely resemble effector/memory cells, rather than naïve B cells, as they can respond very rapidly after antigenic stimulation. A subset of B-1 cells (B-1b cells, which do not appear to produce natural antibodies) may respond similarly to TI-2 antigens. TI-2 responses may not be totally independent of T cell help: cytokines secreted by T cells are required for maximal B cell proliferation to TI-2 antigens and for the limited class switching seen in these responses.

The antibodies made against TI antigens are mainly IgM, although some IgG can be made in response to some TI-2 antigens. This applies particularly to IgG3 in both humans and mice, where a degree of spontaneous switching of the V region gene from the μ to the next downstream γ3 C gene segments appears to occur. This switching may be partially dependent on cytokines derived from T cells since it is less efficient in T cell-depleted mice. The importance of these IgG antibodies in TI responses is not entirely clear. It has also been found that some intestinal IgA, made against commensal bacteria, is apparently synthesized in a TI manner, probably by B-1a cells.

Q6.19. Why might commensal bacteria stimulate IgA responses in a TI manner?

A crucial feature of TI-2 antibody responses is that, unlike TD responses, they show very little if any memory even after repeated immunizations; nor do they exhibit affinity maturation. This has posed a real problem in designing vaccines against pathogens such as Streptococcus pneumoniae and Haemophilus influenzae (not to be confused with the influenza virus) which are important causes of meningitis in young children. The solution to this problem lies in constructing conjugate vaccines. By chemically coupling the microbial carbohydrate to an immunogenic protein, such as tetanus toxoid, the TI-2 antigen is effectively converted into a TD antigen (below) and high-affinity memory responses against these pathogens can now be induced. These conjugate vaccines are proving highly effective clinically for protection against some bacteria such as Haemophilus influenzae Type B (HIB).

6.3.2.1 B Cell Activation

B cell activation first leads to changes in gene expression, an increase in cell size and the construction of cellular machinery for the increased protein synthesis needed for cell division (blasting). The B cell enters the cell cycle and undergoes clonal expansion. Later, patterns of gene expression are altered to permit the synthesis of new components that enable the cell to mediate its effector functions, particularly development into plasma cells and secretion of antibodies. All these events can be induced, at least in part, by BCR signalling. The extent to which signalling through the BCR alone can lead to full B cell activation in response to different types of antigen and to robust antibody responses is not, however, entirely clear, especially *in vivo*. For the most extreme TI-1 responses, to mitogens such as LPS at very high doses, the BCR might not be involved at all (i.e. potentially, strong PRR signalling might suffice alone). At lower doses of TI-1 antigens, it may be that BCR recognition of specific antigenic epitopes may lead to activation if these are coupled with additional signals through PRRs. For TI-2 responses against carbohydrate antigens and other antigens with repetitive structures it may be that the high degree of BCR cross-linking can generate sufficiently high levels of signalling for B cell activation; alternatively, synergy with other molecules, such as complement receptors (CR2; below), may be needed. However for TD responses against protein antigens that do not contain repetitive epitopes, the BCR may not be adequately cross-linked to generate sufficient signalling, which is perhaps why additional help from CD4 T cells is generally needed in this type of response. A common theme, however, is that the BCR may provide Signal 1 for B cell activation, but that full activation may often require additional signals that can be considered the counterpart of Signal 2, or costimulation, for T cell responses. Below we introduce some molecular aspects of B cell activation that may generally apply to both TI (especially TI-2) and TD responses, before considering the latter in more detail in Section 6.3.3.

6.3.2.2 BCR Signalling

As for the TCR, the BCR itself (i.e. the membrane-bound immunoglobulin molecule) has no direct signalling function: it serves purely to recognize cognate antigens. However, the BCR is associated with a molecular complex (CD79) that is responsible for signal transduction and the initiation of downstream signalling, similar to the role of CD3 in T cells. Most information on B cell activation has come from experimental studies where the BCR has been cross-linked, often by using an anti-immunoglobulin antibody as a surrogate antigen. In these situations, the CD79 complex, which possesses immunoreceptor activation motifs (ITAMs) in its cytoplasmic tails, becomes activated by tyrosine kinases that phosphorylate the ITAMs. In turn, another tyrosine kinase, Syk, is recruited and this phosphorylates the scaffold protein BLNK, which serves as a site for the assembly of several signalling components. These include phospholipase Cγ2 and Bruton's tyrosine kinase (Btk). These components then activate several downstream pathways, including those mediated by small G-proteins, calcium and protein kinase C (PKC). These pathways increase cell survival and activate the different transcription factors, including NF-κB, which stimulate the

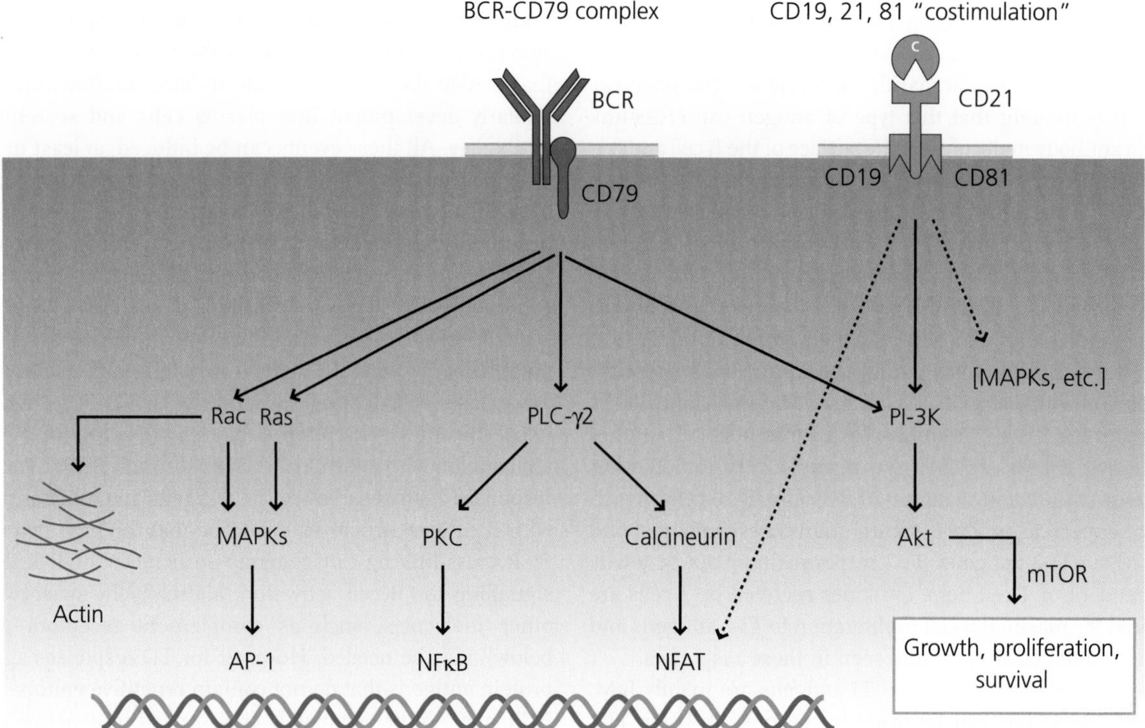

Fig. 6.20 Intracellular signalling pathways in B cell activation. A highly over-simplified scheme of some key intracellular signalling pathways in B cells. When the BCR is ligated or cross-linked by antigen, signals are delivered via CD79 that activate several distinct intracellular signalling pathways. These induce cytoskeletal changes, changes in cell metabolism and the activation and nuclear translocation of several transcription factors, including AP-1, NF-κB and NF-AT. Additional costimulatory signals may derive from the CD19–CD21–CD81 complex (Figure 6.21). In TD activation the activated T cell also delivers signals via CD40 and the secretion of cytokines (Figure 6.22).

transcription of genes needed for activation. While many details differ, some of the general pathways are similar to those induced in T cells after TCR ligation. See Figure 6.20 and compare Figure 5.15.

6.3.2.3 **Positive and Negative Regulation of B Cell Activation**
As mentioned, B cells often need additional signals for full activation and B cells can also be negatively regulated. To provide a couple of examples, here we briefly outline the roles of complement in increasing B cell activation and of antibodies in inhibition. See Figure 6.21.

Positive Regulation of B Cell Activation Similarly to T cells, B cells have co-receptors which can augment or inhibit their responses. CD21 is a complement receptor (CR2) which acts in concert with CD19 and CD81. If the antigen recognized by the BCR is associated with complement – as is the case with many bacteria and other pathogens – the threshold for B cell activation is markedly decreased. Hence, the sensitivity of the B cell to low concentrations of antigen can be dramatically increased. This has been clearly demonstrated from experiments in which the C3d component of complement (the ligand for CD21) was artificially linked to a defined antigen: the concentration of this antigen–C3d complex that was required to activate antigen-specific B cells was found to be three to four orders of magnitude less than when antigen alone was used as a stimulus. Hence, this is an example of an innate immune response (complement activation) modulating an adaptive (B cell) response.

Negative Regulation of B Cell Activation B cells express an inhibitory Fc receptor (FcγRIIB) with immunoreceptor tyrosine-based inhibitory motifs (ITIMs; as opposed to ITAMs) in its cytoplasmic tail. When IgG binds to this receptor on B cells, this molecule recruits SHP phosphatases which inhibit B cell activation. Thus, when IgG has been produced in response to an antigen, antigen–antibody complexes will form. The antigen will bind to the BCR of B cells and the IgG, via its Fc region, to the inhibitory Fc receptor. Thus, B cell activation can be down-regulated by the increased synthesis of specific antibody. This is thought to be a mechanism to regulate the production of IgG by B cells (turning B cells off when a critical level of IgG has been reached) and similar mechanisms may also regulate the production of other isotypes (e.g. via the low-affinity IgE FcR). Other molecules, such as the B-cell specific CD22 (also known as sialoadhesin), can also deliver regulatory signals in both a positive or negative manner.

Q6.20. Why might it be preferable to use $F(ab')_2$ fragments of anti-immunoglobulin antibodies, rather than the intact immunoglobulin molecule to activate B cells experimentally?

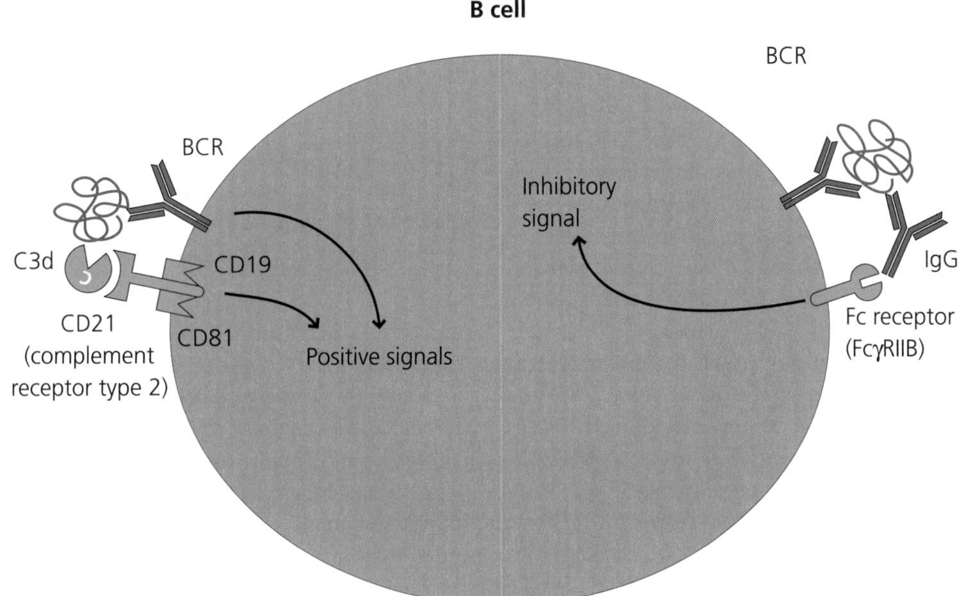

Fig. 6.21 Positive and negative regulation of B cell activation. If an antigen that binds a BCR is itself bound to complement C3d, the complement fragment can bind to CD21 (complement receptor type 2) complexed with CD19 and CD81. Signalling through CD19 then generates positive signals which can increase the sensitivity of the B cell to activation by 100 times or more. In contrast, at sufficiently high levels, IgG can bind to an inhibitory FcR on the B cell, FcγRIIB, which generates negative signals to inhibit further B cell activation and hence to prevent further IgG being produced.

6.3.3 T-Dependent (TD) Antibody Responses

6.3.3.1 Primary and Secondary Antibody Responses

It should now be clear (Section 6.3.2) that if, for example, a mouse is immunized on multiple occasions with a TI (e.g. TI-2) antigen, the antibody response is much the same after each immunization in terms of its speed and the amount and isotype (mainly IgM) of antibody secreted. If, however, a mouse is immunized with a TD antigen, the picture is very different. The first (primary) response takes several days to generate antibodies that are mainly IgM and which disappear rapidly (days) from the circulation. Some IgG may be made late in the primary response, but does not reach high levels. On second and subsequent immunizations the (secondary response) is quite different. IgM is formed with similar kinetics to the primary response, but in secondary responses IgG appears within 1–2 days, reaches high titres and lasts for much longer (weeks or months). This secondary response forms the basis of all vaccinations where antibodies are the protective agent.

6.3.3.2 Molecular Events in TD Responses

6.3.3.3 B Cell Activation

To simplify understanding we will first discuss the initial activation of B cells, involving antigen binding and interactions with CD4 T cells. We will then describe the subsequent events, including isotype switching and the generation of plasma cells and memory cells.

Role of the BCR in TD Responses The BCR has two different functions in B cell activation. First, as outlined above (Section 6.3.2.2), the BCR acts in association with CD79 as a signalling complex for B cell activation. As noted, (Section 6.3.2.1) the contribution of BCR signalling to TD B cell responses is not fully understood. It is, however, clear that in TD responses the BCR additionally serves a crucial function in concentrating antigens to enable processing and presentation to CD4 T cells. Antigens bound to the BCR are endocytosed and processed in the endosomal pathway, enabling peptides to bind to MHC class II and be presented on the cell surface. These steps are crucial in ensuring that B cells can interact with T cells specific for the B cell's cognate antigen (expressed as peptide–MHC complexes at the B cell surface). The importance of the BCR in concentrating antigen has been shown by comparing the efficiency of B cells in presenting their cognate antigen or a non-specific antigen to antigen-specific T cells. B cells can present their cognate antigen at concentrations of 100- to 1000-fold less than needed for the non-specific antigen See Box 6.5.

T Cell–B Cell Collaboration in TD Responses It is convenient to divide TD responses into three main stages (Figure 6.16). First, to provide help, CD4 T cells need to be activated by DCs. This occurs in T cell areas of secondary lymphoid tissues when an antigen-specific T cell recognizes a peptide–MHC complex on an activated DC. These T cells may then become polarized (e.g. to T_h1 or T_h2 subsets). Second, the activated CD4 T cell then recognizes a B cell expressing the same peptide–MHC complex, generated after BCR binding and internalization of the antigen (above). This can lead to activation of the B cell, typically in the T cell area close to its border with the follicle. Here, the B cell can subsequently become committed to isotype switching, it may become a short-lived plasma cell or it may enter the germinal centre.

Once the germinal centre reaction has been initiated, the third stage involves the interaction of germinal centre B cells with T cells. The latter provide essential signals that are needed for class switching, somatic hypermutation and selection of high-affinity B cells in this site. These T cells probably represent a different subset of follicular T cells (below), distinct from conventional T_h1 or T_h2 cells. Follicles and germinal centres do not contain detectable conventional DCs. However, activated B cells can express B7 (CD80, CD86) and CD40, which might be sufficient for activation of T cells expressing CD28 and CD40 ligand. If, for example, CD86 is blocked, the memory B cell response is impaired. Other costimulatory interactions, such as those between CD27 and CD70 (*cf.* CD8 T cell activation; Section 5.4.3.1) are also needed to for the production of plasma cells. It is therefore possible that germinal centre B cells can directly activate follicular helper T cells and, in return, receive the help they need if they are to survive and function further.

Finally, once a primary response involving each of the above three stages has occurred, rapid recall responses can be mounted if the antigen is encountered again (e.g. a secondary response); these are likely to involve interactions between memory B cells and memory T cells (Section 6.4.1). See Figure 6.22.

Q6.21. If the T cell with which the B cell needs to interact has been already been activated by a DC, why might a follicular B cell need to express costimulatory molecules that are involved in T cell activation?

6.3.3.4 Cytokine Synthesis and TLR Expression by B Cells

There are clearly other twists in B cell activation still waiting to be uncovered. Activated B cells do not just secrete antibodies; they can also be potent secretors of cytokines and the pattern of secretion can reflect the manner of their activation. It has been suggested that these patterns reflect different modes of B cell differentiation: B cells activated by T_h1 cells secrete pro-inflammatory cytokines such as tumour necrosis factor (TNF)-α and IL-6, whereas B cells activated by T_h2 cells may secrete IL-10 and IL-4. It is possible that these cytokines act on T cells and/or other B cells involved in the response to increase polarization of the response. It is also possible these cytokines may act on DCs to regulate their functions.

As we have noted, B cells also express several TLRs (Section 6.3.2). The potential roles of these TLRs in TI-1 responses have

Box 6.5 T Cell–B Cell Collaboration in TD Responses: Haptens and Carriers, Adoptive Transfer

How do we know that antibody synthesis to protein antigens needs help from T cells? In the early 1960s it was found that neonatally thymectomized mice generated much reduced antibody responses to some antigens (e.g. sheep erythrocytes) but not others (e.g. bacterial carbohydrates). Soon after, attempts were made to reconstitute irradiated mice, which did not make antibody to sheep RBCs, with spleen thymus or bone marrow cells or mixtures of thymus and bone marrow cells. Spleen cells reconstituted the ability to make antibodies, but thymic or bone marrow cells alone did not. Importantly, however, when thymic and bone marrow cells were given together, antibodies were made, suggesting that cooperation between two different cell types was needed in antibody synthesis. (This was before T and B cells had been identified as different types of lymphocyte. We now know that these tissues contain sufficient, newly-produced mature T and B cells respectively to be able to mount a response.)

A widely used model for examining antibody synthesis is the hapten–carrier system. A hapten is a small molecule (e.g. DNP) that, if it is injected into animals on its own, does not induce antibody synthesis. If, however, the animal is immunized with the hapten bound to a carrier protein, anti-hapten antibody is made. Using this system it was shown that the anti-hapten antibody-forming cells derived from bone marrow cells but that the cells giving carrier-specific help originated in the thymus. Many other experiments have shown that T–B collaboration is essential for antibody synthesis to protein antigens.

How do B and T cells cooperate in an antigen-specific manner? Early experiments showed that carrier and hapten molecules needed to be physically linked in order to stimulate maximal responses. This suggested that the carrier-specific T cell and the hapten-specific B cell needed to be in contact for optimal activation of the B cell. Clearly this would permit the T cell to focus its membrane-bound and secreted activating signals on the specific B cell, but how could this be brought about? Two groups used B cells as APCs for activated antigen-specific CD4 T cells. What they both showed is that if the B cells had been incubated with an antigen for which they expressed cognate (specific) BCRs, they presented that antigen to antigen-specific T cells with around 1000 times greater efficiency than if they were incubated with an antigen for which they did not express specific receptors and which was internalized by non-specific endocytosis. This was interpreted as showing that B cells used their surface immunoglobulin to accumulate and endocytose antigen, which they then processed and presented, via their MHC class II molecules to the CD4 T cell, thus enabling an antigen-specific interaction between the T and B cell. These experiments do, however, raise questions about the absolute need for BCR signalling (as opposed to its antigen concentrating functions) in TD activation of B cells.

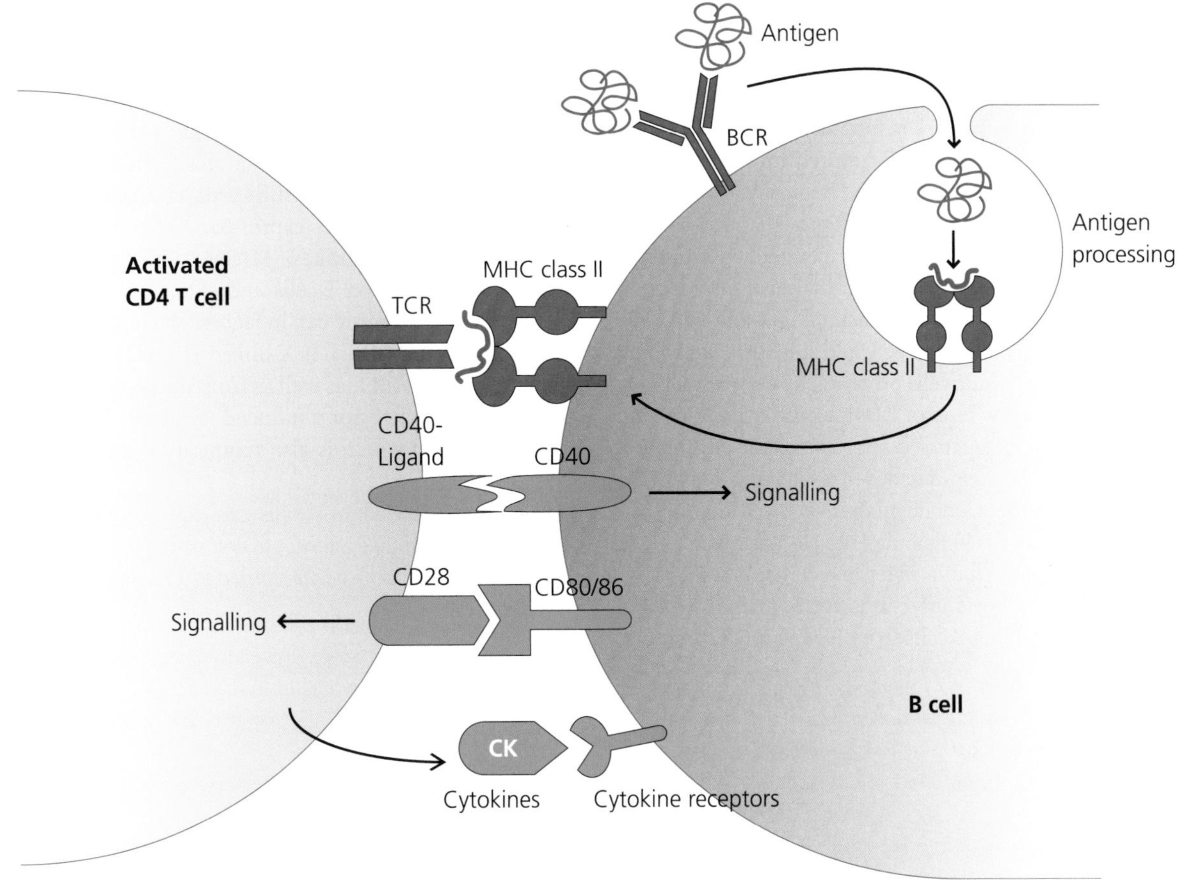

Fig. 6.22 T cell–B cell collaboration. Activated B cells (e.g. in germinal centres) can express co stimulatory molecules such as CD80 and CD86. These may help activate CD4 T cells (e.g. follicular T cells) that recognize peptide–MHC class II on the B cells. Other signals may then be generated (in either or both directions) through CD40 – CD40 ligand and other interactions (not shown) and through cytokines.

been discussed, but some experiments suggest that they may also play a role in TD responses.

6.3.3.5 Differentiation of Activated B Cells

Once activated in a TD response, B cells can switch to express a different immunoglobulin isotype. They can also undergo somatic hypermutation. They can become either plasma cells secreting antibody, or memory cells which have the capacity to become plasma cells on re-stimulation. The early stages of activation, and affinity maturation, are antigen-dependent. Later, the maintenance of plasma cells may be independent of antigenic stimulation. However, the maintenance of memory B cells may need the persistence of antigen – not as a stimulus, but as a reminder not to die.

6.3.3.6 Isotype Switching

Understanding the regulation of isotype switching is crucial to developing new ways of dealing with many immunological problems. In defence against infection it is important that the isotype(s) made are those best able to deal with different infectious agents. Thus, for pyogenic bacteria those IgG isotypes that can mediate opsonization are important, whereas for intestinal infection IgA may prevent pathogen binding, and for parasitic worms IgE may be needed. This is clearly a crucial factor to take into consideration (e.g. in designing new vaccines against infectious diseases). Moreover, as just one more example, if we knew how to regulate IgE synthesis in allergic individuals we could prevent much, sometimes fatal, disease.

It should now be clear that, initially, all B cells express IgM on their surfaces and most also express IgD, and that during TD responses many B cells switch their C region to that of another isotype (IgG, IgA or IgE). As noted earlier immunoglobulin C region genes are arranged in a linear sequence. During isotype switching a loop of DNA is formed with the VDJ gene segments at one end and the required C region gene at the other. The loop then forms a circle and is excised, and the new C region segment is joined to the VDJ segments. Class switching is guided by stretches of repetitive DNA between the J and C regions called switch regions. The result is that all intervening DNA sequences are deleted and a new isotype can ultimately be produced. See Figure 6.23.

How is this differential switching regulated? As mentioned above, cytokines, mainly but possibly not exclusively, secreted by the activated CD4 T cell are key factors responsible for isotype switching in B cells. Thus, interferon (IFN)-γ

Case Study 6.1: Hyper-IgM Syndrome

Clinical: A 4-year-old boy was brought to a paediatric clinic because of recurrent upper respiratory tract infections. A variety of pathogens had been cultured from his sputum during these infections, including *Haemophilus influenzae, Streptococcus pneumoniae* and *Pneumocystis jirovecii*. His numbers of blood neutrophils were elevated, but numbers of B and T cells were normal. Serum analysis showed raised levels of IgM, but a complete lack of IgG and IgA. His peripheral B cells all expressed IgM and IgD, and there were no IgG-expressing cells present. His peripheral blood T cells showed a low level of CD40 ligand expression and expression was not increased after mitogenic stimulation of his T cells. He was diagnosed as having hyper-IgM (HIGM) syndrome. Genetic analysis showed that he had a mutation in his CD40 ligand gene. He was given a bone marrow transplant from a sister and remains well.

Explanation: The boy's infections relate primarily to his inability to switch his B cells from IgM to IgG expressers. His case shows that a functional CD40 ligand gene is essential for this switch. CD40 ligand is encoded on the X chromosome and, thus, in a male child, a single mutation in this gene is sufficient to inactivate and prevent all CD40 ligand expression. This example is therefore one of X-linked HIGM. CD40 ligand is expressed on activated T cells and signals to B cells via CD40. HIGM syndrome can in fact result from defects in both CD40 ligand (which is X-linked HIGM1) and CD40 (which is not; HIGM3) as well as from mutations in other genes such as activation-induced cytidine deaminase (AICD; HIGM2) which is also required for switching.

Q6.22. Cells other than B cells can express CD40 (e.g. DCs). How might we attempt to determine if the lack of CD40 on such cells also contributed to disease in HIGM syndromes?

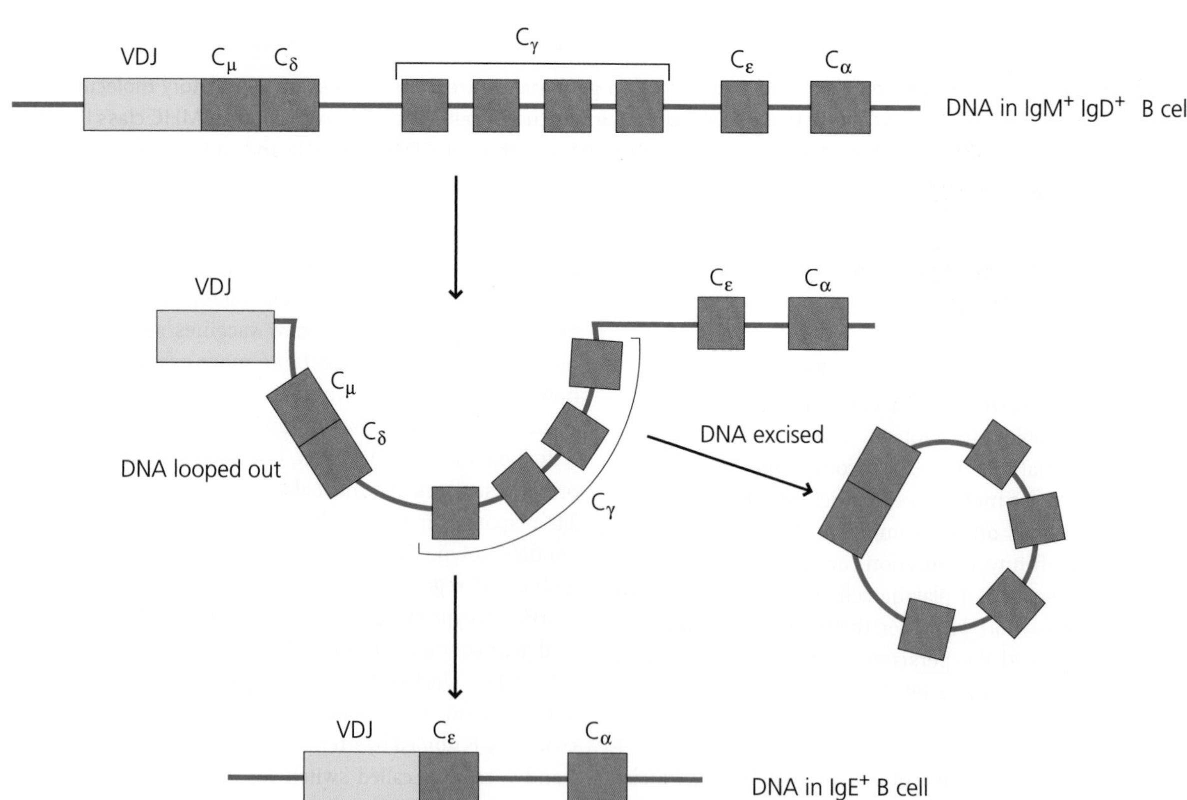

Fig. 6.23 Isotype (class) switching. Isotype switching mainly occurs in TD responses. Naïve B cells express IgM and usually IgD. During activation a B cell may switch to producing a different immunoglobulin class. To bring this about a loop of DNA is formed that contains the intervening C region genes. The loop is excised and the free ends of DNA join so that the particular C region gene is apposed to the pre-assembled V region gene. The cell can now produce a different antibody isotype (here IgE), but of the same antigen specificity. The C region gene segments are shown in a simplified manner.

made by T_h1 CD4 T cells is needed for synthesis of IgG2a, an opsonizing isotype in mice, whereas IL-4, made by T_h2-biased T cells is essential for switching to IgE. IgA switching is less well-understood, but TGF-β, retinoic acid and IL-5 are probably important, at least in mice. However, while cytokines can initiate the synthesis of different classes of antibody, they are not sufficient for class switching. Interactions between membrane-bound molecules on the B cell with those of the helper CD4 T cell are also absolutely crucial, including CD40–CD40 ligand interactions. See Figure 6.24 and Case Study 6.1.

There is some evidence, that as well as T cells, DCs may be involved in isotype switching. Thus, experimentally in the presence of T cell help, naïve (IgM-positive) B cells switched to IgG in the presence of DCs from the spleen, but switched to IgA in the presence of DCs from Peyer's patches. Possibly the DCs may have been acting primarily to instruct the T cells. Recent evidence further suggests that mast cells or possibly basophils recruited to lymph nodes may also be involved since they can secrete IL-4 and IL-13, important in IgE switching.

Follicular T Helper Cells and Isotype Switching It seems clear that T_h1 and T_h2 cells can produce many of the cytokines needed to induce class switching in B cells. Polarization of T cells (e.g. to T_h1/T_h2) occurs in secondary lymphoid tissues, presumably in the T cell areas, but most class switching requires that T cells interact with B cells that are located in compartments anatomically distinct from the T cell areas, in the B cell follicles. As noted (above), secondary lymphoid tissues contain another type of T cell, the follicular T cell. Similar cells can be detected in blood and appear to represent a distinct recirculating T cell subset (or population) that may be specially selected in the thymus. Follicular T cells express a unique transcription factor, Bcl-6, not expressed by T_h1 or T_h2 cells, strongly suggesting that they represent a different T cell subset. Follicular T cells are needed to provide help for B cells and perhaps to trigger the B cell proliferation that is required for somatic hypermutation and affinity maturation. Where, and if, T_h1/T_h2 cells interact with B cells and whether or not follicular T cells are required for class switching is not yet clear.

Q6.23. What might be the relative roles of differentiated T_h1 or T_h2 cells, compared to T_{fh} cells, in B cell responses?

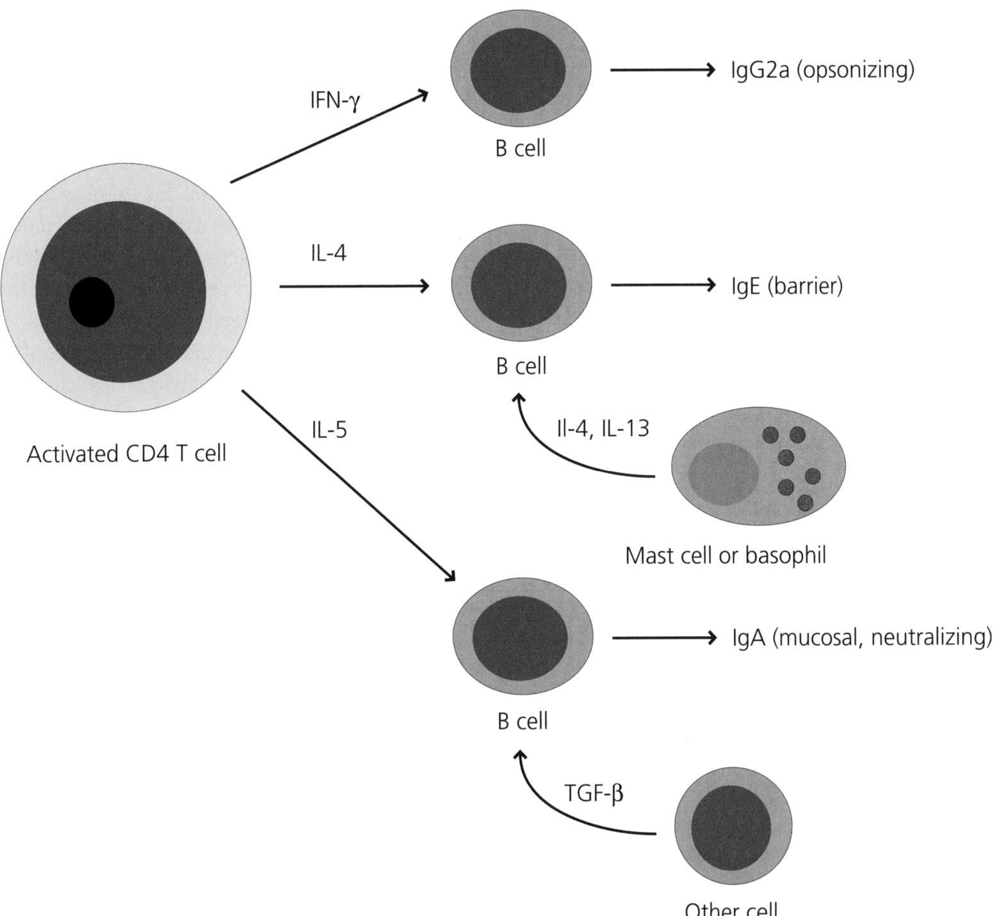

Fig. 6.24 Regulation of isotype switching. Which isotype a B cell expresses after activation is largely determined by cytokines secreted by its activated T helper cell, although cytokines secreted by other cells may also be important. Thus, in mice, IFN-γ induces switching to IgG2a, IL-4 to IgE, and IL-5 plus TGF-β to IgA.

6.3.3.7 Regulation of immune Responses by Fc Receptors

The cytokines produced by polarized subsets of CD4 T cells play a crucial role in determining the class of antibody that is made when they interact with B cells. These T cell subsets recruit or regulate the functions of other types of effector cells, some of which express FcRs for the same class of antibody that is being produced. This enables the antibodies to target antigens to specific types of effector cell and, in turn, to stimulate effector functions that help to eliminate the antigen.

Many different FcRs exist, and many of the biological functions of different antibody isotypes are dependent on specific interactions between antibody Fc regions and their receptors. Most FcRs are members of the immunoglobulin superfamily; the neonatal FcR (FcRn) is the exception being MHC class I-related. They may be single chain molecules or, more commonly, are associated with other molecules. Many, if not all, FcRs are signalling molecules. Most signalling is dependent on activating ITAM motifs in the cytoplasmic tail of the molecule; however, in a few cases, the FcR carries an ITIM motif and is an inhibitory receptor such as the FcγRIIB1 which inhibits B cell activation (Section 6.3.2.3).

Many functions of IgG and IgE, and to a lesser extent IgA, depend on their interactions with specific FcRs. The outcome of binding immune complexes depends on the cell type involved and on the types of FcR it expresses. For example, some FcRs on phagocytes, such as the FcγRI for IgG on macrophages and neutrophils, are opsonic phagocytic receptors, and phagocytosis may or may not be accompanied by the production of reactive oxygen intermediates (ROIs) or pro-inflammatory cytokines. The outcome in these cases may also depend on other signals (e.g. whether or not complement receptors are ligated) and also the state of the cell (e.g. whether it is a resident or activated macrophage). Another example is the single FcR (FcγRIII) for IgG that mediates ADCC by NK cells, stimulating the release of perforin and granzymes into the target cell, and triggering its apoptosis (Figure 6.15). The high-affinity FcR (FcεRI) for IgE is expressed by mast cells, and cross-linking IgE by antigen triggers degranulation with the release of histamine and other inflammatory mediators. It is also possible that FcRs for IgE (and IgA) cause eosinophils to degranulate and kill certain types of parasite that are coated with the respective class of antibody (Figure 2.16).

There is a remarkable elegance in the way that polarized subsets of CD4 T cells can both recruit different sets of cells into different types of immune response and stimulate the production of specific classes of antibody that help to coordinate the overall response. For example, T_h1 cells secrete IFN-γ which, in mice, is needed for class switching to IgG2a. They also activate macrophages which up-regulate expression of an FcR specific for this isotype (FcγRI; others are down-regulated). Thus, antigens, tagged with IgG2a, can be targeted to the intracellular microbicidal mechanisms of these cells. T_h2 cells secrete IL-4 which induces a class switch to IgE. The same cytokine, in concert with IL-5, stimulates growth, differentiation and/or survival of mast cells and eosinophils, which are, of course, the major cells that express the high-affinity receptor for this isotype of antibody (above) (Figure 2.16).

Finally, we will note that this brief description simplifies the complexity of FcRs and the cells that express them, and their roles in normal immunity and potentially in disease. In addition to the multiple types of FcR for different classes of antibody, and their stimulatory and inhibitory forms, some can be cleaved from the cell surface or secreted, some are allelic, and some have multiple isoforms produced by alternative splicing with the splice variants being differentially expressed on different cell types. Almost certainly others remain to be discovered.

6.3.3.8 Affinity Maturation

If blood is taken during an ongoing immune response to a protein antigen and the average affinity of the specific IgG measured it is found to increase over the course of the response. This phenomenon is known as affinity maturation; it may also occur for IgA and IgE. To remain effective in defence, any antibody must remain bound to its antigen. Thus increasing the strength of binding will increase the antibody's efficacy.

As noted earlier, affinity maturation occurs in germinal centres and results from two or more interacting mechanisms. If the amounts of antigen available to the B cells diminish during the course of a response, only B cells with high-affinity receptors will be able to accumulate sufficient antigen to become activated hence there will be selection for higher-affinity B cells and thus higher-affinity antibody. (In addition, because of the relatively limited numbers of follicular T cells in germinal centres, competition for these could also be important.) There is also, however, a quite separate mechanism that is apparently unique to B cells and immunoglobulin genes, and which does not occur in T cells. This is somatic hypermutation. If a mouse is immunized and monoclonal antibodies are generated at different intervals after immunization, the immunoglobulin genes from each hybridoma can be sequenced. It was found that as the interval after immunization increased, increasing numbers of point mutations appeared in the V regions of the immunoglobulin genes. These mutations appear to occur randomly within V regions since they still occur if the hypervariable region bases are replaced by random sequences. We can assume that most of these mutations will give rise to antibodies that are of equal or lower affinity than the germline form. In rare cases, however, the mutation will generate an antibody of higher affinity and, as described above, B cells expressing such antibodies will have a selective advantage as antigen concentrations decrease, and thus the average affinity of the antibody will increase. This process may be iterative with successive rounds of mutation and selection leading to B cells expressing higher and higher affinity receptors. See Figure 6.25 and Box 6.6.

We do not fully understand how mutations are introduced into V regions of B cells during the germinal centre reaction. It is however, know that this process requires both CD40–CD40 ligand interactions as well as a key enzyme, activation-induced cytidine deaminase (AICD), which is found almost

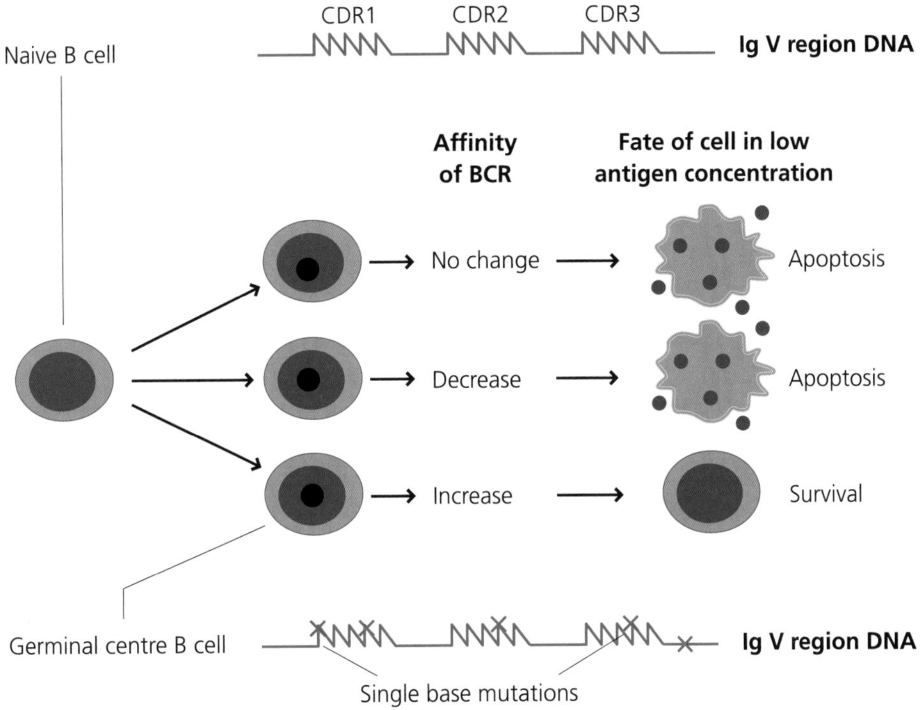

Fig. 6.25 Somatic hypermutation. During TD responses, random mutations are introduced in V region genes of B cells, particularly in the CDR regions (compare top and bottom showing Ig V regions of a naive B cell and one of its germinal centre progeny respectively. This is somatic hypermutation. These mutations may alter the structure of the antigen-binding site. Most mutations will be neutral or decrease binding affinity, and the B cell will apoptose, but occasionally will result in higher affinity. In conditions of low antigen concentration, B cells expressing a higher affinity BCR will accumulate more antigen, and will have a selective advantage and survive. Thus, the average affinity of the secreted antibodies will increase. This is affinity maturation.

Box 6.6 Affinity Maturation *In Situ*: Laser Capture Micro-Dissection

Laser capture micro dissection is a technique that permits analysis of gene expression in single cells in an organ or tissue. Histological sections of tissues are examined and cells of interest identified (e.g. by immunostaining). A laser working through the microscope is used to burn away areas surrounding the cell(s) of interest. The remaining cell(s) can then be micro-pipetted from the section and, using the reverse transcription-polymerase chain reaction (RT-PCR), gene expression examined. In the case of B cells, sections of germinal centres from lymph nodes supporting an active response were used. It proved possible to sequence the immunoglobulin genes being expressed in individual B cells and to identify mutations in their V regions. In some cases it was possible to identify a B cell expressing the germline gene and then to identify other B cells in the same germinal centre that expressed the same V regions, but which had V region mutations. In this way it was possible to follow the life history of the progeny of an individual B cell as they accumulated increasing numbers of mutations, and even to map the migration of the B cells within the germinal centre. A different approach to a similar problem used B cell hybridomas that were generated at different times after the start of the immunization procedure. The result here was that the longer the time after immunization, the more V region mutation were seen in the resulting hybridomas. The use of such techniques which permit analysis of gene expression in individual cells is giving insight into the complexity of immune responses. This complexity is not always seen when the properties of bulk populations of cells are examined, and may lead to us rethinking our ideas of immune response initiation and development.

exclusively in activated B cells. As with CD40–CD40L interactions, AICD is also required for class switching (above). Patients with mutant AICD suffer from the form of HIGM syndrome called HIGM2 See Case Study 6.1.

6.4
B Cell Memory, Antibodies and Long-Term Resistance to Re-Infection

One of the central features of adaptive immunity is that individuals who have recovered from an infection or who have been vaccinated against an infectious disease are usually resistant to re-infection with the same organism. This does not, however, apply to all B cell responses for example vaccination with the TI pneumococcal capsular polysaccharide antigens generates only weak, short-lived resistance. In contrast, individuals immunized against tetanus remain resistant for at least 10 years without any infection or boosting. Long-term resistance to re-infection is often loosely called memory. Mechanistically, however, long-term resistance to infection may result for several reasons. Where resistance to re-infection is dependent on antibodies, the key requirement is to ensure that antibodies are readily available should infection recur. This can be due to (i) populations of long-lived memory B cells that are rapidly re-activated if the same infectious agent is re-encountered; (ii) long-lived plasma cells that produce antibodies continuously over long periods, even in the absence of antigen; and (iii) at the population level, repeated, subclinical infection of different members of the species with the same pathogen, or a cross-reactive antigen. See Figure 6.26.

6.4.1
Memory B Cells

Memory B cells are generated in a B cell response from activated B cells that have not differentiated into plasma cells since there is no evidence that mature plasma cells can convert into memory cells. In mice, memory cells do not express IgM or IgD, but express other isotypes (i.e. they are class-switched); in humans, however, large B cells with a memory phenotype that express IgM can be found. Memory B cells can be studied by adoptive transfer in animals (Figure 2.8), or by stimulating B cells from an immunized animal or human *in vitro*. Early after immunization, memory B cells

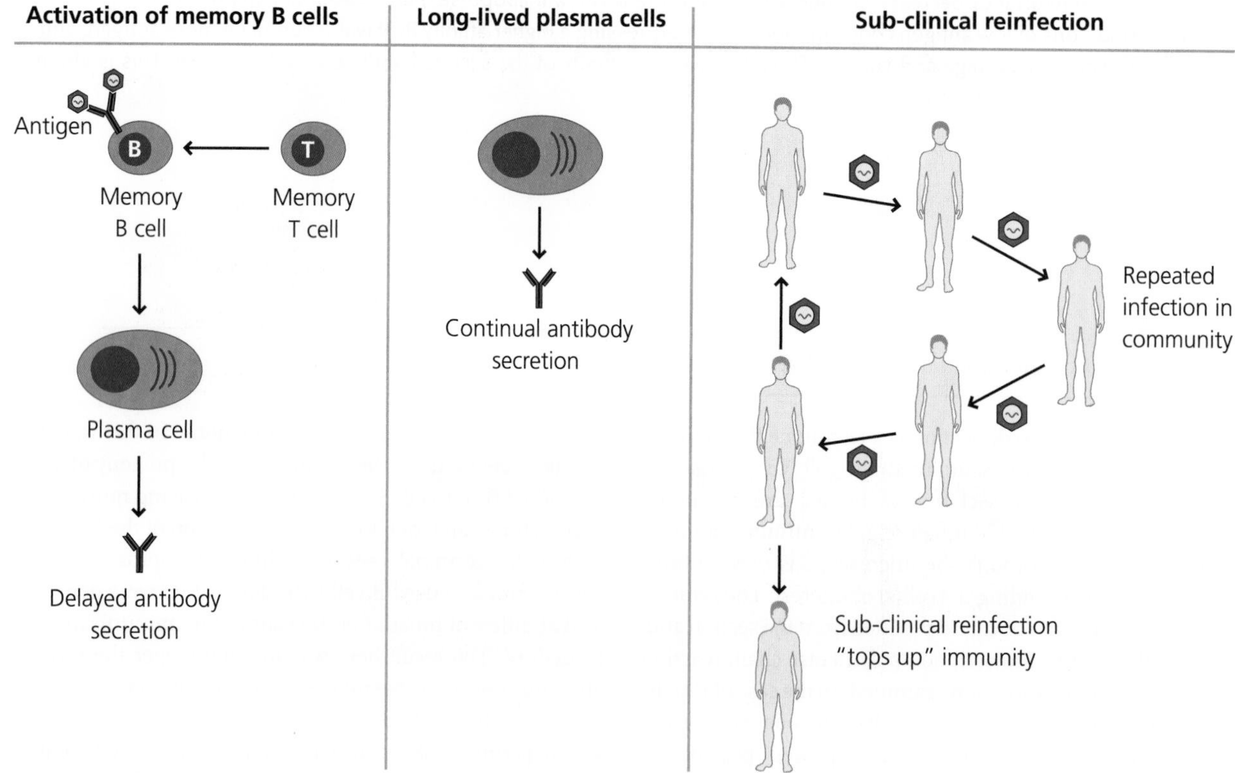

Fig. 6.26 Antibody-mediated resistance to re-infection. At least three different mechanisms can be involved in the long-term maintenance of antibody-mediated resistance to re-infection. In any immunocompetent individual, memory B cells specific for the antigen can be re-activated rapidly following infection. This, however, takes time and resistance to some infections requires pre-formed antibodies. Long-lived plasma cells, particularly in the bone-marrow, can also secrete antibodies for long periods (years), giving immediate protection. At the population level, repeated subclinical infection of immune individuals by an infection endemic in the community can serve to top up immunity without the individual being aware of the infection.

show signs of activation, but with time they come more and more to resemble resting cells. Thus, after immunization there are expanded populations of memory B cells which respond to re-infection more rapidly than naïve B cells. Many will also have switched their immunoglobulin isotype to IgG, IgA or IgE. These cells do, however, need to be re-activated and to differentiate into plasma cells before protective antibody becomes available and re-activation is T cell-dependent.

B cell memory (the ability to make a rapid, isotype-switched secondary response to an antigen) is very long-lasting in immunized mice, well over a year. If B cells from an immunized mouse are adoptively transferred into a non-immune mouse, together with antigen-specific T helper cells, immunization will induce a rapid IgG response. If, however, immunization is delayed following the transfer of B cells, the ability to generate the rapid IgG response disappears within weeks. This suggests that there is something in the immunized mouse that maintains memory B cells and the best candidate is antigen retained on the surface of FDCs. These are long-lived, non-haematopoietic cells present in B cell follicles which can retain antigen–antibody complexes on their surfaces for long periods. Antigenic protein can be detected on FDCs for long periods (months in mice, for most of their life span). It is suggested that memory B cells, recirculating through follicles, recognize their cognate antigen on FDCs. This may serve to prevent the memory B cell apoptosing, but may also enable the B cell to capture some of the antigen and to present peptides from the antigen to antigen-specific memory CD4 T cells. These in turn will stimulate the memory B cells to become plasma cells and secrete antibody. Thus, antibody synthesis may be maintained for as long as antigen is present on FDCs. See Figure 6.27.

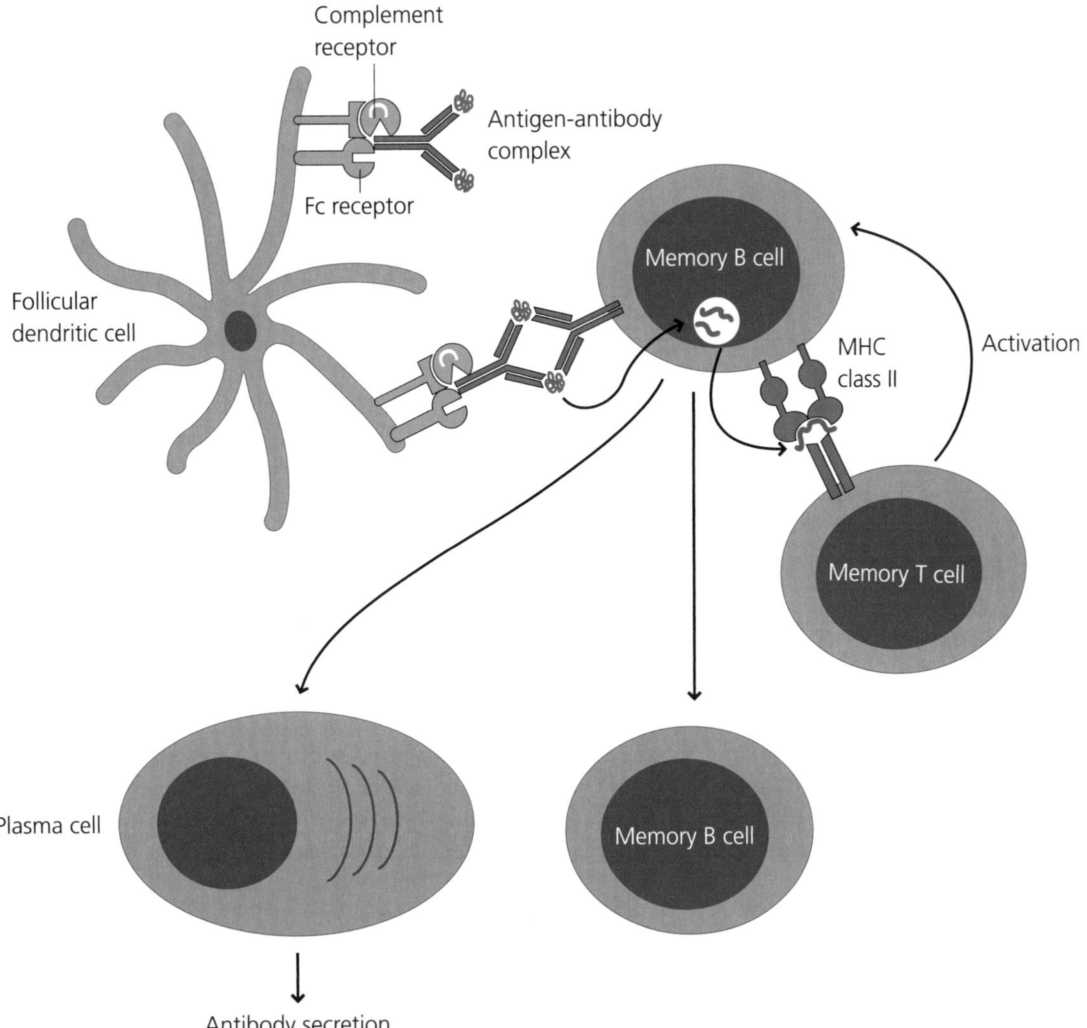

Fig. 6.27 Follicular dendritic cells. FDCs are mesenchymal cells (not bone marrow-derived). They are long-lived resident (non-migratory) cells found in B cell follicles in secondary lymphoid tissues. FDCs express complement and Fc receptors that do not mediate endocytosis. This enables them to retain native (i.e. non-degraded) antigens that are complexed to antibodies on their surfaces for long periods. These complexes may be sampled by memory B cells and used to maintain low-level stimulation of these cells. The B cells may be able to process and present the antigen they have acquired to memory T cells, which in turn activate the B cells. This serves to provide a continual supply of antibody and to maintain the memory B cell pool.

6.4.2
Long-Lived Plasma Cells

It used to be thought that plasma cells lived for only a few days. This seems to be true of splenic plasma cells, but it has been found that the bone marrow is a major source of plasma cells that migrate there following immunization. These plasma cells can live for very long periods – months or a year or more in mice, perhaps much longer in humans, and can continue to secrete antibody over these periods. Plasma cells, purified from mouse bone marrow and transferred into normal mice, continued to secrete antibody for long periods in the apparent absence of the antigen for which they are specific. This will clearly provide long-term resistance to re-infection that is not dependent on memory B cells.

6.4.3
Subclinical Infections

Another important method for maintaining protective levels of antibody is by repeated subclinical infection of immune individuals. As the individual is immune they do not develop symptomatic infection, but the pathogen tops up their immunity. This is illustrated by studies on rubella (German measles). In the United Kingdom, all children are immunized in childhood and immunity to rubella was thought to be life-long. However, when girls approaching reproductive age were tested, their antibody levels were worryingly low, possibly too low to mediate protection, and a programme of re-immunization was introduced. The explanation is that the vaccination of children had been so effective that the virus could not survive in the community and, in the absence of subclinical infection, antibody levels could not be maintained at protective levels.

6.5
B Cell Differentiation and Selection

In this section we will discuss B cell development, and the mechanisms involved in reducing the probability of them reacting to self antigens and inducing autoimmunity.

6.5.1
Development of B Cell Populations

As we have discussed previously, B cells can be divided into three major groups. B-1 cells arise early during foetal development, before B-2 cells are generated, and in adult mice form a self-renewing population in the spleen, intestine and the pleural cavities (in humans, apparently similar B cells have also been identified in the peritoneal cavity). B-2 B cells include a sessile population of marginal zone B cells in the spleen and a recirculating population of follicular B cells which migrate through B cell follicles of secondary lymphoid tissues. These three populations arise from the common lymphoid progenitor (CLP), as do T cells, NKT cells and NK cells (Section 3.5). Once formed, the three B cell lineages are largely independent. Evidence for this comes from studies of cytokines that regulate their differentiation. For example, the cytokine BAFF is required for survival of B-2 cells. Mice lacking the gene for this cytokine or its receptor have greatly decreased numbers of B-2 B cells, whereas the B-1 population is relatively unaffected. See Figure 6.28.

Q6.24. How might we show that there is a bone marrow cell type that can give rise to B and T cells, but to no other cells types?

6.5.2
Differentiation of B-2 Cells

In mice and humans all B cells originate in the bone marrow. Different stages in B cell development in the marrow are identified by the expression of different surface markers, whether or not the cell is dividing and, crucially, the different stages of immunoglobulin gene rearrangement in the DNA of these cells, which occur sequentially during development. Thus if a functional immunoglobulin H chain is produced (at the pro-B cell stage) it pairs with a germline-encoded surrogate L chain and this forms a pre-BCR, which is expressed at the cell surface; the cell is now termed a pre-B cell. Signalling through the pre-BCR stimulates several rounds of proliferation and, importantly, inhibits further heavy (μ) chain gene recombination hence contributing to allelic exclusion (Section 6.2.3.2). It also stimulates L chain gene rearrangement. The pre-B cell then rearranges its L chain genes and, if a productive κ or λ chain is produced, this L chain combines with the H chain to form a competent IgM BCR (immature B cell). Productive rearrangement of a κ chain shuts off further rearrangement of the other κ locus (allelic exclusion) as well as λ chain rearrangement or vice versa, hence leading to isotypic exclusion (Section 6.2.3.2). As κ chain rearrangement occurs first, most B cells express the κ chain. Early B cell development (pro-B cells) appears to require contact with stromal cells, but later (pre-B cell) development, at least in mice, requires IL-7.

Effective signalling through both the pre-BCR of pre-B cells, and the BCR of immature B cells involves many different signalling pathways and requires, amongst many others, the intracellular signalling molecule Btk. If there is a mutation in the BTK gene the developing B cell dies at the pre-B stage. The BTK gene is expressed on the X chromosome so a single mutation can lead to a B cell deficiency in males. This is the origin of Bruton's X-linked agammaglobulinaemia, one of the commoner primary antibody deficiencies which, as you may suppose, leads to an increased incidence of pyogenic infection. There are several other primary B cell immunodeficiencies. These include defects in AICD and the BAFF receptor. All lead to increased susceptibility to extracellular pyogenic bacterial infections.

Immature B cells that have not been identified in the bone marrow as highly autoreactive leave the bone marrow as

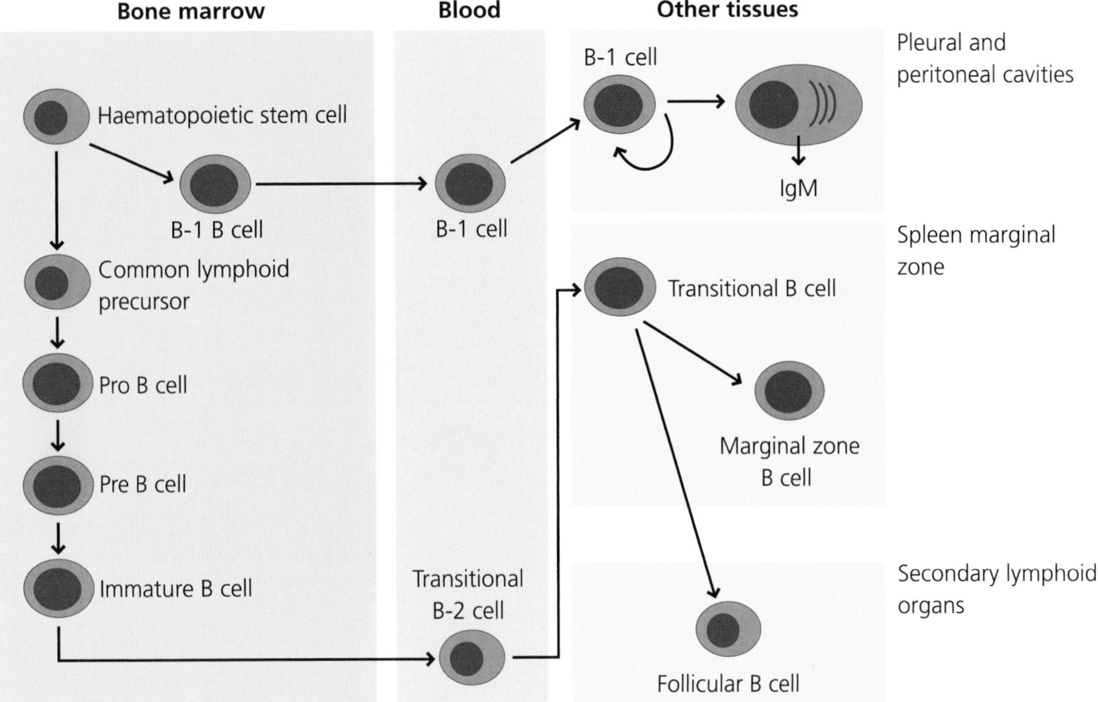

Fig. 6.28 B cell differentiation. In adult mice and humans B cells are formed in the bone marrow from the common lymphoid precursor. B-1 B cells leave the marrow and enter sites such as the peritoneal cavity, where they form a self-renewing population. B-2 B cells develop through pro-B and pre-B stages to become immature B cells. These leave the marrow as transitional B cells that migrate to the spleen. Here, a B-2 B cell may become either a resident marginal zone B cell or leaves to become a recirculating follicular B cell.

transitional B cells and further development occurs in the periphery. Endothelial cells in the marginal zone and red pulp of the spleen may provide a niche for development of transitional B cells into marginal zone B cells, which are primarily responsible for TI-2 responses. The remaining transitional B cells become follicular B cells and recirculate through B cell follicles in secondary lymphoid organs where they are responsible for responses to TD antigens.

6.5.3 B Cell Tolerance

Unlike T cells, B cells are not MHC-restricted, but many developing B cells will express potential anti-self receptors and mechanisms exist to reduce the numbers of such cells. If an immature B cell receives a strong signal from a multivalent self antigen in the marrow (e.g. a cell surface molecule on a stromal cell) the immature B cell either dies (clonal deletion) or receptor editing may occur. In the latter process, there is a short time window where RAG gene expression is maintained and further L chain gene rearrangements can occur. If a BCR is generated that is not autoreactive, the B cell undergoes further maturation and leaves the marrow. If the new receptors do recognize a self antigen, the B cell dies. Alternatively, if the newly formed B cell reacts with a soluble self molecule it is rendered unresponsive (anergized). Such B cells continue to express the relevant chemokine receptors that will enable them to reach T cell areas and interact with T cells. However their anti-apoptotic mechanisms are decreased, so that T cells that express Fas ligand, for example, can kill them thus leading to their elimination. See Figure 6.29 and Box 6.7.

B cell tolerance is incomplete, but it has been estimated that only 2–5% of newly formed B cells escape negative selection in the marrow. These cells are then released as transitional B cells into the circulation (above). Some of these cells will be potentially autoreactive, but self-reactive follicular B cells need CD4 T cell help if they are to become activated and T cell tolerance is more complete. If the follicular B cell recognizes antigen in the absence of help it becomes unresponsive (anergic) and subsequently dies. If, however, a mature, autoreactive follicular B cell does become activated another process can occur. This is sometimes termed receptor revision (to differentiate it from receptor editing that can occur at the immature stage). Under these circumstances the RAG genes (and other components of the recombinase-activating complex) are turned on again, and further immunoglobulin gene rearrangements can occur in a further, and apparently final, attempt to generate a new BCR that is no longer autoreactive. We stress that this is a mechanism to induce peripheral B cell tolerance, that it occurs in mature B cells, and that it is not known to occur in T cells.

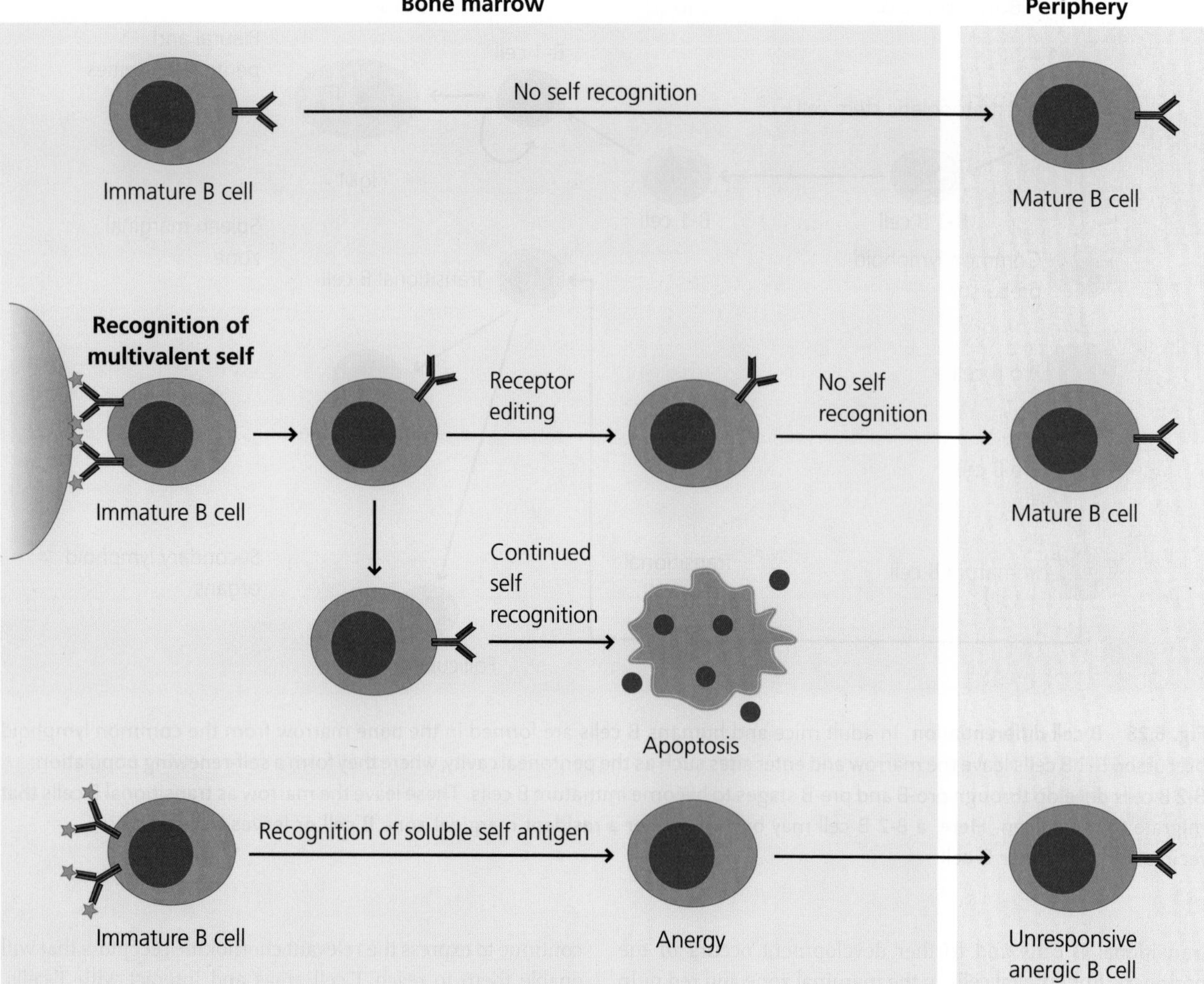

Fig. 6.29 B cell tolerance in the bone marrow. Immature B cells with BCRs that do not recognize self antigens enter the periphery and develop into mature B cells. Immature B cells that recognize multi-valent self antigen in the marrow can, for a short time, attempt to form new BCRs that are non-reactive through receptor editing. If the new receptor does not recognize self antigen, the B cell exits the marrow. If the BCR continues to recognize self antigen, the B cell apoptoses. If an immature B cell recognizes soluble self antigen in the marrow it may become an anergic B cell that is unresponsive to stimulation.

6.6
Therapeutic Antibodies

6.6.1
Antibodies for Immunotherapy

The roles of antibodies in vaccination are generally clear and straightforward and the development of effective vaccines is discussed in Chapter 2. Suffice it to say here that all the wholly effective vaccines currently available depend on antibodies for the protection they give. Antibodies are, however, becoming increasingly important in the immunotherapy of non-infectious disease. We have already mentioned the treatment of Rhesus disease (Section 6.2.6.5). More recently, therapeutic monoclonal antibodies have become available for the treatment of a variety of diseases and these represent a major advance in therapy. One of the first examples was the use of anti-TNF-α antibody in the treatment of some forms of rheumatoid arthritis that were refractory to other available treatments. This treatment was remarkably successful in suppressing the inflammation in these patients. This was followed by the development of monoclonal antibodies for the treatment of many diseases, not all immunological in origin. Over 100 monoclonal antibodies are now being tested in clinical trials and more than 20 have been licensed by the US Food and Drug Administration for clinical use. The target diseases are mainly autoimmune diseases or malignant tumours, but also include macular degeneration (the major cause of blindness in elderly people) and the prevention of transplant rejection (we note that this is not a disease).

6.6.2
Antibody Engineering

A major problem with the use of monoclonal antibodies for therapy is that at present they are derived from mice or rats,

Box 6.7 Mechanisms of Tolerance in B Cells: Studies with Double Transgenics

Tolerance Induction in Immature B Cells by a Membrane-Bound Antigen

Transgenic mice were made in which many B cells were able to express an IgM BCR specific for MHC class I molecules expressed by the H-2^k strain of mice. In the H-2^d strain, transgenic mice (which do not express H-2^k) between 20 and 50% of the B cells expressed the transgenic BCR. In transgenic F_1 mice expressing H-2^k (H-2^k × H-2^d) no detectable B cells expressed the transgenic BCR. The numbers of peripheral B cells was reduced by around 50% in the F_1 mice. These observations suggest that in mice expressing H-2^k, B cells with BCRs specific for that molecule were deleted in the bone marrow.

Soluble Self Antigen Induces Anergy in Mature B Cells

Transgenic mice were made that could express a BCR specific for hen egg lysozyme (HEL). In these mice around 90% of B cells expressed the transgenic receptor and these cells were able to react to HEL. Other transgenic mice were made that expressed a secreted form of HEL (i.e. HEL was present in their serum). F_1 crosses between the two transgenic strains were made. These double-transgenic mice possessed large numbers of anti-HEL B cells but did not make antibody to injected HEL and the amount of IgM expressed on peripheral B cells was much reduced, suggesting a reason for the lack of response (**anergy**). If mature B cells expressing the anti-HEL BCR were adoptively transferred into HEL-transgenic mice, the transferred B cells became anergic and expressed decreased amounts of surface IgM. HEL in these mice represents surrogate self antigen and these observations suggest that soluble self antigens, seen by B cells in the absence of T cell help, can induce anergy and thus tolerance.

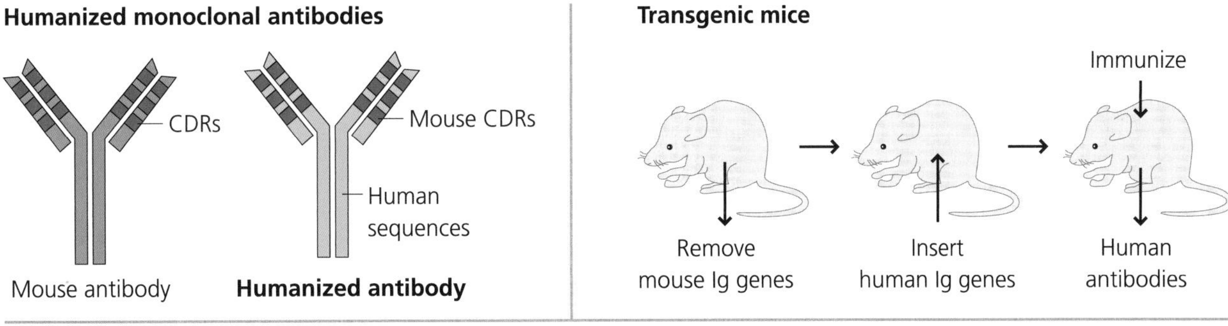

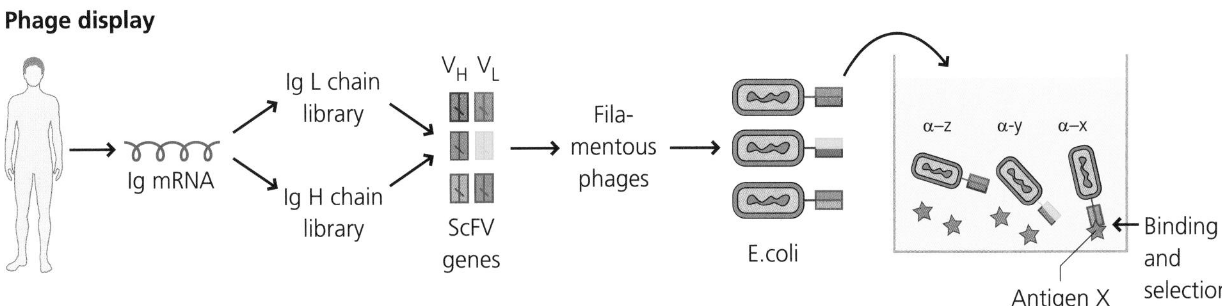

Fig. 6.30 Antibody engineering. Non-human monoclonal antibodies induce antibody responses in humans that render them ineffective. Several approaches are being used to overcome this problem. **Humanized monoclonal antibodies.** Genetic engineering can be used to replace all of a mouse immunoglobulin gene, except the CDRs, with human sequences. These antibodies retain their antigen specificity, but are much less immunogenic **Human immunoglobulin transgenic mice.** A mouse's immunoglobulin genes can be replaced with human genes and on immunization such mice make fully human antibodies. **Phage display.** Libraries of human V_H and V_L gene segments can be produced and the V_H and V_L segments joined to make genes that code for single-chain variable fragment (ScFV) antibody fragments. These ScFV fragments can be expressed on filamentous bacteriophages, and phages expressing antibodies of interest can be selected by binding to antigen-coated plates. The selected ScFV can then be joined to desired C region gene segments.

not humans, and are thus seen as foreign proteins by the human immune system. This results in the synthesis of antibodies against the monoclonal antibody, resulting in its very rapid disappearance and loss of efficacy. To minimize this occurring, some monoclonal antibodies have been humanized: they have been genetically modified so that most of the molecule, apart from the antigen-binding site, has been replaced by human sequences. These antibodies are much less immunogenic than their original forms. More recently transgenic mice have been engineered in which the murine immunoglobulin genes are replaced by human genes so they make human antibodies if they are immunized. Thus, human monoclonal antibodies can be generated from these mice.

Human monoclonal antibodies can also be generated *in vitro* by using bacteriophages. Phages are made that incorporate a random selection of human V region domain (Fv) genes. These combine on the phage surface to form complete antigen-binding domains. Thus, the phage will bind to the cognate antigen, and this binding is used to select and isolate the phage expressing the wanted antibody. The DNA can then be extracted from the phage and used to produce the antibody by standard molecular biological methods.

There is, however, another more generic problem with therapeutic antibodies: they act indiscriminately on cells or cytokines and their activity is necessarily antigen-non-specific. Thus, they can also affect useful responses (e.g. responses to infection). The most dramatic example of this is with the use of anti-TNF-α antibodies to treat rheumatoid arthritis; some patients who had latent tuberculosis re-activated their infection and developed clinical disease infections.

Learning Outcomes

By the end of this chapter you should be able to understand, explain and discuss the following topics and questions; the relevant sections of the chapter are indicated. You should understand some of the evidence from human and animal studies supporting what we know about these topics. You should have some idea of the areas where our understanding is incomplete. You may be able to suggest ways in which our understanding could be advanced.

- B cell populations (Section 6.1)
  - What are the main types of B cells?
- Antibody structure and function (Section 6.2)
  - Why are antibodies multivalent?
  - How is the diversity of antibodies generated?
  - Where are different classes of antibodies generally found, what do they do and how do they do it?
- B cell responses (Section 6.3)
  - How are B cell responses triggered and where does this occur?
  - How do B cells become activated, and what are the roles of different surface molecules in B cell activation and its regulation?
  - What are natural antibodies and what are their functions?
  - What are TI antibody responses?
  - What are the roles of T cells in TD antibody responses?
  - What is isotype switching and why is it important?
  - What is somatic hypermutation and why is it important?
- B cell memory, antibodies and long-term resistance to re-infection (Section 6.4)
  - How is long-term antibody- or B cell-dependent protection against infection mediated?
- B cell differentiation and selection (Section 6.5)
  - How and where do B cells develop
  - What are the different types of B cell tolerance?
- Therapeutic antibodies (Section 6.6)
  - How are monoclonal antibodies made?
  - How can monoclonal antibodies be engineered in different ways and why is this important?
- GENERAL: What types of natural and experimental defect lead to deficiencies in B cell development and function, and what are their outcomes in terms of infection?
- INTEGRATIVE: What are the similarities and differences between T cell and B cell development and/or activation and/or effector functions?
- INTEGRATIVE: How does tolerance in B cells differ from that in T cells?
- INTEGRATIVE: What are the similarities and differences between isotype switching and somatic hypermutation, and why might these not occur for T cells?

Further Study Questions

Qa. How well do we really understand the function of different antibody isotypes in defence against infection? (Section 6.2)

Hint The functions of four of the main human and mouse isotypes (IgM, IgG, IgA and IgE) might seem reasonably well established. However, IgM is not always a pentamer, IgG is comprised of different subclasses and IgA comprises two subclasses (in humans) that can be expressed as monomers or dimers. Presumably, these play specialized roles in different types of defence, but how well do we understand them? What really is the function of IgE in defence against infection, rather than in pathological settings (allergy)? What about more recent insights into the potential functions of IgD in host defence?

Qb. To what extent are different types of B cell responses important in immunity? (Section 6.3)

Hint At the cellular level, we might start by considering the different populations of B cells, how they are activated and where they are found. Then consider which of them preferentially produces natural antibodies or TI versus TD responses, under what conditions and where. At the molecular level, we might consider to what extent these responses change after repeated stimulation. We could also potentially ask which types of stimuli might lead to signalling through BCRs or TLRs, for example? Much of our understanding of these comes from contrived settings (e.g. responses to hapten-carriers or anti-immunoglobulin antibodies). What of real disease settings?

Qc. How well can we explain the different migratory patterns of naïve B cells, plasma cells and memory B cells? (Section 6.4)

Hint The post code principle is likely to be essential (Section 4.3.4.1), but is it sufficient to explain the different anatomical sites to which these cells home? We do know quite a lot about preferential expression of different chemokine receptors, for example, but what about selectins or integrins? Within each site (e.g. secondary lymphoid tissues or the bone marrow), what is known of how different cells are guided to different compartments or what controls their release or retention?

Qd. What is known of the relative importance of different mechanisms in antibody-mediated resistance to infection? (Section 6.5)

Hint We have provided some examples of long-lived plasma cells or memory B cells and resistance to infection. However, how important is each in different types of infection? Why should memory B cells have evolved, if long-lived plasma cells can continue to secrete antibodies? Are the latter really end cells or are they in fact more plastic being able to adopt different functions during immune responses? What is the role of short-lived plasma cells? What actually controls the survival of memory B cells?

Qe. What advances are being made in the development and application of therapeutic antibodies? (Section 6.6)

Hint One could start by thinking about these separately. In terms of development, we might consider what types of technical advances are being developed to enhance their function, such as by preventing host responses to foreign (e.g. mouse) monoclonal antibodies. Or we might ask for which disease settings new antibodies are being developed, and why and to what extent – both in terms of the disease and their potential targets – they have been tested clinically.

7
Immunity, Disease and Therapy

7.1
Introduction

Innate immunity has evolved over several billion years, and has selected recognition systems that can discriminate between molecules expressed by infectious agents and their hosts. Innate immunity rarely causes direct damage to the host. The same is not the case for adaptive immunity, which has been around for about one-eighth of the time of innate immunity and is a much less perfect system. This imperfection has arisen largely because the antigens expressed by pathogens that the adaptive immune system needs to recognize are often not inherently different from the host's own molecules. This means that the antigen receptors of the adaptive system cannot be pre-selected to recognize only pathogen-expressed antigens. The outcome is that both the T cells and antibodies of adaptive immunity can potentially recognize host antigens. On occasion this can result in serious damage and sometimes death to the host. However, the damage is largely caused by recruiting and amplifying components and functions of innate immunity that are otherwise needed to deal with infections. In this chapter we examine the different types of immune-related diseases and other conditions that directly involve the immune system. We also introduce some types of therapy that are being used or developed to treat them, by redirecting immunity to another end and some cases by using components of immunity as therapeutic tools.

We start with a brief overview of the different areas we will discuss (Section 7.1). Then we consider the different ways in which the immune system can cause tissue damage and review what we know of the underlying causes of these harmful responses (Section 7.2). We focus on how lymphocytes are normally regulated to discriminate between infectious agents and host molecules (leading to active responses in the first case, tolerance in the second), under what circumstances they can be triggered to make abnormal responses, and the functions of the genes that control these processes. We next explain how immune responses that are directed against apparently harmless molecules such as components of food, or the body itself, can cause serious disease (Section 7.3). Using case studies to provide examples we then outline what we know, and often still do not know, about the pathogenesis or genetic basis of some of these conditions, what we have learnt from animal models and sometimes how these conditions can be treated (Section 7.4). We then describe how the immune system can cause the rejection of foreign transplants and how it is that some types of transplant can directly attack the recipient (Section 7.5). We end by providing a very brief overview of the potential role of the immune system in helping to defend against not just infectious agents, but perhaps also tumour cells, and how these might fail and lead to cancer (Section 7.6).

By the end of this chapter you will have a basic understanding of the ways in which adaptive immunity can cause harmful responses and disease, and of the reasons that these unwanted responses occur. You will understand why transplants are rejected and what we can do to prevent rejection. You will also begin to appreciate how the immune system interacts with tumours. Overall, you will gain some feeling for what we do and do not know about the roles of immune responses in causing disease.

7.1.1
Immunity, Disease and Therapy

What are the diseases and conditions that relate to the immune system, and what forms of therapy are being developed to deal with them? See Figure 7.1.

7.1.1.1 Immunity and Disease: Immunopathology

Immunodeficiency diseases are conditions resulting from genetic or acquired defects where there is a failure of immune responses. By definition, immunodeficiency diseases result in severe, persistent, unusual or recurrent (SPUR) infections, and the type of infection is usually characteristic of the deficiency in question. These are introduced in Section 1.6.3.1, and are used as evidence for the importance of different cells and molecules in host defence against infection throughout other chapters of this book. Hence, we will not discuss them in detail in this chapter. Likewise other types of genetic defect that lead to dysregulation of immunity and lead to autoimmune and/or lymphoproliferative diseases, such as APECED (Autoimmune Polyendocrinopathy–Candidiasis–

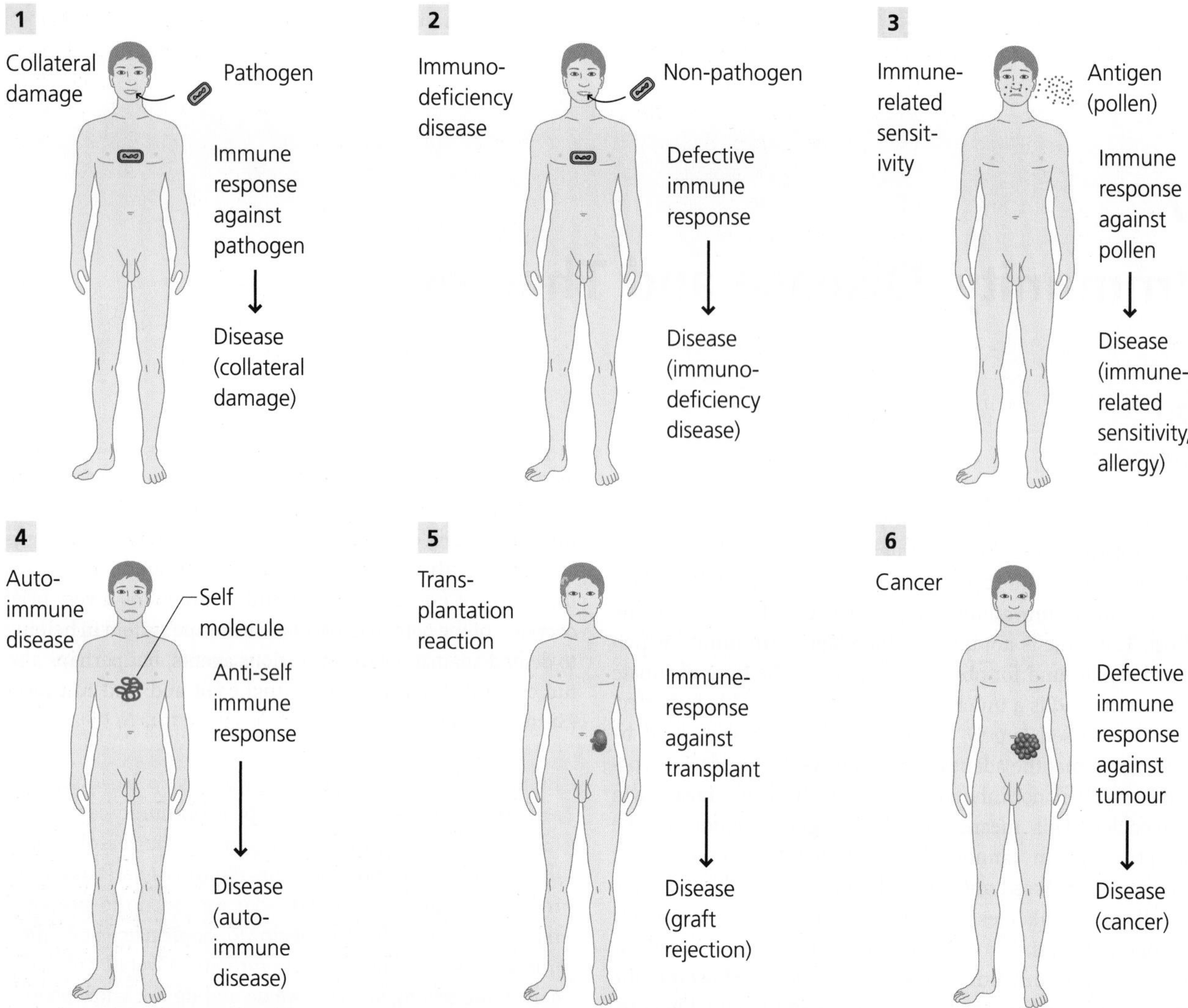

Fig. 7.1 Immune-related diseases. Immune responses are necessary to eliminate pathogens. (1) These responses may, however, cause **collateral damage** to the host. (2) Defects in immunity can result in severe infections by agents that are not normally pathogenic; such **immunodeficiency** diseases can be primary genetic defects or secondary in cause (e.g. because human immunodeficiency virus (HIV) damages crucial components of immunite reactions). (3) Some individuals make responses against normally harmless substances (peanuts, pollen, fungal spores) resulting in **allergies** or other **immune-related sensitivitie**. (4) Other individuals may make immune responses against self components leading to **autoimmune diseases**. (5) Immune responses can also cause **transplant reactions including rejection**. (6) If, however, immune responses are not able to reject abnormal tumour cells, a cancer may develop. (Autoinflammatory diseases that result from abnormalities of the innate immune system are not shown, Section 4.2.2.4.)

Ectodermal Dystrophy), IPEX (Immune dysregulation, Polyendocrinopathy, Enteropathy X-linked) and CTLA-4 deficiency, are discussed in 5.5.5.1 and not considered further in this chapter.

Immunopathology In this chapter we focus initially on aberrant responses against self or harmless antigens that cause immune-related diseases. Collectively these are generally known as immunopathological conditions; they can also result from an aberrant or unwanted adaptive immune response to an infectious agent. The diseases we will consider mostly are mainly allergies and autoimmune diseases.

- Allergies, and other immune-related sensitivities, are sometimes termed hypersensitivity diseases. Well-known examples are hay fever, peanut allergy and, contact sensitivity to metals. These conditions result from an abnormal or exaggerated response to non-infectious, external antigens that are normally harmless to other individuals. Collectively we will refer to these foreign antigens that originate from our environment as extrinsic antigens.
- Autoimmune diseases. Well-known examples of these are rheumatoid arthritis, insulin-dependent (Type I) diabetes and multiple sclerosis. These represent the consequences of an abnormal or exaggerated response against non-infectious, internal antigens to which most individuals do not respond. Collectively we will refer to these self antigens that are part of the body itself as intrinsic antigens. Autoimmune diseases are commonly divided into organ-specific and systemic diseases. In the former case they involve

one main organ in the body, such as the thyroid gland. In the latter case they affect several organs or body systems, such as the blood vessels and joints in systemic lupus erythematosus (SLE) and rheumatoid arthritis, respectively. An autoimmune disease may be precipitated by an extrinsic trigger, such as a viral or bacterial infection, but this is quite distinct from the intrinsic mechanisms that are involved in the persistence of disease once they have been induced.

Q7.1. Why might humans have not evolved in ways that would prevent autoimmune diseases, allergies and other immune-related sensitivities?

In the last part of this chapter we discuss immunity in relation to transplantation and cancer.

- **Transplant and transfusion reactions**: Transplant rejection represents entirely normal, but unwanted, responses against foreign antigens on transplanted tissues (e.g. kidneys) or cells (e.g. bone marrow) that are derived from other people or different species. Such reactions lead to rejection of the transplant and are triggered by foreign antigens that are, in themselves harmless. Collectively these are called transplantation antigens. Conversely, bone marrow transplants can -host cause damage to the host; graft-versus-host disease. Blood transfusion is a special case of transplantation. Responses to non-self antigens such as the ABO or Rhesus blood group antigens expressed on erythrocytes lead to transfusion reactions.
- **Tumour immunity**: New, uncontrolled clonal growth of single cells and development into tumours may be associated with expression of new proteins by the cell. It is possible that the immune system could recognize these new antigens. Such an immune reaction might destroy or control these cells in the same way as infectious agents are dealt with this is called tumour immunosurveillance. Collectively these types of altered or aberrant self antigens are termed tumour antigens and, if they stimulate an immune response that eliminates the mutant cells, they can be termed tumour rejection antigens (by comparison with transplant rejection). In some cases, such as a virally induced tumour, tumour antigens may actually originate from infectious agents. However, in the case of clinical malignancy this form of rejection has clearly not worked, but it might be possible to stimulate the immune system to bring about rejection of the tumour.

A final, general category of immune-related diseases are the autoinflammatory diseases, where the conditions are caused by abnormal responses of the innate immune system and the adaptive system is not involved. These diseases result from genetic defects, but differ from primary immunodeficiencies (above) because these mutations frequently lead to a gain rather than a loss of function and do not lead to SPUR infections; these conditions are discussed in Section 4.2.2.4 and are not consider further in this chapter.

It is very important to bear in mind, throughout this chapter, that any immune response may involve inflammation, though not all immune responses do provoke inflammation *per se*. Also, some immune deficiencies result in inflammation either with or without infection (e.g. the autoinflammatory diseases). It is important to understand which diseases do include inflammation when considering different types of therapies for them.

Q7.2. Is it likely that invertebrates might suffer from any of the immune-related diseases or conditions we are discussing in this chapter?

7.1.1.2 Disease and Therapy: Immunotherapy

Immunotherapy Throughout this chapter we will provide examples of the use of biological components of the immune system to treat ongoing disease, which may or may not have an immunological causation. We will not discuss vaccination in the prevention of disease since, this is covered in Sections 2.5 and 4.6. We will include antigen- and antibody-based therapies, therapy involving adoptive transfer of immune cells, and mention cytokine therapy. We will also mention some pharmacological approaches to therapy including inhibitory or immunosuppressive drugs. See Figure 7.2.

- **Antigen-based therapies**: One of the best known examples of an antigen-based therapy is vaccination (e.g. as used to stimulate protective immunity against an infectious diseases; Section 2.5). In the context of this chapter, however, we will focus on the use of an antigen, or relevant epitopes from the antigen, to decrease or alter the immune response to that antigen. This can be done by administering the antigen in such a way that it will induce tolerance antigen unresponsiveness), often by triggering suppressive mechanisms mediated by regulatory (T_{reg}) T cells or even inhibitory antibodies, or will re-programme the response in a non-harmful direction. To date this approach has been used mainly in the treatment of allergies. For example, desensitization to allergens involves injecting increasing doses of antigenic material, starting with very small doses.
- **Antibody-based therapies**: Pre-formed antibodies can be used in a variety of ways. For example, they can be used to deplete or inhibit cell populations expressing specific molecules or to block the activity of potentially harmful molecules such as cytokines. Particular human antibodies can be used to prevent certain conditions. However, more recently, monoclonal antibodies are being used in a variety of therapeutic settings. These therapeutic antibodies are being used to treat autoimmune diseases, to inhibit transplant rejection, and as therapies for certain cancers. In other cases, autoimmune diseases can be treated by depleting antibodies that cause damage (by plasmapheresis) while in others, paradoxically, they can actually be treated by giving large doses of pooled antibodies from normal donors, intravenous immunoglobulin (IVIG) therapy.

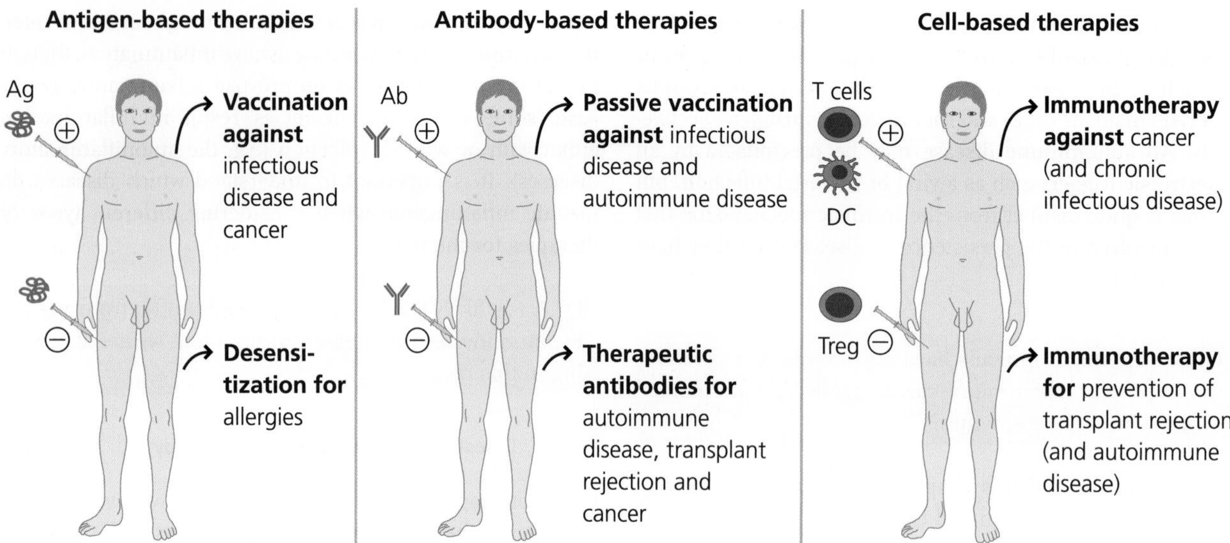

Fig. 7.2 Immune-related therapies. Antigen-based vaccines. A variety of agents ranging from DNA to attenuated living microbes can be used to induce a specific adaptive immune response that protects against subsequent infection. In the future it may be possible to vaccinate against cancers. Antigens can sometimes be used to reduce specific immune responses (e.g. desensitization in allergies). **Therapeutic antibodies.** Monoclonal antibodies that bind to and inhibit a molecule involved in a disease can be used to treat the disease. Such therapeutic antibodies are now used in some autoimmune diseases and cancers and in prevention of organ transplant rejection. **Cell-based therapies.** In the treatment of cancer, antigen-specific lymphocytes can be transferred or dendritic cells (DCs) used to induce specific responses. In the future it may be possible to use regulatory T cells (T_{reg}) to suppress unwanted responses (e.g. in transplantation or autoimmune diseases).

- **Adoptive cell-based therapies**: Key cells of immunity can be transferred to stimulate desired immune responses, or to inhibit damaging or unwanted immune responses. At present these forms of treatment are being explored mainly for the treatment of tumours and chronic infections, but may in the future be used for autoimmune diseases and transplantation. For example, dendritic cells (DCs), expressing the antigen of choice, may be injected in order to stimulate or modulate ongoing responses. Alternatively, tumour-specific T cells recovered from a patient can be expanded *in vitro* and re-injected into the same patient. Experimentally, T_{reg} have also been used to suppress transplant rejection.
- **Immunosuppressive drugs**: This is a huge area and we will not cover it in depth in this chapter. Many pharmacological drugs are anti-inflammatory and are widely used to prevent transplant rejection or to control autoimmune diseases. Others interfere with key intracellular signalling pathways in cells such as T cells and are also widely used in these settings.
- **Cytokine therapy**: Given the potency and selectivity of cytokines, it is not surprising that their use in therapy has been widely explored, primarily in cancer and chronic infections such as hepatitis B and C, and tuberculosis. Type I interferons (IFNs) and various other cytokines (e.g. IL-2) have been trialled in different conditions, but the results to date are not encouraging due to the multiple biological effects these cytokines can stimulate and their sometimes serious side effects. Growth factors, (e.g. granulocyte-macrophage colony-stimulating factor (GM-CSF)), can be used, sometimes successfully, to mobilize cells (e.g. neutrophils) from the bone marrow in cases where the numbers of these cells are reduced.

In some cases, some of the above therapies are needed to treat the side effects of other therapeutic procedures that are themselves immune-related. Examples are the use of therapeutic antibodies to overcome rejection following organ transplantation, or to treat graft versus host disease (GVHD) following bone marrow transplantation.

7.2 What are the Mechanisms of Tissue Damage Caused by the Immune System?

7.2.1 Initiation and Effector Phases of Disease

There is nothing novel or unusual about the effector mechanisms that cause cell or tissue damage during an immune response. They are the same mechanisms used to combat infections and may, for example, involve antibodies, cytotoxic T cells or activated macrophages. It is just that in these pathological conditions, these the responses are inappropriate or unwanted. The great bulk of these diseases are initiated by inappropriate activation of CD4 T cells responses, which of course necessarily also involves the innate immune system. In this section we will concentrate on the effector mechanisms that cause the damage and disease. This area is relatively well-understood. In Section 7.3 we will then approach the difficult question of why such unwanted responses are initiated.

7.2.2
Classification of Immunopathological Mechanisms

In the 1960s, Gell and Coombs introduced a classification of immune-related diseases that is still in use today. They called these hypersensitivity reactions and divided them into four types (I–IV). However, we now know that many diseases usually involve more than one mechanism. It is crucially important, at the outset, to appreciate the key points:

i) These hypersensitivity reactions are simply different mechanisms that can cause cell or tissue damage, whatever the setting.
ii) Some of these mechanisms are common to immune-related sensitivities against extrinsic antigens, to autoimmune diseases against intrinsic antigens, to some types of transplantation reactions and, potentially, to immune responses that might help to eliminate tumours.

The first three types of hypersensitivity diseases depend on antibodies as the effector mechanism. Type I, allergic hypersensitivity or, simply, allergy, depends on the interactions of antigen with IgE that is bound to mast cells. Type II, originally termed cytotoxic hypersensitivity, depends on the ability of antibodies, often IgM or IgG, to kill cells or to modulate their functions. Type III, immune complex-mediated hypersensitivity, depends on the ability of antigen–antibody complexes to initiate inflammation. In contrast, Type IV, delayed-type hypersensitivity (DTH), is mediated by cellular responses such as those including CD8 T cells or activated macrophages that damage the tissue, and is independent of antibodies. Of course we now know that all these mechanisms are dependent, at least in part, on the inappropriate activation of CD4 T cells as the initiating immune event. However, at the time, Gell and Coombs could only classify these types according to the effector mechanisms they could identify. Hence, even though we have gained an increasingly sophisticated understanding of the different mechanisms that can cause tissue damage, and increasingly find that we cannot explain any given disease on the basis of a single mechanism alone, this classification remains in use today. See Figure 7.3.

The types of immune response seen in hypersensitivity reactions are often divided into immediate and delayed responses. This refers simply to the time taken for a response to occur after antigen encounter in a sensitized individual i.e. one who has previously met the antigen and made a response. Immediate responses take only minutes to a few hours to happen because they represent the activity of pre-formed antibodies. In contrast, delayed responses usually take 24–48 h to appear. This is because they depend on re-activation of expanded populations of previously primed T cells, and their recruitment from the secondary lymphoid tissue and migration to the site where the antigen is present; this takes a relatively long time. Once again we stress that these responses are only seen in individuals previously sensitized to an antigen. When an antigen is encountered for the first time there is usually no overt response, although of course if the antigen hangs around for long enough, or is produced by a pathogen resident in the body, effector mechanisms can be activated and a reaction may be seen. See Figure 7.4.

7.2.3
Diseases Caused by IgE Antibodies (Type I: Allergic Hypersensitivity)

IgE is normally synthesized in response to parasitic infections. True parasites can be divided into single-celled protozoa and complex, multi-cellular metazoa such as intestinal worms. IgE is generally synthesized only against the metazoa. In these situations IgE may have a protective effect by helping with worm expulsion, although given the very chronic nature of such infections, it is often clearly not effective.

Allergy, which is wholly IgE-dependent, is thought to be the only form of hypersensitivity reaction that does not have a counterpart in autoimmune diseases. Recently, however, it has been found that some SLE patients do have IgE antibodies to nuclear antigens, although it is not clear if these antibodies play a role in pathogenesis. Allergies are common; hay fever is probably the commonest immune disorder in the Western world. It can be a minor inconvenience (runny eyes and sneezing because of local inflammation) or fatal (death within minutes due to severe, systemic responses). We understand the mechanisms underlying allergy very clearly, but we have very little idea why some of us become allergic and others do not, even when they are in the same family or even identical twins; this is discussed below. Allergic reactions happen very quickly in sensitized individuals, typically within minutes of antigen exposure.

What does IgE do in allergy? The effector mechanisms are well understood. During the sensitization phase B cells are, for unknown reasons, switched to express IgE by activated CD4 T cells (T_h2 cells Section 1.4.5.1). IgE binds to high-affinity IgE Fc receptors (FcRs) on mast cells in the absence of antigen. All of us who are allergic have mast cells coated with IgE specific for the particular antigen(s) to which we are sensitive. In this specific case the antigen(s) is referred to as an allergen(s). If a sensitized individual is re-exposed to that antigen, the antigen can bind to IgE coated mast cells at the mucosal surface or in the tissues. See Figure 7.5.

Q7.3. How might particles such as pollen grains manage to cross the epithelium of the eye to trigger mast cell degranulation in hay fever sufferers?

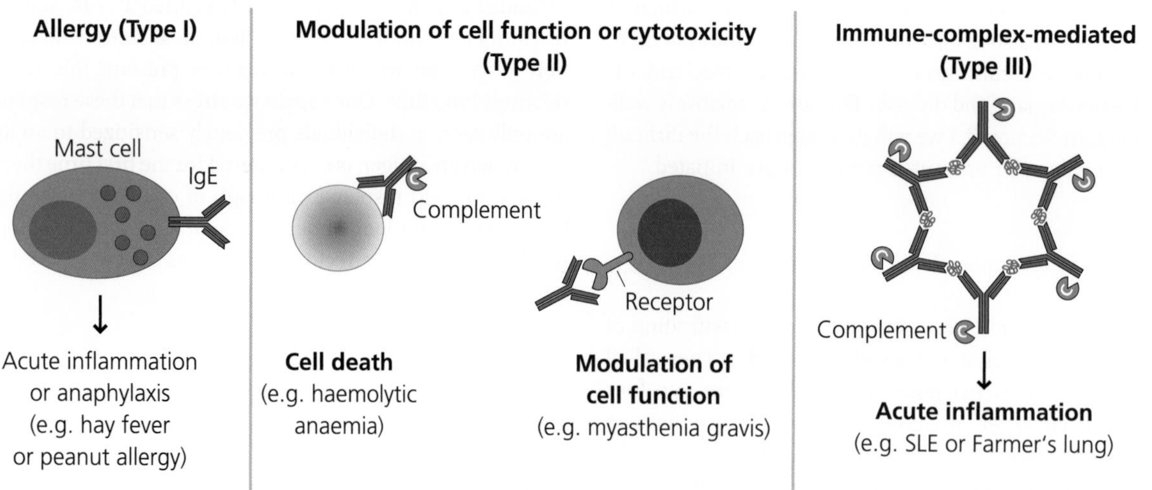

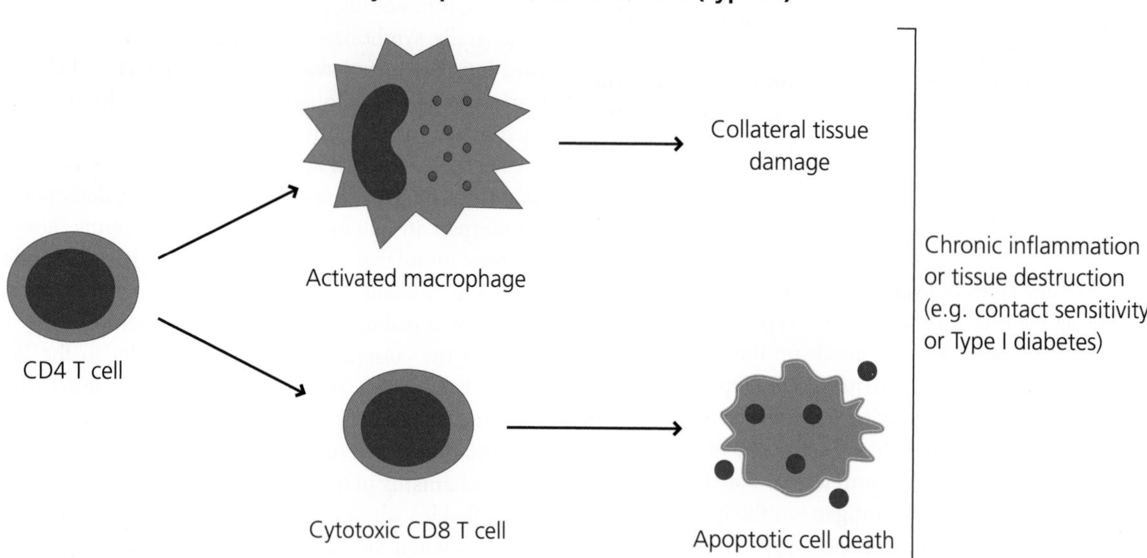

Fig. 7.3 Mechanisms of immunopathology. Tissue damage caused by adaptive immune responses can be due to or independent of antibodies. IgE antibodies (IgE) cause allergy and anaphylaxis by acting through mast cells. IgG and IgM can modulate cell function directly, by killing the cell or by blocking or stimulating receptors. Immune complexes can cause acute inflammation locally or systemically. Cell-mediated (antibody-independent) immunopathology is usually due to T_h1-biased responses involving cytotoxic CD8 T cells and/or activated macrophages, but in some cases T_h2-biased responses are important such as in some forms of asthma.

If the antigen is able to cross-link two or more IgE molecules, a signal is generated by the IgE FcRs to which they are bound. This results in the very rapid activation of the mast cell, within seconds to minutes. Mast cells have large cytoplasmic granules which are storage organelles for preformed mediators of inflammation. Amongst the latter one of the most important is histamine. Activation of the mast cells results in the contents of the granules being released into the extracellular spaces. Locally, histamine causes vasodilatation (e.g. red eyes), increased vascular permeability leading to swelling or oedema (e.g. blocked nose) and the stimulation of sensory nerves (e.g. itching). This is not usually serious, but think what will happen if you are sensitive to wasp stings and are stung in the back of the throat: not a good site for oedema as the swelling will block air entry to the lungs. It is even worse if the effects are body-wide (systemic). Histamine can cause widespread dilatation of veins, leading to a potentially catastrophic fall in blood

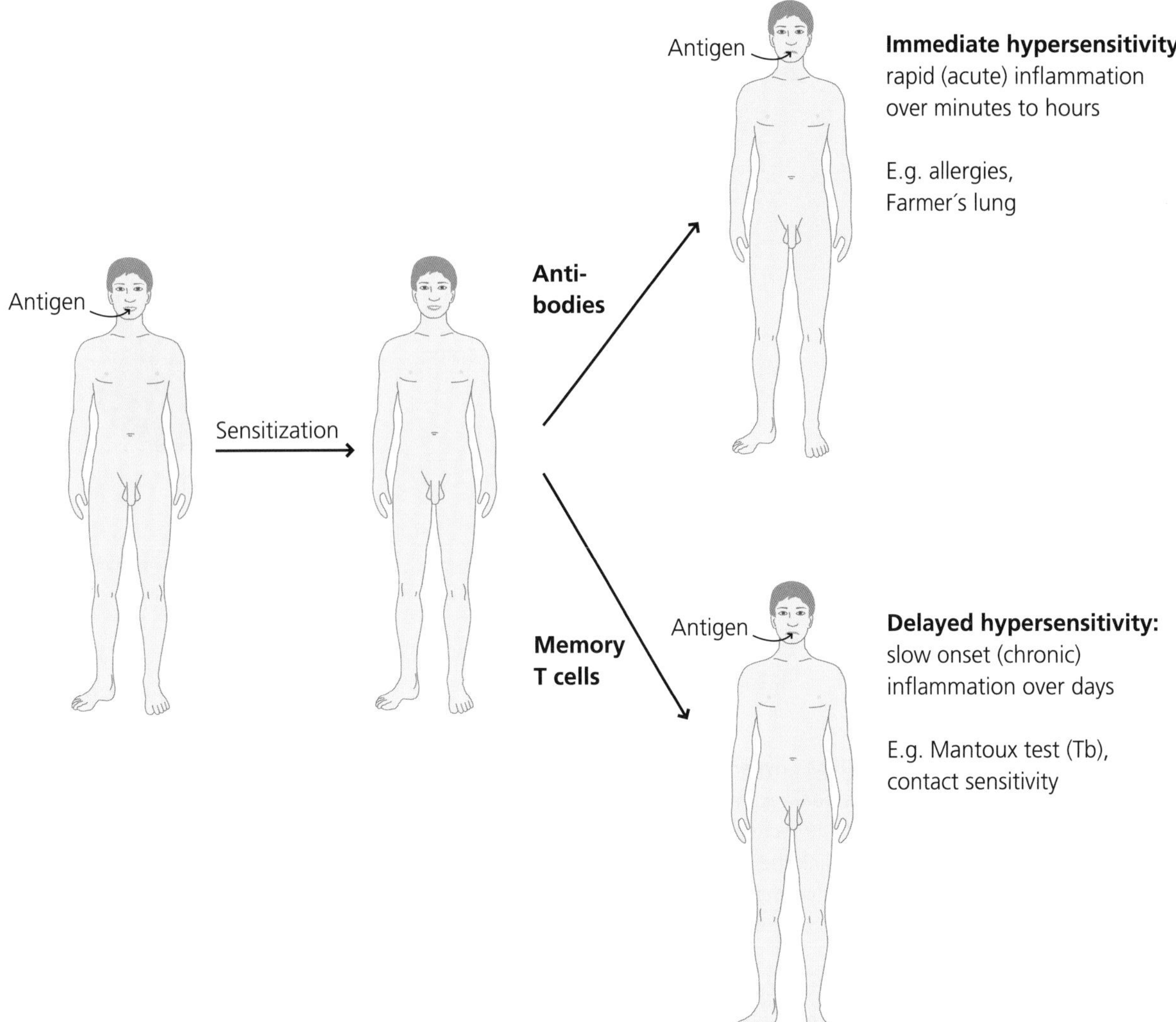

Fig. 7.4 Immediate and delayed-type hypersensitivity reactions. If an individual who has been sensitized to a particular antigen is challenged with the same antigen, an inflammatory response may occur very rapidly, within minutes to hours. This is called **immediate hypersensitivity**, and is seen in allergies and conditions such as Farmer's lung. Immediate hypersensitivity occurs, and can only occur, because there is pre-formed antibody. Other responses take much longer (24–48 h) to occur. This is called **delayed hypersensitivity**, and is seen in the Mantoux test for immunity to tuberculosis and contact sensitivity to substances such as nickel. We call these responses delayed due to time taken to reactivate and expand memory cells, but the response in fact accelerated relative to the primary T cell response that occurs during sensitization.

pressure, and constriction of bronchial smooth muscle, causing respiratory obstruction. Together these features are called anaphylaxis and this is often fatal if not treated with adrenaline to reverse these effects. Mast cells additionally store pre-formed pro-inflammatory cytokines that are also released very rapidly after activation. After this very rapid response, the mast cells later synthesize other mediators such as the lipid metabolites prostaglandins and leukotrienes that also contribute to the oedema and bronchial constriction. Treatment of less severe reactions includes antihistamines and non-steroidal anti-inflammatory drugs (NSAIDs) to inhibit synthesis of the lipid mediators.

7.2.4 Diseases Caused by Antibody-Dependent Modulation of Function (Type II: Cytotoxic Hypersensitivity)

In defence against infection. Antibodies act in several different ways: they can block binding of a pathogen or toxin to its target cell (neutralization), they can kill some pathogens or the cells they have infected through complement mediated lysis or antibody-dependent cell-mediated cytotoxicity (ADCC), or they can opsonize pathogens, leading to their phagocytosis and killing by cells such as neutrophils. All these mechanism have counterparts in some autoimmune diseases. The

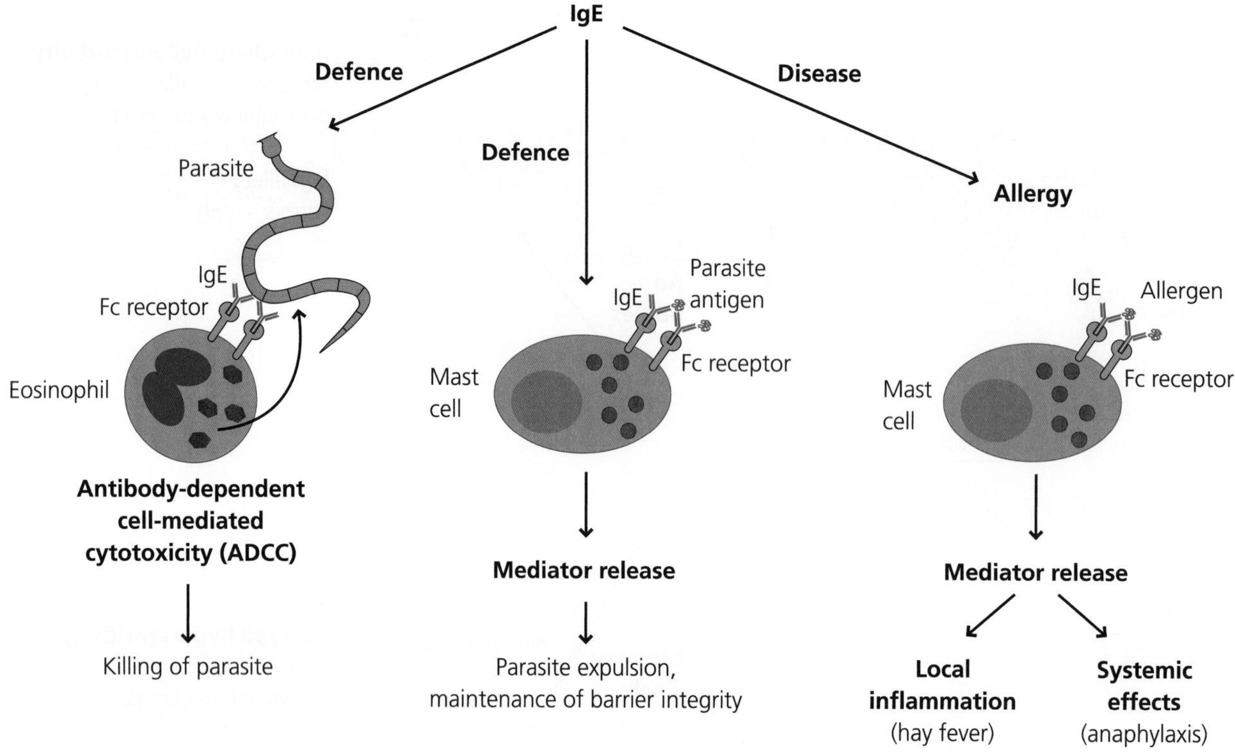

Fig. 7.5 IgE in defence and disease (type I hypersensitivity reactions). IgE is thought to give some protection against parasitic helminth worms. However, some individuals exposed to antigens such as grass pollen or peanuts make adaptive responses in which B cells switch to IgE production. The secreted IgE binds to high-affinity Fc receptors (FcRs) on mast cells. If the individual is re-exposed to the antigen, it can cross-link the IgE on mast cells. This induces the mast cells to release inflammatory mediators. Locally these cause acute inflammation but if they are released into the blood they may cause potentially fatal anaphylaxis. IgE responses are typically controlled by T_h2 cells.

diseases caused by antibodies targeting host cells were called cytotoxic Type II hypersensitivity disease by Gell and Coombs, although in many cases we now know that cells are not actually killed. Importantly, in this type of mechanism, the antibodies are specific for molecules expressed on a cell surface, and are mostly IgM or IgG. See Figure 7.6.

The most direct way of modulating cell function is by killing the cells. Cells that have bound antibodies can be killed in two ways. One is mediated by complement as in haemolytic anaemia and the other by a variety of cells (ADCC) (Section 6.2.6.4). Other than killing cells, antibodies can inhibit cell function by blocking functional molecules on cell surfaces. Thus, in myasthenia gravis, antagonist antibodies against the acetylcholine receptor makes muscle cells unable to respond to acetylcholine, leading to muscle weakness. In some cases (e.g. hyperthyroidism) agonist antibodies can cause abnormal stimulation of the receptor.

7.2.5 Diseases Mediated by Immune Complexes (Type III Hypersensitivity)

Although strictly, any antigen that is bound to an antibody may be called an immune complex, the term is usually used to describe multi-molecular complexes of antibodies and soluble antigens. Such complexes may form during any antibody response to a soluble antigen and are usually disposed of rapidly following binding of complement, which facilitates their uptake by macrophages. In some cases immune complex formation can be beneficial since they can enhance and regulate adaptive immune responses. For example, immune complexes with or without bound complement components, may be taken up by DCs, leading to increased T cell responses, or may bind to receptors on B cells and regulate antibody responses. In some cases, however, large, insoluble complexes are formed, which cannot be cleared by macrophages. These are potentially harmful and lead to Type III hypersensitivity reactions. In these reactions immune complexes act together with complement and/or neutrophils to stimulate acute inflammation, leading to tissue damage. See Figure 7.7.

7.2.5.1 Diseases Caused by Blood-Borne Immune Complexes

Normally, circulating immune complexes bind to red blood cells (RBCs) via a complement receptor (CR1) and are transported to the liver and spleen where macrophages line blood vessels. Here the complexes are removed from the red cells, endocytosed and destroyed, allowing the red cells that transported them to continue to circulate. If, however, complexes form that cannot be removed, they remain in the circulation and are likely to be deposited in small blood vessels where, over time they cause disease. This may occur anywhere in the body, and these diseases are therefore called systemic, but the kidneys and

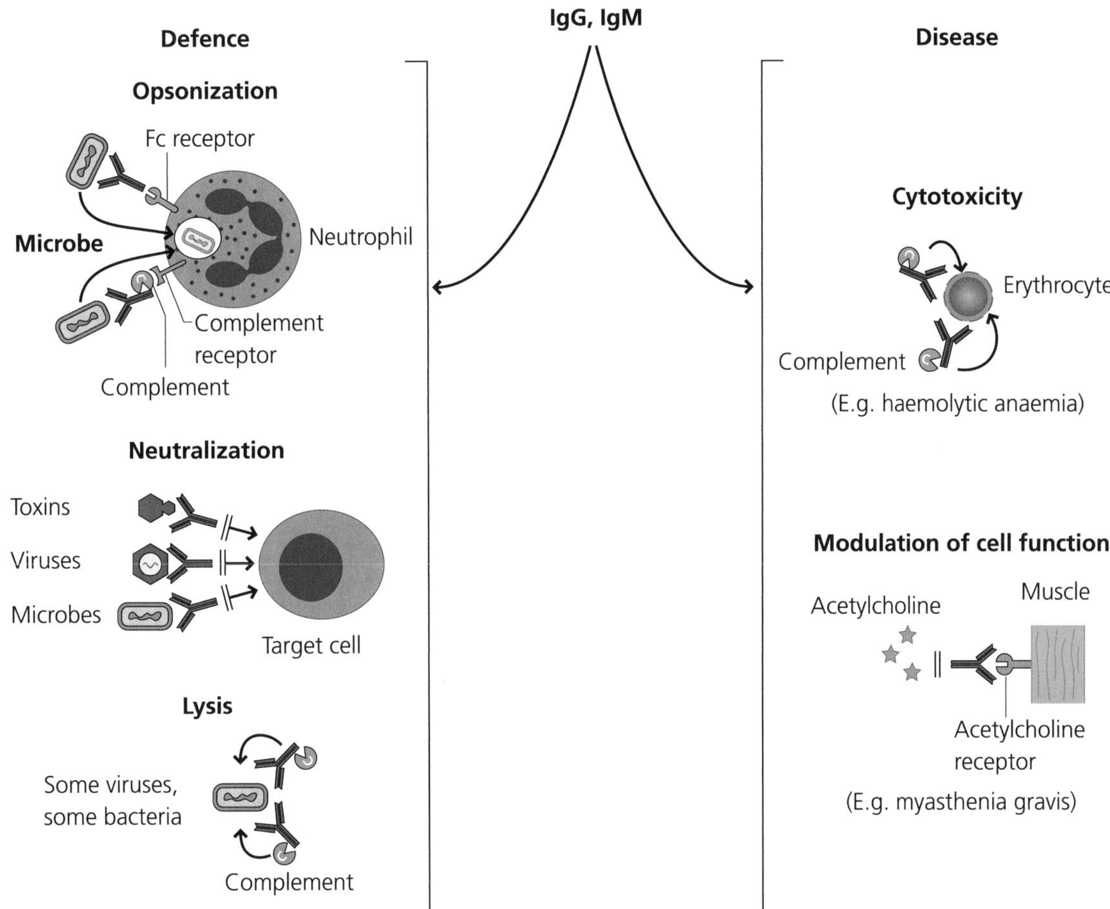

Fig. 7.6 IgM and IgG in defence and disease (type hypersensitivity reactions). Antibodies are crucial in defence against many bacterial and viral infections, largely by opsonizing the microbe, or by neutralization, preventing binding of toxins, viruses or bacteria to cells. Occasionally lysis of bacteria may occur. In disease, antibodies can act on cells to kill them (as in haemolytic anaemia) or modulate their functions. For example, in myasthenia gravis, autoantibodies to the acetylcholine receptor block binding of acetylcholine and cause muscle weakness. IgG responses can be controlled by either T_h1 or T_h2 responses.

joints are often particularly affected. Damaged tissues may also be the sites of deposition. For example, in SLE, complexes may be deposited in UV-damaged skin, leading to the characteristic facial butterfly rash. Wherever they are deposited, immune complexes will inevitably induce acute inflammation at the site of deposition. It is the inflammation that causes the disease. Inflammation results from the activation of complement and recruitment of neutrophils. Neutrophils bind the complexes via FcRs and complement receptors but may not phagocytose them, leading to the release of damaging enzymes and other molecules.

7.2.5.2 Disease Caused by Local Deposition of Immune Complexes

In the above examples the antigen–antibody complexes causing disease were systemic (i.e. the complexes were present in the blood). In some cases, however, complexes may be formed *in situ* at sites where antigens are deposited (e.g. in the lung following inhalation of antigens). As with systemic immune complex disease, the pathology is caused by acute inflammation and the symptoms reflect this local inflammation.

7.2.6 Diseases Caused by Cell-Mediated Immune Responses (Type IV: Delayed-Type Hypersensitivity (DTH)

Activation of CD4 T cells, particularly along the T_h1 pathway, leads to the generation of potent effector mechanisms, (Section 1.2.2) crucial for recovery from infection, including cytotoxic CD8 T cells and activated macrophages. These mechanisms are also all potentially mediators of tissue damage. Collateral damage may occur in an infection if the pathogen is not effectively controlled, as in hepatitis B infection or tuberculosis (Section 2.4.1.1). Damage can also occur if the response to a harmless extrinsic or intrinsic antigen is not regulated, as in Crohn's disease, a chronic inflammation of the intestines. The damage can be direct, for instance if cytotoxic CD8 T cells kill host cells expressing viral or intrinsic antigens. It can also be indirect, for instance if a macrophage secretes enzymes that can break down connective tissue (presumed to be one of the mechanisms of damage in tuberculosis and rheumatoid arthritis). The actual site of the damage may also be crucial to disease severity. Inflammation in the meninges of

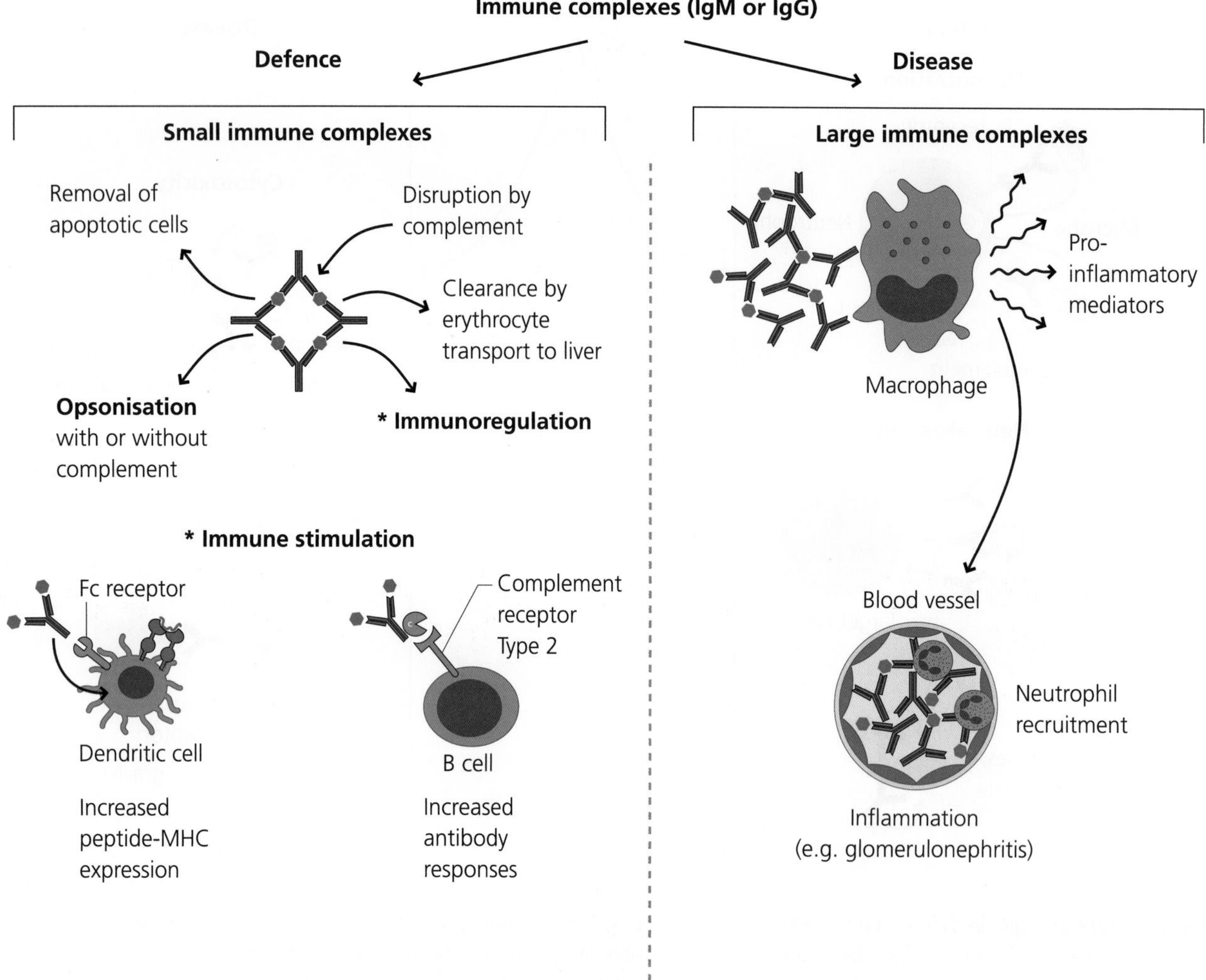

Fig. 7.7 Immune complexes in defence and disease (type III hypersensitivity reactions). Immune complexes of soluble antigens and IgM or IgG antibodies bind complement and are normally cleared by macrophages after transport to liver and spleen by erythrocytes. Such complexes may also regulate immune responses following binding to DCs or B cells via FcRs or complement receptors. Large complexes cannot be cleared this way and are deposited locally in tissues or in small blood vessels in organs such as the kidney. Here, activated complement initiates acute inflammation and recruited neutrophils cause local tissue damage.

the brain may be fatal, whereas in other sites the same degree of inflammation can be relatively harmless. We now understand that many of the features of DTH responses involve aberrant activation of T_h1 or T_h17 cells (it is only more recently that T_h17 cells were discovered and it seems that some responses previously thought to be due to T_h1 cells are really caused by T_h17 cells, Section 1.4.5.1). However, in some cases, Type IV sensitivities may involve aberrant activation of T_h2 cells (e.g. in conditions where eosinophils accumulate in tissues). Hence, we can now discriminate between different types of DTH depending on the cells involved, but they are all thought to be largely or totally independent of See Figure 7.8.

However, in many immunopathological conditions, both B and T cells are activated against extrinsic or intrinsic antigens. Thus, in a patient suffering from a type IV DTH disease, cytotoxic T cells, activated macrophages and antibodies may all be present. It is, however, very important is to know which of these effector mechanism(s) is or are playing a significant role in causing the tissue damage and disease since such information will inform the application and development of therapies. It can be difficult to determine the relative importance of different effector mechanism(s) in human diseases, but in animal models adoptive transfer of antibodies or cells can give clear indications. These aspects are discussed in later sections of this chapter.

7.3 Why Do We Make Harmful Immune Responses to Harmless Antigens?

In this section we will discuss why it is that immune responses can be made to harmless antigens that do not derive from pathogens, including extrinsic foreign and

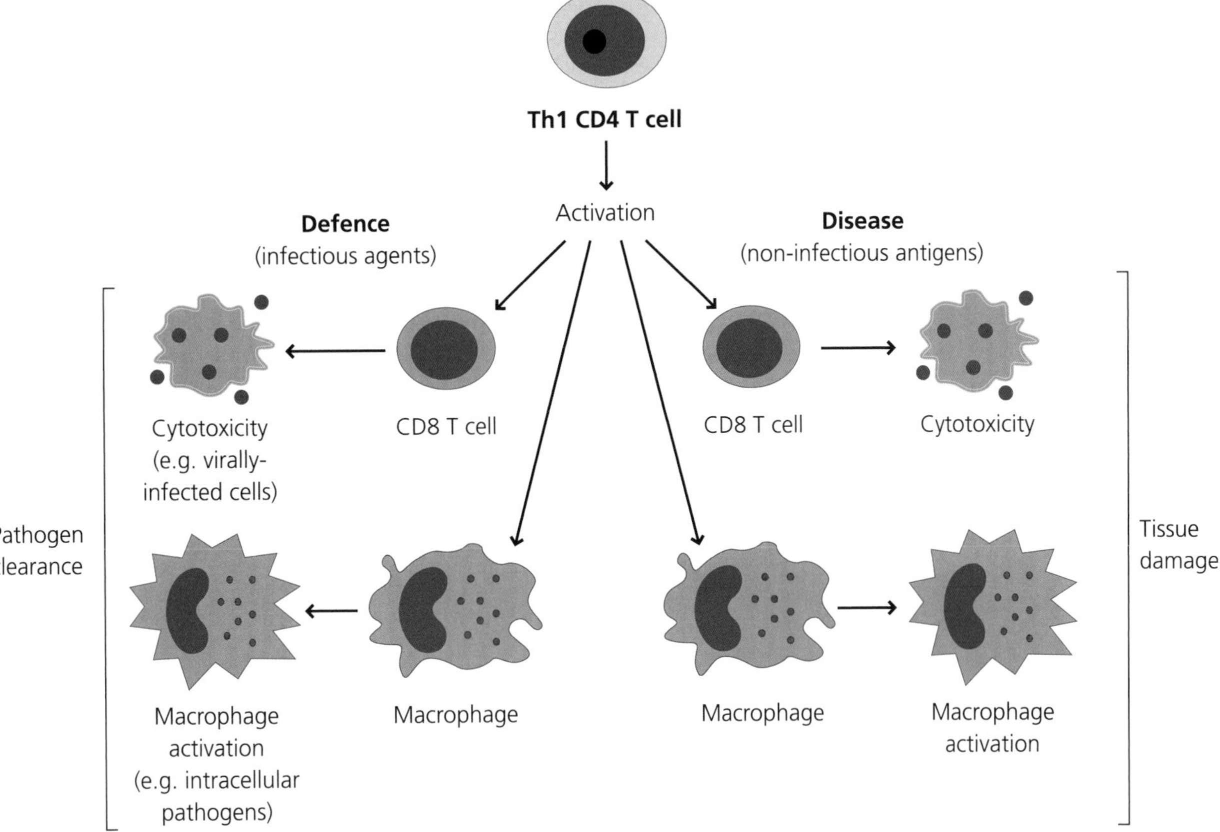

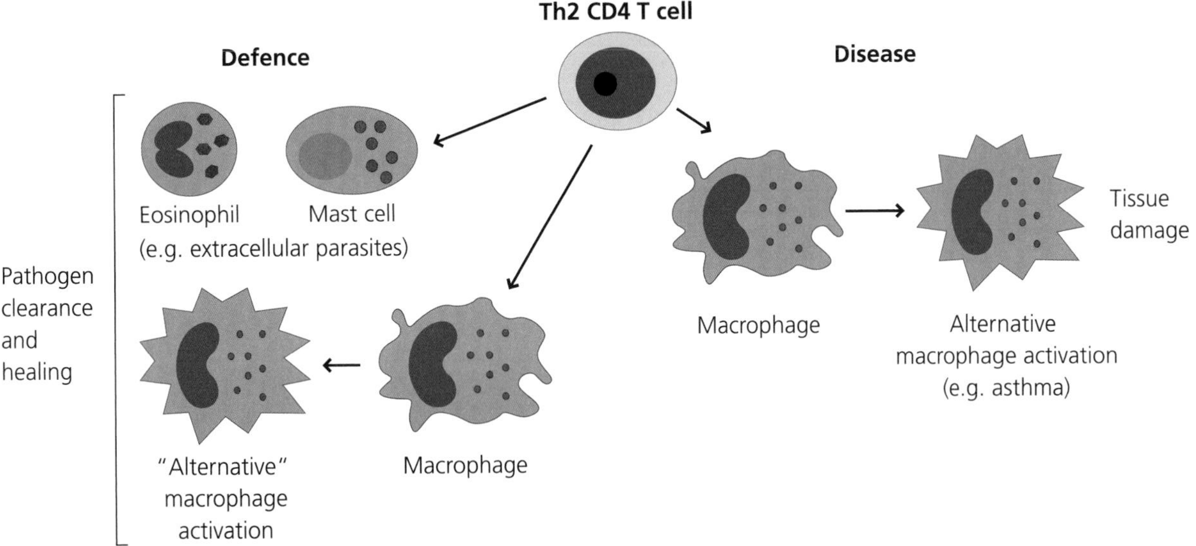

Fig. 7.8 T cell-mediated immunity in defence and disease (type IV hypersensitivity reactions). T_h1-biased CD4 T cells are essential for protection against many microbes via the activation of CD8 T cells or macrophages. If, however, CD4 T cells have been sensitized against a self or harmless foreign antigen, they can stimulate tissue damage, again via CD8 T cells or macrophages. T_h2-biased T cells may be important in defence against some parasites, but are also responsible for the damage seen in conditions such as asthma.

intrinsic self antigens, as well as transplantation and blood group antigens. We will first consider the antigenic repertoire of lymphocytes – how conventional T cell receptors (TCRs) and B cell receptors (BCRs) are generated, and how they can recognize antigens. We will then discuss lymphocyte activation, and why and how it may be that lymphocytes can be activated inappropriately, causing disease. This section will also discuss the polarization of immune responses in relation to immune-mediated diseases. Finally, we will introduce the genetic bases of some of these diseases and the roles that environmental factors may play in their pathogenesis.

7.3.1
Antigen Recognition by Lymphocytes in Normal and Pathogenic States

7.3.1.1 Generation of the Antigenic Repertoire of Lymphocytes

In this section, and this chapter in general, we focus almost exclusively on responses of conventional lymphocytes of humans and mice, which have highly diversified antigen receptors. These are the αβ T cells and the follicular B cells. Generally speaking, the potential roles of other types of lymphoid cells have been much less considered, and certainly much less studied, in most of the areas we will be discussing here. See Figure 7.9.

T cell activation, particularly of CD4 T cells, is crucial to almost all forms of immunopathology, including those where it is antibodies that cause the damage. Why can T cells potentially recognize self and normally-harmless foreign antigens? To understand this we need to re-consider the antigenic repertoires of lymphocytes. These repertoires represent the collective specificities of all the TCRs and BCRs that are expressed in an individual. The repertoire defines what can and cannot be recognized in the external (foreign) and internal (self) worlds. This means the range of antigens, or more precisely, the parts of antigens that can bind to the individual's antigen receptors. In terms of antibody-mediated immunity these are termed B cell epitopes, more or less anything that is the right size and shape to fit into the binding

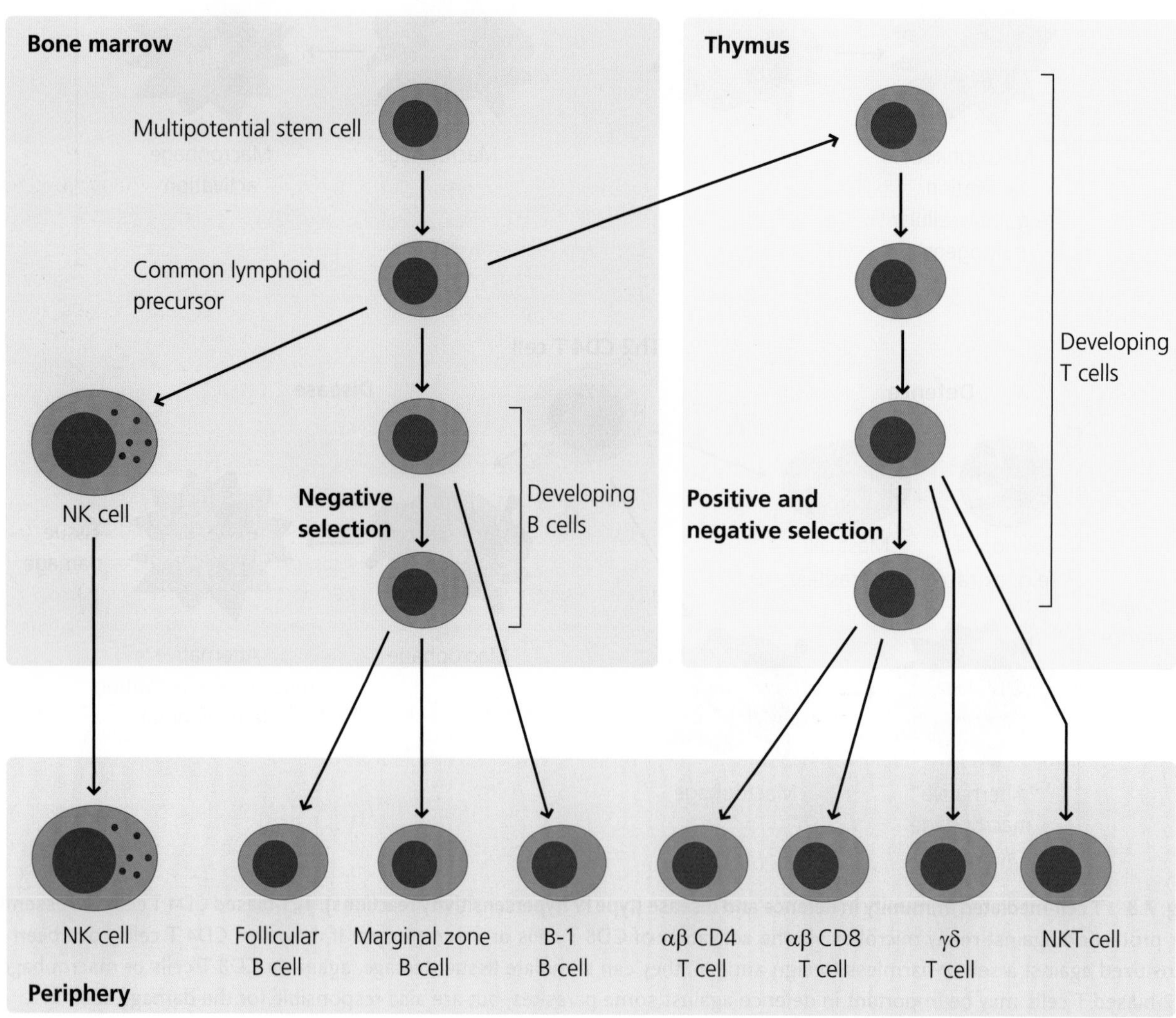

Fig. 7.9 Lymphocyte development. The multi-potential stem cell gives rise to the common lymphocyte precursor. This cell can develop into NK cells, B, T and NKT cells. NK cells and B cells complete their development in the marrow, while T and NKT cells complete theirs in the thymus. Conventional follicular B cells and αβ T cells undergo negative selection to reduce the chance of autoreactive cells being produced, although this process is incomplete (Figure 7.11). αβ T cells also undergo positive selection to ensure that their receptors are MHC-restricted. The interested reader might wish to consider whether or not, and if so to what extent, positive and negative selection might apply to the other cell types shown.

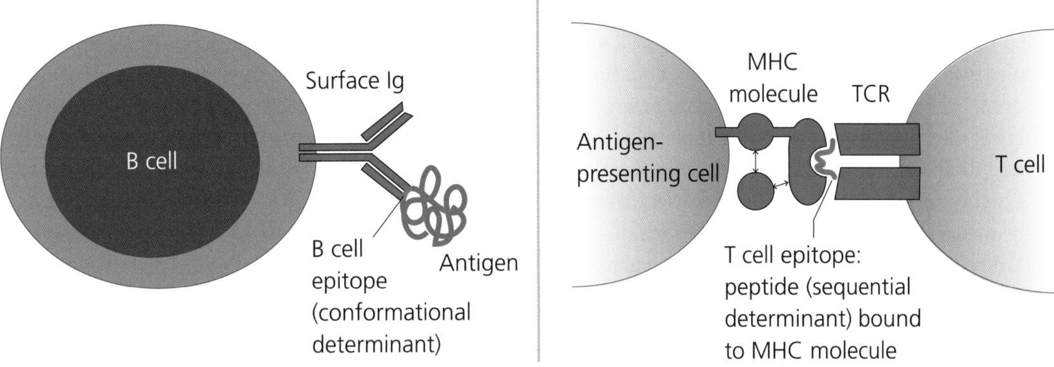

Fig. 7.10 B cell and T cell epitopes. Conventional BCRs (surface Ig) and secreted antibodies recognize regions that are exposed on the surface of macromolecules as well as small molecules. The regions recognized are termed **B cell epitopes** or **conformational epitopes**. If an epitope is to be recognized it must have a relatively rigid three-dimensional structure. Conventional T cells, in contrast, recognize MHC molecules with bound peptides that have been produced by degradation of proteins. These peptides (that can be bound by MHC molecules and recognized by TCRs) are called **T cell epitopes** or **sequential epitopes** since the peptide is a linear sequence from the intact protein. In contrast, conformational epitopes for B cells can be formed by amino acids that are widely separated in the primary sequence of a protein.

site of a BCR. For conventional T cells, the antigens are usually peptide–major histocompatibility complex (MHC) complexes, which are often called T cell epitopes. See Figure 7.10.

TCRs and BCRs are generated by random rearrangement of gene segments and by other mechanisms that further increase diversity (Section 1.5.3.2). As a result of this randomness, the antigenic specificities of the receptors generated cannot be predetermined. Hence, there are specific mechanisms that are designed to ensure that these receptors function in the correct way and do as little damage as possible. However, while these are highly sophisticated mechanisms that work well in the population as a whole, they may be less than perfect at the level of individuals with immune-related disease.

In the case of T cells the peripheral repertoire is shaped by MHC molecules. This is because the randomly generated TCRs are selected so that T cells can only recognize peptides bound to self MHC molecules and not peptides alone. Hence, during their development in the thymus, MHC molecules present self peptides to the developing T cells. Only those developing T cells that weakly recognize a self peptide–self MHC molecule complex are allowed to develop further, while the remainder die within the thymus. This process is called positive selection. What this means is that the T cell repertoire is inherently self-reactive, although normally the strength of recognition of self peptide–MHC complexes by T cells is very weak (i.e. it is of low avidity).

Later in their development, T cells that recognize self peptide–self MHC complexes very strongly (high avidity) undergo apoptosis, so that highly autoreactive T cells do not escape the thymus. This process is called negative selection (Section 5.5.3). These selection processes ensure that those T cells that leave the thymus are generally not highly reactive against self peptide–MHC complexes expressed in the thymus. T cells are thus MHC-restricted, but non-reactive to many self peptides. Because it happens in the thymus, this process is called central tolerance. Negative selection is, however, necessarily incomplete, because it is limited by the range of self peptides that are expressed in the thymus. As it is not possible for all self peptides to be expressed in the thymus, self-reactive (autoreactive) T cells inevitably escape into the periphery. Negative selection of developing B cells also occurs, but this process turns out to be less stringent than for T cells. Hence, it is likely that even more potentially autoreactive B cells exist. It is therefore important to appreciate that self-reactivity of lymphocytes is a normal state in adaptive immunity and that the presence of these autoreactive cells does not necessarily cause us problems. Clearly, however, if such lymphocytes are activated, it is easy to understand how they could cause autoimmune diseases; without them, there could be no autoimmune diseases.

Q7.4. What might be the potential impact of negative selection on the ability to recognize tumour antigens?

7.3.1.2 Promiscuity of Antigen Recognition by Lymphocytes

There is, however, another important consideration in relation to lymphocyte antigen recognition. TCR and BCR recognition is highly discriminatory since both types of receptor can recognize extremely small changes in their respective epitopes. For this reason recognition is sometimes said to be highly specific. However, this term is misleading. TCR and BCR antigen recognition is actually highly cross-reactive and very promiscuous. For B cells there is abundant evidence from structural and functional studies that any given antibody can bind with greater or lesser strength to a wide variety of different epitopes. We know that TCR recognition must also be cross-reactive because of the phenomenon of alloreactivity

(Section 1.6.5.1). Astonishingly, it is estimated that maybe 1–3% of all the T cells in any individual can recognize a single MHC molecule from any other (allogeneic) individual (Figure 1.36 and Section 7.5.1.2, 7.5.5.1). Given that there are hundreds of different alleles of different MHC molecules in the population, and that this of alloreactive T cells frequency remains largely the same however it is tested, there must be a vast amount of overlap that represents an enormous promiscuity in TCR recognition. The only satisfactory explanation for alloreactivity is that it represents a very high frequency of T cells that have been selected to recognize viral or microbial peptides bound to self MHC molecules but which happen to cross-react with foreign MHC molecules, with or without their bound peptides.

Q7.5. How could we estimate the extent of cross-reactivity of a T cell antigen receptor?

So the promiscuity of lymphocyte antigen recognition readily explains why allografts are rejected, and the extent of cross-reactivity in this setting explains why such immune responses are so vigorous. Cross-reactivity also explains why B cells can potentially recognize extrinsic antigens from the external world and make antibodies against food proteins or pollens, for example. Taking this one step further, it may also explain why such a response to an extrinsic antigen might also cross-react with an intrinsic antigen and cause collateral (autoimmune) disease. The same considerations apply to T cells in relation to peptide–MHC recognition. Hence, it is clear that lymphocytes must inevitably be able to recognize harmless antigens. However, to make any sort of response they must first be activated. Thus, the next question to be considered is why they are activated in these pathological settings. See Figure 7.11.

7.3.2 Lymphocyte Activation in Immunological Diseases

Understanding lymphocyte activation in immunopathology is crucial for understanding the pathogenesis of the respective diseases, and for designing appropriate and effective therapies.

7.3.2.1 Signals for Lymphocyte Activation

Lymphocyte activation is essential for any adaptive response against an infectious agent. It is also essential for the development of any of the immunopathological conditions described above. We re-emphasize that all such pathogenesis probably depends on CD4 T cell activation. Activation requires that two different sets of signals are delivered to the T cell, and its subsequent fate (e.g. what type of function it acquires) is controlled by additional signals.

If a TCR recognizes a peptide-MHC with sufficient strength this generates a signal (Signal 1) that is essential for activation. Whether or not a T cell that has received Signal 1 does become activated is regulated by signals from costimulatory molecules, collectively known as Signal 2. Costimulation can be stimulatory (positive) or inhibitory (negative). The generation of costimulatory signals is regulated by signals from the innate immune system in response to the sensing of danger, such as that presented by infection. For example, DCs typically increase their expression of costimulatory molecules in response to pathogen-associated molecular patterns (PAMPs). These are conserved features of pathogens that are recognized by pattern recognition receptors (PRRs) such as Toll-like receptors (TLRs) (Section 1.2.3.1). Hence, they become able to trigger activation of T cells that recognize epitopes from the same pathogens. As another example, activation of complement and deposition of specific complement components on immune complexes, or potentially on surfaces, can also enhance the ability of B cells to make specific antibody responses. See Figure 7.12.

If a CD4 T cell does become activated it can then differentiate to express a particular set of properties, such as its capacity to secrete a selected group of cytokines or migrate to particular tissues. Which properties it does acquire is again determined by external signals coming primarily from innate immune and other non-immune cells; these can be loosely termed Signal 3. In immunity to infection, this control of T cell activation and differentiation is crucial as T cells, particularly CD4 T cells, determine which effector mechanisms are activated because different pathogens require different mechanisms for their elimination.

7.3.2.2 Lymphocyte Activation in Disease

We have seen that, inevitably some peripheral lymphocytes are able to recognize self or harmless foreign antigens. For these antigens, if potentially reactive lymphocytes do not meet their cognate antigen, or in the case of follicular B cells do not receive help from a T cell, harmful responses cannot occur. Even if T cells do recognize such antigens, under normal circumstances these potentially reactive T cells, and B cells, are controlled by regulatory mechanisms outside the thymus; we know this from the consequences of human defects in the genes for FoxP3 and Fas or Fas ligand, for example (Section 5.5.5.1). It is clear, however, that in some diseases these lymphocytes do become activated. Since we produce self-antigens continuously as part of cell renewal, are continually exposed to extrinsic harmless antigens, and as most of us carry persistent viruses, positive control mechanisms (regulation) are important in suppressing autoreactive responses. So how does the body normally avoid harmful reactions? Let us consider this in relation to the two signals for T cell activation (above).

Understanding Signal 1 is relatively straightforward. In immunopathological settings this, for example, requires B cell recognition of epitopes from extrinsic, intrinsic or transplantation (and possibly tumour) antigens, and T cell recognition of MHC-bound peptides derived from the same antigens. Understanding Signal 2 is more complex. In immunity to infection the pathogen expresses PAMPs that stimulate

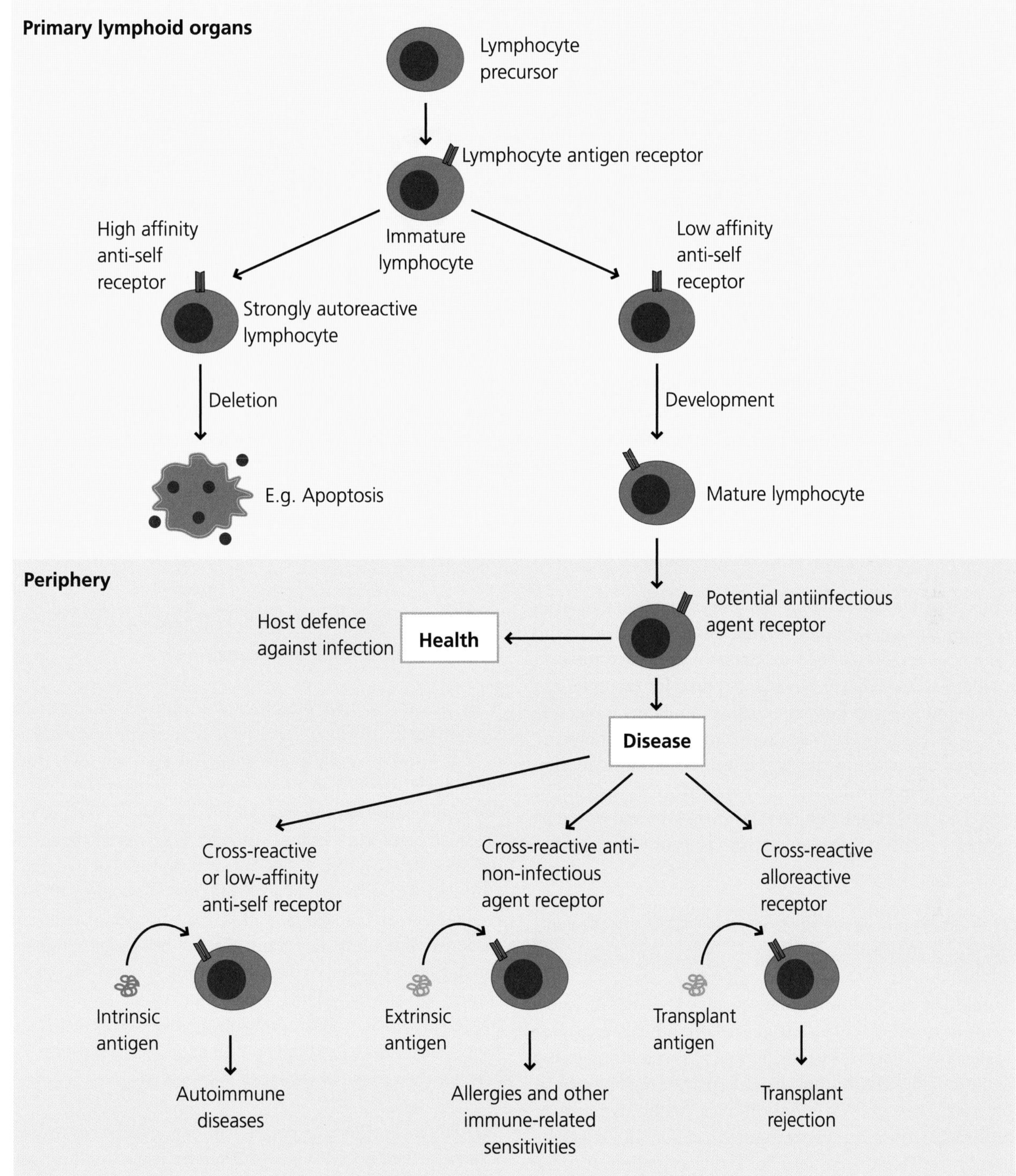

Fig. 7.11 Lymphocytes in health and disease. B and T cells as they develop express a pre-formed, anticipatory repertoire of antigen receptors that can potentially recognize antigens from infectious agents. For both B and T cells, cells that express receptors for self antigens that are present in their environment during development may be eliminated before they enter the periphery, including secondary lymphoid organs. This process is inevitably incomplete and some potentially self-reactive lymphocytes will enter the periphery where, under abnormal circumstances, they may induce autoimmune responses. Lymphocytes that express receptors for harmless self antigens such as pollen or food cannot be eliminated during development and unless peripheral mechanisms work to prevent their activation may induce immune-related sensitivities. A high frequency of T cells can also react against molecules expressed by other (allogeneic) individuals and can be responsible for transplantation reactions (rejection and GVHD).

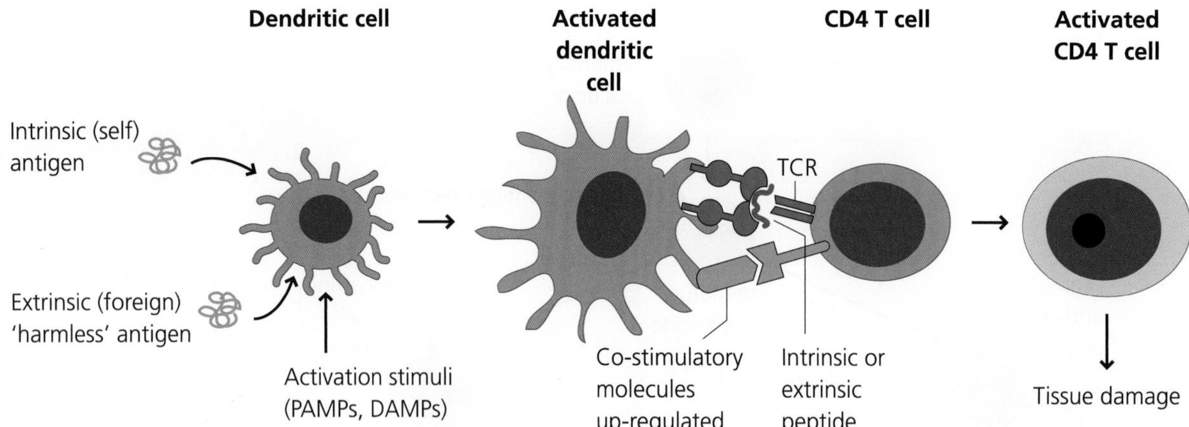

Fig. 7.12 CD4 T cell activation in immunopathology. Normally, DCs express intrinsic (self) or extrinsic (harmless foreign) peptides on their MHC molecules. If these DCs are not fully activated, when if they encounter a T cell specific for these peptide-MHC complexes the T cell will not be activated and may undergo apoptosis. If, however, the same DC has been stimulated by PAMPs or possibly DAMPs in the periphery, it will increase expression of costimulatory molecules and may become be able to activate the T cell, which will then have the potential of mounting a damaging response against the self or harmless antigen.

the innate system. In addition, it has been suggested that cell or tissue damage, whether or not this is associated with infection, generates damage- (or danger-) associated molecular patterns (DAMPS; Section 4.2.2.4) that, for example, can also lead to increased expression of costimulatory molecules by DCs, or perhaps activate complement. What, however, stimulates Signal 2 in responses to self or harmless foreign proteins? Self or harmless foreign antigens are not usually, by themselves, associated with PAMPs or DAMPs and would be predicted to induce active or passive forms of tolerance. See Figure 7.13.

Q7.6. How could we attempt to identify DAMPs involved in immune activation?

Frankly, in most autoimmune diseases we do not yet know why the immune response is triggered. In some there is clear evidence that infection is a trigger and it has been suggested that most or all autoimmune diseases may be precipitated by infections. Hence, PAMPs may induce costimulation that, with direct or cross-reactive recognition of the self antigen by autoreactive TCRs, may lead to T cell activation and the initiation of damage to the tissue. In such settings it is clear that the autoimmune disease is caused by direct recognition of an autoantigen. However, there is a further complication that can seriously hamper our understanding of which antigens are the most important in any given disease. This is because the initial damage to any tissue may then lead to exposure of self antigens to which T cells have not been tolerized (and potentially the generation of DAMPs). Hence, this can lead to the induction of new T cell responses to these antigens. This is termed epitope spreading. Epitope spreading is also seen in pathological B cell responses. See Figure 7.14.

Q7.7. How might evidence for an infectious trigger for autoimmune disease be gathered?

In the case of other allergies and immune-related sensitivities (hypersensitivity diseases) again, in many instances, we do not know why these are triggered and why only some individuals exposed to these antigens develop the disease. Some allergens are proteases, which may modify self proteins or cause damage to tissues. In such cases, even though no infection has occurred, DAMPs may be induced or released to stimulate the induction of costimulation. Together with TCR recognition of the allergen, this may trigger an immune response that ultimately damages the tissue (although the sensitization phase may be silent). This is often the case in contact sensitivity, where a molecule applied to the skin triggers T cell activation, probably against self proteins that have been modified by the agent. In many cases, however, it is just not clear how the response is triggered.

Q7.8. An anti-self response will not necessarily cause disease. Patients who have suffered a myocardial infarct (heart attack) often develop anti-myocardial antibodies, but this is usually a short-lived, self-limited event and the antibodies are harmless. Why might this response be limited? In a few people the heart attack is followed 10–21 days later by further symptoms of heart damage. Why might it be that in these circumstances the response is not self-limiting?

Transplants represent another form of harmless antigen, yet grafts from different members of the same species (allografts) or from different species (xenografts) are rejected by vigorous responses. Any transplant is, however, subject to

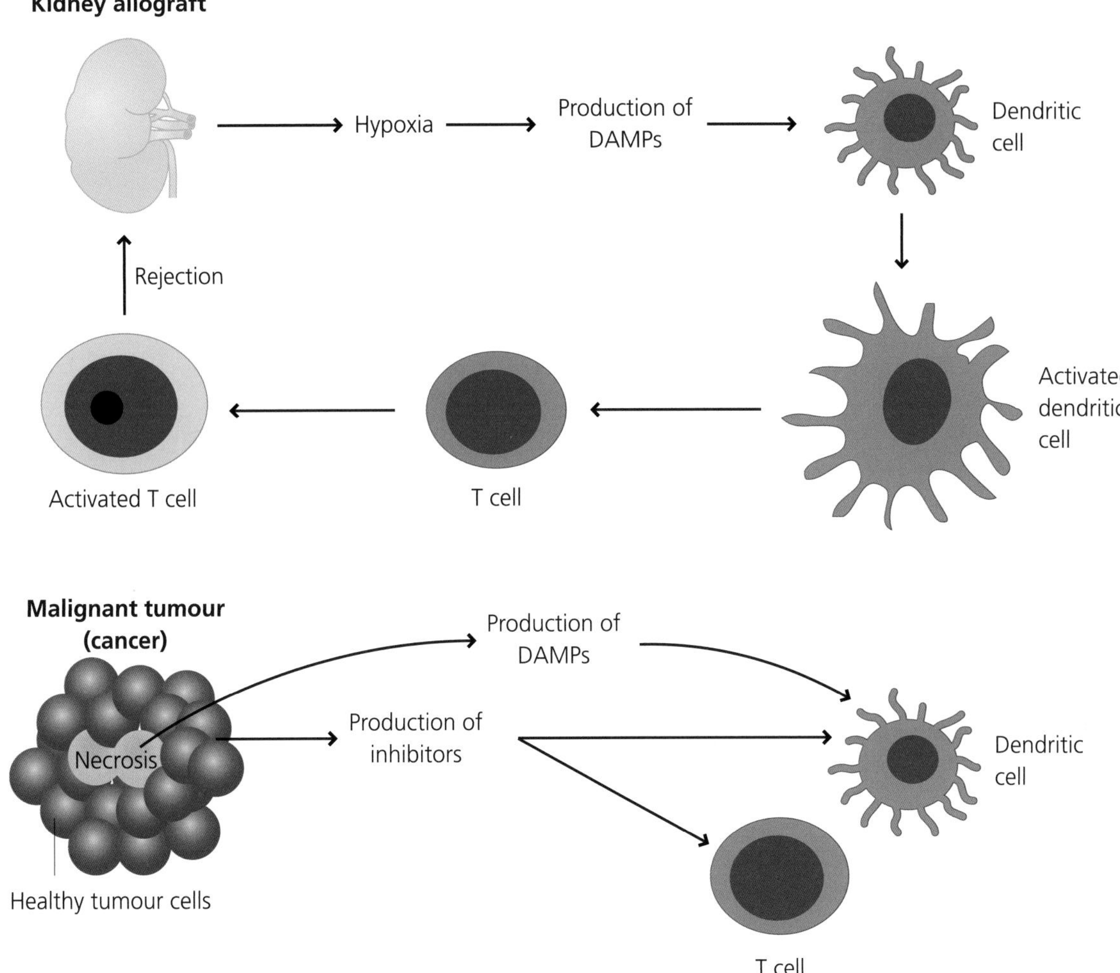

Fig. 7.13 Damage-associated molecular patterns. The initiation of adaptive immune responses to non-microbial antigens (e.g. allograft rejection) may depend on the release of molecules from dead or damaged (e.g. hypoxic) cells that activate innate immune cells such as DCs. These molecules have been termed DAMPs. Some candidate DAMPs such as uric acid and ATP have been identified. Malignant tumours often possess areas of necrosis that could release DAMPs; however, such tumours have evolved multiple ways of preventing adaptive responses from rejecting the tumour.

unnatural stimuli during its retrieval, including an inadequate supply of oxygen, leading to hypoxia for organ grafts, or storage at 4°C under non-physiological conditions for these and bone marrow grafts. Clinically, it has been observed that the shorter the time between removal of the organ from the donor and its transplantation, the more likely it is to be successful. In part this reflects the damage caused to the organ by hypoxia, but it also seems probable that this hypoxia induces the expression of DAMPs in the organ that induce innate immune activation, leading to increased costimulation.

Malignant tumour cells are antigenically very similar to the cells from which they arose. Why should we think that they might potentially induce an active response? Most tumours possess areas of necrosis, often caused by the tumour outstripping or invading its own blood supply, and this may stimulate the expression of DAMPs. The observation that many tumours have developed mechanisms to avoid the adaptive response is itself very good evidence that these tumours do have the potential to induce immunity, and that the immune response is selecting for mutants which can evade the response. Thus, tumour cells can evade T cell recognition by down-regulating MHC expression, and hence not generating Signal 1. In addition, they may produce molecules that inhibit expression of positive costimulatory molecules by DCs, and may also themselves express negative costimulatory molecules. Thus, they avoid generating, or actively modulate, Signal 2.

In immunity to infection, if the infectious agent is successfully eliminated, the immune response is no longer stimulated and a regulated set of negative costimulatory (and other) signals then helps to dampen down the response. If, however, the pathogen is not removed, ongoing immune stimulation can lead to serious collateral tissue damage (e.g. as may happen in tuberculosis). Similarly, in allergy and other immune-related sensitivities, if the antigen is removed the damage stops, for example in hay fever when the pollen season

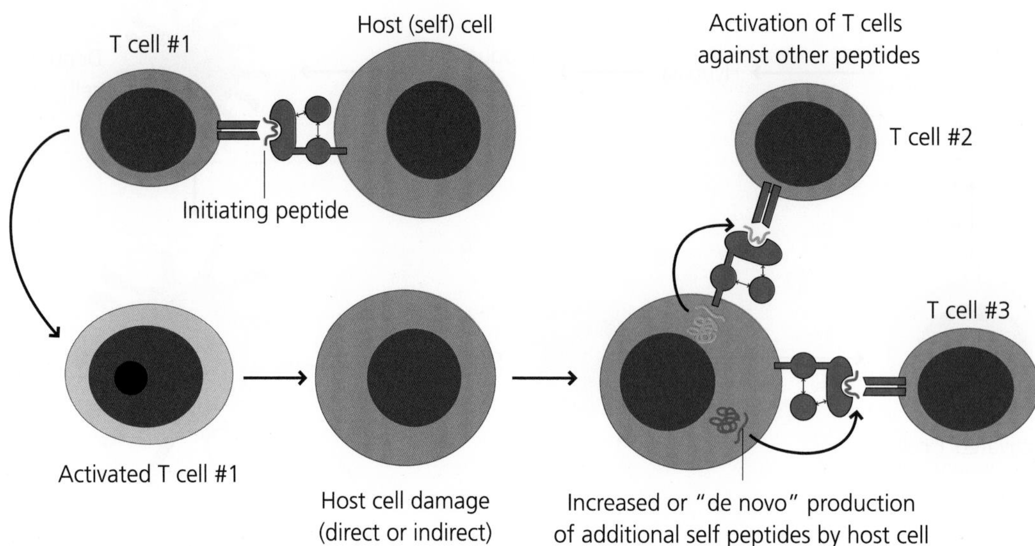

Fig. 7.14 Epitope spreading. In established autoimmune disease, T cells may be found that respond to a variety of antigens from the cells or tissues affected. This does not, however, mean that these antigens were involved in the initiation of the disease. A single peptide could be responsible for the initiation of the response, but subsequent damage to the cells, perhaps with DAMP involvement, means that other peptides may then be presented in an immunostimulatory form to T cells. This is known as epitope spreading and may serve to increase the overall strength of an autoimmune response.

is over or in coeliac disease if gluten is avoided. In autoimmune diseases, however, there is a more or less unlimited source of antigen from self tissues and the immune response may continue indefinitely, as in rheumatoid arthritis, or until the tissue is completely destroyed as is the case for pancreatic β cells in Type I diabetes.

7.3.2.3 Polarization of CD4 T Cells in Immunopathology

CD4 T cell responses to infection show more or less polarization, at least in mice. This is an essential part of the response (being induced by Signal 3 in the case of T cells) and permits the generation of appropriate effector mechanisms (Section 1.4.5.1). Many autoimmune and other immunological diseases also show marked polarization of CD4 T cells. Type I (insulin-dependent) diabetes is an example in which marked T_h1 polarization results in macrophage- and CD8 T cell-mediated damage. In contrast, the rare disease eosinophilic gastroenteritis shows marked T_h2 polarization in terms of the antibody isotypes produced and the marked infiltration of eosinophils into the intestinal submucosa. In many, maybe even in all, cases, however, this polarization is not complete and any immune-related disease can involve a mix of mechanisms. For example, asthma can be viewed as a combined T_h1 and T_h2, and in some cases T_h17, disease. In fact it is now suggested that in many diseases originally thought to result from T_h1 polarization, the CD4 T cells are actually polarized towards T_h17.

Polarization may reflect long-term stimulation of T cells in many immune-related diseases. In mice, polarization needs repeated antigenic stimulation, but we have relatively little understanding of the signals that induce these biases in humans. That the genotype of the individual plays a major role is clear from studies of autoimmune diseases in twins, but at present no genes have been identified which clearly explain polarization in disease (below). Understanding polarization does, however, hold promise for therapy. If a T_h1-polarized disease such as Crohn's disease could be redirected so that the activated T cells became polarized towards T_h2, there would be a real possibility of therapy. In one trial in which Crohn's disease patients were infected with a porcine intestinal worm (Trichuris suis, which induces a T_h2 polarized response) a remarkably beneficial effect was observed. Similarly, allografts are generally acutely rejected by T_h1-biased responses. If we knew how to redirect the response towards a T_h2 type or how to induce T_{reg}, it might be easier to prevent rejection. In the case of tumours, during their development the tumour microenvironment may induce T_{reg} that suppress any otherwise beneficial responses, even if these could be induced. Again, understanding how inhibit or deplete these cells, or to re-programme the response, could lead to effective immunotherapy.

7.3.3 Genetic Basis of Immunopathology

Genes control immune responses in many different ways. We know much about this from analysis of immune responses to model protein antigens and from studies of susceptibility to infection. It is thus not surprising that pathological and unwanted immune responses are strongly influenced by genetic factors (e.g. the MHC in allograft rejection). In this section we will concentrate on autoimmune diseases and discuss other abnormal responses where there is relevant evidence. Whilst the effects of genes are easy to determine in inbred, homogeneous mice, it is much more difficult in outbred humans. Studies in populations of consanguineous families, where gene sharing is common, have been very informative as have those in twins.

7.3.3.1 Concordance in Twin Studies

Studies of the incidence of disease in twins provide valuable evidence about the genetic basis of the disease. If one twin has a disease, the frequency with which the other twin also develops the disease can be estimated. This frequency is termed the concordance. If the concordance is higher in identical (monozygotic) twins than in non-identical (dizygotic) twins this is strong evidence for a genetic influence in susceptibility. In autoimmune diseases and immune-related sensitivities the concordance is frequently much higher in monozygotic twins. However, the degree of concordance in monozygotic twins is often in the range of 20–40%. (Coeliac disease resulting from sensitivity to oral gluten is an exception since the concordance in identical twins is 70–90%). The usual explanation for a low level of concordance is that environmental factors, such infections, play a major role in pathogenesis. See Figure 7.15.

There is, however, another consideration to take into account in relation to the concordance between twins. Although we think that identical twins are wholly identical, this is only true of their genotypes. TCRs and BCRs are not identical in monozygotic twins. The combinatorial nature of receptor gene assembly means that a huge number of different receptors can in theory be generated. Any one individual will generate only a small subset of these possible re-arrangements and thus it is just not possible that identical twins will have identical sets of TCRs. If the possession of a particular TCR predisposes to an immune-related sensitivity or autoimmune disease, it is unlikely that both identical twins will express this particular TCR. It is not clear, however, whether these differences do play a part in explaining low-concordance.

Q7.9. What kind of evidence might support the hypothesis that differences in TCR repertoires may influence susceptibility to autoimmune disease?

Another possible reason for non-concordance relates to epigenetics. Gene expression in an individual cell can be modified in ways that can be inherited by daughter cells when the cell divides (e.g. by DNA methylation or histone modification). This is a new area, but studies suggest that epigenetic modification of immune-related genes can differ between monozygotic twins, and that these differences increase with age and geographical separation. The importance of epigenetic modification is, however, far from fully understood at present.

An important observation from human studies is that it is not uncommon for one individual to suffer from two or more autoimmune diseases concurrently. Thus, of individuals with ankylosing spondylitis (an arthritic disease of the spine), around 15% will have psoriasis (an inflammatory disease of the skin), 3% will have Crohn's disease and 5% will have ulcerative colitis (inflammatory bowel disorders). Of patients suffering from asthma, 7–13% will have Crohn's disease and 8–12% ulcerative colitis. These percentages are much higher than are seen in the general population, and strongly suggest that there are allelic variants of genes that bias those individuals who possess them towards the development of immune-related diseases.

7.3.3.2 Which Genes Confer Susceptibility or Resistance to Immunopathology?

Twin studies can identify a genetic basis for disease, but they cannot tell us which genes are involved. Identifying the actual

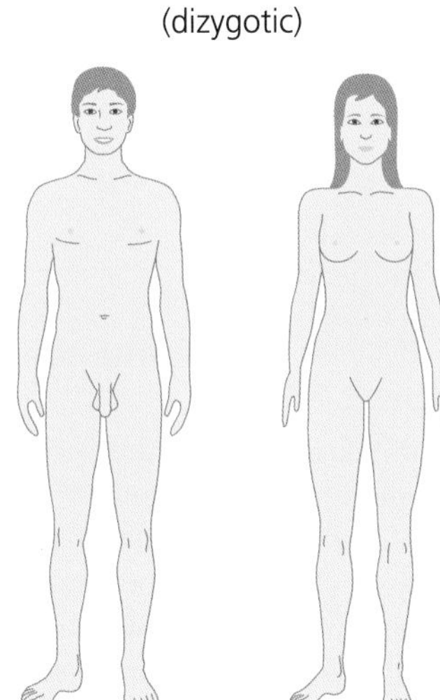
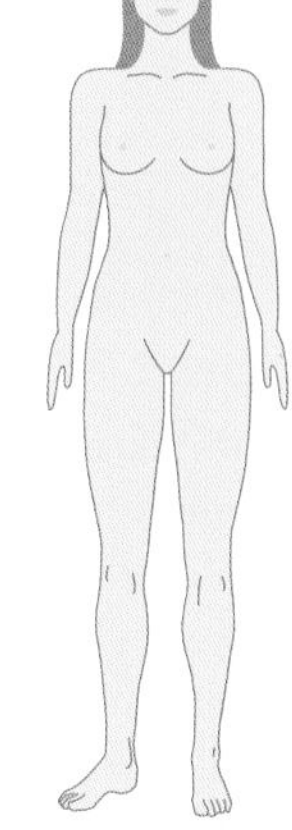

Fig. 7.15 Disease concordance in twins. The genetic contribution to a disease can be estimated by studying sets of identical (monozygotic) and non-identical (dizygotic) twins where one twin has the disease. If the frequency of the disease in the other twin (concordance) is higher in identical than non-identical twins this is strong evidence for a genetic contribution. In most autoimmune diseases concordance is less than 50% in identical twins.

genes involved is a major goal for clinical geneticists. Much information has come from linkage analysis. These are studies in which the association of disease susceptibility with other phenotypic traits can identify the region of a chromosome involved in susceptibility. Now that the human genome has been sequenced, candidate genes in this region can be sequenced and examined for differences (i.e. polymorphisms) that could explain susceptibility. More recently genome-wide association studies have been used. In these, the genomes of large numbers of individuals with and without the disease are scanned very rapidly using a very large number of genetic markers called single nucleotide polymorphisms (SNPs). This technique can enable us to determine if a particular gene(s) shows differences between affected and non-affected individuals.

In the case of autoimmune diseases, the outcome of the above types of these studies to date is that the MHC is by far the most common set of genes associated with susceptibility, in particular the class II genes (HLA-DR, -DQ and -DP). Given the importance of CD4 T cell activation in these diseases, this is perhaps not surprising. Many other genes or genomic regions have been associated with susceptibility. They include variants of CTLA4, an inhibitory costimulatory molecule in rheumatoid arthritis and SLE; the IL-23 receptor, important in T_h17 survival (inflammatory bowel disease, ankylosing spondylitis and psoriasis); the IFN regulatory factor 5 gene, involved in production of Type I IFNs (rheumatoid arthritis and SLE); and the IL-2 receptor, central to T cell activation (Type I diabetes and multiple sclerosis). As a further example, a gene coding for a cytoplasmic viral RNA receptor has been associated with *resistance* to Type I diabetes. (Observations in immune-related sensitivities are discussed below.) It is, however, surprising that those genes which have been identified as risk factors for susceptibility have been estimated to account for only a small proportion of the total genetic basis, often only 10–20%. This shows us that in outbred humans, many genes, or combinations of genes, most as yet unknown, contribute to disease susceptibility.

Q7.10. Which other types of molecule with allelic variants might lead to a generalized tendency to develop autoimmune disease?

Q7.11. Is it possible to suggest ways in which we will be able to fill in the gaps in our understanding of the genetic basis of immune-related disease susceptibility?

7.3.3.3 Animal Studies and Genetic Susceptibility

Different inbred strains of mice show marked differences in susceptibility to spontaneous and induced autoimmune disease. (In some cases this is also true for rats, chickens and domestic animals.) Some of these models have proved very useful in unravelling the pathogenesis of autoimmune diseases and have given information about their possible genetic bases. Where relevant, some of these models are discussed in the next section of this chapter. Some general points do, however, emerge from these studies. Spontaneous disease rarely, if ever, develops in all members of a susceptible strain of laboratory mice. These mice are genetically identical and are reared in identical environments (often in the same cage). This suggests that non-environmental factors may play a role and that differences in TCR repertoires or perhaps epigenetic differences may also prove to be important. In models of inflammatory bowel disease, symptoms only develop if commensal intestinal bacteria are present. Thus, as commensal bacteria are really non-self, there must be environmental influences at work. Surprisingly, the NOD (non-obese diabetic) mouse, which develops insulin-dependent diabetes, suffers more severe disease in a germ-free environment (Section 7.4.4.1).

7.3.3.4 The Increasing Incidence of Immunological Diseases

The frequency of allergies, particularly asthma, is increasing rapidly in the developed world. This is a real clinical problem, yet we do not know why it is happening. Children growing up on farms, or with several siblings, or who are in the company of other children from an early age (e.g. at a nursery), are less likely to suffer from allergies and asthma. These observations have led to the hygiene hypothesis, which suggests that early exposure to multiple antigens and infectious agents biases the immune system against allergic responses. It was originally suggested that these exposures prevented the immune system from being biased towards T_h2-type responses. However, it is not just allergies that are on the increase; autoimmune diseases such as Type I (insulin-dependent) diabetes are also increasing and this is very much a T_h1-type disease. There are also suggestions that allergic individuals are less able to generate T_{reg}, or anti-inflammatory cytokines such as IL-10 and transforming growth factor (TGF)-β, suggesting that, in part, disease may result from defects in immune regulation (see also below). However, we are still a long way away from understanding why allergies are increasing and thus of being able to develop strategies to prevent their occurrence. One clinical trial has, however, started in which dust mite, cat, and grass antigens are being placed under the tongue of children with susceptibility to asthma in attempts to prevent them becoming sensitized. The authors will continue to encourage their respective grandchildren and step-grandchildren to play in the dirt.

7.4 Immunopathology and Therapy in Action

In this section we will use selected clinical and experimental examples of immunopathology to illustrate the principles involved, and to provide some insights into what we do and do not understand about the pathogenesis of these conditions. We will first cover allergy. We will then consider other types of immune-related sensitivities and autoimmune diseases. We are grouping the latter two types of condition (involving extrinsic or intrinsic antigens respectively) into one section because of the mechanistic overlaps between them (Section 7.2.2) and the difficulty in designating many diseases as strictly resulting from responses to extrinsic or intrinsic antigens. In the following sections we will briefly discuss transplantation immunology (Section 7.5) and, finally, introduce the contentious field of tumour immunology (Section 7.6).

7.4.1 Allergic Diseases (Type I Hypersensitivity Reactions)

Allergies result from the production of IgE antibodies against extrinsic antigens. We know this because passive transfer of IgE from a sensitized individual to a normal person transfers the allergy. Allergies were once thought to represent a neurosis, but this was disproved when in 1941 an American doctor transferred allergy by giving normal individuals a blood transfusion from a hay fever sufferer. See Figure 7.16.

7.4.1.1 Anaphylaxis

Allergy is very common. Often this is a mild but irritating condition, as with hay fever, but it can be much more serious. The mechanisms underlying allergy are relatively well understood. Certain individuals, exposed to some antigens (allergens) make adaptive responses in which B cells switch to making IgE. This switch is very probably dependent on IL-4 produced by Th2 cells. The IgE binds to high-affinity FcRs expressed on mast cells. If the allergen cross-links the mast cell-bound IgE, the mast cell is triggered to degranulate, releasing inflammatory mediators such as histamine and several cytokines, resulting in local acute inflammation. Exposure to the antigen (allergen) may, however, lead to a systemic response that can be fatal. This is anaphylactic shock. See Case Study 2.1.

Pathogenesis of Anaphylactic Reactions The reasons that some antigens stimulate IgE synthesis in certain individuals are very unclear (below). For whatever reasons, the girl in Case Study 7.1 did make IgE, which bound to high-affinity FcRs on her mast cells. On the fatal occasion, the antigen she ingested bound to and cross-linked the IgE antibodies, stimulating the mast cells in her mouth and digestive tract to degranulate, releasing histamine and other inflammatory mediators which entered her blood and induced the systemic effects. Histamine caused her respiratory smooth muscle to contract, leading to respiratory obstruction, and her vascular smooth muscle to relax, leading to venous pooling of her blood and preventing blood returning to the heart. These factors

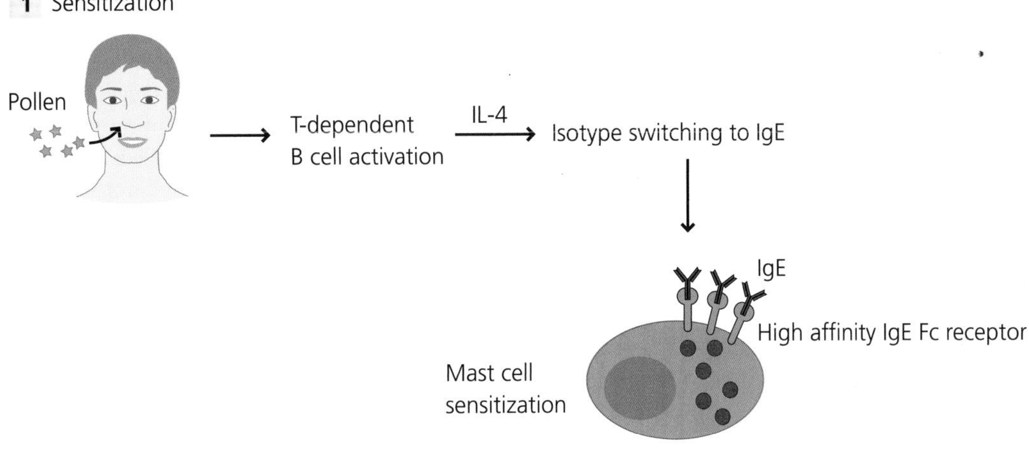

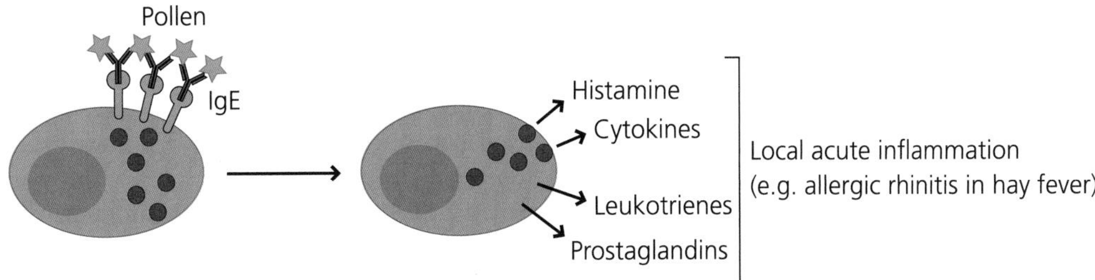

Fig. 7.16 Mechanisms of allergy. Exposure to some antigens may lead to IgE synthesis, which binds to specific Fc receptors (FCRs) on mast cells. If the antigen cross-links mast cell-bound IgE, the mast cell degranulates, releasing pre-formed mediators such as histamine and some cytokines. Other mediators such as the lipd metabolites (eicosanoids) leukotrienes and prostaglandins, are synthesized and secreted. by the mast cell. The result is local acute inflammation as in allergic rhinitis (runny nose) in hay fever.

Case Study 7.1: Peanut Allergy

Clinical: A 12-year-old girl knows she is allergic to peanuts. She buys a meat burger from a stall but is given a veggie burger by mistake. The veggie burger contains peanuts. Within minutes of eating the burger she develops difficulty in breathing. She realizes immediately what has happened but has left her adrenaline syringe in her other jacket. She collapses and dies.

Explanation: On a previous occasion(s) the girl had eaten peanuts. She made IgE to an antigen(s) in the peanuts that bound to mast cells. The binding of the peanut antigen to IgE on mast cells when she ate them again triggered the responses that caused her death.

combined to cause her death. This condition is called anaphylaxis and the alterations in her physiology represent one form of shock. Had she been able to inject herself with adrenaline her death would have been prevented: adrenaline contracts vascular smooth muscle, but induces relaxation in smooth muscle in the respiratory tract by acting on different receptors. See Figure 7.17.

7.4.1.2 Asthma

Asthma is an example of a complex immune-related disease. Asthma is an increasing problem amongst children in the developed world. In the United States, between 8 and 10% of the population suffer from asthma, and the incidence is continuing to increase rapidly. Asthma is a descriptive name that applies to more than one pathological process. Most cases of asthma in children have an initial allergic aetiology, in other words they are Type I reactions. However, maintenance of the asthmatic condition depends on Type IV reactions. Other forms of asthma may not in fact have an allergic basis but for simplicity, here we will only consider asthma that is allergic in origin. See case Study 7.2.

Pathogenesis of Asthma Asthma is not just an allergic disorder. It is also a chronic inflammatory condition. The acute symptoms of broncho constriction are due to the release of mediators by sensitized mast cells, leading to contraction of smooth muscle and excessive secretion of mucus, a Type I reaction. If, however, the immune response is not controlled, structural changes occur in the airways (tissue remodelling). The most important of these are thickening of the bronchial walls, laying down of collagen (fibrosis) under the epithelia and increased goblet cell numbers (leading to increased mucus secretion). Eosinophils are major players in this type of inflammation, as well as tissue remodelling, probably through their secretion of the growth factor TGF-β. These later non-allergic, responses are dependent on cytokines secreted by T_h2-biased CD4 T cells.

In Case Study 7.2 and many other instances of mild to moderate asthma, the initial immune response is a T_h2-biased response to an environmental antigen. In the case study the antigen may well derive from house dust mites. Several different proteins have been identified as allergens and, interestingly, many of these have biological activities that may relate to the induction of immunity. Two such house dust mite antigens are Der P1 and Der P2. Der P1 can act as an innate activator. It has cysteine protease activity that can trigger the release of pro-inflammatory molecules. Der P1 can also signal via TLR4, by mimicking MD-2 (Section 4.2.2.3), while Der P2 can signal via TLR2. It may also be important that dust mite faeces contain high levels of lipopolysaccharide (LPS), since this may supply the initial innate immune activation. Although many cases of asthma show a T_h2 bias, examination of activated T cells from patients with

Case Study 7.2: Asthma

Clinical: A 6-year-old boy is taken to his doctor because he has difficulty with breathing. He has developed this condition over the last few months, and it has recently worsened. On examination he is struggling to breathe and his expirations are particularly prolonged and laboured. There is widespread wheeze on auscultation (listening to the chest). He is treated with inhaled salbutamol (albuterol)– a broncho dilator which eases his symptoms. He is also given an inhaled steroid to use regularly. His symptoms have remained under control since then. Injecting house dust mite antigens into his skin resulted in a very rapid local inflammatory response (redness and swelling, a Type I reaction). His mother and one sister had also been diagnosed with asthma, and all had eczema as babies.

Explanation: The boy had developed allergic sensitivity to a common environmental antigen, probably proteins present in the faeces of the dust mites common in bedding and house dust generally. His symptoms were caused by constriction of his airways due to contraction of the smooth muscle. He was treated with salbutamol which is a β_2-adrenergic agonist that induces smooth muscle relaxation. He was given steroids to inhibit the development of chronic inflammation as a result of later non-allergic immune responses.

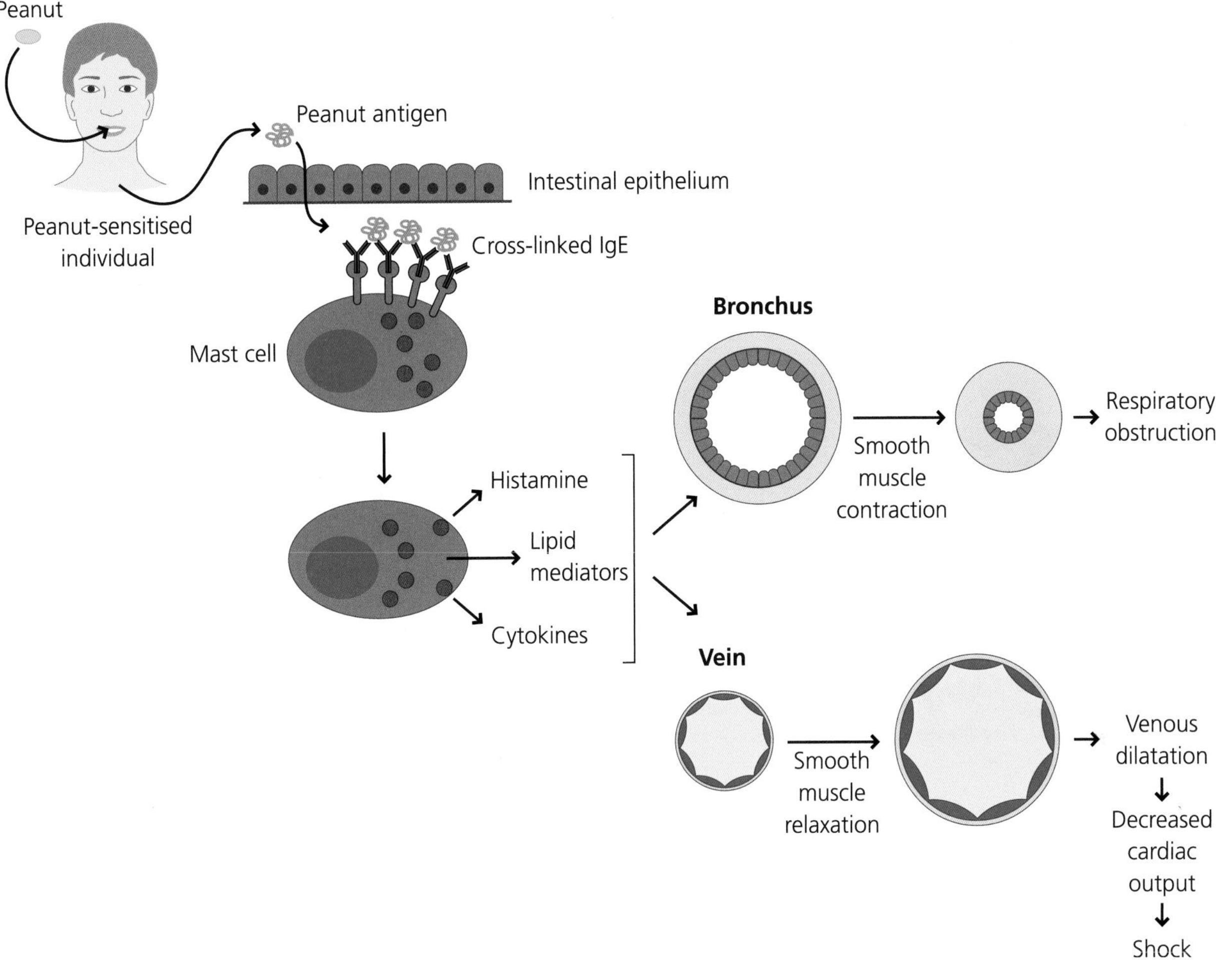

Fig. 7.17 Peanut anaphylaxis. If inflammatory mediators are released in large amounts from sensitized mast cells they may enter the circulation and cause potentially fatal systemic effects. The most important effects are those on smooth muscle. These include broncho constriction, which may cause respiratory obstruction, and venous dilatation, which leads to the pooling of blood in veins, preventing it returning to the heart and leading to potentially fatal hypotension (low blood pressure).

severe asthma shows that they are often of the T_h17 phenotype. See Figure 7.18.

Why Do People Become Allergic? Atopy is defined as a heritable tendency to develop allergy. This is, however, just a descriptive definition and we need to understand the molecular and cellular basis for this susceptibility. There is much evidence for genetic influences on susceptibility to allergies: it often runs in families and if one monozygotic (identical) twin is allergic, there is a 60–80% concordance with the other. If the other twin is not identical there is a 30–40% concordance. This shows that there is a strong genetic influence, but that other factors must also play a role. We have very little idea what these other factors may be: infection is a candidate but there is no hard evidence for its role. Over 100 genes have been identified as contributing to the risk of developing asthma. These include T_h2-related genes such as IL-4 and IL-13, MHC class II alleles, TNF-α, CD14 (involved in LPS recognition), an FcR for IgE, and a number of genes whose roles are unclear. As with other associations between immune-related diseases and the MHC, the precise molecular basis for the asthma association is unclear. However, we may be gaining some insights in the case of diabetes and possibly multiple sclerosis: below.

What Can We Learn from Animal Models of Allergy? It is possible to induce a condition in mice that shows many similarities to human asthma. Initial injection of ovalbumin with alum as an adjuvant (Section 4.6) is followed by repeated airway exposure to ovalbumin by aerosol. Some strains of mice develop symptoms of both early- and late-phase asthmatic responses. The use of this model has given important insights into the disease, but the extrapolation of such results into humans must always be made with caution. Studies with knock-out mice showed that both IL-4 and STAT6 were required for the development of asthma; these are crucial for T_h2 polarization (Section 5.4.2.4). Mouse strains differ in their

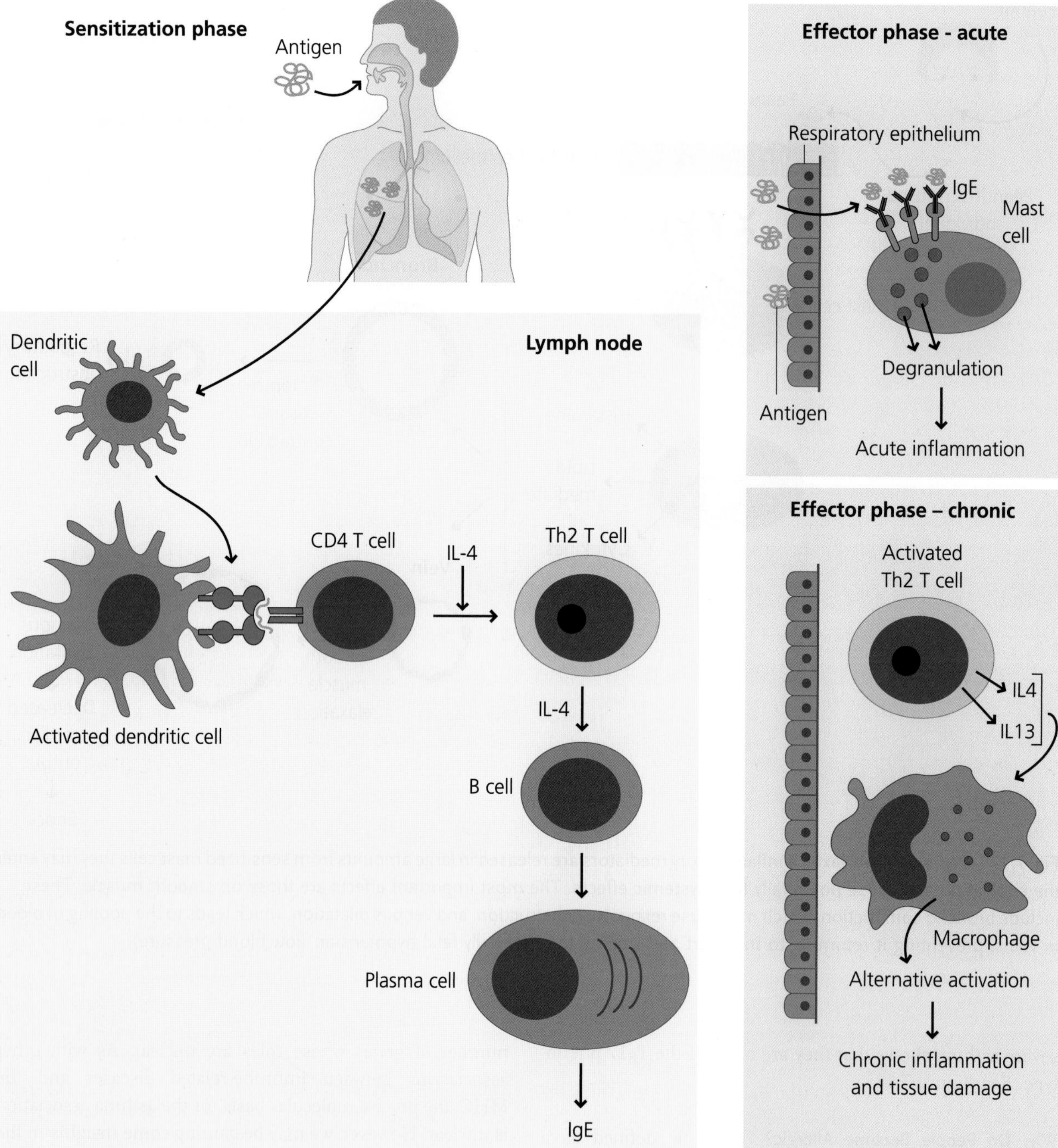

Fig. 7.18 **Asthma.** Immune-mediated asthma (there are other types) involves a T_h2 response to inhaled antigens (allergens). In the presence of IL-4, activated T cells switch responding B cells to make IgE. The acute, IgE-dependent response to the antigen causes mediator release from mast cells that induce broncho-constriction; this is responsible for the respiratory problems (wheezing). A T_h2-biased cell-mediated chronic inflammatory response can also occur that is responsible for the structural changes in the airways known as tissue remodelling, which can become irreversible. In severe asthma, T_h17 T cells may be involved.

ability to make T_h2 responses (e.g. BALB/c mice are more T_h2-biased than C57BL/6 mice). In keeping with this, C57BL/6 mice lacking the IL-5 gene cannot be sensitized to give T_h2 responses to ovalbumin, but BALB/c mice can. T-bet is the CD4 T cell transcription factor typical of T_h1 responses and mice lacking T-bet spontaneously develop an asthma-like condition. This may reflect the regulatory interactions that occur between IFN-γ and IL-4 in determining CD4 T cell polarization (Section 5.4.2.4).

Q7.12. T-bet knock-out mice develop an asthma-like condition in their airways, but other organs and systems do not develop disease. Why might disease be restricted in this way?

Treatment of Allergy Avoidance of allergens is 100% effective in preventing the disease, but is not generally available to patients who have become sensitized; individuals who have to work in contact with allergens to which they have become sensitized may have to use full respiratory protection, or change jobs. Anaphylaxis, which is potentially fatal, demands immediate vigorous treatment and injected adrenaline is the essential first-line treatment. As noted above (Section 7.4.1.1), adrenaline relaxes bronchial smooth muscle, but stimulates contraction of vascular smooth muscle.

Pharmacological approaches are generally successful for more minor allergies. For example sodium cromiglycate inhibits mast cell degranulation though it may not be very effective in practice, maybe due to a short half-life. Systemic corticosteroids are very effective at preventing the inflammatory effects of allergic responses but are associated with multiple side-effects, notably increased risks of infection, bleeding, psychosis, diabetes and high blood pressure. However, inhaled corticosteroids are commonly used in asthma and provide local anti-inflammatory action without causing undue systemic side-effects. They can be important in preventing tissue remodelling.

An alternative approach to the treatment of allergy is desensitization, which has a long history. In this treatment allergic patients are given increasing doses of the allergen, starting with miniscule amounts and working gradually up to large doses. This has proved successful in treating allergies such as those to bee and wasp venom. Why it works is not clear. Possibly the allergic T cells are rendered unresponsive to stimulation (anergy), or perhaps the regime stimulates the formation of IgG antibodies that can diffuse onto mucosal surfaces and mop up the allergens before they get to the mast cells, or maybe antigen-specific T_{reg} are stimulated.

Recently, a novel therapy has been licensed for the treatment of hay fever. Pollen allergy can be treated by giving an extract of Timothy grass pollen sublingually. Oral or mucosal tolerance is a well-recognized phenomenon (Section 5.5.5.3 see Chapter 5). Antigens given by mouth can render individuals non- or hypo-responsive to the same antigen given systemically. Experimentally, oral tolerance is quite easy to induce in non-sensitized animals, but much less so in those that have already been immunized. However, the Timothy grass treatment does appear effective in hay fever sufferers, how it works remains unclear. The real importance of this approach is that it is immunologically specific and hence there are none of the side effects associated with all non-specific treatments. This approach is being tried in several autoimmune diseases (e.g. giving oral bovine collagen in rheumatoid arthritis), but without significant benefits so far. A current trial is using nasal administration of myelin-derived peptides in multiple sclerosis (Section 7.4.4.2).

Q7.13. In what other examples of immune-mediated disease might antigen-based therapy might be appropriate?

7.4.2 Diseases Caused by the Actions of Antibodies on Cells (Type II Hypersensitivity Reactions)

Type II hypersensitivity reactions, also known as 'cytotoxic' hypersensitivity reactions, can be involved in organ-specific autoimmune diseases. As described previously, antibodies can act on cells directly, often by binding to receptors on the plasma membrane and blocking, or in some cases stimulating, the function of the receptor. They can also act, with complement, to cause damage to or death of the cell. Here we will use several examples to illustrate different mechanisms of pathogenesis in type II reactions.

7.4.2.1 Molecular Mimicry

Pathogenesis of Guillain-Barre Syndrome Molecular mimicry refers to the concept that an antigen from a microbe is antigenically cross-reactive with a self antigen, and that the immune response produced against the microbe then targets that self antigen, causing disease. Molecular mimicry is a well-recognized cause of immune-related sensitivities, and we use Guillain-Barre syndrome as one example. See Case Study 7.3.

Guillain-Barre syndrome is an example of molecular mimicry between an infectious agent and an intrinsic self antigen.

Case Study 7.3: Guillain-Barre syndrome

Clinical: A 45 year old man eats some undercooked chicken at a barbecue. A couple of days later he develops severe diarrhoea and vomiting. He recovers from this, but around two weeks later notices a tingling sensation in his legs. Soon after this he finds difficulty in walking. The weakness progresses rapidly until he is almost completely paralysed. He is treated by plasmapheresis and over a period of months he makes a near-complete recovery.

Explanation: The chicken was contaminated with the bacteria *Campylobacter jejuni*, a common cause of food poisoning. This caused the gastro-enteritis. This bacterium contains a lipopolysaccharide (LPS) in its cell wall that stimulated an IgG antibody response. The myelin coat of the patient's nerves contains a ganglioside, GM1, which has structural similarity to part of the LPS molecule. The cross-reaction of this antibody with GM1 in his peripheral nerves caused demyelination, leading to loss of function and paralysis (a cell-mediated response may also be involved). Plasmapheresis was used to deplete his plasma of the damaging antibodies.

Antibodies made against bacterial lipopolysaccharide (LPS) recognize a self molecule, the ganglioside (glyolipid) GM-1 on nerves, and cause damage. This happens because these two quite different molecules happen to express one or more epitopes that are, structurally, sufficiently similar to permit cross-reactive binding of the antibodies. The formation of antibodies against a self molecule (GM-1) tells us that in patients who develop Guillain-Barre syndrome there must be B cells present that are potentially self-reactive. In other words, these autoreactive cells have not been tolerised, presumably because B cells do not have access to GM1 (a component of nerves) during their development in the bone marrow. Normally they would remain quiescent, but the infection provides a sufficiently strong stimulus to activate them (compare with costimulation, Section 1.5.3.3). See Figure 7.19.

One difficulty in understanding Guillain-Barre syndrome (and other similar diseases) is that the antibodies causing the disease are of the IgG class. LPS, as an antigen (rather than a mitogen), typically induces IgM antibodies in a T-independent manner but the synthesis of IgG antibodies will usually need T cell help (Section 6.3.1.2). One likely possibility is that the LPS is linked to a bacterial protein which can therefore act as a "carrier", permitting CD4 T cell activation and the initiation of class switching in B cells which subsequently secrete IgG anti-LPS antibodies (Box 6.5). Guillain-Barre syndrome may be associated with infection by a variety of organisms, including Epstein-Barr virus, but *Campylobacter jejuni* is the most common. It is however very rare for an infected individual to develop Guillain-Barre syndrome: in one study there was not a single case of the syndrome in a sequence of 8000 microbiologically-confirmed infections with *Campylobacter jejuni*.

Molecular mimicry may also be involved in T cell-mediated disease. A host T cell that has not been tolerised against a self peptide-MHC complex (Section 5.5.2) will not cause problems unless it is activated. However, if a pathogen protein can be processed by DCs to form a peptide that is identical to the self peptide, the PAMPs associated with the pathogen may stimulate the activation of the anti-self T cell. The activated T cell may then recognize self peptide presented on host MHC molecules of a tissue, and initiate damage. Whether or not such a peptide can stimulate a T cell will depend on the MHC of the individual who was infected. Clearly the peptide will

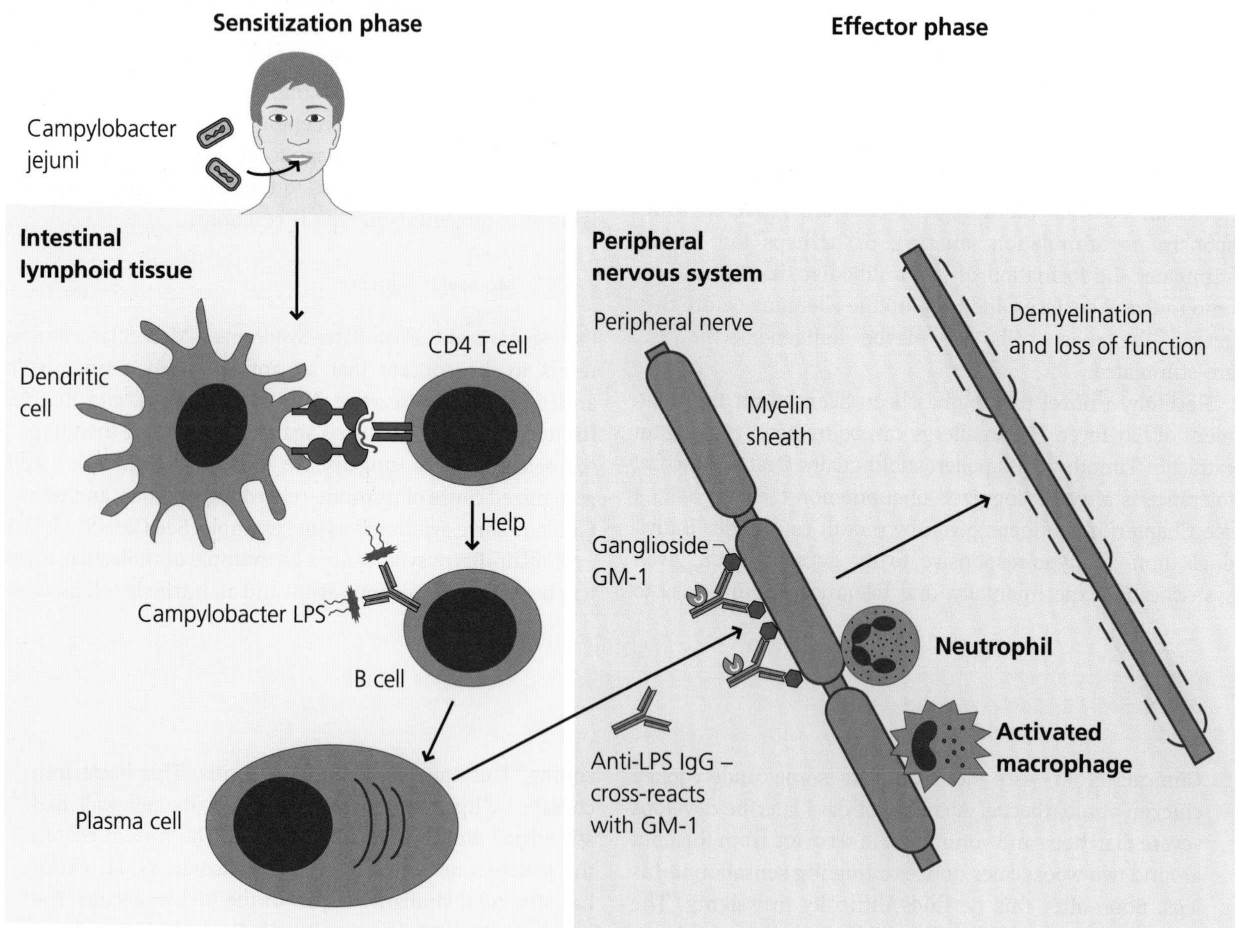

Fig. 7.19 Guillain–Barré syndrome. Some infections, of which *Campylobacter jejuni* is the most common, may be followed by a progressive muscular paralysis. An antibody response is made to bacterial LPS, and the antibody cross-reacts with a ganglioside, GM_1, expressed on the myelin sheath of peripheral nerves. Complement damages the nerve sheath, leading to impaired conduction and paralysis and neutrophils may also be involved. There may also be a T cell-mediated response which results in lymphocytes and macrophages infiltrating the nerves. The disease is usually self-limiting over a long period, but may be treated by plasmapheresis to remove damaging antibodies non-specifically or by giving large amounts of non-specific Ig intravenously (IVIG therapy).

Case Study 7.4: Thyroid Autoimmune Disease

Clinical: An 18-year-old student has become very anxious about his studies and is finding it difficult to sleep. His friends have noticed that his eyes have a staring appearance. On examination he has bulging eyes (exophthalmos), a fine tremor of his hands, and his pulse is fast and irregular. Blood tests reveal increased levels of thyroid hormones. He is diagnosed as having hyperthyroidism (Graves' disease) with atrial fibrillation (his heart is being stimulated to contract abnormally by aberrant signals generated in the atrium in response to the increased secretion of thyroid hormone). He is treated with drugs to treat his cardiac arrhythmia and with a drug to block the excessive secretion of thyroid hormones. He remains healthy 2 years later.

Explanation: The student had developed antibodies to the thyroid-stimulating hormone (TSH) receptor. These antibodies, on binding to the receptor deliver an activating (agonistic) signal, resulting in unregulated secretion of thyroid hormones. The disease can be treated by inhibiting hormone synthesis. Previously it was treated by removing the thyroid gland surgically and giving the patient replacement thyroid hormone orally.

need to possess appropriate amino acids at positions important for binding to the 'anchor residues' in the peptide-binding groove of the respective MHC molecule (Section 5.2.2.1). This may be one explanation for the MHC associations seen in so many of these diseases

A separate question is, of course, how widely infections are actually involved in initiating autoimmune disease. Some researchers suggest that they may be a major stimulus in many, if not all, cases of autoimmune diseases, but there is as yet no substantive evidence that this is the case.

7.4.2.2 Anti-Receptor Antibodies

In the above case the antibodies were damaging the nerves, but it is also possible for antibodies to cause disease by acting directly on cells in ways that alter their functions. See Case Study 7.4.

Pathogenesis of Thyroid Autoimmune Diseases In Case Study 7.4 (hyperthyroidism in Graves' disease), the patient has developed antibodies against the thyroid-stimulating hormone (TSH) receptor that act as an agonist to stimulate increased production of thyroid hormones (mimicking the natural TSH). We can be certain it is these antibodies that cause this disease for two reasons:

i) If a pregnant woman has Graves' disease, it is possible for her new-born child also to be hyperthyroid. The hyperthyroidism is cause by IgG antibodies that have crossed the placenta. As these antibodies are catabolized, the symptoms disappear from the child.
ii) If serum from such a patient is injected into animals it can cause long-term thyroid stimulation. The factor responsible was shown to be an anti-TSH receptor antibody originally called long-acting thyroid stimulator (LATS). See Figure 7.20.

In another condition, myasthenia gravis, it is again clear that antibodies are both necessary and sufficient to cause the disease. In myasthenia gravis the patient develops muscle weakness (often first noticed as drooping of the eyelids) that is associated with blocking antibodies to the acetylcholine receptors. If a pregnant woman has myasthenia gravis, the baby may be born with the same symptoms due to the placental transfer of these antibodies, but these symptoms also disappear as the maternal antibody is catabolized.

In contrast to the above case study, some patients suffer from autoimmune diseases in which thyroid hormone secretion is insufficient (hypothyroidism, which is associated with myxoedema). In Hashimoto's thyroiditis, for example, the thyroid gland shows chronic inflammation with marked T cell infiltration and loss of hormone-secreting cells. There may also be autoantibodies to thyroid antigens and blocking antibodies to TSH receptors in the blood. Hence, it is not completely clear whether the antibodies or the chronic T cell-mediated inflammation (or both) cause the decreased hormone secretion. However, new-born babies delivered by hypothyroid mothers do *not* show symptoms of hypothyroidism even though the anti-thyroid antibodies can be detected in the babies' blood, suggesting that these antibodies are not the causative agent. It is therefore more likely that the damage to the thyroid is initially caused by a cell-mediated response. This damage could then lead to epitope spreading (Figure 7.14), which stimulates antibody synthesis. This illustrates a very important point relating to many autoimmune diseases as well as allergy and other immune-related sensitivities: often a variety of immune effector mechanisms are present and it is difficult to determine which is or are responsible for the actual damage.

7.4.2.3 Drug Sensitivities

Some Type II diseases are due to a immune-related sensitivity to a drug. An interesting example is the now-superseded anti-hypertensive drug, methyldopa, which caused loss of erythrocytes (haemolytic anaemia) in some patients. One explanation for this is that the drug bound to a protein on the surface of erythrocytes, creating a new CD4 T cell epitope (a neo-epitope) to which the patient was obviously not tolerant. CD4 T cells were activated and provided help to a B cell which synthesized IgG antibodies against the new epitope. The erythrocytes might then be destroyed directly by complement-mediated lysis or though phagocytosis by macrophages in the spleen or liver. The result is haemolytic anaemia. This represents one mechanism for antibody-mediated autoimmune disease. A similar scenario applies to other drugs which may, for instance, cause destruction of

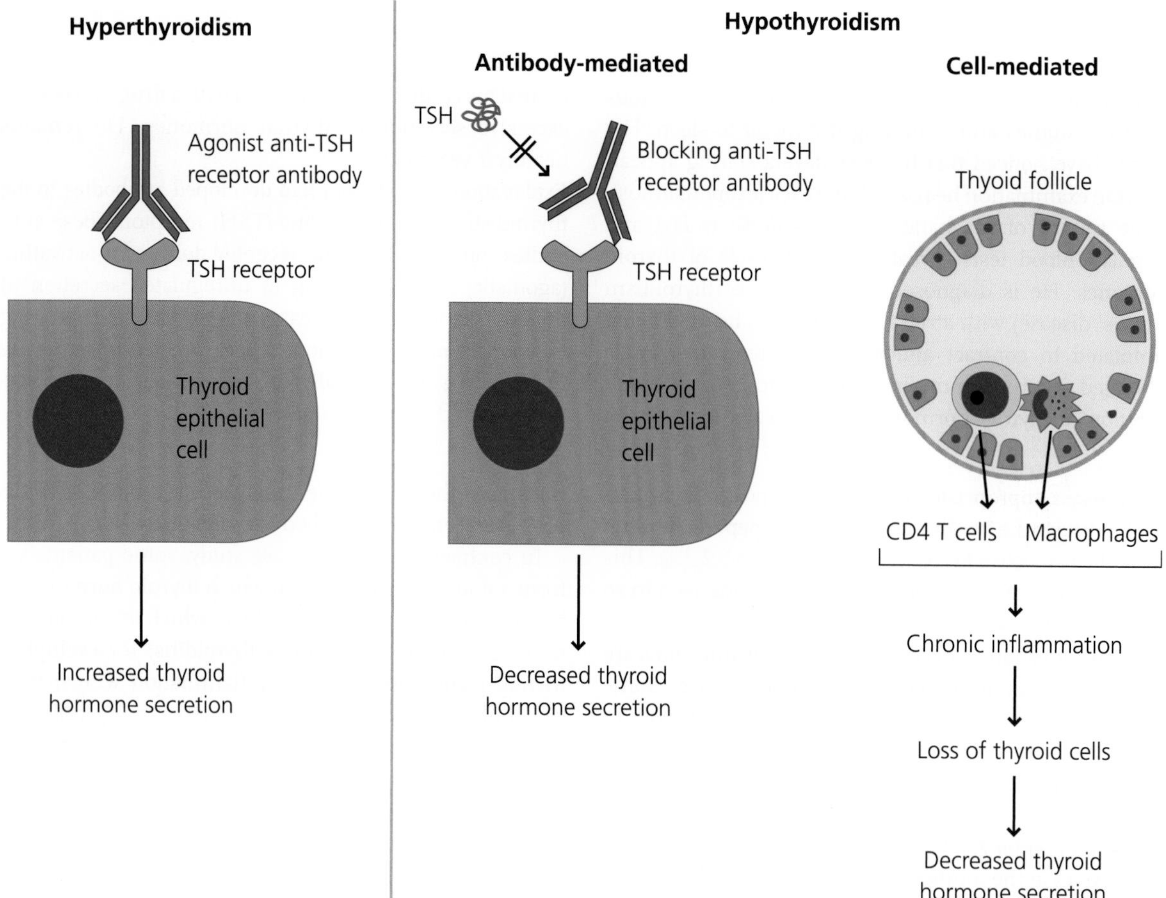

Fig. 7.20 Thyroid autoimmune disease. Hyperthyroidism (Graves' disease) may be caused by agonist antibodies to the TSH receptor which induce long-lasting stimulation of thyroid hormone secretion. Hypothyroidism involving decreased secretion of thyroid hormones, may be caused by blocking antibodies to the TSH receptor, but may also be due to antibody-independent cell-mediated immunity involving cytotoxic T cells and activated macrophages a type IV response, very different to the other two types shown.

platelets (thrombocytopenia). In many cases, however, we have no idea why patients develop antibodies against their own cells. See Figure 7.21.

Q7.14. Where might the innate activation danger signal needed to stimulate T cell responses come from? arise in drug sensitivities?

It is important to realize that drug sensitivities can be mediated by IgG, as above, or by IgE, as in reactions to penicillin metabolites. The latter (but not the former) can lead to fatal anaphylaxis, but this is fortunately rare.

7.4.2.4 Blood Group Antibodies and Disease

Many of us possess preformed antibodies to our own erythrocytes that can potentially cause problems if we need a blood transfusion. These are the blood group antibodies. The ABO blood groups are the major barrier to universal blood transfusion. Individuals possess IgM antibodies to those AB antigens that they do not express. In other words A-positive individuals have anti-B antibodies; B-positives have anti-A; those expressing neither (O) have anti-A and anti-B; while AB-positives have neither. These antibodies may have been made originally against antigens expressed on intestinal bacteria, but which cross-react with the respective blood group antigens on erythrocyte RBCs. The result is that if a Group A person (whose blood contains anti-B antibodies) receives Group B blood, the donor red cells will be destroyed rapidly by complement-mediated lysis or opsonization. This does not just cause anaemia, but can induce potentially fatal shock, probably due to massive cytokine release. This is why blood groups are always identified and matched before transfusions. Additionally, the recipient's serum is always tested (cross-matched) against the potential donor's red cells in case there are other anti-RBC antibodies present.

Q7.15. A Group O mother will possess anti-A and anti-B blood group antibodies. If she becomes pregnant with an A-positive foetus will the foetus develop haemolytic anaemia?

7.4.2.5 Haemolytic Disease of the New-Born

The Rhesus (Rh) blood groups cause a different kind of problem. People are classified as Rh-positive or Rh-negative depending on whether or not they express the strongest

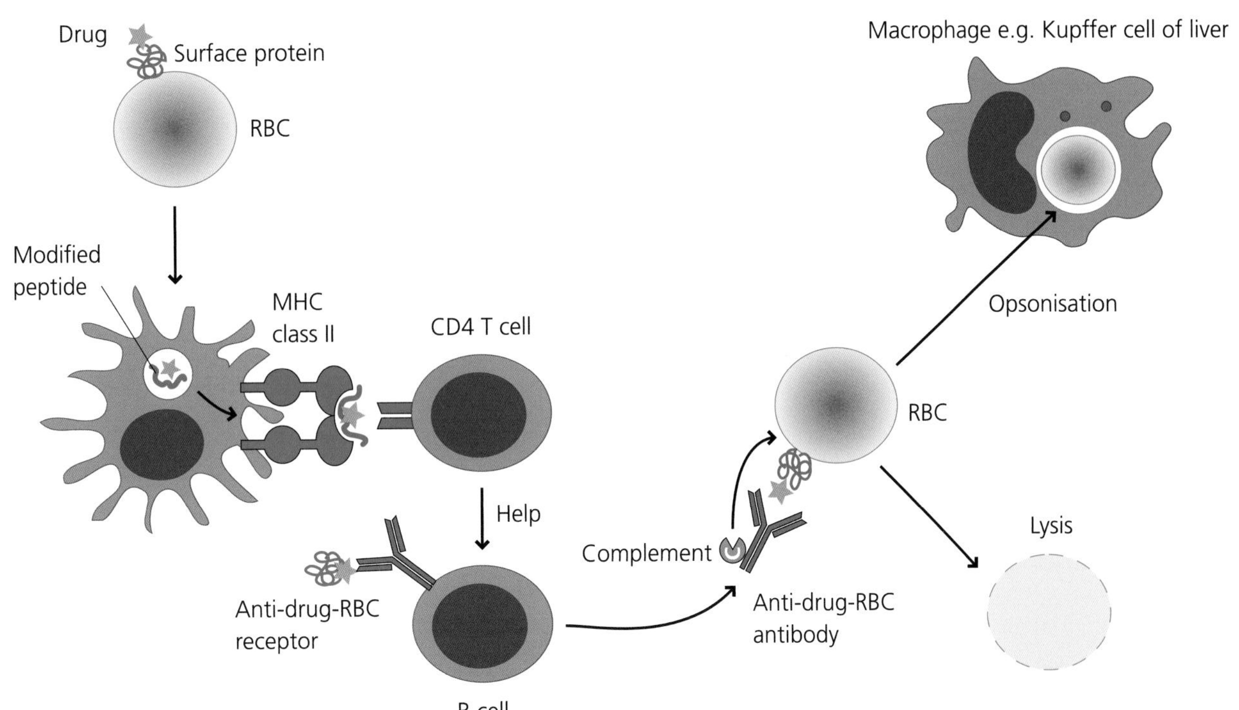

Fig. 7.21 Drug-induced haemolytic anaemia. Some drugs, such as methyldopa previously used as a treatment for hypertension, on rare occasions may induce a haemolytic anaemia. It is suggested that the drug modifies an erythrocyte (RBC) protein, possibly generating modified peptides that can activate CD4 T cells. B cell tolerance is less complete than T cell tolerance and an antibody response is induced that leads to RBC destruction by lysis or opsonization. Stopping the drug resolves the problem.

Rhesus system antigens (called D) on their red cells. We do not naturally possess antibodies to Rh antigens (unlike antibodies to ABO antigens; above). However, if a mother is Rh-negative and the father Rh-positive, the foetus may be Rh-positive. If this is the case, all will be well during the first pregnancy. However, during birth there is always some leakage of foetal blood into the maternal circulation and foreign red cells are potent antigens. Thus, the mother will make IgG anti-Rh antibodies. If she becomes pregnant again with another Rh-positive foetus, the antibodies, being IgG, will cross the placenta into the foetal circulation and mediate haemolysis of the foetal red cells. This can cause death *in utero* or severe haemolytic anaemia (haemolytic disease of the new-born). See Figure 7.22.

Q7.16. The activation of adaptive immunity against Rh antigens should require activation of the innate system by danger. Where might this danger come from when foetal RBCs enter the maternal circulation?

Q7.17. A Rh-negative mother has a Rh-positive child and is not given anti-Rh antibodies. She becomes pregnant again but the child does not develop haemolytic anaemia. What reasons might account for this?

If it occurs, this form of anaemia can be treated by giving the foetus *in utero* transfusions of Rh-negative blood. However, this is a totally preventable disease. If, as soon as the first Rh-positive baby is born, the mother is given an injection of anti-Rh antibodies, this is very effective in preventing her becoming sensitized, so she does not make anti-D antibodies. How this works is not completely clear. The antibodies may mop up the foetal red cells in the maternal blood, targeting them to macrophages for destruction and possibly preventing their access to DCs. Alternatively the anti-Rh antibodies may have more subtle feed-back effects on the immune response, perhaps involving the inhibitory FcγRIB on maternal B cells (Section 6.3.2.3). Whatever the mechanism, this is one the first examples of (passive) antibody-mediated immunotherapy and represented a major immunotherapeutic advance. The treatment is also of course antigen-specific.

7.4.2.6 Treatment of Type II Sensitivities

In many cases the treatment is straightforward (e.g. blood transfusion for haemolytic anaemia). In other cases an effective treatment can be to remove the antibodies causing the disease from the blood. This involves plasmapheresis in which blood is taken from the patient, plasma (containing the antibodies) is removed and the depleted blood (without

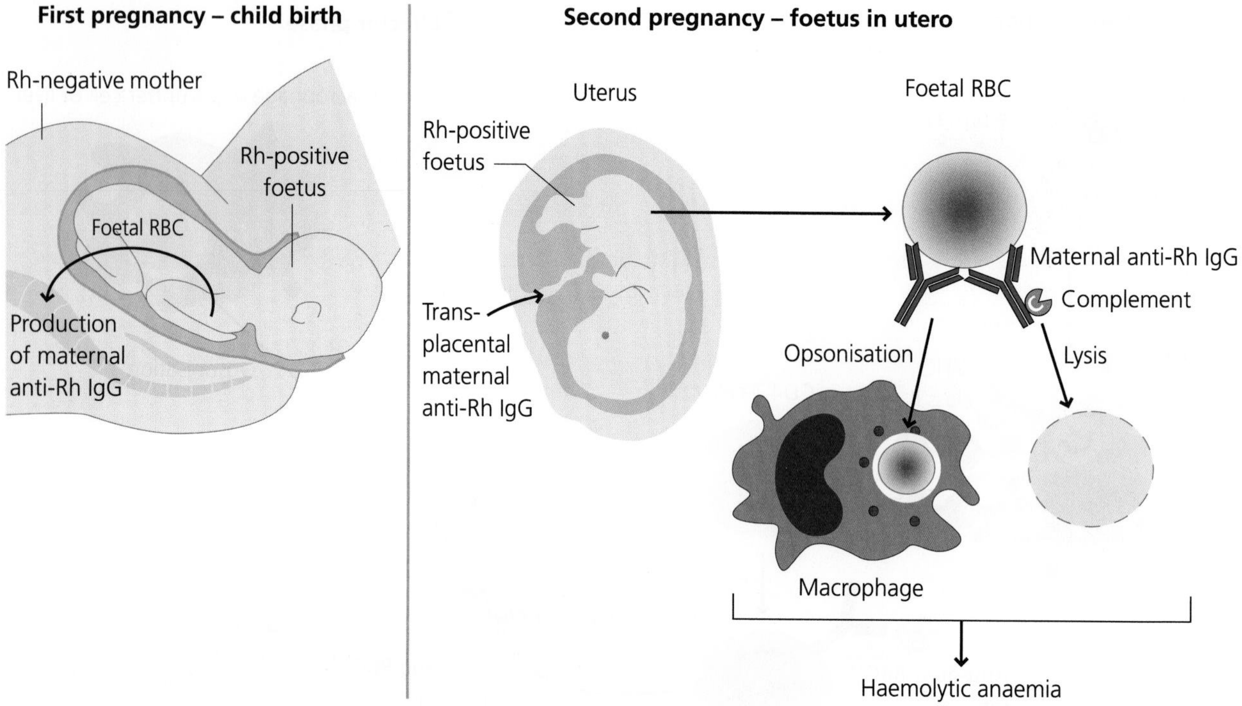

First pregnancy – child birth
Rh-negative mother
Rh-positive foetus
Foetal RBC
Production of maternal anti-Rh IgG
Second pregnancy – foetus in utero
Uterus
Foetal RBC
Rh-positive foetus
Maternal anti-Rh IgG
Complement
Trans-placental maternal anti-Rh IgG
Opsonisation
Lysis
Macrophage
Haemolytic anaemia

Fig. 7.22 Haemolytic disease of the new born. If a Rh-negative mother has a Rh-positive baby, during birth some foetal RBCs escape into the maternal circulation and the mother makes an IgG anti-Rh response. If she has a second pregnancy with a Rh-positive foetus, the anti-Rh antibodies will cross the placenta and destroy the foetal RBCs. This can be prevented very effectively if, immediately after the first and subsequent births, the mother is given anti-Rh antibodies; these this prevents her becoming sensitized and she does not make an active anti-Rh response.

the causative antibodies) is returned. Again, this is a non-specific treatment, but can be very effective in diseases such as Guillain–Barré syndrome and myasthenia gravis. Alternatively, large amounts of non-specific human polyclonal IgG can be administered (IVIG therapy). It is not fully clear why IVIG is successful, and there is some controversy about its efficacy. It may act by stimulating inhibitory FcRs on B cells or compete in some way with pathological antibodies for activating receptors. The holy grail of immunotherapy is to render therapy antigen-specific, but such therapy is a long way off.

Q7.18. Why might the administration of large amounts of pooled human immunoglobulins (IVIG) bring benefit in antibody-mediated diseases?

7.4.3 Diseases Caused by Immune Complexes (Type III Hypersensitivity Reactions)

Type III hypersensitivity diseases result from the formation of antigen–antibody complexes that activate complement and lead to inflammation. They can be systemic or localized to particular tissues or organs, and they can include both autoimmune diseases against intrinsic antigens and immune-related sensitivities against extrinsic antigens.

Systemic antigen–antibody complexes are formed in the blood. As described earlier (Section 7.2.5), under normal circumstances, systemic complexes are removed rapidly following transport by RBCs and cleared by sinusoid-lining macrophages in the liver and spleen. There are, however, circumstances where, for whatever reason, this uptake is inefficient (often this is due to the large size of the complexes) and these complexes may be deposited in small blood vessels and lead to local inflammation. There are many clinical conditions associated with systemic immune complex deposition (e.g. SLE; below and Case Study 7.5). Similarly there are many diseases associated with local immune complex deposition. If antigens are inhaled and becomes deposited in the mucosa, immune complexes form in the lungs and lead to local acute inflammation (e.g. Farmer's lung below and Case Study 7.6). In many cases of immune complex disease in the kidneys, such as in SLE (below), circulating complexes are deposited in glomerular vessels, leading to focal areas of inflammation (glomerular nephritis). In other forms of glomerular nephritis (Goodpasture's syndrome), specific antibodies are made against glomerular antigens. These bind to antigens in the glomeruli, forming immune complexes that induce more diffuse inflammation.

Case Study 7.5: Systemic Lupus Erythematosus

Clinical: A 24-year-old woman develops a facial rash under her eyes and across her nose (butterfly rash). Shortly afterwards she develops painful, swollen joints in her hands. She is found to have protein in her urine. Blood tests show that she has a very low level of the complement C4 component and high levels of antibodies to double-stranded DNA. She is diagnosed as suffering from SLE. She is treated with steroids and immune suppression (as her kidneys are involved). She finds that her symptoms show alternating periods of remission and exacerbation over the next decades, although she is reviewed carefully to ensure that overt renal failure does progress.

Explanation: Her symptoms are caused by the deposition of antigen–antibody complexes in small blood vessels in skin damaged by UV light joints and in the kidneys. These complexes bind and activate complement, and as a result of this activation, local acute inflammation is induced. Neutrophils are recruited to the site and are responsible for the local tissue damage. Steroids reduce the inflammation and immune suppression reduces autoantibody production.

7.4.3.1 Diseases Caused by Systemic Immune Complexes

Pathogenesis of SLE. See Case Study 7.5 Why do SLE patients deposit immune complexes in their blood vessels? The antigens in the complexes are largely derived from cell nuclei, and typically include DNA and histones. It cannot be that the antibodies are damaging the cells directly since they do not have access to the nuclei of intact cells. Might it be that cells undergoing apoptosis are not dealt with efficiently? The concordance for SLE in identical twins is estimated to be between 30–70% in different studies. Defects in genes involved in apoptosis have been shown to be involved and possibly also those involved in autophagy (a specialized process in which a cell can engulf and eliminate its own cytoplasmic organelles). Complement components are major risk factors. Crucially, complement is involved in the aiding clearance of apoptotic cells, and defects in C1, C2 and C4 components are strongly associated with SLE. For example, 90% of individuals with a C1q deficiency and 75% with a C4 deficiency develop SLE; this is a hugely strong association. Other genes have been identified as risk factors for SLE. The pro-inflammatory cytokine TNF-α is implicated and it is suggested that SLE patients may make more TNF-α than normal subjects. Not surprisingly, the MHC is another major player. Whether this is due to the genes for TNF being close the MHC genes, or if the involvement of the MHC genes, is an independent factor, remains to be defined. See Figure 7.23.

Q7.19. C1q is a crucial component in complement activation by immune complexes via the classical pathway. Might we not predict that a C1q deficiency would lead to less effective and thus inflammation?

Animal Models of SLE Some strains of mice such as the NZB (New Zealand Black) strain develop an SLE-like disease spontaneously. In these mice three genetic loci have been identified as contributing to the disease. One is involved in tolerance, one is expressed in B cells and one in T cells, but the functions of the genes encoded by these loci are not completely understood. Mice in which the genes for complement C1q or C4 have been deleted also develop spontaneous SLE-like disease, clearly paralleling many examples of the human disease. Several other mutant mouse strains develop SLE-like disease, including those with the mutations related to defects in apoptotic cell clearance and the regulation of BCR signalling.

7.4.3.2 Diseases Caused by Local Deposition of Immune Complexes. See Case Study 7.6.

7.4.3.3 Pathogenesis of Occupational Hypersensitivities

The disease in Case Study 7.6 (below) is classified as a hypersensitivity pneumonitis and particularly affects the lung alveoli. The patient inhaled fungal spores and spore antigens were transported via lymph to his local (mediastinal) lymph nodes. These initiated an immune response leading, on

Case Study 7.6: Farmer's Lung

Clinical: A 54-year-old farmer notices that some hours after he has been working in his hay barn he becomes wheezy and short of breath. A chest X-ray shows widespread small areas of opacity. A blood test reveals that he has high levels of IgG antibody to antigens found in fungal spores. He was treated with corticosteroids and advised to avoid working in the barn if possible. He followed this advice and remains symptom-free.

Explanation: Repeated exposure of his immune system to the fungal spores present in mouldy hay led to him producing high levels of anti-spore IgG antibodies. Inhaled spores lodge in the lung mucosa, bind to the IgG and complement is activated, leading to localized acute inflammation. The steroids are non-specific anti-inflammatory agents.

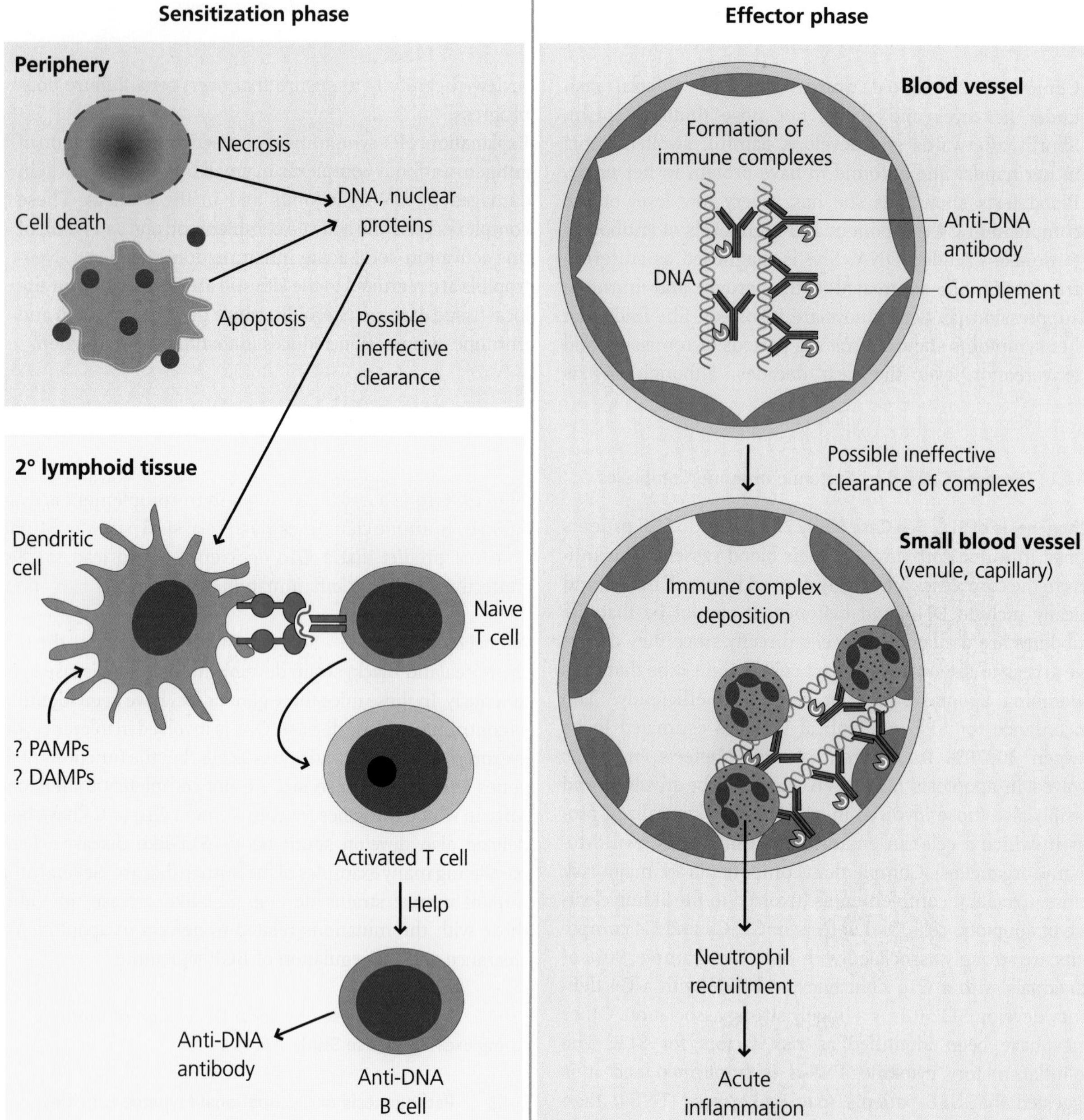

Fig. 7.23 Pathogenesis of systemic lupus erythematosus. SLE is an inflammatory condition resulting from the deposition of circulating immune complexes in small blood vessels such as those in the skin, joints and kidney. These complexes activate complement, which induces acute inflammation, including the recruitment of neutrophils and macrophages, at the site of deposition. In SLE the antibodies are specific for DNA or DNA-associated proteins. Two suggestions for pathogenesis are that there may be inefficient removal of apoptotic cells, leading to the release of DNA and associated proteins, and/or that clearance of circulating immune complexes is inefficient.

repeated exposure, to the production of anti-spore IgG antibodies. The IgG antibodies spread throughout the body and diffused out of blood vessels into the lung connective tissues. When the patient next inhaled the spores they were bound by local IgG, complement was activated, and acute inflammation was initiated. As this is a multi-step process it takes several hours before the inflammation builds up in the alveoli sufficiently to interfere with breathing. This pathogenesis contrasts with, and is quite different from, allergic (Type I) hypersensitivity where the symptoms appear in minutes. Many other examples of this type of disease exist and some are quite exotic bird fancier's lung (avian proteins), crack lung (crack cocaine), bagassosis (pressed sugar cane) and suberosis (Penicillium in mouldy cork dust) are just a few examples. See Figure 7.24.

7.4.3.4 Treatment of Immune Complex-Related Diseases

If the antigen causing the disease is environmental (e.g. fungal spores), avoiding exposure is a guaranteed remedy. In most cases, however, particularly if the antigen is self as in

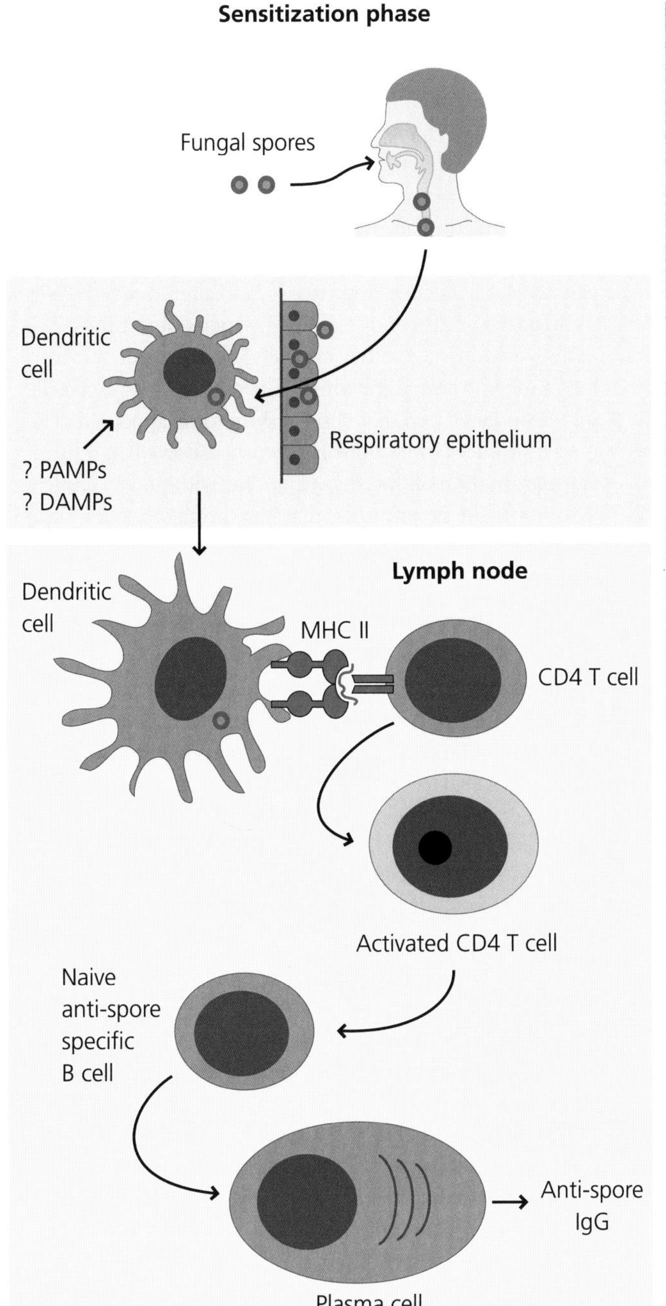

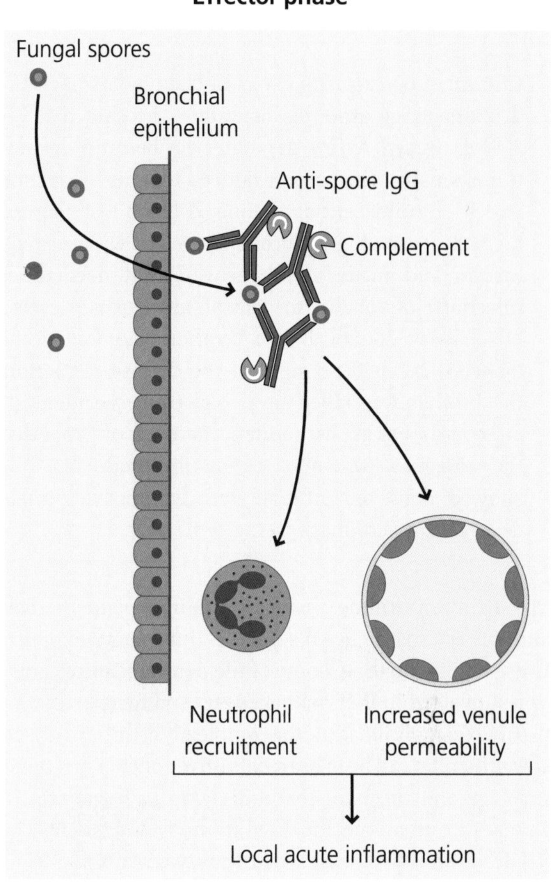

Fig. 7.24 Farmer's lung. Some individuals who inhale fungal spores over time make an IgG response to spore antigens. The IgG diffuses into the lungs and, when spores are next inhaled, local immune complexes form. These activate complement and local acute inflammation follows. Symptoms appear a few hours after the spores are inhaled.

SLE, avoidance is not possible and treatment is directed at suppressing the inflammation that is the cause of the damage. Drugs used include anti-inflammatory drugs; NSAIDs (such as ibuprofen), corticosteroids) and immunosuppressants. The problem with anti-inflammatory and immunosuppressant drugs is that they are not immunologically specific and can also suppress immune responses to pathogens, leading to increased susceptibility to infection or re-emergence of a latent infection.

7.4.4 Cell-Mediated (Type IV) Hypersensitivity Reactions

In many cases of autoimmune disease, cell-mediated responses are involved and antibodies appear to play no part in the pathogenesis of the disease; these are the type IV reactions. Antibodies are often present in these diseases, but the evidence generally suggests that they are not causing the damage. We will use four common and important examples of

Case Study 7.7: Type I Diabetes

Clinical: A 6-year-old boy was noticed by his parents to be drinking large amounts of water – he said that he was always thirsty. A few days later he became drowsy and unresponsive. His parents noticed that his breath smelled sweet – reminiscent of acetone. A blood test showed that he had a very high glucose level. He was treated with insulin and made a full recovery, but needed regular injections of insulin to control his glucose levels. The boy had an identical twin brother who was tested for diabetes, but had no signs of the disease. Unfortunately the brother was killed in a road traffic accident shortly afterwards and at post-mortem the Islets of Langerhans in the pancreas (the sites of insulin-producing β cells) showed a marked infiltrate with lymphocytes (mainly T cells) and macrophages.

Explanation: For unknown reasons the affected child had activated CD4 T cells that reacted with antigens expressed by β cells in the Islets of Langerhans. These CD4 T cells activated effector mechanisms (cytotoxic CD8 T cells and activated macrophages) that brought about the destruction of all his β cells, making him unable to secrete insulin. This caused blood glucose levels to rise, and altered metabolism induced the production of ketones responsible for the smell of his breath. The attack on his islets had been going on for a long time and it was only when all, or almost all, of his β cells had been destroyed that clinical symptoms appeared. The post-mortem examination of his brother showed that chronic inflammation of the islets can occur without any clinical symptoms. It is probable that the brother would have developed diabetes had he lived.

type IV autoimmune disease: insulin-dependent diabetes, multiple sclerosis, Crohn's disease and rheumatoid arthritis. These will to illustrate some of the central features and principles involved in their pathogenesis and treatment; they will also increasingly illustrate the complexity of these diseases, and how difficult it can be to draw definitive conclusions regarding aetiology and mechanisms underlying pathogenesis. These diseases are organ specific. As in antibody-dependent (Types I–III) autoimmune diseases, there are two main phases: initiation and maintenance of the disease process.

7.4.4.1 Type I Insulin-Dependent Diabetes

Pathogenesis of Type I Diabetes. See Case Study 7.7 In diabetes we have not identified the initiating antigen but there are several candidates including insulin itself. It has been suggested that a viral antigen, expressing a cross-reactive peptide, could be the trigger. Coxsackie virus is a popular candidate but there is no firm evidence for this pathogenesis. Whatever the trigger, the damage is dependent on the initial activation of CD4 T cells. Whether it is effector cytotoxic T cells or activated macrophages, or both, that cause the actual damage is unclear. Knowing the actual effector mechanism may, however, help to design more effective therapies. See Figure 7.25.

Q7.20. The Islets of Langerhans contain a mixture of insulin-secreting β cells and α cells that secrete glucagon (a hormone that raises blood glucose levels). Given that in typical autoimmune Type I diabetes only the β cells are affected, what might this suggest about the likely immunological effector mechanism?

Genetic Basis As for the other examples in this section, the MHC is the major identified locus affecting the risk of developing diabetes. Particular MHC class II alleles are associated with an increased risk or, in some cases, decreased risk of developing the disease. The molecular basis of this association is unknown, but it is intriguing that in humans and mice one of the high-risk MHC alleles has the same amino acid substitution, Asp57, which is predicted to alter the shape of the peptide-binding groove, giving it a more open conformation (Section 5.2.2.1). The MHC could regulate immune responses at two main levels. First, at the level of the antigen-presenting cell (APC, e.g. a DC) it could control binding of a disease-associated peptide epitope for specific T cells. Second, at the level of the T cells it could control which self peptides bound to the MHC in the thymus and thus change the repertoire of T cells that were positively or negatively selected relative to the wild-type allele. There is in fact evidence for both.

In both humans and mice, other multiple (more than 20) gene loci have been identified as being associated with increased risk of diabetes, but in only a few cases have the actual genes been identified. One of these is the insulin gene itself. More recently, a genome-wide association analysis has implicated the IF1H1-MDA5 gene that gene codes for a cytoplasmic receptor for viral RNA, and the rare gene variant identified was associated with strong protection against diabetes. Viral infection has long been suggested as a trigger for diabetes (as in many other autoimmune diseases) and this study give indirect evidence in support of this hypothesis. The protective variant could in principle control a more effective immune response to clear an RNA virus, such as a retrovirus, for example.

Animal Models of Diabetes Strains of mice (NOD), rats (BB) and chickens develop Type I diabetes spontaneously. The disease that develops in NOD mice shows similarities to human Type I diabetes. They both start with a chronic inflammatory infiltrate of lymphocytes and macrophages into the Islets of Langerhans, and the structural similarity (Asp57) of high-risk MHC alleles in NOD mice that are humans is of interest, but as yet unexplained. It may relate to the peptide(s) involved in disease initiation. Experiments with NOD mice have defined clearly the role of CD4 T cells in disease initiation. CD4 T cells taken from a diabetic NOD mouse that are transferred into a young NOD mouse that is

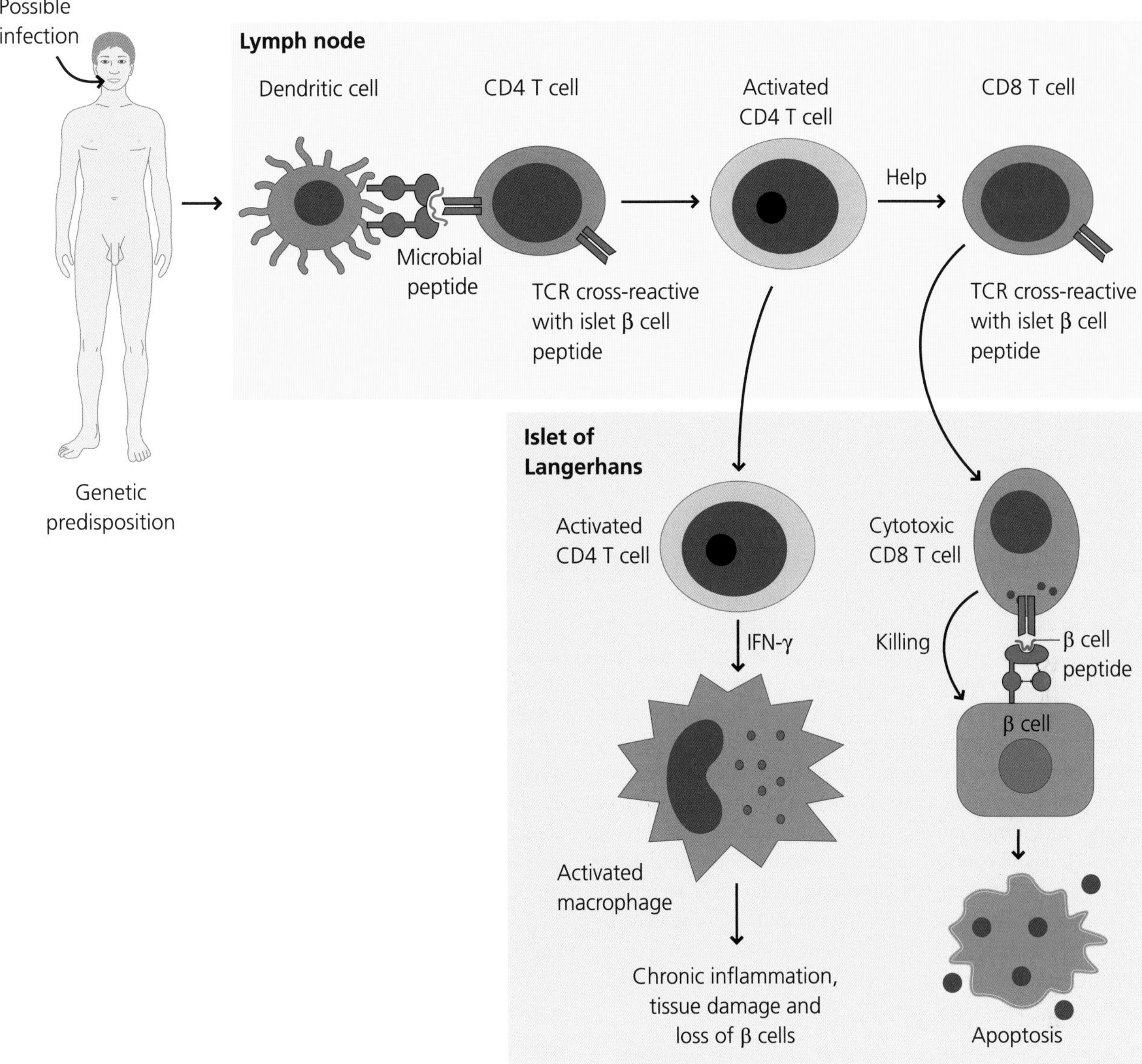

Fig. 7.25 Type I (insulin-dependent) diabetes. For reasons that are largely unclear, some individuals make an adaptive immune response that destroys their insulin-secreting β cells in their pancreatic Islets of Langerhans. It is suggested that an infection (possibly Coxsackie virus) activates CD4 T cells that cross-react with a β cell peptide. This activates CD4 and CD8 T cells that invade the Islets and induce destruction of the β cells. Infiltrating macrophages are also present and may play a part in causing the damage.

not diabetic can speed up the onset of the disease. As with humans, multiple gene loci contribute to disease susceptibility. As mentioned above, over 20 susceptibility loci have been identified in NOD mice, but few actual susceptibility genes have been identified. Apart from the MHC, the IL-2 gene may be important. Defects in T_{reg}, which are very IL-2-dependent, are prime candidates for influencing diabetes pathogenesis in NOD mice, but results to date are not clear cut. One feature shared by BB rats and NOD mice is that they show a T cell lymphopoenia (deficit) and this is also a feature of several other animal models of autoimmune disease. There is, however, no evidence that diabetes-prone humans are lymphopoenic. Another intriguing feature of diabetes in NOD mice is that it is more frequent and severe in mice kept in a germ-free environment. This contrasts with models of inflammatory bowel disease, where the disease is dependent on commensal intestinal bacteria and does not occur in germ-free mice.

Q7.21. Why might diabetes be more severe in NOD mice that are germ-free?

Treatment of Diabetes By the time Type I diabetes is clinically apparent, virtually all the insulin-secreting cells have been destroyed and they are not capable of significant regeneration (see Case Study 7.7). At present the only widely available treatment is replacement insulin by injection. Transplantation of Islets of Langerhans is being carried out in several centres and with some success, but as with all allografts prevention of rejection is a major challenge. In the future, stem cells may be used to generate β cells, but there will remain the problem of the autoimmune response still being present.

If individuals who are likely to develop diabetes could be identified before they had lost all their β cells, there would be a much better chance of inhibiting the disease process. We cannot, however, at present predict who will and will not develop the disease; even with identical twins the concordance is only in the order of 30%. If, however, individuals with a high risk of disease could be identified, it might be possible to test them repeatedly for markers of disease (e.g. autoantibodies or activated T cells specific for β cell antigens) and potentially those with disease markers could be given preventative treatment. None of the potential treatments are without risk and we would need to be sure that the benefits of the treatment in preventing disease outweighed the chances of damaging side effects occurring.

7.4.4.2 Multiple Sclerosis

Pathogenesis of Multiple Sclerosis. See Case Study 7.8 As for most of the diseases discussed in this chapter, we are largely unable to explain how the disease is initiated. Oligodendrocytes are the cells in the CNS that produce myelin and it is these cells that appear to be the main targets in multiple sclerosis. Analysis of T cells (both CD4 and CD8) from multiple sclerosis patients suggests that the major antigen is myelin basic protein, the most abundant protein in the myelin sheath. However, other antigens are also involved, possibly via epitope spreading, and IgG antibodies are present in the lesions. It is still controversial as to whether these antibodies play a role in pathogenesis or whether they are a secondary phenomenon. The predominant tissue damage is likely to be caused by cytotoxic CD8 T cells and/or chronic inflammation with the secretory products of activated macrophages causing the damage. See Figure 7.26.

Q7.22. How might we determine if antibodies do play a significant role in the pathogenesis of multiple sclerosis and why might this information be helpful?

Genetic Basis of Multiple Sclerosis One of the most striking features of multiple sclerosis is its geographical distribution. Multiple sclerosis is significantly more frequent in Northern latitudes and the Western hemisphere than in Southern latitudes and the Eastern hemisphere (coeliac disease shows a rather similar distribution). The concordance rates for multiple sclerosis in twins from these different areas are variable but are at most around 30%, again suggesting strong environmental influences are involved.

Q7.23. Why might an autoimmune disease be more common in particular geographic regions?

The MHC is the major genetic risk factor: in multiple sclerosis possession of a particular pair of alleles at the DR and DQ loci is associated with a very high risk of the disease. Two other genes have been associated with increased risk that encode for the IL-2 receptor α chain and the IL-7 receptor α chain. It is, however, clear, from studies examining the frequency of multiple sclerosis in families, that these genes account for only a small part of the total genetic basis for multiple sclerosis in humans.

Evidence from Animal Studies Mice or rats injected with myelin basic protein in the presence of an adjuvant can develop a demyelinating CNS disease with similarities to multiple sclerosis, called experimental autoimmune encephalomyelitis (EAE). Unlike human disease it is often however self-limiting. The disease can be adoptively transferred to healthy mice by transferring activated CD4 T cells from an affected mouse. Studies using blocking anti-cytokine antibodies have shown that IL-23 is a major cytokine in EAE pathogenesis. As with multiple sclerosis, the MHC plays a major part in determining susceptibility to EAE, but no other specific genes have been identified to date.

Another model of multiple sclerosis has been adopted by some scientists. Humanized mice are mice in which a mouse's immune system has been replaced by part of human immune system. This is done by taking mice that have no adaptive immune system (e.g. RAG-deficient, Section 5.5.1.3) and reconstituting them with human haematopoietic stem cells. When a HLA-DR2 human MHC class II gene that is associated with susceptibility to multiple sclerosis was engineered into these mice, together with genes

Case Study 7.8: Multiple Sclerosis

Clinical: A 32-year-old woman noticed that her vision was blurred. A few weeks later she developed weakness in one leg. A brain scan showed focal lesions in the white matter and an oligoclonal IgG band was present in her cerebrospinal fluid. She was diagnosed as having multiple sclerosis (MS). Despite a variety of treatments her disease progressed although at times she had periods of remission, but she repeatedly relapsed and became weaker – finally having difficulty with breathing. Twelve years after her initial diagnosis she died of pneumonia. At post-mortem she was found to have scattered areas in her brain and spinal cord where the myelin coat of her nerves had been destroyed.

Explanation: For unknown reasons she had activated CD4 T cells that were reactive with her central nervous system (CNS) myelin. These activated T cells migrated into her CNS and initiated localized areas of chronic inflammation. This inflammation caused the demyelination of her nerves, leading, over time, to complete loss of function in those particular areas. Although auto-antibodies can often be detected in the lesions, it is not clear if they play a significant role in the pathogenesis of MS.

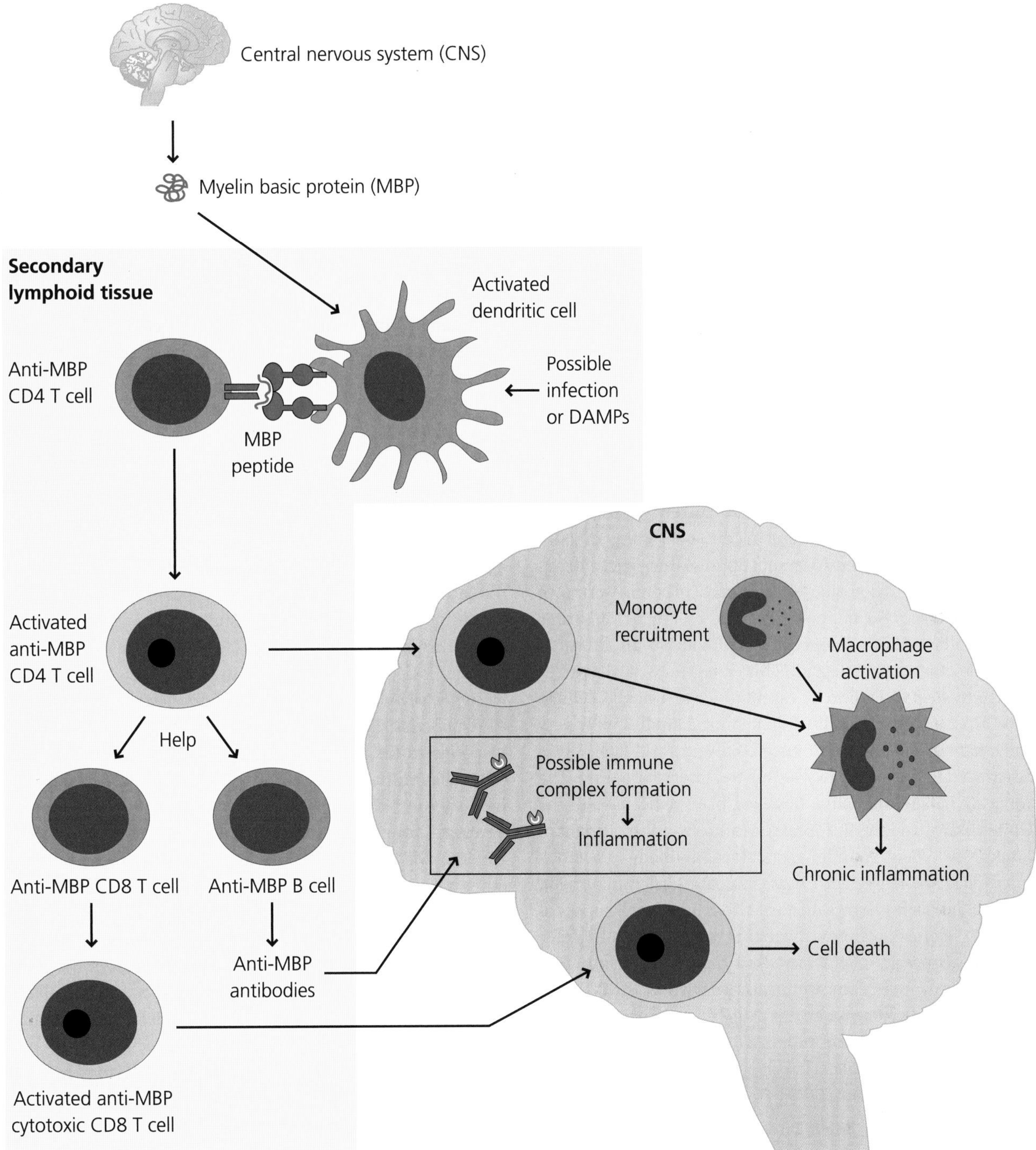

Fig. 7.26 Multiple sclerosis. For reasons that are largely unclear, some individuals make adaptive responses to myelin proteins expressed in the CNS; often myelin basic protein is the major antigen. Activated CD4 and CD8 T cells migrate to the CNS and recruit macrophages into the area. Effector mechanisms generated by these cells result in damage to the myelin sheaths of nerves, resulting in apparently random patches of demyelination, which can cause a variety of different symptoms. Antibodies specific for myelin may also be present but their role in pathogenesis is unclear.

encoding human TCR specific for a myelin basic protein peptide, and human CD4, they developed a multiple sclerosis-like disease. Not only this, but, in human multiple sclerosis, different class II alleles are associated with different clinical patterns of disease and, when these alleles were transferred into the humanized mice, the respective pattern of disease resembled that seen in human multiple sclerosis. This illustrates very directly the central role of the MHC in this disease and this approach will permit analysis of the molecular basis of this association. See Figure 7.27.

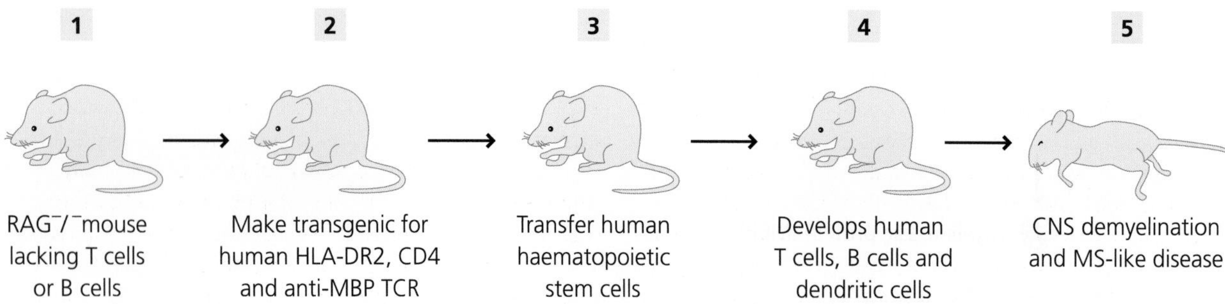

Fig. 7.27 Multiple sclerosis-like disease in humanized mice. Mice that have no T or B cells ($RAG^{-/-}$) can be made transgenic for human MHC class II molecules, human CD4 and T cells that express a TCR for a human myelin basic protein peptide (MBP). These mice are given human haematopoietic stem cells and develop human lymphocytes and DCs. The T cells will have been selected in the thymus on human MHC class II molecules. If the mice have been given MHC class II genes associated with susceptibility to multiple sclerosis, they may develop a multiple sclerosis-like disease with clinical features very similar to those found in human patients expressing the same MHC alleles.

Treatment of Multiple Sclerosis The current treatment of multiple sclerosis is very non-specific, consisting of anti-inflammatory drugs and immunosuppressants. These drugs have of course major side effects, particularly in terms of increasing the risks of infection. One of the newer drugs involves the use of a monoclonal antibody to an adhesion molecule, the $\alpha_4\beta_1$ integrin. This integrin was shown in mouse models of EAE to be essential for activated T cells to cross CNS endothelial cells. Initial clinical studies were very encouraging, but the occurrence of a very rare, fatal, brain disease led to the withdrawal of this treatment for a time. The antibody, natalizumab, has, however, been re-licensed in the United States and the European Community for treatment of multiple sclerosis. Another monoclonal antibody to a molecule, CD52, expressed on T cells Campath-1 (alemtuzumab), has had dramatic effects when used to treat a small subset of multiple sclerosis patients. Since this drug affects all T cells, there are concerns about increased risks of infection and, more recently, infections with cytomegalovirus (CMV) have proved to be a problem when this drug is used. Other forms of pharmacological immune suppression remain the mainstay of therapies for multiple sclerosis.

7.4.4.3 Inflammatory Bowel Disease

Several different diseases come under the heading of IBD, but the two most important are Crohn's disease and ulcerative colitis. We will focus primarily on Crohn's disease. Although this disease predominantly affects the small and large intestines, there can be widespread systemic symptoms, suggesting that the underlying immune mechanisms are complicated. Here, we will discuss Crohn's disease to illustrate this complexity See Case Study 7.9.

Pathogenesis of Crohn's Disease Crohn's disease is a chronic inflammatory condition, initiated by activated CD4 T cells, with the damage being caused primarily by activated macrophages and possibly CD8 T cells. It seems unlikely that antibodies play a significant role in causing tissue damage.

What do we know of the factors leading to the onset of Crohn's disease? Genetic susceptibility is important. Concordance for Crohn's is between 50–60% for identical twins and much less for non-identical twins. It appears that the MHC is a less important risk factor in Crohn's disease than in many other immune-related diseases. However, at least 30 separate genetic loci are associated with increased susceptibility. Recently, several important risk factors have been identified. A major risk factor is NOD2, a cytoplasmic PRR which responds to the muramyl dipeptide component of bacterial cell walls (Section 4.2.2.2). Individuals who are homozygous for a variant NOD2 allele are around 20 times more likely to develop Crohn's disease. However, only 20% of Crohn's patients are homozygous for this variant so there must be other unidentified genetic risk factors. Two other genes associated with increased susceptibility to Crohn's are involved in autophagy, a process in which a cell phagocytoses part of its own cytoplasm. Autophagy normally forms part of the intracellular recycling mechanisms involved in recycling cellular components, and may also play a part in destroying intracellular bacteria such as Listeria monocytogenes and Mycobacterium tuberculosis which can enter the cytoplasm of infected cells. However, in mice in which one of the above autophagy genes (ATG16L1) has been inactivated, there are defects in Paneth cells, a specialized cell type which resides at the bottom of small intestinal crypts and which plays a role innate defence. These mice appear to be more susceptible to intestinal inflammation.

Q7.24. To what extent might Crohn's disease be considered an autoinflammatory disease? (Section 4.2.2.4).

Apart from genetic factors there is strong evidence that environmental influences play an important role in Crohn's pathogenesis. Much interest centres around the roles of commensal intestinal flora, largely because of animal studies (below). It is intriguing that the majority of patients with Crohn's disease possess antibodies (IgG and/or IgA) specific

Case Study 7.9: Crohn's Disease

Clinical: A 23-year-old woman developed persistent diarrhoea and abdominal pain. On occasion blood was present in her stools. Stool cultures excluded an infective cause. Colonoscopy showed a patchy inflammation in the colon with some deep ulceration and bleeding. A biopsy from an inflamed site showed a mononuclear cell infiltrate (mainly activated T cells) with granuloma formation. The inflammation penetrated deep into the colonic wall. She was diagnosed with Crohn's disease and was treated with an anti-inflammatory drug, sulfasalazine, which reduces secretion of inflammatory mediators in the intestine. This caused some improvement in her symptoms, but over a period of time they worsened. After other medications, including steroids and immunosuppressants failed to bring about relief, she was treated with infliximab, an anti-TNF-α monoclonal antibody. This caused most of her symptoms to disappear and she is maintained on sulfasalazine.

Explanation: She had developed a chronic inflammatory disease affecting mainly her large intestine. This disease typically occurs in young adults but the aetiology is largely unknown. Treatment is anti-inflammatory and is therefore not immunologically specific. Sulfasalazine is poorly absorbed from the intestine and thus its actions are mainly localized to the intestinal wall. Steroids cause generalized immunosuppression with an increased risk of infection and other serious side-effects. Infliximab neutralizes one of the major cytokines produced by activated T cells and macrophages, TNF-α, and has been a major advance in the treatment of this disease. However, since TNF- α is a major cytokine involved in suppressing spread of *Mycobacteria tuberculosis*, patients treated with infliximab who have latent tuberculosis are at risk of re-activating the disease.

for the yeast Saccharomyces cerevisiae, but the significance of this observation remains obscure.

Animal Studies Many mouse and rat models exist which bear important similarities to human inflammatory bowel disease, especially Crohn's disease. These include knock-out mice for IL-2 (critical for T_{reg} development and survival) and IL-10 (a key anti-inflammatory cytokine). One striking feature of all these models is that they depend on the animal having commensal gut flora. No disease develops in germ-free animals, but if such mice are given commensal bacteria the disease does develop.

One the most informative models available at present depends on the adoptive transfer of CD4 T cells into T cell-deficient rats or mice. If such mice are given unfractionated CD4 T cells they remain healthy, but if they are given only naïve T cells they develop a Crohn's-like inflammatory bowel disease. If are given a mix of naïve T cells and T cells expressing CD25 (the IL-2 receptor), they remain healthy. Thus, the CD25-expressing T cells inhibit the ability of the naïve T cells to cause disease: these are one type of T_{reg}, previously called suppressor cells. See Figure 7.28.

7.4.4.4 Rheumatoid Arthritis

Rheumatoid arthritis is the most important of a group of autoimmune disease in which joints are particularly involved, affecting around 1: 200 of the general population. It is quite different from osteoarthritis in which the joint destruction appears to be due largely to mechanical factors and for which as yet there is no evidence for immune involvement. While rheumatoid arthritis has many features of Type IV hypersensitivity, the roles of antibodies in its pathogenesis are controversial (below) See Case Study 7.10.

Pathogenesis of Rheumatoid Arthritis Histological examination of affected joints shows an infiltration with macrophages. These are thought to be the major secretors of pro-inflammatory cytokines such as TNF-α and IL-1 that are present in the joint tissues. Activated CD4 and CD8 T cells and B cells are also present. As in other examples of chronic inflammation, tertiary lymphoid structures may appear (Section 3.5.3) and these may contain many plasma cells. Whether antibodies or immune complexes play a role in the pathogenesis of the joint damage is, however, still controversial. Two types of antibody are of interest. Rheumatoid factor is a group of antibodies, mainly of the IgM isotype, that are specific for host IgG. These antibodies are not, however, only present in rheumatoid arthritis and can be found in other inflammatory conditions. Antibodies specific for citrullinated polypeptides appear to be much more specific for rheumatoid arthritis, and these polypeptides are present in inflamed joints. Although antibody levels correlate with disease severity, this may be a secondary phenomenon and may not relate to a role in pathogenesis. These antibodies may, however, have a potentially important role in diagnosis: examination of serum samples taken from patients who subsequently developed rheumatoid arthritis has shown that the antibodies may be present long before clinical symptoms appear. Thus it may be possible to identify patients at high risk of developing rheumatoid arthritis, enabling treatment to be started before there is any actual joint damage. See Figure 7.29.

Once again we are unsure why anyone initially develops rheumatoid arthritis. There is a genetic susceptibility and concordance in identical twins is 15–20% (dizygotic twins concordance is 4%), but this is lower than in many other autoimmune diseases. A large number of genes or genetic loci have been identified as potential contributors to susceptibility.

Recipient mouse	Reconstituting cells	Outcome	Interpretation
 T^-	Bulk T cells	Healthy	Normal T cell populations are non-pathogenic
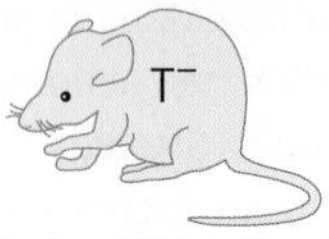 T^-	Effector/memory $CD25^+$ T cells	Healthy	$CD25^+$ T cells are non-pathogenic
 T^-	Naive $CD25^-$ T cells	IBD	Naive T cells contain potentially pathogenic cells
 T^-	$CD25^+$ plus $CD25^-$ T cells	Healthy	$CD25^+$ T cells can suppress potentially pathogenic naive T cells
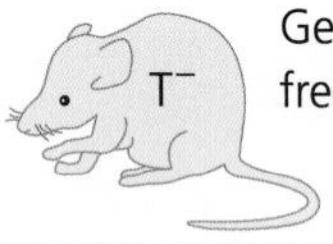 T^- Germ-free	Naive $CD25^-$ T cells	Healthy	Naive T cells become pathogenic only if gut commensals are present

Fig. 7.28 Regulatory T cells in murine inflammatory bowel disease. T cell-deficient mice can be reconstituted by giving them T cells from a normal animal. When CD4 T cells were separated into naïve ($CD25^-$) and effector/memory ($CD25^+$) populations using surface molecule expression, it was found that the transfer of naïve T cells induced an inflammatory disease in the large intestine (inflammatory bowel disease), somewhat similar to human Crohn's disease. If, however, the supposed effector/memory T cells were also transferred, the inflammatory bowel disease was suppressed. The effector/memory population was found to contain CD25-expressing CD4 T cells that had this suppressive function, T_{reg}. If, however, naïve T cells were transferred into germ-free mice, no inflammatory bowel disease developed.

Many of these are involved with the regulation of immune responses that are likely to lead to inflammation. One chromosomal region strongly linked with susceptibility encodes two genes of potential importance: complement C5 and TNF-α. Although we know from therapy that TNF-α is crucial in the disease process, we cannot tell from genetic data whether the risk factor is TNF-α or C5.

7.4.5 Conclusions

The examples of different immune-related diseases given in this section will serve to illustrate what we do and do not know about how these diseases are caused. In some cases, such as myasthenia gravis, Guillain–Barré syndrome and

Case Study 7.10: Rheumatoid Arthritis

Clinical: A 35-year-old woman developed stiffness in the joints of her hands and shoulders. This was worst when she woke up but eased during the morning. When her finger joints began to swell she went to her doctor. A blood test showed the presence of antibodies to cyclic citrullinated peptides. She was treated with a NSAID, which gave her partial relief for a time. Her symptoms worsened, however, and X-rays showed erosion of several joints. She was treated with an immunosuppressant, but her disease continued to progress. She was then treated with an anti-TNF-α therapy (etaenercept. To make Etanercept, the gene for a soluble TNF receptor was joined to the Fc part of a human Ig gene. The gene is expressed *in vitro* as a fusion protein, and following injection will bind secreted TNF and target it to macrophages for destruction). Her symptoms improved dramatically and she is maintained on this therapy with no further deterioration.

Explanation: This woman had developed rheumatoid arthritis. Chronic inflammation in the synovium of her joints led to erosion of the cartilage and bone. Antibodies to cyclic citrullinated peptides are specifically found in rheumatoid arthritis and not present in other autoimmune or inflammatory conditions. She was treated with anti-inflammatory drugs, and finally with soluble TNF-α receptors to mop up TNF-α secreted by activated T cells and macrophages, thus preventing damage. This form of therapy has resulted in dramatic improvements in cases of rheumatoid arthritis that were refractory to all other treatments.

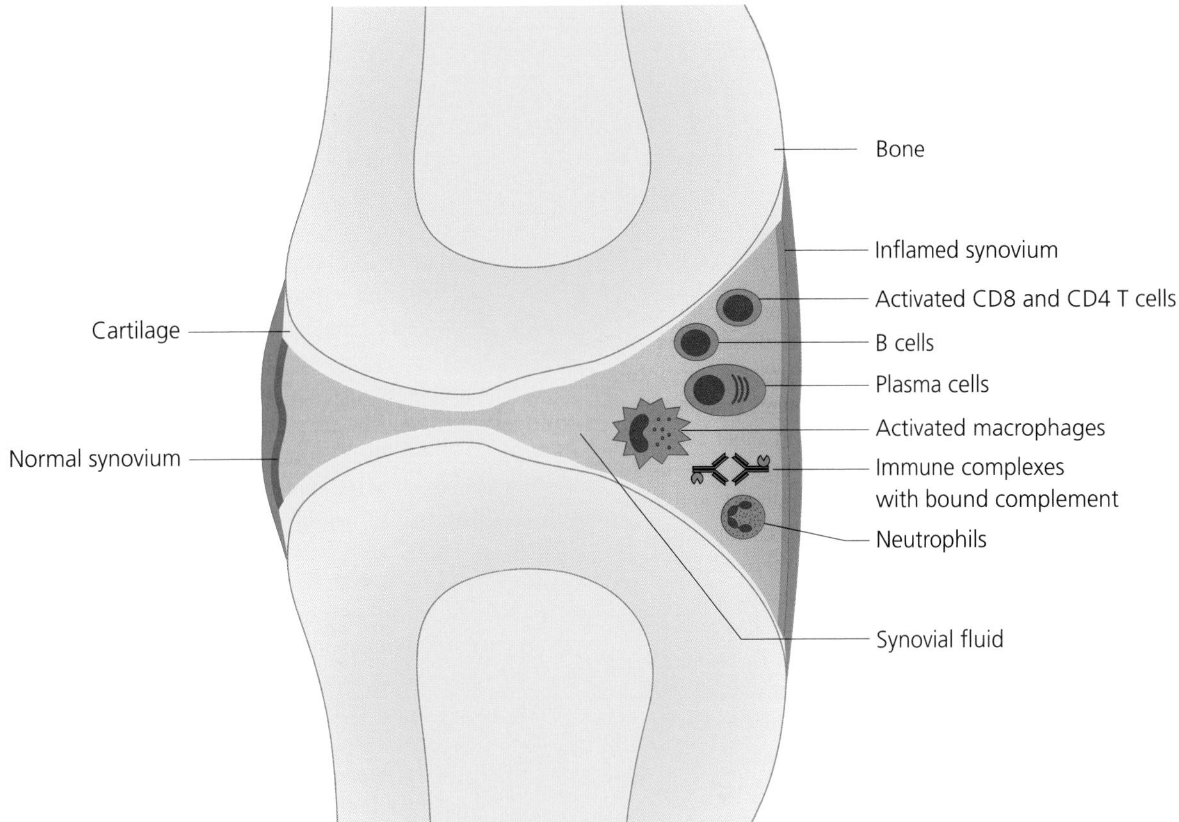

Fig. 7.29 Pathological mechanisms in rheumatoid arthritis. In rheumatoid arthritis, a chronic inflammatory response in the synovium of joints leads to destruction of cartilage and bone. Examination of the affected synovium shown an infiltration with cells which include T and B lymphocytes, plasma cells, activated macrophages and some neutrophils. It thus appears that all the possible effector mechanisms of adaptive immunity are present and determining which is or are the most important in the pathogenesis of the disease is proving difficult.

hyperthyroidism, the effector mechanism is clearly defined antibodies in these cases – and this is crucial in deciding how to treat these patients. In most of these diseases we are, however, still unclear as to the relative importance of antibodies, cytotoxic T cells and activated macrophages in causing tissue damage, but defining the actual roles of these effector mechanisms is crucial for designing new therapeutic strategies.

When it comes to understanding why only some individuals develop these diseases we are on even more shaky ground. In most cases, only a very small percentage of the individuals who possess known risk factors do develop disease. As noted, a recent study examined around 8000 individuals with proven Campylobacter jejuni infection and not one had developed Guillain–Barré syndrome. In the case of ankylosing spondylitis, an arthritic disease of the spine, around 95% of the sufferers express the MHC class I allele, HLA-B27, but only 1–5% of individuals who carry the gene will actually develop ankylosing spondylitis. We also know that concordance for these diseases in identical twins is usually less that 40% (coeliac disease is one exception). Is it infection, epigenetic DNA modification, differences in T cell repertoires or some other factor(s) that are responsible? We just do not know.

You will see that our lack of understanding represents a major problem for scientists attempting to understand these diseases and for clinicians who are treating affected patients. In some cases treatment is simple and effective. A relative of one of the authors who had her thyroid removed because of hyperthyroidism has been taking thyroid hormones for around 70 years, and remains fit and well. In most cases, however, treatment is not so simple and carries potential dangers. Corticosteroids were a wonderful therapy when first introduced for rheumatoid arthritis, but have very serious side effects. The ultimate aim is to identify treatments that are antigen-specific. As we have mentioned, a sublingual pollen extract is an effective treatment for hay fever, and this type of approach is being developed in many animal models and some clinical trials in other diseases. In many diseases, however, we do not know the identity of the antigens that are important in the initiation or progression of the disease and, without this knowledge, antigen-specific therapy is largely guesswork.

Animal models of disease are crucial in these areas. Potentially they can permit the identification of both the effector mechanisms and the antigens important in pathogenesis. Animal models do, however, have many limitations. They often do not resemble the human disease very closely, and the immune systems of mice and man are far from identical. Animal studies can, however, inform human studies very directly. We have used the examples of diabetes and multiple sclerosis in this respect, and this interaction between clinical and experimental studies may represent the most hopeful strategy for the future understanding and treatment of immune-related diseases.

7.5
Transplantation Immunology

Kidney transplantation for renal failure is one of the most successful treatments for a potentially fatal disease and the success rates are much better than most treatments for cancer. Overall, around 70% of renal transplant recipients are still alive 5 years after surgery. There are of course many other situations where a transplant can or could be curative: organs such as the heart, lungs and liver are transplanted routinely, and bone marrow transplants can be very successful in curing immunodeficiency diseases and leukaemias. Thus, there are multiple possibilities for using transplantation in the treatment of disease. The real problem lies in understanding why transplants are rejected and to how to prevent graft rejection. The holy grail is to prevent graft rejection in an immunologically-specific way.

In this section we will discuss the immunology of transplant rejection, focusing on allografts. We will then examine the ways in which transplant rejection can be suppressed or prevented. We will also very briefly consider the foetus since because it is semi-allogeneic, half its MHC genes having come from the father, it is a natural transplant.

7.5.1
Immune Responses to Transplants

7.5.1.1 The Language of Transplantation

Before we discuss the biology of transplantation, there is some terminology (jargon) to clarify.

- A transplant within one individual, e.g. skin from the thigh to cover a serious burn to the face is called an autograft.
- If you need a transplant, and you are lucky enough to have an identical twin, a transplant from your twin is called an isograft. The relationship between the two of you is termed isogeneic or syngeneic. Transplants between genetically identical members of an inbred strain of mice are also isografts.
- Most common transplants are allografts, i.e. they are transplants between genetically non-identical individuals of the same species. In these situations the relationship between the donor and the recipient is termed allogeneic.
- In some cases transplants are made between species, e.g. transplantation of heart valves from pigs into humans and are termed xenografts.
- The phase of the response when the immune response is being initiated is known as sensitization and is closely followed by the effector phase, often resulting in rejection.

Untreated allografts and (most) xenografts are usually rejected because they are incompatible, whereas isografts and autografts are not because they are compatible. See Figure 7.30.

7.5.1.2 Major Histocompatibility Antigens

MHC molecules are the major antigens that determine whether allografts are accepted or rejected. MHC antigens were identified in humans (HLA) and mice (H-2) at about the same time. In mice, the genetic loci responsible for stimulating allograft rejection were identified by classical genetic methods (Section 5.2.1).

Alloreactivity It is not surprising that organs and tissues from a non-identical, allogeneic individual of the same species can be recognized by TCRs because they are foreign. The antigens that stimulate transplant rejection are called transplantation antigens, and MHC molecules are by far the strongest of these. The surprising reason for the strength of the anti-MHC response is the very high frequency of T cells that reacts to other MHC molecules from the same species; it is estimated that 1–5% of all T cells from any individual can respond to an allogeneic MHC (the frequency of T cells reactive against a standard protein is in the order of 1: 10^5 – 10^6). This is the phenomenon of alloreactivity. The only satisfactory explanation for this is that it represents a very high frequency of cross-reactive T cells (Section 7.3.1.2). In other words, many T cells that would normally react with self MHC–foreign peptide complexes are also able to recognize foreign MHC molecules (with or without their bound peptides). T cell recognition of an allogeneic organ can thus lead to a very broad and highly potent attack against the organ – leading to rejection.

7.5.1.3 Minor Histocompatibility Antigens

Multiple minor histocompatibility antigens exist (probably hundreds) and some of these have been identified at a molecular level. In theory, any protein that exists in allelic forms could act as a minor antigen. All that is needed is for antigen processing to be able to generate a peptide from the allelic variant of the donor that is different from the peptide expressed by the graft recipient. Recently some minor antigens have been identified as being mitochondrial in origin, and others are coded for by endogenous retroviruses (all mammals carry many of these viruses quite silently). Males express molecules encoded by genes on the Y chromosome, these of course are not present in females and, male grafts can therefore be rejected by female recipients.

In general, minor antigens stimulate weak rejection compared to the MHC. However, minor antigens can summate to give a strong response. Mouse strains have been generated that possess identical MHCs, but differ for the rest of their genomes. Skin grafts between these strains are rejected vigorously. Minor histocompatibility antigens are important clinically. Even were it possible to match a donor and recipient for all their MHC antigens, it would be impossible to match all the minors.

7.5.2
Transplant Rejection

Transplant rejection is classified into three types, based largely on the time taken for rejection to occur. Hyperacute rejection

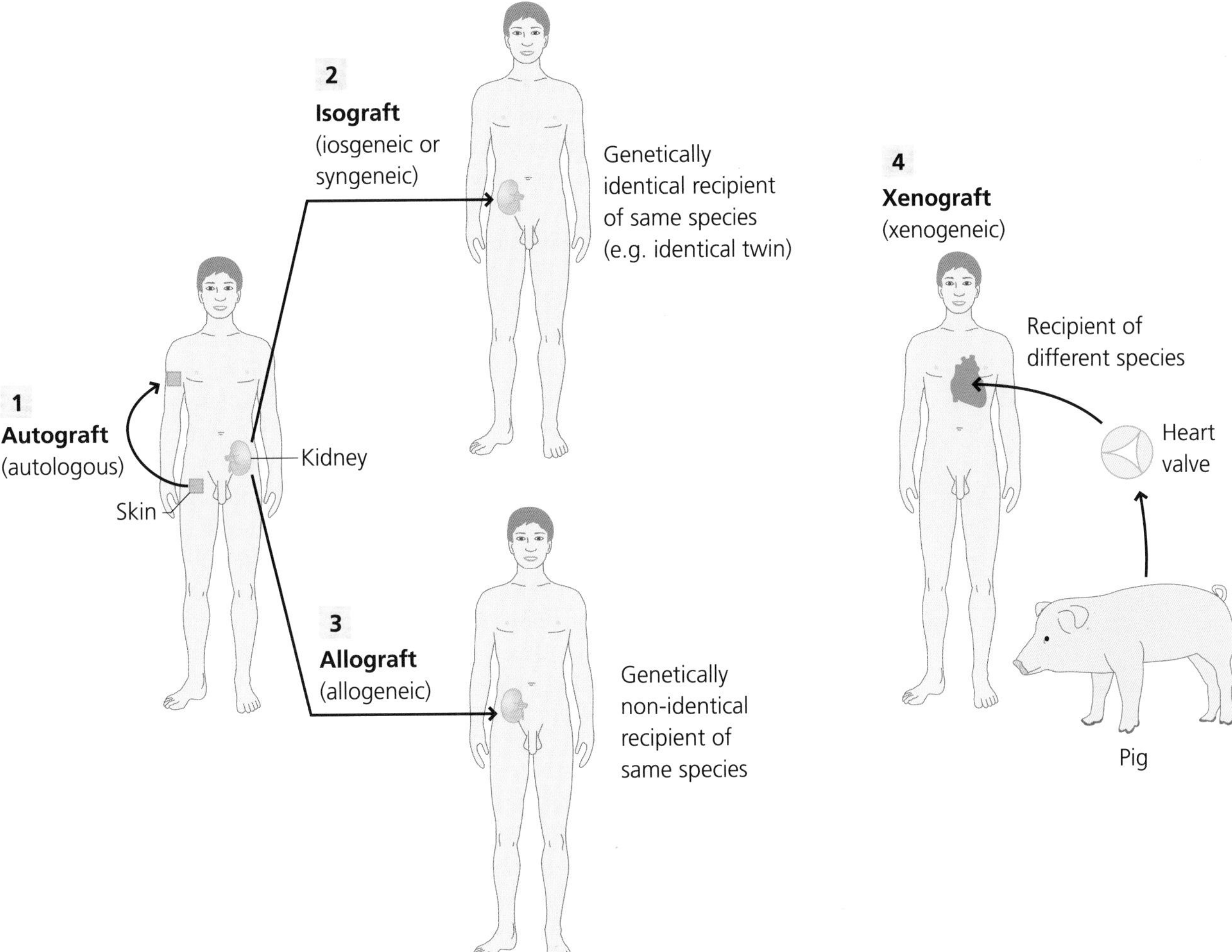

Fig. 7.30 Transplantation terminology. An **autograft** is a transplant made within the same individual. An **isograft** is a transplant between two genetically identical isogeneic or syngeneic individuals, such as identical twins or inbred strains of animals. An **allograft** is a transplant between genetically different, allogeneic members of the same species. A **xenograft** is from one species to another xenogeneic species.

occurs in minutes or hours. You will recall that any immune-mediated phenomenon that occurs so rapidly depends on preformed antibody (Section 7.2.2) the same is true for hyperacute rejection. Acute rejection occurs in days or weeks and represents the standard type of rejection seen in allografts because of the time taken to trigger and execute the immune response. Chronic rejection takes place over months or years: its pathogenesis is only partly understood, but probably involves both antibody- and cell-mediated immunity.

7.5.2.1 Hyperacute Rejection

When a kidney is transplanted and the vascular clamps are removed, the kidney, which was pale because it had been perfused with clear salt solutions to preserve it in a healthy state, becomes darker and redder as blood flows though it. Normally it will stay in this happy state. If, however, a kidney is transplanted by mistake into a recipient who has preformed antibodies against antigens expressed on the endothelial cells of the graft, within a short time (minutes or hours) the graft become dark and urine flow ceases because the kidney has died though lack of a blood supply. Histology shows that the small vessels in the kidney are completely blocked by thrombi, aggregates of blood platelets and fibrin. What has happened? The antibodies in the blood bound to antigens on endothelial cells, which can include MHC molecules, minor histocompatibility antigens or ABO blood group antigens. Complement was activated and there was a rapid accumulation of neutrophils. Neutrophils and/or the products of complement activation caused damage to the endothelial cells, stimulating platelet adhesion and aggregation, and blood coagulation, forming solid masses of thrombus that completely blocked blood flow. This response is similar to the Type II hypersensitivity responses described in Section 7.2.4. There is nothing that can be done at this stage: the kidney is dead and has to be removed. Hopefully, this type of rejection will now never be seen clinically because all recipients are tested for antibodies against donor antigens. Hyperacute rejection is often seen in xenografts and is generally due to preformed natural antibodies which in many cases are specific for certain carbohydrates. See Figure 7.31.

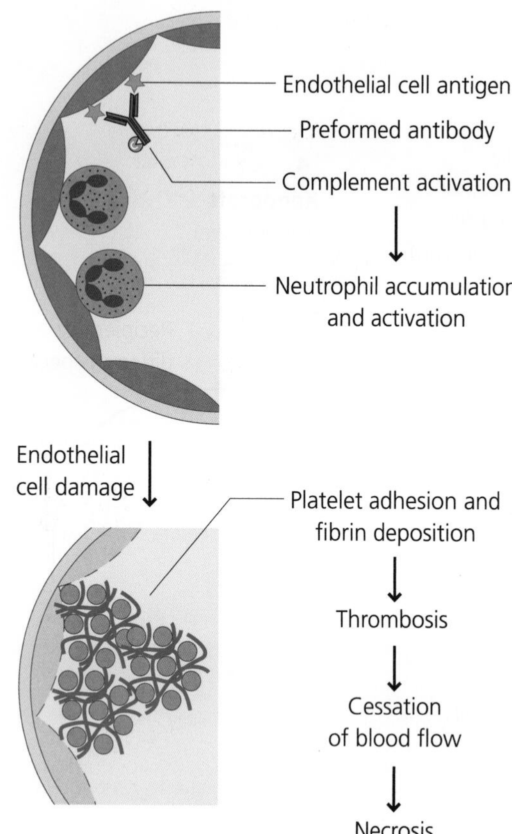

Fig. 7.31 Hyperacute allograft rejection. If the recipient of a vascularized graft such as a kidney possesses preformed antibodies against antigens expressed on the kidney endothelium (e.g. MHC molecules or blood group antigens), the antibodies may bind and activate complement. This leads to endothelial damage (neutrophils may be involved in causing damage), which in turn induces platelet aggregation and fibrin deposition, leading to thrombus formation and occlusion of the kidney blood vessels. The kidney dies of anoxia.

7.5.2.2 Acute Allograft Rejection

If a piece of skin is transplanted onto an allogeneic recipient, it initially heals as if it were syngeneic, and blood vessels and lymphatics grow into the graft over a period of days. By around 8–9 days, however, the graft starts to look abnormal, and by 12–14 days it is shrunken and dead and sloughs off because it has been rejected. Histological examination shows that the first difference between syngeneic and allogeneic grafts is an accumulation of mononuclear cells (lymphocytes and macrophages) around the venules in the latter. With time the numbers of infiltrating cells increase and they become activated. Then the epidermal cells and blood vessels begin to show signs of damage, which increase until the whole graft is necrotic. This pattern of rejection is called acute rejection (for skin grafts it used to be called first-set rejection).

The sequence of events is similar in solid organ grafts except that is it quicker – kidney grafts in rodents can be rejected within 6–7 days. This speed reflects the way in which organ grafts become vascularized since blood flow is established immediately the vascular clamps are removed. This picture of acute rejection occurs when an allograft is transplanted to a naïve recipient and represents the situation with most clinical transplants (of course human transplant recipients are always given immunosuppressive treatment and hopefully the picture of complete acute rejection will rarely be seen). See Case Study 7.11.

Sensitization of the Host to the Transplant The major cell type that can activate naïve T cells efficiently is the DC. Donor-derived DCs are present in the transplant (they are sometimes called passenger cells as they are carried over to the recipient), and they can directly activate alloreactive T cells of the host; this is called the direct pathway). Host DCs can, however, pick up antigens from the graft and activate host antigen-specific T cells; this is called the indirect pathway. The direct pathway is the most important: if thyroid glands are cultured *in vitro* the DCs migrate out of the gland, and if the glands are then transplanted into allogeneic recipients their rejection time is markedly prolonged. Such grafts are not, however, accepted indefinitely, and this must mean that host DCs are acquiring and presenting donor MHC molecules and/or multiple minor transplantation antigens. See Figure 7.32 and Box 7.1.

Effector Mechanisms in Acute Allograft Rejection Nude mice are hairless but also, because of a developmental branchial arch defect, they lack a thymus. Such mice will accept allogeneic skin grafts indefinitely (they will also accept skin from other species – a mouse with feathers is a most unusual sight). The ability to reject skin grafts can be restored by transferring T cells. This shows that T cells are essential for acute allograft rejection but it does not, however, show that T cells are sufficient for rejection. Much work has gone into trying to define the essential rejection mechanisms but there is no unifying answer. Different types of allograft (e.g. skin, solid organ or bone marrow cells) in different species, and even between different strains of the same species (e.g. mice), appear to be rejected by different mechanisms. For example, in T cell-depleted rats and mice, transferred CD4 and CD8 T cell are both able to bring about rejection of skin grafts, but antibodies do not appear to play a significant part even though anti-MHC antibodies are synthesized during rejection. Thus, if a T cell-deficient mouse bearing a healed skin allograft is injected with antibodies to the MHC molecules on the graft, it is not rejected. The same may be true for skin grafts in humans. In experiments that would be difficult to justify ethically today, children with B cell deficiencies, unable to make antibodies, were given allogeneic skin grafts to see if they were rejected. These were rejected as efficiently as those on normal children. What this shows is that antibody is not essential for acute skin graft rejection. It does not however, show that antibody can or does not play a part in such rejection in normal individuals. Acute allograft rejection is therefore generally considered to be a typical Type IV hypersensitivity reaction (Section 7.4.4) or at least this type of mechanism is a strong component of the rejection process. See Figure 7.33.

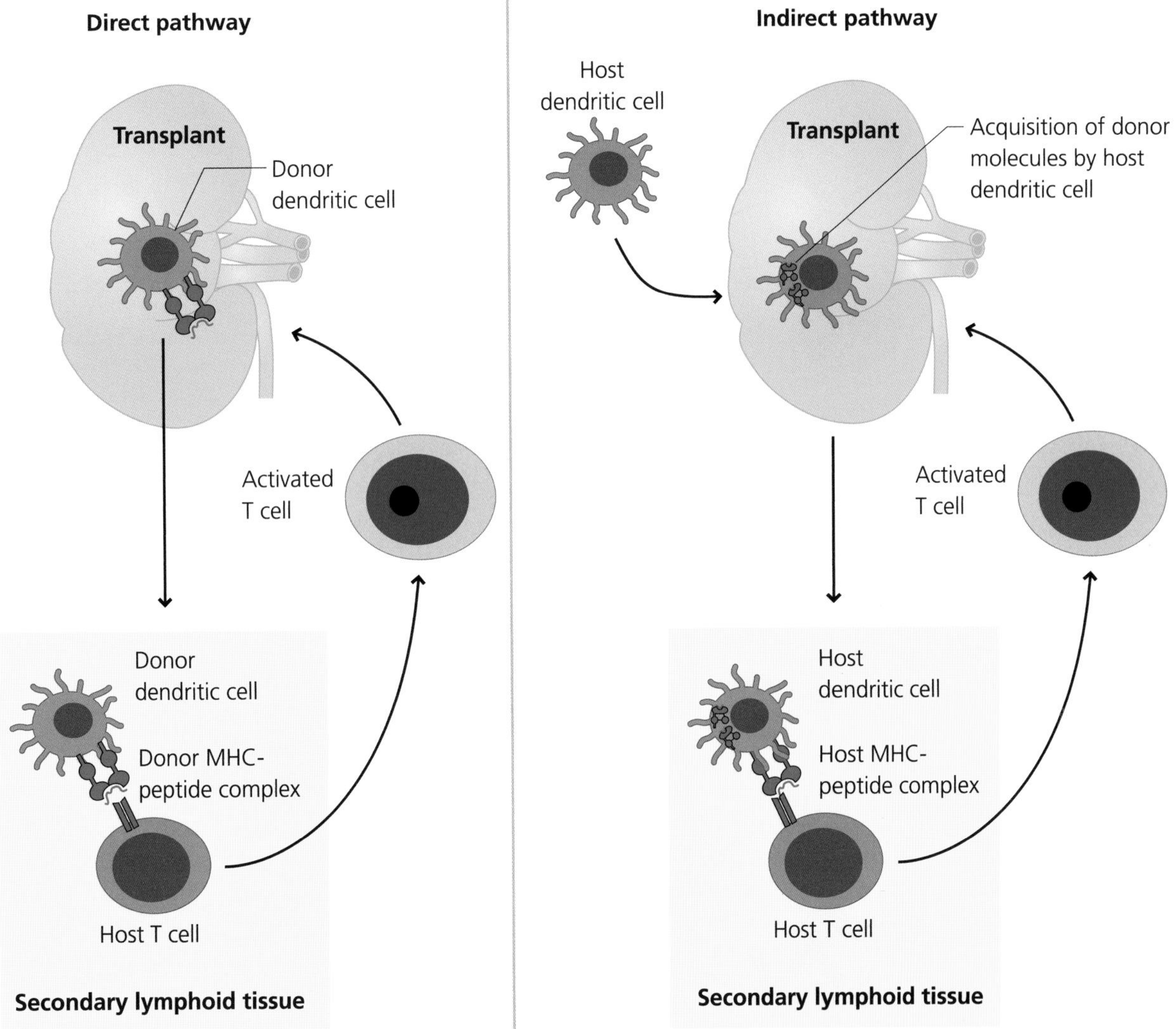

Fig. 7.32 Sensitization to allografts. Direct pathway. DCs (passenger cells) in the transplant migrate from the graft to secondary lymphoid tissues where they stimulate alloreactive host T cells. **Indirect pathway.** Host DCs or their precursors migrate into the transplant from the blood and acquire donor antigens, including MHC antigens. The DCs then migrate from the graft to secondary lymphoid organs where they present peptides from the donor antigens, bound to host MHC molecules, to host T cells. The indirect pathway is the one used for the induction of responses to "normal" antigens, whereas the direct route is unique to the setting of transplants.

Case Study 7.11: Acute Allograft Rejection

Clinical: A 34-year-old woman develops end-stage renal failure and is maintained by kidney dialysis. She has four sibs. She and her sibs are tested to determine which MHC alleles they express. One of her brothers has a closely matched set of MHC alleles, and volunteers to give her a kidney. The operation is successful and she starts to pass urine. She is given steroids and tacrolimus to suppress rejection. Ten days after the operation she develops a fever and the transplanted kidney is tender. Her urine flow decreases. She is treated with high-dose methylprednisolone and her urine flow returns to normal. She is maintained on immunosuppressants and 5 years later remains well.

Explanation: The plumbing needed to replace the kidney is relatively straightforward. The nascent immune response to the graft antigens, which is initiated rapidly after the vascular clamps are released, is partially suppressed by the drugs she is given; tacrolimus inhibits T cell proliferation while methylprednisolone has anti-inflammatory properties. The response is, however, so strong that it starts to attack the kidney, leading to early acute rejection. This requires an extra layer of immunosuppression, mediated by the steroids. As time passes after transplantation, the amount of immunosuppression can be reduced, but in practice it is almost never completely stopped.

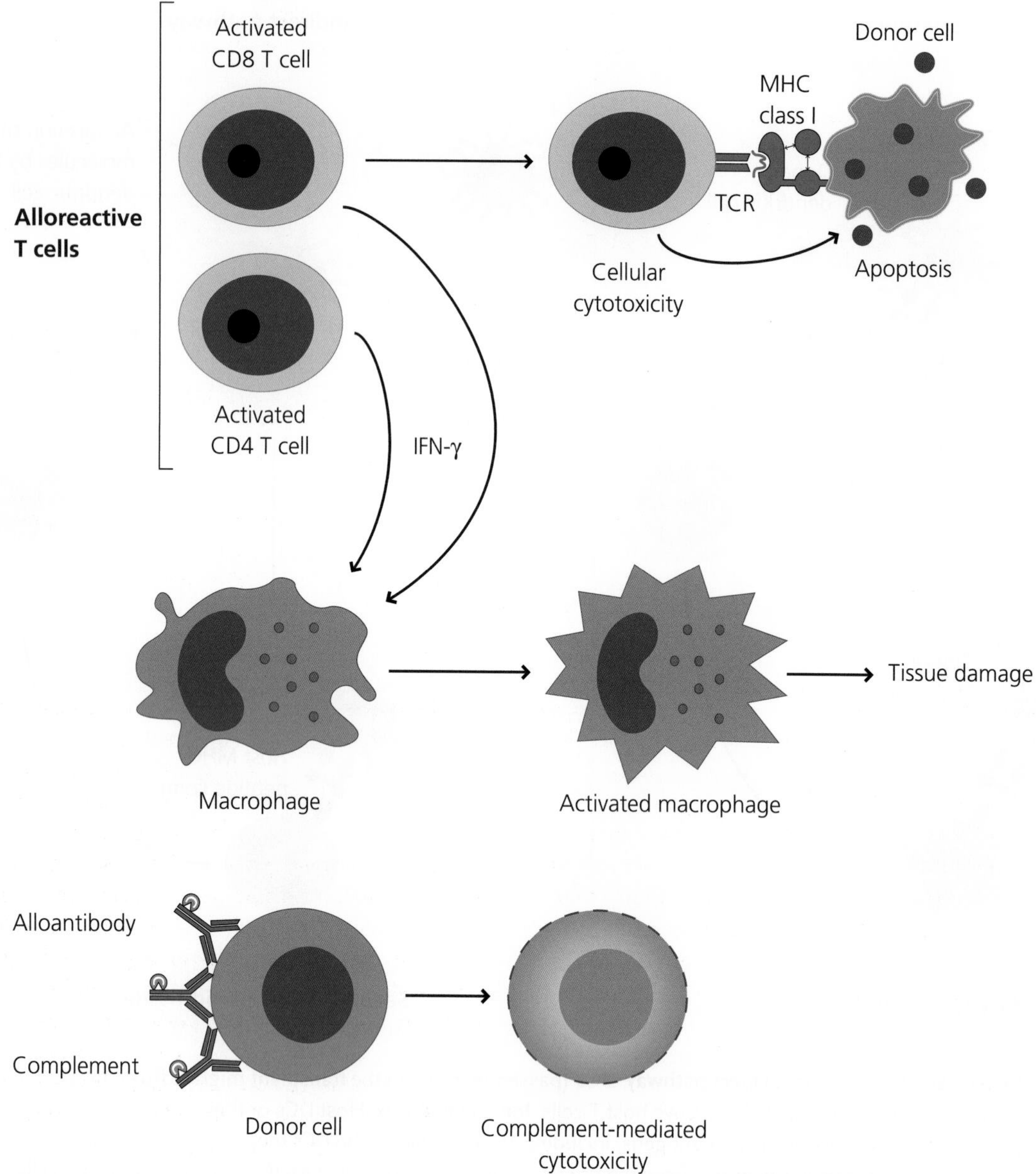

Fig. 7.33 Effector mechanisms in acute allograft rejection. When tissues or organs are transplanted between normal individuals of the same species, rejection takes place over days (organs) or 1–2 weeks (skin). The immune response activates all effector mechanisms of the immune system including CD8 and CD4 T cells, macrophages and antibodies. All of these may be present within the graft and it is not clear which is/are the most important. Antibodies are probably less important than cell-mediated mechanisms, and the strength of the response is such that either cytotoxic T cells or activated macrophages may be sufficient to cause rejection.

Treatment of Acute Rejection HLA matching. Prevention is better than cure. The best way to prevent a transplant being rejected is to have it from an identical twin, who will have identical MHC and minor genes. Sibs are also good donors: if two sibs share a single MHC allele they are likely to have identical MHCs and will also share 50% of the minor antigens. For outbred donors it is best to find ones who share as many MHC alleles as possible, and current evidence suggests that matching for MHC class II gives better survival of the transplanted organ than matching for MHC class I. Thus, the 2-year survival of MHC class II-matched kidney grafts is around 90%, whereas for class I-matched grafts it is around 70%.

Immunosuppression. Drugs which have a selective effect on T cells are the mainstay of immunosuppression. Cyclosporin, a fungal product, binds to cyclophilin in the cytoplasm of T cells and the complex inhibits the action of calcineurin, whose activity is needed for the activation of the NFAT transcription factor, which is required for synthesis of IL-2 and other cytokines (see T cell signalling in Section 5.3.3.2 and Figure 5.15). This activity is much more restricted to T cells, but still leaves the problem of general immunosuppression. Tacrolimus (originally known as FK506) is another fungal product that binds to a different target, called FK506 binding protein, but which also inhibits calcineurin and activation of

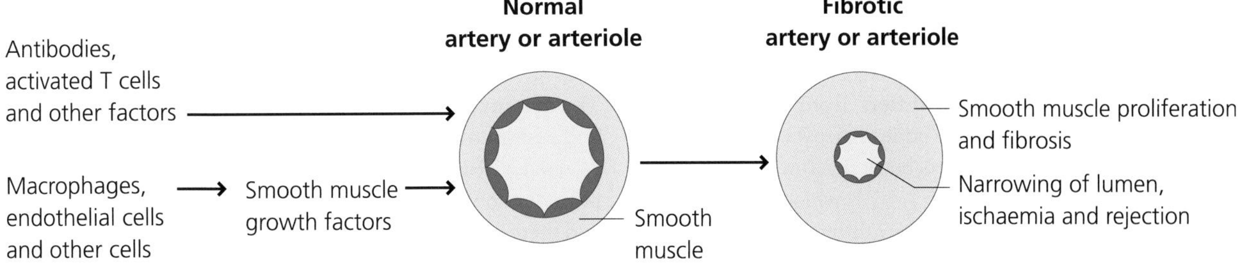

Fig. 7.34 Chronic allograft rejection. Sometimes, an organ graft that has been performing efficiently for a considerable time – even years – can start to fail insidiously. This often represent chronic rejection. The main pathological change in chronic rejection is smooth muscle proliferation and fibrosis (collagen deposition) leading to thickening of the walls of small arteries, diminished blood flow (ischaemia) and organ failure. Both antibody- and cell-mediated immune responses may play a role in pathogenesis, and chronic rejection may represent an excessive healing response. Secretion of growth factors for smooth muscle cells by macrophages and endothelial cells may play a role in the arteriole wall thickening.

NFAT. Another drug, rapamycin (also called sirolimus), also binds FK506 binding protein, but this complex inhibits a different target called mTOR (mammalian target of rapamycin) and reduces cellular proliferation. Nevertheless, in the absence of immunologically-specific immunosuppression, in treating allograft recipients we will be always struggling to maintain a balance between efficacy and toxicity. Moreover, by preventing T cell activation or proliferation in general, some or all of these drugs may in fact also prevent the induction of potentially beneficial T cell responses, e.g. T_{reg}.

Other drug treatment includes corticosteroids, which have non-specific anti-inflammatory properties. Often these are used to treat acute rejection episodes and are not in general used for long-term therapy. Cellular methods focus on depleting defined T cell subsets using monoclonal antibodies. The problem with all these methods of immunosuppression is that they are non-specific since generally all immune responses are inhibited, this leads to greatly increased risks of infection and some cancers (Section 7.6).

In experimental models of allograft rejection, transferring T_{reg} can dramatically increase graft survival. If ways of increasing T_{reg} activity in humans can be devised there is a real possibility of therapeutic advances. Similarly, re-programming the predominantly Th1 allogeneic response towards T_h2 could be beneficial.

7.5.2.3 Chronic Rejection

Not infrequently, a kidney that has been transplanted successfully, and has been functioning well for months and years, starts to deteriorate and over a long period it loses function completely. This is chronic rejection. Histologically the most prominent feature of chronic rejection is thickening of the walls of small arteries and arterioles. This is due to proliferation of smooth muscle cells in the media of the vessel, and the secretion by these cells of collagen and other connective tissue components. The result is narrowing of the lumen of the vessels, and the kidney become ischaemic and eventually dies. The causes of chronic rejection are incompletely understood. However, antibodies may play a role, as may infiltrating cells such as macrophages that can secrete smooth muscle cell growth factors, as also can endothelial cells. Chronic rejection is very difficult to treat. See Figure 7.34.

Box 7.1 Mechanisms of Sensitization to Skin and Vascularized Organ Allografts

For skin grafts, sensitization occurs primarily in the draining lymph nodes. This is shown clearly by experiments done in rats. Skin pedicles were constructed in which a piece of skin was isolated and left connected to the rat only by its blood vessels but all lymphatic connections were disrupted. Allogeneic skin grafts placed on pedicles were retained for very long periods, often more than 100 days, before being rejected. This showed the importance of lymphatics for skin graft rejection. We now explain this as being due to the necessity of DCs migrating from the skin graft into the recipient's lymph nodes to initiate the response.

Q7.25. How might the importance of migrating DCs in skin graft rejection be investigated? Could there be other possible explanations for the non-rejection of alymphatic skin grafts?

In other experiments kidneys were transplanted between outbred (allogeneic) dogs, but the blood vessels of the donor kidney and recipient were connected by plastic tubes and the kidneys were kept in a box on the dog's back to prevent any lymphatic connections. Surprisingly, the kidneys were rejected with the same speed as kidneys transplanted normally. The original explanation put forward for this was that rejection of kidney transplants is triggered within the donor tissue itself, whereas rejection of skin grafts is triggered in draining lymph nodes. We now believe that the former explanation is incorrect.

Q7.26. What alternative explanations might explain why alymphatic kidneys transplants in dogs were rejected normally?

7.5.3 Bone Marrow and Stem Cell Transplantation

Bone marrow transplantation differs from skin and solid organs in that a suspension of bone marrow cells is injected intravenously. The transplanted bone marrow contains haematopoietic stems cells (Section 1.4.1.1) which migrate to the recipient's bone marrow where they differentiate into different blood cells. As with other allogeneic transplants, without immunosuppression, bone marrow transplants are rejected vigorously. Bone marrow transplantation is, however, a well-established, highly successful way of treating a variety of diseases. Rejection of bone marrow transplants is primarily T cell-mediated, as for acute rejection of skin and kidney transplants (above). Unlike solid grafts, however, there is evidence from experimental studies that NK cells, recognizing the absence of self MHC class I antigens on the donor cells, can kill them (Section 1.4.4). Thus, F_1 hybrids between two mouse strains (call them A and B) will accept skin grafts from either parental strain (A or B) because the F_1 T cells have been tolerized to both A and B during their development. Bone marrow transplants from the parental strain A into the (A × B) F_1 may, however, be rejected. It is thought that this represents the F_1 NK cells recognizing the lack of B MHC class I antigens on the bone marrow cells and becoming activated; this phenomenon is called hybrid resistance.

7.5.3.1 Graft-Versus-Host Disease (GVHD)

In some cases when a patient has received a bone marrow transplant, the bone marrow cells are not rejected but the patient becomes unwell: fever, skin rashes, liver damage and diarrhoea are common features. This represents GVHD. In most cases of bone marrow transplantation the recipient is allogeneic to the donor and it is therefore crucial to avoid rejection of the graft. If the transplant is for an immunodeficiency the recipient may be able to accept it without immunosuppression, but in other cases the recipient will need to be immunosuppressed. Bone marrow, however, contains mature T cells from the donor; these can recognize host MHC antigens and will be activated to them. The activated T cells can then cause the tissue damage associated with rejection, but in this case it is the grafted T cells that are effectively rejecting the host. There is also evidence that donor NK cells in the bone marrow can have a beneficial effect since they appear to recognize host haematopoietic cells selectively and may kill host APCs, inhibiting the activation of donor T cells. See Figure 7.35.

7.5.3.2 Graft Versus Leukaemia Effect

To prevent GVHD T cells can be removed from the bone marrow cells before they are transferred. Bone marrow transplantation is frequently used as part of the treatment of leukaemia, and clinical data shows that survival of the transplanted cells may be increased if T cells are not removed. In these cases the donor T cells may act to kill remaining leukaemia cells this phenomenon is called the graft versus leukaemia effect.

7.5.4 The Foetus as an Allograft

The mammalian foetus is an allograft because half of its MHC and minor antigens are from the father, yet it is not usually rejected. Clearly this non-rejection is crucial for successful reproduction, but why the foetus survives is still a major puzzle and there are no clear-cut answers. Many hypotheses have been put forward, but none has yet proved sufficient to fully explain foetal survival. These hypotheses include the following. There could be a mechanical barrier blocking cell and molecular traffic between the mother and foetus? This is unlikely because maternal cells can be found in the foetal circulation and vice versa. Additionally IgG can cross the placenta freely. There could be suppression of maternal immunity. This is unlikely because pregnant women do not generally show an increased incidence of infectious disease. It may however be significant that some autoimmune diseases are alleviated during pregnancy. Access of foetal MHC antigens to the mother could be prevented? Some foetal tissues such as the trophoblast, which is most directly in contact with the mother – do not express MHC antigens. Could there be active inhibitory mechanisms? There is a population of NK cells in the placenta that may inhibit the activity of maternal immune cells. Recently, it has been shown that indoleamine deoxidase (IDO) is expressed by placental cells. IDO depletes tryptophan from local environments and causes cell starvation and this may inhibit T cell activation and effector function. Overall, although many mechanisms have been suggested, it is still not clear why foetuses are not rejected. It may be that because non-rejection is so vital to survival of the species, many different mechanisms cooperate to bring this about, so that the absence of any one mechanism will not lead to the foetus being rejected.

Q7.27. How may further elucidation of mechanisms preventing foetal rejection be helpful in other settings?

7.6 Tumour Immunity

As is true for all immunology, tumour immunology is a highly complex and developing field, but the possibility of using immune responses to reject growing tumours is a very attractive possibility and is being widely researched at present. We will try to introduce some basic concepts that we feel are helpful and informative.

Tumours are clonal expansions of cells that can divide indefinitely (they are immortal) and whose growth is unregulated, their growth is independent of external growth factors; they are autonomous. Tumours can be solid masses (e.g. carcinoma of the breast, sarcoma of bone) or diffuse (as in leukaemias, tumours of leukocytes). Tumours may be benign or malignant (cancer). Benign tumours are localized and

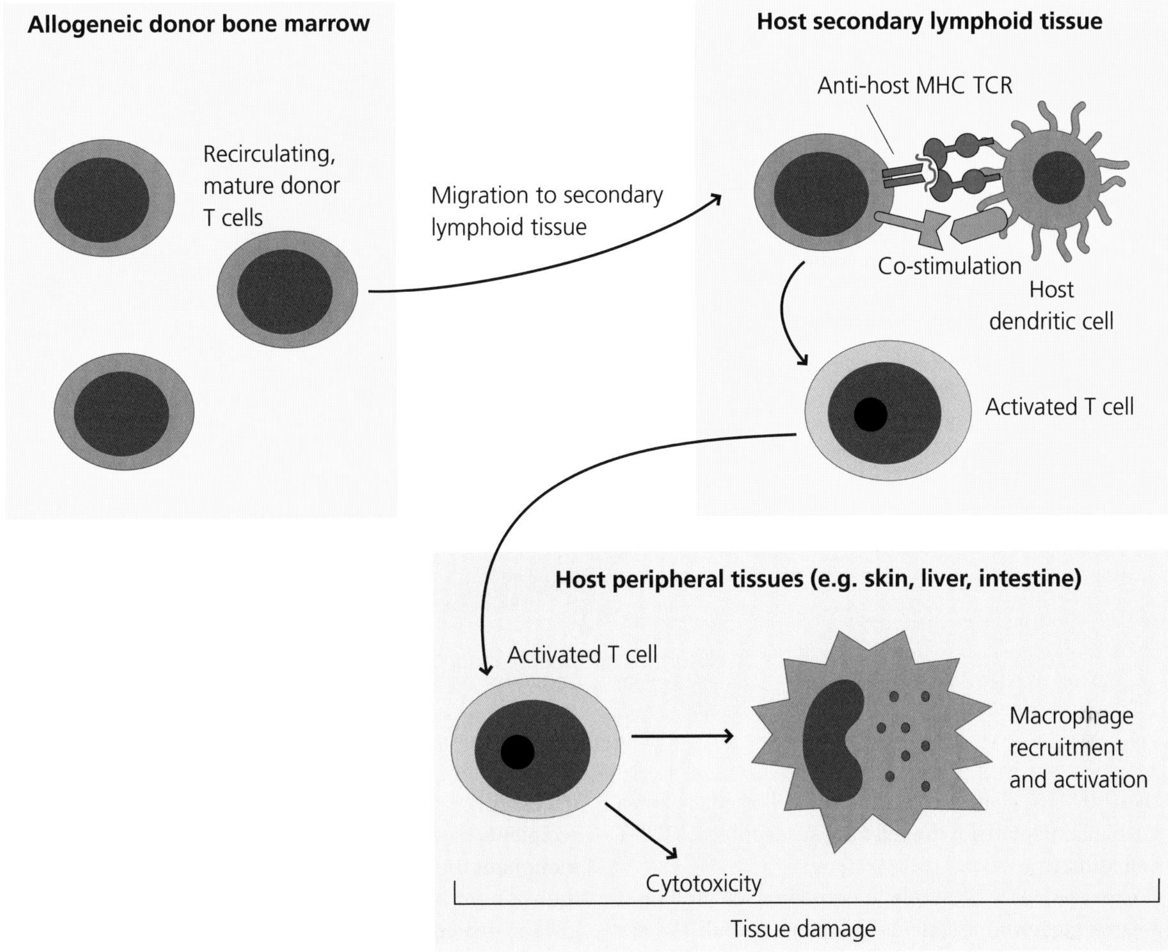

Fig. 7.35 Graft versus host disease (GVHD). Transplanted allogeneic bone marrow contains donor-derived mature T cells that are not rejected by the host. Some of these T cells are alloreactive and recognize and are activated by host MHC molecules. The activated T cells then migrate to tissues such as the skin, liver and intestines, where they induce T cell-dependent tissue damage. Because they are in effect rejecting the host this phenomenon is called GVHD.

because they cannot invade normal tissues, do not spread outside the local site. In general, benign tumours are not a serious clinical problem (a benign tumour deep in the brain is however not, good news). Malignant tumours are able to infiltrate (invade) normal tissues, and because they can invade into lymphatic and blood vessels, they can spread to other sites (metastasize). It is metastases that kill patients.

7.6.1 Tumour Antigenicity

Tumours are essentially self so should we not perhaps be tolerant to all the molecules expressed by tumours? It is, however, clear that tumours can be antigenic. Thus, patients and tumour-bearing animals may express antibodies that recognize tumours, and T cell clones specific for tumours can be grown from tumour-bearing animals. Why should tumours be antigenic? There are several ways in which this may happen.

i) Mutations. All tumours carry many mutations, maybe hundreds or thousands, some of which are for them to escape growth control. If a mutation in one of these genes leads to a novel peptide being generated by antigen processing, the tumour may become immunogenic.
ii) Some tumours in adults re-express molecules that are normally only found in the foetus (and adult testis). These are the oncofoetal antigens. Immunological tolerance depends on a continuous supply of antigen being available to the immune system; new T cells are being formed continually and can only be tolerized if they see antigen. Thus, adults may not be tolerant to onco-foetal antigens.
iii) Some tumours are induced by viruses. Even if the viral genome is incorporated into the host's, viral proteins may be made which could serve as sources of foreign peptide.
iv) We are not tolerant to all cellular proteins, for example some are present in too small an amount to induce tolerance. If, however they are expressed in tumours and, T cells could be activated by immunizing tumour-bearers with these proteins, there may be enough peptide expressed on tumour MHC molecules to permit an activated T cell to respond. In some cases less

Fig. 7.36 Tumour antigens. Tumours may express antigens derived from at least three different sources. If the tumour is associated with viral infection, antigens from the virus may be expressed. All tumours require multiple mutations to develop their typical characteristics. These mutations may generate novel peptides that can be recognized by T cells. In some tumours, genes that are normally only active during early development of a tissue or organ may be re-activated e.g. synthesizing "oncofoetal" antigens. (In other cases, there is dysregulation of normal cell antigens, e.g. leading to over-expression; not shown.)

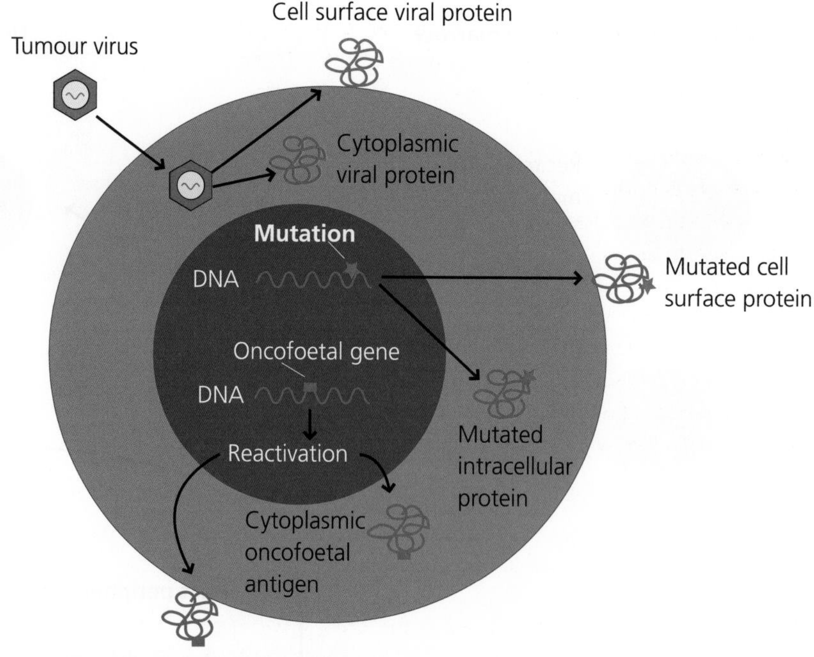

than 10 MHC class I molecules on a cell need to carry a particular peptide for the cell to be susceptible to CD8 T cell killing.

v) We may be able to break tolerance to some self proteins if we give the immune system appropriate stimuli. If these proteins are not essential for life, such as proteins associated with melanin synthesis in melanocytes, these may serve as targets for T cells to destroy melanomas; they may also kill the normal cells, but it may be judged better to lose these cells if there is a chance of eliminating the tumour. Thus, melanoma patients given immunotherapy may lose pigmentation of their skin (vitiligo). See Figure 7.36.

Thus, there are multiple ways in which tumours may be antigenic. What is crucial is to identify those antigens that might be targets for an immune response that will lead to rejection of the tumour, and to determine how to maximize the effectiveness of that response.

7.6.2
Tumour Immunosurveillance

It has long been suggested that the immune system may recognize individual tumour cells at an early stage and prevent them developing into clinical tumours. This is termed tumour immunosurveillance. If immunosurveillance is important in defence against tumours, we would predict that certain immunodeficiencies would lead to an increased incidence of the disease. Certainly there is an association between some immunodeficiency states and an increase in tumours. However, an association does not necessarily mean a causal relationship. Kidney transplant recipients have a much higher incidence (up to 100 times) of malignant tumours, but in most cases these are tumours of the skin or lymphoid system. Transplant recipients are always treated with immunosuppressive drugs which are designed to inhibit the immune response and which may also induce mutations in lymphoid cells. Acquired immunodeficiency syndrome (AIDS) patients show a high incidence of Kaposi's sarcoma. This tumour, however, is very strongly associated with a human herpes virus. Are the tumours arising because of a lack of tumour surveillance or because without CD4 T cells, the viral infection cannot be cleared? Another example is squamous cell carcinoma of the skin, which is seen much more frequently in recipients of solid organ transplants than in the normal population. In many cases there is evidence that such tumours are associated with the human papilloma virus (HPV) or EBV. It seems quite possible that all tumours seen in immunosuppressed transplant recipients could be of viral origin – if this is the case, then the defect is that the immune system cannot get rid of the virus rather than not being able to reject the tumour.

7.6.3
Tumour Immunity and Darwin

Tumours have high mutation rates. Any mutation that increases a tumour cell's ability to survive will give that particular mutant and its progeny a selective advantage, and the mutant clone will come to dominate the tumour. If the tumour is immunogenic, any immune response against the tumour will act as a selective pressure, favouring the survival of mutants that can avoid or evade that particular response. Thus, over time, tumour variants will be selected that have become resistant to everything that the immune system can throw at them. Exactly the same principles apply to tumour treatment with chemotherapy. This is one reason that tumours are often treated with multiple drugs since the probability of a cell generating sufficient mutations to resist multiple drugs is much less than for single drugs given sequentially. One implication of this selection by the immune system is that a late stage tumour will be resistant to immune attack and, if so what chance is there then for immunotherapy? If immunotherapy is to succeed, patients should be treated at as early a stage as possible, before the immune system has selected for highly-resistant variants. See Figure 7.37.

7.6.4 Evasion of the Immune Response by Tumours

To survive, potentially-immunogenic tumours must have acquired the ability to evade or avoid the immune response in a variety of ways, not surprisingly, these are remarkably similar to those used by pathogens to avoid the response. Many mechanisms of evasion have been described in animal models. These include, amongst others, the secretion of immunosuppressive cytokines such as IL-10 and TGF-β, decreased expression of MHC class I or costimulatory molecules, induction of T_{reg} and interference with TCR signalling. An important question is, however, to what extent these tumours resemble tumours arising spontaneously in humans. It is of course more difficult to do such studies in humans but there is evidence, for instance, of human tumours secreting TGF-β and recently, CD4 $CD25^+$ T_{reg} specific for tumour antigens have been isolated from human melanomas. That these tumours do develop evasion mechanisms is in itself very strong evidence for the existence of anti-tumour immunity and for the selection of variants by the immune response. See Figure 7.38 and Box 7.2.

Q7.28. Might the observation that tumours down-regulate immune responses to their own antigens suggest approaches to the treatment of tumour-bearing patients?

7.6.5 Immunotherapy for Cancer

In animal models it has proved relatively easy to immunize mice to a tumour if this is done before the tumour is transplanted and this may prevent the tumour from developing. This can be done by a variety of ways, but its relevance to human tumour is distant there are few if any conditions in which we can predict with certainty that an individual will develop a tumour. There are some however: for example women with mutations in the BCRA-1 and BCRA-2 genes have such a high risk of developing breast cancer that many such women have their breasts removed surgically to prevent the disease. In such cases an effective prophylactic vaccine would be hugely beneficial. In another context, vaccination against hepatitis B, which causes hepatic carcinoma, has been very successful in preventing this tumour. Another situation where prophylactic vaccination is being used is in prevention of cervical carcinoma by vaccinating young women against HPV that is strongly associated with this tumour. It is important to realize, however, that in both the latter cases the vaccination is preventing infection by the virus, not stimulating immunity directly against the tumour.

The major goal of tumour immunotherapy is therapeutic vaccination, immunizing in such a way that a pre-existing tumour is rejected. Many trials using different approaches are underway. These include the use of antibodies, the expansion of anti-tumour T cells *in vitro* and their re-transfusion, and the injection of DCs that express tumour-derived antigens. None of these trials are at present showing general efficacy. In many cases a proportion of patients show some clinical improvement, and in rare cases the tumours have regressed completely, but the overall success rate is low. In all these trials, it is end-

Box 7.2 Evasion of Immune Responses by Tumours

The Meth A sarcoma is a mouse tumour induced by treating mice with methylcholanthrene, a potent carcinogen. If normal mice are injected with Meth A tumour cells, a tumour grows that will kill the mouse. If a second injection of tumour cells is given soon after the first, the second injection does not cause a second tumour: this is known as concomitant immunity. Over time after inducing the first tumour, however, concomitant immunity becomes weaker and eventually disappears altogether, and the second injection gives rise to a tumour very efficiently.

If tumour cells are injected into T cell-deficient mice, again the tumour grows progressively and kills the animal. In one early series of experiments normal mice were made immune to the tumour by injecting endotoxin, and this caused complete regression of the tumour, and subsequent injections of tumour cells into these mice did not induce tumours. T cells from these endotoxin-treated mice were then injected into T-deficient mice bearing large Meth A tumours. The T cells induced complete rejection of the tumour. If, however, tumour-bearing T cell-deficient mice were injected with a mixture of T cells that could induce rejection, together with T cells from a mouse in which concomitant immunity had waned, the tumours were not rejected but continued to grow and killed the mice. Thus, T cells from the tumour-bearing mice were able to suppress the ability of cells from the immunized mice to reject the tumour. This was a very convincing demonstration of suppressor T cells – now called regulatory T cells (T_{reg}).

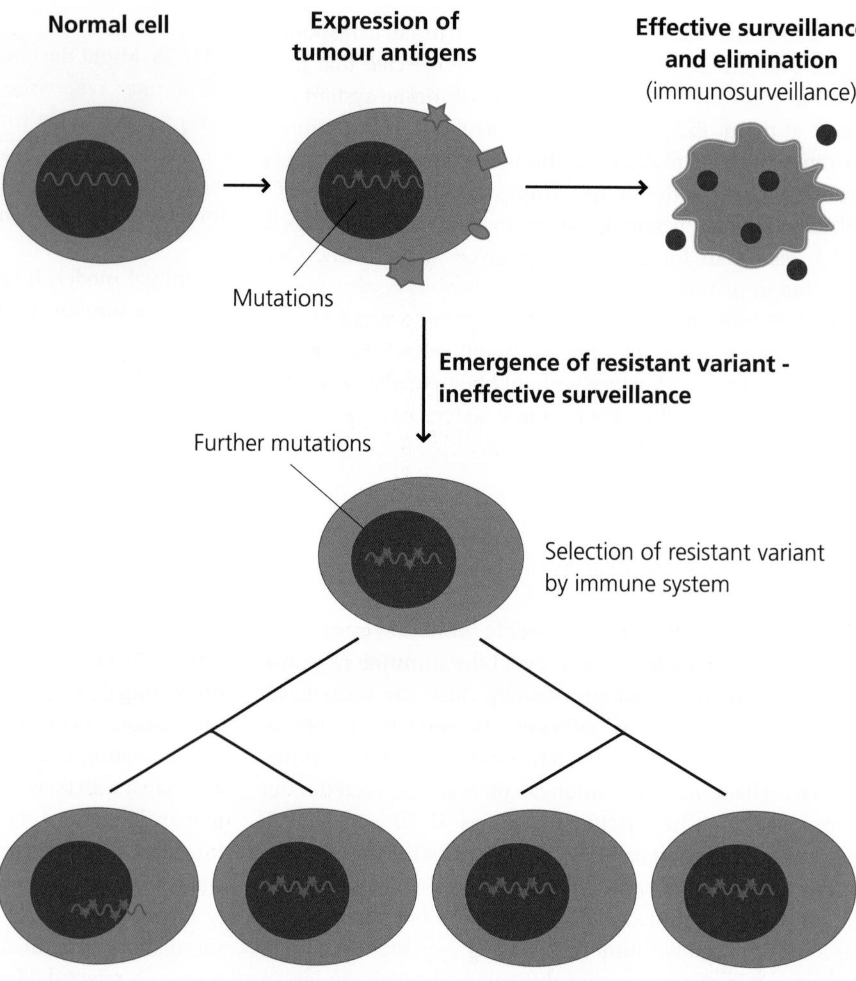

Fig. 7.37 Stages in tumour immunity. As tumours progress they can express tumour antigens that induce T cell and or NK cell responses for example. These may destroy tumour cells before a tumour is clinically apparent; this is sometimes called the elimination phase. The tumour cells will, however, continue to mutate. At a certain stage, it is possible that the immune system can kill mutated cells as they arise and an equilibrium phase is reached. However, any mutation that enables a mutant cell to evade or avoid an immune response that would kill the tumour cell will give the mutant cell a selective advantage, and the mutant clone will outgrow the other tumour cells, this is now the escape phase. Over time a tumour will have been selected by the immune response to become resistant to all effector mechanisms.

stage patients who are being treated and, as discussed above, in such patients the tumours will have evolved by selection against immune responses to become resistant. If these therapies are to succeed they will have to take into account all the evasion strategies that the tumours have evolved during their development.

Q7.29. What might be some limitations of cell-based therapies?

7.7 Conclusions

Lymphocyte antigen receptors in the adaptive immune system cannot inherently distinguish between what is potentially harmful and what is not. As a result of this, the immune system has had to evolve indirect mechanisms to avoid reactivity to non-harmful antigens, e.g. by getting rid of potentially harmful lymphocytes (tolerance) or by preventing their activation (regulation). These mechanisms are not, however, 100% effective. The results are the immune-related sensitivities and autoimmune diseases, both representing major clinical problems. Although there are hints that antigen-specific immunotherapy may be a realizable goal, at present there are very few examples of this in clinical practice. Many current therapies are not antigen-specific and carry important and potentially life-threatening side-effects, such as increased risks of infection. The design of effective antigen-specific therapies that may involve the effective induction of antigen-specific tolerance or regulation will require a much better understanding of immune response initiation and regulation than we have at present. We are of the opinion that such understanding will only come from integrated approaches involving both experimental and clinical studies.

Transplantation is not a disease, but a most effective form of therapy for many diseases. Rejection is of course the major problem and, as with autoimmune diseases, the development of antigen-specific suppression of rejection is a major goal. Developing such strategies for transplantation will inform the treatment of autoimmune diseases and vice versa. Malignant tumours pose an opposing set of problems. If human tumours are immunogenic – and the evasion strategies they induce during their development is very strong evidence that they are

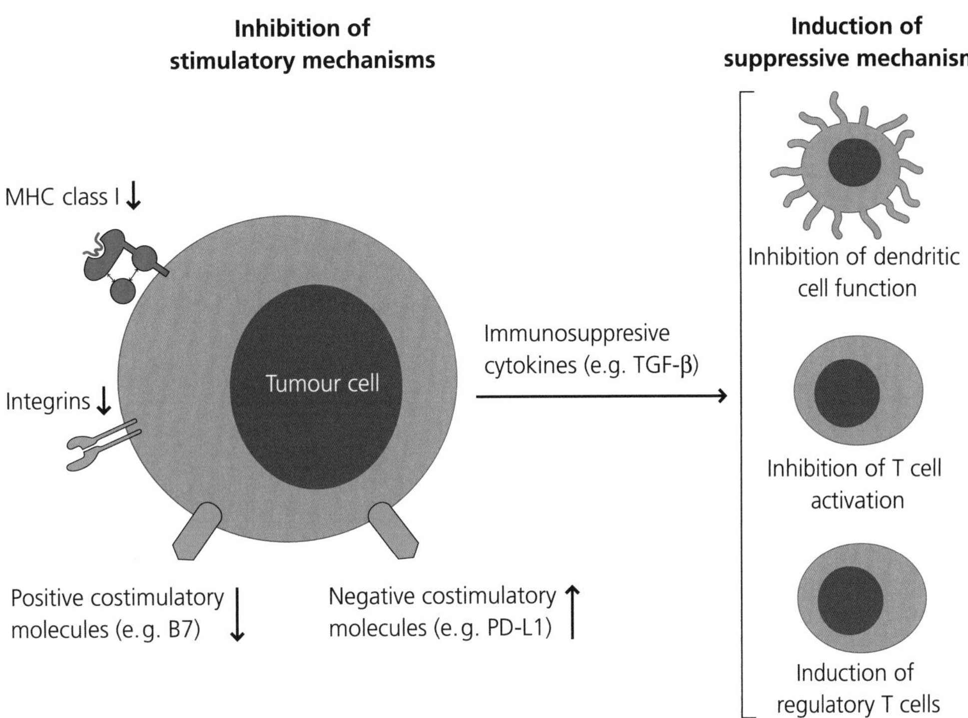

Fig. 7.38 Tumour immune escape mechanisms. Many mechanisms have been identified in tumours that assist their evasion of immune response. Tumours may decrease expression of MHC class I and other molecules involved in immune activation (e.g. adhesion and costimulatory molecules) and increase expression of negative costimulatory molecules. Tumours may secrete inhibitory cytokines and other molecules that inhibit DC functions, induce the formation of T_{reg} and inhibit T cell activation.

Learning Outcomes

By the end of this chapter you should be able to understand, explain and discuss the following topics and questions – the relevant sections of the chapter are indicated. You should understand some of the evidence from human and animal studies supporting what we know about these topics. You should have some idea of the areas where our understanding is incomplete. You may be able to suggest ways in which our understanding could be advanced.

- Immunity, disease and therapy (Section 7.1)
  - In what general ways can adaptive immune responses cause tissue damage?
  - What general types of antigen are involved in harmful immune-mediated responses?
  - What general types of therapy can be used to treat immune-related diseases?
- What are the mechanisms of tissue damage caused by the immune system? (Section 7.2)
  - What are the different types of hypersensitivity reactions and what types of effector mechanisms are involved in each?
- Why do we make harmful immune responses to normal harmless antigens? (Section 7.3)
  - What do we know about roles of lymphocyte antigen recognition in harmful responses?
  - What do we know about the roles of lymphocyte activation in harmful responses?
  - What do we know about the identity and functions of genes that predispose or protect against harmful responses?
- Immunopathology and immunotherapy in action (Section 7.4)
  - What are the similarities and differences between harmful immune responses against extrinsic and intrinsic antigens
  - How do allergies differ from other immune-related sensitivities?
  - What types of disease can be caused by antibodies?
  - What types of disease can be caused by immune complexes?
  - What types of disease can be caused by cell-mediated reactions?
  - What specific examples are there of immune-related diseases where the causes and/or mechanisms of damage seem clear?
  - What specific examples are there of immune-related diseases where the causes are unknown and/or where mechanisms of damage overlap?
  - What specific examples are there of immune-related diseases that can or cannot be treated effectively?
- Transplantation immunology (Section 7.5)
  - How is acute rejection initiated and how does it damage transplants?

  - What are the types of rejection and how do they cause damage?
  - How might graft rejection be treated or prevented?
  - What are graft versus host reactions and how are they similar to or different from host versus graft reactions?
  - What might prevent rejection of foetuses?
  - Can the hypersensitivity mechanisms identified by Gell and Coombs be used to explain transplant rejection?
- Tumour immunity (Section 7.6)
  - What kind of immune responses are made to tumours?
  - What are the different types of tumour antigen?
  - Does tumour immunosurveillance exist?
  - How might tumours escape immunity?
  - What new approaches are being tried for cancer immunotherapy?
- INTEGRATIVE. How useful is the Gell and Coombs classification of hypersensitivity reactions in different settings? Where would we be now if they had never lived?
- GENERAL. To what extent might infections play a role in triggering different types of immune-related diseases?

immunogenic – the problem is overcoming these evasion strategies. This might be done, for example, by suppressing the activity of regulatory cells or by boosting or redirecting the host immune response against the tumour, such as by therapeutic DC vaccination.

It is clear that the evolution of the adaptive immune system has been crucial for defence against infection, even though it has also driven the evolution of avoidance or evasion mechanisms in pathogens – an arms race. The imperfections of the adaptive system are apparent to all of us. We cannot tell if there might have been other ways of evolving adaptive immunity, or even if we really needed to evolve such as system. After all, the great majority of animals do very well without an adaptive system. Our adaptive system is, however, the one we have to live with. Continuing to increase our understanding of how it works is absolutely crucial, both for the prevention of infectious disease and for the treatment of the diseases that involve malfunctioning of this system.

Answers to the Questions

Answers to the Questions in Chapter 2

Q2.1 How far is it valid or useful to think of the immune system as a sixth sense?

A2.1 It has receptors for external stimuli, responds adaptively and appropriately to these stimuli, and its responses include central processing of information. There is also memory for a particular external stimulus. We do not know if this helps our understanding of immunology!

Q2.2 If influenza only infects the respiratory epithelium, why do we feel so generally unwell when we have flu?

A2.2 The influenza virus activates the innate immune system via its PAMPs. This results in the systemic release of cytokines such as IL-1, Type I IFNs and TNF-α, which cause the general symptoms. You might want to ask yourself at this point how these systemic symptoms might contribute to defence against infection.

Q2.3 If immunodeficiencies lead to so much death and disease, why has evolution not selected for their elimination?

A2.3 Many of these diseases are X-linked or autosomal recessive disorders. In these cases the majority of female or other individuals carrying a single mutant gene respectively will not show symptoms and will breed as successfully as those not carrying the mutant gene, passing it on to the next generation.

Q2.4 Another important function of complement is to clear immune complexes from the bloodstream. Individuals lacking some particular complement components often present with skin rashes and, in some cases, severe kidney damage. Why might this be?

A2.4 If immune complexes are not cleared effectively they may be deposited in small blood vessels and initiate inflammation. Specific complement components needed for immune complex clearance are often defective in these conditions. That such diseases develop in the absence of these components is strong evidence for their importance in normal clearance. This is discussed further in Section 7.2.5.

Q2.5 Why might neutrophils play a lesser role in host defence against *intracellular* than *extracellular* bacteria?

A2.5 Bacteria such as mycobacteria replicate intracellularly and are very resistant to killing in this niche. Neutrophils are short-lived cells that probably cannot generate sufficiently strong killing mechanisms over a sustained time period to kill mycobacteria that they might have phagocytosed. There is, however, evidence from studies in mice that neutrophils can be important in the early stages of defence against another intracellular bacterium, *Listeria*.

Q2.6 Why might macrophages play a lesser role in defence against *extracellular* bacteria?

A2.6 This is a numbers game. Monocytes are present at a much lower frequency in the blood than neutrophils and thus cannot be recruited at the same rate. Additionally, there are not large additional stores of monocytes (some may be stored in the spleen) available for rapid deployment, and there are fewer precursor in the bone marrow, preventing rapid increases in their production.

Q2.7 Why might NK cells play lesser roles in host defence against viruses *other than* the herpes viruses?

A2.7 This is not really clear (at least to us). Presumably there are other mechanisms that can mediate effective defence against the early stages of infection with other viruses.

Q2.8 Might you expect NK cells to play a role in host defence against *tumour cells* as well as viruses?

A2.8 Probably yes. Tumour cells that express new antigens (Section 7.6.1) may be susceptible to attack by cytotoxic CD8 T cells. Any tumour cell that has decreased expression of MHC class I molecules will have a selective advantage with regard to T cells, but may become susceptible to NK cell attack since these cells monitor the levels of MHC class I expression, and can kill their targets if these are reduced or absent (Section 1.4.4).

Q2.9 Why might it be advantageous to the host not to polarize fully all CD4 T cell responses?

Exploring Immunology: Concepts and Evidence, First Edition. Gordon MacPherson and Jon Austyn.

A2.9 If a particular pathogen always induced a fully polarized response, any mutations in that pathogen that enabled it to evade that response would give the pathogen a selective advantage. By not having full polarization, such mutants could presumably still be attacked by other arms of the response.

Q2.10 What might be the selective advantage to Salmonella of inducing intestinal inflammation?

A2.10 This profuse diarrhoea will greatly increase the probability of contaminating drinking water and increase the probability of spreading infection to other hosts.

Q2.11 In contrast to most infections, patients who recover from tetanus are usually not immune to re-infection and need to be actively immunized. Why might this be?

A2.11 The amounts of tetanus toxin required to kill the patient are very small and there is not enough released to stimulate an effective, protective immune response.

Q2.12 Normal individuals possess IgM and IgG antibodies against capsular polysaccharides of *Streptococcus pneumoniae.* Given that such polysaccharides are T-independent (TI) antigens (Section 1.4.5.3), how might IgG be synthesized?

A2.12 If the some of the polysaccharides were linked to proteins from the bacteria, this would permit the generation of the T cell help needed for IgG synthesis. This is the basis of successful conjugate vaccines against some bacteria (e.g. *Haemophilus influenzae* Type B, an important cause of meningitis).

Q2.13 If for some reason, the antigen that stimulates such an immune response is not derived from a pathogen, but is a self antigen (e.g. a molecule in β cells of the Islets of Langerhans in the pancreas that produce insulin), what might ensue?

A2.13 The immune response, once activated, will continue until the supply of antigen is exhausted. If the effect of the immune response is to kill the cells expressing the antigen, e.g. islet β cells the supply of insulin will dry up, resulting in diabetes. Diabetes will appear, however, only when almost all the β cells have been killed; thus the disease process must have started long before clinical symptoms appear. If we could identify patients at this preclinical stage we would have a much better chance of preventing the disease.

Q2.14 Can you suggest a hypothesis to explain how, following a primary infection, the tubercle bacteria are maintained in a dormant state?

A2.14 The reactivation suggests that the dormant state is maintained by an active, on-going immune response. One hypothesis would be that there is a cycle of events: the bacteria start to replicate and this re-stimulates the immune response, which then contains the growth until they start to replicate again, and that this can be repeated over many years.

Q2.15 Robert Koch – he of the postulates fame (Section 2.4.1.2) – recognized the importance of DTH responses in tuberculosis and attempted to develop a cure for tuberculosis by inoculating diseased patients with killed mycobacteria. Unfortunately a significant number rapidly became seriously ill and died. Why might this have happened?

A2.15 Patients with active tuberculosis who are given tuberculosis antigens generate a very strong, potentially harmful immune response, probably resulting from excessive cytokine secretion (a "cytokine storm"). This is thought to occur because there are many activated tuberculosis-specific T cells in these patients, and giving them extra, exogenous antigenic stimulation results in rapid release of large amounts of cytokines.

Q2.16 What might be the selective advantage to Salmonella of inducing intestinal inflammation?

A2.16 It may permit more easy penetration of the intestinal epithelium and further dissemination within the underlying tissues.

Q2.17 Although diarrhoea and vomiting can lead to serious dehydration these responses are often considered also to be beneficial to the host. Why?

A2.17 It may assist in getting rid of the bacteria from the intestine more rapidly.

Q2.18 How might you try to show that *Helicobacter pylori* is the cause of gastric cancer, rather than just being associated with it?

A2.18 This is difficult. It is not ethical to do the definitive experiment of experimentally infecting a human and seeing if gastric cancer eventually develops, and there are no suitable animal models available at present. It may be that the inflammation induced by the bacteria increases the rate of division of gastric epithelial cells, increasing the probability of cancer-related mutations being introduced into the dividing cells. Situations of chronic inflammation are often associated with an increased risk of malignancy perhaps for this reason.

Q2.19 Polio virus infects humans by the oral route. However, most humans infected with polio virus do not develop paralysis. What might determine whether an infected individual does develop paralysis?

A2.19 It could relate to the strain of polio virus since some strains are more virulent (able to cause disease) than others. Or to the dose of virus that has infected them. It might also relate to the ability of the host to resist the virus, and this would most likely be determined genetically. If it was determined genetically we would expect to see a higher incidence of paralytic polio in families where one member was affected. In particular, where one of a pair of twins was affected, we would expect to see a higher incidence in the other twin if they were monozygotic than if they were dizygotic (Section 7.3.3.1).

Q2.20 Why are colds and influenza more common in the winter months?

A2.20 Colds and flu are spread by droplets emitted from patients by coughs and sneezing, at least partially; hand-hand contact, or contact with surfaces containing droplets (e.g. on public transport) is also a very efficient way of spreading a cold or flu. If a droplet dries out the virus is inactivated. Droplets will survive longer in cold, damp conditions than in hot, dry conditions.

Q2.21 What might be the consequences of the early stages of influenza being asymptomatic?

A2.21 If you do not know that person is infected (asymptomatic) then you cannot institute procedures to isolate that individual and prevent the spread of infection. Thus an infected individual may infect many others before symptoms appear. This was crucial to the spread of "Spanish flu" in the 1914–1918 war and is a major reason to be concerned about bird flu mutating to be able to spread from person to person.

Q2.22 Why do you feel cold and shiver while you are developing a fever?

A2.22 Because in order to increase the body's core temperature (fever), heat needs to be generated. Shivering (repeated muscle contractions) generates extra heat. When fever is remitting, heat is lost by sweating and evaporation.

Q2.23 What might be the evolutionary function of fever in defence against infection?

A2.23 Some microbes reproduce less efficiently at higher temperatures. A direct experiment used lizards infected with a bacterium. At their normal environmental temperature, the lizards were able to resist an infection. When kept at a suboptimal temperature they died of the infection.

Q2.24 If it were proven that a massive production of cytokines in response to H5N1 infection causes mortality, what new therapeutic approaches might be developed?

A2.24 Strategies to prevent the systemic effects of the cytokines could be investigated. The first question would be to determine which cytokine(s) are important in causing the systemic effects; TNF-α would be a prime candidate. The effects might be blocked by using inhibitory antibodies to the cytokines, or by using soluble (decoy) cytokine receptors, or by using small molecules which block the cytokine-binding site on the receptor. It would, however, be important to ensure, if at all possible, that the beneficial effects of the cytokines are not affected.

Q2.25 Does the finding that vaccination against papilloma virus prevents the development of cervical carcinoma prove a causal relationship between the two?

A2.25 It does not prove a causal relationship. The viral infection could be generating a permissive environment in which the mutated cells can grow. As with gastric carcinoma and *Helicobacter*, (Q2.18) we cannot do the definitive experiments. In practice this is probably not a significant concern if the tumours are prevented by vaccination, although clearly it is of great scientific interest.

Q2.26 How might the discovery of such viral immune evasion mechanisms affect our thinking about vaccine design?

A2.26 If a prophylactic vaccine is being developed, full efficacy would mean that we would not need to worry about viral evasion sonce the virus would not be able to infect. However, for therapeutic vaccines we would need to consider developing vaccines, which include molecules, or genes coding for molecules, that can counteract the effects of the viral evasion molecules. At a very blunt level this may be achieved by the use of suitable adjuvants (Section 4.6). A more sophisticated approach would be to identify the crucial evasion molecules and design ways of counteracting these.

Q2.27 Until very recently, essentially all humans co-existed from neonatal stages with intestinal parasites. These parasites generate T_h2-biased immune responses in their hosts. In the developed world we no longer co-exist with these parasites. How might this change have affected our overall immune responsiveness?

A2.27 In general metazoan parasites bias adaptive immunity towards T_h2 responses. These responses have to be carefully regulated to avoid host damage and thus parasites may also induce the development of regulatory mechanisms, such as T_{reg}. These mechanisms may act in a strictly antigen-specific manner. Might it be that if the immune system develops in the absence of chronic infection, these regulatory mechanisms do not develop effectively, and that this results in defective regulation of responses to normally harmless environmental antigens such as food proteins or house dust mite faeces? This might for example result in an increase in allergies, which we know is happening. This type of thinking forms the basis for the "hygiene hypothesis" to explain the increased incidence of such conditions (Section 7.3.3.4).

Q2.28 Can you think of any examples when it might be to a pathogen's benefit to kill its host?

A2.28 Perhaps if a dead host provides an environment where the pathogen can reproduce successfully. Bacteria such as the clostridia (the causes of tetanus, botulism and gas gangrene) can only reproduce in strictly anaerobic conditions. Hence, these bacteria will be able to grow in a dead but not living tissue (they are saprophytic) and a dead host might provide an almost limitless source of nutrients for them. Such bacteria then form spores which pass into the environment, ready to infect other individuals.

Q2.29 What would be the outcome if the protein was injected without an adjuvant?

A2.29 There may be at best a very weak response or no response at all. The injection itself causes some tissue damage which might release DAMPs to act as adjuvants. There is however also a possibility that the protein injected on its own might induce tolerance. Protein, free of aggregates, can induce tolerance if injected i.v.

Q2.30 Why are HIV and malaria somewhat similar, in terms of the challenge they represent to designing a vaccine against them?

A2.30 The main similarity is that they both display massive antigenic variation and this makes it very difficult to generate a vaccine that will protect against all strains of the pathogen. The main difference between HIV and malaria in this context lies in the ways they generate variation. For HIV it is the result of a very high mutation rate that arises from inaccurate RNA copying during viral replication. For malaria, it arises from the very large number of genes coding for surface proteins that each malaria parasite possesses, and which can be expressed sequentially.

Answers to the Questions in Chapter 3

Q3.1 Can you think of some other examples that illustrate the barrier role of skin?

A3.1 The need of insects to bite in order to transmit infection is one example. The microbes causing malaria and leishmaniasis (protozoans), plague (bacteria), and yellow fever and rabies (viruses) cannot infect through the skin unless it is penetrated by biting.

Q3.2 What other mechanisms might protect the respiratory tract against infection?

A3.2 Coughing: patients who are bedridden often cannot cough effectively. This allows secretions to pool in the lower respiratory tract and to serve as culture media for bacteria. Physiotherapists strive to make post-operative patients cough efficiently.

Q3.3 The avian H5N1 virus is an enveloped virus. In birds, viral transmission is via the oro-faecal route. How might the virus be able to survive in the bird intestine?

A3.3 This is difficult for us to explain. The avian digestive tract may differ from the mammalian, but birds do have gall bladders and secrete bile containing the bile salts that can disrupt viral envelopes. One suggestion is that the envelope proteins are so closely packed that the lipid membrane is protected from the detergent action of the bile salts.

Q3.4 A technician working in a bacterial laboratory developed a severe upper respiratory tract infection and was treated with a broad-spectrum antibiotic. On his return to work he set up some large-scale cultures of a normally harmless bacterium, Haemophilus influenzae. He developed a very severe respiratory tract infection with the Haemophilus influenzae and nearly died. Why? What might have been done to prevent the infection?

A3.4 The likely explanation is that the antibiotics killed the commensal bacteria lining his respiratory tract, allowing the *Haemophilus* bacteria to colonies. He should not have been permitted to work with potential pathogens for a longer period after stopping antibiotic treatment. Had he followed good practice in the handling of the bacteria his respiratory tract would not have become colonized.

Q3.5 How might we show that mast cells are important in the initiation of acute inflammation following infection with a pyogenic bacterium such as Staphylococcus aureus?

A3.5 Mast cells degranulate and produce histamine which is a largely or completely mast cell-specific product. There are also drugs that block mast cell degranulation which might be useful tools. There are also strains of mice that are deficient in some mast cells and these might be used.

Q3.6 Can you suggest some situations in which oedema may not be beneficial to the host?

A3.6 Any situation where swelling of an organ or tissue can cause obstruction. For example, a bee sting in the upper respiratory tract of an allergic individual may cause respiratory obstruction. Oedema in the brain can cause raised intracranial pressure, which may force the brain stem into the narrows at the site of entry of the spinal cord leading to loss of CNS function.

Q3.7 How might children with LAD be treated?

A3.7 First, any infection is treated with antibiotics. In some cases, neutrophils from a normal individual can be transfused since these can migrate normally and kill the bacteria. The most effective long-term treatment at present is bone marrow transplantation, but this carries significant risks. In the future, genetic modification of the patient's own stem cells using gene therapy to replace the faulty gene(s) with normal versions may become available.

Q3.8 How might we attempt to identify a novel molecule involved in neutrophil adhesion to endothelial cells?

A3.8 The approach most widely used is to attempt to make a monoclonal antibody that will block adhesion, and then to use the antibody to isolate and characterize the molecule. The first thing needed is an assay for adhesion. We could start *in vitro* by using preparations of cultured endothelial cells to which neutrophils adhere. Next, we need to make monoclonal antibodies to neutrophils (Section 6.2.5). The resulting candidate antibodies could then screened for their ability to block neutrophil adhesion in our assays. We could then administer our antibody to mice and test its effect *in vivo*: a local site of inflammation could be monitored for inhibition of neutrophil adhesion; this could even be done directly by using multi-photon imaging with fluorescent neutrophils. If an antibody does block adhesion, it can then be used to isolate the molecule it binds to from a preparation of neutrophil membranes. The nucleotide sequence of the protein can then be determined by reverse genetics.

Q3.9 Why may lobar pneumonia resolve with little or no permanent tissue destruction, whereas a pyogenic skin infection resulting in an abscess leads to permanent scarring?

A3.9 In lobar pneumonia the inflammation is very much confined to the alveolar spaces in the lungs and the connective tissues are surprisingly unaffected. The inflammatory exudate in the spaces can be rapidly and effectively expelled by coughing, leaving the lung structure intact. This anatomical disposal route is not available in an abscess. By the time it discharges there has been much connective tissue inflammation. A healing response is initiated that is needed to repair the local tissue damage that leads to scarring (See Section 2.4.4.2).

Q3.10 If 90% of a rat's liver is removed surgically, the remaining hepatocytes start dividing and continue until the liver has regained its original mass. They then stop dividing. What mechanisms might enable the liver to regenerate to its previous mass?

A3.10 We really do not know the answer to this. It illustrates one of the fundamental problems in biology, the regulation of cell population sizes, which applies to the cells of immunity as much as to any other cell type. How is the mass of the liver "measured" by the body? Could it be that a secretory product of hepatocytes regulates growth in an autocrine, but inhibitory manner so that when a certain concentration is reached it stops hepatocyte division? If an external growth factor is involved might serum from the liver-deficient rat contain a growth factor? Might this serum stimulate hepatocyte growth if transferred into a normal rat?

Q3.11 What might be the reason for having separate T cell and B cell areas in secondary lymphoid organs?

A3.11 This is surprisingly difficult to answer in functional terms. The general, rather vague answer usually offered is something like "it enables lymphocyte responses to be tightly regulated". Certainly we are finding out about *how* different lymphocytes localize to different areas of these organs and what they do there, but exactly *why* this occurs in different compartments is not entirely clear at all.

Q3.12 Why might the examination of the thoracic cavity of a long-term smoker illustrate the filtration function of lymph nodes?

A3.12 The mediastinal lymph nodes, which drain the lungs, would be dark grey or black. Macrophages in the nodes are filled with carbon particles derived from tobacco smoke.

Q3.13 DCs also migrate from normal, steady-state tissues in the absence of infection, although at a lower rate. Why might this be important?

A3.13 We do not think there is an absolutely definitive answer to this as yet. However, these DCs would be transporting self antigens from peripheral sites. If they had not been stimulated by infection, they may have reduced levels of costimulatory molecules. Hence when they got into lymph nodes, for example, they might interact with weakly autoreactive T cells which could become tolerized. This has therefore been suggested to be one possible mechanism for the induction and maintenance of peripheral tolerance.

Q3.14 The investigators noticed that at later time points, there were fewer labelled DCs present remaining in the node if the DCs carried the antigen than if they did not. What might be the significance of this observation?

A3.14 It is possible that following their activation, T cells were able to induce apoptosis of the DCs with which they had interacted. This could be a means of preventing the activation of more T cells and thus of regulating the immune response.

Q3.15 If labelled chemokines are injected subcutaneously into a mouse, within 30 min the chemokines are expressed on the luminal surface of HEVs. How might this phenomenon aid immune responses.

A3.15 Chemokines form part of the postcode mechanism for directing leukocyte migration. By expressing chemokines on HEVs it is possible that novel populations of lymphocytes – perhaps effector memory cells may be recruited into the node.

Q3.16 How might we be able to start sorting out the relative importance of different peripheral influences on T cell differentiation.

A3.16 This is a difficult but important problem. You might like to consider first whether these peripheral factors could influence T cell differentiation? This could be approached by *in vitro* experiments, for example activating T cells in the presence of different cytokines. If we did find that for example, a specific cytokine did affect differentiation it would be important to determine its *in vivo* relevance. This could be attempted by using antibodies to block the potential effector molecule or to deplete specific cell types that could be the source of the factor, or use mice in which the gene coding for the factor or its receptor had been knocked out. It is important to remember that there may be more than one factor causing the same effect; if blocking a factor did not alter polarization this need not mean that the factor is not involved.

Q3.17 As splenectomized children grow older it appears that their susceptibility to infection with encapsulated bacteria decreases. Why might this happen?

A3.17 We wish we knew. There are many examples of age-related changes in susceptibility to infection and the severity of infection; some decrease with age, others increase. For example, chicken pox in a child usually causes only mild symptoms, but in adults can be much more serious.

Q3.18 How, experimentally, might we test the hypothesis that the production of monocytes is under feedback control during normal, steady-state conditions?

A3.18 If the hypothesis is correct, depleting monocytes from the periphery should lead to an increase in monocyte production in the bone marrow. In principle a specific monoclonal antibody might be used to deplete monocyte, in mice for example (it would need not to bind to monocyte precursors in the marrow). The numbers of peripheral monocytes could be counted at different times after depletion, and the numbers of monocytes precursors in the marrow assessed by using flow cytometry to enumerate the numbers of cells expressing surface markers that define these precursor subsets (Section 3.5.1.2).

Q3.19 Why might it be that all animals that have evolved T lymphocytes have also evolved a thymus in which these cells develop?

A3.19 We are not entirely clear about this. Why could T cells not develop in the bone marrow as B cells mostly do? Presumably the thymus provides a relatively closed, isolated site which is perhaps crucial for tolerizing T cells to self antigens before they are released. However, it is known that antigens injected intravenously can get into the thymus. So what happens if there is an infection, isn't this potentially dangerous, tolerizing T cells to these antigens? Or doesn't it matter, because we have already developed mature T cells specific for the infectious agent? On a related point, which we also cannot answer, why did birds evolve a specialized Bursa of Fabricius for B cell development?

Q3.20 A researcher in Australia removed thymi from foetal sheep while they were still in the uterus. When these sheep were born they showed normal cell-mediated immunity. The worker concluded that the thymus was not important in the development of lymphocytes involved in cell-mediated immunity. Why was he wrong to make this conclusion?

A3.20 Different mammalian species are born at different times in their development. Mice are for example born relatively early; they are blind and need full-scale maternal care for some time. Sheep and other herd animals are born much later in development since they need to be able to get up and run very soon after birth to avoid predators. The same is true of the development of the immune system in different species. Thus, at the time the thymi were removed from the foetal sheep, large numbers of T cells had already left and entered the periphery. Humans are also born relatively late in development and in fact thymi can be removed from very young children during heart surgery with no apparent subsequent ill effects.

Q3.21 What might be the functional significance of having tissues resembling secondary lymphoid tissues present in an inflamed organ?

A3.21 We do not know of any for certain. Given that lymph from the organ will drain to local nodes there would not seem to be any real reason for having an extra site for lymphocyte activation actually in the organ. The phenomenon may represent an "accident" resulting from the secretion of a particular set of chemokines and cytokines in the inflamed area. It may, however, also give us clues as to how normal secondary lymphoid tissues develop, and we might yet discover a "true" purpose for the development of tertiary lymphoid tissues.

Answers to the Questions in Chapter 4

Q4.1 Is it in any way surprising that the Toll family of molecules can be involved in functions as diverse as development and defence against infection?

A4.1 One of the authors was surprised to learn this some years ago. The toll molecule that is involved in development is expressed on embryonic cells where it binds a molecule known as dorsal protein; involved in the generation of a protein gradient. In immunity, however, toll is expressed by different cells and binds a molecule called spaetzle which is activated after the binding of a soluble PRR to fungal or some bacterial products; this ultimately results in the production of anti-fungal or -bacterial peptides . Thus the same molecule can mediate very different effects depending on which cell expresses it, and which downstream molecules interact with it. Therefore, in the mammalian immune system, it is not surprising that a particular TLR expressed on dendritic cells, for example, may lead to a different set of responses than the same TLR expressed on macrophages.

Q4.2 How might the relative roles of TLR versus IL-1 receptor signalling be dissected experimentally?

A4.2 We need to find ways in which the activity of the two classes of receptor could be inhibited selectively. One possibility might be to use a blocking anti-IL-1 receptor antibody. Might there also be ways of interfering with genes specific to the respective pathways?

Q4.3 Why might the effects of defective TLR3 signalling be seen only with herpes viruses and only in the CNS?

A4.3 There is no clear answer. Cells with this deficiency produce little Type I IFN following viral stimulation, but not all children with these deficiencies develop encephalitis after herpes simplex infection. What this does demonstrate, however, is that a single TLR can be crucial for resistance to a viral infection in a specific.

Q4.4 Why might recurrent pyogenic disease be seen only in childhood in people with defective Myd88 signalling?

A4.4 There is no clear answer. Possibly there is a gradual build up of resistance due to antibody synthesis. There are many examples of susceptibility to infectious and other disease changing with age, and in most cases the underlying reasons are not understood.

Q4.5 The autoinflammatory diseases are not associated with an increased risk of infections. Why might this be?

A4.5 The autoinflammatory diseases reflect over-activity of the innate system. Very often they develop because of mutations that involve loss of regulation of innate activation through mutations in regulatory genes or by mutations which cause increased activation ("gain of function"). Thus, responses to infection are if anything likely to be maintained, posibly even enhanced.

Q4.6 How might we be able to determine the relative contributions of TLR activation in intestinal epithelial cells and macrophages to innate immunity?

A4.6 In mice one possibility is to use bone marrow chimaeras. First, mice are given lethal irradiation, which destroys their bone marrow stem cells. The mice are then rescued by giving them bone marrow cells from a normal mouse. By using mice in which Myd88 or other TLR-related signalling molecules are knocked out, as bone marrow donors or recipients, chimaeras can be made in which only bone marrow-derived cells can or cannot signal effectively. Experimentally, infection of these mice showed that effective defence depended on TLR signalling in their bone marrow-derived cells, but not host cells such as epithelial cells.

Q4.7 How could we show that the triple response is caused by histamine?

A4.7 The local injection of histamine induces a triple response, and treatment of individuals with an anti-H_1-histamine receptor drug abolishes the triple response.

Q4.8 How might mast cells be able to "sense" mechanical damage?

A4.8 Probably by themselves being directly damaged by mechanical stress. Some neuropeptides released by stimualted nerves can stimulate mast cell degranulation and it is possible that local release of these mediators is involved. However, the early phase of the response still occurs in tissues that have been denervated, suggesting that it is a direct effect on the mast cell.

Q4.9 If Kupffer cells are isolated from a mouse's liver, and tested for their ability to synthesize reactive oxygen intermediates, they are very inefficient compared to recently recruited macrophages. Why might Kupffer cells be adapted in this way?

A4.9 Kupffer cells line liver sinusoids and are primarily involved in "house-keeping" rather than having major defence roles (e.g. removal of old red cells and immune complexes). If they generated reactive oxygen intermediates each time these events occurred there would be a real possibility of collateral tissue damage to the liver.

Q4.10 How might we attempt to discover a novel phagocytic macrophage receptor?

A4.10 The conventional way would be to attempt to make a monoclonal antibody that blocked phagocytosis. Thus, a rat could be immunized with mouse macrophages, its B lymphoblasts fused with a B cell lymphoma and the resulting hybridomas screened for a possible blocking antibody (Section 6.2.5). This would need an *in vitro* phagocytosis assay. Western blotting (Figure 4.6) could be used to estimate the molecular size of the molecule recognized by the antibody; if this differed from the sizes of known receptors there is a good chance it is a novel molecule. A partial amino acid sequence could be identified, and a partial gene sequence predicted from this could be used to "fish out" the mRNA from macrophages. This would give the full gene sequence which is the definitive way of identification. Alternatively, by using the DNA sequences of known receptors, the mouse genome could be searched for genes that are related, but which differ from known receptors.

Q4.11 L-selectin is the molecule used by lymphocytes to adhere loosely to endothelial cells in high endothelial venules of lymph nodes. Neutrophils and monocytes also express L-selectin, but do not emigrate into lymph nodes under normal conditions. Why might this be?

A4.11 Selectins are just the first of (usually) three major molecular interactions between leukocytes and endothelial cells. If the requisite combination of chemokines and integrins and their respective receptors is not present, transmigration will not take place. Unlike lymphocytes, neutrophils and monocytes do not express the required combination.

Q4.12 Can you think of some situations where understanding the "post code" principle may lead to innovations in the treatment of disease?

A4.12 Most obviously in the treatment of inflammatory disorders such as rheumatoid arthritis (Section 7.4.4.4). Blocking the entry of cells such as monocytes into the synovial tissues could prevent further inflammatory damage. Less obviously in perhaps preventing metastasis of malignant tumours. Blood-borne tumour cells need to arrest in small blood vessels to set up metastases. This is thought to be facilitated by the tumour cells binding to platelets and leukocytes, and this is mediated by the tumour cells expressing ligands for selectins. Integrins may then promote adhesion of tumour cells to endothelium. Blocking these interactions may help to prevent metastasis. The selectivity of these interactions may also help explain why certain tumours metastasise to particular tissues and organs.

Q4.13 Where do all the molecules that enter inflamed tissues go to?

A4.13 Mainly into lymph which will eventually enter the blood, where they will be diluted. If bioactive molecules are present at high enough concentrations, they will have the potential to cause systemic effects such as IL-1 causing fever.

Q4.14 Suppose we have identified a novel small protein that is secreted by intestinal epithelial cells when they are cultured with *Salmonella* bacteria. How could we determine if this protein is chemotactic for neutrophils?

A4.14 There are several chemotaxis assays available. The most widely used is perhaps the Boyden chamber. This has two compartments separated by a membrane through which cells can migrate. The potential chemotactic protein is placed in the bottom chamber and neutrophils in the upper. The protein will form a concentration gradient across the membrane. The assay counts the number of neutrophils that have migrated across the membrane in a given time. In well-controlled experiments, one control is to place the potential chemotactic protein in both chambers. This is because some molecules may stimulate movement generally, without any directional component (chemokinesis) and this would also increase the number of cells migrating without there being any chemotaxis. If increased migration happens with the protein in both chambers this suggests it is not due to chemotaxis.

Q4.15 If neutrophils are treated *in vitro* with low concentrations of ethanol, their ability to generate ROIs is reduced. Might this observation have clinical significance?

A4.15 Chronic alcoholics have an increased risk of many infections, including pyogenic infections. Given the importance of neutrophils in defence against infection, it is a tenable hypothesis that alcohol, by damaging neutrophils, could increase susceptibility to pyogenic infections. Chronic alcoholics do however, have many other problems that could also contribute.

Q4.16 Can you suggest some other ways in which bacteria might have evolved to avoid being killed by neutrophils?

A4.16 Some bacteria, such as the mycobacteria, possess cell walls that are very resistant to degradation. These bacteria can also prevent phago-lysosomal fusion, while other bacteria, such as *Listeria*, can escape into the cytoplasm of the infected cell (Figure 2.13). These are just a few examples and you should be able to find many more.

Q4.17 Can you suggest two ways in which people affected by the radiation leakage at Chernobyl might present clinically?

A4.17 Because neutrophils have a very short life span, their numbers will decrease rapidly after irradiation. Thus, pyogenic infection is a likely presentation. Blood platelets are also short-lived and a decrease in their numbers will lead to bleeding, often noticed as bleeding from the gums. In the longer term we would expect to see an increase in the frequency of cancers due to the mutagenic effects of radiation. In fact, in the areas around Chernobyl, apart from a large increase in the rates of thyroid cancer (related to the release of radioactive iodine), there has been very little, if any increase.

Q4.18 How might patients with CGD be treated? What might be the problems with these approaches?

A4.18 These should include antibiotics to treat any ongoing infections, and bone marrow transplantation which can be curative but which does have significant risks (Sction 7.5.3). In the future, gene therapy may become available, in which the defective gene is replaced *in vitro* by a normal gene in bone marrow stem cell, which are then given back to the patient. The problems with bone marrow transplantation are primarily the increased risk of infection that accompanies the immunosuppression that is essential to prevent the graft being rejected, and graft versus host disease (GVHD), in which T cells in the marrow attack the recipient. The problems with gene therapy are less well-defined, but one is that it is difficult to ensure that the gene is inserted in the correct site, and another is that if it is inserted close to a growth control gene might cause uncontrolled proliferation of cells, similar to leukaemia.

Q4.19 Some workers have claimed that peritoneal macrophages can secrete IFN-γ. Others, however, have suggested that this secretion may come from a small population of contaminating NK cells. What sort of experiments could we do to help sort out this controversy?

A4.19 You might like to think about the difficulties of preparing really pure populations of different cell types. Selective removal of specific cell types using monoclonal antibodies and cell sorting or magnetic beads is very efficient, but could you detect the presence of less than 1% contaminating cells? Might the use of macrophages generated *in vitro* from bone marrow or blood monocytes be useful? It is now possible to do reverse transcription-polymerase chain reaction (RT-PCR) on single cells – might this also help?

Q4.20 How might we assess whether NK cell defence against herpes virus infection is due to their cytotoxic activity or their ability to secrete cytokines?

A4.20 In humans this is very difficult since at present there are no identified defects in cytotoxicity or cytokine secretion that are limited to NK cells. In mice it would be possible to use strains of mice in which both perforin and Fas ligand genes are knocked out to abolish cytotoxicity, but this would of course also affect CD8 T cells. However, if there were effects in these mice early in infection, before cytotoxic T cells could be detected, this might suggest a role for NK cell cytotoxicity. Similarly if specific cytokines were knocked out or blocked with monoclonal antibodies and had an early effect on infection this too might suggest a role for NK cells. Beware, however plasmacytoid DCs are very rapid cytokine secretors (Section 5.4.1.4).

Q4.21 How experimentally might it be possible to dissect the respective roles of IL-1 and TNF-α in systemic inflammatory responses?

A4.21 The aim is to inhibit selectively the actions of one of these proteins in an inflammatory response. The protein itself could be inhibited by using a specific antibody. In an experimental animal, the gene could be knocked out or protein synthesis inhibited by the use of small interfering RNA (siRNA), which blocks mRNA. An alternative approach would be to administer the protein directly to a normal animal and assess its effects. The problems with all approaches are to know whether the effects seen are directly and solely due to the protein itself, or whether there are indirect effects involving other proteins: might giving IL-1 stimulate IL-6 secretion, for example?

Q4.22 Can you suggest two ways in which the functions of acute-phase proteins such as serum amyloid A could be explored?

A4.22 This is a real problem, and there is a remarkable lack of understanding of the real functions of these proteins. Some approaches are similar to those described in Question 4.21: trying to identify changes following the inhibition of the protein or after administration of the protein. Another approach is to investigate the proteins properties *in vitro* if it can act as an opsonin or whether it modulates the functions of leukocytes. This illustrates how very difficult it can be to know where to begin investigation without having a clear hypothesis.

Q4.23 In pyogenic infections, the numbers of circulating neutrophils are greatly increased. How could we design an experiment to test the hypothesis that this increase is related to the presence of a blood-borne growth factor?

A4.23 The most direct test would be to prepare serum from an animal where neutrophil production is starting to increase, such as in the early stages of a pyogenic infection. The presence of a growth factor could be tested directly by injecting the serum into a normal animal and monitoring neutrophil production. Alternatively, the serum could be added to cultures of bone marrow cells and the development of neutrophil colonies monitored. By adding inhibitory antibodies specific for individual growth factors to the cultures, clues as to the nature of a putative factor might be obtained.

Q4.24 How likely is it that adjuvants will be developed for use in vaccination with absolutely no side effects at all?

A4.24 This would be an admirable goal, but is it realistic? An effective adjuvant almost certainly needs to activate the innate immune system to be effective, and such activation will inevitably cause side effects such as local inflammation. If it were possible to dissect the innate response to understand how to stimulate only the parts needed for induction of adaptive immunity, side effects might be minimized.

Answers to the Questions in Chapter 5

Q5.1 How might we determine the life span of T cells in the mouse?

A5.1 The most direct method would be to take T cells from a donor mouse that expressed a different allele of a particular gene (e.g. Thy-1) and transfer these cells to mice expressing the other allele. At intervals, sample the recipient mice and count the numbers of surviving donor cells. We would need to be careful to define which T cells we are examining, because CD4 and CD8 T cells could differ, so separation into subsets on the basis of these markers would be important. With appropriate markers this could also be done for subsets of effector and memory cells. We need to be careful however, that the allele used to detect the transferred cells is stably expressed, and does not itself induce rejection of the cells.

Q5.2 How many different MHC class II molecules is any individual human likely to express?

A5.2 Humans have three loci coding for MHC class II molecules, in humans DP, DQ and DR. Each locus contains structural genes for the α and β chains which are codominantly expressed. Given the very high degree of polymorphism at most of these loci, most individuals will be heterozygous for many of the structural genes. Thus you might expect that most individuals will express six MHC class II molecules, three encoded by each chromosome. It is, however, more complicated for MHC class II. For example, DR actually contains four structural β genes. It is also possible, for example, for a DP α chain made on the paternal chromosome to pair with a DP β chain made on the maternal chromosome to form a heterodimer. Thus the actual number of MHC class II genes expressed could be considerable higher that at first might seem to be the case.

Q5.3 Might there be any problems in using fluorescently labelled lymphocytes for long-term adoptive transfer experiments?

A5.3 Lymphocytes divide many times when they are activated and at each division the amount of label is halved. It does not take many divisions before label becomes undetectable. This is why genetic labels are crucial. If the transferred cells do not divide, or if the experiments are short-term, fluorescent labelling is an effective technique.

Q5.4 Why might it be that some CNS neurons do not express MHC class I?

A5.4 Most CNS neurons cannot be regenerated if they die, so to kill them would be very serious. It is therefore presumably crucial that they continue to live, even with a chronic viral infection, rather than to be killed by cytotoxic T cells. You might also like to ask yourself why, if they do not express MHC class I molecules, they are not killed by NK cells (Section 4.4.5).

Q5.5 In epithelial cells, MHC class II molecules are present, not at the cell surface, but in intracellular vesicles. What might be the significance of this finding?

A5.5 There are no clear cut answers, but it is known that a variety of cells, including epithelial cells, can release small vesicles called exosomes, which can express MHC class II and costimulatory molecules. The physiological functions of exosomes are unclear but it has been claimed that they can for instance stimulate anti-tumour immunity in clinical trials. It is also possible that they are involved in tolerance induction.

Q5.6 How might the antigen-processing mechanisms distinguish between normal self proteins and those derived from pathogens?

A5.6 In general they cannot. We know this because if peptides are separated from MHC molecules, most peptides are derived from the host cell and in the case of MHC class II, its environment. This does not normally present a problem, because many potentially self-reactive T cells are removed in the thymus and periphery and are thus not available to recognize self peptides. That we develop autoimmune diseases (Section 7.3) shows that this process of removal is not always effective.

Q5.7 If peptides are needed to get MHC class I molecules to the cell surface, how can MHC class I molecules bind peptides that are simply added to the cultures?

A5.7 β_2-Microglobulin is not covalently attached to the MHC class I heavy chain and is continually exchanging with free β_2-microglobulin. When β_2-microglobulin dissociates, the affinity of the heavy chain for peptide decreases markedly and the peptide may dissociate. If the added peptide is at high concentration, it is likely that it will replace the original peptide.

Q5.8 Chloroquine, added to cultures of APCs and protein, strongly inhibits antigen presentation. Are there possible reasons that might account for this other than inhibition of proteolysis? How might we attempt to set up control experiments to exclude the other possible reasons?

A5.8 It is possible that chloroquine is toxic to APCs; in fact, this actually is the case at higher concentrations. Thus, the lack of presentation may reflect death or damage to the APCs rather than inhibition of processing. We would need to show that the APCs are able to function in other ways. One way would be to use the allogeneic mixed leukocyte reaction in which the APC stimulate allogeneic T cells to proliferate. This stimulation does not depend on the antigen-processing mechanisms being intact. Using this assay we could directly test the capacity of chloroquine-treated DCs to stimulate naïve T cells. With macrophages, however, we would need to test responses of T cell clones or lines, because these APCs are usually unable to activate naïve T cells.

Q5.9 Why might it be important to have mechanisms that increase the likelihood of peptides binding with high affinity to class II molecules in the acidic conditions of the endosomal pathway?

A5.9 The binding of peptides to MHC class II molecules is generally very pH-dependent. It is important that peptide–MHC class II complexes are stable at the neutral pH present at the cell surface, so they need to be bound with high affinity. In order to permit stable binding of these same peptides in the acid conditions of the endosome other mechanisms such as those dependent on HLA-DM and perhaps-DO may be important to select for these higher-affinity peptides.

Q5.10 How might we attempt to show that the cytoplasmic tail of the invariant chain was responsible for targeting MHC class II molecules to the endosomal system?

A5.10 By using cells in which the part of the gene coding for the cytoplasmic invariant chain has been genetically modified. By generating molecules in which individual amino acids are changed, (e.g. site-directed mutagenesis) it is then possible to dissect the molecular requirements for selective localization.

Q5.11 How might we determine if cross-presentation requires TAP for the delivery of peptides to the RER?

A5.11 The main cross-presenting cells in mice are DCs (particularly those expressing CD8. A mouse with one of the TAP genes knocked-out could be used and CD8 DCs isolated from the mouse and tested *in vitro*. Alternatively, CD8 DCs could be isolated from a normal mouse and small interfering RNA (siRNA) specific for TAP mRNA used to block TAP synthesis by transfecting it into the DC.

Q5.12 How might we investigate different molecules that may be associated with MHC class I molecules at different stages in their life cycle within a cell, such as in the RER?

A5.12 It is generally not possible to use microscopy, since the resolution even of electron microscopy is insufficient. Biochemical approaches are needed. Cell fractionation can be used to isolate different cellular compartments (e.g. the RER). Molecules present in the different compartments can then be identified, but this does not tell us if they are associated with MHC class I molecules. To do this, immunoprecipitation might be needed. Antibodies to MHC class I molecules could be used to purify the MHC molecules which might remain associated with other components. These could therefore be isolated and identified. By combining this approach with cell fractionation and pulse–chase labelling, it would be possible to determine the different associations that occur throughout the life history of the MHC molecule. Box 4.2 in Chapter 4.

Q5.13 How might we attempt to identify a natural ligand for a CD1 molecule?

A5.13 Many are trying, but with little success to date. CD1 knock-out mice show altered susceptibility to infection with the α-proteobacteria suggesting involvement of iNKT cells. In an infection with these bacteria it might be possible to isolate APCs from the site of infection, purify CD1 molecules from the APCs and identify the CD1-bound molecules.

Q5.14 Endothelial cells express MHC class I molecules. Might this have a role in directing the migration of antigen-specific CD8 T cells, in addition to the usual post code signals (Section 3.3.3.2)?

A5.14 This might happen if the endothelial MHC class I expressed the relevant peptides for CD8 T cell recognition. It has been shown that the injection of antigen into the brain can lead to such expression and that this can help direct antigen-specific CD8 T cells to cross the endothelium into the brain.

Q5.15 How might we attempt to estimate the concentration of a particular secreted molecule in the immunological synapse?

A5.15 This is an important but difficult question. For example, the effects of cytokines on cells in culture are generally concentration-dependent. If the gene coding for the molecule was coupled with a gene coding for a fluorescent protein such as Green Fluorescent Protein (GFP), it might be possible to use the degree of fluorescence to estimate the protein concentration. In the case of a cytokine that might be secreted by one cell at the site of contact with another, the absolute concentration is likely to be miniscule, but the effective concentration could be vastly increased, because it could be secreted onto a very small surface area.

Q5.16 Might the immunological synapse have roles in infection other than being involved in T cell activation or function.

A5.16 It has been suggested that it could be a route for virus transmission. DCs in the periphery are ideally sited to be infected by viruses and then contact T cells that, particularly after activation, are in some cases, favoured cells for viral replication. Such a route has been suggested for measles virus and HIV, but direct evidence is lacking.

Q5.17 Why does robust activation of IL-2 synthesis require strong signalling?

A5.17 Perhaps in order to prevent CD4 T cells being fully activated unless there is a real infection, with the presence of both the viral or microbial antigen and strong activation of the innate immune system.

Q5.18 Why might we need so many different types of DC?

A5.18 We are starting to understand some of the specializations of different DCs, including their patterns of cytokine secretion, ability to cross-present, differences in localization and so on, however it is not yet clear why these functions need to be carried out by different cell types. Understanding this is however, important if we are to target DCs for vaccination and immunotherapy (Section 5.6.2).

Q5.19 How might we begin to identify accurately the cell type which is the crucial secretor of IL-4 in the initiation of T_h2 polarization?

A5.19 The definitive answer may come by generating mice in which IL-4 synthesis is inhibited only in specific cell types such as mast cells or basophils. This could theoretically be done *in vivo* by knocking out genes in specific cell types or *in vitro* by targeting siRNA to specific cell types. Of course, we should also consider whether different cell types might be crucial for secreting IL-4 in different settings.

Q5.20 What might be the importance of IFN-γ stimulating MHC class II expression on epithelial cells?

A5.20 It is important that the initial activation of CD4 T cells is tightly regulated to avoid autoimmune disease (Chapter 7). Thus, only DCs are able to activate naïve CD4 T cells under normal circumstances. However, activated CD4 T cells need to maintain their activation as long as there is infection to be dealt with and it is likely that any cell expressing MHC II, such as these epithelial cells, can present antigen to activated CD4 T cells. In turn, the CD4 T cells may be able to help the epithelial cells acquire specialized functions that may enable them to resist or eliminate the infection.

Q5.21 Why might perforin not polymerize in the CTL membrane and induce suicide?

A5.21 CTLs can certainly be killed by other CTLs if the former express the appropriate peptide–MHC class I complex. CTLs do, however, express a protease, cathepsin B, on the membrane of their intracellular granules which becomes attached to the plasma membrane during granule exocytosis. Cathepsin B can degrade perforin and thus presumably helps to prevent polymerization in CTL membranes. Hence, if CTLs are used in a cytotoxicity assay in the presence of a cathepsin B inhibitor, they themselves may die by apoptosis.

Q5.22 What might be the benefits to the host of inducing apoptosis in target cells rather than straightforward lysis?

A5.22 Apoptosis is generally an immunologically "silent" event and does not induce inflammation because apoptotic cells are rapidly taken up by cells such as macrophages. Additionally, apoptosis induces breakdown of nucleic acids, both of the infected host cell and probably also of viruses, thus preventing production of new infectious virus.

Q5.23 How might one be able to assess the relative importance of CTL killing versus secretion of cytokines for eliminating infectious agents?

A5.23 One way would be to construct mice in which both the granule-dependent (e.g. perforin or granzymes) and -independent (Fas) pathways are inhibited. This could be done by making double knock-out mice which could then be tested for resistance to infection. Alternatively, if we suspected the involvement of a particular cytokine, we could attempt to inhibit the secretion of the cytokines or the expression of its receptor (e.g. by using siRNA technology *in vitro* or *in vivo*). We would need to be aware that some cytokines such as IFN-γ may however, by altering expression of MHC molecules, also alter the susceptibility of cells to killing.

Answers to the Questions in Chapter 6

Q6.1 Why can we not make antibodies to gelatine? (denatured collagen)

A6.1 Antigenic epitopes for B cells need to be structurally stable, three-dimensional structures. Gelatine is denatured collagen, which has lost its tertiary structure and the whole molecule is floppy so there are no stable epitopes.

Q6.2 If allelic and isotype exclusion did not occur, how many different antibodies could be generated in any given B cell?

A6.2 We have maternal and paternal alleles of heavy (H), κ and λ (L) chains. If any given antibody was only comprised of chains with L chains of either type (as they normally are), a B cell could potentially produce up to eight different antibodies. If, however, an antibody could be formed from a hybrid of a H chain with a κ chain, plus a H chain with a λ L chain, this number would be considerably increased.

Q6.3 What might be the biological relevance of the extremely different typical affinities of antibodies and TCRs for their respective antigens?

A6.3 In most circumstances, antibodies can only function in defence against infection if they remain bound to their antigens. Thus, the higher the avidity, the better. In contrast the TCR only appears to be involved in signalling (via CD3). Once it has delivered its signal it does not need to remain bound to its cognate peptide–MHC complex. Indeed, if it did so with a very high affinity it might even prevent the T cell detaching easily from an APC (such as a DC) and migrating to the site where it is needed.

Q6.4 How likely it is that natural antibodies are produced in the complete absence of any antigenic stimulation?

A6.4 This is probably impossible to answer definitively. Even germ-free mice lacking commensal organisms are fed on proteins. Some of these cross intact into the circulation and could potentially be recognized by B cells, thus providing an antigenic stimulus. (Typically, however, this source of antigen induces tolerance).

Q6.5 What might be the role of commensal organisms in stimulating the production of the intestinal IgA antibodies found in normal individuals?

A6.5 Germ-free animals have very few IgA plasma cells in their intestinal lamina propria and very low levels of intestinal IgA. It is impossible to exclude the possibility that a very small amount of IgA is made independently of commensal bacteria, but the important point is that the great bulk of IgA present in uninfected animals is in fact commensal-dependent. Researchers are now interested in the potential role of this IgA in regulating oral tolerance.

Q6.6 If the poly-immunoglobulin receptor can bind both IgA and IgM, why might it be that IgA is the most abundant isotype in mucosal secretions?

A6.6 IgM is typically secreted by plasma cells in non-mucosal sites. Under normal, non-inflamed conditions, IgM is confined to the blood and does not enter extravascular tissues. In contrast, IgA is secreted by plasma cells lying in extravascular connective tissues in mucosae and has direct access to the epithelial cells across which it must be transported. Any IgM that does enter mucosal tissues can be, and is, also transported onto the luminal surface.

Q6.7 The relationship between breast feeding and protection is a correlation, not an explanation. What other interpretations might be possible?

A6.7 There are many: perhaps the breast-fed children were generally better looked-after, were fed better diets and so on. However, a good clinical trial will control for these alternative explanations by using groups where the children are matched for all potential confounding factors, leaving the only significant variable as whether or not they were breast-fed. The correlation is, however, very strong and has been found in many independent studies.

Q6.8 How might the concept of a common mucosal immune system be relevant for the generation of vaccines against mucosal infections?

A6.8 It raises the possibility that, for instance, a vaccine against a respiratory or uro-genital tract infection might be delivered orally, or intranasally to avoid intestinal digestion of the vaccine. From a practical point of view this is highly desirable because of the ease of delivery and the lack of any potentially distressing effects of administration by injection. In addition, mucosal vaccination also tends to provide some degree of systemic (non-mucosal) immune defence, whereas vaccination into the skin, for example, does not provide mucosal defence.

Q6.9 Why might genetically identical animals make different antibody responses to the same antigen?

A6.9 Genetically identical animals do not have identical B cell (or T cell) antigen receptor repertoires (Section 7.3.3.1). Any individual animal will only make a small percentage of all the possible receptor gene rearrangements, largely at random, and their receptor repertoires will inevitably differ. Additionally, the antibodies generated in an immune response will depend on which B cells happen to be present in the secondary lymphoid tissue at the time the antigen reaches it. Finally, large antigens express many different epitopes and the antibodies made to the antigen will be a selection of antibodies to different epitopes these may differ between individuals.

Q6.10 What might be the limitations of using antibodies from another species to treat clinical disease (e.g. for passive immunization against tetanus)?

A6.10 The antibodies will be viewed by the immune system as a foreign protein, and host antibodies will be made against them. (This does happen but, you may ask, where is the "danger" in an intravenous injection of a soluble protein? We are also puzzled). An antibody made against the injected antibodies will lead to them being cleared very rapidly from the circulation and losing its efficacy. The antibody response to the injected antibodies might also lead to serum sickness in which immune complexes between the host's and the injected antibodies are deposited in small blood vessels, leading to complement activation and local acute inflammation (Section 7.2.5).

Q6.11 In what ways might hybridomas might be screened for production of a monoclonal antibody against a given antigen?

A6.11 This depends very much on the nature of the antigen you wish to detect. If it is a cell- or tissue-associated molecule, using immunohistology on frozen sections can be very informative. If it is expressed by cells in suspension, flow cytometry might be the method of choice. If you are looking for a functional molecule you could try to block or stimulate that function. For example, to find a new Fc receptor, you might try to block binding of an opsonized particle to a macrophage. If you wish to make an antibody to a soluble molecule you could try to adhere your molecule to a tissue culture surface and detect binding of a monoclonal antibody to it using an ELISA assay (Box 4.3). There are of course many other approaches, each being tailored to the individual need in question (See also Box 6.4).

Q6.12 In what areas other than immunology might monoclonal antibodies might be of use?

A6.12 There are just too many to make a comprehensive list. How about identifying cell adhesion molecules in plants, the detection of contaminating proteins in foods or uses in forensic science?

Q6.13 Monoclonal antibodies to CD4, CD8 and CD25 have been used to deplete T cell subsets *in vivo* in mice. Why might it be of relevance that other cells also express these molecules?

A6.13 CD4 and CD8 are expressed on some DCs as well as on T cells. CD25 is expressed on activated CD4 T cells and DCs as well as on T_{reg}. The important point is to note that such expression may potentially lead to misleading results because other cells could also be depleted. This concern generally applies to any study that tries to draw firm conclusions based on use of any single marker.

Q6.14 Given that the human immunodeficiency virus (HIV) can live and replicate within macrophages, how should this make us think about vaccine design?

A6.14 Should we be concerned that a vaccine that induced opsonizing antibodies might just target HIV to macrophages more efficiently, as may be case with yellow fever and Dengue? This emphasizes the point that it is crucial in vaccine design to understand the nature of protective immune responses to a pathogen.

Q6.15 How might it come about that mast cells can be coated with IgE, yet serum levels of IgE are barely detectable?

A6.15 It could be important to know if the IgE is synthesized in lymphoid tissues or locally (e.g. in mucosal tissues). Immunocytochemistry of lympoid and peripheral tissues, staining for IgE-containing plasma cells would be informative. If IgE is mainly secreted locally, it could bind directly to mast cells in the vicinity. Might local synthesis of IgE and capture by mast cells be the explanation for the very low levels found in the blood?

Q6.16 How might we determine if ADCC is actually carried out by NK cells, monocytes or indeed any other specific type of cell present in a mix of leukocytes?

A6.16 Either we can attempt to purify a specific cell population (positive selection) or deplete a specific population (negative selection) from the total pool. We could use separation techniques such as cell sorting or magnetic bead separation to isolate relatively pure populations of cells. We need to have antibodies that are specific for each population of cells. The purity of each population needs to be checked by FACS. The separated populations can then be assessed for their ability to mediate ADCC. Think about ways of separating specific cell types from mixed populations. The principles involve negative and positive selection. How can such selection be carried out and what are the advantages and disadvantages of each type of selection? Would we want to know if the different cells are capable of mediating ADCC *in vitro*?

Q6.17 What consequences may follow if the mutated V region generated during somatic hypermutation makes the antibody autoreactive?

A6.17 Potentially it could of course cause autoimmune disease. This, however, would require T cell help and T cells are tolerized much more stringently than B cells. If the B cell recognizes self antigen without T cell help it is possible that the BCR may itself change by reactivation of the RAG genes that mediate rearrangement (receptor revision; Section 6.5.3). If this is not successful the B cell is likely to die by apoptosis.

Q6.18 What might be the evolutionary advantage of a TI-1 response to LPS?

A6.18 TLR stimulation can assist in the activation of B cells and this may be the physiological function of the LPS response in helping to produce antibodies against Gram-negative bacteria. Normally, LPS will only be present at low concentrations. High concentrations of LPS can be lethal *in vivo*, so the polyclonal response to high LPS concentrations may well be found only *in vitro*.

Q6.19 Why might commensal bacteria stimulate IgA responses in a TI manner?

A6.19 This is a recent observation and the underlying mechanisms are not fully clear. It has been suggested that the IgA is secreted by B-1 cells that migrate to the gut but which generally make IgM in the absence of apparent antigenic stimulation. Another possibility is that pathogen-associated molecular patterns (PAMPs) in the commensal bacteria may provide activation signals for the B cell, through B cell PRRs, that bypass the need for T cell help. Another is that antigens on the bacterial surface may be presented as "arrays" that can cross-link the BCR efficiently, which again might bypass T cell help. These hypotheses could explain B cell activation, they do not, however, explain the switch to IgA.

Q6.20 Why might it be preferable to use $F(ab')_2$ fragments of anti-immunoglobulin antibodies, rather than the intact immunoglobulin molecule to activate B cells experimentally?

A6.20 Because the B cell expresses an inhibitory FcR. If the anti-immunoglobulin antibodies bind to this receptor activation may be suppressed.

Q6.21 If the T cell with which the B cell needs to interact has been already been activated by a DC, why might a follicular B cell need to express costimulatory molecules that are involved in T cell activation?

A6.21 There is no good evidence that DCs migrate into follicles after activating the CD4 T cell and, to maintain T cell activation, signalling from the B cell may well be required.

Q6.22 Cells other than B cells can express CD40 (e.g. DCs). How might we attempt to determine if the lack of CD40 on such cells also contributed to disease in HIGM syndromes?

A6.22 We do not know the answer to this question and the literature is very sparse. We do, however, think it possible that it may be an important determinant of the types of infections seen in these patients. In mice it is possible to inhibit the expression of a particular molecule in a particular cell type by selective gene targeting. Thus it may be possible to inhibit CD40 expression in DCs but not B cells, or vice versa, and test susceptibility to different types of infection that are typically seen in clinical cases of HIGM.

Q6.23 What might be the relative roles of differentiated T_h1 or T_h2 cells, compared to T_{fh} cells, in B cell responses?

A6.23 The problem of how different polarized CD4 T cells interact with B cells is not completely resolved. T_{fh} cells do not appear to secrete the cytokines needed for B cell class switching, but it is not clear that T_h1 and T_h2 T cells migrate into follicles. T_h1 and T_h2 cells that have been polarized *in vitro* do migrate into follicles following adoptive transfer, but it is not clear how this relates to T cells activated *in vivo*. It is not impossible that T_h1 and T_h2 cells regulate class switching, perhaps even outside the follicles whereas T_{fh} cells drive the proliferation of B cells in follicles and germinal centres.

Q6.24 How might we show that there is a bone marrow cell type that can give rise to B and T cells, but to no other cells types?

A6.24 One approach requires the use of adoptive transfer: (i) identify phenotypic differences between subsets of bone marrow cells; (ii) use these differences to separate the subsets, e.g. by fluorescence-activated cell sorting (FACS); (iii) transfer the subsets into irradiated mice (whose own stem cells have been killed); and (iv) determine if there is a subset that can give rise only to T and B cells. Alternatively, use a technique that will introduce a heritable marker into individual bone marrow cells. This could be mild irradiation, which will induce unique chromosomal abnormalities into individual cells, or infection with a retrovirus, which will show unique patterns of integration into the genome of individual cells. Then transfer these bone marrow cells into irradiated mice and determine if, in any mice, there are T and B cells which share the same marker, and that this particular marker is not present in any other cell type.

Answers to the Questions in Chapter 7

Q7.1 Why might humans have not evolved in ways that would prevent autoimmune diseases, allergies and other immune-related sensitivities?

A7.1 All these diseases are caused by adaptive immune mechanisms which are important in defence against infection, so we cannot do without them. Additionally, pathogens are continually mutating and being selected for variants that can avoid or evade immune responses, and we cannot evolve rapidly enough to deal with pathogen mutations. Hence, adaptive immunity needs to be anticipatory, and this involves the generation of receptors at random. It is almost impossible to ensure that every lymphocyte that could possibly react against an intrinsic antigen is deleted and inevitably these receptors also have a chance of cross-reacting with extrinsic antigens (see Figure 7.11). Finally, many of these diseases are associated with multiple genetic factors, often caused by mutations in genes important in defence, but which do not apparently affect our ability to counter infection. Hence these genes will not have been selected against during our evolution.

Q7.2 Is it likely that invertebrates might suffer from any of the immune-related diseases or conditions we are discussing in this chapter?

A7.2 Most of these diseases involve the adaptive immune system which is not present in invertebrates. Autoinflammatory diseases and some immunodeficiencies do involve the innate system (see Section 4.2.2.4 and elsewhere in Chapter 4). Any animal in the wild that has such defects is however likely to die, and is unlikely to be found. Invertebrates can, however, develop tumours that share many features with mammalian examples.

Q7.3 How might particles such as pollen grains manage to cross the epithelium of the eye to trigger mast cell degranulation in hay fever sufferers?

A7.3 This is not clear. A study in Finland used electron microscopy of biopsies from the conjunctiva of the eyes of medical students to show that, in individuals sensitized to birch pollen (a major cause of hay fever), pollen grains applied to the conjunctiva crossed several layers of epithelial cells in a few minutes. This did not happen in non-sensitized students. The mechanisms of transport are quite unknown.

Q7.4 What might be the potential impact of negative selection on the ability to recognize tumour antigens?

A7.4 Many tumour antigens are derived from normal self molecules. If the relevant peptides from these molecules are expressed in the thymus or in peripheral tissues in the absence of danger signals, it is likely that T cells recognizing such molecules will be tolerized. Hence, the frequency of lymphocytes could recognize this type of antigen is likely to be very low. Tumours are, however, highly mutated and there is the possibility that they will express mutant peptides on their MHC molecules (Section 7.6). T cells will not be tolerant to such peptides, so some anti-tumour reactivity can be expected.

Q7.5 How could we estimate the extent of cross-reactivity of a T cell antigen receptor?

A7.5 This is a complex, but very important immune problem. Most experiments have used a single T cell clone, generated against a particular peptide-MHC combination, and have examined the frequency with which different peptides can activate the clone. The problem is, of course that it is impractical to synthesise and test all possible combinations and permutations of different peptides. However from other approaches we have gained the surprising answer is that an individual T cell can react to about 10^8 different 11-mer acid peptides. It has however, been estimated that the total number of peptides that could be generated from proteins, and which expressed the appropriate anchor residues for a particular MHC class II molecule, is in the order of 10^{12}. This implies that the chances of a T cell activated against a particular peptide actually meeting a cross-reactive peptide must be very low.

Q7.6 How could we attempt to identify DAMPs involved in immune activation?

A7.6 We first need an assay for innate immune activation. One study showed boosting of CD8 T cell activation by co-injection of antigen mixed with dying cells into mice. The dying cells were solubilized and fractions were separated on the basis of molecular size. Mass spectrometry was used to characterise the molecule in the active fraction, revealing it to be uric acid (Section 4.6.1). Similar principles could be used to identify other potential DAMPs. These experiments do not, however, tell us how uric acid or other possible DAMPs might act *in vivo*.

Q7.7 How might evidence for an infectious trigger for autoimmune disease be gathered?

A7.7 In experimental situations evidence can be obtained directly. In one model, for example, a viral protein is expressed as a transgene under the control of a tissue-specific promoter. Thus, a viral envelope gene, under the control of the insulin promoter, can be expressed in pancreatic β cells. In some experiments, if the mouse is infected with the virus, the β cells are destroyed and diabetes ensues. This shows proof of principle, but it does not show that this happens in human disease. In some human diseases the association with an infectious aetiology is very strong (e.g. streptococcal infection and rheumatic fever), but in most cases there is only weak statistical evidence at best. In others there is suggestive evidence but no proof. Thus, a comparison of patients with Type I diabetes and controls showed an association between disease and both a polymorphism in an antigen-processing gene and having been infected with avian mycobacteria.

Q7.8 An anti-self response will not necessarily cause disease. Patients who have suffered a myocardial infarct (heart attack) often develop anti-myocardial antibodies, but this is usually a short-lived, self-limited event and the antibodies are harmless. Why might this response be limited? In a few people the heart attack is followed 10–21 days later by further symptoms of heart damage. Why might it be that in these circumstances the response is not self-limiting?

A7.8 The damage caused to the heart by the myocardial infarct could lead to the exposure of self antigens to which we are not tolerant (e.g. because these are normally sequestered in the tissue and are not adequately represented in the thymus to enable reactive lymphocytes to be eliminated). In addition, the hypoxia may have led to the release of DAMPs, which resulted in these lymphocytes being activated, including B cells that secreted the anti-myocardial antibodies. There are two main possibilities for the limited response. The supply of antigen may not be sustained because the heart muscle heals and the immune response is no longer capable of damaging the muscle to release more antigens. Another possibility is that regulatory mechanisms operate to shut off the response. Why in some cases it is not limited is unclear. Might the anti-myocardial response be stronger or more sustained in individuals with risk factors for autoimmune disease?

Q7.9 What kind of evidence might support the hypothesis that differences in TCR repertoires may influence susceptibility to autoimmune disease?

A7.9 It is possible to analyze differences in the CDR3 region of TCRs (Section 5.1.2, compare with Figure 6.2) in individual T cells, and thus gain an idea of the potential T cell repertoire by spectratyping. If it could be shown, for instance, that identical twins who both suffered from the same autoimmune disease possessed clones of T cells using the same TCR, but that this did not occur in non-concordant twins, this would provide some evidence for the role of particular TCRs in pathogenesis. However, to our knowledge, there is no substantial evidence for this occurring.

Q7.10 Which other types of molecule with allelic variants might lead to a generalized tendency to develop autoimmune disease?

A7.10 We already know of such molecules involved in the generation of T_{reg}, such as genes involved in the regulation of expression of the transcription factor FoxP3 (Section 5.5.5). From first principles we might also expect these to include molecules involved in the polarization of CD4 T cell responses (e.g. cytokines such as IFN-γ, IL-4, IL-12 or IL-23). Cytokines with regulatory functions such as IL-10 or TGF-β might also be important. Additionally, and probably more important than allelic variants within structural genes, may be variants in the regions controlling gene expression such as promoters.

Q7.11 Is it possible to suggest ways in which we will be able to fill in the gaps in our understanding of the genetic basis of immune-related disease susceptibility?

A7.11 If professional geneticists are having difficulty in doing this, perhaps it is not surprising that immunologists are also finding it difficult. It may be that there are variants of different genes that in isolation are not significant risk factors, but if they are combined in an individual may summate to increase the risk of disease. An analogy might be being dealt a bad hand of cards in a poker game.

Q7.12 T-bet knock-out mice develop an asthma-like condition in their airways, but other organs and systems do not develop disease. Why might disease be restricted in this way?

A7.12 Such responses may require other concomitant external or internal factors, such as infection with a specific type of microbe or the presence of particular tissue microenvironments that may only be present in the respiratory tract. The take-home message is, however, that we do not really understand this observation.

Q7.13 In what other examples of immune-mediated disease might antigen-based therapy might be appropriate?

A7.13 The first requirement is that we have identified the antigen that is most important in causing the disease. This may be the actual initiating antigen, but it may be an antigen that has induced an immune response though epitope spreading (Section 7.3.2.2). Antigens that have been tried in clinical therapy include collagen for rheumatoid arthritis and a uveal antigen for autoimmune uveitis (the uvea is that part of the eye containing the iris). We must of course know that the antigen is itself harmless before using them. Might beef collagen given in the 1990s have been contaminated with bovine spongiform encephalomyelitis (BSE), for example?

Q7.14 Where might the innate activation danger signal needed to stimulate T cell responses come from? arise in drug sensitivities?

A7.14 This is difficult and there are no definitive answers. If there is a new peptide-MHC epitope created by the drug binding to a protein, T cells will not have been tolerized to that epitope. If a DC carrying that modified peptide is activated by a PAMP or DAMP, perhaps by a concurrent infection, it could activate T cells specific for the modified peptide.

Q7.15 A Group O mother will possess anti-A and anti-B blood group antibodies. If she becomes pregnant with an A-positive foetus will the foetus develop haemolytic anaemia?

A7.15 The foetus will not develop haemolytic disease. The ABO antibodies are IgM and cannot cross the placenta. If, however, a mother has been sensitized to ABO antigens – perhaps by a mis-matched transfusion – and has developed anti-A or -B antibodies that are IgG, these can cross the placenta and the foetus will develop very severe haemolytic disease.

Q7.16 The activation of adaptive immunity against Rh antigens should require activation of the innate system by danger. Where might this danger come from when foetal RBCs enter the maternal circulation?

A7.16 As with drug sensitivities, this is again difficult and no clear answers are available. There is of course trauma during childbirth, and perhaps this releases DAMPS which generate the innate signalling. However, if the foetal RBCs get into the maternal circulation, the response will be primarily in the spleen and it is difficult to see how DAMPs could also be delivered in significant quantities to this organ (although this is possible). An additional question is why is IgG synthesized, since the foetal RBCs represent a primary immunization and there is no secondary challenge?

Q7.17 A Rh-negative mother has a Rh-positive child and is not given anti-Rh antibodies. She becomes pregnant again but the child does not develop haemolytic anaemia. What reasons might account for this?

A7.17 Two possibilities that come to mind are the following. (i) If the father is heterozygous for the Rh antigen gene, the baby may not carry the Rh gene, and thus will not be susceptible to the anti-Rh antibody. (ii) Different father.

Q7.18 Why might the administration of large amounts of pooled human immunoglobulins (IVIG) bring benefit in antibody-mediated diseases?

A7.18 Despite the efficacy of high dose IVIG therapy in many conditions (and not just antibody-mediated diseases), there is as yet no clear answer as to how it works. Most likely it may act in different ways, perhaps depending on the actual condition that is being treated. Mechanisms of action may include the administered antibodies binding via their Fc regions to inhibitory Fc receptors on B cells (Section 6.3.2.3), preventing their synthesis of damaging antibodies; binding of sialic acid to a receptor(s) that reduces macrophage activation and the production of inflammatory mediators; and binding via F(ab')2 to damaging antibodies and forming immune complexes which are then cleared. Other mechanisms have also been suggested. By following this question up, the interested reader may gain a deeper insight into what we think we know, and what we probably don't know, about the regulation of immunity.

Q7.19 C1q is a crucial component in complement activation by immune complexes via the classical pathway. Might we not predict that a C1q deficiency would lead to less effective and thus inflammation?

A7.19 Inflammation is a central feature of SLE and it is a puzzle that SLE is so common in C1q-deficient individuals. Given the central role of C1q in activating complement vial the classical pathway, we might have predicted that immune complexes found in SLE would cause less inflammation in C1q-deficient patients. However analysis of SLE lesions shows that the complexes do contain complement, so we suggest that in SLE, immune complexes can activate complement in other ways, not involving the classical, C1q-dependent pathway. Conceivably this happens through the other two pathways (Section 4.4.2.1).

Q7.20 The Islets of Langerhans contain a mixture of insulin-secreting β cells and α cells that secrete glucagon (a hormone that raises blood glucose levels). Given that in typical autoimmune Type I diabetes only the β cells are affected, what might this suggest about the likely immunological effector mechanism?

A7.20 This selectivity in destruction of β cells is difficult to explain if a macrophage-dependent DTH mechanism is responsible. Destruction by cytotoxic T cells would be much more cell-specific and the fact that only β cells can be affected could suggest that CTL are involved, (assuming that β cells are accessible to cytotoxic T cells). It may, however, be relevant that in some very severe cases of Type I diabetes both α and β cells are destroyed, suggesting a non-specific mechanism such as might be caused by macrophages causing collateral damage.

Q7.21 Why might diabetes be more severe in NOD mice that are germ-free?

A7.21 We are unaware of a definitive answer to this. Might it be that commensal bacteria have a role in generating non-antigen-specific T_{reg}? If this is the case, then we are also faced with the question of why commensals are essential in mouse models of inflammatory bowel disease (Section 7.4.4.3) and, again, this is not clear.

Q7.22 How might we determine if antibodies do play a significant role in the pathogenesis of multiple sclerosis and why might this information be helpful?

A7.22 It clearly is important to find out if antibodies do play a role or if this is just a secondary phenomenon, as this will inform therapy. Animal models are useful but not definitive in this respect. EAE differs in many ways from multiple sclerosis. Treatments designed to deplete antibodies (e.g. plasmapheresis) may give information. For example if they alleviate the disease this would suggest that pathogenic antibodies do have a role. A major and important point, however, is that many autoimmune diseases typically show a pattern of remission and relapse. This means that, in a proportion of patients, remission will be associated with any form of treatment and may lead to a misleading interpretation.

Q7.23 Why might an autoimmune disease be more common in particular geographic regions?

A7.23 This is a major puzzle and there are no definitive answers. Diseases which show geographical variations are often infectious in origin. Perhaps there is an infectious trigger for a disease such ultiple sclerosis and the infection is more prevalent in some regions. Multiple sclerosis is, however, more common in Northern latitudes and it has so far not been possible to link it to known endemic infections.

Q7.24 To what extent might Crohn's disease be considered an autoinflammatory disease? (Section 4.2.2.4).

A7.24 Autoinflammatory diseases are a large group of conditions in which there is abnormal activation if the innate immune system (Section 4.2.2.4). In many cases, these diseases are apparently independent of the adaptive system, and the tissue damage is due to T cell-independent inflammation. In Crohn's disease, at least in some patients, variants of the NOD2 gene may lead to abnormal activation of the innate system. However, there is also strong evidence for the involvement of T cells, since T_h1 and T_h17 cells are present in Crohn's lesions. Evidence from animal studies also supports the involvement of T cells, particularly T_{reg}, and responses to commensal flora. Thus, Crohn's disease involves both the innate and adaptive systems and it may not be particularly helpful to classify it as an autoinflammatory disease *per se*.

Q7.25 How might the importance of migrating DCs in skin graft rejection be investigated? Could there be other possible explanations for the non-rejection of alymphatic skin grafts?

A7.25 It is now possible to deplete DCs selectively in mice. This can be done by using transgenic mice which express the human diphtheria toxin receptor (DTR) selectively in DCs, and then treating the mice with the toxin (Box 5.7). If only DCs express the toxin, there is a selective depletion. Tissues from such a treated mouse could be transplanted into a normal allogeneic recipient and rejection monitored. The experiment could be done in reverse, using toxin-treated DTR mice as recipients to investigate the roles of recipient DCs. It could also be done in situations where both donor and recipient were toxin-treated DTR transgenic mice to ask if DCs are needed at all. The non-rejection of alymphatic skin grafts might involve the passage of immature DCs from the graft in blood to the spleen, where they could induce tolerance or T_{reg}; possibly, soluble graft MHC antigens in blood could have similar effects.

Q7.26 What alternative explanations might explain why alymphatic kidneys transplants in dogs were rejected normally?

A7.26 Kidneys may be able to release DCs directly into the blood, and these cells could migrate to the spleen and initiate potent alloreactive responses. There is clear evidence, for example, that this might occur during rejection of vascularized heart transplants in mice; after transplantation, donor-derived DCs were found in the recipient spleen for example. The relative importance of different modes of sensitization may also differ between skin and kidney grafts (from other experiments, they probably do) and even between mice and dogs.

Q7.27 How may further elucidation of mechanisms preventing foetal rejection be helpful in other settings?

A7.27 If we understood how a semi-allogeneic foetus survives, we might be much better placed to suppress aberrant responses in autoimmune diseases or unwanted responses in transplantation.

Q7.28 Might the observation that tumours down-regulate immune responses to their own antigens suggest approaches to the treatment of tumour-bearing patients?

A7.28 Here are some possibilities. If MHC class I expression is down-regulated, in tumours could NK cell activity be stimulated? If DC activation is inhibited by tumours could exogenous, activated DCs be used to stimulate an effective response? If the tumour stimulates the generation of T_{reg}, could these be removed by anti-CD25 treatment?

Q7.29 What might be some limitations of cell-based therapies?

A7.29 The primary problem is that of rejection if the cells did not originate in the recipient, because they are likely to be rejected as is any other allogeneic transplant. This means that, in these circumstances, usually the recipient is immunosuppressed, either as a result of the disease or by using active immunosuppression. These recipients are at danger of infection, and of generating virally-induced tumours. Hence autologous cell-based therapies are preferred, but these come with the added complication that it is necessary to obtain cells from each and every individual patient, which can be laborious and time consuming, before they are re-administered.

Further Study Questions

Chapter 3

Qa. What types of constitutive versus inducible defences might the natural barriers have? (Section 3.2)

Hint These barriers are external (skin), or situated at topologically external sites (e.g. lungs, gut, genital tract), lined by epithelia. Epithelial cells would be in a good position to secrete some basic defence molecules constitutively, and perhaps to secrete more of these, or new ones, if infection occurs. Perhaps think also about how much the secretion of mucus, or the acidity of these sites, might be controlled.

Qb. To what extent is it possible to identify different types of inflammation and their purposes? (Section 3.3)

Hints It might be helpful to start with the settings of sterile trauma and infection in general. Then we might consider different types of viruses, microbes or larger parasites – and the sorts of tissues they might infect. This then brings us on to inflammatory responses in different organs. We know, for example, that those in the brain can be very different to those in other non-lymphoid tissues such as skin. What about responses in different types of lymphoid tissues?

Qc. How much do we know about the regulation of adaptive immune responses in lymphoid tissues by events that occur in peripheral tissues? (Section 3.4)

Hints We find the conduits fascinating structures and are sure we have much to learn about how much, and what type, of information they might be able to transmit directly into the core of a lymph node, for example. It also appears that leukocytes recruited to inflammatory sites might then traffic to such tissues (some could be recruited directly from the blood). Perhaps start by considering neutrophil swarms or if there is anything known about granulocytes such as basophils in lymph nodes.

Qd. To what extent can different secondary lymphoid organs be considered also to function as primary lymphoid tissues, or vice versa? (Section 3.5)

Hints Perhaps start by thinking about different settings. Are we talking about the normal steady state in the absence of infection or after an adaptive immune response has occurred. What about if an organ is removed (e.g. a ruptured spleen in a road traffic accident) – might the functions of other lymphoid organs be affected? What about different species or different stages of maturity (humans versus mice, adults versus neonates)?

Qe. What are the indications and complications of stem cell therapy (and/or the potential for gene therapy) particularly for the treatment of immune-related diseases? (Section 3.6)

Hints We need to define what sort of stem cell therapy we mean – transplantation of bone marrow (e.g. HSCs) or of organs grown from ES cells? The complications of the first are in fact very well known, those of the latter rather more theoretical at present. How about gene therapy? What sort of immune-related diseases can we think of where replacement of a defective gene would be advantageous? What are the potential risks of messing about with a person's DNA? (Do we want to get into ethics as well?)

Chapter 4

Qa. To what extent can altered, damaged or stressed self components act as agonists for PRRs? (Section 4.2)

Hints We do know that some types of PRRs, such as the scavenger receptors, can bind modified self components such as oxidized low-density lipoprotein. Others, such as some of the TLRs, are also thought to respond to components of the host, including heat-shock proteins that are produced by stressed cells (e.g. after infection). We do, however, need to be cautious because some apparent responses might actually be due to traces of contaminants in the preparations used, LPS being a prime example.

Qb. How much do we understand about the regulation of inflammatory responses? (Section 4.3)

Hints Inflammation is dangerous and needs to be carefully controlled. Do we want to consider local or inflammatory responses, or regulation at the tissue, cell or molecular level? There is, for example, increasing understanding of the nervous and endocrine control of inflammation (e.g. the

Exploring Immunology: Concepts and Evidence, First Edition. Gordon MacPherson and Jon Austyn.

hypothalamus–pituitary–adrenal axis). How about specific types of cells making anti-inflammatory response to counter-balance it? What about the kinetics of secretion of different types of molecules, including cytokines, that might dampen it down?

Qc. To what extent can we define the different stages of macrophage activation, the function of different NK cell receptors and their ligands or the roles of acute-phase reactants in defence? (Section 4.4)

Hints Yes, we apologise (a little) – this is three questions in one. There is in fact a great deal known about each of these. However, for macrophages, are these really defined stages or is there more of a continuum? For NK cells, are these receptors expressed differently by different subsets, and what types of responses might they control? For acute-phase reactants, in some cases the answer is very clear, but in others it is probably true to say we still have absolutely no idea of why these molecules are produced.

Qd. How much do we know about the control of production of different types of immune cells? (Section 4.5)

Hints The best place to start is probably to choose one cell lineage that takes your fancy. (For us this would probably be DCs; we also find mast cells interesting because their origins are still relatively obscure.) Where are the precursors of these cells first produced? What types of stromal cells might they contact there, what types of cytokines might be produced to help them develop and what types of transcription factors might regulate development? Do they get into the blood as precursors, or at a mature stage? Where do they go? How do these populations change if infection or inflammation occurs?

Qe. To what extent is it possible to define different types of DAMPs and their receptors?

Hints This is a minefield at present. Perhaps start by defining what we mean by a DAMP. Is it an "altered self" component or something produced by cells or tissues in response to damage, and to what extent might it overlap with a PAMP? Then perhaps find out about how much we do or do not yet know about the mechanism(s) of alum as an adjuvant, how inflammasomes are activated, and such like. But tread carefully!

Chapter 5

Qa. How much is known about the functions of non-classical MHC molecules, their ligands (if any), and their cell or tissue distributions?

Hint We have provided a few examples of non-classical MHC molecules, but many others have been identified in the human and mouse MHC region. Some have ligands such as peptides, but is antigen processing needed to generate these? Some are restricted to certain cell types or tissues, such as the placenta. For some, it is clear that in some cases they can be recognized by non-conventional lymphoid cells, but why?

Qb. To what extent do we understand the functions of different DC subsets in immunity?

Hint Perhaps start by defining what we mean by a subset. Multiple subsets of classical DC have been identified in mouse lymphoid tissues, for example, and we believe that some have specialized functions. What of pDCs in normal and pathological settings, and to what extent can they regulate lymphocyte responses? Do we want to consider FDCs as well?

Qc. How far can we define different polarized subsets of effector or memory T cells?

Hints The ability of CD4 T cells to adopt specialized functions (T_h1, T_h2, etc.) seems clear, at least in mice. However, additional subsets with different functions, such as T_h9 and T_h21, have also been postulated. In some cases CD8 T cells also seem to adopt specific functions (T_c1, T_c2, etc.) and this has even been suggested to be the case for iNKT cells. Might this also apply to other non-conventional T cells such as γδ T cells? To what extent is this apparent polarization retained or lost if the CD4 or CD8 T cells develop into central or effector memory cells?

Qd. To what extent might mechanisms of positive and/or negative selection apply to development of non-conventional lymphocytes?

Hint We know that, during development, follicular B cells and αβ T cells are subject to negative selection, and that the latter are also subject to positive selection. Some have postulated that positive selection also applies to B cells. It certainly applies to iNKT cells, but what about γδ T cells? If we wish to extend this to NK cells, it is clear these cells are selected to express receptors for specific MHC class I alleles, for example, but how? Does negative selection apply here as well?

Qe. What advances are being made in cell-based adoptive therapies for different immune-related diseases?

Hint Perhaps start by considering diseases in which one might wish to stimulate protective immunity or in which one might wish to suppress aberrant or unwanted responses. Then, ask which immune cell types might be best suited for one or the other. We have provided a few examples. However, new approaches are being tried even for these, such as transfection of tumour-specific TCRs into T cells, exposure of DCs to different types of stimuli to modulate their functions or different ways of generating antigen-specific regulatory T cells. To what extent do we really want to repolarize or redirect immune responses rather than simply turn them on or off?

Chapter 6

Qa. How well do we really understand the function of different antibody isotypes in defence against infection? (Section 6.2)

Hint The functions of four of the main human and mouse isotypes (IgM, IgG, IgA and IgE) might seem reasonably well established. However, IgM is not always a pentamer, IgG is comprised of different subclasses and IgA comprises two subclasses (in humans) that can be expressed as monomers or dimers. Presumably, these play specialized roles in different types of defence, but how well do we understand them? What really is the function of IgE in defence against infection, rather than in pathological settings (allergy)? What about more recent insights into the potential functions of IgD in host defence?

Qb. To what extent are different types of B cell responses important in immunity? (Section 6.3)

Hint At the cellular level, we might start by considering the different populations of B cells, how they are activated and where they are found. Then consider which of them preferentially produces natural antibodies or TI versus TD responses and under what conditions and where. At the molecular level, we might consider to what extent these responses change after repeated stimulation. We could also potentially ask which types of stimuli might lead to signalling through BCRs or TLRs, for example? Much of our understanding of these comes from contrived settings (e.g. responses to hapten-carriers or anti-immunoglobulin antibodies). What of real disease settings?

Qc. How well can we explain the different migratory patterns of naïve B cells, plasma cells and memory B cells? (Section 6.4)

Hint The post code principle is likely to be essential (Section 4.3.4.1), but is it sufficient to explain the different anatomical sites to which these cells home? We do know quite a lot about preferential expression of different chemokine receptors, for example, but what about selectins or integrins? Within each site (e.g. secondary lymphoid tissues or the bone marrow), what is known of how different cells are guided to different compartments or what controls their release or retention?

Qd. What is known of the relative importance of different mechanisms in antibody-mediated resistance to infection? (Section 6.5)

Hint We have provided some examples of long-lived plasma cells or memory B cells and resistance to infection. However, how important is each in different types of infection? Why should memory B cells have evolved, if long-lived plasma cells can continue to secrete antibodies? Are the latter really end cells or are they in fact more plastic being able to adopt different functions during immune responses? What is the role of short-lived plasma cells? What actually controls the survival of memory B cells?

Qe. What advances are being made in the development and application of therapeutic antibodies?

Hint One could start by thinking about these separately. In terms of development, we might consider what types of technical advances are being developed to enhance their function, such as by preventing host responses to foreign (e.g. mouse) monoclonal antibodies. Or we might ask for which disease settings new antibodies are being developed, and why and to what extent – both in terms of the disease and their potential targets – they have been tested clinically.

Index

Exploring Immunology: Concepts and Evidence, First Edition. Gordon MacPherson and Jon Austyn.

b

j

k

l

n